11/23/76
FINAL EXAM

TABLE 34

Drill Sizes

Letter drills begin where **number drills** end.

Number and Letter Drills	Fractional Drills	Decimal Equivalents	Number and Letter Drills	Fractional Drills	Decimal Equivalents	Number and Letter Drills	Fractional Drills	Decimal Equivalents	Number and Letter Drills	Fractional Drills	Decimal Equivalents
80	...	.0135	42	...	.0935		13/64	.2031		13/32	.4062
79	...	.0145		3/32	.0937	6	...	.2040	Z	...	.4130
	1/64	.0156	41	...	.0960	5	...	.2055		27/64	.4219
78	...	.0160	40	...	.0980	4	...	.2090		7/16	.4375
77	...	.0180	39	...	.0995	3	...	.2130		29/64	.4531
76	...	.0200	38	...	.1015		7/32	.2187		15/32	.4687
75	...	.0210	37	...	.1040	2	...	.2210		31/64	.4844
74	...	.0225	36	...	.1065	1	...	.2280		1/2	.5000
73	...	.0240		7/64	.1094	A	...	.2340			
72	...	.0250	35	...	.1100		15/64	.2344		33/64	.5156
71	...	.0260	34	...	.1110	B	...	.2380		17/32	.5312
70	...	.0280	33	...	.1130	C	...	.2420		35/64	.5469
69	...	.0292	32	...	.1160	D	...	.2460		9/16	.5625
68	...	.0310	31	...	.1200	E	1/4	.2500		37/64	.5781
	1/32	.0312		1/8	.1250	F	...	.2570		19/32	.5937
67	...	.0320	30	...	.1285	G	...	.2610		39/64	.6094
66	...	.0330	29	...	.1360		17/64	.2656		5/8	.6250
65	...	.0350	28	...	.1405	H	...	.2660			
64	...	.0360		9/64	.1406	I	...	.2720		41/64	.6406
63	...	.0370	27	...	.1440	J	...	.2770		21/32	.6562
62	...	.0380	26	...	.1470	K	...	.2810		43/64	.6719
61	...	.0390	25	...	.1495		9/32	.2812		11/16	.6875
60	...	.0400	24	...	.1520	L	...	.2900		45/64	.7031
59	...	.0410	23	...	.1540	M	...	.2950		23/32	.7187
58	...	.0420		5/32	.1562		19/64	.2969		47/64	.7344
57	...	.0430	22	...	.1570	N	...	.3020		3/4	.7500
56	...	.0465	21	...	.1590		5/16	.3125			
	3/64	.0469	20	...	.1610	O	...	.3160		49/64	.7656
55	...	.0520	19	...	.1660	P	...	.3230		25/32	.7812
54	...	.0550	18	...	.1695		21/64	.3281		51/64	.7969
53	...	.0595		11/64	.1719	Q	...	.3320		13/16	.8125
	1/16	.0625	17	...	.1720	R	...	.3390		53/64	.8281
52	...	.0635	16	...	.1770		11/32	.3437		27/32	.8437
51	...	.0670	15	...	.1800	S	...	.3480		55/64	.8594
50	...	.0700	14	...	.1820	T	...	.3580		7/8	.8750
49	...	.0730	13	...	.1850		23/64	.3594		57/64	.8906
48	...	.0760		3/16	.1875	U	...	.3680		29/32	.9062
	5/64	.0781	12	...	.1890		3/8	.3750		59/64	.9219
47	...	.0785	11	...	.1910	V	...	.3770		15/16	.9375
46	...	.0810	10	...	.1935	W	...	.3860		61/64	.9531
45	...	.0820	9	...	.1960		25/64	.3906		31/32	.9687
44	...	.0860	8	...	.1990	X	...	.3970		63/64	.9844
43	...	.0890	7	...	.2010	Y	...	.4040		1	1.0000

Metalwork
Technology and Practice

Metalwork
Technology and Practice

Oswald A. Ludwig
Formerly Department Head
Vocational Education
Henry Ford High School
Detroit, Michigan

Willard J. McCarthy
Professor
College of Applied Science
and Technology
Illinois State University
Normal, Illinois

Victor E. Repp
Associate Professor of
Industrial Education
and Technology
Bowling Green State University
Bowling Green, Ohio

McKnight Publishing Company, Bloomington, Illinois

SIXTH EDITION
Lithographed in U.S.A.

Ronald E. Dale, Vice President - Editorial, wishes to acknowledge the skills and talents of the following people and organizations in the preparation of this publication.

Wesley D. Stephens
President
Instructional Structure and Sequence

Donna M. Faull
Production Editor

Ann Urban
Copy Editor

Elizabeth Purcell
Interior Design/Layout Artist

Sue Whitsett
Proofreader

Eldon Stromberg
University Graphics Incorporated
Carbondale, Illinois
Cover Design

Gorman's Typesetting
Bradford, Illinois
Composition

R. R. Donnelley & Sons
Chicago, Illinois
Preproduction and Printing

William McKnight III
Manufacturing

Library of Congress
Card Catalog Number: 74-21561

SBN: 87345-117-1

First Edition Copyright 1943
Second Edition Copyright 1947
Third Edition Copyright 1955
Fourth Edition Copyright 1962
Fifth Edition Copyright 1969

Preface

Metals are basic engineering materials that are essential to our industrial development and technological progress. The purpose of Metalworking Technology and Practice is to provide a comprehensive introduction to the technology of metalworking, both its theory and practice. This task has become increasingly difficult with each revision because of the accelerating rate of new developments in the processing of metals. In most cases, the new processing methods do not replace existing methods, resulting in an ever-expanding body of knowledge which must be structured and covered in a single volume.

In order to accommodate the new material, it was necessary for this sixth edition to be a complete revision. Content was reorganized along conceptual lines, requiring combination of some of the previous units. Obsolete information was removed and new information was added to bring each unit up to date. Many units had to be completely rewritten in order to integrate the large quantity of new material. New units on metric measurement, adhesive bonding, and nontraditional machining processes were added. However, the simple language and abundant illustrations traditionally associated with this book have been retained. Many of the old illustrations were redrawn to reflect current equipment and practices, and many new illustrations were added.

A comprehensive vocabulary list is included at the end of each unit. It is important to learn the meanings of technical words in each unit, especially the words which are boldface. Review questions also are included at the end of each unit. The questions may be used as an aid in guiding your study or for review purposes after demonstrations or during class discussions. A revised Study Guide (workbook) is also available for use with this new edition.

This has resulted in a book which provides a truly comprehensive introduction to metalworking technology, making it an excellent introductory text for use in metalworking technology and materials processing curriculums, as well as a valuable reference for manufacturing curriculums.

Victor E. Repp

Acknowledgments

Acknowledgment and appreciation are expressed by the publishers and the present coauthor to the following people who provided helpful criticism, information, and suggestions used in the preparation of previous editions: Mr. F. R. Kepler, Mr. Harry M. Dextor, Mr. Floyd C. Allison, and Mr. Carnot Iverson. Mrs. Oswald A. Ludwig participated unstintingly in the preparation of previous editions, and Mr. Earl A. Ludwig provided assistance in the fourth edition.

Mr. Willard J. McCarthy authored the fifth edition and expressed appreciation for the extensive help received from many industrial firms and individuals, including Mrs. McCarthy, who handled correspondence and manuscript typing.

Mr. Howard T. Davis prepared new drawings for both the fifth and the present edition, and Mr. William DuBois provided a number of photographs for the present edition.

Sincere appreciation is also expressed to the following industrial firms and organizations which provided technical information and numerous illustrations used in the book:

Acme-Gridley
Adjustable Clamp Company
Aerospace Industries Association of America, Inc.
Ajax Electric Company
Alcoa
Aluminum Company of America
American Foundrymen's Society
American Gas Furnace Company
American Iron and Steel Institute
American Machinist
American Screw Company
American Society of Mechanical Engineers
American Welding Society
Ames Precision Machine Works
Anaconda Company
Armstrong Brothers Tool Company
Atlas Press Company
Barber-Colman Company
Bausch & Lomb, Inc.
Bendix Corporation
Bethlehem Steel
Bridgeport Machines, Inc.
Brown & Sharpe Manufacturing Company
Chevrolet Division, General Motors Corporation
Chrysler Corporation
Cincinnati Incorporated
Cincinnati Milacron/ Cintamatic Division
Cincinnati Milling Machine Company
Cincinnati Milling Machine Company, Cimtrol Division
Clausing Division, Atlas Press Company
Cleveland Twist Drill Company
Clipper Belt Lacer Company & The Bristol Company
The Craftool Company
Cyril Bath Company
Di-Acro Corporation
Eugene Dietzgen Company
Henry Disston & Sons, Inc.
DoAll Company
Eclipse Counterbore Company
Federal Products Corp.

Fellowcrafters, Inc.
Fenn Manufacturing Company
Fisher Body
The Foot-Burt Company
Forging Industry Association
Foxboro Company
Friden, Inc.
G. A. Gray Company
General Electric Company
Gorham Company
Great Lakes Machinery Corporation
Great Lakes Screw Corporation
Greenfield Tap & Die Corporation
Hamilton Standard Division United Aircraft
Hobart Brothers Company
Hossfeld Manufacturing Company
Hughes Aircraft Company, Industrial System Division
IBM
Inland Steel Company
International Silver Company
I. S. U. Photographic Service
Johnson Gas and Appliance Company
C. E. Johansson Gage Company
Kearney & Trecher Corporation
King Machine & Manufacturing Company
Le Blond
Linde Air Products Company
Link-Belt Company
McEnglevan Heat Treating & Manufacturing Company
McGraw-Hill Book Company
Miller Electric Manufacturing Company
Morse Twist Drill & Machine Company
National Bureau of Standards
National Cash Register Company
National Cylinder Gas Company
National Machine Tool Builders Association
National Twist Drill & Tool Company
Niagra Machine & Tool Works
Norton Company
Peck, Stow & Wilcox Company
Pontiac Division, General Motors Corporation
Charles Porker Company
Pratt & Whitney Company
Precision Tool & Manufacturing Company
Rex Charnbelt, Inc.
Reynolds Metals Company
George Scherr Company
Sheldon Machine Company
Shore Instrument & Manufacturing Company
Sikorsky Aircraft
Simonds Saw & Steel Company
South Bend Lathe, Inc.
L. S. Starrett Company
Supreme Products — Rigid Tool Company
Taft-Pierce Manufacturing Company
Teledyne Ryon Aeronautical
Tempil°
Thermolyne Corporation
TOCCO Division, Park-Ohio Industries, Inc.
Today, International Harvester Company
Union Drawn Steel Company
Union Twist Drill Company
United Chromium, Inc.
Up-Right Scaffolds
U. S. Electrical Manufacturing Company
U.S.M. Fastener Company
U. S. Steel Corporation
Waldes Kohimoor, Inc.
The Walton Company
J. H. Williams Company
Wilson Division, ACCO
World, International Harvester Company

Content

List of Tables

PART I

Introduction

UNIT 1

Introduction to Metalwork

1-1. The Importance of Metals

Metals are essential to the conduct of our daily lives and to the industrial society in which we live. Metals are everywhere around us. They are widely used in the production and construction of transportation vehicles, including aircraft and spacecraft, automobiles, buses and trucks, railroad cars, bicycles and motorcycles, and ships and submarines. Structural steel and other structural metals are used in the construction of roads, bridges, tunnels, and buildings. See Fig. 1-1.

Many metals are used in the construction of home appliances. These include familiar labor-saving devices—such as stoves, washing machines, clothes dryers, and dishwashers—and convenience items such as toasters, refrigerators, furnaces, and air conditioners. Also included are entertainment items such as radio and television sets, record-playing equipment, tape recorders, cameras, and projectors. Metals are used extensively in the construction of hand and portable power tools, machine tools, farm and manufacturing machinery, and road-building equipment. Sports equipment—such as fishing reels, pleasure boats, outboard motors, golf clubs, and guns—also makes use of metals. Valuable or precious metals are often used in making coins, jewelry, tableware, and cutlery.

Fig. 1-1. Metals Are Everywhere Around Us

A large proportion of the labor force is employed in metalworking careers. This labor force necessarily includes workers with various levels of education and skills. It includes semiskilled workers, craftsmen (a level of skill identified by trade unions), technicians, technologists, and engineers in many areas of specialization.[1] Careers in metalworking are discussed further in Unit 2.

1-2. Properties of Metals

Metals have different characteristics, and these characteristics are called **properties.**[2] When an engineer or product designer selects the metals which are to be used in a

[1]A very good source of information concerning all types of careers, including metalworking careers, is U.S. Department of Labor, Bureau of Labor Statistics, **Occupational Outlook Handbook** (Washington, D.C.: U. S. Government Printing Office, 1972-73 Edition), 880 pages. A revised edition is printed every two years.

[2]The student should learn the meaning of all **boldface** words.

modern metal product, he selects them on the basis of their properties. Six common properties of metals are described below.

Density refers to the weight of the metal. The density may be indicated in pounds per cubic inch (lb/in^3) or pounds per cubic foot. The density of steel is 0.284 lb/in^3. Aluminum, which is much lighter in weight, has a density of 0.097 lb/in^3. Hence, steel is nearly three times as dense or as heavy as the same volume of aluminum.

Corrosion resistance is the ability to resist rusting or other chemical action. Aluminum, stainless steel, and copper are far more corrosion-resistant than steel.

Hardness means resistance to penetration. Steel is much harder than lead or pure aluminum. Some steels can be made even harder by heat-treatment processes. They can be made so hard that they will cut other metals. Examples of products which are made possible by the heat treatment of steel are files, hacksaw blades, drills, and other metal-cutting tools.

Toughness in metal refers to its ability to withstand shock or heavy impact forces without fracturing. A metal ranking high in toughness will generally bend or deform before it fractures or breaks. When steel is extremely hardened by heat treatment, it loses some of its toughness. Files or drills, which are hardened by heat treatment, will usually break before they bend significantly. Toughness is very often more important than hardness in steel. It is more important for steering knuckles, bumpers, springs, and other parts of an automobile chassis to rank high in toughness than to rank too high in hardness. The durability of a product is related to the combination of its toughness and hardness properties.

Brittleness refers to the ease with which metals will fracture without bending or deforming greatly. Glass is very brittle. Hardened tool steels and gray cast iron are relatively brittle when compared to ordinary unhardened steels.

Tensile strength means resistance to being pulled apart. It is the force necessary to pull apart a piece of metal which has one square inch of cross-sectional area. Tensile strength is generally expressed in terms of thousands of pounds per square inch. For example, pure soft aluminum has a tensile strength of about 13,000 psi. Soft low-carbon steel has a tensile strength of about 69,000 psi. The tensile strength of certain kinds of tool steels can be increased to about 200,000 psi by heat-treatment processes. See No. 1095 steel in Table 13, page 158.

Metals have a number of other important properties. You will probably be studying these later.

1-3. Classification of Metals

Pure Metals

Metals are classified as **pure metals** or as **alloys**. A pure metal is a single chemical element which is not combined with any other chemical element. Examples of metals which are available as pure metals are iron, aluminum, copper, lead, tin, and zinc. Pure metals are generally too soft, lack high strength, or rank low in some other desired property. Thus, their use in the pure state is limited to laboratory experiments and a few construction applications.

Alloys

The properties of a pure metal may be changed by melting and mixing one or more pure metals with it; this procedure produces a new metal which is called an **alloy**. An alloy may have characteristics very different from either of the original pure metals from which it was formed. Stainless steel is a familiar alloy composed of steel, nickel, and chromium. It is strong, tough, and much more corrosion-resistant than plain steel.

An alloy, therefore, may be defined as a metallic substance composed of a combination of two or more metallic elements, one of which must be intentionally added to the base metal. Nonmetallic elements may also be included in alloys.

Metals may be further classified as either **ferrous** or **nonferrous** metals. The word ferrous is derived from the Latin word **ferrum**, which means iron. The principal element in all steel is iron. Thus, all steels are called ferrous metals. Examples of nonferrous metallic elements are aluminum, copper, lead, tin, and zinc.

The many kinds of alloys may also be classified as either **ferrous alloys** or **nonferrous alloys**. Alloys are named after the principal metal, called the **base metal**. Thus, steels which are intentionally alloyed with nickel, chromium, or tungsten are called **alloy steels**. Metallic-alloying elements are alloyed with aluminum to form **aluminum-base alloys**. They may also be added to copper to form **copper-base alloys** or with zinc to form **zinc-base alloys**. Aluminum-base alloys, copper-base alloys, and other alloys of nonferrous base metals are called nonferrous alloys. Whether a metal is a pure metal or an alloy, it is still called a metal.

1-4. Selection of Metals

Engineers, industrial product designers, technicians, and skilled craftsmen in metal-working occupations must know about various metals and their properties. There are hundreds of different grades of structural steels, alloy steels, tool steels, and special steels available for selection. More than 60 different aluminum-base alloys and more than 60 copper-base alloys are available.

Fig. 1-2. The Electric Furnace Is Widely Used for Producing Steel Alloys

Many other kinds of nonferrous alloys are also available for industrial use today. There are more than 100 different metals used in the manufacture of a modern automobile. There are about 12,000 metal parts used in the average modern automobile. In fact, more than 20,000 different kinds of metals or alloys are available for the design and production of metal products in modern industry. See Fig. 1-2.

You will be learning more about the different metals, their properties, and how they are produced in other chapters in this book. At this time, however, it is important to know that there are many different kinds of metals. Each kind of metal was developed to obtain certain properties which were demanded in the design and production of various metal products. You will want to know the principal properties of the common metals when you design, work with, and construct metal products in your metalworking course.

The reasons for selecting metals on the basis of properties can be understood when you consider several modern industrial products. Aluminum is used in the construction of items such as aircraft, small engines, lawn furniture, storm windows, and fishing boats. It is a desirable metal for those items because of its light weight, its corrosion resistance, and its strength. Special grades of aluminum have a tensile strength rating which is as great as that for ordinary low-carbon steel. See Table 14.

Steels which can be hardened by heat-treatment processes must be selected for making cutting tools which are used to cut other metals. Steels of this type are called **tool steels**. Files, hacksaw blades, drills, chisels, and threading dies are made of tool steels.

1-5. How to Profit from the Study of Metalwork

The following are a few of the many ways in which you can profit from the study of metals and metalworking processes:

1. You will learn how to plan, construct, inspect, and evaluate the quality of the metal products which you produce.
2. You will have opportunities to work with common metalworking tools, machines, materials, and processes. The range of experience which you acquire in metalwork will, of course, depend on the length of the course or the amount of time spent in studying metals.
3. You will apply your knowledge of mathematics, science, and drawing to practical metal products or exercises. You will be solving problems such as planning products, calculating cost of materials, and calculating speeds and feeds for machine tools. You will apply scientific theory in understanding the internal changes which take place in steel when it is hardened and tempered during heat-treatment processes.
4. You will learn to work safely with tools, machines, and materials. In the technical and mechanical age in which we live, these experiences will always be valuable. You can apply these kinds of experiences to the development of mechanical hobby interests and do-it-yourself jobs.
5. Metalworking will provide information and experience which will enable you to understand, care for, and repair metal products around the home. You will be spending a sizeable portion of your lifetime income on metal products which must be maintained and cared for.
6. Metalworking will provide the knowledge and skill necessary in many careers which involve the care, servicing, and maintenance of metal products. These types of careers include auto mechanics, appliance service workers, millwrights, factory maintenance workers, and maintenance workers for many kinds of machines and equipment.
7. Metalworking will provide experiences valuable in other kinds of careers where mechanical ability and a knowledge of metals are important. Examples of such careers include most engineering careers, dentistry, industrial designers, laboratory technicians, technologists, and scientists.
8. You will have the opportunity to develop an understanding of the metalworking industry, its workers, its processes, and many of its products. This study should provide knowledge and experience helpful in determining whether you wish to pursue a metalworking career.

1-6. Developing a Basic Knowledge of Metalwork

Basic knowledge and skill in metalworking may be acquired through systematic study. **Metalwork technology** is sometimes called **general metalwork** or just **metalwork**. It is a broad subject which covers many different areas of metalworking materials, processes, and activities. Following are some of the important areas of activity:

1. Designing and planning metalwork products or operations, which includes reading drawings, making sketches, and making a bill of materials.
2. Sawing and bench work, which includes use of hand tools and power sawing.
3. Study of various metals, alloys, and their properties.
4. Care of equipment.
5. Drill press work and drilling operations.
6. Threads, taps, dies, and threading.
7. Fitting and assembling metal parts.
8. Fabricating sheet metal.
9. Hot metalworking processes such as soldering, brazing, welding, forging, heat treatment of steel, molding, and casting.

10. Tool sharpening.
11. Finishing and inspecting, including precision measurement.
12. Machine tools such as the lathe, shaper, milling machine, and grinding machine.

1-7. Getting the Most from Metalwork

The following hints will help you to get the most out of your metalworking course:

1. Learn the rules and procedures established for working in your metalworking laboratory as soon as possible, and follow them faithfully.
2. Do the assigned reading and other work when it is assigned by your instructor. By doing so, you will be better prepared to profit from the classroom lessons and demonstrations.
3. One of the distinguishing characteristics of a good craftsman is the knowledge of the principal parts of the machines which are operated. Learn the correct name for each tool and machine, then learn the principal parts of each.
4. Learn the properties of the common metals. You will be using some of these metals in the construction of products or exercises in the laboratory. Knowledge of metal properties will enable you to select the best material for each part of each product you produce.
5. Make an effort to learn as much as possible about different kinds of metalworking industries. Study the different career opportunities available in metalworking industries. Visit metalworking plants whenever the opportunity occurs.
6. Your textbook includes basic information concerning many kinds of metals, tools, machines, and metalworking processes. For further information, or for more advanced information, you will want to study other metalworking books.

Words to Know

alloy
aluminum
career
corrosion
density
ferrous metal
metallic element
nonferrous metal
properties of metal
tensile strength
tool steel

Review Questions

1. List ten important items made from metals.
2. Name five industries which employ metalworkers.
3. Explain what the word **properties** means, as applied to metals.
4. What is meant by the **density** of metals?
5. Explain the meaning of **corrosion resistance.**
6. Explain the importance of hardness in metals.
7. List three metal products where toughness is more important than hardness.
8. Explain the meaning of the term **tensile strength** as applied to metals.
9. List the approximate tensile strength for the following:
   a. Pure aluminum
   b. Low-carbon steel
10. List five metals which are commercially available as pure metals.
11. What is an alloy?
12. What is the principal metal used in producing ferrous metals?
13. List several nonferrous metals.
14. What is meant by the **base metal** in an alloy?
15. List several occupations in the metalworking industries in which the worker must know about various metals and their properties.

16. How many different kinds of metals and alloys are available for the design and construction of metal products?
17. Why were so many different kinds of metals developed?
18. List several common metal products made of aluminum.
19. What kind of steels are used in making tools which cut metals?
20. List several ways in which you can profit from the study of metals.
21. List several occupations, other than metalwork occupations, in which the study of metalwork is important.
22. List several different areas of metalworking processes which are included in the study of the broad area of **metalwork technology.**

Mathematics

1. Calculate the tensile strength of a C1018 cold-drawn steel rod which has a cross-sectional area of $3/8$ths of a square inch.
2. A skilled worker loses 20 days' work due to a serious industrial accident. If his usual wage were $4.50 per hour, 8 hours per day, how much money has he lost in wages?

Social Science

1. Write a story in which you describe the kind of society we would probably be living in if metals had not been developed for useful purposes in making machinery and consumer goods.
2. Write a story in which you describe the effect that development of metalworking industries had on the location of major centers of population.

Career Information

1. If you have ever visited a manufacturing plant which produces metal products, write a story about it. Explain how the product was made and the kinds of work you saw being done.
2. Select some common metal product with which you are familiar. Identify the kinds of metals and processes used in its construction and list the metalworking occupations which were necessary in order to produce it.

UNIT 2

Careers in Metalworking

2-1. Factors to Consider in Selecting a Career

It is entirely possible that you will live and work for 30 to 50 years after completing your formal education. Your living, home, pleasures, happiness, and success will depend to a large extent on the amount of income your job provides. Before making a career choice, it is important to know everything you can about the occupation which you plan to make a career. For example, if you were interested in becoming a machinist, you would want to get answers to the following questions:

1. Do you know what a machinist does?
2. How much does a machinist earn?
3. How many hours a day are usually worked?
4. Is year-round work available?
5. How long does it take to learn to be a machinist?
6. Can you learn all of it in school?
7. Is the machinist's work dangerous?
8. What is the difference between a machinist, toolmaker, diemaker, diesinker, millwright, etc.?
9. What are your chances of getting a job after you have learned the trade?
10. Will you still be able to work at that job when you are 40 years old; will you still want to do it then?
11. What are the chances of being promoted to supervisory jobs?
12. What is the possibility of starting your own machine shop?
13. What other occupational opportunities, such as teaching, are open to the machinist?

These are questions for you to answer. Ask your friends, parents, teachers, and relatives about them. Read about them. Talk to a machinist if you can.

It is important that you know all about the job that you select and that you check on whether or not you will be able to do the work. This unit should help you to get this information and thus give you happiness and success in your work.

2-2. Meaning of Skill and Knowledge

Skill is the training of your hands to do certain things. **Knowledge** comes from the word **know.** To learn a trade or to become a technician (see § 2-10) you must acquire both **skills** and **knowledge**. A worker in a skilled trade must be able to apply **knowledge** effectively to the tasks which he performs. To learn a trade, to become a **technician**, or to become an **engineer** (see § 2-8), you must **know** something about mathematics, drawing, science, and other subjects. You must know **why** things are done in a certain way. You also need to know much of the information in this book for the purpose of communicating with and understanding the people that work in various metalworking industries.

2-3. Hours and Wages

Employees in metalworking jobs usually work seven to eight hours each day. It is difficult to state here how much workers get paid, because it changes from time to time and is different for various jobs and in the various sections of the country. Remember that the most highly skilled workers, or those with the most education and training, generally get the highest pay. Extra pay is usually received for overtime. A highly skilled worker in any occupation or trade is seldom without work, but it may require constant study to keep up with the changes in machines and methods.

2-4. Classification of Occupations

An **occupation** is the kind of job or work at which one is employed in order to earn a living. An **occupation** has the same meaning as a **job** or **vocation**. There are many kinds of occupations or jobs which can become rewarding careers. There are actually more than 20,000 different occupational titles, or job titles, described in the **Dictionary of Occupational Titles.**[1] Hundreds of the occupational titles described are metalworking occupations, trades, or jobs.

The amount of education and training required for different metalworking occupations may vary considerably. The training period may range from a period of several days to five years or longer. For example, a drill press operator may be trained to perform simple drilling operations with a small drill press during a period of several days in a school shop or in an industrial plant. On the other hand, a period of four years is generally required for training an all-around machinist. See § 2-11.

General Classifications

Metalworking occupations and the workers employed in these occupations are often broadly classified under categories which are based upon the knowledge, skill, and length of training needed to perform the required work. The following are several common broad occupational classifications used:

1. Unskilled workers
2. Semiskilled workers
3. Skilled workers
4. Technicians
5. Technologists
6. Engineers

These very broad and general occupational classifications will be explained in the following sections.

2-5. Unskilled Workers

This classification includes workers who require little or no special training for the tasks they perform. Examples of jobs in this classification include laborers who handle and move materials manually. Other examples include floor sweepers, dish washers, and domestic workers who perform tasks requiring little thought or little application of knowledge. The percentage of unskilled workers in the total labor force is decreasing and will probably continue to decrease in the years ahead.

2-6. Semiskilled Workers

This classification includes workers in occupations or jobs requiring some special training for the tasks they perform. A broad range of occupations or jobs is included in this classification. Thousands of different kinds of jobs may be classified as semiskilled jobs. The training period for semiskilled jobs, in most instances, may range from several days to about one year. The training may be provided in a school shop, or it may be provided on the job by the employer. In some instances, however, the training period may require as long as two years. Examples of metalworking jobs in the semiskilled classification include assembly-line workers in factories, machine tool operators, inspectors, maintenance mechanics,

[1] U. S. Department of Labor, Bureau of Employment Security, **Dictionary of Occupational Titles,** Vol. 1, **Definition of Titles** (Washington, D.C.: U.S. Government Printing Office, 1965), 809 pages.

painters, spot welders, and punch press operators.

Machine tool operators are generally included in the semiskilled classification. This group includes drill press operators, lathe operators, milling machine operators, planer and shaper operators, and operators of nearly every kind of specialized production machine tool. Machine tool operators are generally employed to operate one kind of machine tool. As the operators become skilled, they can perform all of the operations which can be performed on the machine.

Because of the variation in skill and training required, semiskilled workers, such as machine tool operators, are further classified for job promotion and pay purposes. It is common practice to classify machine tool operators as **Class A**, **Class B**, or **Class C** operators. The Class A operator generally possesses more knowledge, skill, and experience than the Class B or Class C operator.

Hourly pay rates vary considerably for machine tool operators in various parts of the United States — from the Federal minimum wage to nearly the rate for toolmakers. During 1973, a survey indicated that a large group of Class A machine tool operators averaged 30-40¢ an hour more than Class B operators and 60-80¢ an hour more than Class C operators.

Highly competent machine tool operators are able to make all necessary machine setups on the machines they operate. They are able to make all calculations and adjustments, including the determination of cutting speeds and feeds. Frequently they must be able to read blueprints and use precision-measuring tools and machine parts accurately.

Experience in high school or vocational school metalwork or machine shop classes is valuable for securing employment and advancing more rapidly as a machine tool operator. This kind of experience is also valuable in securing employment in many kinds of semiskilled jobs in the metalworking industry.

2-7. Skilled Workers

Workers under this classification include those employed in the skilled trades. A **trade** is a job or work which generally requires from two to five years to learn. It requires both **knowledge** and **skill**. One generally learns a skilled trade through a combination of shop instruction, classroom instruction, and on-the-job training. The classroom instruction generally includes mathematics, blueprint reading, technical theory, science, and any other necessary instruction required in the trade.

Examples of skilled trades included under metalworking occupations include the following: machinist, layout artist, tool-and-die maker, instrument maker, boilermaker, welder, sheet metalworker, molder, and heat treater. Descriptions of these trades, the nature of the work involved, and educational training requirements are explained later in section 2-11.

A **tradesman** (a level of skill identified by trade unions), also called a **craftsman**, must be able to perform all of the jobs or tasks which are common in the trade or craft. For example, a **machinist** is a skilled worker who must be able to set up and operate all of the machine tools used in the trade. The machinist must know shop mathematics, how to read blueprints, and how to use precision-measuring tools.

Apprenticeship Method

This is one highly recommended method for learning a skilled trade. An **apprentice** (a level of skill identified by trade unions) is one who is employed to learn a trade in a systematic order under a master of the trade or under the direction of a company. The length of the apprenticeship training period may vary for different trades, anywhere from two to six years. The apprenticeship period for becoming a machinist or tool-and-die maker, for example, is usually four or five years. Upon completion of an apprenticeship, one becomes a **journeyman** (a level of skill identified by trade unions), a worker who has met minimum qualifications for entrance into the trade.

When an apprentice completes the term of training, a written document is generally received which shows a satisfactory completion of the apprenticeship training program. The document specifies the trade or occupation in which the worker is qualified. This document is recognized by many employers and labor unions throughout the country as qualification for entrance into the trade. The new journeymen in a trade must continue studying the tools, processes, and procedures in the trade if they wish to become highly skilled and to advance more rapidly in the trade.

To qualify for apprenticeship training in a skilled metalworking trade, one must generally be a high school graduate or must have equivalent trade or vocational school education. One must have better-than-average mechanical ability. High school or vocational school graduates with a good background in science, mathematics, English, drafting, and metalwork or machine shop courses are frequently sought after as apprentices for skilled metalworking trades or occupations.

An apprentice earns while learning a trade. The wage scale is graduated so that earnings increase periodically with an increase in experience.

Pickup Method

This is a second way in which many workers acquire the broad knowledge and experience required for employment as a skilled tradesman. This method involves working on different kinds of semiskilled jobs within one occupational area until one acquires sufficient broad knowledge and experience to gain employment as a skilled worker in that occupational area.

For example, many workers have learned the machinist's trade by the pickup method. They acquired the broad knowledge and skill required of the machinist by working at different semiskilled jobs, on many different machine tools, until they could set up and operate all of the common machine tools.

Workers who choose to learn the machinist trade or any other skilled trade by the pickup method will find that it generally takes longer than serving an apprenticeship. Frequently they must attend vocational or technical schools to learn blueprint reading, shop mathematics, and technical theory relating to the trade which they are learning.

It is becoming increasingly difficult to learn and enter skilled metalworking occupations by the pickup method. The apprenticeship method is generally recommended and recognized as a more efficient method for learning and entering a skilled trade.

2-8. Engineers

Engineers plan, design, and direct the building of roads, bridges, tunnels, factories, office buildings, waterworks, dams, mines, automobiles, aircraft and spacecraft, ships, railroads, power plants, electrical appliances, electronic equipment, radio and television stations, machinery, and engines. Engineers must have college degrees, must like mathematics and drawing, and know a lot about physics and chemistry.

Engineers frequently specialize in some particular phase of engineering such as mechanical, electrical, or chemical engineering. Many kinds of engineers design metal products, develop and supervise the manufacturing procedures and processes for metal products, or utilize metals in the products they design. The following are some phases of engineering in which engineers work directly or indirectly with metals in their work; therefore, they must know and understand the properties of metals and metalworking processes:

1. **Architectural engineers** design all types of buildings ranging from small homes constructed largely of wood, to factories and large buildings constructed largely of structural metals and masonry.
2. **Aeronautical and aerospace engineers** develop new designs for aircraft, missiles, and space vehicles.

3. **Civil engineers** design highways, bridges, dams, waterways, and sanitary systems.
4. **Electrical and electronics engineers** design and develop electrical machinery, electrical switches, controls for machines and appliances, radios, television, automatic controls for industrial machinery, electric power generators, electric distribution systems, and many other electronically controlled products.
5. **Marine engineers** develop new designs for commercial and military ships, submarines, and other types of marine equipment.
6. **Mechanical engineers** design many different kinds of machines, appliances, and mechanical equipment, including both industrial and consumer products.
7. **Metallurgical engineers** develop processes and methods of extracting metals from their ores, refining them, and preparing them for practical use. They develop new alloys with improved properties, as required for many kinds of metal products.
8. **Tool and manufacturing or industrial engineers** generally start with the model of an industrial or consumer product created by a product engineer. They plan and organize workers, materials, and machines for the complete and economical mass production of the product. This involves analyzing and planning the industrial processes involved. It involves designing the special manufacturing machines, equipment, assembly line, and any packaging system required. It involves supervision of the construction and installation of the equipment needed for the entire production of the product. And it involves the supervision and control of production through all phases, from the raw material to the finished and packaged product.

Other Kinds of Engineers

Another kind of engineer is one who runs a railroad locomotive, an engine on a steamship, or one who runs the machinery for heating, ventilating, and supplying power for a factory or large building. Some engineers work indoors while others work outdoors. These kinds of engineers should have at least a high school education or better, depending upon the work. It takes from two to five years to learn the work. They must pass an examination before being qualified for the job.

2-9. Technologists

The occupational classification of **technologist** is a new category of professional worker, ranking between the engineer and the technician. Graduates of four-year college programs, technologists receive academic preparation in mathematics, physics, chemistry, engineering, graphics, computer programming, business organization, and management. Many of the technologist degree programs require one or more periods of on-the-job work experiences which are called **internships**. The internships provide valuable professional experiences impossible to obtain in the classroom.

Engineering technologists are prepared to serve in positions of engineering support. With the work of the engineer becoming more and more theoretical and scientific in character, technologists are taking over much of the practical, or applied work. The technologist may be called upon to do the on-site surveys, mathematical computations, design studies, laboratory experiments, and other tasks once performed routinely by engineers. With technical and managerial abilities beyond those of the two-year technician, the engineering technologist may also supervise a staff of technicians.

Industrial technologists are employed either in technical positions or, more typically, in positions of middle management. Some types of positions open to industrial technologists include department head, per-

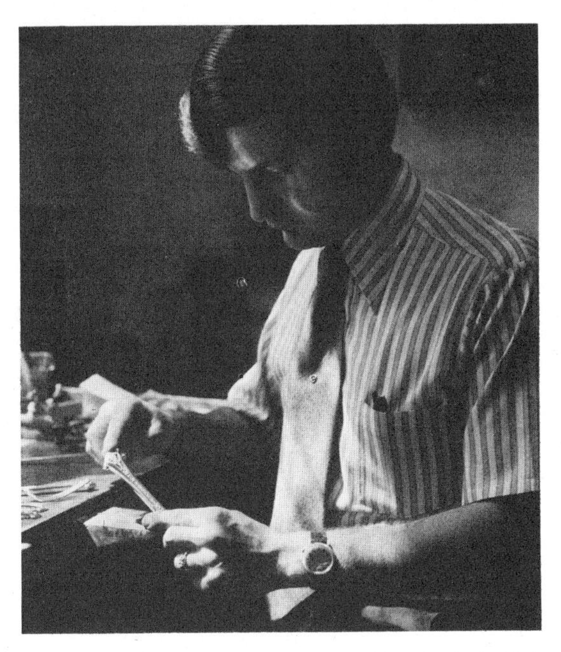

Fig. 2-1. This Silver Designer Utilizes Artistic Talents and Technical Knowledge

sonnel manager, training director, technical writer, labor relations director, production manager, plant and product designer, and as a specialist in quality control, facilities planning, cost estimating, and time studies. See Fig. 2-1. They may also work in the area of sales as a manufacturer's representative, sales engineer, product analyst, sales manager or analyst, and in various capacities in advertising, purchasing, and sales and service.

2-10. Technicians

Technician occupations are among the fastest growing occupational groups in the United States. Technicians include workers whose jobs generally require the application of scientific and mathematical theory. Their work generally involves helping to translate scientific ideas into useful products or services. They frequently work directly under scientists, engineers, or industrial managers. It has been estimated that industry needs

Fig. 2-2. Technician Checking Final Program of Numerically Controlled Machine Tool

several technicians for every professional engineer.

In general, the educational and training requirements for technicians include high school graduation and two years of post-secondary school training. This kind of technical training is available in various types of schools, including technical institutes, junior colleges, community colleges, area vocational or technical schools, armed forces schools, technical-vocational high schools, private technical schools, and extension divisions of colleges and universities.

The term **technician** does not have a generally accepted definition. It is used by different employers to include many kinds of workers, in a variety of jobs, with various backgrounds of education and training. In some instances it is used to designate workers performing routine technical work confined within a limited sphere. More frequently, however, the term is used to designate employees who perform tasks requiring greater breadth of technical knowledge and experience, such as assisting engineers and scientists. See Fig. 2-2.

Generally, technician occupations in modern industrial establishments require technical education and training which rank between that required of the skilled tradesman and the engineer. Technicians usually must possess more theoretical knowledge of drafting, mathematics, science, and technical writing than a skilled craftsman or tradesman. However, they are not expected to know as much about these subjects as the engineer.

Often, technicians train in only one area of technology, such as the following common areas of specialization: mechanical technology, tool technology, industrial or manufacturing technology, aeronautical technology, automotive technology, electrical technology, chemical technology, civil engineering technology, metallurgical technology, instrumentation technology, and safety technology. Technicians of one specialty may require greater familiarity with one or more of the skilled trades than technicians of other specialties. However, they are not required to perform as skilled craftsmen. Some technicians are expected to know about many kinds of industrial machines, tools, and processes. This is particularly true in the case of **industrial** or **manufacturing technicians** who are employed to supervise skilled and semiskilled workers in modern metals manufacturing establishments.

Aeronautical Technicians

Technicians in this area work with engineers and scientists by assisting with problems involving the design, production, and testing of aircraft, rockets, helicopters, missiles, and spacecraft. They aid engineers by preparing layouts of structures, by collecting information and making calculations, checking drawings, and performing many other duties.

Chemical Technicians

Chemical technicians work mainly with chemists or chemical engineers in the development, production, utilization, and sale of chemical products. The chemical field is so broad that technicians in this area specialize in the chemical problems involved in one industry, such as the food processing, electroplating, or paper industry.

Civil Engineering Technicians

Technicians in this area assist civil engineers with the many tasks involved in the planning and construction of highways, bridges, dams, viaducts, and other structures which civil engineers design. The technician helps in the planning stage of a structure by assisting with the surveying, drafting, and preparation of specifications for materials. When a structure or project is under construction, the technician works with the engineer and the contractor in scheduling construction activities. He also inspects the construction to determine that the workmanship and materials conform to blueprint specifications.

Electronics Technicians

This field includes work in the areas of radio, television, communications equipment, computers, automated machines, electric motors, and electronic controls on many types of electrical machines. The field also involves many kinds of electronic measuring and recording devices, such as missile and spacecraft guidance devices. In the broad field of electronics technology, a technician generally specializes in one phase of the work, such as television, electric motors, or electronic controls.

Heating, Air-Conditioning, and Refrigeration Technicians

Technicians in this field often specialize in one of the areas of work, such as air conditioning. They may further specialize in one activity such as the design of layouts for air conditioning. In the manufacture of air-conditioning equipment they may be assigned the problem of analyzing production procedures. They may also be assigned problems such as devising methods for testing air-conditioning equipment.

Industrial Technicians

Technicians in this area are often called **industrial technicians, manufacturing technicians,** or **production technicians.** Industrial technicians assist industrial engineers, manufacturing engineers, and tool engineers in a wide variety of problems in many different industries. Industrial technicians are particularly important in metals manufacturing industries. They assist engineers who generally start with a product which has already been designed and which must be mass produced.

Their work involves assisting engineers with problems which involve efficient use of employees, materials, and machines in the production of goods and services. These problems involve plant layout, development and installation of special production machinery, planning the flow of raw materials or parts, developing materials handling procedures, and controlling inventories. They are also concerned with problems involving time-and-motion studies, analysis of production procedures and costs, quality control of finished products, and packaging methods.

Industrial technicians often acquire experience which enables them to advance into specialized areas of work. These areas include industrial safety, industrial job supervision, and industrial personnel work which involves interviewing, testing, hiring, and training employees.

Instrumentation Technician

The **instrumentation technician** assists engineers in designing, developing, and making many different kinds of special measuring instruments and gages. Such instruments and measuring devices are used for automatic regulation and control of machinery; measurement of weight, time, temperature, and speed of moving parts; measurement of volume, mixtures, and flow; and the recording of data.

Mechanical Technician

This is a broad term which often includes such areas as tool design technology, machine design technology, production technology, automotive technology, and diesel technology.

Technicians employed in the above areas of technology assist engineers with problems involved in the design and development of machine tools, production machinery, automotive engines, diesel engines, and other kinds of machinery. Technicians assist engineers in making sketches and drawings of machine parts; estimating costs for materials and estimating production costs; solving design problems involving surface finish, stress, strain, and vibration; and developing and performing test procedures on machines or equipment.

When making performance tests on machines and equipment, technicians use many kinds of measuring instruments and gages. They also prepare written reports of test results, including graphs, charts, and other data concerning the performance and efficiency of the equipment. See Fig. 2-3.

The **tool designer** is a well-known specialist included under the area of mechanical technology. The **tool design technician** de-

Fig. 2-3. Checking Accuracy of Diamond Grinding Wheel

signs and draws tools, jigs, fixtures, and holding devices for use on many kinds of production machines and may also supervise others in making the tools which he or she designs.

Metallurgical Technician

The metallurgical technicians generally work with metallurgists, assisting them with various metallurgical problems. They assist the metallurgists in testing samples of metals for their chemical content, hardness, tensile strength, toughness, corrosion resistance, durability, and machinability. They work with the metallurgists on problems which involve the development of improved methods of extracting metals from their ores and also assist in the development of new metals which have properties that are different and not yet established.

Other Kinds of Technicians

In addition to specialization in the above areas of technology, technicians also specialize in other fields. **Mathematical technicians** work with and assist scientists, engineers, and mathematicians by performing many kinds of computations. **Agricultural technicians** work with agricultural scientists in the improvement of farm products, foods, and soils. Other kinds of technicians include medical and X-ray technicians, dental technicians, optical technicians, and petroleum technicians.

2-11. Descriptions of Occupations

The following descriptions of many skilled and semiskilled metalworking jobs or occupations are arranged in alphabetical order.

Aircraft-and-Engine Mechanic

An aircraft-and-engine mechanic repairs, inspects, and overhauls airplanes. He may do either emergency repairs or major repairs and frequent inspections. He should be able to measure with **micrometers**, use **hand tools**, and run such **machine tools** as the **drill press** and the **grinder**. The work is greasy, dirty, and some work has to be done outdoors.

Aircraft-and-engine mechanics should have a high school education and should attend a school or technical institute. They will learn drafting, electricity-electronics, physics, and other subjects which will be needed in the trade. If a trade school is not attended, these may be learned in night school. It takes a minimum of four years to learn the trade.

An **auto mechanic** may wish to become an aircraft-and-engine mechanic. Tests must be passed to get a license from the **Federal Aviation Agency**; without this license, the mechanic may be a **helper**.

The airplane mechanic must guard against gasoline fumes, explosions, and the danger of inhaling poisonous carbon monoxide gas.[2]

Assembler

An assembler puts together the finished parts of a machine, automobile, engine, typewriter, lock, watch, etc. Such work may be very simple or it may call for much skill.

Auto Mechanic

An auto mechanic services and repairs mechanical, electrical, and body parts of not only cars but buses, trucks, and various gasoline-powered vehicles, too. The mechanic should be able to measure with **micrometers,** use many **hand tools,** and run the **drill press** and the **grinder.** The work is greasy and dirty, and some work may be done outdoors.

An auto mechanic should have at least a high school education. The trade will be learned and progress made quicker if he or she is a technical school or trade school graduate and is good at mathematics and drafting (see § 4-2.) If a technical school or

[2]**Carbon monoxide** is a gas made of one part **carbon** and one part **oxygen**. It is the same gas which comes from the exhaust pipe of an automobile. Only 2% carbon monoxide in the air of a room will make a person unconscious in half an hour.

trade school is not attended, these may be learned in night school. It takes three to four years, and sometimes even longer, to learn the trade. A person who has done metalwork such as can be learned from this book may learn the trade more quickly and would be a much better auto mechanic if some of the training of a **machinist** were acquired.

The auto mechanic must guard against gasoline fumes and explosions and the danger of inhaling the poisonous **carbon monoxide gas.**

Bench Mechanic

A bench mechanic works at a bench with hand tools and must read **blueprints** (see § 4-3.) The mechanic's primary function is the repair of parts that have been disassembled from a machine or vehicle such as carburetors, transmission, and engines.

Boilermaker

The boilermaker assembles prefabricated parts of boilers, tanks, and machines out of iron and **steel plates** and can also repair them. The boilermaker must drill and punch holes, use machines for cutting and bending the plates, drive hot **rivets**, and read **blueprints**. Boilermakers must be strong because the work is heavy and hot. They work at the site where the boiler, tank, or machine is to be assembled and must be skilled in using tools and equipment for installation and repair. There is the danger of being burned by hot boilers and rivets. A young person should be 18 years old when beginning to learn this trade. It takes about four years as a helper to become a boilermaker.

Coremaker

A coremaker makes **cores** used to form holes or hollow parts in **castings** which make them lighter so that less metal will have to be cut away afterwards. The coremaker should learn mathematics and also learn how to read **blueprints**. A young person may learn the trade in a **foundry** during an apprenticeship of about four years.

The work is usually steady. The coremaker's job is closely related to the **molder's** trade but is lighter work than molding.

Diemaker

The diemaker makes metal forms or patterns, called **dies**, which are used in **punch presses**[3] to stamp out forms in metal. Automobile fenders are made with such dies. These dies must be exact or the many hundreds or thousands of pieces made with them will be wrong. Diemakers can set up and run any machine and use any tool in the shop. They must read **blueprints**, make **sketches,** use **layout tools,** and measure with **micrometers**. See Fig. 2-4.

The diemakers use most of the information given in this book. They must have good eyesight to make fine measurements. Diemakers should have at least a high school education but will learn the trade and progress faster if they are technical school or

Fig. 2-4. Diemaker Honing the Detail of a Major Die

[3]A **punch press** is a powerful machine which punches holes of different shapes in metal as easily as you punch a ticket. It may also be used to press a flat sheet of metal into a certain shape or form, such as an automobile body or fender.

trade school graduates and are good at mathematics and drafting. A prospective diemaker can learn mathematics and drafting in night school if a technical school or trade school is not attended. Technical training may bring a promotion to the rank of supervisor or superintendent. It takes at least four to five years to learn the trade as an **apprentice**.

A diemaker is seldom without work and is one of the last persons to be laid off. The diemaker's, **diesinker's**, and **toolmaker's** trades are closely related.

Diesinker

The diesinker makes the metal dies which are used in **drop hammers** to hammer hot steel into the form. The object thus made is called a **drop forging** and the dies are called **drop forging dies**.

A **drop hammer** is a large, powerful machine which has a heavy weight that acts as a hammer. The weight has a **die** fastened to its bottom side. The hot, soft metal is laid upon another die which is fastened on the base, or anvil. The weight then drops on the hot metal and hammers it into both dies. This forms the metal into the shape of the dies. A drop hammer which is run by steam is a **steam hammer**.

Fig. 2-5. Bank of Drill Presses for Production Work

Besides tools, many automobile and machine parts which must have great strength are hammered out this way. Automobile axles and many wrenches are drop forgings. The dies must again be exact or the many hundreds or thousands of pieces made with them will be wrong.

The diesinker must have good eyesight, read **blueprints,** make **sketches**, use **layout tools**, and measure with **micrometers**. A diesinker uses most of the information given in this book to set up and run any machine and use any tool in the shop. A diesinker should have the same educational preparation as the **diemaker**.

A diesinker is seldom without work and is one of the last to be laid off. The diesinker's, **diemaker's**, and **toolmaker's** trades are closely related.

Drill Press Operator

A drill press operator earns a living by running a **drill press**. Units 50-53 give complete information about drill presses and how to run them. A drill press operator can do many different operations on the drill press, set up the work, and sometimes sharpen **drills.** A person can learn to run a drill press in a week or, at most, in a few months. See also **machine operator**. See Fig. 2-5.

Electrician

This is often called the **Age of Electricity and Electronics**. Electrical devices are in the home, office, factory, store, on ships, automobiles, farms, streets, and elsewhere. Electricity furnishes light, heat, and power; it carries messages and gives us entertainment; it controls machines, does difficult mathematical jobs, and keeps records. Electrical things are installed and repaired by the electrician, who uses much of the information given in this book.

An electrician should be at least a high school or trade school graduate before the trade may be learned as an **apprentice**, which takes four years. An **electrician** must know mathematics and **physics**, how to read **blueprints**, know something about metals, **drills** and **drilling, assembly tools, pipe, pipe-fitting tools, tubing,** and **soldering**. An electrician must know the **National Electrical Code** which states the correct way to install wires, switches, etc. and must guard against electrical shock.

Some electricians work indoors while others work outdoors. The electrician who works outdoors putting up poles, wires, and cables and maintaining them, is called a **line operator**. The electrical field is large and includes many kinds of work. A good electrician is one who may feel lost in another type of electrical work. For example, an electrician doing **house wiring** may not be able to work on motors in a factory without additional training. The electrician's and **radio service worker's** trades are closely related.

Electroplater

The electroplater covers metal articles with a protective, attractive surface coat of chromium, nickel, silver, gold, or another metal by using **electricity**. The worker must know something about electricity and **chemistry**, and, if possible, should be a high school or trade school graduate. Some of the work is damp, there is danger of getting **acid burns**, and the worker must guard against poisonous fumes. See Fig. 2-6.

Forge Operator

A **forge operator**, also called a **hammer operator**, runs a **drop hammer** on which automobile axles, wrenches, etc. are forged. The difference between **hand forging** and **drop forging** is explained in section 37-1. The forge operator must have a knowledge of **iron** and **steel** and must be able to read **blueprints**. One should have at least a high school education and must be strong and healthy. The work is hot, heavy, dirty, noisy, and the danger of being burned by hot metal is ever present.

Founder

See **molder**.

Gagemaker or Instrument Maker

The gagemaker, sometimes called an **instrument maker,** makes and repairs all kinds of **gages** and is a type of **toolmaker**. To do this one must be able to set up and run any machine in the shop, must also make accurate measurements and therefore, must have good eyesight. The gagemaker must read **bueprints**, make **sketches**, **lay out** the work, and use most of the information given in this book.

The worker will learn the trade much quicker and progress faster as a technical school or trade school graduate with good mathematics and drafting skills. If a technical school or trade school is not attended, these skills may be learned in night school. Technical school or trade school training will help one get promoted to a supervisory post. It takes at least four years to learn the trade.

A gagemaker is seldom without work and is one of the last to be laid off. However, the danger of being cut by sharp tools is ever present.

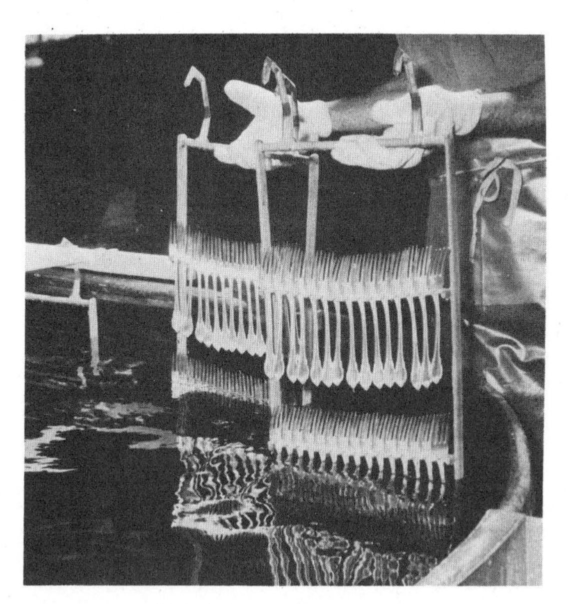

Fig. 2-6. Silverplating

Fig. 2-7. Precision Grinding

Gas Fitter

A gas fitter measures, cuts, fits, and connects **gas pipes** and **pipe fittings** and installs gas stoves, gas heaters, gas meters, and other gas equipment. The gas fitter's trade is closely related to the **plumber's** trade.

Grinding Machine Operator

A grinding machine operator operates precision-grinding machines such as the surface grinder, cylindrical grinder, tool and cutter grinder, and centerless grinder. They machine parts to very close tolerances, which requires great care and skill. Grinding machine operators are usually trained machinists who specialize in the operation of grinding machines. The work is often damp and dirty. See Fig. 2-7.

Gunsmith

A gunsmith makes and repairs firearms.

Heat Treater

A heat treater is one who performs heat-treatment operations on steel and other metals. He must know how to **harden, temper, case harden, anneal,** and **normalize** metal. This work is usually learned by working in the heat-treating department of a factory. A heat treater should have at least a high school education. The work is hot and there is danger of being burned by hot metal and hot liquids. The heat treater must sometimes guard against poisonous fumes.

Inspector

An inspector checks or examines materials, parts, or articles while they are being made or immediately after they are finished. Inspectors must read **blueprints**, know the different kinds of **fits**, and use all kinds of measuring tools. An inspector should have at least a high school or trade school education.

Jeweler

The jeweler makes high-grade jewelry of platinum, gold, and silver. The quality of this work requires good eyesight, even though most of it is done while looking through an eye **loupe** or magnifying glass. The jeweler must know how to run a **jeweler's lathe** and other small hand and machine tools. If possible, the jeweler should have a high school or trade school education and learn the trade as an **apprentice**.

Journeyman

A journeyman is one who has learned a **skilled** trade.

Lathe Operator

A lathe operator uses the **lathe** described in Unit 56. The operator can do different operations on it, set up the work, and sharpen the needed tools. A young person can learn to run a lathe in three months to a year and should have a high school or trade school education. Lathes are more abundant in this country than any other metalworking machine tool. See also **machine operator**. See Fig. 2-8.

Fig. 2-8. Operating a Turret Lathe

Fig. 2-9. Machine Operator Milling a Face on a Crankshaft

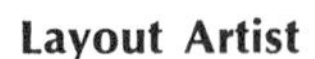

Layout Artist

A layout artist reads the dimensions given on the **blueprint** and then with fine measuring and marking tools draws lines and marks on the metal surface to show where to cut or form the metal. Knowledge of mathematics, how things are made in the shop, and the properties of various metals is essential.

The layout artist is a **diemaker, diesinker, machinist, toolmaker,** or **sheet metalworker** who is chosen to do all the layout work. More information about **layout work** is given in sections 6-1, 7-1, 31-4, and 53-7.

Machine Operator

A machine operator earns a living by adjusting and running only one machine, therefore becoming **specialized** in the operation of that machine. If the machine operators run a **drill press,** they are **drill press operators**; if they run a **lathe,** they are **lathe operators**. Thus, there are also **shaper operators, planer operators, milling machine operators**, and **grinding machine operators.**

A young person can learn to run any one of these machines in three to six months. It may take a year to specialize on some of them. A machine operator should have a high school education and should be able to read **blueprints** and **micrometers**. There is the danger of being cut by sharp tools. Some of the work is greasy and dirty. See also **machinist**. See Fig. 2-9.

Machine Setup Workers

The machine setup worker specializes in getting machine tools ready for operation, instructs **machine operators** in their use, keeps an eye on the machine run by the machine operators, and keeps them adjusted. The machine setup worker may be a fully qualified machinist or may have learned how to set these machines by observing them while employed as a machine operator. An ability to read blueprints and to use all kinds of measuring tools and gages is required.

Machinist

Machinists repair and construct machine tools. They do **fitting**, set up and run any machine, are able to use any tool in the shop, must read **blueprints**, make **sketches**, use **layout tools**, and measure with **micrometers**. A machinist uses most of the information given in this book.

Good eyesight is essential, because the machinist must make fine measurements, have superior judgment of depth and distance, and also have good coordination. A high school education is preferred. The trade will be learned quicker and progress made faster if the machinist is a technical school or trade school graduate and is good at mathematics and drafting. If a high school or trade school is not attended, these skills may be learned in night school. Technical school or trade school training will help the machinist to be promoted to supervisor and superintendent. It takes four years to learn the trade as an **apprentice**.

A machinist is seldom without work. There is the danger of being cut by sharp tools, and some of the work is greasy and dirty. See also **machine operator**. See Fig. 2-10.

Mechanic or Machine Repairman

A mechanic is skilled in working with machines and in shaping and uniting materials by using tools and instruments. **Preventive maintenance** is a major part of the job. Mechanics inspect the equipment, oil and grease the machines, and clean and repair any parts, thus **preventing** any breakdowns or work delay.

If major breakdowns do occur, the machine must be completely disassembled in order to make the necessary repairs.

Fig. 2-10. A Basic Tool for the Machinist Is the Lathe

Metallurgist

The metallurgist has studied the art and science of separating metals from the rocks and earth in which they are found and of preparing them for use. The metallurgist knows how to mix different metals and make new metals (the meaning of **alloys** is explained in section 19-1). Progress is aided with every new metal discovered. The metallurgist must study chemistry and have a college education.

Metal Patternmaker

The metal patternmaker makes the **metal patterns** which are used to make molds in the **foundry**. These patterns are prepared from metal stock, rough castings of the original work pattern, wax, or ceramics. The patternmaker must read **blueprints**, make **sketches**, and use the same hand tools and machines that are in a **machine shop**.[4] Precision and accuracy are the keynotes of the job.

A patternmaker should complete high school or trade school. It takes about five years to learn the trade as an **apprentice**. The work is interesting and usually steady, although there is a danger of being cut by sharp tools.

Metal Spinner

A metal spinner forms bowls, cups, trays, saucers, vases, pitchers, etc. by pressing flat pieces of soft sheet metal over forms

[4]A **machine shop** is a place where pieces of metal are formed, cut, polished, and finished by machines into tools, parts of machines, etc.

which turn in the **lathe. Metal spinning** is explained in Unit 33. The metal spinner must have skill, read drawings, and should have a high school education. Wages are usually high. There is danger of being cut by sharp tools.

Milling Machine Operator

A milling machine operator can do many different operations and set up the work on a **milling machine.** A young person can learn to run a milling machine in three months to a year and should have a high school education. See also **machine operator**. See Fig. 2-11.

Millwright

A millwright moves and installs heavy machines and equipment in shops (see Fig. 26-15), and constructs any special foundation for them. The millwright must read **blueprints**, and lubricate, dismantle, or repair the machinery installed. A young person with at least a high school education may learn the trade as an **apprentice**. The millwright's and **erector's** trades are closely related.

Molder

The molder makes **molds** and **castings** by packing sand around the pattern of a metal part to be copied. Health, strength, and skill are a necessity. The molder needs a high school education and should learn mathematics and know how to read **blueprints**. It takes about four years to learn the trade as an **apprentice.**

Molders get higher pay than workers in most other trades. The work is hot, dusty, dirty, and much of it is heavy lifting and working with damp sand. There is the danger of being burned by hot metal and sparks, but it is usually steady work. The molder's and **coremaker's** trades are closely related.

Parts Programmer

A parts programmer analyzes and schedules on a program sheet the operations involved in machining many kinds of metal parts which are to be machined on **numerically controlled** machine tools. Numerically controlled machine tools process parts automatically, with little effort or control on the part of the machine tool operator. The operator loads parts into the machine, replaces worn tools, and removes parts from the machine.

The parts programmer reads and interprets the blueprint or drawing of a part to be machined and analyzes and lists on a program sheet, in proper sequence, the machining operations involved in machining the part. The operations are listed in coded form on the program sheet. See Fig. 2-12.

Fig. 2-11. Three-Spindle Bridge Miller Trimming Helicopter Rotor Hub

Fig. 2-12. Programming a Tape-Controlled Automatic Machine Tool Is a Complex Task

The programmer must indicate the kinds of operations to be performed, the tools to be selected and used, the correct cutting speeds and feeds, and when cutting fluids are to be used. Hence, a parts programmer must have a good knowledge of mathematics, blueprint reading, and machine shop operations. With large, complex, multipurpose machine tools, knowledge of mathematical computers and how to use them is necessary.

Pipe Fitter

A pipe fitter measures, cuts, fits, and connects pipes and **pipe fittings** (see § 27-5) for gas, air, oil, or water. The pipe fitter should know something about **steam fitting** and read blueprints. The pipe fitter's, **plumber's,** and **steam fitter's** trades are closely related.

Planer Operator

A planer operator can set up work on the planer and make precision parts. A planer operator can be trained in six to eight months, but most are trained machinists specializing in planer operation. See also machine operator.

Plumber

The plumber installs and repairs sewer and drain pipes, water pipes, gas pipes, meters, sinks, bathtubs, showers, faucets, tanks, etc. The plumber helps to keep our homes, schools, and other buildings clean and free from sickness by keeping the sewers, water systems, and heating in order.

Plumbers should have a high school or trade school education, must learn mathematics, read **blueprints**, and know something about **steam fitting** and building in general. They go from job to job; the work is often dirty and disagreeable, but quite steady. It takes five years to learn the trade as an **apprentice**. State examinations must be passed before a license can be obtained. The plumber's, **pipe fitter's**, and **steam fitter's** trades are closely related.

Polisher and Buffer

A polisher and buffer uses abrasive belts, discs, or wheels to smooth metal, and buffing wheels for polishing. The work is usually learned on the job. It is dusty and dirty work.

Riveter

A riveter spreads the small end of a hot **rivet** into the form of a head by hammering. This is done with a hammer that works with air forced through a hose and a trigger to start and stop. It is called an **air hammer**. The riveter works in a shipyard, boiler shop, or on the steel frame of a railroad car, bridge, big building, or skyscraper. It is heavy work. The riveter who works on ships, bridges, and skyscrapers must work outdoors, winter and summer, and must travel from job to job. There is the danger of falling from high places. It is necessary to be careful and strong. The riveter's work is closely related to the **boilermaker's** and **structural ironworker's** trades.

Machine, Tool, or Materials Salesperson

A salesperson of machines must know how to demonstrate the use of various machines, must be able to answer questions about the goods being sold, and show others how to use them. It is also necessary to know how they are made.

A salesperson of machines or tools has often worked at a trade such as **diemaker, diesinker, machinist**, or **toolmaker** and often is a college graduate. Sales personnel must also know something about advertising, how to show or display goods, and how to write contracts for the things they sell. They must have a knowledge of the goods made by their company's competitors in order to compare these goods with those they are trying to sell. There are courses in **salesmanship** which can be studied. The salesperson must be neat and pleasant, must know what people need and want, must speak good English, and must know when to talk and when to be silent. Some sales personnel travel over a large territory and are away from home for long periods.

Sheet Metalworker

A sheet metalworker makes and repairs such things as furnace pipes, furnaces, ventilators, signs, eave troughs, metal roofs, metal furniture and lockers, automobile and airplane bodies, etc., which are made out of **sheet metal**. Work may be done in a factory, on buildings, or on ships.

Sheet metalworkers must know how to **rivet, solder**, and read **blueprints**. They should have a high school or trade school education, know **geometry**,[5] drafting, and how to make **patterns**. This trade takes at least three to four years to learn.

Spring and fall are the busy seasons for outdoor work. Factory work is more steady. The work is noisy at times, and there is the danger of falling from high places and of being cut by sharp edges of metal. The sheet metalworker must also guard against **lead poisoning** caused by solder made of **lead** and **tin**. The sheet metalworker's and **tinsmith's** trades are closely related. See Fig. 2-13.

Fig. 2-13. Many Jobs Require Working on Scaffolds and at Considerable Heights

Steam Fitter

A steam fitter measures, cuts, fits, and connects pipes and **pipe fittings** for steam and hot water heating systems. One must read **blueprints** and know how to **weld**. The work is dirty and at times requires heavy lifting. A young person should have a high school or trade school education. It takes four to five years to learn the trade. The steam fitter's, **pipe fitter's**, and **plumber's** trades are closely related.

Structural Ironworker

You have seen the great steel frames which form the skeletons of large buildings or skyscrapers. Fastening the many **steel beams** and frames together is called **structural ironwork**. You may have seen a structural ironworker stand on a steel beam and swing high in the air. The structural ironworker also builds bridges and ships on which the big parts are fastened together with welding or hot **rivets**. Blueprints must be read, and it is necessary to be a careful and strong outdoor worker and a good climber. There is the danger of falling from high places. See also **riveter**. See Fig. 2-14.

[5] **Geometry** is the branch of mathematics which deals with points, lines, surfaces and angles.

Toolmaker

The toolmaker makes and repairs all kinds of special tools, cutting tools, **jigs**, and **gages**. These must be measured with fine instruments and therefore one's eyesight must be good. The jigs must be exact,

Fig. 2-14. Structural Work Is Usually Welded

Fig. 2-15. Welding Is Frequently Used in Small Job Shops

or the many hundreds or thousands of pieces held in them, to be drilled or otherwise cut, would be wrong. The toolmaker can set up and run any machine and use any tool in the shop, must read **blueprints**, make **sketches**, use **layout tools**, and measure with **micrometers**. The toolmaker uses most of the information given in this book.

Toolmakers should have at least a high school education, but they will progress faster if they are technical school or trade school graduates. Knowledge of mathematics and drafting is necessary; these subjects may be learned in night school. Technical school or trade school training will assist promotion to the positions of supervisor and superintendent. It takes four to five years to learn the trade as an **apprentice**.

A toolmaker is seldom without work and is one of the last persons to be laid off. There is the danger of being cut by sharp tools if care is not taken. The toolmaker's, **diemaker's, diesinker's,** and **gagemaker's** trades are closely related.

Welder

The welder joins metal parts by melting the parts together with the use of the oxy-acetylene welding process, the electric-arc welding process, or with any of the other different kinds of welding processes used in industry. The welder also cuts metal by burning it with a cutting torch. See Fig. 2-15. A highly skilled welder knows how to utilize a number of the different welding processes and is able to weld many different metals.

The welder should know about metals and how to read **blueprints**. A beginner can learn to do simple welding jobs in a month. To become skilled in welding with several different processes may take from six months to several years, or longer. Welders must be licensed in many states.

The welder is very important in industry today and will become increasingly important in future years. There is some danger of being burned by the torch or hot metal if one is not careful.

Words to Know

apprentice
assembler
auto mechanic
bench mechanic
boilermaker
carbon monoxide
career
coremaker
diemaker
diesinker
drop forging
drop forging die
electrician
electroplater
engineer
forge operator
foundry
gagemaker
geometry
gunsmith
hammer operator
heat treater
inspector
instrument maker
jeweler
journeyman
layout artist
lead poisoning
line operator
machine operator
machine setup worker
machine shop
machinist
mechanic
metallurgist
metal patternmaker
metal spinner
millwright
molder
pipe fitter
plumber
polisher
riveter
salesperson
semiskilled worker
sheet metalworker
skilled worker
steam fitter
structural ironworker
technician
technology
toolmaker
trade
unskilled worker
welder

Review Questions

1. List several factors which you should consider in selecting your occupation.
2. Explain the meanings of **skill** and **knowledge**.
3. List several school subjects which you should know if you wish to learn a trade, become a technician or technologist, or an engineer.
4. Define the meaning of occupation.
5. List several kinds of unskilled jobs.
6. List several kinds of semiskilled metalworking jobs.
7. What is the length of the training period for most semiskilled metalworking jobs?
8. List several skilled metalworking trades.
9. In what book can one find further information about skilled trades or occupations?
10. Describe the apprenticeship method of learning a trade.
11. What is a journeyman?
12. Of what significance is the written document which an apprentice receives upon completion of the training program?
13. What qualifications must one have in order to qualify for an apprenticeship?
14. Describe the pickup method of learning a trade.
15. What are the educational qualifications for becoming an engineer? A technologist? A technician?

Science

1. Of what importance is physics in industrial occupations?
2. Of what importance is chemistry in industrial occupations?

Mathematics

1. Of what importance is mathematics in industry?

Drafting

1. Of what importance is drafting in industry?

Career Information

1. Write a story entitled, "Why I Picked the Metalworking Course."
2. Select an occupation in which you are interested, then find out the following things about it:
   (a) What things are made by doing that work?
   (b) Is the work seasonal?
   (c) What are the wages earned?
   (d) What are the number of hours of work per day?
   (e) How long does it take to learn the trade?
   (f) Where can you learn the trade?
   (g) Is the work dangerous?

(h) Are good health, strength, and good eyesight needed for the work?
(i) What chances are there of getting a position?
(j) What possibilities are there for promotion?
(k) What chances are there for you to keep the job after you are 40 or 50 years old?
(l) What education is required?

3. If you are interested in several metalworking jobs, use the chart below to compare them.

Summary of Selected Metalworking Jobs

Job	Wages	Kind of Work	Hours	Training Needed	Dangers Involved	Where Can It Be Learned
1.						
2.						
3.						
4.						
5.						

UNIT 3

Safety in Metalworking

3-1. Dressing Safely for Work

Be Clean and Neat

A worker who is clean and neat is usually a **safe worker** and ordinarily does clean, neat work. Loose or torn clothing should not be worn, especially around machines. Loose clothing may be caught by a moving part of the machine and pull the worker into gears or blades.

Apron

Wear a clean apron made of heavy **canvas**. It should hang down to the knees as shown in Fig. 3-1. **Apron strings** should be strong and long enough so that they can be tied properly at the back. Strings tied in front may be caught in a machine. The apron should fit snugly on the chest and fit closely around the waist; apron strings that cross on the back fit best because they pull the apron up close to the chest and snugly around the body. **Pockets** should be sewed on properly; a torn pocket may get caught in a machine.

Aprons should be washed regularly. An apron is for keeping clothing clean, not for wiping dirty hands. It is best to own two aprons; one may be worn while the other is being washed. A sweater should be removed because it overheats the worker.

Sleeves

It is safer to wear short sleeves than long sleeves. Long sleeves should either fit the arm closely or be rolled up above the elbow. Loose or torn sleeves, or sleeves that are not rolled up may get caught in a machine.

Necktie

The long or **flowing necktie** should be removed entirely or tucked into the shirt between the first and second buttons; the first button is the collar button (see Fig. 3-1). This will keep it from getting tangled in a machine.

Gloves

Gloves should never be worn around a machine because they may get caught. However, workers must wear gloves while

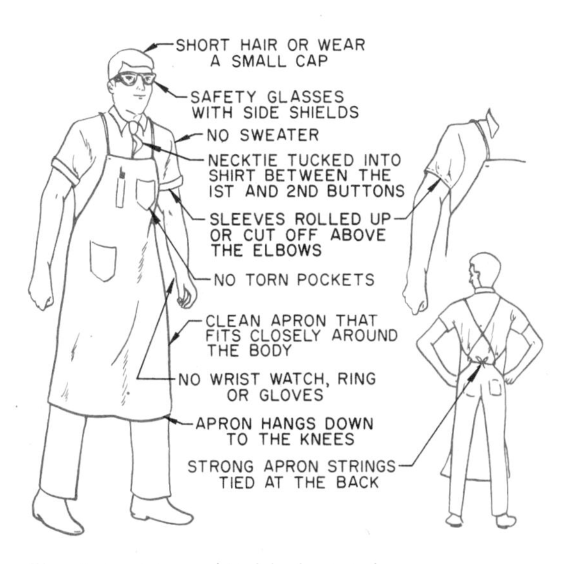

Fig. 3-1. Dressed Safely for Work

handling hot materials or containers and when arc welding.

Wristwatches and Rings

Wristwatches and rings are a definite hazard to safety. They have a history of causing serious and even fatal injury, so they should be removed while a person works around machinery or works with electricity.

Shoes

Open-toed sandals, canvas, and other soft shoes do not provide adequate foot protection. Ordinary leather dress and work shoes offer considerable protection and should be worn while working in shops and laboratories. Industrial quality safety shoes provide even greater protection against toe stubbing and crushing-type accidents. See Fig. 3-2.

Eye Protection

Many states require teachers and students to wear safety glasses while working in school shops and laboratories. Visitors should also be so protected. Safety glasses with side shields, goggles, and face shields should all meet or exceed State and Federal safety specifications. PROTECT YOUR EYES. Be sure you learn and observe the eye protection rules at your school.

Fig. 3-2. Safety Shoe with Steel Reinforced Toe

Other Safety Clothing

Some metalworking operations, such as pouring molten metal and arc welding, require additional safety clothing to be worn. The units covering these topics also include further explanation and illustration of the additional safety clothing required.

3-2. Safe Work Practices for Metalworking

Safe work practices and safe work habits result when you use machines, equipment, tools, and materials correctly. You must follow commonly recognized safety rules and safety practices in order to avoid accidents and personal injury. The following safety rules, precautions, and procedures should be observed:

1. Always notify the instructor immediately when you are injured in the shop or laboratory, no matter how slight the injury.
2. Always have proper first aid applied to minor injuries. Always consult a physician for proper attention to severe cuts, bruises, burns, or other injuries.
3. Safety glasses, goggles, or a face shield of an approved type should be worn at all times in the shop or laboratory.
4. Oil or grease on the floor is hazardous and can cause slipping; hence, it should always be cleaned up immediately.
5. Place oily rags or other flammable wiping materials in the proper containers.
6. Keep aisles and pathways clear of excess stock, remnants, or waste. Store long metal bars in the proper storage area.
7. Return all tools or machine accessories to the proper storage areas after use.
8. Operate machines or equipment only when authorized to do so by your instructor.
9. Avoid needless shouting, whistling, boisterousness, or play when in the shop

or laboratory. Give undivided attention to your work.

10. Never touch metal which you suspect is hot. If in doubt, touch the metal with the moistened tip of your finger to determine whether it is hot.
11. When approaching people who are operating a machine, wait until they have finished that particular operation or process before you attract their attention.
12. Avoid touching moving parts of machinery.
13. Do not lean on a machine which someone else is operating.
14. Do not operate a machine until the cutting tools and the workpiece are mounted securely.
15. Be sure that all of the safety devices with which a machine is equipped are in the proper location and order before using the machine.
16. If more than one person is assigned to work on a certain machine, only one person should operate the controls or switches.
17. Never leave a machine while it is running or in motion.
18. A machine should always be stopped before oiling, cleaning, or making adjustments on it.
19. Always use a brush or a stick of wood to remove metal chips from a machine. Otherwise you may be cut by sharp chips.
20. Do not try to stop a machine, such as a drill press spindle or a lathe spindle, with your hands.
21. Do not touch moving belts or pulleys.
22. Always be certain that the machine has stopped before changing a V-belt.
23. Before starting a machine, be sure that it is clear of excess tools, oil, or waste.
24. Request help from a fellow worker when it is necessary to lift a heavy machine accessory or other heavy objects. Always lift with your legs, not your back.
25. Do not work in restricted areas which are marked off as safety zones.
26. Be sure that everyone around you is wearing approved safety glasses, goggles, or a safety shield if you are permitted (by your instructor) to blow metal chips from a machine with compressed air.
27. Know the location of the nearest fire alarm in the building in case of fire. Also learn the location of the nearest fire extinguisher and ask your instructor to explain how it is operated.
28. Always place flammable materials, such as paint thinners, lacquers, and solvents, in a metal cabinet away from open flames.

3-3. Safe Use of Hand Tools

Use of hand tools frequently results in personal injury. See Fig. 3-3. The following safety rules or safe work practices should be followed when using hand tools:

1. Use the right tool for the job to be performed.
2. See that tools and your hands are clean and free of grease or oil before use.
3. Cutting tools should be sharp when using them. Dull tools cause accidents because of the greater forces required to use them.

Fig. 3-3. Many Accidents Happen in Using Hand Tools

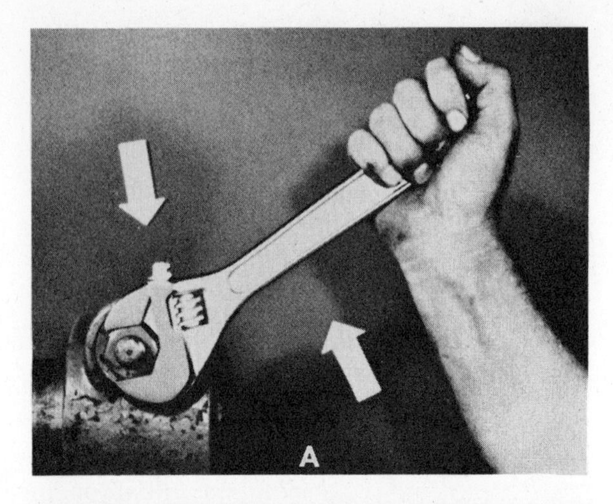

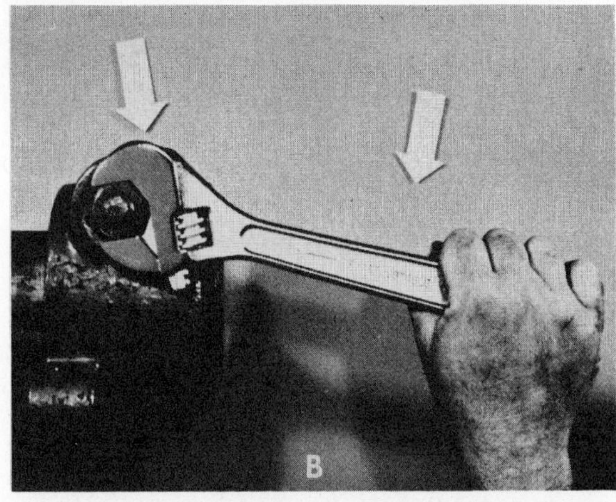

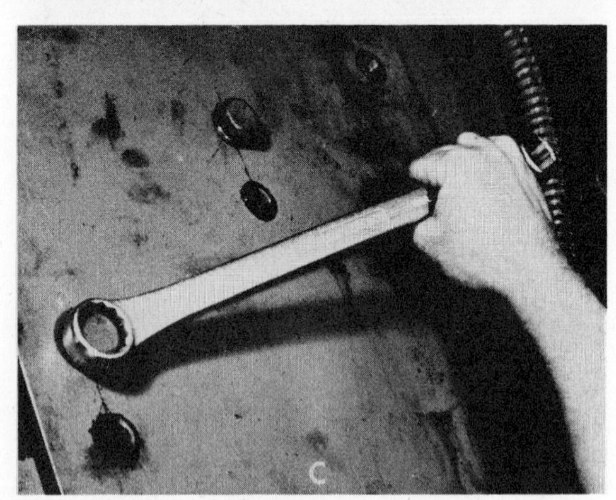

Fig. 3-4. Use a Wrench Properly
A. Wrong: Do not push on the handle.
B. Right: Pull on the handle.
C. Use offset box wrench when possible.

4. Sharp-edge tools should be carried with their points and cutting edges pointing downward.
5. Heads of cold chisels and punches should not be allowed to mushroom or crack; they should be properly dressed or repaired.
6. When using a chisel, always chip in a direction which will prevent flying chips from striking others.
7. Choose the correct type of wrench for the job and use it properly. You can injure your knuckles or hand if the wrench slips. See Fig. 3-4.
8. When using a file, be sure that it is equipped with a snug-fitting handle. Otherwise, the sharp tang on the file could injure your hand.
9. When you hand tools to others, give them with the handle first.
10. Always report damaged tools to the instructor. Damaged tools can cause injuries.
11. Tools should always be wiped free of grease or dirt after use and then returned to the proper storage location.

3-4. Other Safety Rules

Other safety rules are given throughout the book as each machine, tool, or operation is explained.

Review Questions

1. Why shouldn't sweaters or loose clothing be worn while a person is around machinery?
2. If a worker always wears a clean apron or work clothes, what does this tell you about how good a worker the person is?
3. Under what conditions should gloves be worn in the shop or laboratory?
4. Why is long hair a potential hazard around machinery?
5. Describe several types of accidents which can be caused by wearing rings or wristwatches.

6. List two safety rules or safe practices regarding first aid and the reporting of injuries.
7. Why should oil or grease be removed from the floor or cleaned from hand tools?
8. Why is it important that oily rags be stored in metal containers?
9. Describe a simple way to test whether a piece of metal is too hot to pick up with bare hands.
10. Why is it a good rule not to try to get the attention of a person who is operating a machine until that person has finished the particular operation or process being worked on?
11. What are the state and Federal laws on eye protection in school shops and laboratories? In industry?
12. Why is a careless worker a hazard to others?

Mathematics

1. If a worker who earns $30.80 a day loses 13½ days' work due to an accident, how much pay does the worker lose?

Social Science

1. What are the goals or purposes of the Social Security Law?

Career Information

1. How should a person who works around machinery be dressed to do work well and safely?
2. Why is it important that a worker be clean and neat?

PART II

Planning and Organizing Work

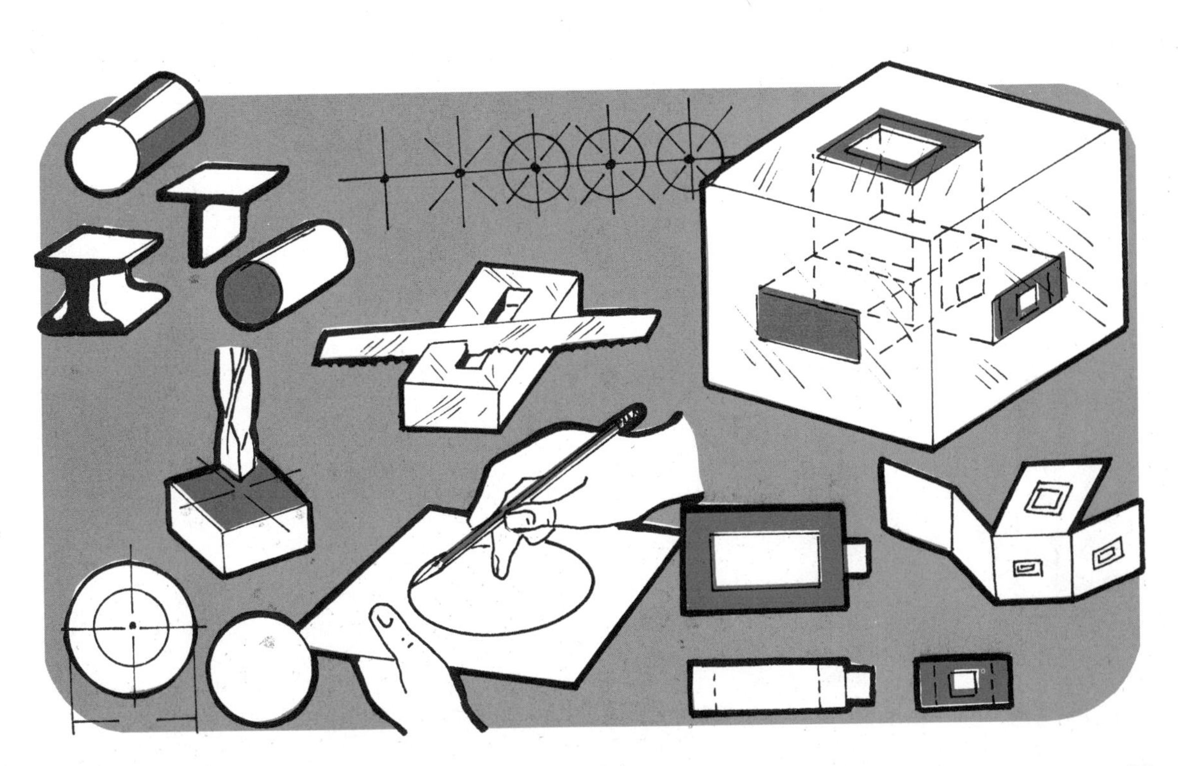

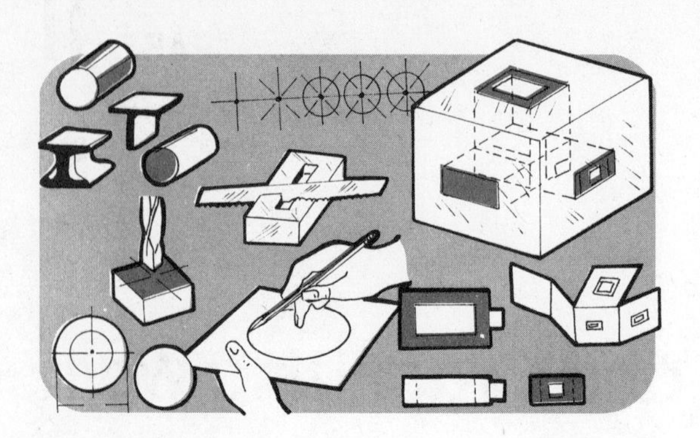

UNIT 4

Reading Drawings and Making Sketches

4-1. Why Are Drawings Needed?

Drawings of several types are used by product designers, engineers, and skilled craftsmen. **Pictorial sketches** and **pictorial drawings** are used at the design stage because the appearance of the product is very important to its successful sale. See Fig. 4-1. The designer usually makes many pictorial sketches of a product before finding a preferable design. Pictorial drawings and sketches, however, usually do not provide enough information to enable workers to make the product.

Fig. 4-1. Pictorial Drawing

Engineers and skilled craftsmen use **working drawings,** also known as **orthographic drawings** and **blueprints,** from which to work.

4-2. What Is a Working Drawing?

A working drawing gives all the information needed by workers to make the product or object; it is a drawing from which the workers work, Fig. 4-8. It enables workers to make objects exactly alike, even in different factories, so that the parts are **interchangeable.** Automobile parts are made in different parts of the country; yet, when assembled, they fit perfectly. The working drawing makes this possible. It must show the following items:

1. Shape of every part of the object.
2. Sizes of all parts.
3. Kind of material.
4. Kind of finish.
5. How many pieces of each part are wanted.

A **mechanical drawing** is a working drawing made with mechanical drawing instruments. It is usually made by the **draftsman** in the **drafting room.**

Drafting includes mechanical drawing and **sketching.**

4-3. What Is a Blueprint?

The draftsman often makes a tracing of a working drawing on **translucent**[1] paper or cloth so that several copies can be made quickly. The **diazo printer** in Fig. 4-2 makes an exposure through the tracing to light-sensitive paper using a bright light. The paper is developed using ammonia fumes to make a copy called a **blueprint,** or simply "print." Diazo copies are on white paper rather than the blue used for blueprints.

4-4. Reading Working Drawings

A working drawing is the language of all mechanical occupations — the drafting room, the shop, the manufacturing plant, and, in fact, the industrial world.

To **read** a working drawing one must know what certain kinds of lines, signs, and **abbreviations** mean. To become a **mechanic,** a person must first learn to read this language.

4-5. Parts of a Circle

It is important that the beginner learn the names of the different parts of a **circle.** See Fig. 4-3.

The **circumference** is the length of the curved line which forms the circle.

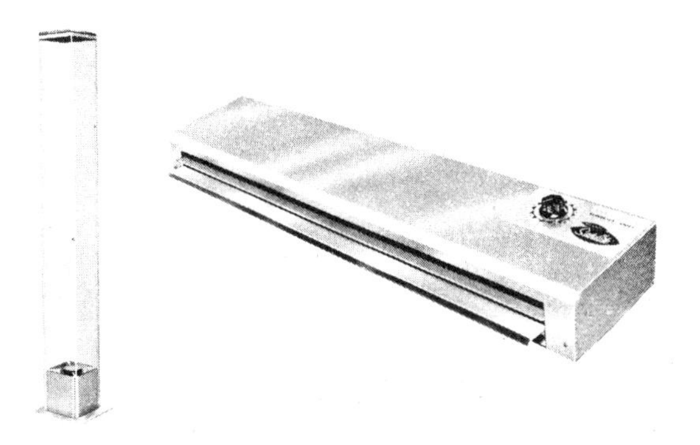

Fig. 4-2. Diazo Printer and Developing Tube

[1]**Translucent** is something through which you can see outlines, such as frosted glass or tissue paper.

The **diameter** is a straight line drawn through the **center** of a circle, both ends touching the circumference. When we speak of a 2″ circle, we mean that 2″ is the diameter of the circle.

The **radius** is one-half of the diameter.

An **arc** is any part of the circumference of a circle.

A **chord** (pronounced cord) is a straight line connecting the two ends of an arc or connecting two points on the circumference of a circle.

A **semicircle** is a half circle.

Tangent (pronounced **tanjent**) means touching at only one point; thus a **tangent line** touches an arc or the circumference of a circle at one point.

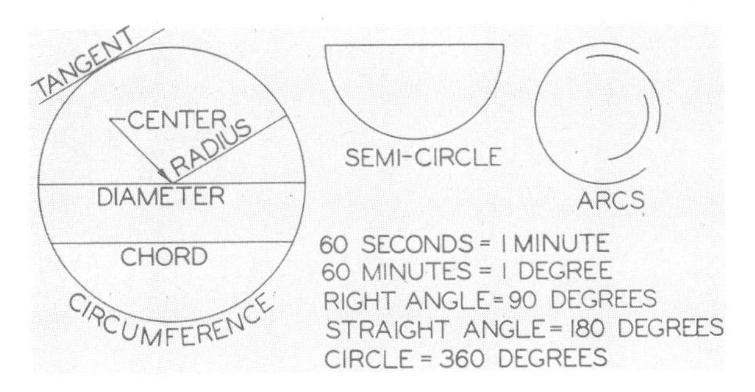

Fig. 4-3. Parts of a Circle

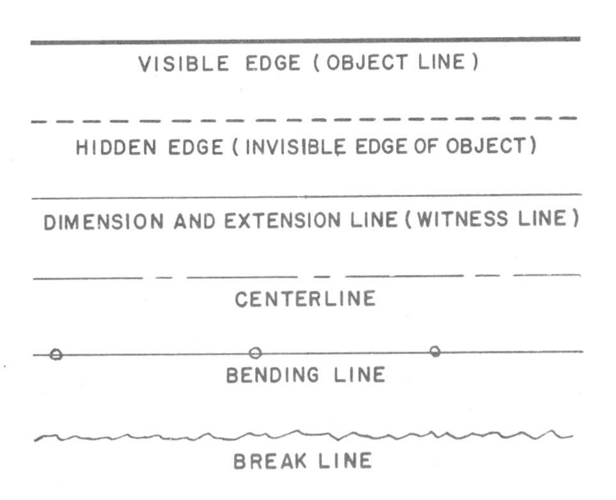

Fig. 4-4. Lines Used on Working Drawings

There are 360 **degrees** in a circle, hence a **degree** is 1/360 of a circle. The **area** is the amount of surface inside the circumference.

4-6. Lines on Working Drawings

The lines on working drawings which the beginner must know are shown in Fig. 4-4.

A **visible line** is a thick line used to show all edges that can be seen. See Figs. 4-4 through 4-8.

A **hidden line,** sometimes called an **invisible edge** or a **dotted line,** is made of 1/8″ dashes 1/16″ apart. It is used to show hidden edges, just as if one could look right through the object. See Figs. 4-4 and 4-7.

An **extension line** is a thin line. It is drawn from the edge from which the measurement is to be made. Note that the extension line does not touch the object; it should be about 1/16″ from the object and extend about 1/8″ past the **arrowhead** (see § 4-9). See Figs. 4-4 and 4-8.

A **dimension line** is also a thin line; it is drawn between the extension lines about 3/8″ from the object. The **dimension** is in the opening in the dimension line. See Figs. 4-4 and 4-8.

Whenever we talk about the distance from one hole to another, we mean the distance from the **center** of the one hole to the **center** of the other. Dimensions between holes should always be given from **center to center** because the worker in the shop measures that way. See Fig. 4-8. Measurements are always made from the center of a circle or **arc.**

Centerlines, Figs. 4-4 and 4-8, locate the centers of circles or arcs. They are the foundation lines for measuring and must be drawn before the circle or arc is drawn. The centerline is also used to show the **axis** of an object and to locate slots. A centerline is a thin line made up of a 5/8″ line, then a 1/16″ dash, a 5/8″ line, a dash again, with 1/16″ space between the lines and dashes. It is drawn horizontally and vertically through the center of every circle and arc. The 1/16″ dashes should cross at the center of the circle or arc as in Fig. 4-8. If a rounded metal object is to be made, the layout on the metal surface is started with centerlines. Dimensions given between the centerlines on the drawing are thus used; they should be located with reference to centerlines or finished surfaces. See also how centerlines are used in layout work in Figs. 6-6 and 7-8. Centerlines are also used to show the centers of slots and grooves. See Fig. 4-8.

Bending lines are shown in Figs. 4-4 and 31-1; a small circle is drawn by hand at each end of the line.

A **break line** is a wavy line; it is used to show that a part is broken off as in Figs. 4-1, 9-1, and 49-9.

4-7. Lettering

It is often necessary to add short notes, words, or **abbreviations** of words to give all the necessary information on a working

TABLE 1

Abbreviations and Symbols Used on Drawings

(See section 4-7.)

Symbol	Meaning
′	Feet, or Minutes
″	Inches, or Seconds
°	Degrees
±	Plus or minus; more or less
℄	Centerline
D or Dia	Diameter
R or Rad	Radius
P	Pitch
RH	Right Hand
LH	Left Hand
USF	United States Form
USS	United States Standard
SAE	Society of Automotive Engineers
USAS	United States American Standards Institute
Thds.	Threads
NC	National Coarse
NF	National Fine
UNC	Unified National Coarse
UNF	Unified National Fine
CS	Carbon Steel
HRS	Hot-Rolled Steel
CRS	Cold-Rolled Steel
HSS	High-Speed Steel
√	Finish
Csk	Countersink

drawing. See Fig. 4-13. All these must be **lettered,** frequently in freehand, neatly and plainly because lettering can be read more easily by everyone. Table 1 gives a list of standard abbreviations and **symbols** used on drawings. Abbreviate only when there is not enough space for the whole word.

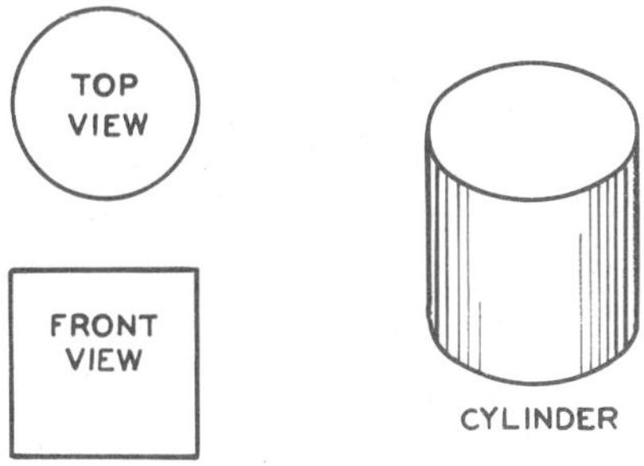

4-8. Views

On a working drawing, each view shows the outline or shape of the object as seen from that side. To show an object com-

Fig. 4-5. Front and Top Views of Cylinder

Fig. 4-6. Front and End Views of Cylinder

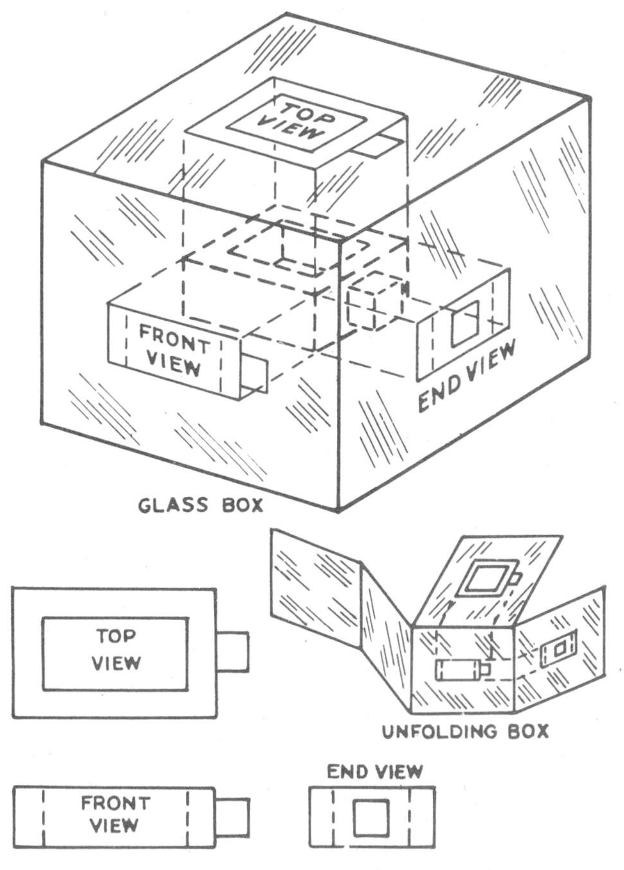

Fig. 4-7. Front, Top and End Views of a Hollow Block

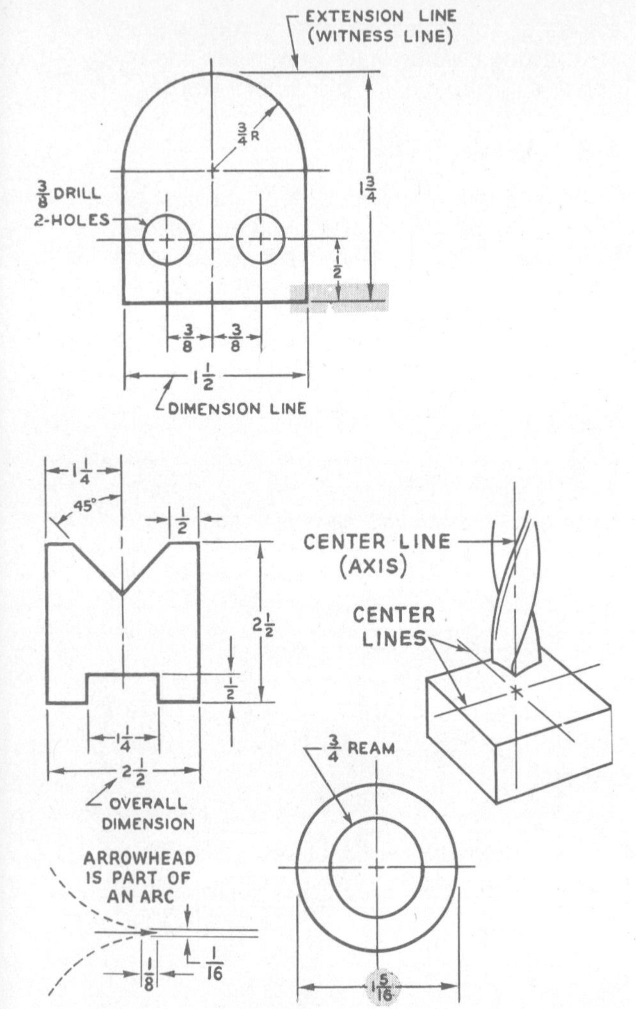

Fig. 4-8. Centerlines and Dimensions

pletely, two or more views are usually necessary. Some objects can be described with only one view. No more views than are necessary should be drawn. A working drawing of a **cylinder** needs only two views, the **front view** and **top view,** Fig. 4-5, or the **front view** and **end view** when the cylinder lies on its side as in Fig. 4-6. Some objects need three views; a front view, a top view, and an end (or side) view. It will help you

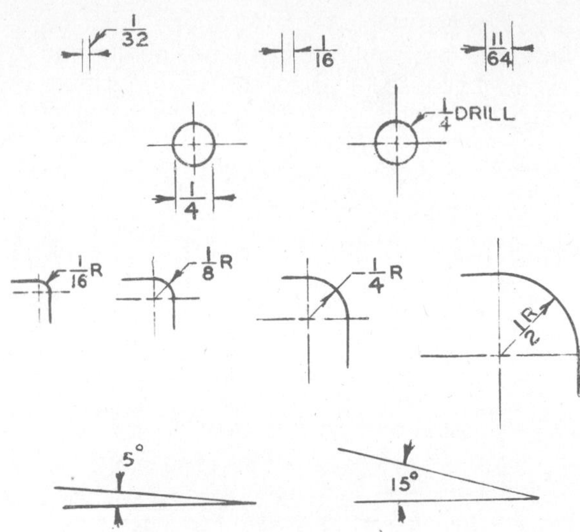

Fig. 4-9. Dimensions for Small Spaces

to draw these views if you can imagine that the object is in a glass box, Fig. 4-7, with the views drawn on the front, top, and end of the box. When the box is unfolded, the views will be in their correct positions.

Dimensions are most often read from the bottom or right side of the sheet. Some drafting rooms make all dimensions read from the bottom of the sheet as in Figs. 4-8 and 4-9. **Fractions** must be made with a **horizontal line,** as $\frac{1}{2}$″, $\frac{3}{4}$″. When all dimensions are in inches, the **inch marks** (″) can be omitted.

4-9. Dimensions

Dimensions are the most valuable part of a working drawing because they give the size of the object, the location and size of holes, and other features. The **dimension lines** and dimensions are drawn as in Fig. 4-8. Draw long, narrow **arrowheads** that touch the **extension lines** which extend from the object to show the distances given by the dimensions.

Give the **diameter** of a circle rather than the **radius.** The diameter should be marked **D** or **Dia** except when it is plain that the dimension is the diameter. The radius of

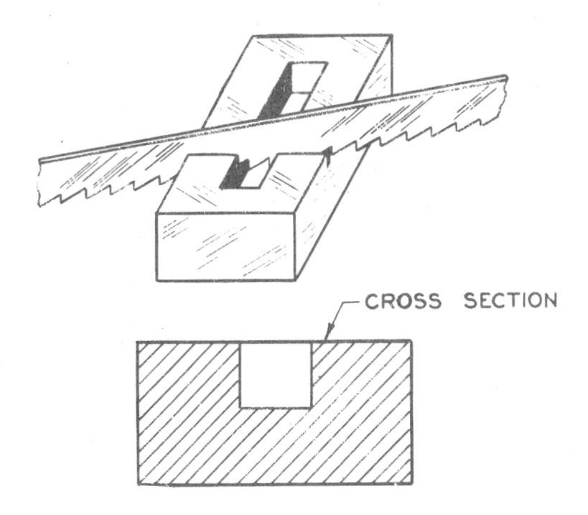

Fig. 4-10. Cross Section of Metals

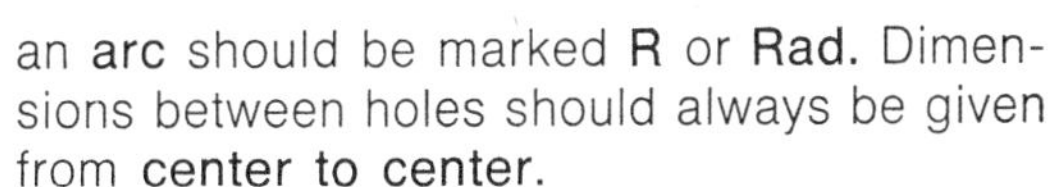

an **arc** should be marked **R** or **Rad.** Dimensions between holes should always be given from **center to center.**

A dimension should be repeated only when there is a special reason for doing it. If the space is too small for a dimension, use one of the ways shown in Fig. 4-9. Always give **overall dimensions** which are the dimensions that give the total length, width, and height or thickness of the object. See Fig. 4-8.

4-10. What Is Meant by the Scale of a Drawing?

Oftentimes, an object is too large to draw **full-size** on a sheet of paper. It is, therefore, drawn **half-size, quarter-size,** etc., or it is drawn **to scale** which means that it is drawn so that ⅛″ equals 1 foot, ¼″ equals 1 foot, etc. The dimensions are placed on such a drawing the same as if the drawing were full-size. See Fig. 31-1. If the drawing is other than full-size, the scale that is used must be given in a note on the drawing, as "Scale: ¾″ = 1 Foot."

4-11. What Is a Cross Section?

The part of a working drawing that shows the object as if part of it had been cut away is called a **cross section** or simply **section.** It shows the inside shapes of holes and the thickness of parts. See Figs. 4-10 and 23-8. The cut metal is often shown by **parallel lines** called **section lines.**

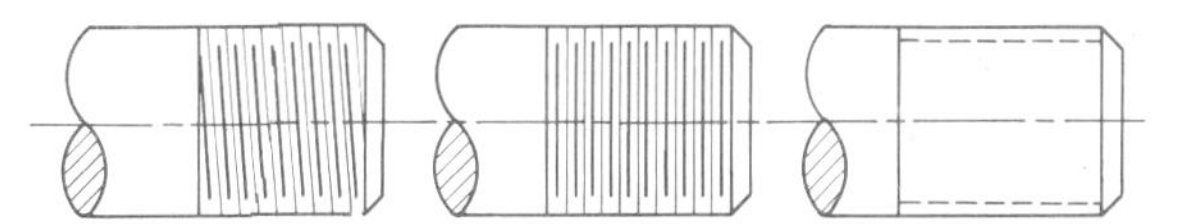

Fig. 4-11. Quick Ways to Draw Screw Threads

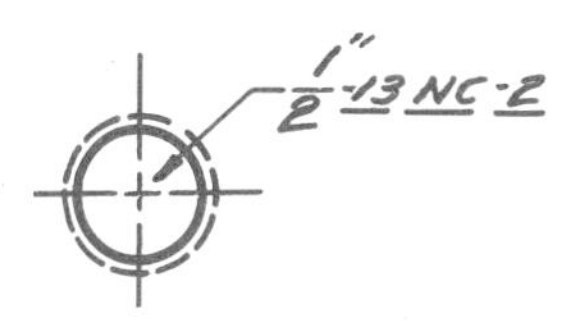

Fig. 4-12. Looking Into a Hole that Has Threads in It

4-12. Quick Way to Draw Screw Threads

Section 21-1 describes **screw threads.** A quick way to draw screw threads is shown in Figs. 4-11 and 4-12. The lines may be made either slanted or straight across.

A **hole** that has threads in it is a **tapped hole** and it may be shown as in Fig. 4-12. The note, ½″-13NC-2, means that the screw or bolt which must screw into the hole is ½″ in diameter and has 13 threads per inch, the kind of thread is **American (National) Coarse Thread,** and has a **Class 2, Free Fit.**

Sections 21-15, 21-16, 22-7, and 23-1 give more information about **threads** and **taps.**

4-13. Meaning and Reasons for Sketching

Often the step to a better job for a young person is the ability to **read drawings** as well as make them. See § 4-4. The worker

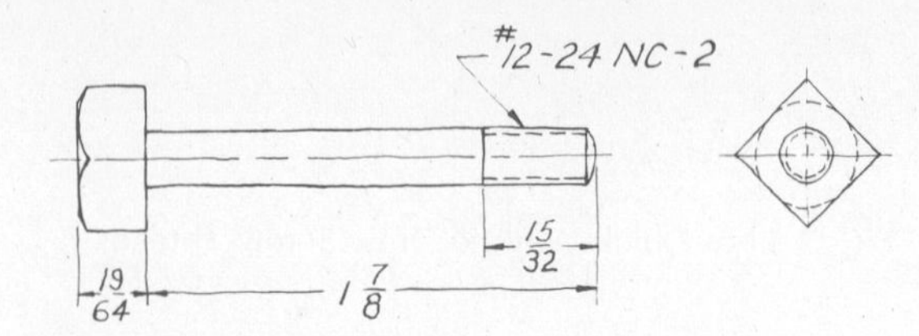

Fig. 4-13. Sketch of a Bolt

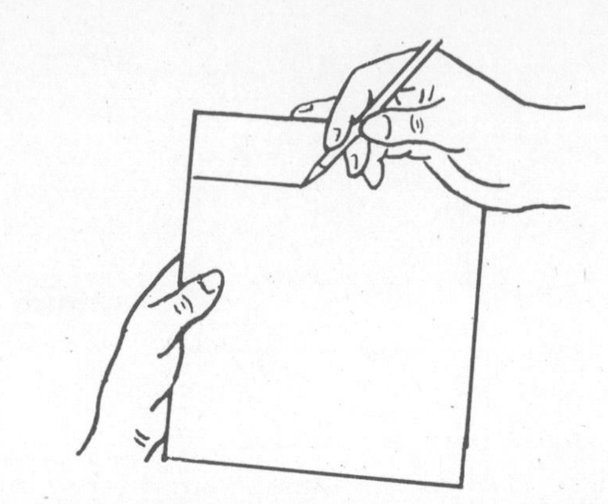

Fig. 4-16. Sketching a Line Along Upper Edge of Paper

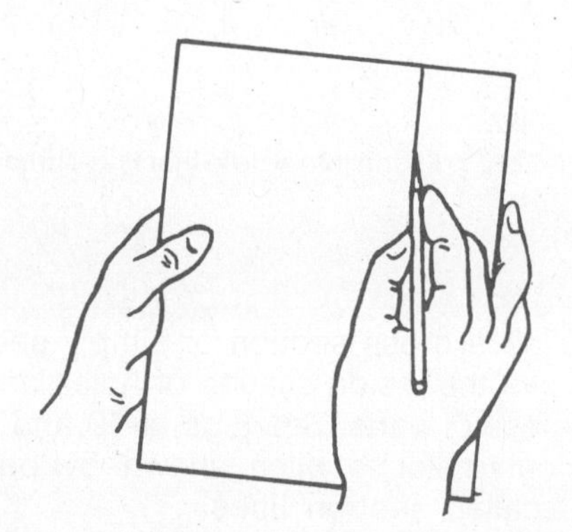

Fig. 4-14. Sketching a Line Near Edge of Paper

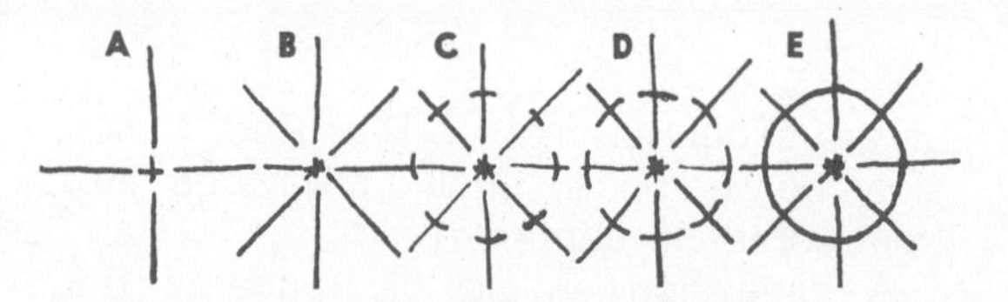

Fig. 4-17. Steps in Sketching a Circle

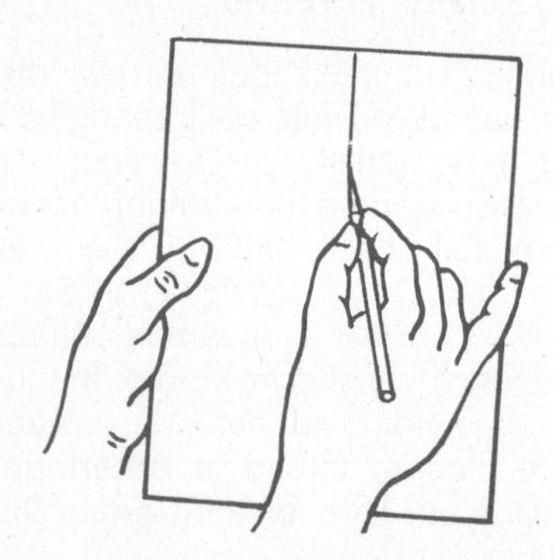

Fig. 4-15. Sketching a Line Near Center of Paper

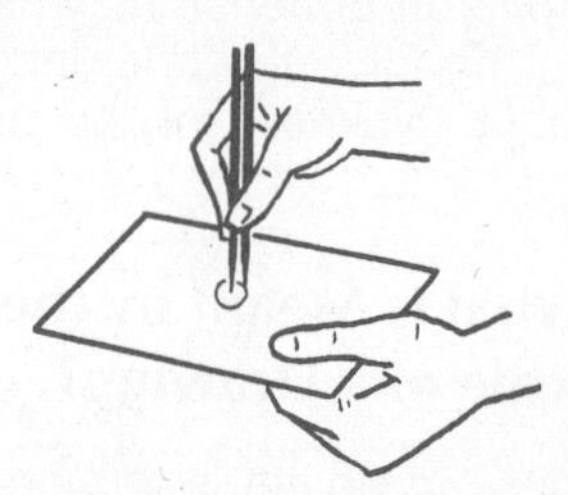

Fig. 4-18. Sketching a Small Circle with Two Pencils

must not only know how to read the language of the industrial world but must know also how to write it, giving ideas and describing an object to someone else. It is often necessary to make a part for a machine for which there is neither time to make a drawing with instruments nor to have a draftsman make a drawing.

Sometimes only one piece of a machine part is to be made and, therefore, it would be too expensive to have the draftsman make a mechanical drawing. Very often the worker in the shop has ideas about the **design** of an object. Then too, in the shop there are usually no mechanical drawing instruments handy. In such cases the worker must make a sketch or freehand drawing in the shop with the tools at hand. These are usually only paper and pencil. With these

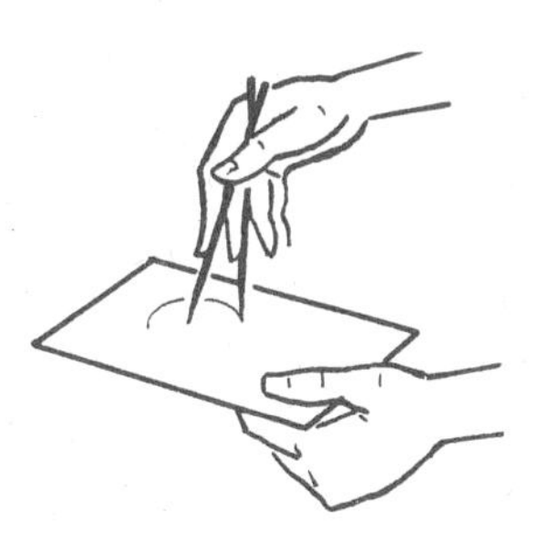

Fig. 4-19. Sketching a Large Circle with Two Pencils

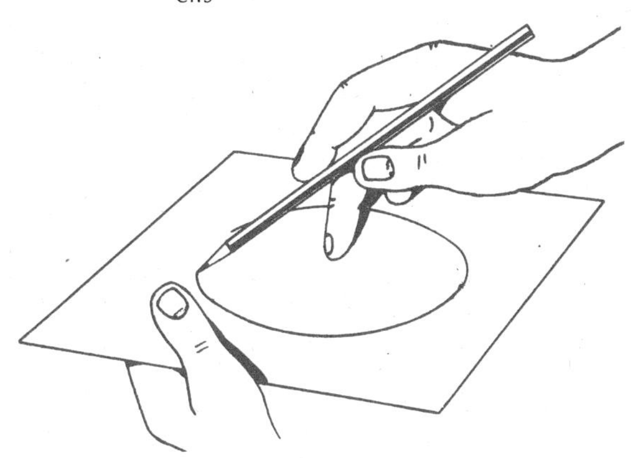

Fig. 4-20. Sketching a Circle with Pencil and Middle Finger

tools the worker can, in a very short time, make a working drawing that will answer the purpose. Even when a rule is handy, the experienced worker often does not use it. A sketch of a bolt is shown in Fig. 4-13.

Sketches are valuable because they show ideas and dimensions of a part. The cleverest idea, however, may result in serious mistakes if the sketch is poorly made. The greatest joy in making things lies in planning and making them yourself. The beginner should practice making clean sketches that give all the information needed and that can be read easily by others.

4-14. Sketching Lines

The straight edges of a paper pad may help to sketch straight lines near the edge of the paper my moving the third finger as a guide along the edge of the pad as in Fig. 4-14. Straight lines near the center of the paper may be sketched by moving the little finger as a guide along the edge of the pad as in Fig. 4-15. Straight lines along the upper edge of the paper may be sketched by bending the wrist as shown in Fig. 4-16. Straight lines near the lower edge of the paper may be sketched by using the back of the little finger as a guide.

4-15. Sketching Circles

One way to sketch circles is shown in Fig. 4-17.

A. Draw **centerlines** through the point that is to be the center of the circle.
B. Draw two **diagonal** lines.
C. Mark the **radius** of the circle on each of the lines.
D. Draw short **arcs** through the marks.
E. Finish circle.

Another easy way to sketch circles is to use two pencils like a **compass** and turn the paper under the pencils. Fig. 4-18 shows how to sketch a small circle this way. The pencils should be held as in Fig. 4-19 for larger circles. Large circles may also be sketched by turning the paper under the pencil and middle finger, Fig. 4-20.

Words to Know

American (National) Coarse Thread
axis
bending line
blueprint
break line
centerline
chord
circumference
diameter
diazo printer
dimension
dimension line
end view
extension line
front view
full-size drawing
hidden line
invisible edge
mechanical drawing
overall dimension
parallel lines
pictorial drawing
radius
scale drawing
screw thread
semicircle
sketch
symbol
tangent line
top view
visible line
working drawing

Review Questions

1. What is a pictorial sketch?
2. What is a pictorial drawing?
3. Who uses pictorial sketches and drawings?
4. What is a working drawing?
5. Who uses working drawings?
6. What is a mechanical drawing?
7. What is a blueprint?
8. What views of an object are shown in a working drawing?
9. What is an overall dimension?
10. What is a cross section?

Mathematics

1. What is the diameter of a 1½″ circle?
2. What is the circumference of a 3½″ circle?
3. What is the diameter of a circle that has a circumference of 7.346″?
4. What is the radius of a circle that has a circumference of 10.479″?
5. How long is an arc that equals 1/12 of the circumference of a 4″ circle?
6. Which has more degrees, a small circle or a large circle?
7. How many degrees are there in a quarter of a circle? In half a circle? In three-quarters of a circle?
8. What is the area of a 3″ circle?
9. What is the scale of a drawing that is 1/4 size?
10. What is the difference between "quarter-size" and 1/4″ = 1′?

Drafting

1. Sketch a working drawing of the first project that you are expected to make.
2. Make a pictorial sketch of the first project you are expected to make.

Career Information

1. Explain in writing why working drawings are the language of the shop, the drafting room, and the industrial world.
2. Of what use is a draftsman in an industrial plant? Explain in writing.
3. In which of the jobs discussed in Unit 2 is a knowledge of reading working drawings and sketching important?

UNIT 5

Product Planning

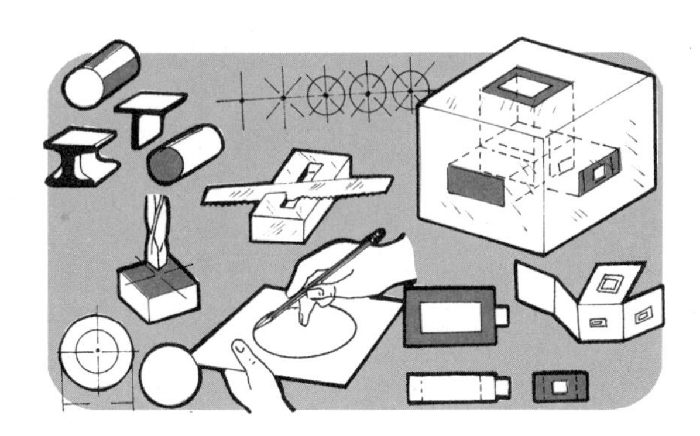

5-1. Making a Product Plan

Before attempting to construct a product of your own design, you should carefully prepare a product plan. A product plan contains all the information essential for successful construction of the product. A good plan includes the following information:

1. A working drawing of the product; this could be a carefully sketched freehand drawing, a pictorial drawing, or a sketch.
2. A bill of materials.
3. A list of the steps of procedure, in proper order, for making the product.
4. Approval of your instructor (if required) before making the product.

It is often easier to prepare the product plan if you use a form sheet designed for the purpose. A form sheet of the type shown in Fig. 5-1 may be used, unless your instructor provides you with one. The working drawing should be attached to the completed form sheet, thus completing the product plan, as shown in Figs. 5-2 and 5-3.

When you prepare a plan for a product you wish to construct, you will be using procedures similar to those used in industry. Every item made, whether it is a tin can or an automobile, must be carefully planned before it is produced. Working drawings must be made for each part of the product. The proper materials must be selected for each part, and the costs must be calculated. Tools, equipment, and machinery must be provided, then the manufacturing procedures must be carefully planned for the most economical production. Your plan, therefore, involves many of the kinds of activities which are performed by manufacturing engineers, industrial technologists, industrial technicians, and skilled workers.

5-2. What Is a Bill of Materials?

You must have the correct metal before you can make a metal product; therefore, it is necessary to know how to specify and order metals. The working drawing gives all of the information needed to make a bill of materials. See Fig. 5-2. The bill of materials should be made in the form shown in Fig. 5-1.

A bill of materials should show:

1. The parts of the product, identified by numbers or letters.
2. The number of pieces needed for each part.
3. The size of the material.
4. The shape and the kind of material.
5. The standard parts used in the product.
6. The unit cost of the material: the cost per pound, per foot, per square foot, etc.
7. The total cost of the materials.

5-3. What Is a Standard Part?

A standard part is a part that is made by several companies and is the same no matter who makes it. Hardware such as bolts, nuts, rivets, screws, washers, etc., that are made to standard sizes and shapes are

Product Plan Form Sheet

Name ______________________________ Grade ________
Product ______________________________ Hour ________
Source of product idea, if other than your own design ______________________
____
Estimated Time __________ Actual Time__________ Approved__________

BILL OF MATERIALS

Part No.	No. of Pieces	Size			Material	Unit Cost (Per sq. ft., lb., etc.)	Total Cost
		T	W	L			

Total Cost ________

Manufacturing Procedure

1. Select standard stock
2. Mark stock to length or overall size
3. Cut standard stock
4. Make part A
   a.
   b.
   c.
   d. etc.
5. Make part B
   a.
   b.
   c.
   d. etc.
6. Assemble the product
   a.
   b. etc.
7. Inspect
8. Apply finish
9. Inspect

Fig. 5-1. Product Plan Form Sheet
A form like this may be used for planning your product.

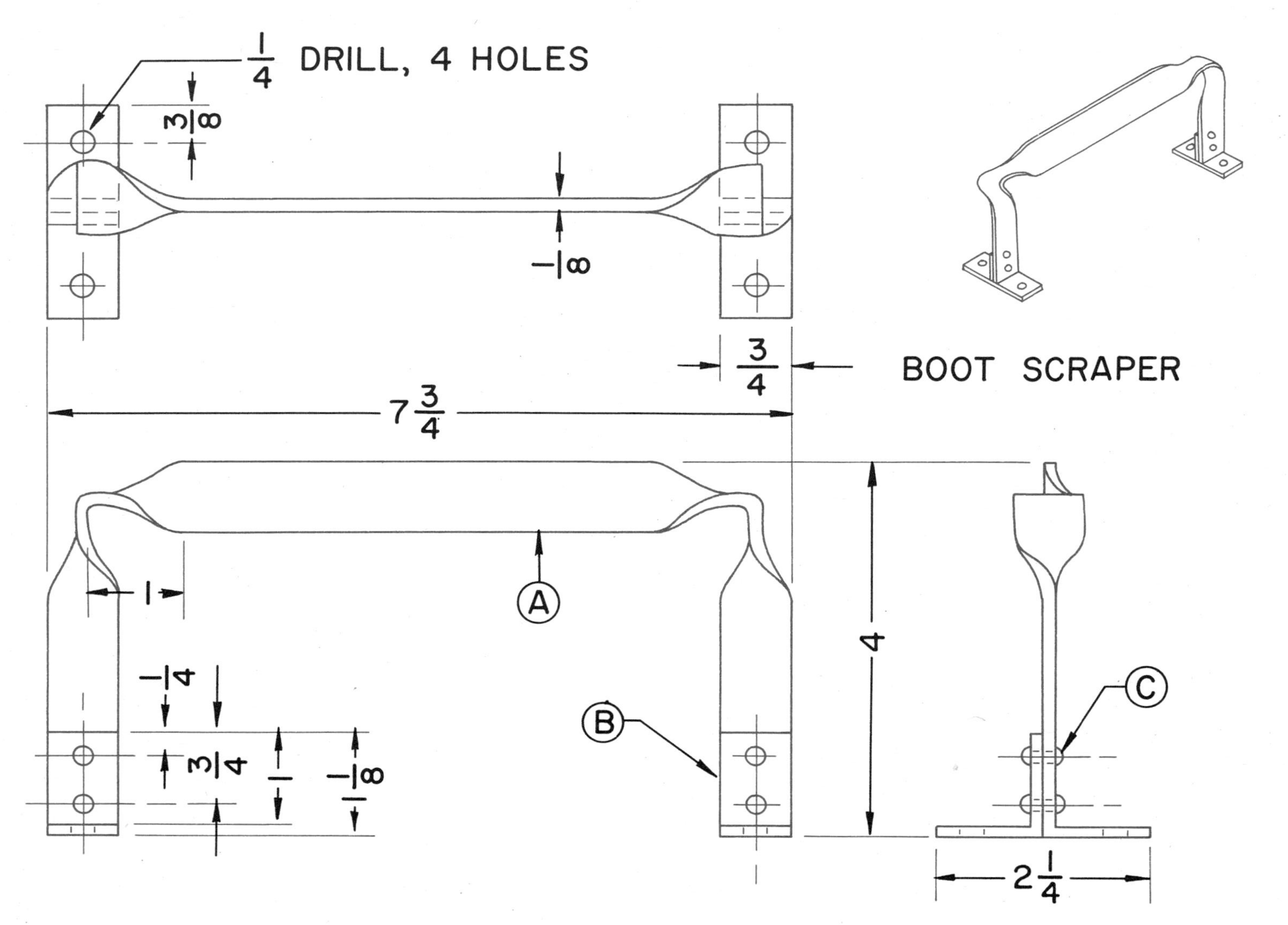

Fig. 5-2. Working Drawing of a Boot Scraper

Completed Product Plan Form Sheet

Name Jackson, Frank Grade 9

Product Boot Scraper Hour 3rd Period

Source of product idea, if other than your own design Shop product idea file

Estimated Time 3 hours Actual Time 2 hours Approved HESS

BILL OF MATERIALS

Part No.	No. of Pieces	Size			Material	Unit Cost (Per sq. ft., lb., etc.)	Total Cost
		T	W	L			
A	1	⅛	¾	15⅝	H.R. Steel	8¢/ft.	13¢
B	2	⅛	¾	2	"	"	3¢
C	4	⅛		⅜	R.H. Rivet	1¢	4¢
					Paint		4¢

Total Cost 24¢

Manufacturing Procedure

Part A

1. Cut one piece of ⅛″ x ¾″ x 15⅝″ hot-rolled steel.
2. File or belt grind one end square, removing as little metal as possible.
3. Mark overall 15½″ length with a combination square and scriber.
4. File or belt grind to layout line.
5. Remove sharp edges with a flat mill file.
6. Lay out bend and twist lines according to working drawing.
7. Make 90° bends at top of scraper, using vise and ball peen hammer.
8. Make 90° bends at bottom of scraper using vise and ball peen hammer.
9. Make 90° twists using vise and adjustable wrench.
10. Check bends and twists for correct alignment.

Part B

1. Cut two pieces of ⅛″ x ¾″ x 2⅛″ hot-rolled steel.
2. File or belt grind one end of each piece square, removing as little metal as possible.
3. Mark overall 2″ length with square and scriber.
4. File or belt grind to layout line.
5. Remove sharp edges with a flat mill file.
6. Lay out bend line and rivet hole centers using a combination square and scriber.
7. Center punch rivet hole centers.
8. Drill ⅛″ rivet holes; remove burrs from holes.
9. Make 90° bends using vise and ball peen hammer.

Assembly

1. Clamp parts A and B together.
2. Drill holes in part A, using holes in part B as guides.
3. Rivet parts A and B together, using a ball peen hammer and rivet set.

Finishing

1. Paint with flat black enamel.

Fig. 5-3. Completed Product Plan Form Sheet for a Boot Scraper

standard parts. Use catalogs to get information about standard parts and materials.

5-4. Meaning of Standard Stock

Standard stock is the material that is used in the manufacture of finished products. Steel as it comes from the steel mill is standard stock. Standard stock is purchased from metal wholesalers who stock each shape in many sizes. Common shapes of standard stock are shown in Fig. 5-4.

5-5. Measuring Standard Stock

Remember that the size given on the bill of materials is the size of the standard stock that you will order. The size given on the working drawing is the finished size. The size of any part, as given on the working drawing, must have added to it the extra metal which is needed for finishing the object to size. Standard stock is specified or described as follows:

Flat Sheet or Strip
- Thickness x width x length, as follows:
  - 1/8″ x 1 3/4″ x 4 1/4″

Square Bar
- Thickness x width x length, as follows:
  - 1″ x 1″ x 4 1/4″

Round Bar
- Diameter x length, for example:
  - 2″ Dia. x 4 1/4″

Hexagonal and Octagonal Bar
- Distance across flat sides x length, for example:
  - 1 1/4″ x 4 1/4″

Tubing
- Outside dimensions x wall thickness x length, as:
  - 7/8″ Dia. x .049 wall x 12″, and
  - 1″ x 1″ x .062 wall x 18″

Structural Shapes
- Overall cross-sectional dimensions x shape x wall thickness x length, for example:
  - 1 1/2″ x 1 1/2″ angle x 3/16″ wall x 36″

The length of metal needed to make a scroll or spiral may be measured by first making the shape out of soft wire, then straightening it out and measuring the length. It may also be measured by using a divider as explained in section 7-10.

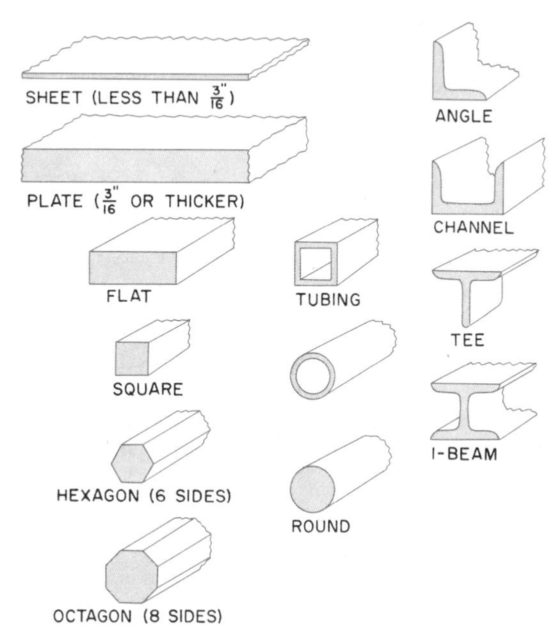

Fig. 5-4. Shapes of Standard Metal Stock

5-6. How Standard Stock Is Priced

Metal wholesalers normally sell all metals by weight. Prices are quoted as so many dollars and cents per hundred pounds. School suppliers, however, tend to sell bar stock by the lineal foot, sheet materials by the square foot, and casting metals by the pound. The prices charged for materials in your shop or laboratory very likely follow the pricing policies of school suppliers.

Words to Know

distance across flats
bill of materials
product plan
scroll
angle
channel
hexagon
I-beam
octagon
plate
sheet
tee
spiral
standard part
standard stock
steel mill

Review Questions

1. What is a product plan, and what information is generally included on it?
2. What is a bill of materials and what information should be on it?
3. Where may the information be found for making out a bill of materials?
4. What is meant by standard stock?
5. What are standard parts?
6. Should the size given on the bill of materials be that of the standard stock or of the finished part?
7. Explain how the length of a curve or scroll can be measured.

Planning

1. Develop complete plans for several products you have at home and which could be made in your shop or laboratory.
2. Make a complete plan for a product you expect to make.

Mathematics

1. What is the cost of 16″ of ½″ dia. steel rod which is priced at 15¢ per foot?
2. Figure the cost of 8¾ pounds of ¼″ steel plate which is priced at 12¢ per pound.
3. Find the cost of 126 square inches of 22 gage sheet metal which is priced at 24¢ per square foot.

PART III

Layout and Measurement

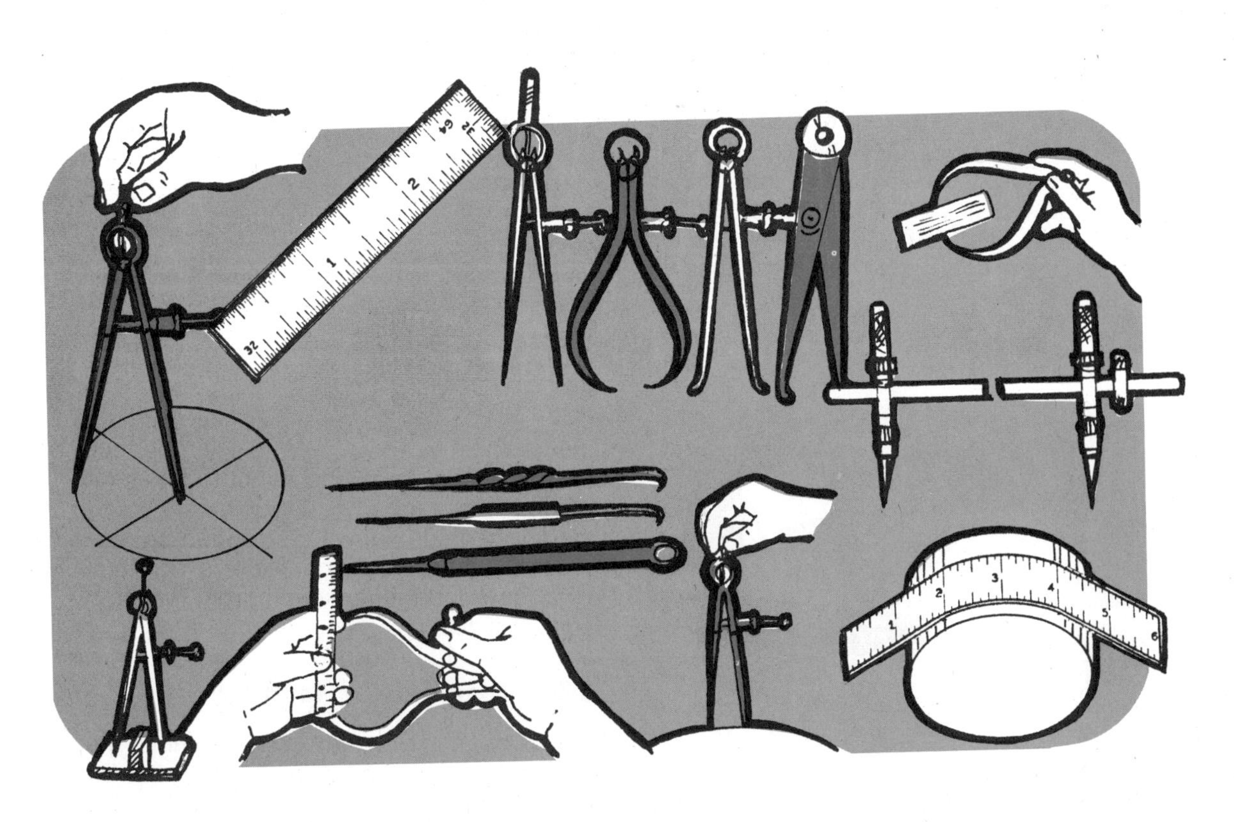

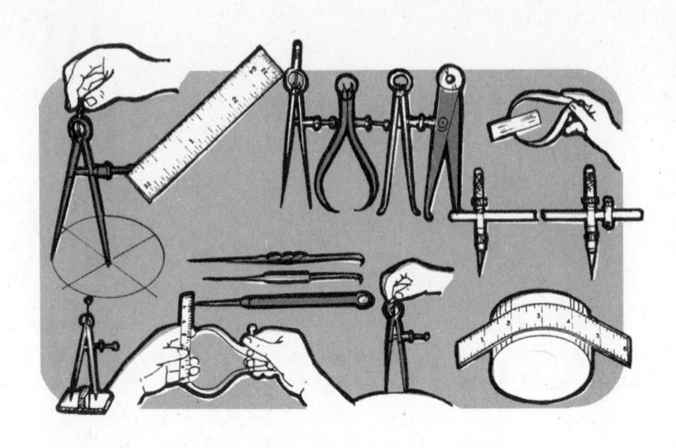

UNIT 6

Layout Tools

6-1. What Is Meant by Laying Out?

Laying out is the marking of lines, circles, and **arcs** on metal surfaces; such work is called **layout work.** It is the transferring of information from a **working drawing** to metal surfaces to show the worker at the machine or bench the location and amount of metal to be cut away. In many ways, laying out on metal is the same as making a **mechanical drawing** on paper.

6-2. Layout Tools

The tools used for making a drawing on metal are **layout tools.** They are different from those used for making a mechanical drawing because lines must be scratched or cut lightly into the metal. The layout tools that are most often used are described in the sections that follow.

Fig. 6-1. Surface Plate

6-3. Who Uses Layout Tools

The layout tools described in this unit are used in many trades. They are used by the **auto mechanic, bench mechanic, diemaker, diesinker, gage maker, inspector, layout artist, machine operator, machinist, metal patternmaker, ornamental ironmaker,** and **toolmaker.** Some of them are used by the **sheet metalworker.**

6-4. Surface Plate

The surface plate is a large iron or granite plate with a very flat surface, Fig. 6-1. It is cut on a machine and then carefully **ground** or **scraped.** The surface plate is an expensive tool. It may be placed on the bench and used as a foundation or table upon which to rest the work, gages, and other tools used to lay out work as shown in Figs. 6-21 and 6-22.

The surface plate should be handled with care so that the finished surface will not be nicked or scratched. A small nick will keep the work or layout tools from laying perfectly flat and might cause the layout to be wrong. When the surface plate is not in use, protect the finished surface with a **wooden cover** or case. If an iron surface plate is not to be used for a long time, keep it from rusting by coating it with oil.

6-5. Machinist's Hammer

The **head** of a machinist's hammer has a ball-shaped **peen**; thus it is called a **ball**

peen hammer, Fig. 6-2. The **face** and peen are **hardened;** the middle contains the **eye,** which is the hole and is left soft. The **face** should be rounded a little so that it will not make marks. Machinist's hammers weigh from 1 ounce to 3 pounds, without the handle. A hammer weighing ¾ to 1¼ pounds is used for ordinary work around the shop. A light hammer is better for layout work; one weighing from two to six ounces can be handled easily since only very light blows are necessary. The peen end is sometimes used for riveting and for making decorative marks on wrought iron products.

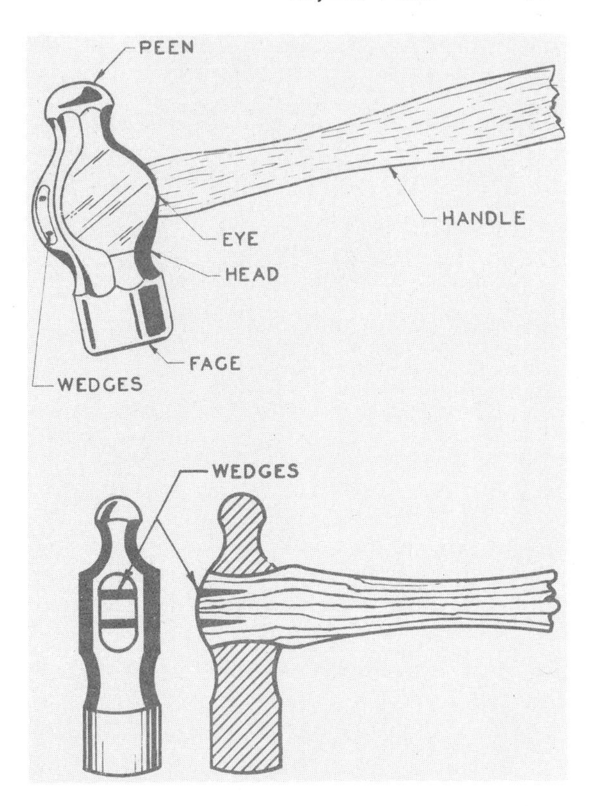

Fig. 6-2. Machinist's (Ball Peen) Hammer

6-6. Hammer Handle

The **hammer handle** is made of a kind of wood called **hickory.** It is thinner near the middle to make it springy so that the shock of the blow will not hurt the wrist. The length of the handle depends upon the size of the hammer. The handle should fit the eye, which is smaller in the middle than at the outside. This helps to keep the head from slipping off the handle after it is fastened with steel **wedges.**

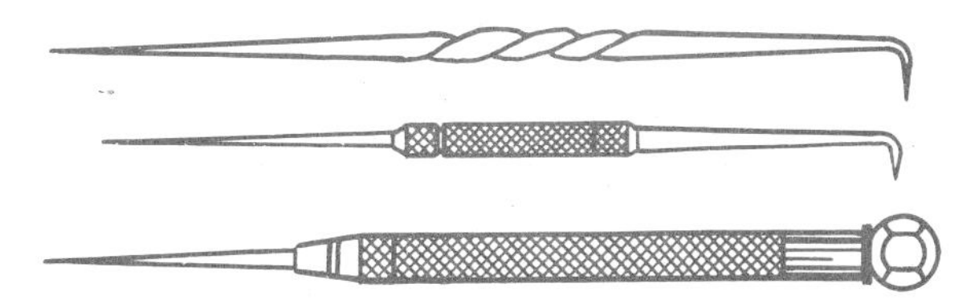
Fig. 6-3. Scribers

6-7. Scriber

The scriber is a piece of **hardened steel** about 6″ to 10″ long, pointed on one or both ends like a needle, Fig. 6-3. It is held like a pencil to scratch or **scribe** lines on metal. The bent end is used to scratch lines in places where the straight end cannot reach. Sharpen the points on an **oilstone.**

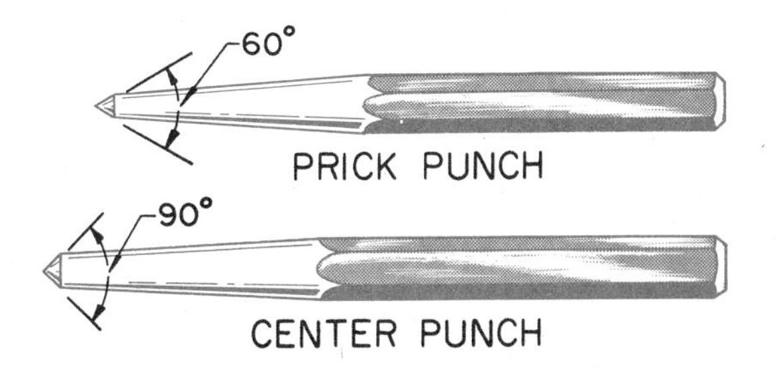

Fig. 6-4. Difference between Prick Punch and Center Punch

6-8. Prick Punch

The prick punch is a sharply pointed tool of **hardened steel,** Fig. 6-4. It is used to accurately locate points to be center punched as well as to make small **punch marks** on layout lines in order to make them last longer.

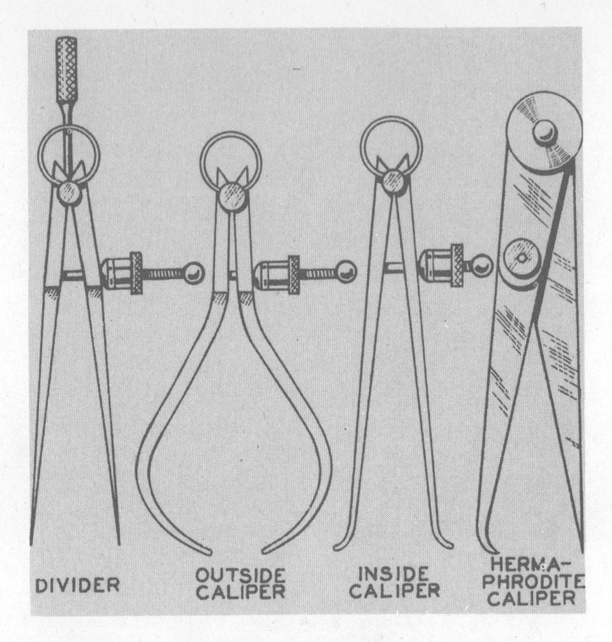

Fig. 6-5. Divider and Calipers

6-9. Center Punch

The center punch looks like a prick punch, Fig. 6-4. It is usually larger than the prick punch and has a 60° or 90° point, while the prick punch has a sharper point. It is also made of **hardened steel.**

The center punch is used only to make the prick-punch marks larger at the centers of holes that are to be drilled, hence the name center punch.

6-10. Divider

The divider is a two-legged steel instrument with hardened points, Fig. 6-5. Its size is measured by the greatest distance it can be opened between the two points. Thus, a 4″ divider opens 4″ between the points, a 6″ divider opens 6″ between the points, etc. Dividers are used to scribe circles and parts of circles (Fig. 7-8), to lay off distances (Fig. 7-9), and to measure distances. See § 7-10 and Figs. 7-10 and 7-11. Both points should be even in length; sharpen on an **oilstone.** The wing divider is shown in Fig. 31-4.

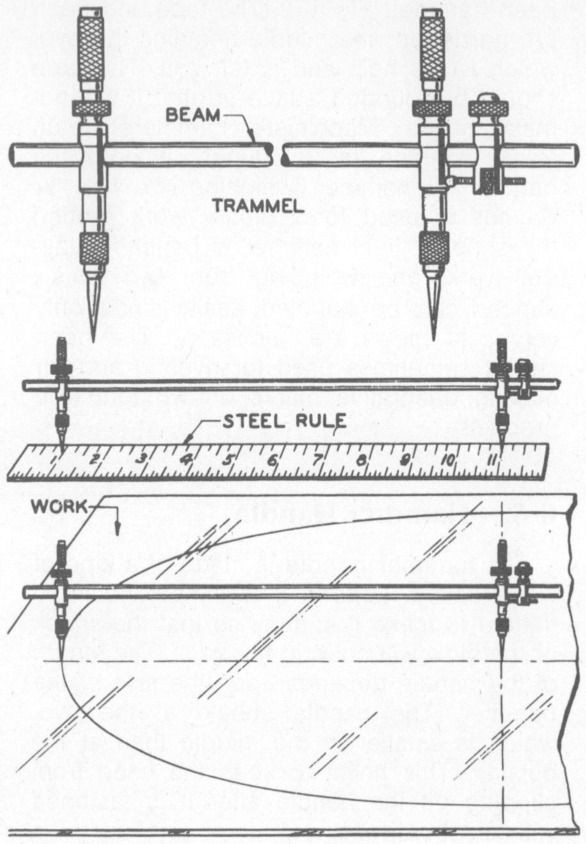

Fig. 6-6. Trammel

6-11. Trammel

A large circle or an **arc** having a large **radius** may be made with a tool called a **trammel,** Fig. 6-6. It is sometimes called a **beam compass** or **beam trammel** and may also be used to measure distances in the same way that a divider is used. The size of the circle or arc that may be made or the distance that may be measured depends on the length of the **beam.**

6-12. Outside Caliper

An outside caliper is a two-legged steel instrument with its **legs** bent inward. See Fig. 6-5. Its size is measured by the greatest

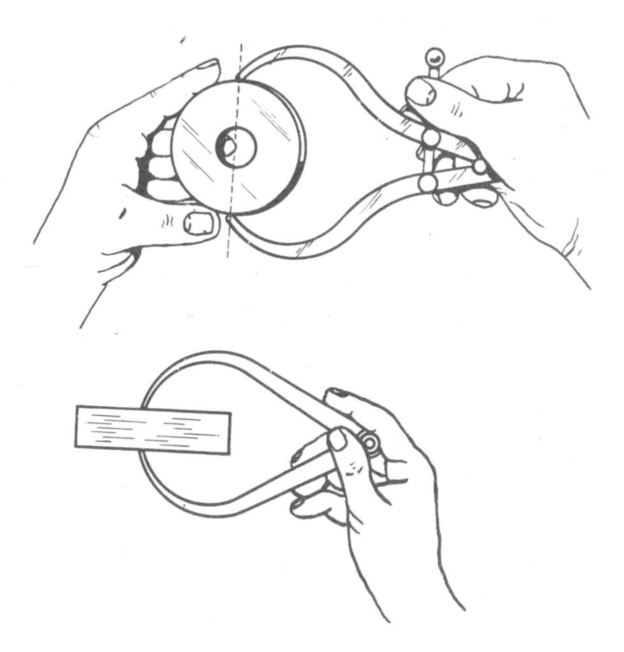

Fig. 6-7. Uses of Outside Caliper

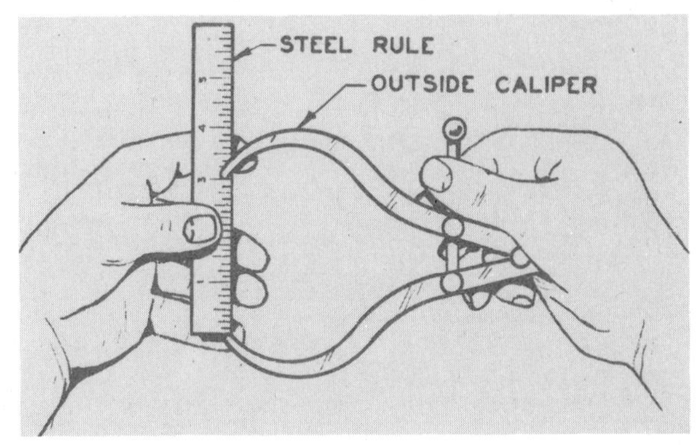

Fig. 6-8. Setting Outside Caliper to Steel Rule

distance it can be opened between the legs. It is used to measure the outside diameters of round objects and to measure widths and thicknesses as is shown in Fig. 6-7. The distance between the legs is then measured with a rule as in Fig. 6-8.

6-13. Fine Measuring with Outside Caliper

Fine measurements with an outside caliper depend upon the sense of **touch** and **feel** in the fingertips. The caliper should, therefore, be held gently with the fingertips. It should be moved back and forth over the work and set until both legs just touch the sides. It is very easy to force the legs over the work and get a wrong measurement. The correct touch or feel is obtained only through much practice.

6-14. Setting Outside Caliper

To set an outside caliper to a certain size, hold a steel rule in the left hand and, with the right hand, place one leg of the caliper against the end of the steel rule, Fig. 6-8. Place the finger behind the end of the steel rule to keep the leg from slipping off while

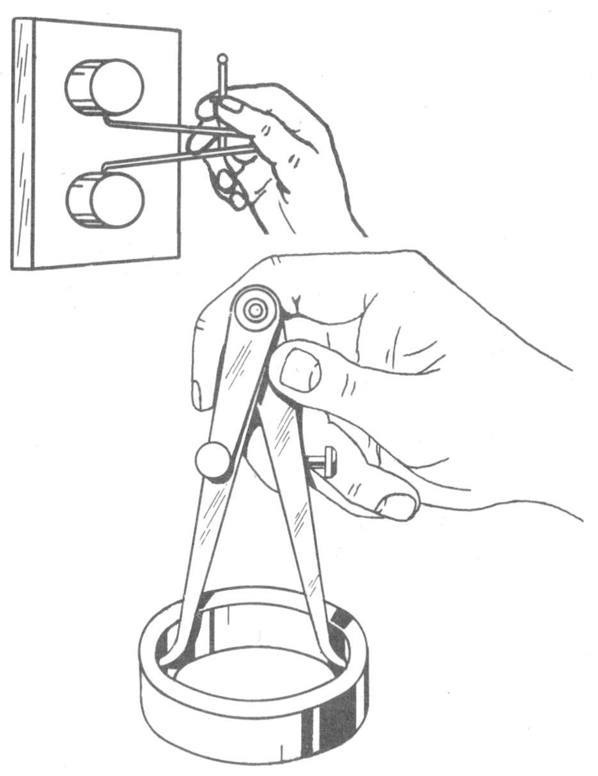

Fig. 6-9. Uses of Inside Caliper

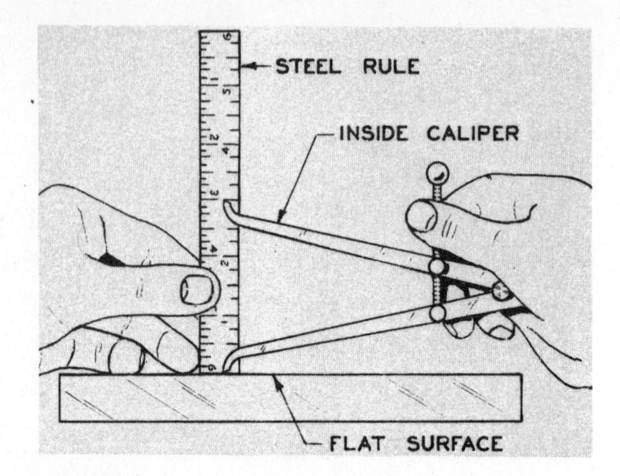

Fig. 6-10. Setting Inside Caliper to Steel Rule

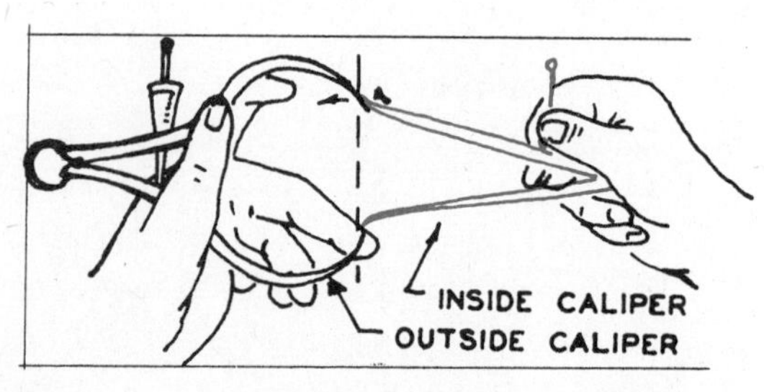

Fig. 6-11. Transferring a Measurement from One Caliper to Another

setting the other leg of the caliper to the size wanted.

6-15. Inside Caliper

An inside caliper is a two-legged steel instrument with its legs bent outward, Fig. 6-5. It is used to measure the diameters of holes or to measure spaces as in Fig. 6-9; it is then measured on the rule as in Fig. 6-10.

6-16. Fine Measuring with Inside Caliper

Fine measurements with an inside caliper depend upon the sense of **touch** or **feel** in the fingertips. The caliper should, therefore, be held gently with the fingertips and moved back and forth in the hole until both legs just touch the sides. It is very easy to spring the legs, thus obtaining a wrong measurement.

6-17. Setting Inside Caliper

To set an inside caliper to a certain size, hold the end of the steel rule against a flat metal surface and then set the caliper as in Fig. 6-10.

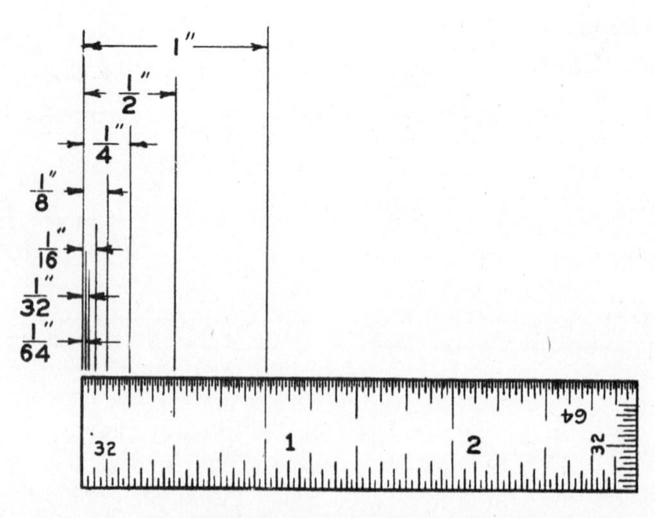

Fig. 6-12. Divisions of One Inch (Actual Size)

6-18. Transferring a Measurement from One Caliper to Another

To transfer a measurement from an inside caliper to an outside caliper, or from an outside caliper to an inside caliper, hold the caliper which already has the measurement in the left hand and the one which is to be set in the right hand. Set the caliper in the right hand to the size of the one in the left hand as in Fig. 6-11. You should feel the points touching lightly.

6-19. Hermaphrodite Caliper

The hermaphrodite caliper has one pointed leg like a divider and one bent leg, Fig. 6-5. It is used to find the center on the end of a round bar (see Fig. 7-14) and to scribe lines **parallel** to the edge of a piece of work. See § 7-15 and Fig. 7-18.

6-20. Steel Rule

The steel rule (Fig. 6-12), also called **machinist's rule,** is made in many thicknesses, widths, and lengths from ¼″ to 4′. The most commonly used steel rule is 6″ long. Because a steel rule is used to measure, it should be handled with care to keep the edges from becoming nicked or worn round.

The edges of steel rules are divided by fine lines into different parts of an inch, such as 8ths, 16ths, 32nds, and 64ths of an inch. See § 9-1. The smallest division is $\frac{1}{64}$″, the next larger is $\frac{1}{32}$″, the next is $\frac{1}{16}$″, then ⅛″, ¼″, and ½″. Some rules are divided into 10ths and 100ths of an inch. The divisions are called **graduations.** Some rules have divisions on the ends to make measurements in small spaces. Steel rules with metric measurements are available. See Unit 10.

A thin, springy rule is called a **flexible steel rule** and is used to measure curves, Fig. 6-13. The use of a **hook rule** is shown in Fig. 7-2. It is also useful in setting an **inside caliper.** See Fig. 6-10. Section 8-4

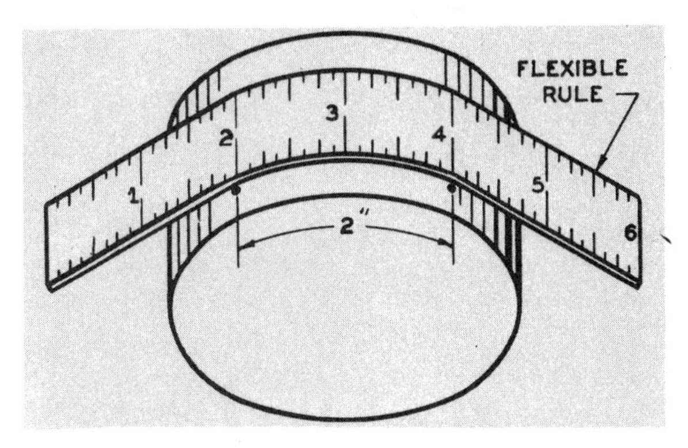

Fig. 6-13. Measuring with a Flexible Rule

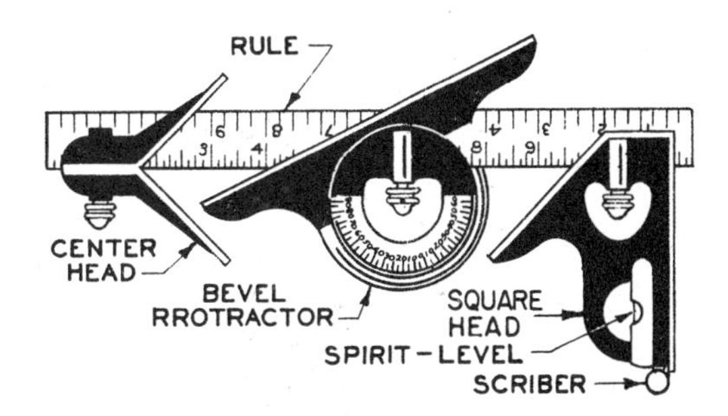

Fig. 6-14. Combination Set

explains the **decimal rule,** and the **shrink rule** is explained in section 39-2. Section 7-4 gives more information about measuring with a steel rule.

6-21. Combination Set

The combination set, Fig. 6-14, is the most commonly used set of tools in the **machine shop.** The set includes a **square head, center head, bevel protractor, spirit level, steel rule,** and **scriber.** The rule or blade may be fastened quickly to each of the first three; the beginner should ask how it is done so that the small parts will not be lost or damaged.

Figures 6-15, 7-4, 7-15, and 7-23 show a few of the uses that can be made of the steel rule when the square head, bevel protractor, or center head are fastened to it.

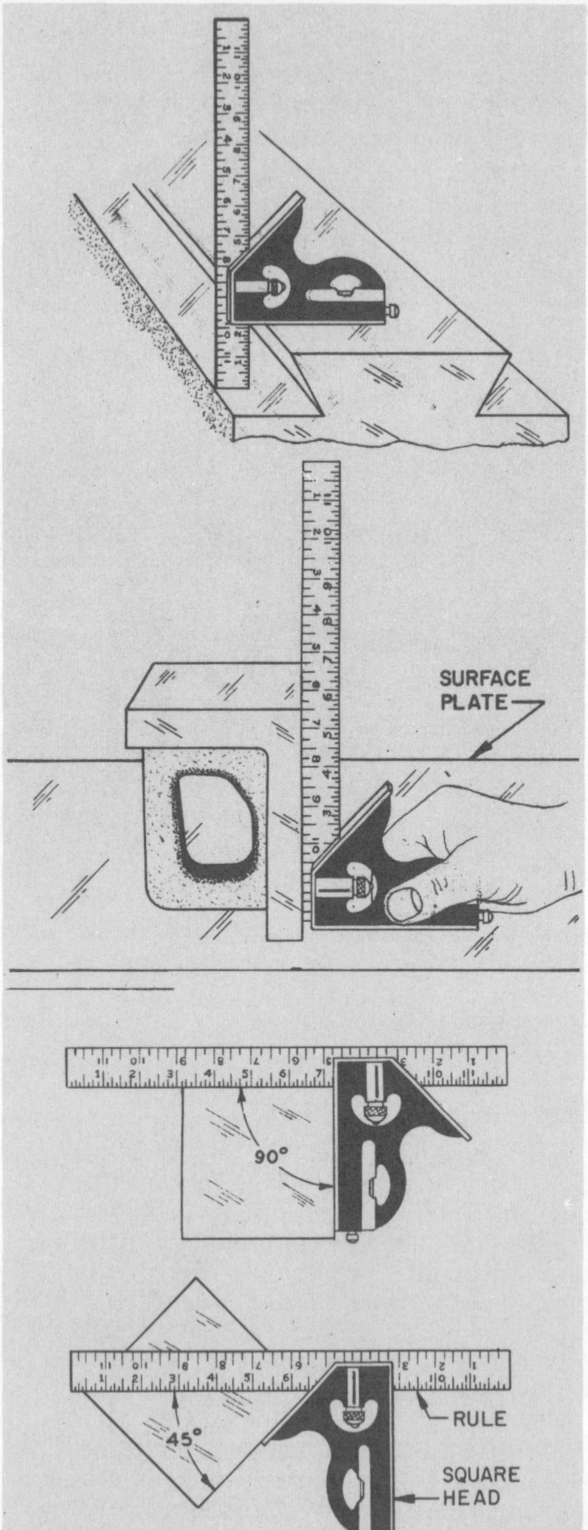

Fig. 6-15. Uses of a Combination Square

6-22. Combination Square

The combination square has many uses, Fig. 6-15. See Figs. 6-14, 6-21, and 7-5. Note that 45° and 90° can be measured with it. Its **squareness** should be tested as explained in section 6-27.

6-23. Center Head

The center head may be used to extend a line around a corner, Fig. 6-16.

The center head with the steel rule fastened to it is called a **center square.** It

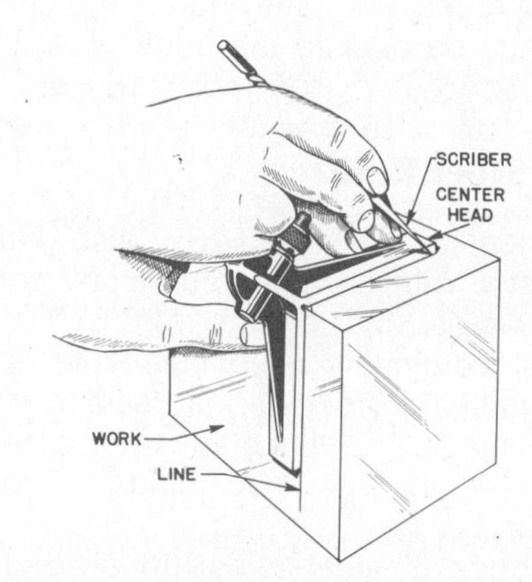

Fig. 6-16. Extending a Line Around a Corner with a Center Head

Fig. 6-17. Setting Work at an Angle with a Spirit Level on a Bevel Protractor

is used to find the center of a round piece. See Fig. 7-15.

6-24. Bevel Protractor

The bevel protractor, Fig. 6-14, is divided into **degrees** and with the rule fastened to it, any **angle** can be measured. See Figs. 7-4 and 15-10.

6-25. Spirit Level

A **spirit level** is usually fitted into the **bevel protractor**, Fig. 6-17, and the **head** of the combination square to help in **leveling** the work or setting it at an **angle.**

6-26. Solid Steel Square

The solid steel square is made in one piece, both **blade** and **beam,** Fig. 6-18. It is made of **hardened steel** and is more exact than the **combination square** with its sliding parts which may become worn, dirty, or nicked. The uses of the solid steel square are shown in Fig. 6-19.

6-27. Testing Squareness of a Square

The squareness of any square may be tested by placing the **beam** of the square against a **straightedge** with the **blade** resting on a smooth surface, Fig. 6-20. While holding the square in this position, scribe a line along the edge of the blade. Then turn the square over as shown by the dotted lines and see if you have the same line. Both inside and outside edges of the blade should be tested this way.

6-28. Surface Gage

The surface gage has a heavy, flat **base** carrying a **spindle** which may be set at any angle, Fig. 6-21. A **scriber,** which may also be set at any angle or at any height, is clamped to the spindle. The surface gage scribes a line **parallel** (see § 7-15) to a surface or to another line and is often used as a **height gage.** It is often placed on a **surface plate** and used as shown in Figs. 6-21 and 6-22. The **adjusting screw** is used for fine setting. The height of the point of the scriber may be measured with a rule as in

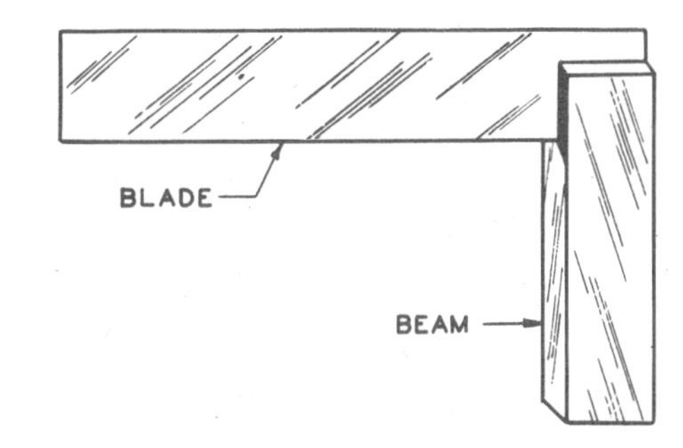

Fig. 6-18. Solid Steel Square

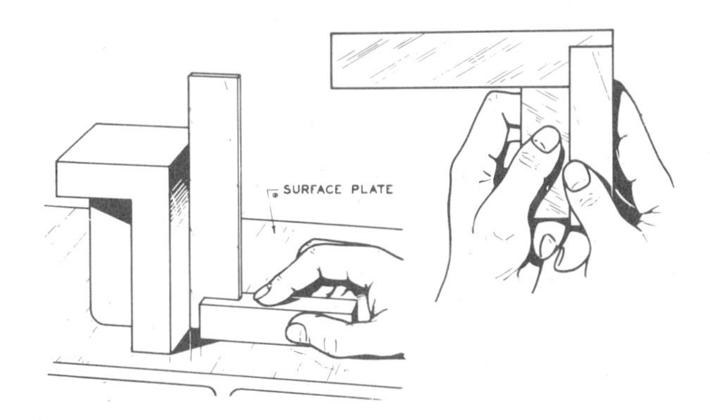

Fig. 6-19. Uses of Solid Steel Square

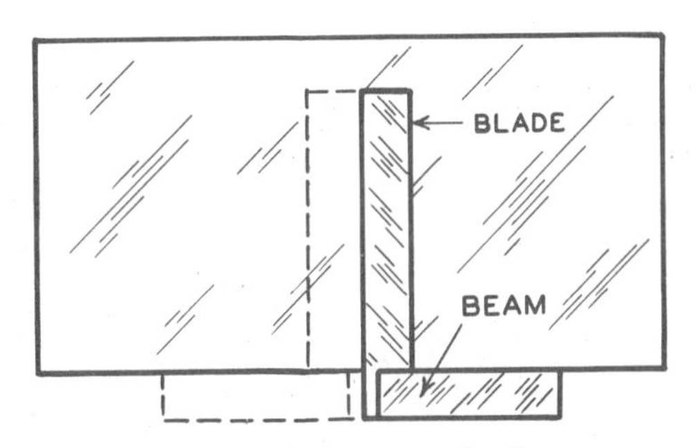

Fig. 6-20. Testing Squareness of a Square

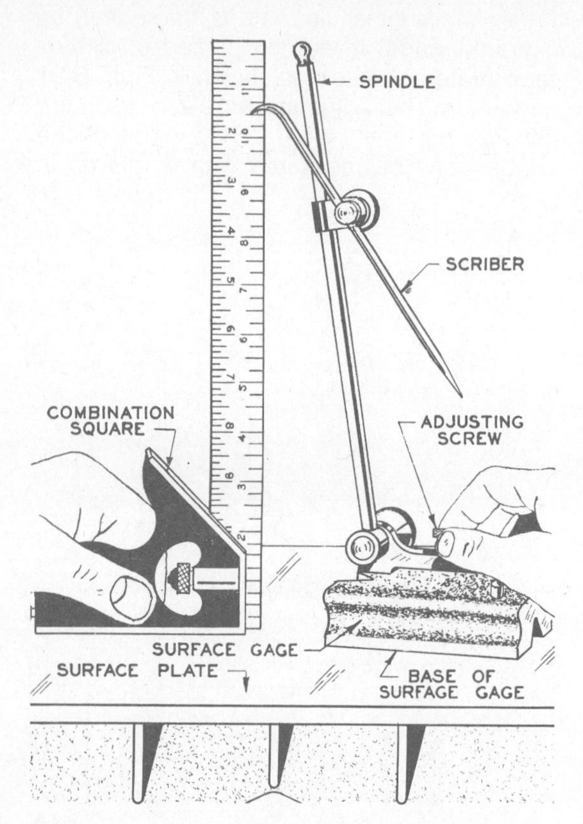

Fig. 6-21. Setting the Height of a Surface Gage

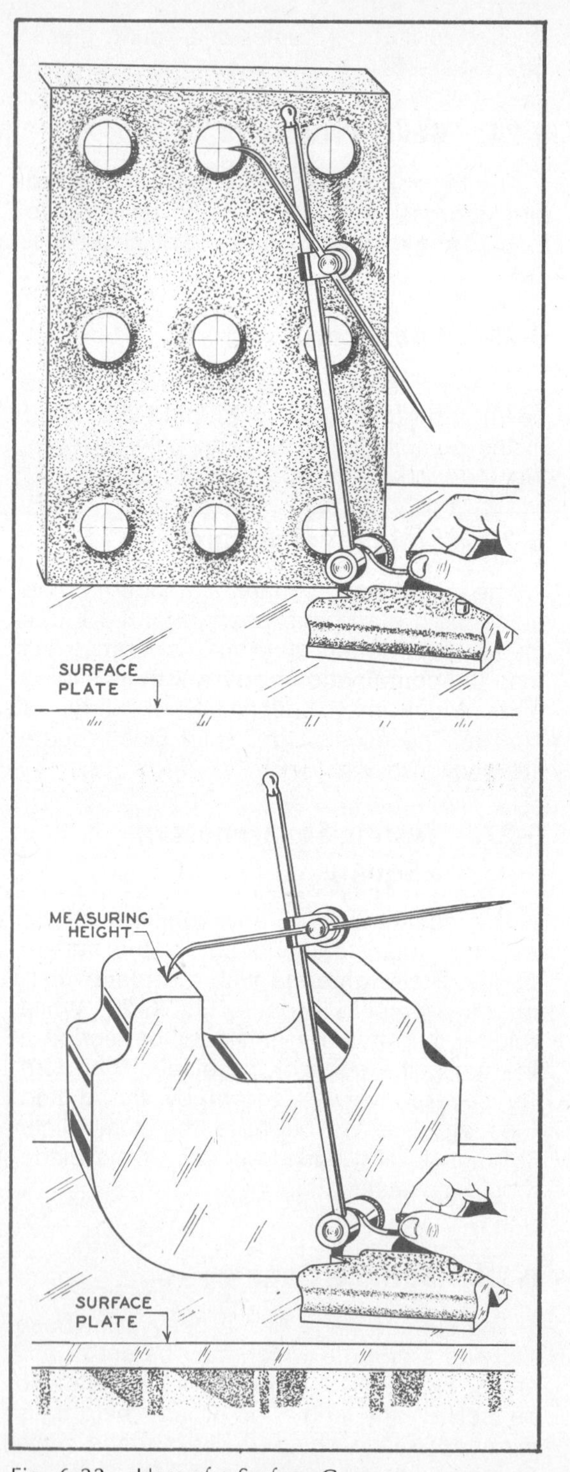

Fig. 6-22. Uses of a Surface Gage

Fig. 6-21. The V-shaped grooves at one end and at the bottom make it useful on or against round work.

6-29. Vernier Height Gage

The vernier height gage, Fig. 6-23, may be used for doing the same kinds of layout work as the surface gage. It is a precision layout tool which utilizes beam graduations and a vernier scale to provide positioning of the scriber to an accuracy of one-thousandth of an inch. Unit 9 explains how to read vernier scales.

6-30. Angle Plate

The angle plate has two **surfaces** at a **right angle** to each other, Fig. 6-24. It is

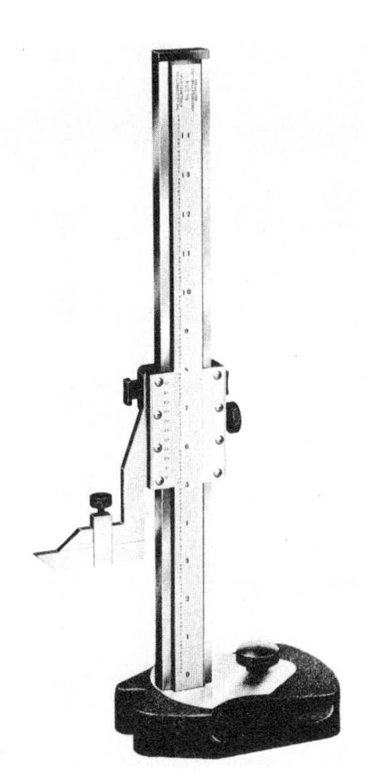

Fig. 6-23. Vernier Height Gage

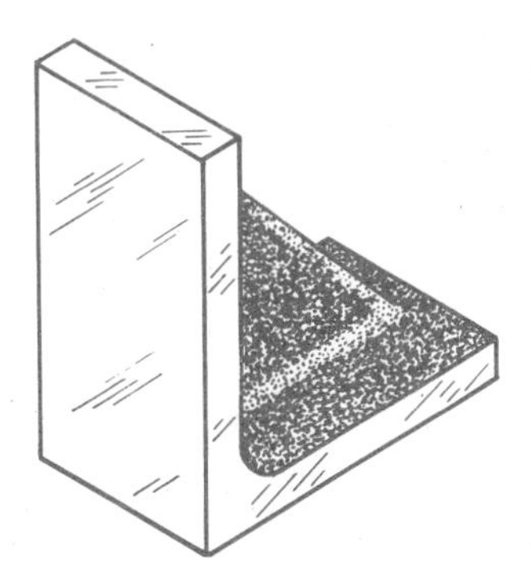

Fig. 6-24. Angle Plate

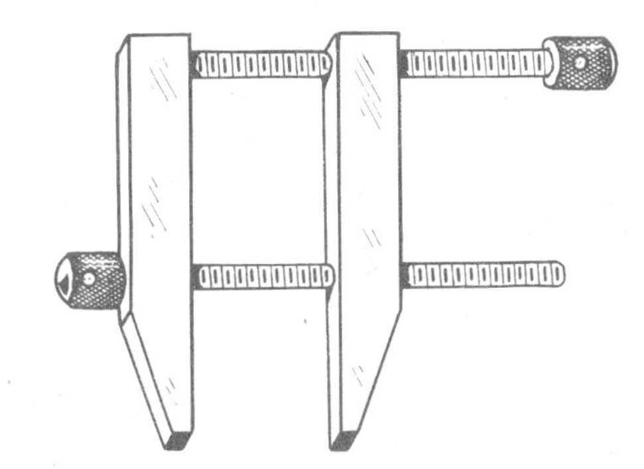

Fig. 6-25. Parallel Clamp

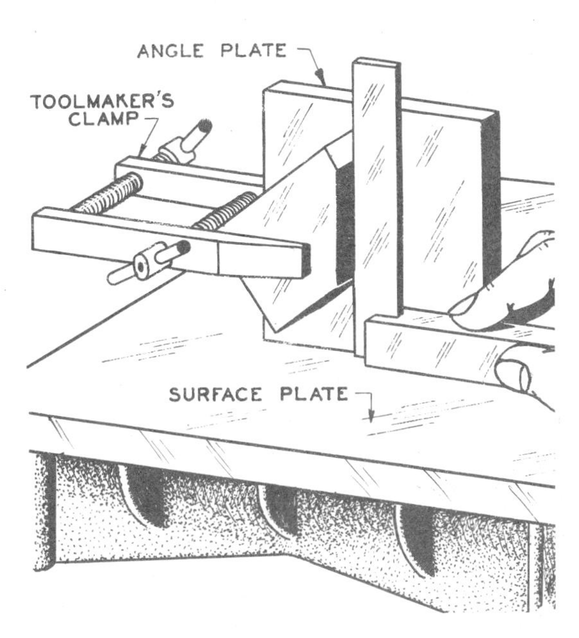

Fig. 6-26. Uses of Parallel Clamp and Angle Plate in Layout Work

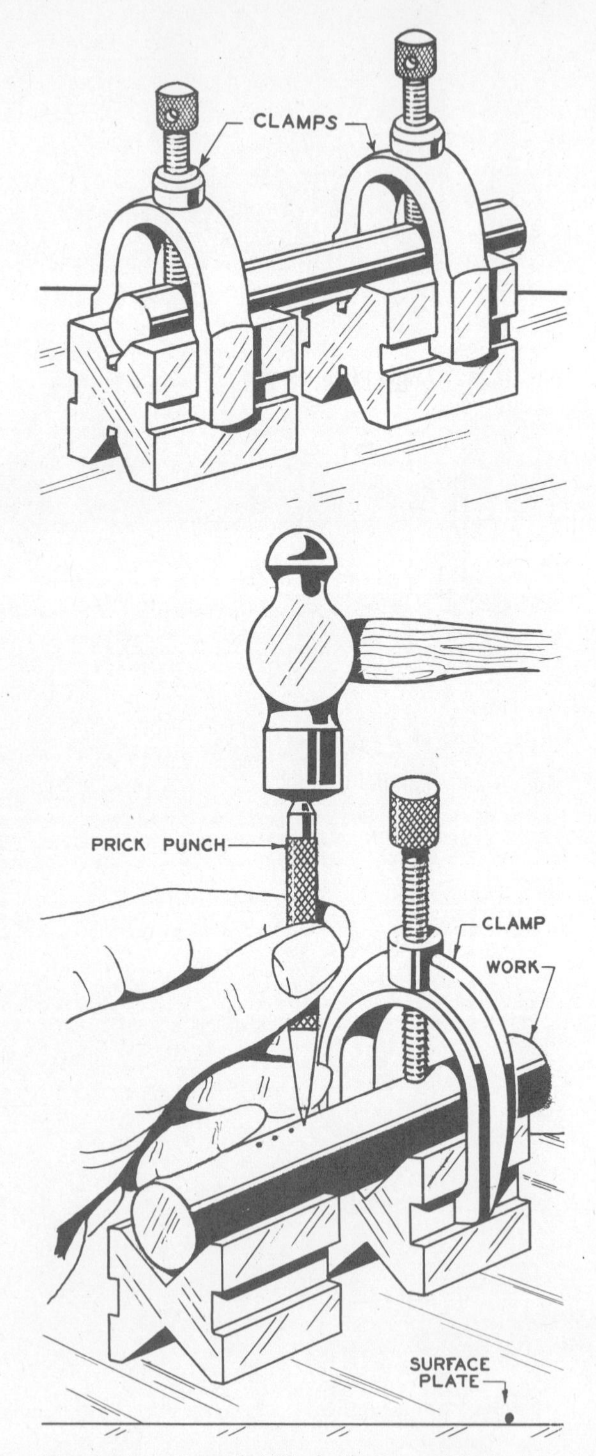

Fig. 6-27. Uses of V-Blocks in Layout Work

often necessary to clamp the work that is to be laid out to the angle plate, Fig. 6-26.

6-31. Parallel Clamp

The **parallel** clamp, also called a toolmaker's clamp, Fig. 6-25, is often used to hold parts together or to hold the work against an **angle plate** while laying out as in Fig. 6-26. The jaws should always be kept **parallel.**

6-32. V-Block

The V-block is a block of steel with V-shaped grooves, Fig. 6-27. It is used to hold round work for laying out.

Words to Know

angle plate
ball peen hammer
beam compass
center punch
combination set
combination square
graduation
hermaphrodite caliper
hook rule
inside caliper
machinist's hammer
outside caliper
parallel clamp
prick punch
scriber
solid steel square
surface gage
surface plate
toolmaker's clamp
trammel
V-block
vernier height gage

Review Questions

1. Why are layout lines put on metal?
2. For what is a surface plate used?
3. Why is the surface plate an expensive tool?
4. How should the surface plate be protected when not in use? Why should it be protected?
5. What kind of hammer is a machinist's hammer? In what sizes are they made?
6. What is a scriber? For what is it used?
7. How may a scriber be sharpened?
8. What is the difference between a prick punch and a center punch? For what is each used?
9. What is a divider? For what is it used?

10. How far can a 4″ divider be opened?
11. What is the diameter of the largest circle that can be made with a 4″ divider?
12. Describe a trammel. For what is it used?
13. Describe an outside caliper. For what is it used?
14. How is the size of an outside caliper measured?
15. For what is an inside caliper used?
16. Describe the hermaphrodite caliper. For what is it used?
17. Write in numbers: six feet, seven and three-quarters inches.
18. Name the parts of the combination set.
19. What two angles can be made with a combination square?
20. Why is a solid steel square more exact than a combination square?
21. Describe the surface gage. How is it used?
22. How are fine adjustments made on the surface gage?
23. What is an angle plate? For what is it used?
24. Describe a parallel clamp. For what is it used?
25. Describe the V-block. Name several uses for it.
26. For what is a vernier height gage used?

Drafting

1. What instrument is used in the drafting room as the scriber is used in the shop? The divider? The surface plate?
2. What tool or instrument is used in the drafting room as the steel rule is used in the shop? The bevel protractor? The steel square?
3. Draw circles of the following diameters: 1½″, 3¼″, 2⅝″, 3 9/16″, 1 19/32″.

Mathematics

1. How many 64ths are there in ⅜″? In ⅝″? In 15/16″?
2. How much larger is ⅛″ than 1/16″?
3. How much is one-half of ¾″?
4. How much is one-half of ½″?
5. How much is one-half of ¼″?
6. How much is one-half of ⅛″?
7. Draw lines of the following lengths: 5½″, 4¾″, 2⅝″, 1 11/16″, 3 27/32″, 2 49/64″.
8. What fractional part of a circle is 90 degrees? 15 degrees? 315 degrees?

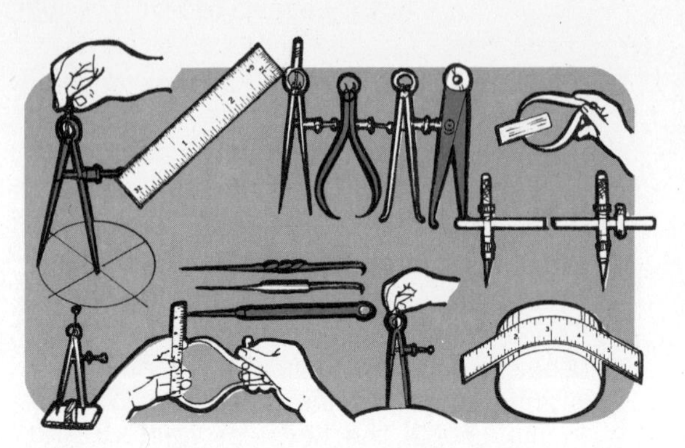

UNIT 7

Layout Techniques

7-1. Layout Work Is Important

The ability to do accurate layout work is a greater skill than the ability required to run some simple machine tools. The layout artist must take information from working drawings and use layout tools to transfer it accurately to the material being worked on. Any mistake in the layout means that the workpiece may be made incorrectly. If the mistake cannot be corrected, then the workpiece is spoiled. In either case, mistakes made in layout cause waste, either in labor or materials, and are therefore very expensive. Mistakes are easiest and least expensive to correct at the layout stage. All layout work should be checked and rechecked against the working drawings before cutting or other work begins.

7-2. Coloring Metal for Layout

The first step in layout work is to color the surface upon which the lines are to be made. Some surfaces are rough while others are smooth and bright. To see the lines on the surface easily, it is necessary to color it. Rough surfaces may be covered with white chalk or a white paint. The chalk should be rubbed in with the fingers until smooth.

Iron and steel surfaces that are smooth and bright may be coated with **layout fluid.**

7-3. Layout Fluid

A **layout fluid** that comes in different colors may be bought for coloring metal surfaces in the laying out process. The metal is first rubbed clean and then the fluid is put on with a brush or sprayed on, Fig. 7-1. It dries quickly. Use steel wool or alcohol to remove the dried layout fluid.

Fig. 7-1. Coloring Metal with Layout Fluid

7-4. Measuring with Steel Rule

To measure with a steel rule, stand the rule up on its **edge** on the work so that the lines on the rule touch the work, Fig. 7-2. The **division line** on the rule has a certain thickness. Always measure to the center of the line. Measure from the 1″ mark because the end of the rule may be worn. Figure 6-13 shows how to use a **flexible steel rule.**

7-5. Scribing Lines

After the layout fluid has dried on the metal surface, the layout can be made. To draw a straight line, place the steel rule, square, or **bevel protractor** in the correct position and hold it against the work with the left hand, Fig. 7-4. Hold the **scriber** with

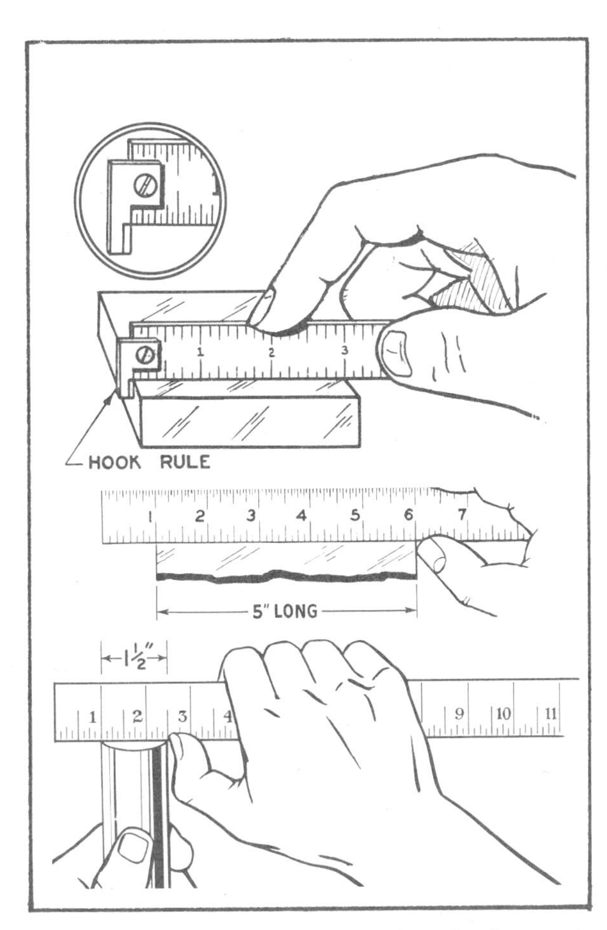

Fig. 7-2. Measuring with Edge of Steel Rule

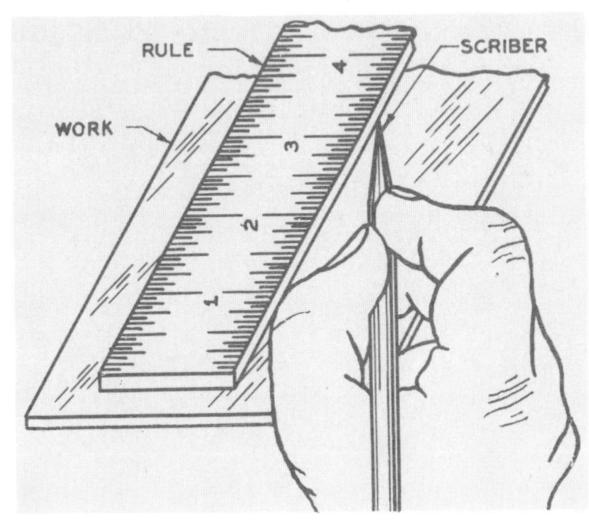

Fig. 7-3. Slant the Scriber so the Point Follows the Lower Edge of the Rule

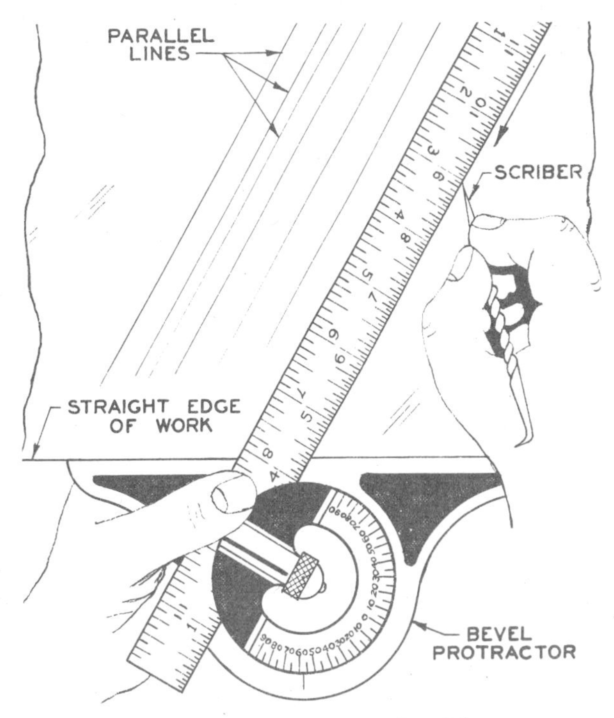

Fig. 7-4. Scribing 57° Lines, Using a Bevel Protractor

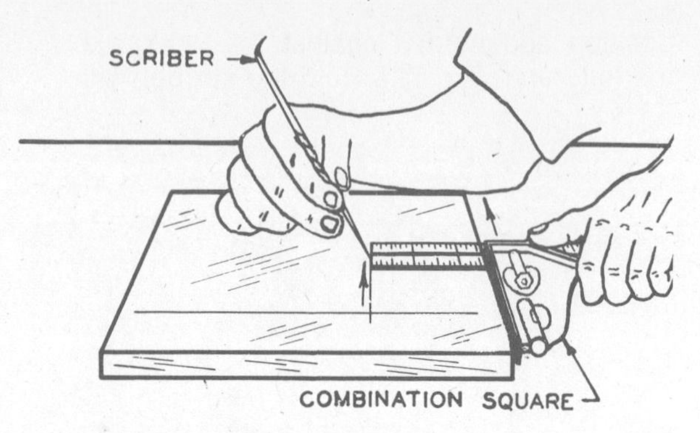

Fig. 7-5. Scribing a Line, Using a Combination Square

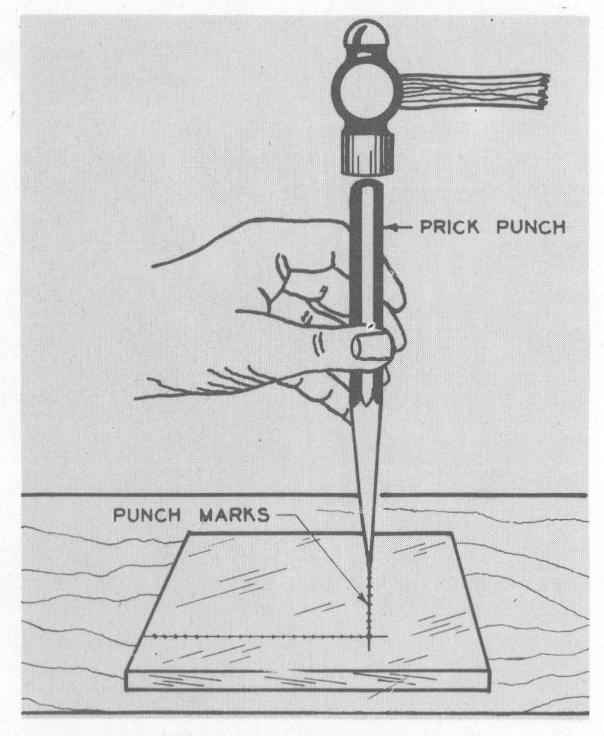

Fig. 7-6. Prick Punching a Line

the right hand just as you hold a pencil and lean it to one side so that the point will draw along the lower edge of the rule, Figs. 7-3 — 7-5, 7-15, 7-23, and 7-28. Scratch one line. Use a sharp scriber; sharpen it on an oil-stone. See § 44-12.

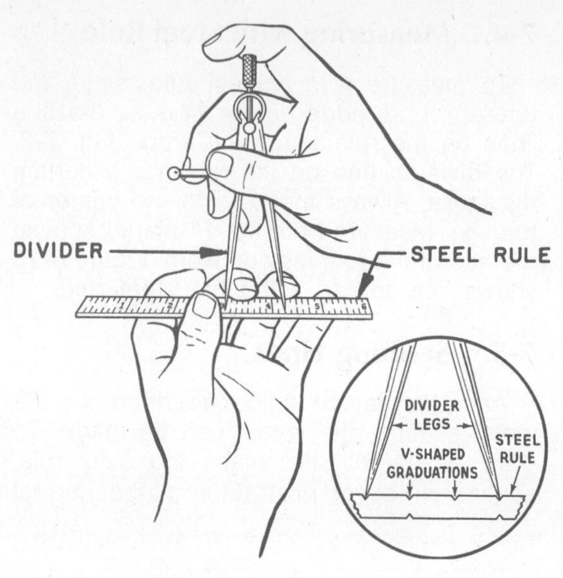

Fig. 7-7. Setting Divider

7-6. Prick Punching

The **coloring** (see § 7-2) and the scribed lines of a layout wear off in the handling of the work. The lines should be **prick punched** to make them last longer, Fig. 7-6. The point of the prick punch should be placed exactly on the line, held squarely on the surface, and struck lightly with the hammer. The **prick-punch marks** should be closer together on curved lines than on straight lines.

7-7. Setting a Divider

To make a circle, set the divider to the size of the **radius,** half of the **diameter** (see § 4-5). For example, if the circle is to be 2″ in diameter, the divider should be set at 1″. Setting a divider to a certain size is shown in Fig. 7-7. Place one point of the divider into one of the lines of the steel rule, usually the 1″ line, then move the other point until it exactly splits the other line at which the divider is to be set. The V-shape of the lines helps to set the divider points correctly by feeling. Handle carefully to avoid springing.

7-8. Scribing Circle with a Divider

To scribe a circle with a divider, hold it by the **stem,** Fig. 7-8. Place one point in the prick-punch mark at the **intersection** of the lines to keep it from slipping and swing it to the right or left. Scratch a sharp circle.

7-9. Laying Out Equal Distances with a Divider

The divider may be used to divide a line or circle into a number of equal parts. This is shown in Fig. 7-9.

7-10. Measuring with a Divider

The divider may be used to measure distances, Fig. 7-10; the distance between the

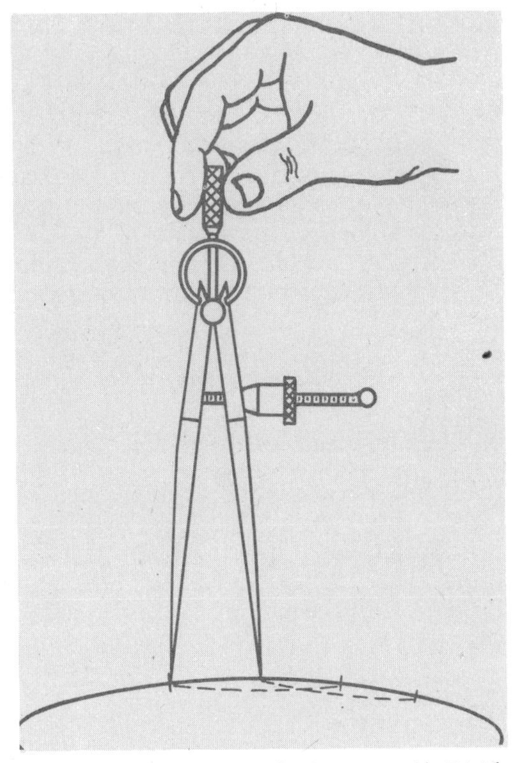

Fig. 7-9. Laying Out Equal Distances with Divider

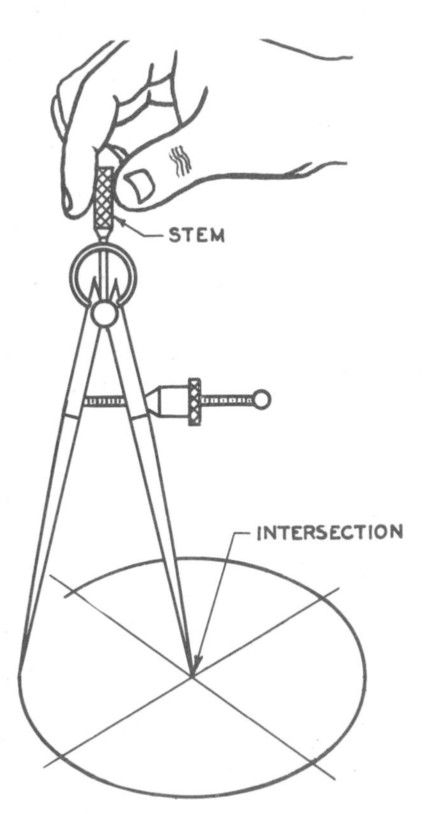

Fig. 7-8. Scribing a Circle with Divider

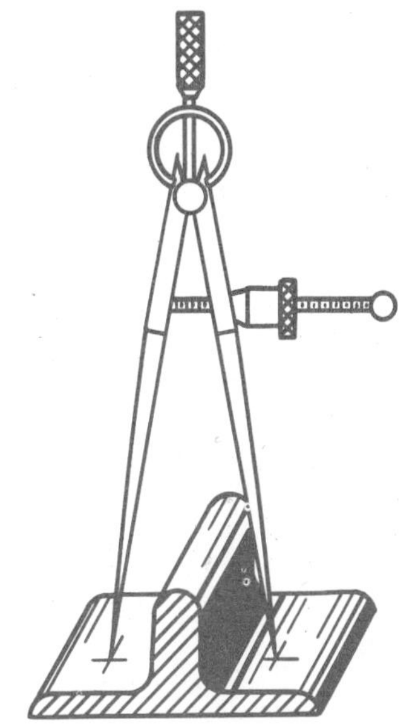

Fig. 7-10. Measuring Distances with Divider

divider points may then be read on the rule. See Fig. 7-7.

Figure 7-11 shows how the divider may also be used to measure the length of a curve or a **scroll.** This is done by setting the divider to a certain distance, ½″ approximately and then **stepping off** along the **measuring line** or **neutral line** of the curve or scroll and counting the number of steps. When a metal curve is bent, the outside of the metal stretches and the inside squeezes together, but the measuring line does not change.

7-11. Finding Center of Circle

To find the center of a circle, put four small prick-punch marks on the **circumference.** Then, with the divider set to the **radius** of the circle, scribe four **arcs** as in Fig. 7-12. The center of the circle is between the four arcs, Fig. 7-13.

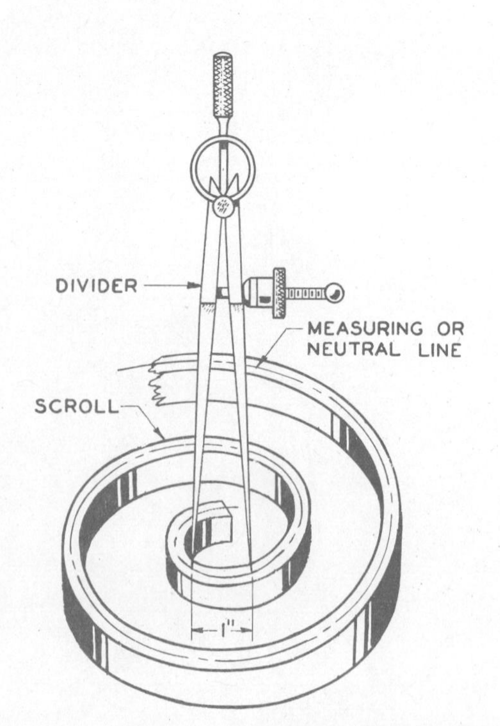

Fig. 7-11. Measuring Length of Curve with Divider

7-12. Finding Center on the End of a Round Bar

The center on the end of a round bar may be found in several ways. The **hermaphrodite caliper** may be used to scribe four arcs while the bar is held in a vise, Fig. 7-14.

The **center square** may be used to find the center of a round piece of metal, Fig. 7-15. Any two lines drawn across the end

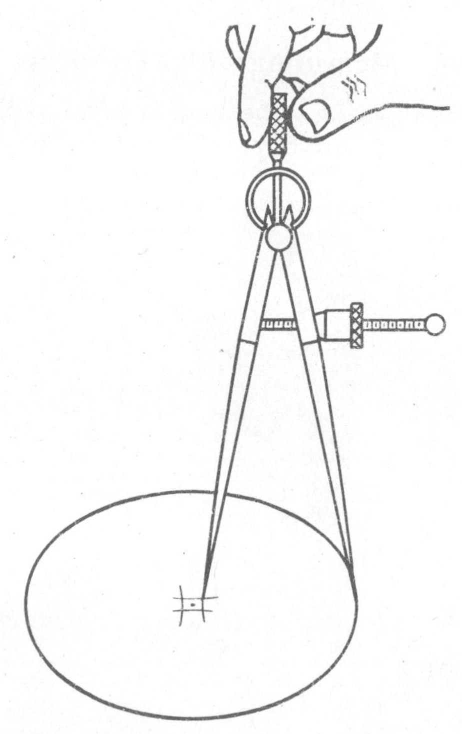

Fig. 7-12. Finding the Center of a Circle with Divider

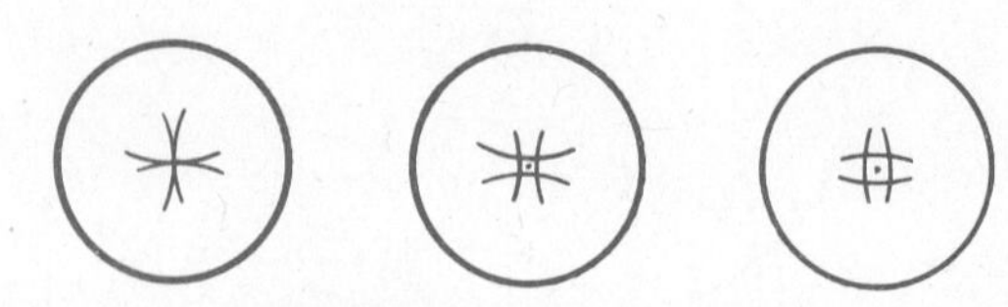

Fig. 7-13. Center of the Circle Is between the Four Arcs

of a round piece of metal with a center square will cross at the center.

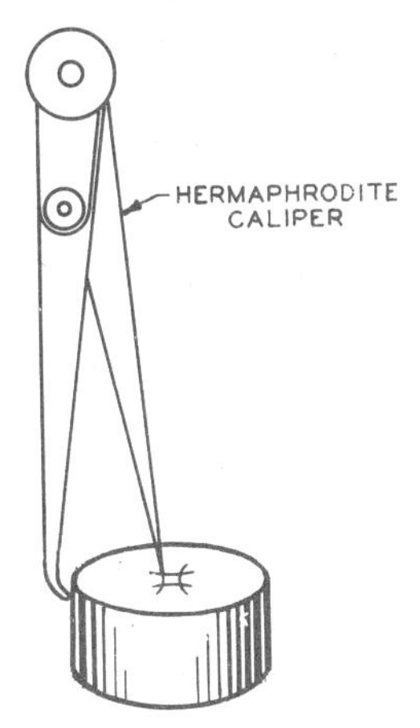

Fig. 7-14. Finding Center of a Round Bar with a Hermaphrodite Caliper

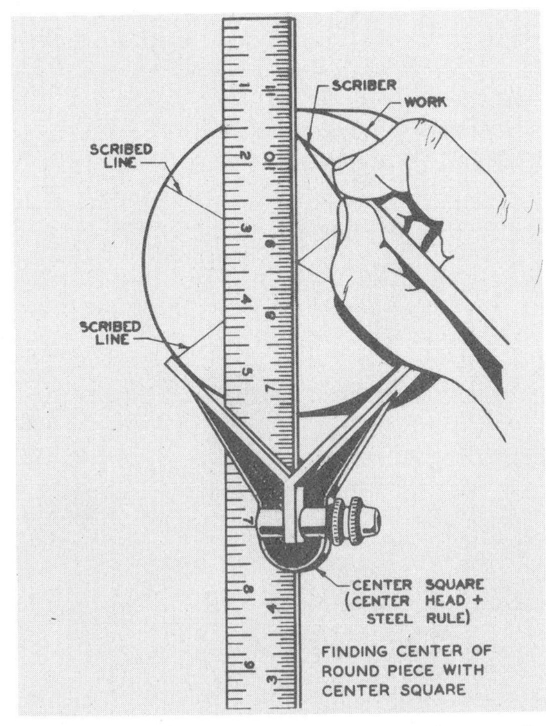

Fig. 7-15. Finding the Centers of Round Pieces

The center of a round bar may also be found with a **surface gage** by following the same procedure as that used for the hermaphrodite caliper.

7-13. Finding Center of an Arc

Suppose that the arc **XY** in Fig. 7-16 is already on the metal surface and that it is necessary to locate its center. To find the center of the arc:

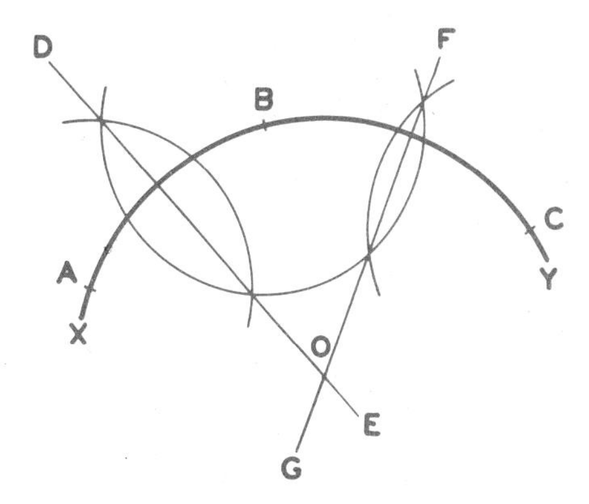

Fig. 7-16. Finding the Center of an Arc

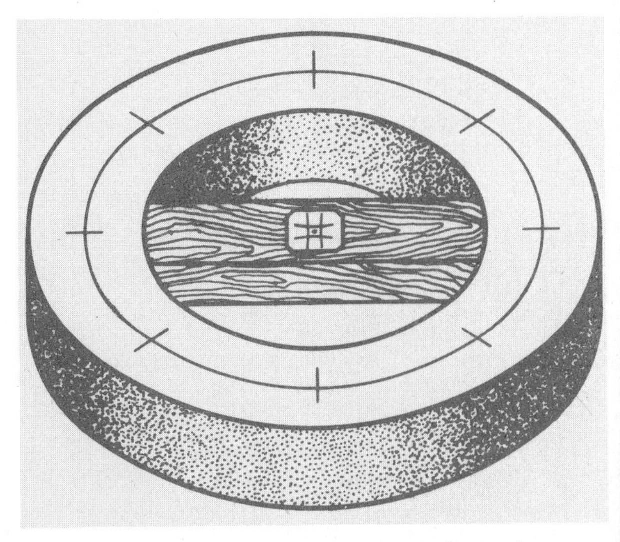

Fig. 7-17. Bridging a Hole to Locate Its Center for Layout

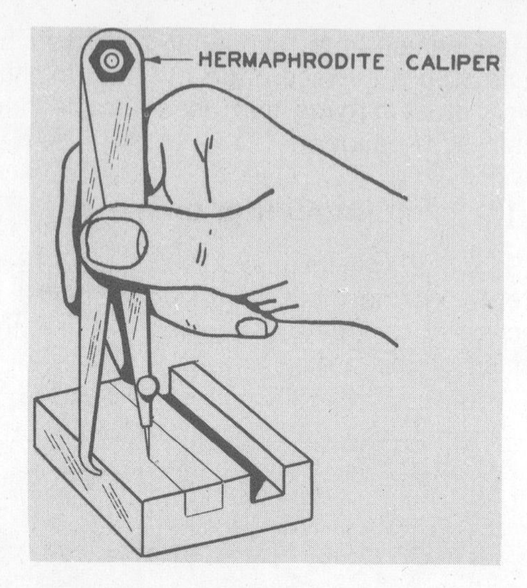

Fig. 7-18. Scribing Lines Parallel to Edge with Hermaphrodite Caliper

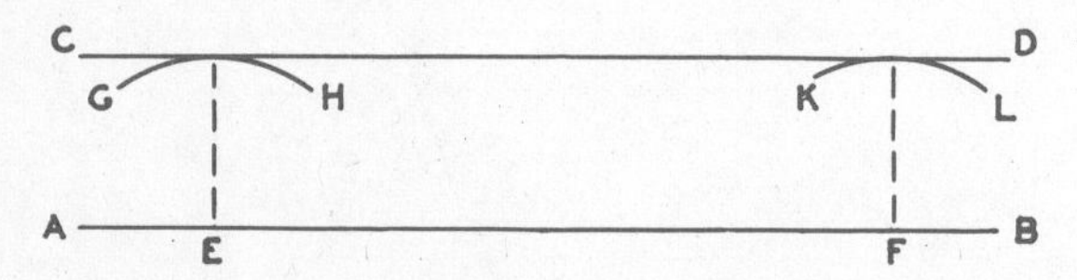

Fig. 7-19. Laying Out Parallel Lines

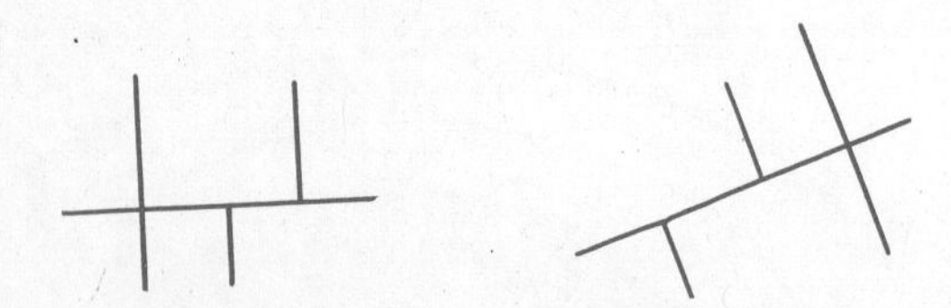

Fig. 7-20. Perpendicular Lines (All Angles Are 90°)

1. Place three prick-punch marks **A, B,** and **C** anywhere on the arc **XY.**
2. With the divider, scribe three arcs of the same size from points **A, B,** and **C.** See Fig. 7-16.
3. Scribe lines **DE** and **FG** through the **intersections** of the arcs. See § 7-8. The intersection at **O** is the center of the arc **XY.**

7-14. Finding Center of Hole

Sometimes the center of a circle comes in a hole. To find the center of a circle, whittle a bridge of soft wood to fit across the hole, Fig. 7-17. A large-headed tack or a small piece of tin, with corners bent down like spurs, is then driven into the wood. The center is then located on the tin by one of the procedures explained in section 7-11 or 7-12.

7-15. Laying Out Parallel Lines with Hermaphrodite Caliper

The **hermaphrodite caliper** may sometimes be used to lay out parallel lines. The bent leg is held against the edge of the work and as the caliper is moved along, the pointed leg draws a line on the surface parallel to the edge of the work, Fig. 7-18. It is very useful when many pieces have to be laid out alike (duplicate parts).

7-16. Laying Out Parallel Lines with Dividers

Suppose that the line **AB**, Fig. 7-19, is already on the metal surface and that it is necessary to draw the line **CD** parallel to **AB** and at a given distance from it:

1. Place small prick-punch marks at **E** and **F** any place on line **AB.**
2. Set the divider to the distance that the two lines are to be apart. See Fig. 7-7.
3. With one leg of the divider set at **E,** draw the **arc GH.**
4. With the same setting of the divider and with **F** as a center, draw the arc **KL.**
5. Draw the line **CD tangent** (see § 4-5) to the arcs **GH** and **KL.**

The line **CD** is parallel to the line **AB** and the given distance from **AB.**

7-17. Laying Out a Perpendicular Line from a Point to a Line

Lines which form **right angles** (90 degrees) with each other are called **perpen-**

dicular lines, Fig. 7-20. Suppose that the line AB and the point C in Fig. 7-21 are already on the metal surface and that it is necessary to draw a **perpendicular** to the line AB from point C:

1. Prick punch point C lightly.
2. Set the divider and make an **arc** so that it crosses the line AB as at D and E.
3. Prick punch lightly at D and E.
4. With D and E as centers, scribe arcs which cross at F.

The line drawn through the points C and F is perpendicular to the line AB.

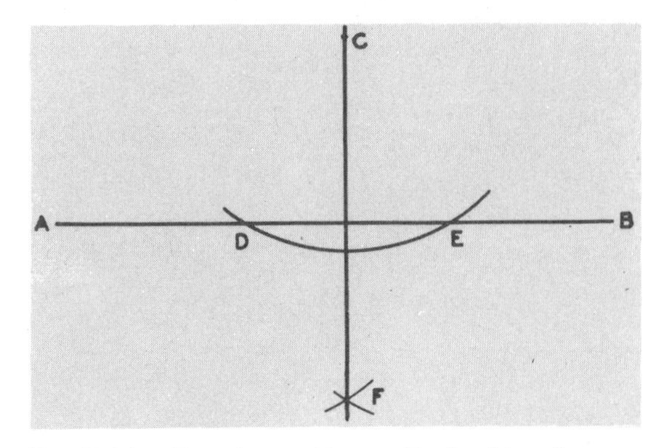

Fig. 7-21. Drawing a Perpendicular Line from a Point to a Line

7-18. Laying Out a Perpendicular Line through a Point on a Line

Suppose that the line AB in Fig. 7-22 is already on the metal surface and that it is necessary to draw a **perpendicular** through the point C which is on line AB:

1. Prick punch point C lightly.
2. With the divider set at any distance and with C as a center, scribe **arcs** crossing the line AB as at D and E.
3. Prick punch lightly at D and E.
4. With D and E as centers and with the divider set at a greater distance than before, scribe arcs which cross at G and H.

The line drawn through the points G and H also passes through the point C and is perpendicular to the line AB.

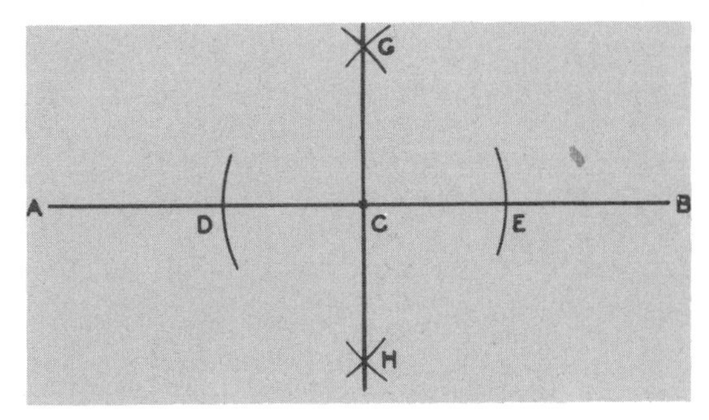

Fig. 7-22. Drawing a Perpendicular Line through a Point on a Line

7-19. Laying Out Parallel and Perpendicular Lines with a Combination Square, the Work Having Only One Straight Edge

If the work has one **straight edge,** parallel lines and perpendicular lines may be laid out as in Fig. 7-23.

7-20. Laying Out Parallel and Perpendicular Lines with a Vernier Height Gage

Since the vernier height gage is always used on a precision flat surface such as a surface plate, all lines drawn by its scriber as it is moved across the flat surface will be parallel to that surface, Fig. 7-24. To scribe parallel lines on a workpiece accurately, use the following procedure:

1. Make certain the workpiece rests firmly on the surface plate. If it "rocks," shim it, hold it in a vise, or clamp it to an angle plate.

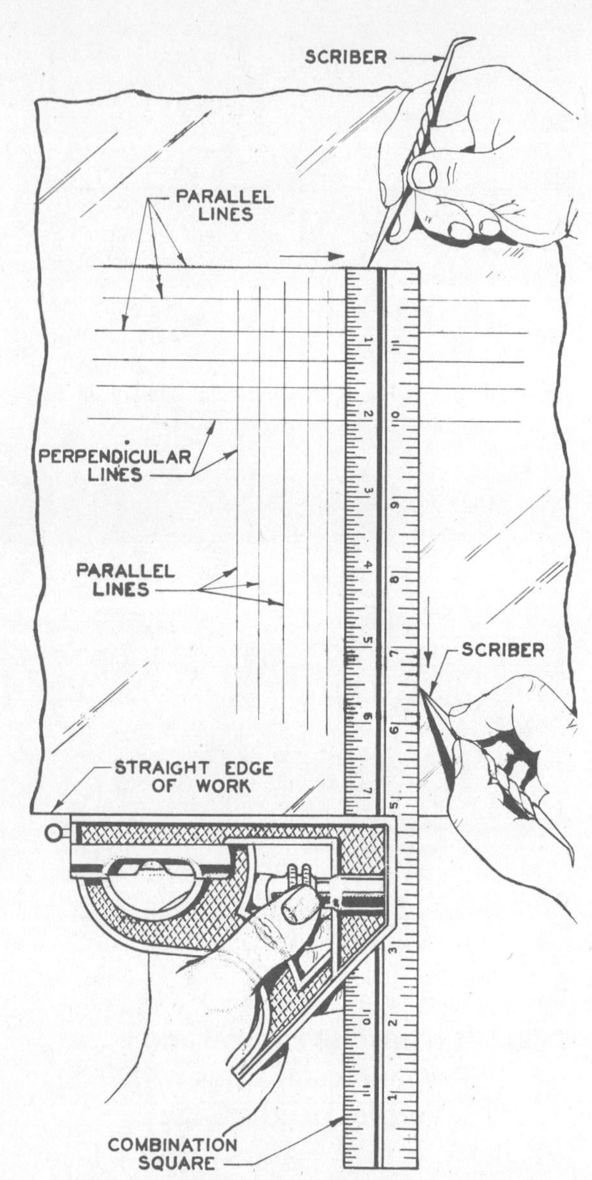

Fig. 7-23. Laying Out Parallel and Perpendicular Lines with a Combination Square When the Work Has Only One Straight Edge

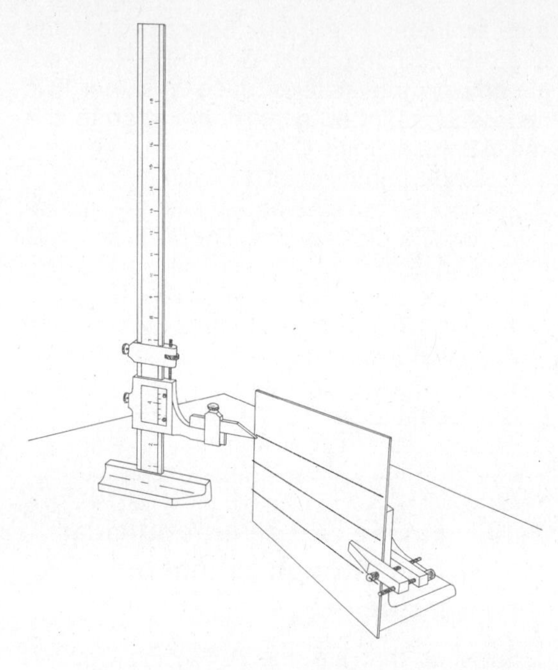

Fig. 7-24. Scribing Parallel Lines with a Vernier Height Gage

2. Set the vernier to the correct reading for the first line and scribe it. (Vernier reading is covered in Unit 9.)
3. Reset the vernier to the reading required for the next line and scribe it.

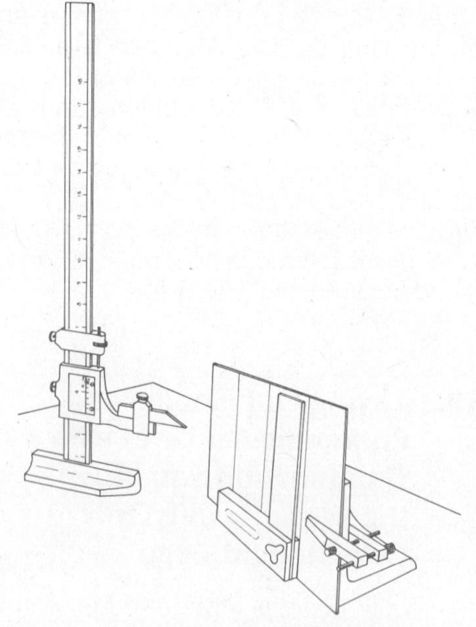

Fig. 7-25. Aligning Lines Vertically with a Precision Square

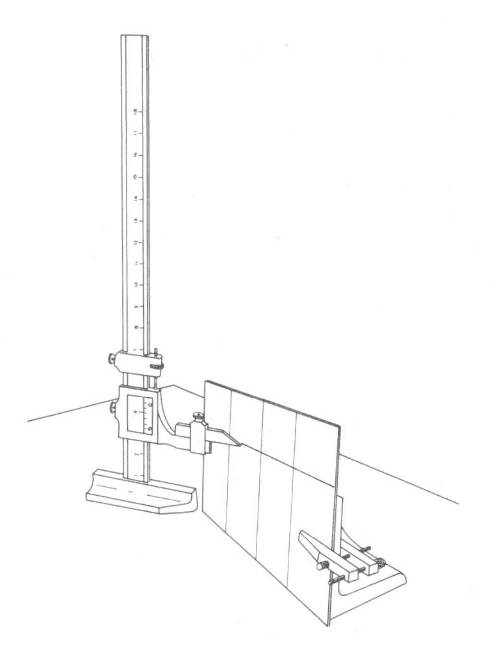

Fig. 7-26. Scribing Perpendicular Lines with a Vernier Height Gage

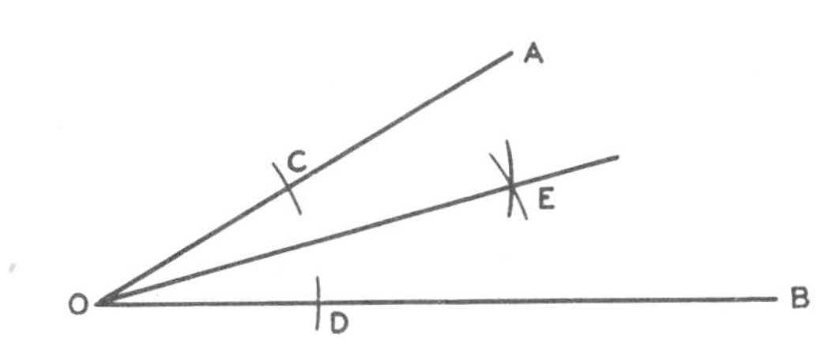

Fig. 7-27. Bisecting an Angle

It will automatically be parallel to the line previously drawn. If additional parallel lines are needed, continue as in Step 3.

To lay out perpendicular lines with a vernier height gage:

1. First lay out the lines required in one direction, following the procedure outlined for parallel lines above.
2. Using a precision square, align these lines vertically, Fig. 7-25, and clamp the workpiece in this position. Any line now drawn parallel to the surface plate will be perpendicular to the lines already scribed, Fig. 7-26.

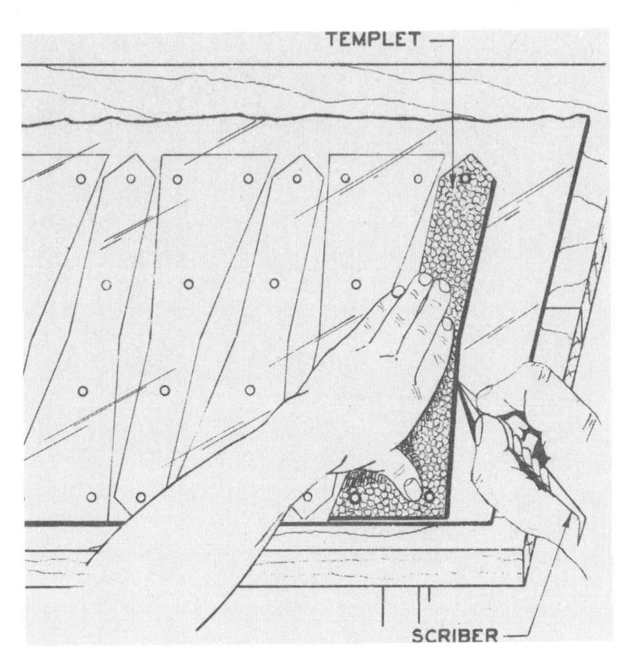

Fig. 7-28. Laying Out Work with a Template

7-21. Bisecting an Angle

Bisect means to cut or divide into two equal parts. Thus, to bisect an angle means to divide the angle into two equal parts or angles.

Suppose that the angle **AOB** in Fig. 7-27 is on the metal surface and it is necessary to bisect it:

1. Prick punch lightly at **O.**
2. With the divider set at any distance and with **O** as a center, draw **arcs** cutting the lines as at **C** and **D.**
3. Prick punch lightly at **C** and **D.**
4. Then with **C** and **D** as centers, draw arcs which cross each other as at **E.**

The line drawn through the points **E** and **O** bisects the angle **AOB.**

7-22. What Is a Template?

A **template,** also spelled templet, is a **pattern** for marking the shape of pieces of work or for marking holes, etc. It is usually made of plastic or sheet metal and is useful when many duplicate pieces have to be laid out, Fig. 7-28. The template is laid on the work; the lines, circles, etc., are then marked with a scriber. Fewer layout tools are thus needed and time is saved because it is not necessary to do a lot of measuring.

Words to Know

align
bisect
layout fluid
perpendicular line
right angle
scroll
template or templet

Review Questions

1. Explain why accurate layout work is so important.
2. Why should the surface for a layout be colored with layout fluid?
3. When is chalk used on a surface instead of layout fluid?
4. After it has dried, how is layout fluid removed from a metal surface?
5. How can lines be made to last longer on a layout?
6. How should a steel rule be placed on the work when measuring?
7. Name three methods of locating the center of the end of a round rod.
8. How can you find the center of the hole in a plate?
9. What does bisect mean?
10. Describe the procedure for scribing parallel and perpendicular lines with a combination square on work having only one straight edge.
11. Describe the procedure for scribing parallel and perpendicular lines with a vernier height gage.
12. What is a template? When is it used the most?

Drafting

1. Draw an arc with a circular object, then locate the center of the arc, using drawing instruments.
2. Construct parallel lines ¾″ apart.
3. Construct a perpendicular line from a point outside the line.
4. Construct a perpendicular line through the center of a 3″ line.
5. Bisect a 45-degree angle.

Career Information

1. Tell why a layout artist needs more education than a worker who runs a simple machine.
2. Explain why layout work is necessary.
3. Name four occupations which require an ability to do layout work.

UNIT 8

Decimal Equivalents

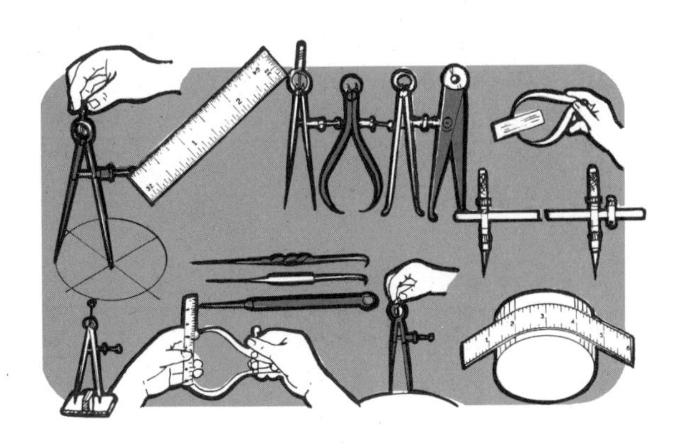

8-1. Decimal Equivalents

You must know how to convert common fractions to decimal equivalents in order to learn how to read and use precision-measuring tools like micrometers and verniers, Fig. 8-1. Measurements made by micrometers and verniers are in decimal fractions. Decimal fractions, which are decimal equivalents of common fractions, are obtained by dividing the numerator of a common fraction by its denominator, as $\frac{1}{2} = 1 \div 2$, or

$$2\overline{)1.0} = .5$$

Since micrometers and verniers are read in thousandths of an inch — which is three places to the right of the decimal point (.001″, .003″ — all decimal fractions must be carried out to three places; thus, ½″ = .500, Fig. 8-2.

The decimal equivalent of any fraction is found in the same way as above. Two other examples follow:

1. $\frac{3}{8}'' = 3 \div 8 =$

$$\begin{array}{r} .375 \\ 8\overline{)3.000} \\ 2\,4 \\ 60 \\ 56 \\ 40 \\ 40 \end{array}$$

A

B

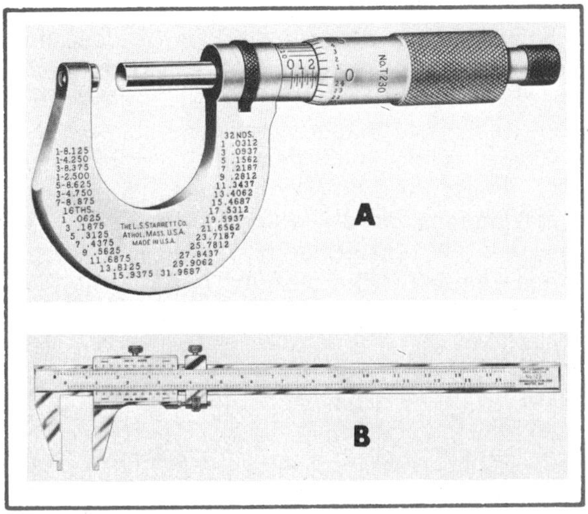

Fig. 8-1. A—Micrometer Caliper; B—Vernier Caliper

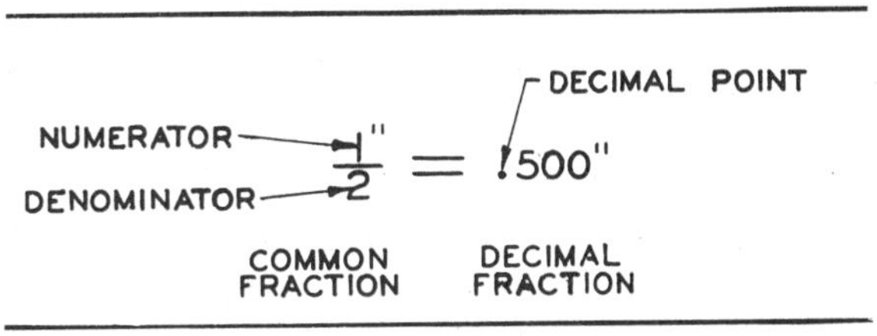

DECIMAL POINT

NUMERATOR — 1″ / DENOMINATOR — 2 = .500″

COMMON FRACTION — DECIMAL FRACTION

Fig. 8-2. Decimal Equivalent

The decimal equivalent of 3/8″ is therefore .375″ and is read three hundred seventy-five thousandths.

2. 9/16″ = 9 ÷ 16 =

```
     .5625
16)9.0000
   8 0
   1 00
     96
      40
      32
       80
       80
```

Therefore, .5625″ is the decimal equivalent of 9/16″ and is read five hundred sixty-two and one-half thousandths or five hundred sixty-two and five ten-thousandths.

When parts are to be made to very exact sizes, decimal fractions are used on working drawings instead of common fractions. When common fractions appear on working drawings of metal parts, the actual size may vary 1/64″ either way from the stated size. However, when decimal fractions are used, the size can vary only .005″ either way from the stated size. The decimal fraction for 1/64″ is .0156″.

8-2. Table of Decimal Equivalents

All common fractions are changed to decimal fractions and put in the form of a table to save the time of figuring. This table is called a **Table of Decimal Equivalents.** It is placed on large cards; every metal shop and drafting room has one or more of these cards hanging on the walls. There is a table on this page.

To save time, decimal equivalents are usually stamped on the frame of the micrometer. The young mechanic should memorize the decimal equivalents of 1/2″, 1/4″, 1/8″, 1/16″, 1/32″, and 1/64″. If these are known, most of the decimal equivalents which must be used can be figured out.

A shop problem often has a fraction such as 3/5 or 7/11 that cannot be measured with a **steel rule,** which has eighths, sixteenths, thirty-seconds, and sixty-fourths. Such fractions must be changed to the nearest eighths, sixteenths, etc., by using the decimal equivalents table. For example: 3/5 = .600 and the nearest fraction that can be measured with the steel rule is 19/32″, which equals .593″.

8-3. How Thick Is One-Thousandth of an Inch?

To get an idea of .001″ or 1/1000″, this paper is about 3/1000″ thick and thin tissue paper is about 1/1000″ thick. A hair on your head is about 3/1000″ thick.

8-4. Decimal Rule

The inch on a decimal rule is divided into 10 and 100 parts, Fig. 8-3, instead of quarters, eighths, sixteenths, etc. It is thus unnecessary to change from a common fraction to a decimal fraction. Measurements are made quicker with this rule when decimal dimensions are wanted.

The smallest divisions on some of these rules are 50 parts to an inch instead of 100 so that they can be read easily.

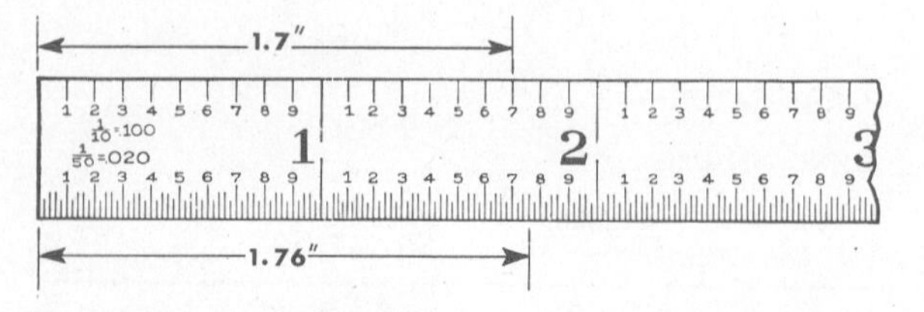

Fig. 8-3. Decimal Rule

Words to Know

common fraction
decimal equivalent
decimal equivalent table
decimal fraction
denominator
equivalent
numerator
thousandths of an inch

Review Questions

1. Write the decimal equivalents of ½″, ¼″, ⅛″, $\frac{1}{16}$″, $\frac{1}{32}$″, $\frac{1}{64}$″. Say them.
2. Write the decimal equivalents of $\frac{47}{64}$″, $\frac{13}{32}$″, $\frac{19}{64}$″, $\frac{15}{16}$″, ⅞″, $\frac{57}{64}$″. Say them.
3. Write the following numbers: six hundred twenty-five thousandths; three hundred ninety-three thousandths; seven-thousandths; sixteen thousandths; one-half thousandth; one-quarter thousandth; one ten-thousandth.
4. The answer to a problem is .225″. What is the nearest decimal equivalent that can be measured with a steel rule?
5. How thick is the paper in this book? Write it. Say it.
6. How thick is a hair on your head? Write it. Say it.
7. How thick is .001″? Make a comparison.
8. What is the allowable variation in a part dimension that is given in the form of a fraction?
9. What is the allowable variation in a part dimension that is given in the form of a decimal?

Drafting

1. When should dimensions on drawings be given in common fractions and when should they be given in decimal fractions?

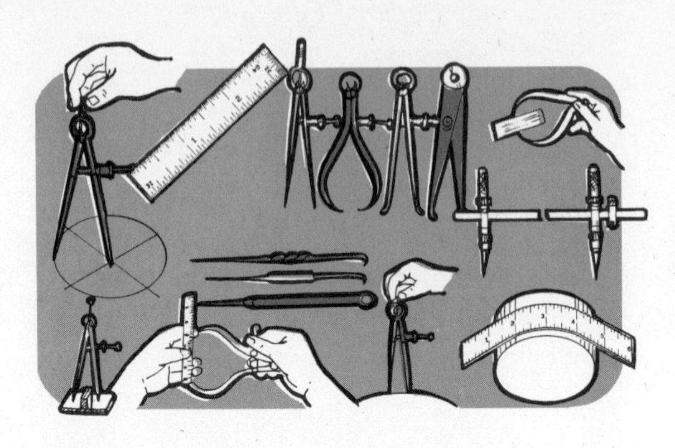

UNIT 9

Micrometers and Verniers

9-1. What Is a Micrometer?

A micrometer, called **mike** for short, is an instrument that measures in thousandths of an inch (1/1000″). An inch on a steel rule is usually divided into 64 parts and sometimes into 100 parts. It is impossible to stamp 1000 lines per inch on the steel rule. Even if it were possible, the measurements would have to be made with a magnifying glass and even then they would be difficult to read. The micrometer, because of the way it is made, divides an inch into a thousand parts and makes fine measuring easier.

9-2. Parts of a Micrometer

A micrometer, Fig. 9-1, has a **frame, sleeve, thimble, spindle,** and **anvil.** The inside of the micrometer is shown in Fig. 9-1. The opening between the anvil and the spindle is made smaller or larger by turning the thimble. The size of the opening is read on the sleeve and the thimble. The **lock nut**

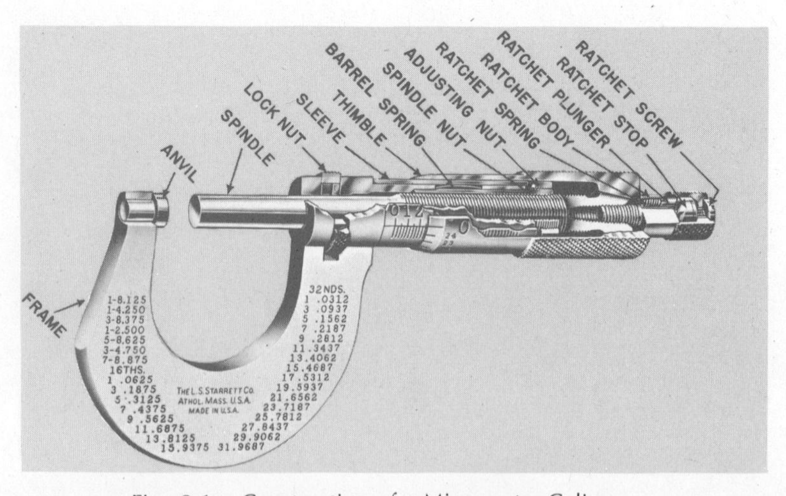

Fig. 9-1. Construction of a Micrometer Caliper

locks the spindle so that it will not turn. The **ratchet stop** is explained in section 9-6.

9-3. Kinds and Sizes of Micrometers

The **outside micrometer** is used the most often. See Fig. 9-1. It is used to measure the outside diameters of round objects and the widths and thicknesses of flat pieces.

The **inside micrometer**, Fig. 9-2, is used to measure the diameters of openings as in Fig. 9-3.

The **depth micrometer** is used to measure the depths of holes, grooves, and slots, Fig. 9.4.

The **screw thread micrometer**, Fig. 9-5, is used to measure the pitch diameter of screw threads.

Outside micrometers are made in many sizes, Fig. 9-6. They are identified by the maximum size they can measure; thus:

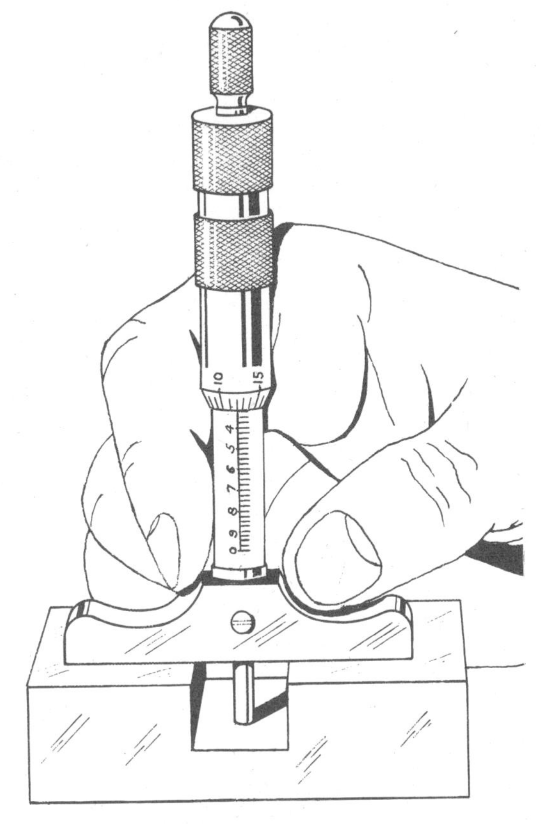

Fig. 9-4. Measuring with a Depth Micrometer

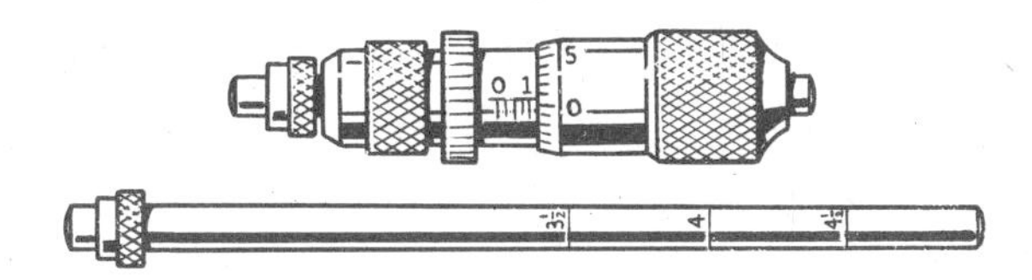

Fig. 9-2. Inside Micrometer

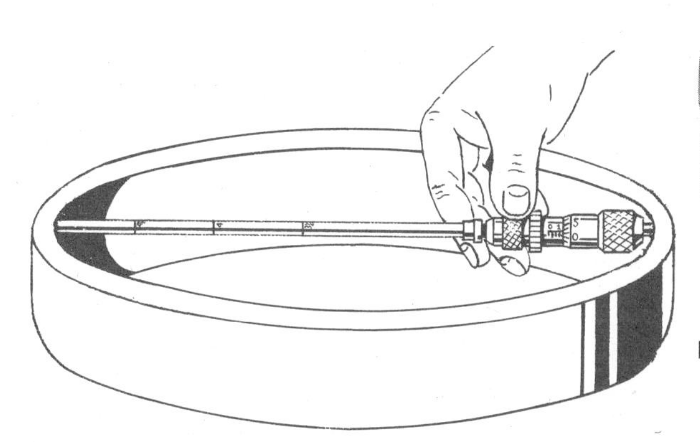

Fig. 9.3. Measuring with an Inside Micrometer

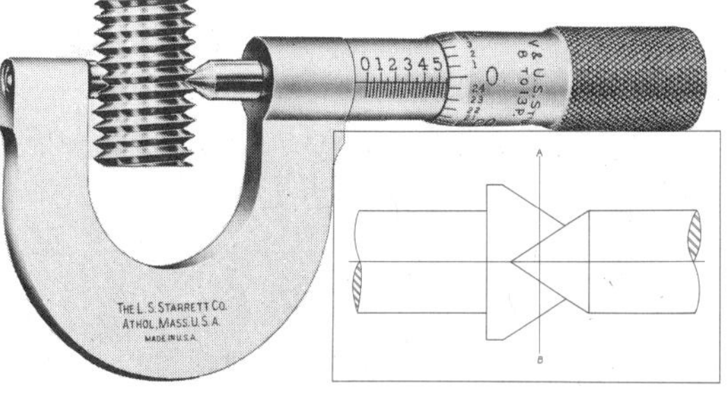

Fig. 9-5. Thread Micrometer Measuring the Pitch Diameter of a Screw Thread Directly

Insert shows anvil and spindle position at line AB which corresponds to zero reading.

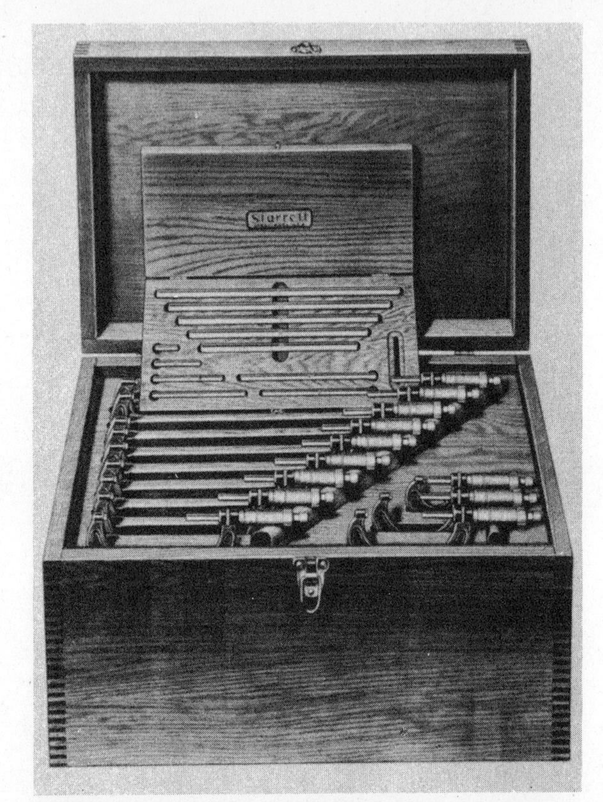

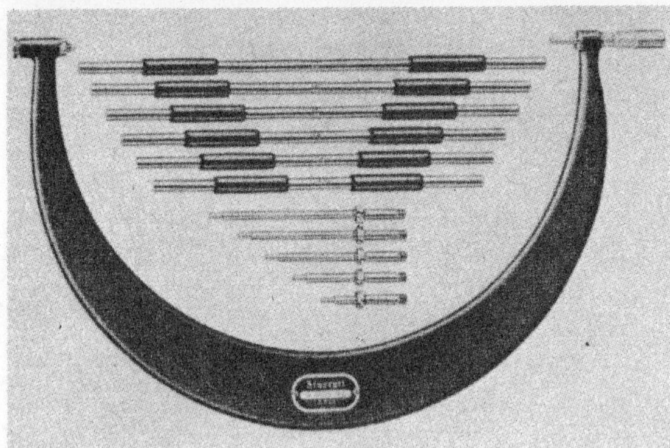

Fig. 9-6. Micrometer Sets, 1″ to 24″

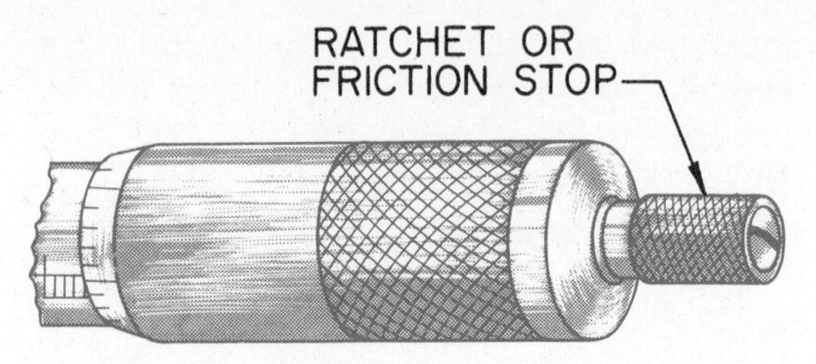

Fig. 9-7. Ratchet Stop on Micrometer

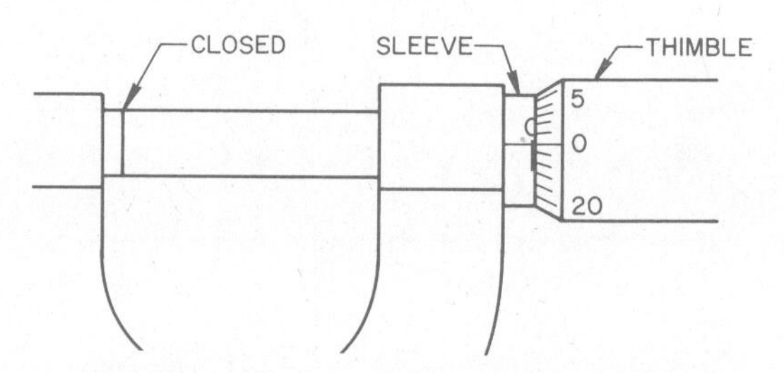

Fig. 9-8. Testing the Exactness of a 1″ Micrometer

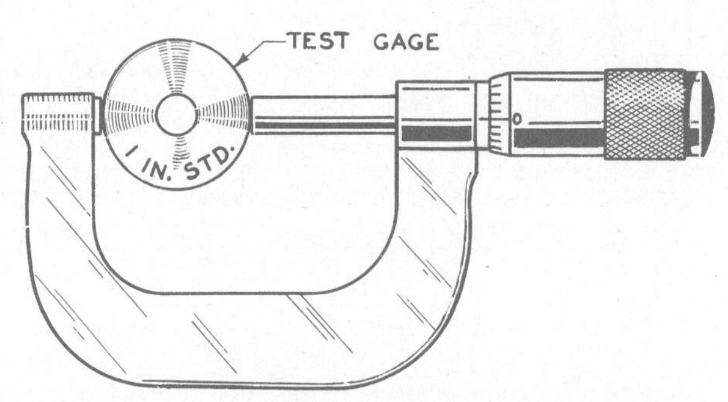

Fig. 9-9. Testing the Exactness of a 2″ Micrometer

1. a 1″ micrometer measures from 0″ to 1″.
2. a 2″ micrometer measures from 1″ to 2″.

9-4. Care of Micrometers

The micrometer is a fine instrument and should be handled as carefully as a watch. Dropping it on the floor or bench may damage its fine parts and make it useless. Keep it away from dust, grit, and grease. It should never be twirled. The spindle should not be screwed down to the anvil (as in Fig. 9-8) when not in use because changes in temperature cause it to **expand** (see § 26-35), and shrink and thus strain it so that it will not measure correctly. See § 46-26.) Its exactness should be tested now and then, as explained in sections 9-7 and 9-8. Micrometers should be oiled before storing away.

9-5. Touch or Feel

The exactness of most measurements depends upon the sense of touch or feel. No two persons' touches are alike. If one mechanic's touch is heavy and another mechanic's touch is light, there will be a difference in the two measurements of the same piece. The sense of touch is most delicate in the fingertips. Among skilled mechanics, it is highly developed. A skilled mechanic using the proper measuring tool can feel 0.00025″. It can only be done, however, by holding the measuring tool delicately with the finger tips. See Fig. 9-10.

Fine measurements cannot be made where there is vibration from machinery, etc., as it destroys the sense of touch. Avoid working with a measuring tool that has been set by another man; set it yourself. The different touches of different mechanics in putting the tool on the work will cause different measurements. In making very fine measurements, even the warmth of the hand will change the size of the measuring tool.

9-6. Ratchet Stop on Micrometer

A **ratchet stop**, or a friction stop, is placed at the end of the thimble on some micrometers so that the same tightness or **pressure** is always used, Fig. 9-7. The **ratchet** or friction stop, slips at the proper pressure for taking measurements.

9-7. Testing Exactness of 1″ Micrometer

Close the 1″ micrometer completely with the correct amount of **touch**. If the micrometer is set correctly, the **0** mark on the sleeve will be in line with the **O** mark on the thimble, Fig. 9-8. Wear on the **anvil** and the

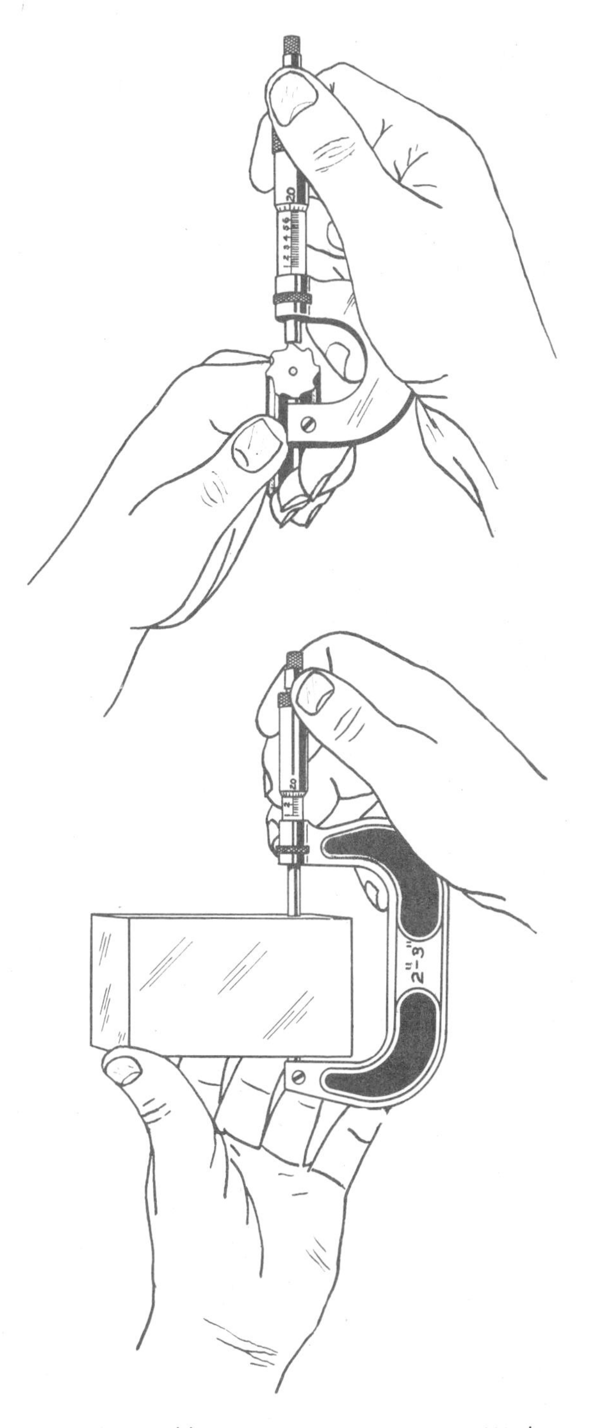

Fig. 9-10. Holding a Micrometer to Measure Work Held in the Hand

screw sometimes makes it necessary to reset the micrometer. This is done by various methods, depending upon the make of the micrometer. It is best to see the manufacturer's catalog for information on how it is done.

9-8. Testing Exactness of 2″ (or Larger) Micrometers

For testing 2″ or larger micrometers, a **gage** of known size must be used and the micrometer then set to this size, Fig. 9-9.

9-9. Mastering the Micrometer

To master the micrometer, you must learn to do the following:

1. Set the micrometer to any number.
2. Measure articles:
   a. Hold micrometer correctly (see § 9-10).
   b. Use correct touch or feel (see § 9-5).
3. Say the number.
4. Write the number.

9-10. Holding Micrometer to Measure Work Held in Hand

Hold the work to be measured in the left hand. Hold the micrometer in the right hand with the third or little finger pressing the frame against the palm, Fig. 9-10; the thumb and first finger are free to turn the thimble. Place the work between the anvil and the spindle; then with the thumb and first finger turn the thimble until a very slight pressure is felt.

It is necessary for beginners to study their touch carefully because the measurement will not be the same when the pressure is light as when it is heavy; a micrometer may be sprung as much as .003″ by too much pressure. This gives the wrong measurement and damages the instrument. A **ratchet stop** on a micrometer always gives the same pressure.

9-11. Micrometer Screw

All micrometers are read alike. The end of the **spindle**, inside the **thimble**, is called the screw spindle. The sleeve is fastened to the end of the screw. The screw has 40 **threads** to an inch; that is, the screw must turn 40 times to move one inch. Thus, each turn of the screw equals 1/40″. Since the micrometer is read in thousandths of an inch instead of 40ths of an inch, 1/40 of an inch must be changed to thousandths of an inch by dividing the **numerator** by the **denominator:**

$$40\overline{)1.000}\quad = .025$$

There are also 40 lines to an inch on the **sleeve**, the same as the number of threads on the screw; these lines show how many times the screw has turned. Each turn equals .025″. If one turn of the screw equals .025″, then 1/25 of a turn equals 1/25 of .025″, or .001″. Thus, by dividing the edge of the thimble into 25 parts, it is possible to make exactly 1/25 of a turn which is .001″, or 2/25 of a turn which is .002″, etc. It makes it possible to measure in thousandths of an inch. One-half and quarters of a thousandth can be judged as nearly as possible.

9-12. Explanation of Marks on Sleeve and Thimble

Turn the thimble until the **0 mark** on the thimble and the **0 mark** on the sleeve come together. See Fig. 9-8. This will give the smallest size that the micrometer will measure; on a 1″ micrometer it is 0; on a 2″ micrometer it is 1″; on a 3″ micrometer it is 2″.

The **marks** on the thimble are .001″ each. Turn the thimble to the next line on the thimble in the direction of the arrow, Fig. 9-11, and on a 1″ micrometer the **spindle** will be .001″ from the **anvil**. Turn the thimble to the line marked 5 on the thimble, Fig. 9-12, and the spindle will be .005″ from the anvil.

Note that one complete turn of the thimble is .025″ on the sleeve, Fig. 9-13. For each turn of the thimble, the thimble moves over one more mark on the sleeve. This means that each mark on the sleeve is .025″. Every fourth line on the sleeve is a little longer than the others and is stamped 1, 2, 3, etc., which stand for .100″, .200″, .300″, etc. See Fig. 9-15. The micrometer is set at .100″ in Fig. 9-14.

9-13. Reading a Micrometer

To find out how much the micrometer is opened (the distance between the anvil and the spindle), the marks on the sleeve may be read like any ordinary rule. Remember that the numbers 1, 2, 3, 4, etc., mean .100″, .200″, .300″, .400″, etc. To this add the thousandths that show on the thimble. For example, the readings in Fig. 9-15 are:

(A) .200″ even
(B) .250″ (.200″ + .025″ + .025″ = .250″)
(C) .562″ (.500″ + .050″ + .012″ = .562″)
(D) .787½″ (.700″ + .075″ + .012½″ = .787½″ or .7875″)

Be sure to add correctly.

9-14. Reading a Ten-Thousandth Micrometer

Some micrometers have a **vernier**, named after the inventor, **Pierre Vernier**. A **vernier micrometer** can be read to a ten-thousandth of an inch (1/10000″ or .0001″). Again, each mark on the thimble is .001″.

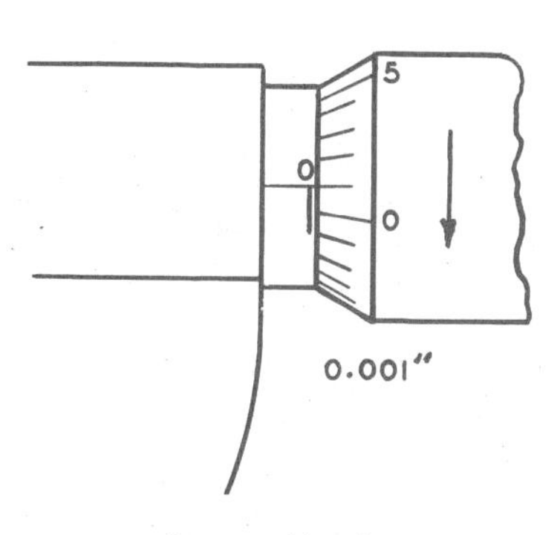

Fig. 9-11. One Thousandth of an Inch (.001″)

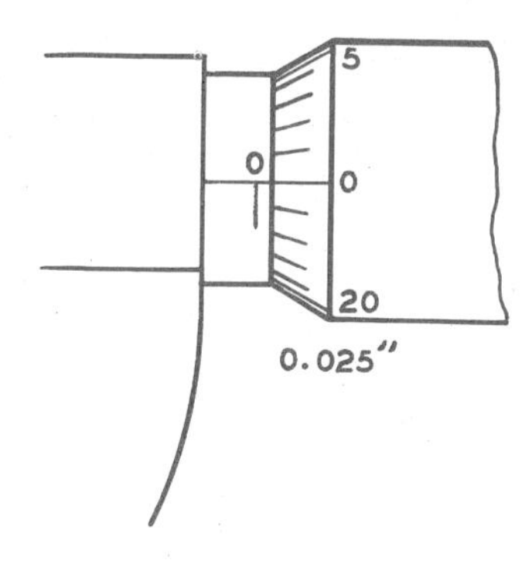

Fig. 9-13. Twenty-five Thousandths of an Inch (.025″)

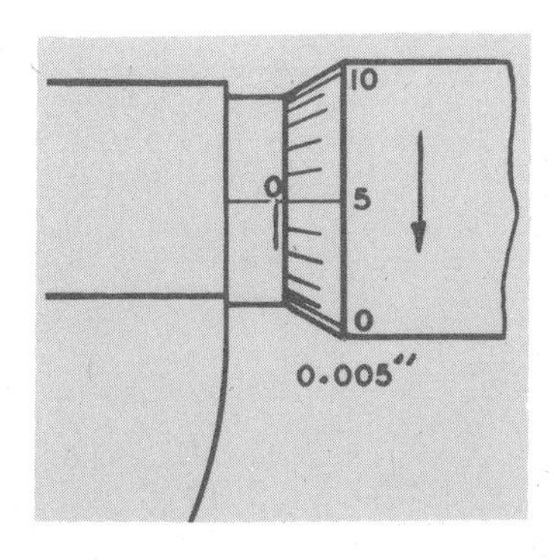

Fig. 9-12. Five Thousandths of an Inch (.005″)

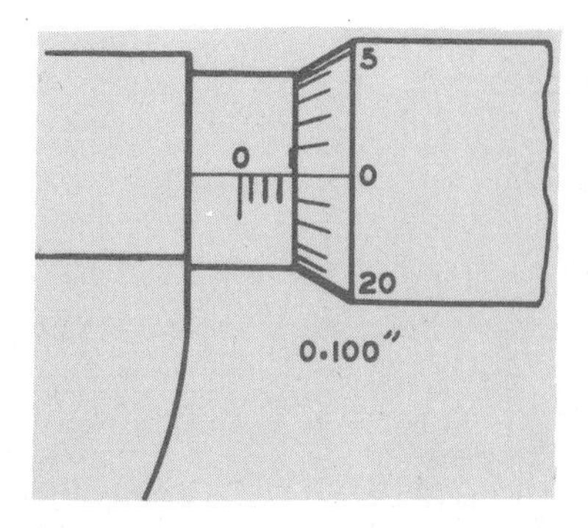

Fig. 9.14. One Hundred Thousandths of an Inch (.100″)

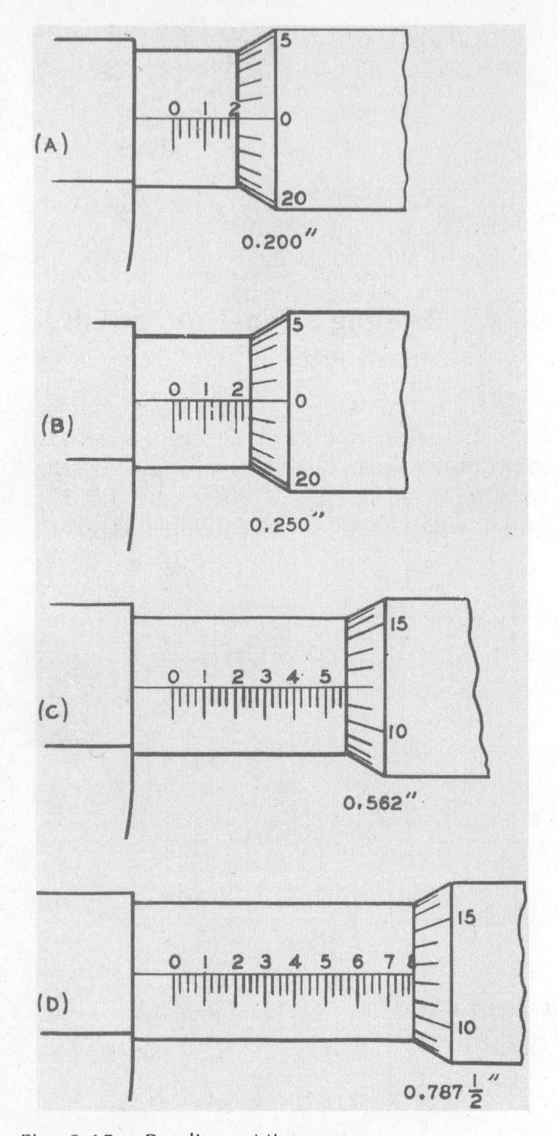

Fig. 9-15. Reading a Micrometer

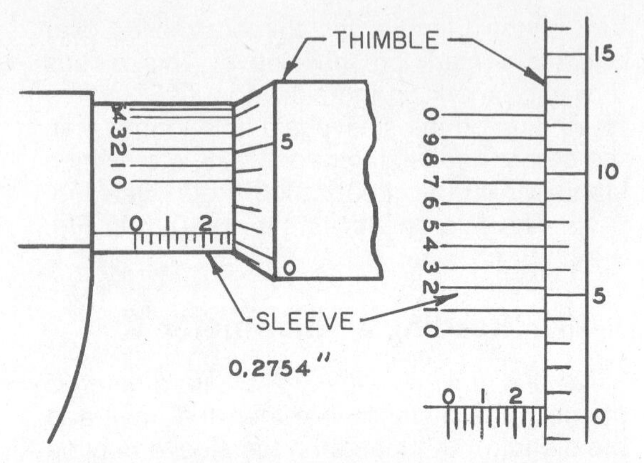

Fig. 9-16. Vernier Micrometer

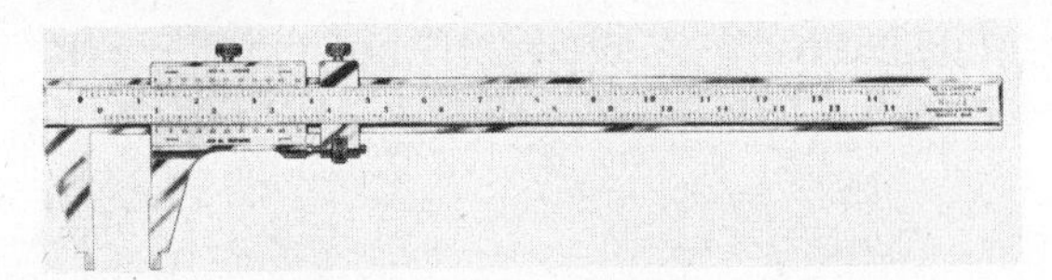

Fig. 9-17. Vernier Caliper

Fig. 9-18. Making an Outside Measurement with a Vernier Caliper

The ten divisions on the **back of the sleeve** are the vernier; the lines are numbered 0, 1, 2, 3, 4, 5, 6, 7, 8, 9, 0 in Fig. 9-16. These lines have the same space as nine divisions on the thimble.

To read the vernier micrometer, first read the thousandths as on any ordinary micrometer. Suppose that this number is between .275 and .276, as in Fig. 9-16. To find the fourth **decimal place**, find the line on the vernier that matches exactly a line on the

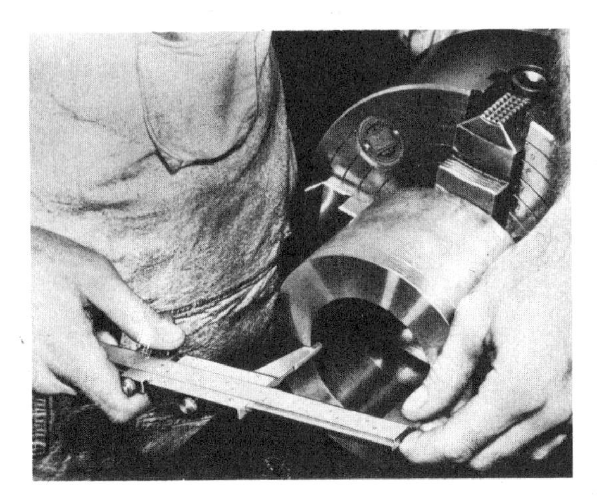

Fig. 9-19. Making an Inside Measurement with a Vernier Caliper

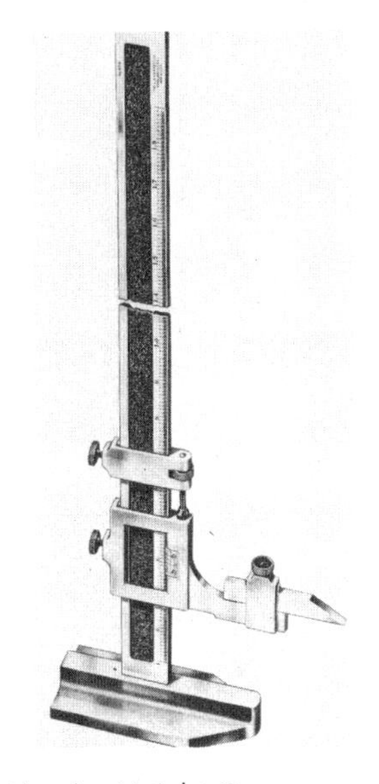

Fig. 9-20. Vernier Height Gage

thimble. In this case it is the line 4; this number 4 means 4 ten-thousandths inches (.0004″) and added to the .275″ gives the fourth decimal place. Thus, the complete reading is .2754″ (.275″ + .0004″ = .2754″).

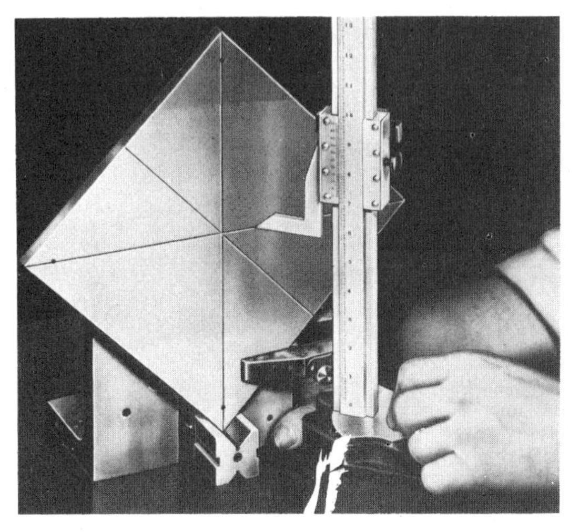

Fig. 9-21. Precision Layout with Vernier Height Gage

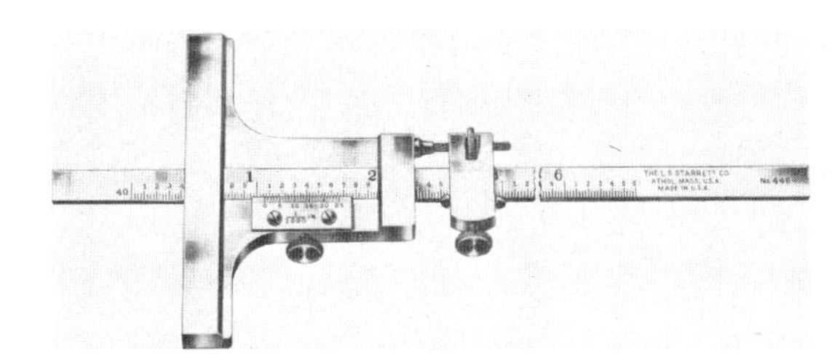

Fig. 9-22. Vernier Depth Gage

9-15. Vernier Measuring Tools

Linear vernier measuring tools also make measurements accurately to one-thousandth of an inch. Unlike the micrometer, a vernier caliper, Fig. 9-17, can make both outside and inside measurements, Figs. 9-18 and 9-19, throughout the entire range of the instrument. Vernier calipers are made in standard lengths of 6″, 8″, 12″, 24″, 36″, and 48″.

The vernier height gage (Fig. 9-20), equipped with a straight or offset scriber, is often used for precision layout work, Fig. 9-21. It is also used in conjunction with a dial indicator for inspecting completed work. Vernier height gages are made in standard sizes and of 12″, 18″, 24″.

A vernier depth gage, Fig. 9-22, like the depth micrometer, is used for measuring hole and slot depths and the distance between flat parallel surfaces.

The vernier gear tooth caliper is used for measuring the accuracy of gear teeth. It may also be used to measure some form cutters and threading tools, Fig. 9-23.

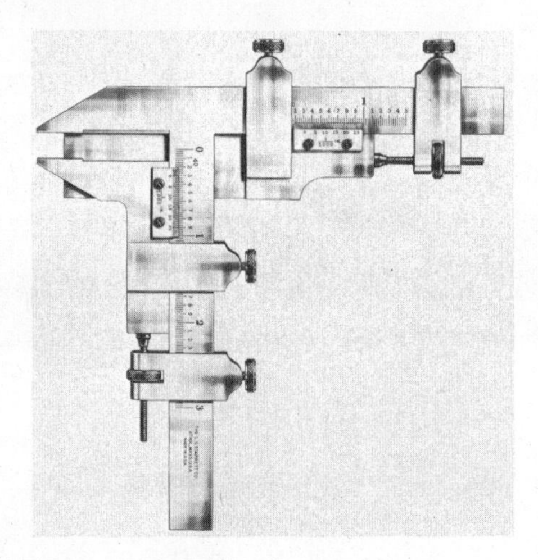

Fig. 9-23. Vernier Gear Tooth Caliper

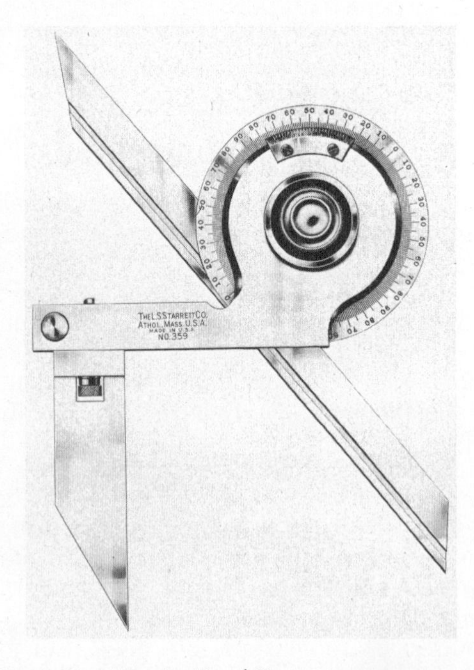

Fig. 9-24. Vernier Bevel Protractor

The vernier bevel protractor, Fig. 9-24, is a precision tool used for layout and measurement of precise angles.

9-16. Care of Vernier Tools

Vernier tools are fine instruments and deserve the finest care. Keep them free of grime and grit by wiping after each use with a soft lint-free cloth. Never force them when making measurements. They should be periodically checked for accuracy against measurement standards of known size. When not in use, they should be stored in a suitable case.

9-17. Using the Vernier Caliper

The vernier caliper is made up of a graduated beam with a fixed measuring jaw, a movable jaw which carries a vernier scale, and a mechanism for making fine adjustments, Fig. 9-17. One side of the caliper beam is graduated for reading outside measurements, the other for reading inside measurements. The outside measurements are read on the scale reading from left to right.

Making accurate measurements with vernier tools requires the same care and delicate touch as when using micrometers. First, the slide assembly should be moved along the beam until the jaws almost contact the work. Then the part of the slide assembly which carries the fine adjusting screw should be locked to the beam. The jaws are now brought into contact with the workpiece by moving the fine adjusting nut. The jaws should make definite contact with the workpiece but should not be tight. The main slide assembly is then locked to the beam, and the caliper is carefully removed from the workpiece to prevent springing the jaws. The reading may now be made.

9-18. Reading a 25-Division Vernier

Reading a vernier is very much like reading a micrometer. Each inch of the beam is graduated into 40 equal parts, Fig. 9-17, just like the sleeve of a micrometer. You will remember that the decimal equivalent of 1/40th of an inch is .025″. Again like the micrometer, every fourth division line is a little longer than the others and is marked 1, 2, 3, etc., which stands for tenths of an inch or .100″, .200″, .300″, etc., Fig. 9-25. The vernier plate corresponds to the micrometer thimble and is graduated into twenty-five equal divisions, representing thousandths of an inch. Every fifth line on the vernier plate is numbered for convenience in making readings. A four- or five-power magnifying glass will aid in making accurate readings.

To read a vernier, first note how many inches, tenths (.100, .200, etc.), and 40ths (.025) the zero on the vernier plate is from the zero on the beam. Then add to this amount the number of thousandths indicated by the matching lines on the vernier plate and beam.
(Note starred lines in Fig. 9-25).

Example: (Fig. 9-25)

(A)	The zero on the vernier plate is between the 3″ and 4″ marks, thus the number of whole inches is	3.000
(B)	The zero on the vernier plate has passed six 1/10th graduations, or	.600
(C)	The zero on the vernier plate has not passed any 1/40th graduations, thus	.000
(D)	The twelfth line on the vernier plate matches a line on the beam, adding	.012
	Total reading	3.612

9-19. Reading a 50-Division Vernier

Some of the new vernier tools are equipped with a 50-division vernier plate, Fig. 9-26.

They are read in the same manner as the 25-division vernier plate. However, they are easier to read with the unaided eye, because the division lines are spaced farther apart on the vernier plate, and the 1/10th divisions on the beam have only two divisions representing 1/20th of an inch (.050) instead of the four divisions representing 1/40th of an inch (.025).

Example: (Fig. 9-27)

(A)	The zero on the vernier plate is between the 1″ and 2″ marks on the beam, making	1.000
(B)	The zero on the vernier plate has passed the 6 on the beam, indicating 6/10ths, or	.600
(C)	The zero on the vernier plate has passed the midpoint division between the sixth and seventh marks, adding 1/20th", or	.050
(D)	The fifteenth line on the vernier matches a line on the beam, adding	.015
	Total reading	1.665

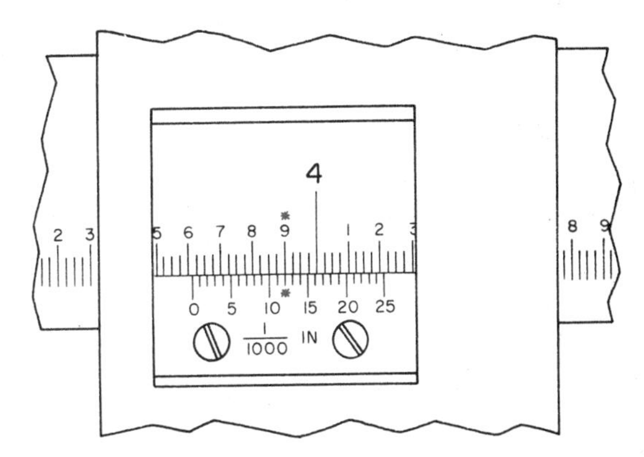

Fig. 9-25. The Vernier Scale

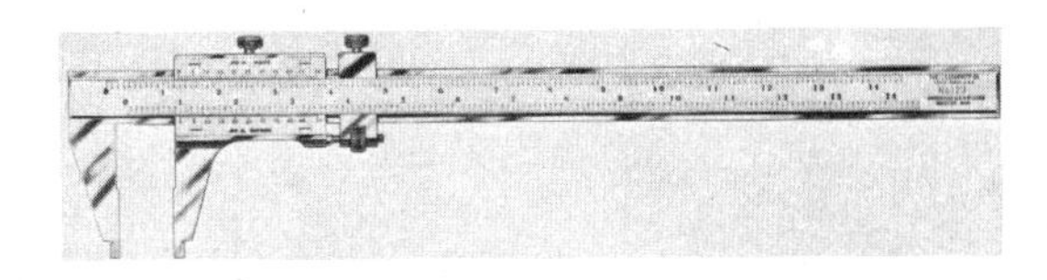

Fig. 9-26. 50-Division Vernier Caliper

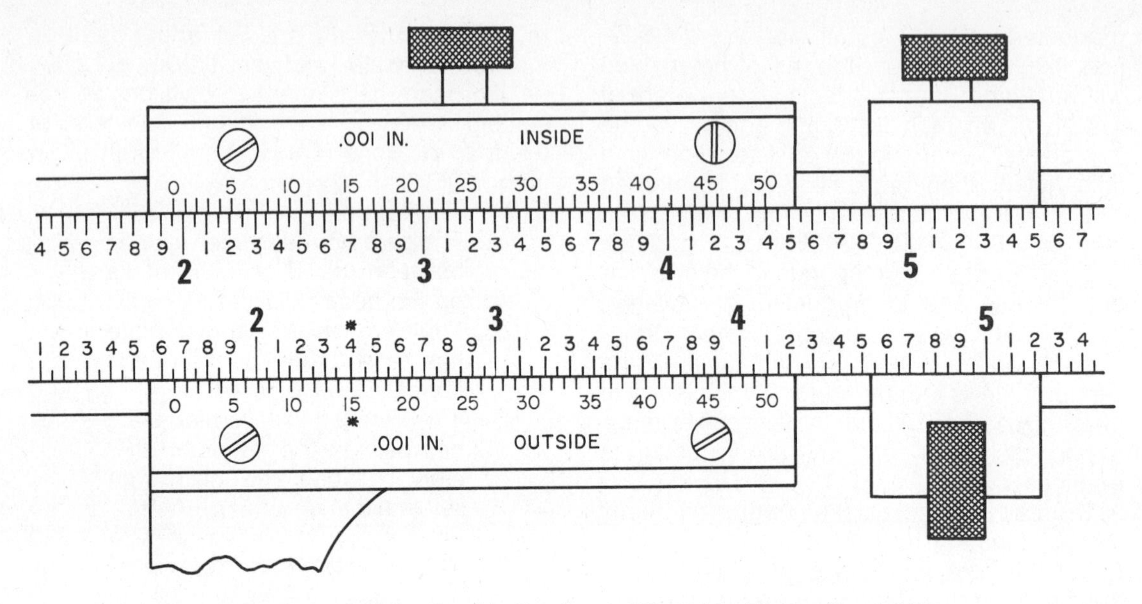

Fig. 9-27. Vernier Outside Reading 1.655 and Inside Reading 1.965

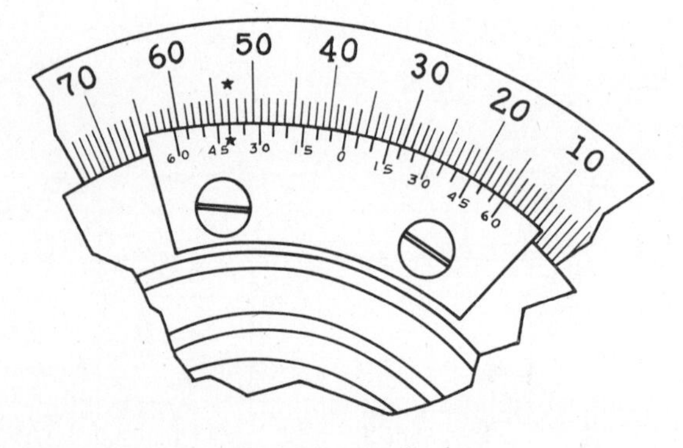

Fig. 9-28. Vernier Protractor Scale

9-20. Reading the Vernier Bevel Protractor

The vernier bevel protractor, Fig. 9-24, measures angles accurately to one-twelfth of a degree, or five minutes. The protractor dial is graduated into 360°, but they are marked 0°-90°, 90°-0°, 0°-90°, and 90°-0°. Each five-degree line is a little longer than the one-degree lines; the ten-degree lines are numbered, as are the longest lines. The vernier plate is divided into twelve equal divisions on either side of the zero, Fig. 9-28. Since there are 60 minutes in each degree, each line represents 1/12th of one degree, or 1/12th of 60 minutes, or five minutes.

To read the vernier protractor, first note the number of whole degrees between the zero line on the dial and the zero line on the vernier plate. Then, reading in the same direction, add the number of minutes represented by the line past the zero which matches a line on the dial.

Example: (Fig. 9-28)

(A)	Number of whole degrees between the zero on the dial and the zero on the vernier plate	37°
(B)	Number of minutes indicated by the line on the vernier plate which matches a line on the dial	00° 40′
	Total reading	37° 40′

Words to Know

depth micrometer
friction stop
inside micrometer
micrometer
  anvil
  frame
  sleeve
  spindle
  thimble
outside micrometer
ratchet stop
ten-thousandths
  micrometer
vernier
vernier caliper
vernier depth gage
vernier protractor
vernier height gage
vernier micrometer

Review Questions

1. On what two parts of a micrometer are the graduations located?
2. On what two parts of a vernier caliper are the graduations located?
3. Name four kinds of micrometers and tell the kind of measurements for which they are used.
4. Name four kinds of verniers and tell the kinds of measurements for which they are used.
5. What is the largest size you can measure with a 1″ micrometer?
6. What is the smallest size you can measure with a 2″ micrometer?
7. What is the largest size you can measure with a 6″ micrometer?
8. Can you measure ½″ stock with a 2″ micrometer? Why?
9. Why should micrometers and verniers be handled with great care?
10. The sense of touch is most delicate in what part of the hand?
11. How is it possible that two people measuring the same object with the same micrometer or vernier would get different readings?
12. What effect does the warmth of the hand have on the measuring tool?
13. What is the use of a ratchet or friction stop on a micrometer?
14. How many threads per inch are there on the micrometer screw?
15. How is the accuracy of a 1″ outside micrometer tested?
16. How is the accuracy of a 2″ or larger, outside micrometer tested?
17. How is the accuracy of a vernier caliper tested?
18. What two advantages does a vernier caliper have over a micrometer caliper?
19. How does a vernier micrometer differ from a regular micrometer?
20. Ask your instructor to give you several objects of known size so that you can practice using the micrometer and vernier calipers. Have him check your work.

Mathematics

1. What is the decimal equivalent of 1/64″, 1/32″, 1/16″, 1/8″, 1/4″, 1/2″, 3/64″, 5/32″, 7/16″, 5/8″?
2. Add .600″ + .050″ + .013″.
3. Add .900″ + .075″ + .011½″.

Social Science

1. Tell how the development of fine measuring tools has helped the manufacturing industry and thereby benefited humanity.

Career Information

1. What kinds of workers need to know how to use micrometers and verniers?
2. Who are the skilled craftsmen who make precision-measuring tools? What kind of education must they have?

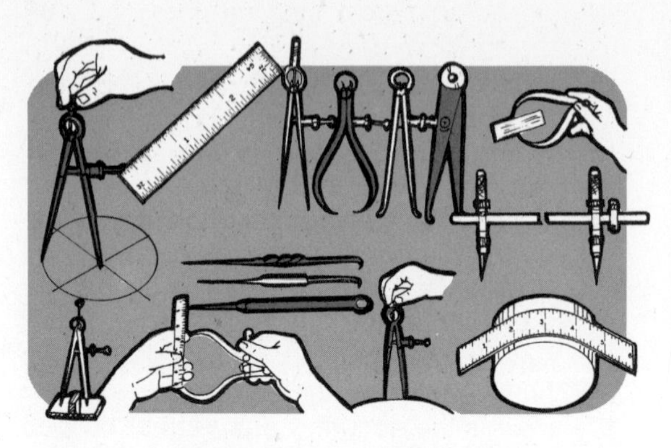

UNIT 10

The Metric System of Measurement

10-1. Development of the Metric System of Measurement

The **metric system of measurement** originated in France about 1790, was adopted there about 1795, and was made mandatory in 1840. It was called the **CGS system** and was based on the **centimetre** for length, the **gram** for mass (weight), and the **second** for time. In 1902 a more practical system of units was developed based on the meter for length, **kilogram** for mass, and **second** for time; this was called the **MKS system**. The **ampere** was added in 1950, creating the **MKSA system**. In 1954, the **kelvin** for temperature and **candela** for light intensity were added, and in 1960 the system was named the **Systeme International d'Unites,** abbreviated **SI** in all languages.

10-2. Importance of the Metric SI System

The metric SI system has been, or is being, formally adopted by every industrial nation except the United States, which is now considering it for adoption. Interestingly, use of the metric system in the United States was authorized, but not made mandatory, by an act of Congress in 1866. Until recently, however, it has only been used extensively in the fields of science and medicine. Many United States manufacturers engaged in international trade have already begun conversion to the metric SI system. One major manufacturer has already announced that all its new products will be made with metric measures and standards. At Lima, Ohio, a major auto manufacturer is already producing, for domestic use, an engine built to metric dimensions. It is expected that Congress will soon recommend a program for adoption of the metric SI system. Handbooks for machinists, engineers, and drafting rooms provide information about metric standards, usage, and methods.

10-3. The Metric SI System

The metric SI system is a **base ten**, or **decimal system.** Its units of measurement are related to each other by **powers of ten.** Each of the main units are changed to units of more convenient size, when desired, by multiplying or dividing by powers of ten. Table 2 lists the names of several basic SI units, their **symbols**, and their relationship to each other. It is important to learn the metric symbols because they are used whenever metric measurements are written or printed. The same symbol is used for singular and plural.

We are primarily concerned here with the basic metric SI units of linear measurement. The **metre** is the basic unit of length. It roughly corresponds to the yard, measuring 39.37 inches, Fig. 10-1. The metre is divided into **decimetres** (0.1 metre), **centimetres** (0.01 metre), and **millimetres** (0.001 metre).

Most metric rulers and tapes are divided into millimetres and are numbered at every 10 mm mark, Fig. 10-2. 30 cm rulers are used rather than the familiar 12″ rulers, and

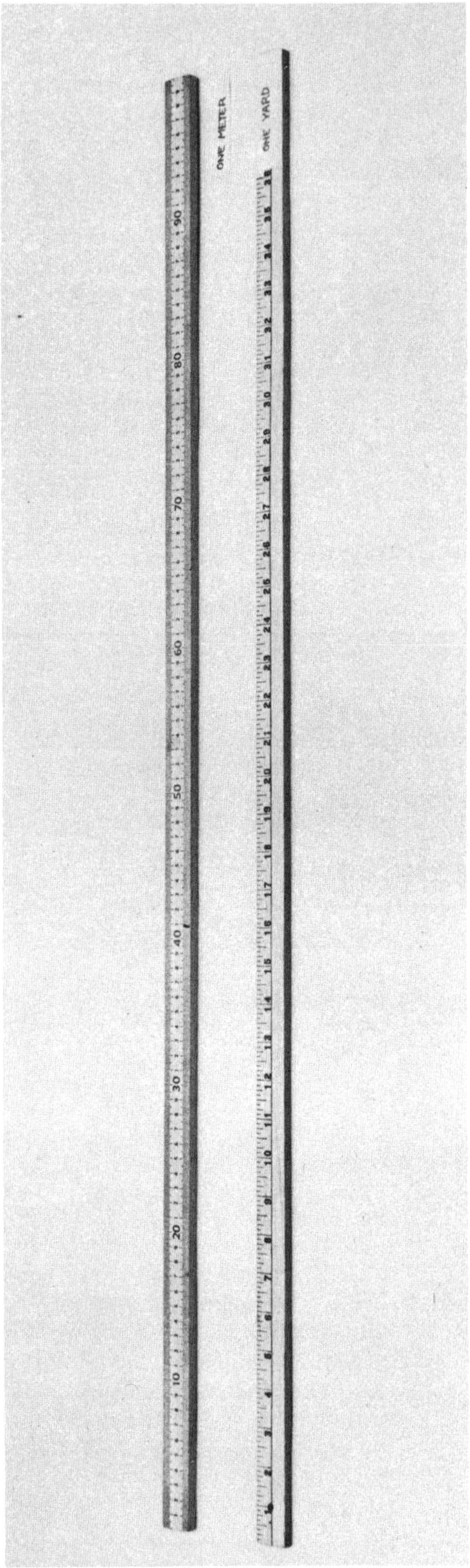

Fig. 10-1. Comparison between One Metre and One Yard

15 cm scales replace the 6″ pocket scale. A 6′ tape is replaced with a 2 m tape, etc. Precision scales are available with graduations as fine as 0.5 mm (½ mm). Fig. 10-3.

10-4. Precision Measurement with Metric Micrometers and Verniers

A full range of metric micrometers, verniers, and gage blocks is available for making linear measurements finer than 0.5 mm. The metric micrometer, Fig. 10-5, is read in much the same manner as the inch micrometer. See Unit 9. Its spindle has 50 threads per 25 mm. Therefore, two revolutions of the spindle are required to move the spindle one millimetre. Since the thimble of the metric micrometer is graduated in 50 equal divisions, and two revolutions of the thimble are required to move one millimetre, the degree of precision obtained is 0.01 mm (1/100th of a millimetre). Note that the sleeve of the metric micrometer has both an upper and a lower set of graduations. The upper set consists of 25 graduations, one millimetre apart, with each fifth millimetre being numbered. The lower set of graduations divides each of the upper set in two, providing 0.5 mm graduations. The

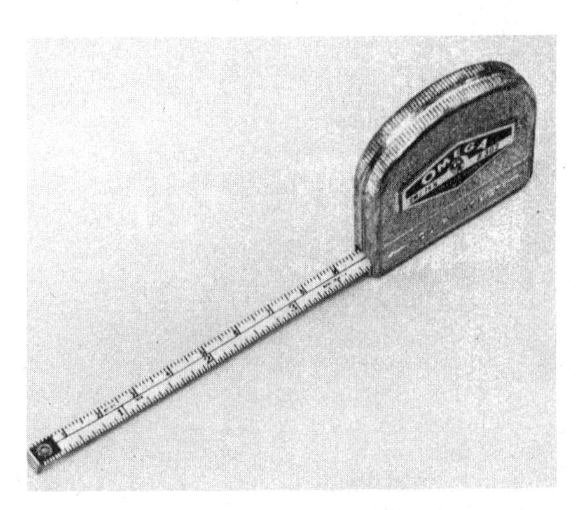

Fig. 10-2. Metric Graduations on Steel Tape

TABLE 2

Metric SI Units

Property		Unit Name	Symbol	Relationship of Units
LINEAR MEASURE	Length	millimetre	mm	1 mm = 0.001 m
		centimetre	cm	1 cm = 10 mm
		decimetre	dm	1 dm = 10 cm or 100 mm
		metre	m	1 m = 100 cm or 1000 mm
		kilometre	km	1 km = 1,000 m
	Area	square centimetre	cm^2	1 cm^2 = 100 mm^2
		square decimetre	dm^2	1 dm^2 = 100 cm^2
		square metre	m^2	1 m^2 = 100 dm^2
		are	a	1 a = 100 m^2
		hectare	ha	1 ha = 100 a
		square kilometre	km^2	1 km^2 = 100 ha
	Volume	cubic centimetre	cm^3	1 cm^3 = 0.0001 litre
		millilitre	ml	1 ml = 0.0001 litre
		cubic decimetre	dm^3	1 dm^3 = 1,000 ml
		litre	l	1 l = 1,000 ml
		cubic metre	m^3	1 m^3 = 1,000 litres
MASS		milligram	mg	1 mg = 0.001 g
		gram	g	1 g = 1,000 mg
		kilogram	kg	1 kg = 1,000 g
		metric ton	t	1 t = 1,000 kg
TIME		second	s	
ELECTRIC CURRENT		ampere	A	
TEMPERATURE		kelvin	K	
LUMINOUS INTENSITY		candela	cd	
AMOUNT OF SUBSTANCE		mole	mol	

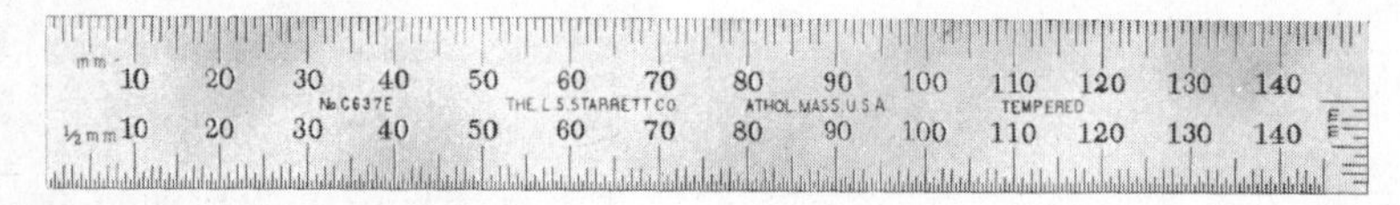

Fig. 10-3. Precision 15 cm Scale with 1 mm and 0.5 mm Graduations

metric micrometer reading of 5.78 mm illustrated in Fig. 10-5, is obtained as follows:

1. Upper sleeve reading (whole millimetres) — 5.0 mm
2. Lower sleeve reading (half millimetres) — 0.5 mm
3. Thimble reading (hundredths of a millimetre) — 0.28 mm

Total reading — 5.78 mm

The vernier metric micrometer is designed for measuring to 0.002 mm (two-thousandths of a millimetre). The vernier is placed on the sleeve above the usual graduations just as with vernier inch micrometers. See Unit 9. Note that the millimetre and half-millimetre graduations are marked on the lower part of the sleeve, to allow the vernier to be read without having to twist the micrometer, Fig. 10-6.

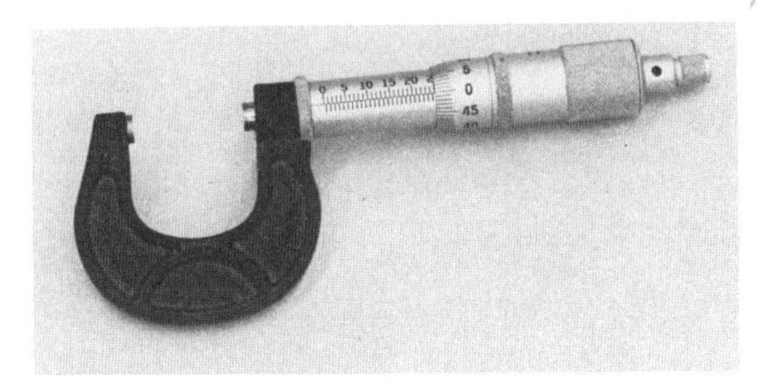

Fig. 10-4. 25 mm Outside Micrometer

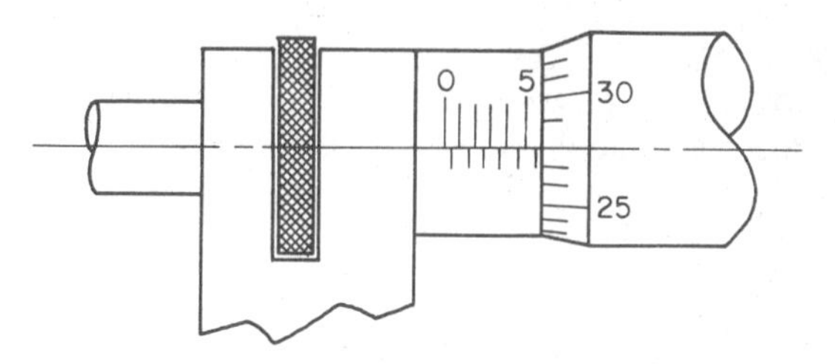

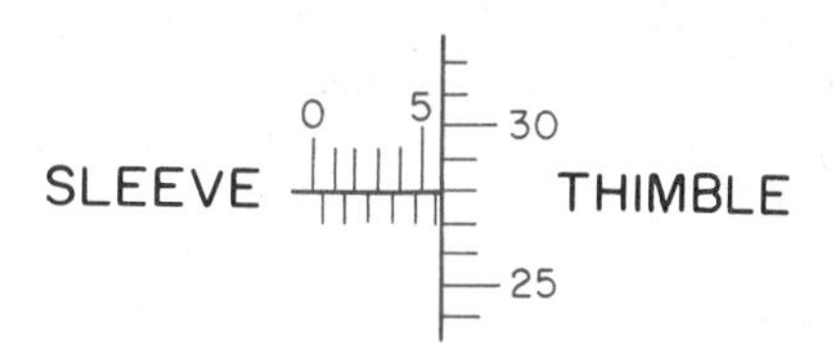

Fig. 10-5. Reading to 5.78 mm

The example in Fig. 10-6 is read as follows:

1. Reading from sleeve (to nearest 0.5 mm)	2.5 mm
2. Reading from thimble (to nearest 0.01 mm)	0.36 mm
3. Reading from vernier (to nearest 0.001 mm)	0.008 mm
Total reading	2.868 mm

Metric vernier calipers, Fig. 10-7, are made to measure to a fineness of either 0.05 mm (.002″) or 0.02 mm (.0008″), depending on how they are calibrated. They are read in exactly the same manner as vernier calipers that are graduated in inches. See Unit 9. The numbers on the main scale mark each centimetre, and the numbers on the vernier scale mark each tenth of a millimetre. As an example, Fig. 10-8 is read:

1. Main scale reading (to nearest millimetre)	47.0 mm
2. Vernier scale reading (to nearest .05 millimetre)	0.75 mm
Total reading	47.75 mm

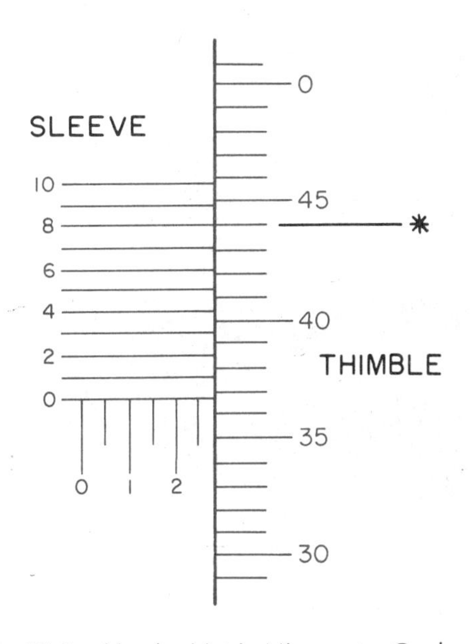

Fig. 10-6. Vernier Metric Micrometer Graduations

10-5. Dimensioning of Engineering Drawings

All dimensions on engineering, or working, drawings will eventually be given in millimetres. Indeed, a few companies have already completely changed to use of metric SI dimensioning. A number of other companies have used **dual dimensioning** for some time. This is the practice of providing both metric and inch dimensions on the same drawing, Fig. 10.9. The metric dimension is either shown above the dimension line and the inch size below, as

$\frac{150 \pm 0.4}{5.906 \pm .016}$, or a diagonal line (/) is used, as 150 ± 0.4/5.906 ± .016. Dual dimensioning makes it equally convenient to make parts on machines calibrated in millimetres or in inches.

10-6. Use of Conversion Tables

Conversion tables are included for convenience in converting dimensions in inches to dimensions in millimetres. Table 3 lists fractions of an inch, their decimal equivalent in inches, and their equivalent in millimetres. With this table, the metric equivalent of a common fraction, for example 5/16″, can quickly be found to be 7.938 mm. (In practice, this would probably be rounded off to 8 mm.)

Table 4 converts decimal parts of an inch to millimetres. Thus, if the metric equivalent of a decimal figure such as .835 inches is desired, it may be found by using the following procedure:

0.8 inches = 20.32 mm
0.03 inches = 0.762 mm
0.005 inches = 0.127 mm
0.835 inches = 21.209 mm

By using Table 3 and Table 4 together, the metric equivalent of any measurement in inches may be calculated.

Words to Know

candela
centimetre
conversion tables
decimetre
dual dimensioning
gram
kelvin
kilogram
kilometre
mass
metre
metric SI system
metric micrometer
metric vernier caliper
millimetre
vernier metric micrometer

Review Questions

1. Why is it becoming more important to know the metric system of measurement?
2. What is the basic unit of length in the metric system?

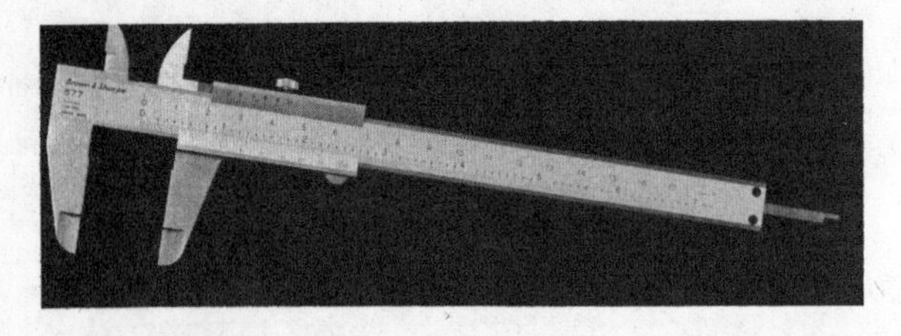

Fig. 10-7. Metric Vernier Caliper

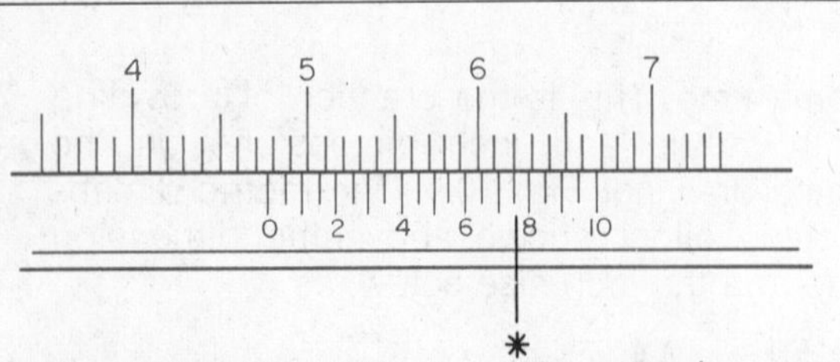

Fig. 10-8. Metric Vernier Caliper Reading 47.75 mm

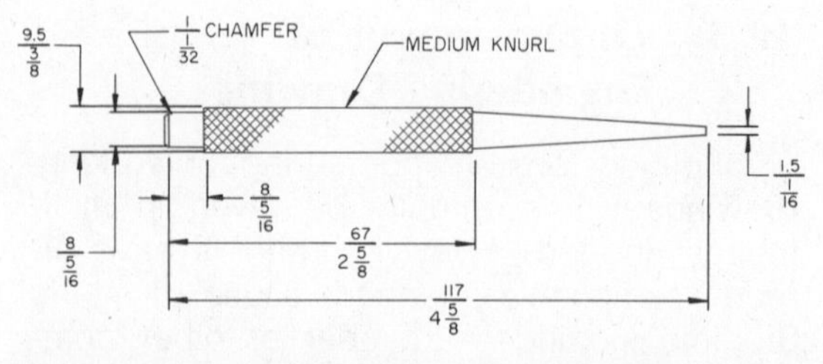

Fig. 10-9. Example of Dual Dimensioning

TABLE 3

Conversion Table
(Inches in Fractions to Decimal Inches to Millimetres

Fraction	Inches	Millimetres	Fraction	Inches	Millimetres
1/64	0.016	0.397	33/64	0.516	13.097
1/32	0.031	0.794	17/32	0.531	13.494
3/64	0.047	1.191	35/64	0.547	13.891
1/16	0.063	1.588	9/16	0.563	14.288
5/64	0.078	1.984	37/64	0.578	14.684
3/32	0.094	2.381	19/32	0.594	15.081
7/64	0.109	2.788	39/64	0.609	15.478
1/8	0.125	3.175	5/8	0.625	15.875
9/64	0.141	3.572	41/64	0.641	16.272
5/32	0.156	3.969	21/32	0.656	16.669
11/64	0.172	4.366	43/64	0.672	17.066
3/16	0.186	4.763	11/16	0.688	17.463
13/64	0.203	5.159	45/64	0.703	17.859
7/32	0.219	5.556	23/32	0.719	18.256
15/64	0.234	5.953	47/64	0.734	18.653
1/4	0.250	6.350	3/4	0.750	19.050
17/64	0.266	6.747	49/64	0.766	19.447
9/32	0.281	7.144	25/32	0.781	19.844
19/64	0.297	7.541	51/64	0.797	20.241
5/16	0.313	7.938	13/16	0.813	20.638
21/64	0.328	8.334	53/64	0.828	21.034
11/32	0.344	8.731	27/32	0.844	21.431
23/64	0.359	9.128	55/64	0.859	21.828
3/8	0.375	9.525	7/8	0.875	22.225
25/64	0.391	9.922	57/64	0.891	22.622
13/32	0.406	10.319	29/32	0.906	23.019
27/64	0.422	10.716	59/64	0.922	23.416
7/16	0.438	11.113	15/16	0.938	23.813
29/64	0.453	11.509	61/64	0.953	24.209
15/32	0.469	11.906	31/32	0.969	24.606
31/64	0.484	12.303	63/64	0.984	25.003
1/2	0.500	12.700	1	1.000	25.400

TABLE 4

Conversion Table
Decimal Parts of an Inch to Millimetres

Inches	Millimetres	Inches	Millimetres	Inches	Millimetres
0.001	0.025	0.01	0.254	0.1	2.54
0.002	0.051	0.02	0.508	0.2	5.08
0.003	0.076	0.03	0.762	0.3	7.62
0.004	0.102	0.04	1.016	0.4	10.16
0.005	0.127	0.05	1.270	0.5	12.70
0.006	0.152	0.06	1.524	0.6	15.24
0.007	0.178	0.07	1.778	0.7	17.78
0.008	0.203	0.08	2.032	0.8	20.32
0.009	0.229	0.09	2.286	0.9	22.86

3. What metric unit of length will be used on all engineering drawings? How is it related to the other metric units of length?
4. How are metric rulers and tapes usually graduated and numbered?
5. What is the smallest unit of metric measurement usually found on precision scales?
6. To what degree of accuracy does a plain metric micrometer measure? Vernier metric micrometer?
7. To what degree of accuracy do metric vernier calipers measure?
8. What two methods are used for dual-dimensioning engineering drawings?
9. For what are inch/metric conversion tables used?

Mathematics

1. Change the following fractions of an inch to millimetres:
   A. 3/16
   B. 1/4
   C. 3/8
   D. 19/32
   E. 45/64
2. Convert the following decimals to millimetres:
   A. .720
   B. .315
   C. .019
3. To make a sheet metal tube 250 mm long and 75 mm in diameter, what size should the metal blank be?
4. What diameter hole, to the nearest fraction of an inch, will a 10 mm drill make?

PART IV

Bench Work

UNIT 11

Hand Sawing

11-1. How the Hacksaw Got Its Name

It is supposed that when man first needed a cutting tool he chopped or **hacked** notches in any hard material he could find. Such a saw would make a very rough cut and this may be how the **hacksaw** got its name. The hacksaw is now a greatly improved tool and is used for sawing all common metals except hardened steel. The cutting tool must always be harder than the material to be cut.

11-2. Parts of Hand Hacksaw

The parts of the hand hacksaw, Fig. 11-1, are the **frame, handle, prongs, tightening screw** and **wing nut,** and **blade**. The frame is made to hold the blade tightly. Frames are made in two styles: the **solid frame** in which the length cannot be changed and the **adjustable frame** which has a back that can be shortened or lengthened to hold blades of different lengths.

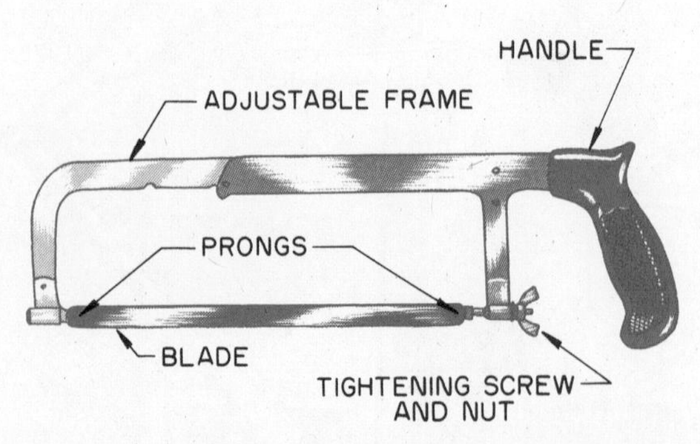

Fig. 11-1. Hand Hacksaw

11-3. Hand Hacksaw Blade Selection

Hand hacksaw blades are made of thin high-grade steel which has been hardened and tempered. Some blades are all **hard** and are therefore quite brittle. Other kinds of blades have hardened teeth and a softer back; these are classified as **flexible** blades. The softer back makes them springy and less likely to break. You should know the following facts about hand hacksaw blades:

Size

They are made in 8″, 10″, and 12″ lengths. The length is the distance between the centers of the holes at each end, Fig. 11-2. The blades are ½″ wide and 0.025″ thick.

Material

Blades are available in several kinds of material including carbon steel, molybde-

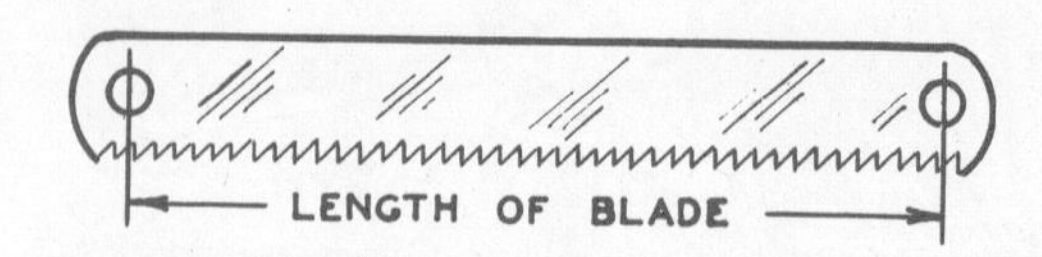

Fig. 11-2. Length of Blade

num alloy steel, tungsten alloy steel, molybdenum high-speed steel, and tungsten high-speed steel.

Number of Teeth Per Inch

A blade always has one more **point** than the number of complete teeth in 1″, Fig. 11-3.

(A) 14 teeth: for cutting soft steel, aluminum, brass, bronze, copper alloys, and other materials 1″ or more in thickness.

(B) 18 teeth: for cutting machine steel, angle iron, drill rod, tool steel, aluminum, copper alloys, and other materials ¼″ to 1″ in thickness; for general-purpose work.

(C) 24 teeth: for cutting materials 1/16″ to ¼″ thickness, iron pipe, metal conduits, light angle iron, etc.

(D) 32 teeth: for cutting materials up to 1/16″ thickness, sheet metals, thin wall tubing, and thin angles or channels.

Tooth Set

Set of saw teeth refers to the way the teeth are bent to one side or the other to provide a **kerf** (the cut made by the saw) which is wider than the thickness of the saw. The wide-saw kerf prevents the saw from binding, Fig. 11-4 (A). Two kinds of set are provided, a **raker** set and **wavy** set, Fig. 11-4 (B). The raker set has one tooth bent to the right, one to the left, and one straight tooth in between. The wavy set has several teeth bent to the right and several teeth bent to the left, alternately. Blades with 14 and 18 teeth have the raker set. Those with 24 or 32 teeth have the wavy set.

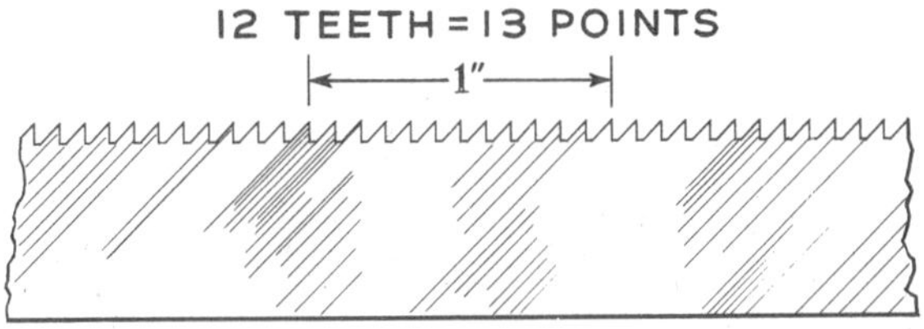

12 TEETH = 13 POINTS
1″

Fig. 11-3. Points per Inch

11-4. Putting Blade in Frame

When putting a new blade in the hand hacksaw, place it so that the teeth of the blade point away from the handle. Fasten one end of the blade to the hook at one end of the frame and the other end of the blade over the hook at the opposite end of the frame, Fig. 11-5. Strain the blade well as you tighten the **tightening screw**; a loose blade makes a crooked cut and is likely to break.

11-5. Holding Metal for Sawing

The work should be held tightly in the vise. The part to be sawed should be near

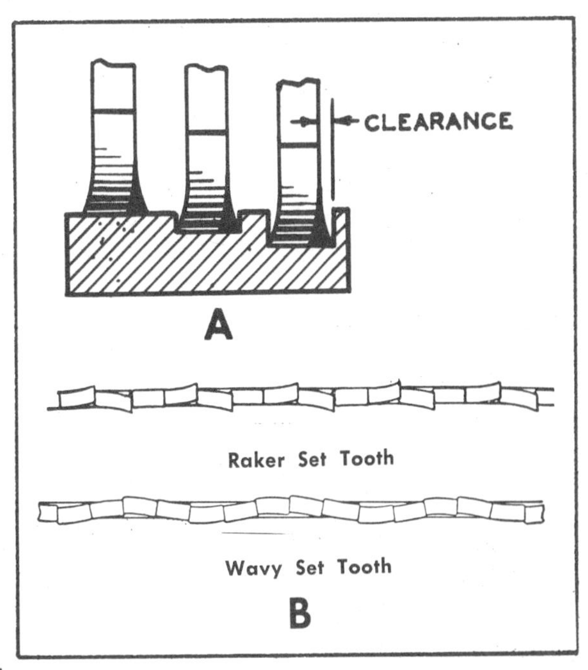

Fig. 11-4. Profile of Blade and the Tooth Set
A. Thickness of Blade and Teeth (Enlarged)
B. Tooth Set on Metal-Cutting Saws: Upper Is Raker Set; Lower Is Wavy Set

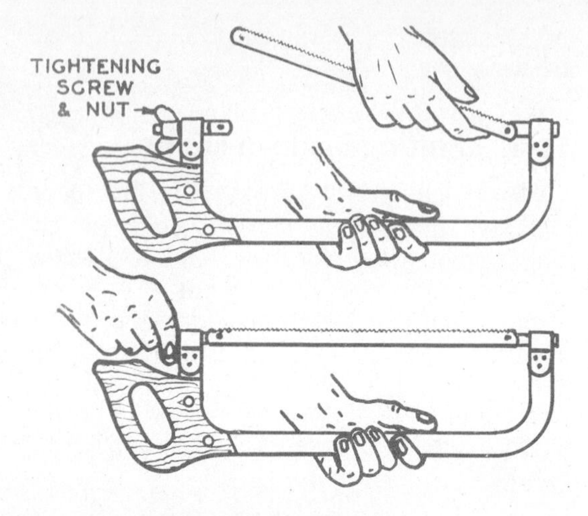

Fig. 11-5. Putting Blade in Frame

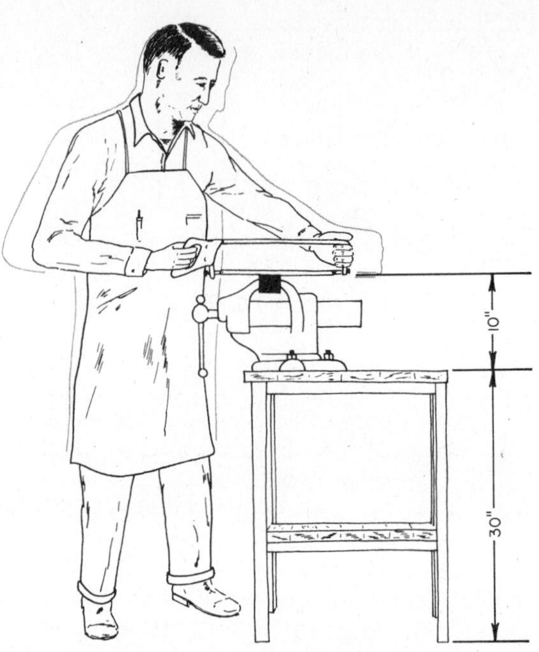

Fig. 11-6. Body Position When Sawing

the vise jaws to keep the work from **chattering.**[1] Protect polished surfaces from the rough, steel vise jaws by covering them with **vise jaw caps**. See Fig. 25-5.

11-6. Body Position for Sawing

When sawing, stand with one foot ahead of the other. Sway forward and backward on the feet in addition to moving the arms, Fig. 11-6.

11-7. Holding a Hand Hacksaw

Grasp the handle firmly with your dominant hand (right- or left-hand). Use the other hand to grip the other end of the saw frame as in Fig. 11-7.

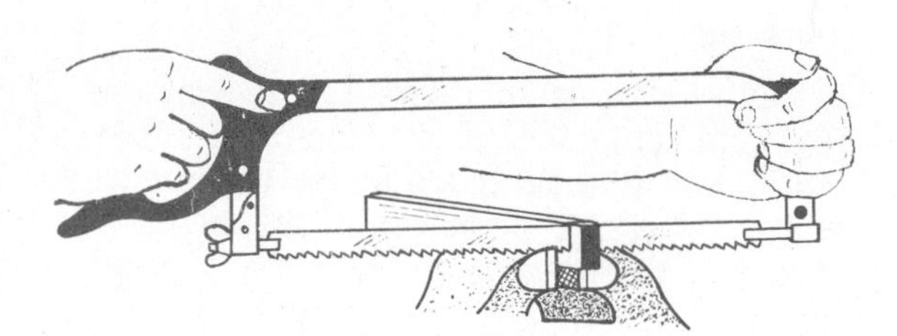

Fig. 11-7. Holding the Hacksaw

11-8. Cutting Stroke

When starting a cut, it helps to notch the starting place with a file. Place the saw on the work and begin with a **backward stroke.** Press down on the **forward stroke** and lift a little on the **return stroke** (Fig. 11-8) because the blade cuts only on the forward stroke. Try to imitate the stroke of the **power hacksaw** (see Fig. 12-1); see how it lifts on the return stroke. The first few strokes should be short ones with only a little pressure. When the saw kerf has been well established so that the blade can't jump out, increase the stroke to the full blade length and add pressure. Tighten the blade again after a few strokes since it will stretch when it becomes warm.

[1]**Chattering** is a fast, rattling noise made by the jarring, vibrating, or springing of the work or the cutting tool. It may cause nicks and notches on the work.

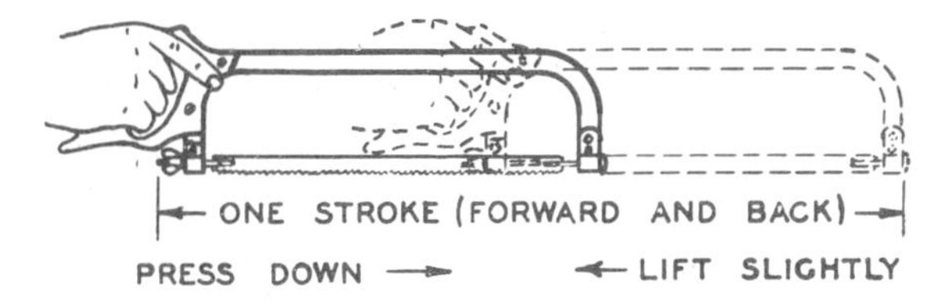

Fig. 11-8. One Stroke

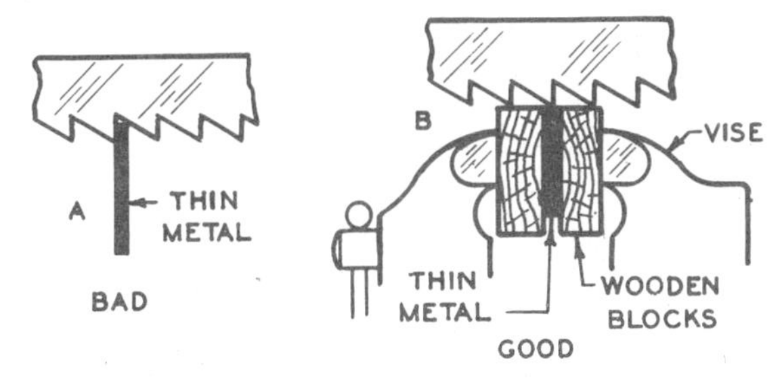

Fig. 11-9. Holding Thin Metal for Sawing

11-9. Speed

Make about 40 **cutting strokes** per minute. See Fig. 11-8. Sawing faster does not appreciably speed cutting. Make long, steady strokes.

11-10. Pressure

Press down on the **forward stroke** and lift a little on the **return stroke**. Large pieces need more pressure than small pieces. A worn blade needs more pressure than a new blade. Rubbing, slipping, or sliding over the metal without cutting makes the cutting edges of the teeth smooth, bright, and shiny like glass. It makes the teeth blunt and dulls them.

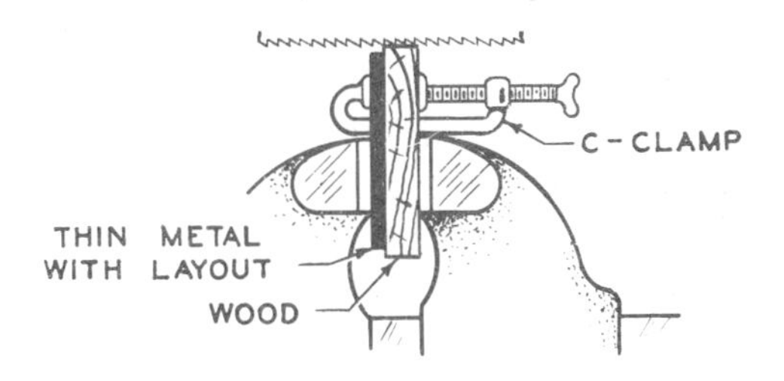

Fig. 11-10. Holding Thin Metal Having a Layout on One Side

11-11. Wear on Sides of Teeth

The sides of the teeth wear down by the rubbing between the blade and the metal which is cut. A used blade makes a narrower cut than a new blade. A new blade, if placed in an old cut, wedges and sticks and is ruined with the first stroke because of the difference in the width of the saw kerf made by the two blades. If a cut cannot be completed with the old blade, start a new cut with a new blade.

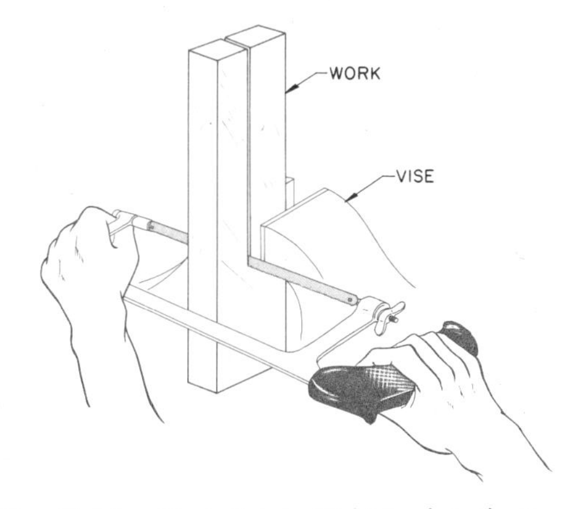

Fig. 11-11. Blade Set at a Right Angle to the Frame

11-12. Sawing Thin Metal

A thin piece of metal to be sawed should be placed in a vise with the metal gripped close to the line of the cut, to prevent it from vibrating, springing, or **chattering**.

If the metal is thinner than the space from one tooth to the next tooth on the blade, Fig. 11-9(A), it should be placed between two boards and clamped in the vise. Saw through both wood and metal at once as in **B** in Fig. 11-9. Remember that at least three teeth should be cutting at any one time. If there is a layout on one side of the thin metal and it is necessary to saw near a line, a **C-clamp** may be used to fasten the metal to the board, so the layout can be seen, Fig. 11-10.

Fig. 11-12. Silhouettes

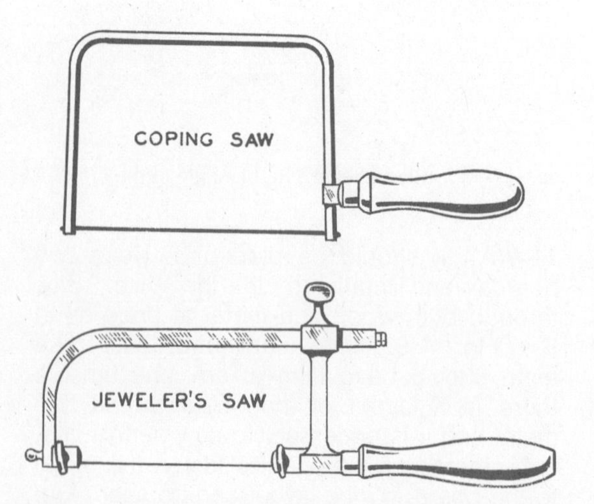

Fig. 11-13. Coping and Jeweler's Saws

11-13. Sawing Wide Pieces of Metal

To saw wide pieces of metal, it is better to set the blade at **right angles** (90°) to the frame, Fig. 11-11.

11-14. Cutting Curved Outlines

Objects with curved shapes — such as bookends, letter openers, bracelets, and jewelry — may be cut out of any attractive sheet metal. Silhouettes of animals, vehicles, and other objects of interest, Fig. 11-12, are ornamental and may be used as pins, buckles, wall decorations, parts of signs, or weather vanes. A small pin may be

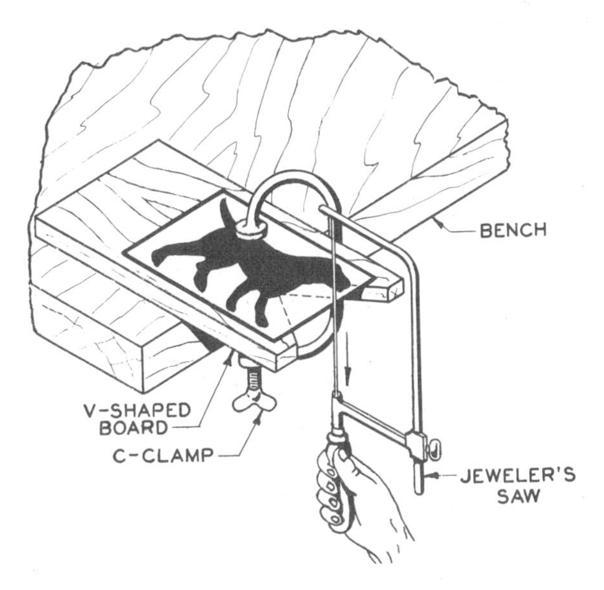

Fig. 11-14. Using Jeweler's Saw

soldered to the back of a silhouette. This makes a nice ornament to be worn on a dress or on a coat.

Cutting curved outlines is done with a **coping saw** or with a **jeweler's saw**, (also called a **piercing saw**), Fig. 11-13, because irregular holes are cut or **pierced** with this saw.

Metal-cutting blades for the jeweler's saw blades have fine teeth and must be very hard in order to cut metal. They are 5″ long, have from 32 to 76 teeth per inch, and are sold by the dozen. The teeth of the blade should point toward the handle.

The work should be held on a board with a V-shaped notch. The board is laid on the edge of the bench and held down with the hand or with a clamp. The sawing should be done with the favored hand, Fig. 11-14. The **cutting stroke** is downward. For internal cuts, a hole must be drilled and the blade put through the hole and then fastened to the saw frame.

11-15. Broken Blades

Saw blades are usually broken by:

1. Loose blade
2. Loose work
3. Too much pressure
4. Blade sticking

Words to Know

adjustable frame	jeweler's saw
backward stroke	piercing saw
chatter	return stroke
clearance	saw kerf
coping saw	saw set
forward stroke	silhouette
hand hacksaw	solid frame

Review Questions

1. Can hardened steel be cut with a hand hacksaw? Why?
2. What materials are used to make hacksaw blades? Which are hardest and will tend to last longest?
3. Which way should the teeth of a hand hacksaw blade point? Why?
4. Why should a new blade never be used in a cut made by an old blade?
5. What should be the cutting speed of the hand hacksaw?
6. What are standard hand hacksaw blade lengths?
7. How should sheet metal be held for sawing?
8. What kinds of saws are used to cut curved outlines?
9. Which way should the teeth of a jeweler's saw blade point?
10. Does the jeweler's saw blade cut on the upward or downward stroke?
11. Name four causes of blade breakage.
12. List two kinds of tooth set used on hand hacksaw blades.

Mathematics

1. Using a 32-tooth blade, what is the thinnest metal that can be safely cut?

UNIT 12

Power Sawing

12-1. Power Hacksaw and Its Parts

The power hacksaw, Fig. 12-1, is a machine for sawing all kinds of metal, except **hardened steel.** The **blade** must always be harder than the material to be cut. The **frame** and blade move forward and backward. It is driven by an **electric motor** and **feeds**[1] and stops by itself.

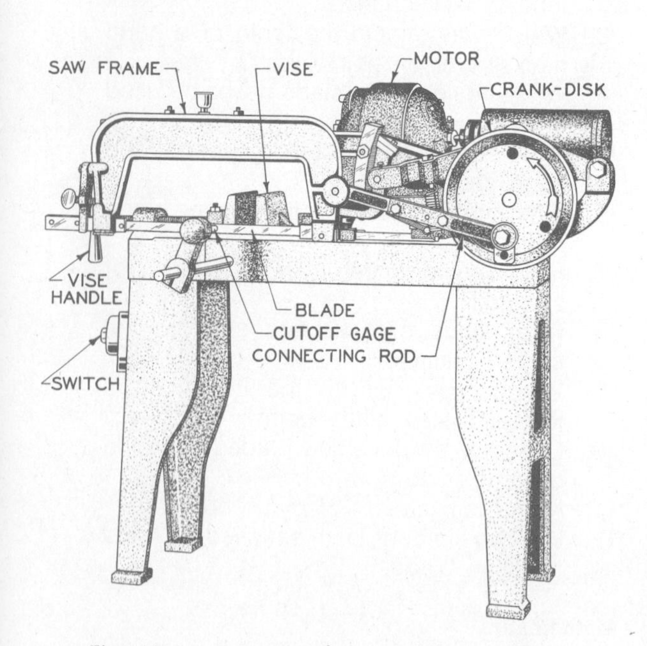

Fig. 12-1. Power Hacksaw and Parts

[1]**Feed** is the movement of a cutting tool into the work on each stroke or revolution.

12-2. Power Sawing Machines

Three common types of power sawing machines are used for sawing metals: (1) power hacksaw, (2) horizontal band saw, (3) vertical band saw. These three power sawing machines are discussed in this unit.

12-3. Care of Power Hacksaw

The power hacksaw should be serviced with the correct lubricants according to the schedule recommended by the manufacturer. It should be cleaned after being used.

12-4. Putting Blade in Frame

When putting a blade in the power hacksaw, check the operator's manual or ask your instructor which way the teeth should point. Some power hacksaws require the teeth to point towards the front of the machine, others towards the rear of the machine. Tighten the blade properly and fasten securely. A loose blade will break more readily than a blade properly tightened.

12-5. Holding Metal to Be Sawed

To cut a piece of metal in the power hacksaw, clamp the stock to be cut tightly in the **vise.** See Fig. 12-1. If the bar from which a piece is to be cut is so short that it will not reach across the full width of the vise, put a piece of metal the same width as the bar at the other end of the vise, Fig. 12-2. The pressure on both ends of the vise must be

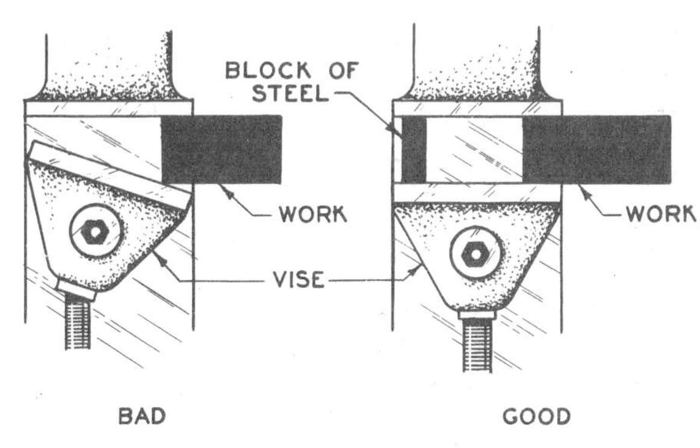

Fig. 12-2. Holding Short Piece of Metal in Power Saw Vise

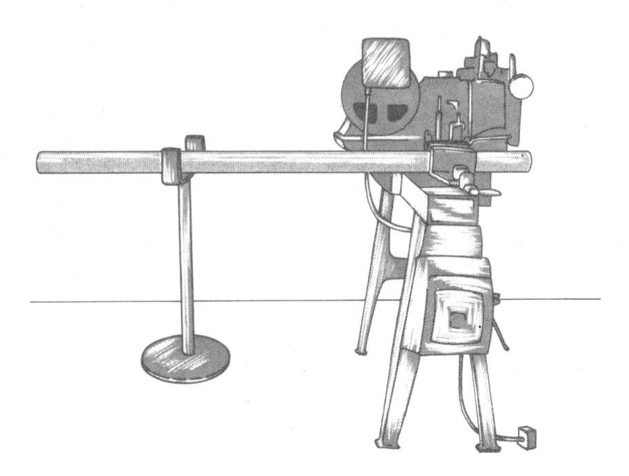

Fig. 12-4. Stock Support

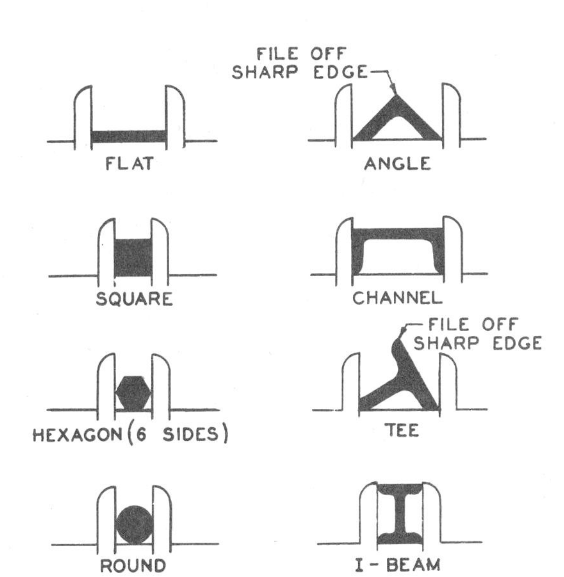

Fig. 12-3. Holding Different Shapes of Metal in Power Saw Vise

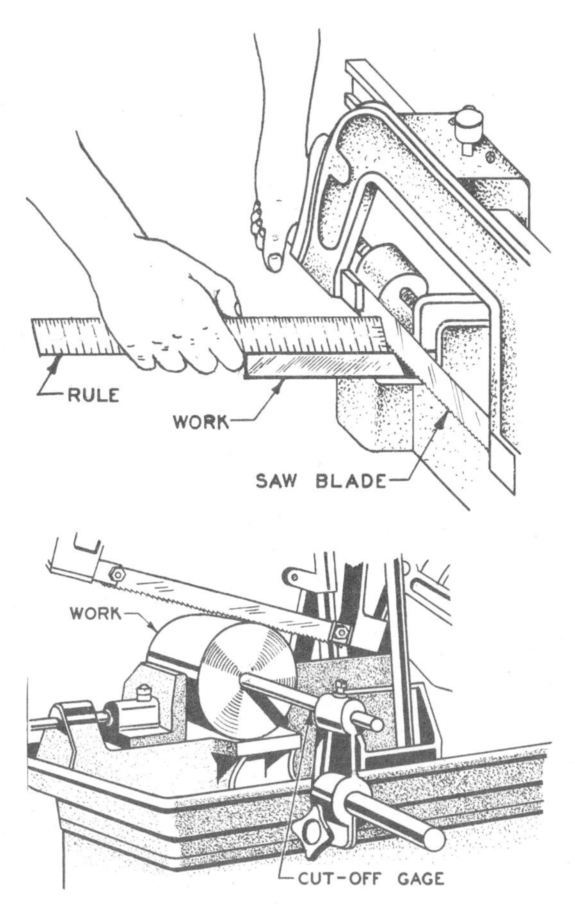

Fig. 12-5. Measuring Metal to Be Cut
Top: Holding Rule Against Saw Blade When Measuring
Bottom: Using Cutoff Gage

the same. Figure 12-3 shows how to hold various shapes of metal in the power saw vise. On some saws the vise may be swiveled up to 45° to make angular cuts.

One end of a long bar may be supported on a stock support, Fig. 12-4. Both ends of the long bar should be the same height from the floor.

12-6. Measuring Metal to Be Cut

When measuring a piece of metal to be cut in the power hacksaw, place the **edge** of the rule alongside the metal to be cut (see Fig. 7-2) and at the same time hold the end of the rule against the side of the saw blade, Fig. 12-5. While it is in this position, move the metal to the length wanted, then tighten the vise against the metal.

12-7. Using Cutoff Gage

When many pieces of the same length are to be cut, the first piece may be measured as described in section 12-6. Additional pieces of the same length may be cut without measuring every piece by setting the **cutoff gage** against the end of the metal after the first piece is measured, Fig. 12-5.

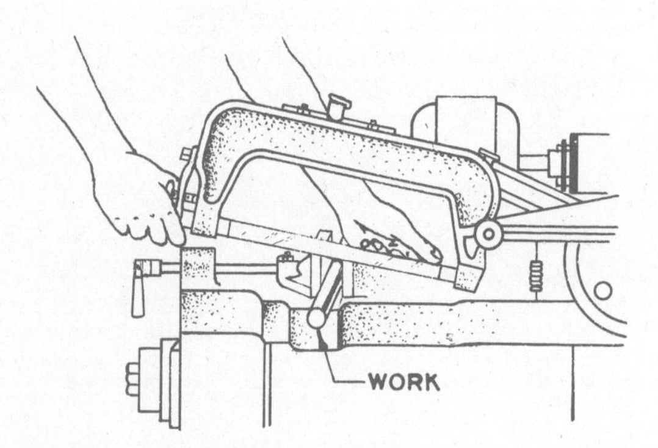

Fig. 12-6. Lowering Saw by Hand to Keep Blade from Breaking

Fig. 12-7. Cutting Fluid Flowing on Cut

12-8. Starting Power Hacksaw

For the type of saw shown in Fig. 12-6, let the saw down by hand so that the blade just touches the work. (Some saws require the blade to be at least 1/4″ away from the work at the start of the cut.)

Next, set the saw in motion by turning on the power. If the power saw is the kind that uses **cutting fluid** (see § 48-6), see that the cutting fluid flows on the cut, Fig. 12-7.

12-9. Hacksaw Speed and Feed Pressure

Power hacksaws are of two basic types, **dry cutting** and **wet cutting.** With the wet cutting machines a **cutting fluid** is used (Fig. 12-7). With the use of cutting fluid the saw operates at higher cutting speeds, it cuts faster, and the blade lasts longer.

The cutting speed for power hacksaws may range from about 35 to 150 **cutting strokes** per minute, depending on the make and type of machine. Ordinarily the speeds range from about 60 to 120 cutting strokes per minute. Note that the feeding pressure is applied during the cutting stroke. The blade lifts slightly and does not cut on the return stroke. Large work needs more feeding pressure than small work. A worn blade needs more feeding pressure to make it cut than a new blade, and so more pressure must be added as the blade becomes worn. Most power hacksaws are equipped with a knob or control device which may be adjusted to increase or decrease the feeding pressure exerted on the saw during the cutting stroke.

Power hacksaws may be single-speed, 2-speed, 3-speed, or 4-speed. Hard and tough metals, such as high-carbon steel or tool steel, should be cut at lower cutting speeds than ordinary low-carbon steels. Table 5 gives recommended cutting speeds for power hacksaws.

TABLE 5
Cutting Speeds Recommended for Power Hacksaws

Material	Strokes Per Minute Dry	Strokes Per Minute Wet
Low-carbon Steel	60-90	90-120
Medium-carbon steel	60	90-120
High-carbon steel	60	90
High-speed steel	60	90
Drill rod	60	90
Alloy steel	60	90
Cast iron	60-90	(cut dry)
Aluminum	90	120
Brass	60	90-120
Bronze	60	90

12-10. Power Hacksaw Blades

If the machine is to operate efficiently, care must be taken in the selection of the proper hacksaw blade for a power hacksaw. Five factors which should be understood in blade selection include: (1) length, (2) thickness, (3) width, (4) tooth coarseness, and (5) the kind of material from which the blade is made.

Length

Blades are available in lengths ranging from 12″ for small machines to 30″ for very large machines. Machines of the size used in school shops generally use blades 12″ or 14″ in length.

Thickness

Blades are available in thicknesses of 0.032″, 0.050″, 0.062″, 0.075″, 0.088″, 0.100″.

Width

Blades are available in widths of 5/8″, 1″, 1 1/4″, 1 1/2″, 1 3/4″, 2″, or 2 1/2″.

Tooth Coarseness

The coarseness refers to the number of teeth per inch of blade length. The number of teeth per inch is also called the **pitch.** Blades are available with 3, 4, 6, 10, 14, and 18 teeth per inch. However, for most applications in the school shop, blades with 6, 10, or 14 teeth per inch are selected. Blades with 14 or 10 teeth are used for the majority of light-duty sawing operations in school shops. The 14-tooth blades are used for cutting thin-wall tubing, thin angles, and other metals of less than about 1/4″ thickness. The 10-tooth blades are used for general-purpose sawing applications of metals from 1/4″ to about 3/4″ thickness. A 6-tooth blade should be selected for efficient cuts on workpieces from about 1″ to 2″ thickness and a 4-tooth blade for workpieces greater than 2″ in thickness.

Tooth Set

Power hacksaw blades have the **raker** tooth set, which sometimes is called the **alternate** or **regular** set. See Fig. 11-4 (**B**) and also § 11-3.

Blade Material

Blades are available in the following kinds of material: flexible high-speed steel, all hard high-speed steel, all hard molybdenum alloy steel, all hard tungsten alloy steel, and the **welded and composite blade** which has a hardened high-speed steel cutting edge welded to a tough alloy steel back.

Blade Selection

Blades 12″ in length are available with 10 or 14 teeth per inch, 1″ width, and 0.050″ thickness. Blades 14″ in length are available in a wide variety of specifications.

Blades commonly used in school shops are 1″ width, 0.050″ thickness and with either 10- or 14-teeth per inch for light-duty cutting applications. However, for general-purpose cutting applications, the 14″ blade is available in 1 1/4″ width, 0.062″ thickness, and with either 10- or 6-pitch teeth. For very heavy-duty applications other pitches and widths are available. As a general rule,

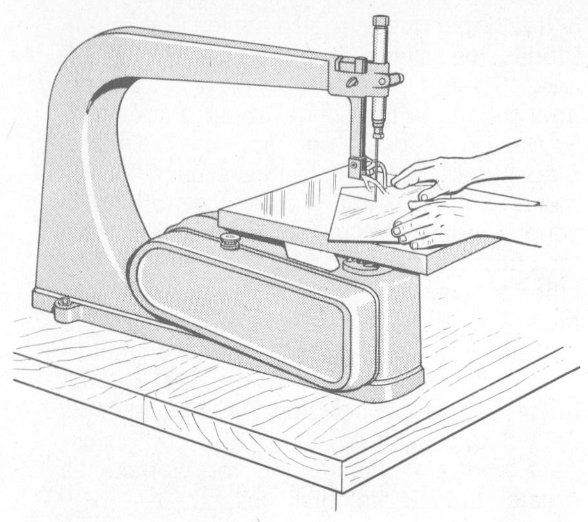

Fig. 12-8. Scroll Saw

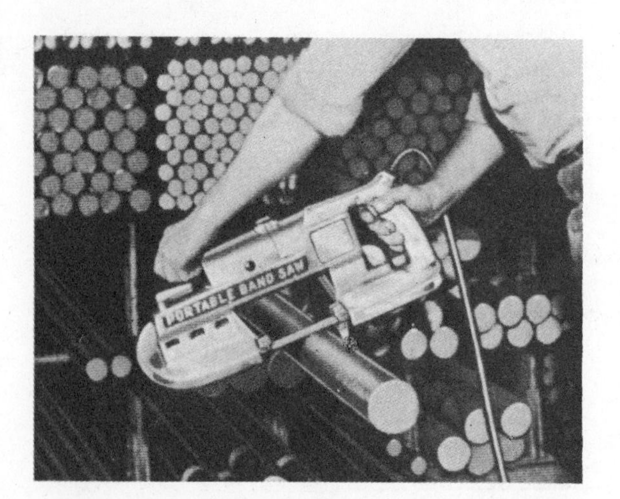

Fig. 12-9. Portable Band Saw

blades with finer teeth should be used for cutting hard stock and thin sections.

12-11. Scroll Saw

The scroll saw, Fig. 12-8, also called **jig-saw,** is used to cut thin, **soft metals** as well as wood, plastics, and other soft materials. It cuts curves in thin materials easily and is a useful tool in the home workshop.

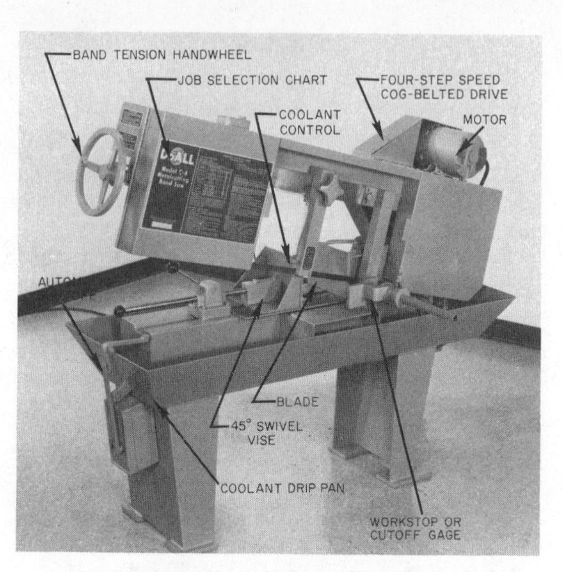

Fig. 12-10. Horizontal Band Saw

12-12. Portable Band Saw

Some metal that must be sawed cannot be easily moved to the saw, so a **portable band saw,** Fig. 12-9, may be carried to the job. This tool has a **band saw blade** that runs continuously in one direction because the blade is a thin, flat strip of steel with a saw edge. It cuts faster than the hacksaw blade that moves back and forth, cutting only on the forward stroke. It can be operated from most electrical outlets.

12-13. Horizontal Band Saw

The **horizontal** band saw, Fig. 12-10, has a blade that runs in only one direction. It is used for cutting off metal pieces. It cuts faster than the power hacksaw which cuts only on one stroke and then lifts and returns for the next cutting stroke. The frame and the blade of the horizontal band saw can be raised and lowered because it has a hinge on one end. The position of the saw blade is horizontal when in the cutting position.

Horizontal band saws may be either the **dry cutting** or the **wet cutting** type. Wet cutting machines, Fig. 12-10, are equipped with a cooling pump which circulates the cutting fluid. Some band saws are equipped with a **spray mist** system which sprays a fine mist of cutting fluid on the blade as it cuts. The use of cutting fluids permits higher cutting speeds, faster cutting, and increased saw blade life.

Kinds of Cuts

The horizontal band saw can cut off stock square or at an angle. For angular cuts, the vise may be swiveled up to 45°.

Feed Rate

The feed rate on most horizontal band saws is hydraulically controlled; therefore, the danger of too rapid feeding is eliminated when the **feed rate** is properly adjusted. The feed rate should be set so that the blade may enter the work steadily and produce a steady flow of metal chips. If the feed rate is too rapid when the blade enters the work, the blade may break. When the feed rate is too slow, the blade does not cut efficiently and it will become dull more rapidly.

12-14. Cutting with a Horizontal Band Saw

1. See that the proper band saw blade is installed for the job.
2. See that the blade is tensioned properly. Check this with your instructor.
3. Determine that the vise is set for the type of cut desired. For a square cut it should be set at a 90° angle to the saw blade; check with a square, make a trial cut, and check the accuracy of the cut.
4. Select the stock, then measure, and mark the point at which the cut is to be made.
5. Mount the workpiece in the saw vise in proper position for the cut. See Fig. 12-3. If more than one piece is to be cut, locate and set the **work stop** or **cutoff gage** at the end of the part so that additional parts may be cut to the same length without measuring each part. See Fig. 12-5. If the stock being cut is a long bar, support the long end on a stock support, Fig. 12-4.
6. Determine the correct cutting speed and set the machine accordingly.
7. Raise the saw frame with the handle provided, start the saw running, and adjust the feed rate; this is the rate at which the blade advances toward the work.
8. Allow the saw to make a very slight kerf, raise the blade slightly, and stop the machine. Measure the length of the stock; adjust if necessary.
9. Again start the machine, turn on the cutting fluid, and complete the cut.
10. Adjust the stock for the next cut and proceed in the usual manner.

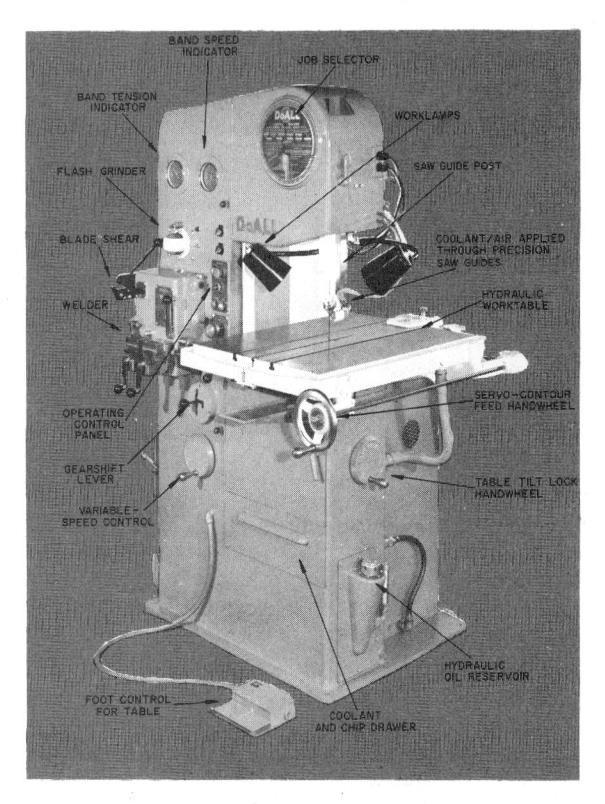

Fig. 12-11. Vertical Band Saw

12-15. Vertical Band Saw

A vertical band saw, Fig. 12-11, is also called a **contouring machine.** Like the horizontal band saw in Fig. 12-10, it has a blade which runs continuously in one direction. The blade of the vertical band saw is in a vertical position. A blade shear and a blade welder are often mounted on the saw for cutting and welding band saw blades, Fig. 12-11. Some manufacturers make a band saw which can be converted for use as either a horizontal- or vertical-type band saw.

Fig. 12-12. Table Tilted for Angular Cut

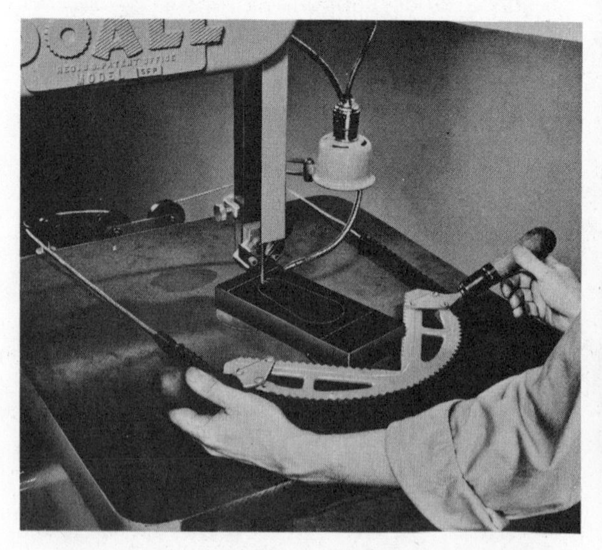

Fig. 12-13. Band Saw Inserted through Drilled Hole for Internal Contour Sawing

Kinds of Cuts

Vertical-type band saws may be used for making either straight line cuts, angular cuts (as in Fig. 12-12), or curved line cuts (as in Fig. 12-13). Curved line cuts are called **contour** cuts. Hence, vertical band saws are also called **contouring machines.** The saw table may be tilted at any desired angle up to 45° for making angular cuts, Fig. 12-12.

When internal contour cuts are made, a hole must be drilled in the workpiece first, Fig. 12-13. The blade is then cut with the blade shear, inserted through the drilled hole in the workpiece, and rewelded with the blade welder. Next, the blade is installed on the machine and properly tensioned. The contour cut may then be made. Of course,

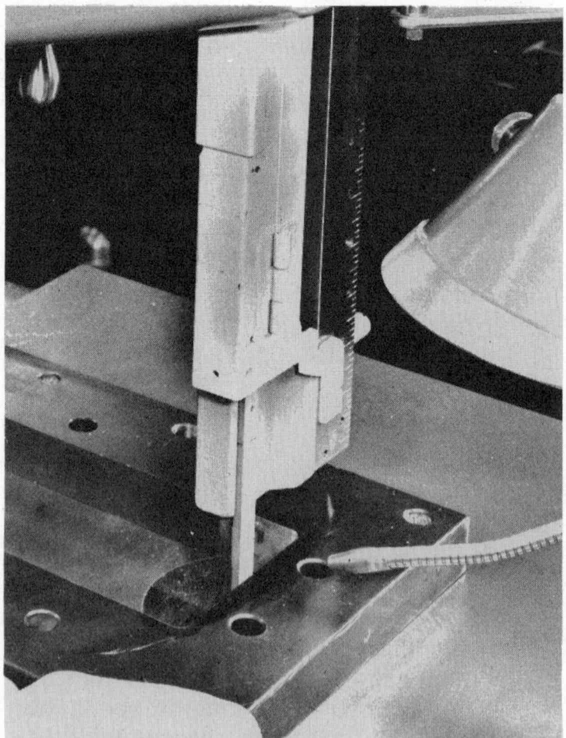

Fig. 12-14. Filing with a Band File

the blade must again be removed from the machine and sheared in order to remove the finished workpiece.

Band Filing

Vertical band sawing machines may also be used for **band filing,** Fig. 12-14. A **band file** is made up of short segments of file blade which are riveted to steel tape and hooked together, Fig. 12-15. Band filing is much faster than hand filing. A workpiece may be band filed by simply holding it against the moving file band until the desired surface and contour are obtained.

Band Polishing

An abrasive belt may be installed on a vertical band sawing machine for band polishing. For this purpose, a **guide** is mounted on the saw guide post, Fig. 12-16. The abrasive belt is installed on the machine and is held in position by the guide provided. Polishing is done by holding the workpiece against the moving abrasive band as shown in Fig. 12-16.

12-16. Band Saw Cutting Speeds

The cutting speed for band saws, including either the horizontal or vertical types, may be varied in several ways, depending on the type of motor drive system on the machine. On many horizontal band saws the speed is changed by changing the position of the V-belt on the motor and machine drive pulleys. The cutting speed on some vertical band saws may be changed in a similar manner. Many vertical-type band saws have a **variable-speed drive** mechanism. With this type drive the speed of the machine may be changed to any desired speed within the speed range for the machine. However, **with the variable-speed drive mechanism, the speed of the ma-**

Fig. 12-15. Uncoupling the Band File

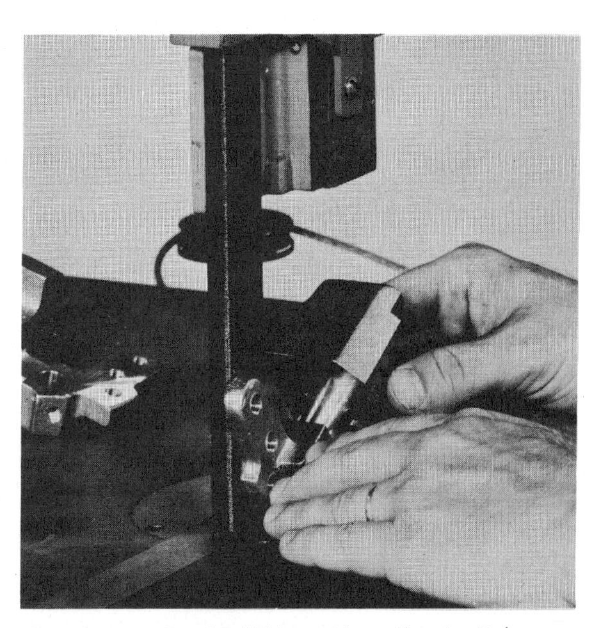

Fig. 12-16. Band Polishing with an Abrasive Belt on the Band Saw

chine can be changed only while the machine is running.

Recommended cutting speeds vary considerably according to the following:

1. Kind of material being cut.
2. Hardness of the material.
3. Thickness of the material.
4. Whether the cutting is wet or dry.

Slower cutting speeds are generally used when harder metals are cut. Slower cutting speeds are also used for cutting thick materials rather than thinner materials. Higher cutting speeds may be used for wet cutting rather than dry cutting.

The following (Table 6) suggests average cutting speeds in surface feet per minute (sfpm), for cutting material ½″ to 1″ thick ness with either horizontal- or vertical-type band saws:

TABLE 6
Cutting Speeds (sfpm)
for Band Saws

Material	Cutting Speed (sfm)	
	Dry	Wet
Alloy Steel (tough)	125	175
Aluminum	250	800
Bakelite	300	
Brass (soft)	500	800
Brass (hard)	200	300
Bronze	200	300
Copper	250	400
Drill Rod (annealed)	75	125
Gray Cast Iron (soft)	125	
Hard Rubber	200	
High-Speed Steel	50	75
Low-Carbon Steel	125	175
Malleable Iron	125	175
Medium-Carbon Steel	100	150

The cutting speeds in Table 6 may be increased approximately 25% for cutting materials ¼″ or less in thickness. They should be decreased by approximately 25% for materials which are 2″ or more in thickness.

12-17. Blade Tension on Band Saw

The band saw blades on both horizontal- and vertical-type machines should be installed with the teeth pointing the direction in which the blade travels. The blade should be tightened with the proper tension, as recommended by the manufacturer of the machine. Some band saws are equipped with a **blade tension indicator** which detects when the proper tension is applied for the width of the blade installed. Ask your instructor to explain how proper blade tension is determined and applied on the band saws in your shop.

12-18. Band Saw Blade Selection

Blade Material

Metal-cutting band saw blades are made from a variety of steels, including high-carbon steel, special alloy steel, and high-speed steel. The alloy steel blades and high-speed steel blades generally are designed for heavy-duty production work. The high-carbon steel blades have hardened teeth and a softer flexible back. They generally are the least expensive to purchase. Carbon steel blades are probably the most widely used on both horizontal and vertical machines in school shops, small machine shops, and maintenance machine shops.

Length

The length of band saw blades varies for different machines. Blades may be purchased in welded bands for all standard machines. Blades also may be purchased in coils 100 feet in length or longer in a special stripout container. The blades may be cut to length and welded in the shop with a band saw blade welder, as shown on the machine in Fig. 12-11.

Tooth Set

Metal cutting band saw blades commonly have two kinds of set, **raker** set and **wavy** set. See **B** in Fig. 11-4. Both types may be

TABLE 7

Metal-Cutting Band Saw Blade Sizes

Width	Thickness	Teeth Per Inch
3/32	0.025	18
1/8	0.025	14-18-24
3/16	0.025	10-14-18
1/4	0.025	10-14-18-24
3/8	0.025	8-10-14-18
1/2	0.025	6-8-10-14-18-24
5/8	0.032	8-10-14-18
3/4	0.032	6-8-10-14-18
1	0.035	6-8-10-14

TABLE 8

Cutting Speeds (sfpm) for Skip-Tooth Band Saw Blades

Material	Speed in Feet Per Minute
Aluminum	2000 to 3000
Aluminum Alloys	300 to 2000
Asbestos	800 to 1500
Bakelite	2500 to 3500
Brass	400 to 1000
Copper	1000 to 1500
Formica	800 to 2000
Lucite-Plexiglas	2000 to 3000
Magnesium	3000 to 5000
Wood	2000 to 3000

used in either horizontal or vertical machines. The raker set generally is recommended for most contouring cuts in vertical machines. Generally, the wavy set is recommended for most operations on horizontal machines and for cutting thin metals or thin wall tubing.

Blade Width

Blades generally are available in the widths shown in Table 7. A narrow width blade cuts a smaller radius than a wider blade; however, narrow width blades will break more easily when the feed rate is too great. The following shows the minimum radius which can usually be sawed with blades of various widths:

Blade Width	Radius
1/2″	2 1/2″
3/8″	1 1/4″
1/4″	5/8″
3/16″	3/8″
1/8″	7/32″
3/32″	1/8″

Blade Thickness

Blades are available in the thicknesses shown in Table 7.

Teeth Per Inch

Also called **pitch,** the "teeth per inch" refers to the coarseness of the teeth. Blades are available with various degrees of coarseness, as shown in Table 7. A coarse pitch generally should be selected for sawing large sections or thicknesses and soft metals. A finer pitch usually should be selected for sawing thinner thicknesses and harder metals. There should be a minimum of **two teeth** in contact with the work at all times, except for very thin sheet metal which is cut at higher speeds. For general-purpose metal-cutting operations, band saw blades of the following pitches will produce good results:

Teeth Per Inch	Metal Thickness
18	up to 1/4″
14	1/4″ to 1/2″
10	1/2″ to 2″
8	2″ to larger

Skip-Tooth Blades

Metal-cutting band saw blades of the skip-tooth type have a wide space between each tooth, as though every second tooth were removed. The wide space provides extra chip clearance. These may be used for sawing nonferrous metals such as aluminum, brass, copper, lead, zinc, etc. They are also used for sawing wood, plastic, asbestos, and other nonmetallic materials. They can be used at higher operating speeds than standard blades with regular teeth and, therefore, cut many times faster. Table 8 gives general skip-tooth sawing speed recommendations.

Words to Know

blade tension
contour sawing
cutoff gage
cutting fluid
cutting speed
feed rate
horizontal band saw
jigsaw
portable band saw
power hacksaw
scroll saw
skip tooth blade
tooth pitch
vertical band saw

Review Questions

1. What is meant by the feed of a saw?
2. In which direction should the teeth of your power hacksaw point?
3. What advantages are achieved in using a cutting fluid on a power hacksaw?
4. On which stroke does your power hacksaw cut?
5. How can the feed pressure be increased or decreased on your power hacksaw?
6. List several factors which must be considered when selecting the proper power hacksaw blade for a job.
7. List the types of tooth set on power hacksaw blades.
8. What harm can result if the proper feed rate is not used on a band saw?
9. Why are vertical band saws often equipped with a shear and a blade welder?
10. What kinds of cuts or other operations can be performed on a vertical band saw?
11. How can the cutting speed be changed on your vertical band saw?
12. What is the necessary precaution to be observed when changing the speed on a machine equipped with a variable-speed drive mechanism?
13. Recommended cutting speeds for band sawing vary according to four important factors. Name them.
14. What is the recommended band saw cutting speed for low-carbon steel ¾″ thick without cutting fluid? With cutting fluid?
15. What recommendation should be followed concerning blade tension?
16. What factors should be considered in the selection of a band saw blade for a vertical band saw?

Mathematics

1. How many 4″ pieces can be cut from a 5-foot bar, allowing ⅛″ for each saw cut?
2. By what % will the cutting efficiency be increased if cutting fluid is applied to the blade of a band saw while cutting dry medium-carbon steel?

UNIT 13

Chisels and Chipping

13-1. Meaning and Importance of Chipping

Chipping is the art of shaping metal by removing small pieces with a cold chisel and a hammer. Most of the work once done by chipping is now done by machine tools. However, chipping is still done in maintenance and repair work.

13-2. Cold Chisels

Cold chisels are used to cut cold metal, hence the name. They are made of high-grade steel in a number of sizes and shapes. The cutting ends are **hardened** and **tempered** because the cutting tool must be harder than the material to be cut.

13-3. Sizes and Names of Cold Chisels

Cold chisels are usually made of steel 3/8″ to 1″ thick and from 6″ to 8″ long. The thickness and length depend upon the use of the chisel. For fine work, the thinner and shorter chisel is used. For large work, the thicker, and usually the longer one, is used.

Cold chisels are known by the shapes of their cutting edges. Thus, we have the **flat chisel, cape chisel, diamond-point chisel,** and **round-nose chisel,** Fig. 13-1.

13-4. Flat Chisel

The flat chisel has a wide **cutting edge,** Fig. 13-1. It is used for chipping flat surfaces, cutting off **sheet metal,** cutting bars and **rivets** (see Fig. 13-9), and for most of the ordinary chipping around the shop. It is the most commonly used chisel.

13-5. Cape Chisel

The cape chisel has a narrow **cutting edge** and is used for cutting narrow grooves. See Figs. 13-1 and 13-5. It should be widest at the cutting edge to keep the chisel from sticking in the groove; it is thus less likely to break.

13-6. Diamond-Point Chisel

The diamond-point chisel has a **cutting edge** shaped like a **diamond,** Fig. 13-1. It

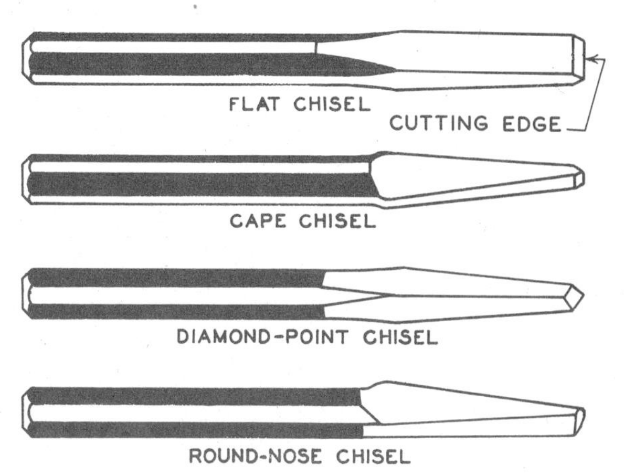

Fig. 13-1. Cold Chisels

is used to cut V-shaped grooves and to chip square corners.

13-7. Round-Nose Chisel

The round-nose chisel (see Fig. 13-1) has a rounded **cutting edge** and is used for chipping round corners and grooves.

13-8. Cutting Angle of Cold Chisels

The **cutting angle** is a wedge which cuts into the metal as shown in Fig. 13-2.

The cutting angle on most cold chisels is from 50° to 75°, depending upon the kind of metal to be cut. The softer the metal, the nearer the angle should be to 50°. For cutting hard metal, the cutting angle should be more blunt. For average work, it should be about 60°.

13-9. Corners of Cold Chisels

The corners of a cold chisel are weak. They are less likely to break off or to dig into the metal while chipping if they are ground so they are a little rounded. See Fig. 13-8.

13-10. Heads of Cold Chisels

The head of the cold chisel should be **tapered**[1] a little and rounded, as **A** in Fig. 13-3, but not **hardened.** See § 38-4. After a cold chisel has been used for some time, the head becomes flattened like a mushroom with rough, ragged edges, as **B** in Fig. 13-3; it is then called a **mushroom head.** This is dangerous, because one of the ragged edges may fly off and injure you or another worker when the mushroom head is struck with the hammer. The mushroom head should be ground on the grinding wheel to the shape shown in **A** in Fig. 13-3.

13-11. Sharpening Cold Chisels

Good work can only be done with sharp chisels. Sharpen the **cutting edge** on a **grinding wheel** as explained in section 44-10. The beginner may use a **center gage** to test the **cutting angle,** Fig. 13-4.

13-12. Hammers for Chipping

The **machinist's hammer** is used for ordinary chipping. It should weigh from ¾ of a pound to 2 pounds, depending upon the

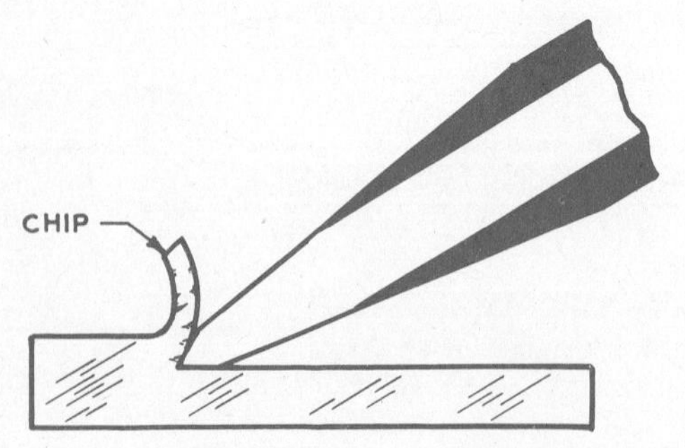

Fig. 13-2. Wedge Action of a Cold Chisel

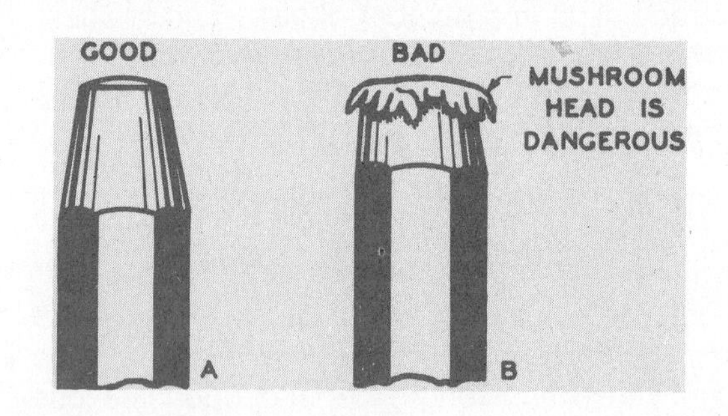

Fig. 13-3. Alternating current.

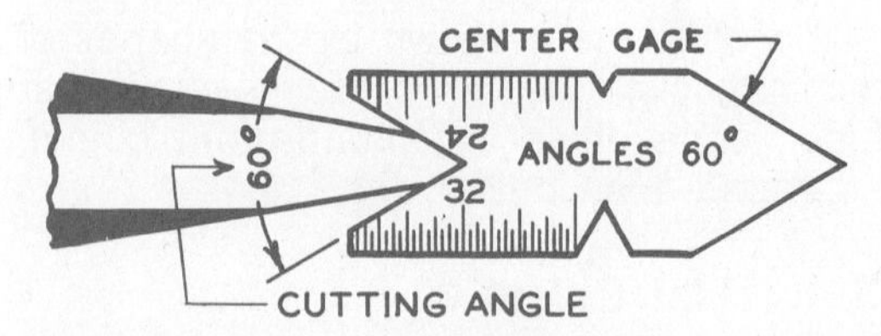

Fig. 13-4. Testing the Cutting Angle of a Cold Chisel with Center Gage

[1]**Tapered** means gradually narrowed toward one end.

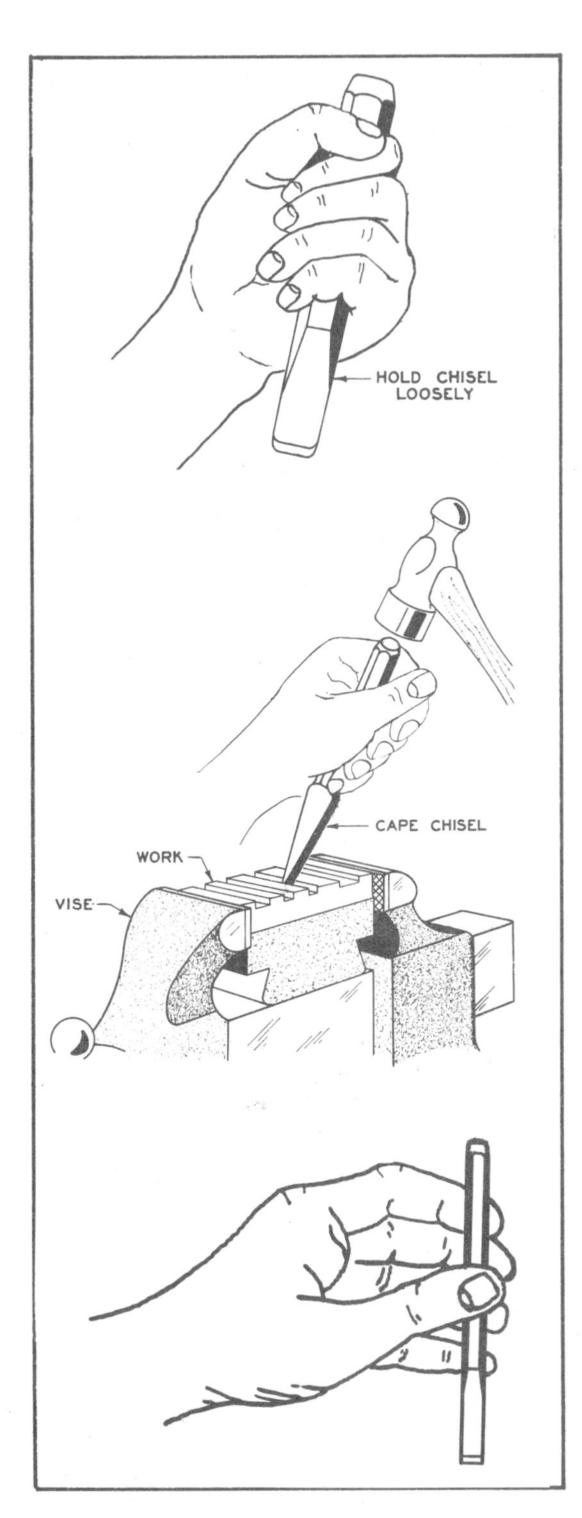

Fig. 13-5. Holding Chisel

work; use a light hammer for small work and a heavy hammer for large work. A one-pound hammer is about right for a ¾″ chisel for ordinary chipping.

13-13. Holding Chisel and Hammer

The middle of the chisel should be held loosely in one hand, Figs. 13-5, 13-7, and 13-9. For small, fine work it may be held in the fingers, as shown in Fig. 13-5. The hammer handle should be grasped near the end with the favored hand and held loosely to keep from tiring the arm.

13-14. Goggles

While you are chipping, wear **goggles** to protect your eyes, Fig. 13-6; flying **chips** cause serious injuries. Know where the chips are landing when you are chipping; eyes have been lost by chips striking another worker. See that the chips fly against a wall or set up some kind of a **shield.**

13-15. How to Chip

Hold the chisel in a position that will make a chip of the right size. Keep your eyes on the cutting edge of the chisel. Swing the hammer over the shoulder with an easy arm movement, Fig. 13-7. Strike the head of the chisel squarely with a good, snappy blow and reset the chisel after each blow for the next cut. Cut a little at a time. The life of the cutting edge will be lengthened, the chisel will be less likely to break, and more metal will be cut in the long run.

Fig. 13-6. Goggles

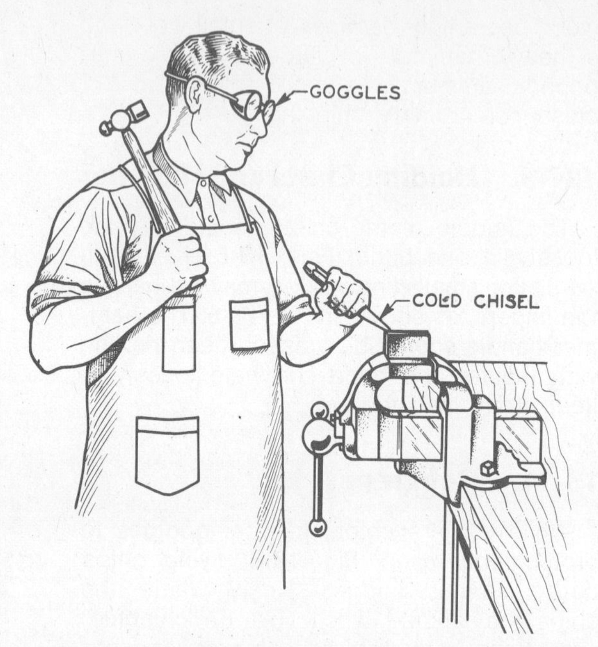

Fig. 13-7. Body Position, also Holding Hammer and Chisel

13-16. Shearing

The **vise jaw** and **flat chisel** may be used together to act like a pair of scissors or **shears.** This is called **shearing.**

Thick **sheet metal** may be cut in a vise with a flat chisel. First, a line which is used as a guide should be made with a **scriber.** Then, clamp the metal tightly in a vise that has good edges so that the line will be just below the tops of the vise jaws. This leaves a little metal for filing to the line afterwards. Continue as in Fig. 13-8.

1. Lay the **cutting edge** of the chisel on top of the vise jaws and against the metal.
2. Slant the chisel a little. Make sure that the **face** of the chisel is **horizontal,** as in Step 1, Fig. 13-8, because it is important to make a square cut.
3. Start to cut.

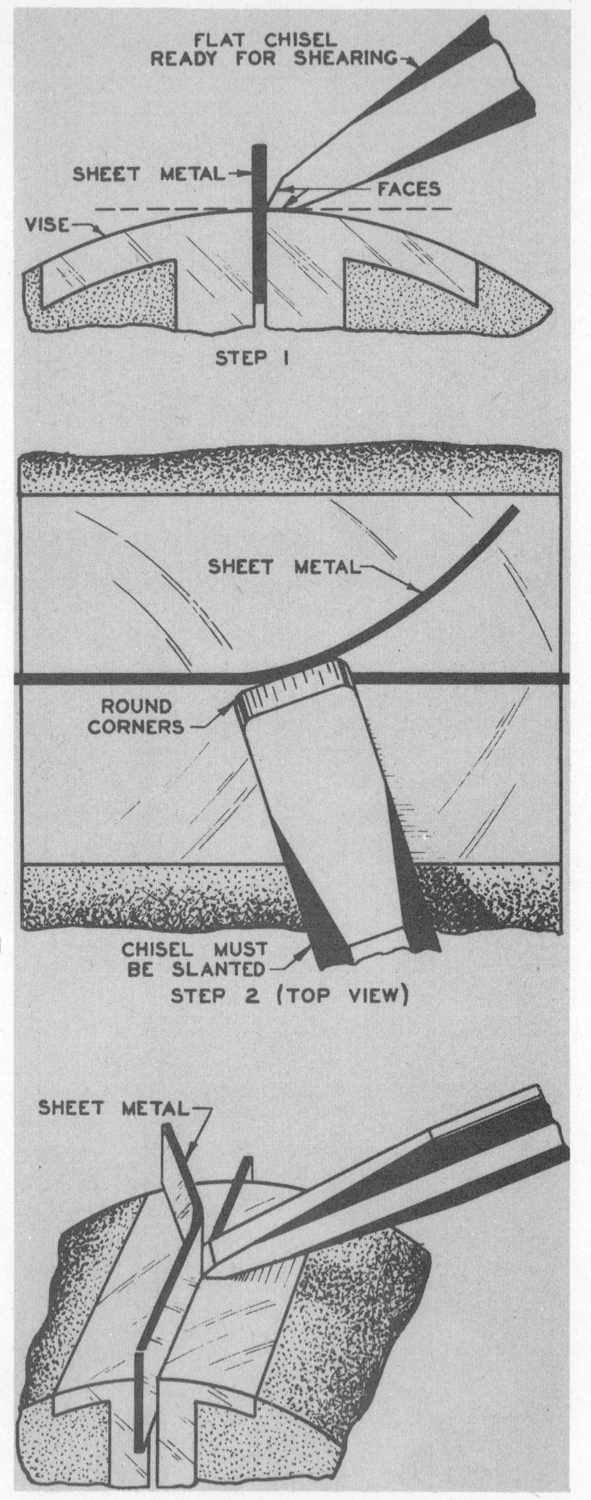

Fig. 13-8. Shearing Sheet Metal with a Flat Cold Chisel

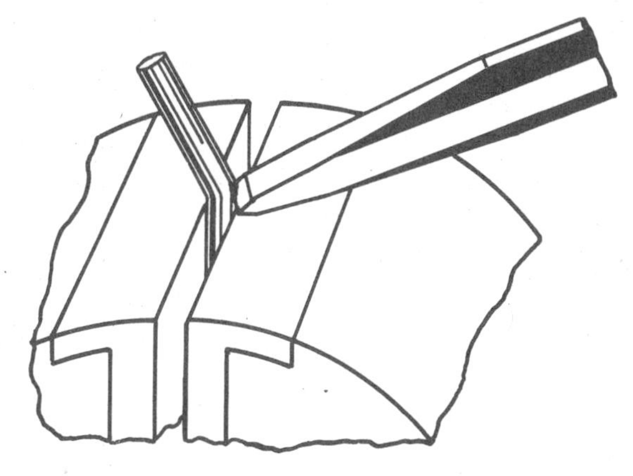
Cutting Stock in Vise with Flat Cold Chisel

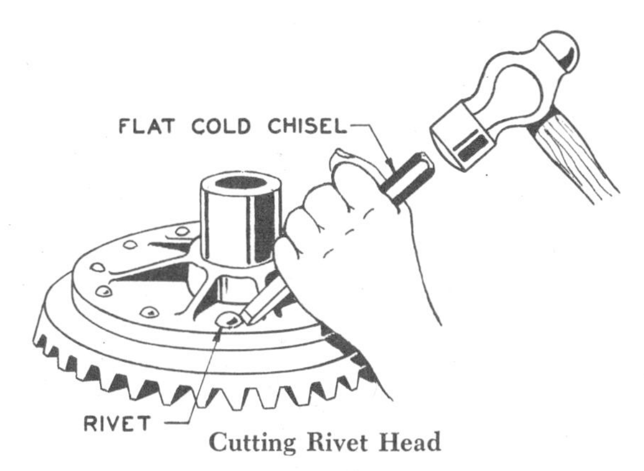

Cutting Rivet Head

Fig. 13-9. Cutting a Rod and Rivet with a Chisel

13-17. Cutting Rods and Rivets

A small rod or bar may be roughly cut by nicking it on opposite sides with the flat chisel, Fig. 13-9, and then bending it until it breaks. Figure 13-9 also shows a **rivet head** being cut off with a flat chisel.

Words to Know

cape chisel
center gage
cold chisel
diamond-point chisel
flat chisel
mushroom head
round-nose chisel
shearing
tapered

Review Questions

1. What is meant by chipping?
2. What part of the cold chisel is hardened and tempered?
3. Give the names and uses of cold chisels.
4. What is the best angle of the cutting edge for average work?
5. What gage may be used to test the cutting angle of a cold chisel?
6. What size hammer should be used for chipping?
7. What part of the chisel should you watch while chipping?
8. What precautions should you take to protect yourself and others while chipping?

UNIT 14

Files

14-1. What Is a File?

A file is a **hardened** piece of high-grade steel with slanting rows of teeth. It is used to cut, smooth, or fit metal parts. It cuts all metal except **hardened steel.**

14-2. Parts of File

The **tang** of a file is the pointed part which fits into the handle, Fig. 14-1. The **point** is the end opposite the tang. The **heel** is next to the handle. The **safe edge,** or side, of a file is that which has no teeth.

14-3. File Handles

All files should be fitted with handles, Fig. 14-2. Higher quality handles have a metal insert which is self-threading on the soft tang of the file, enabling the file to be hung by its handle. Large files should have large handles, and small files should have small handles. Very small files usually are made with a smooth, knurled handle, eliminating the need of attaching a wooden one.

The metal ring on the file handle is called a **ferrule.** It keeps the handle from splitting.

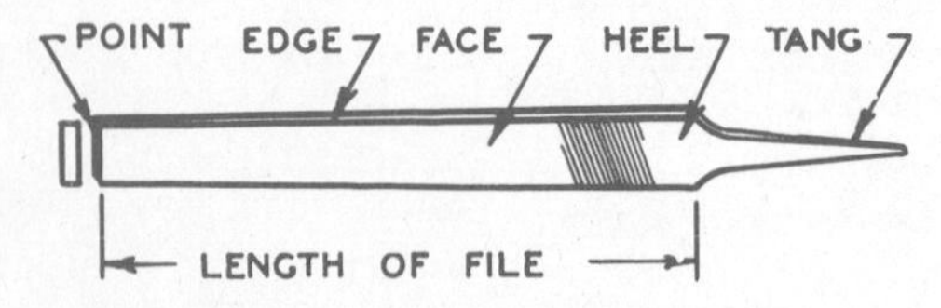

Fig. 14-1. Parts of a File

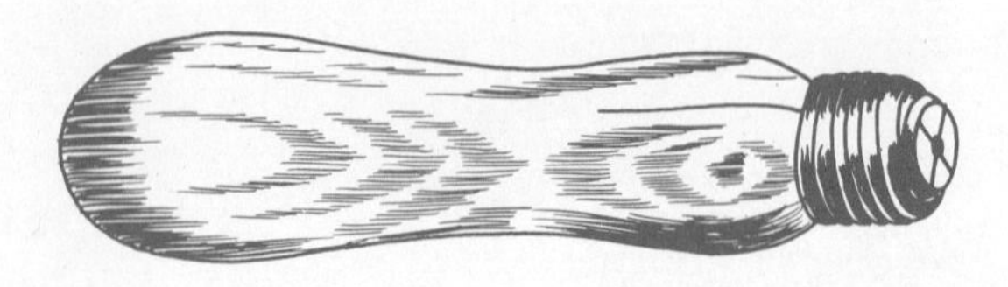

Fig. 14-2. File Handle

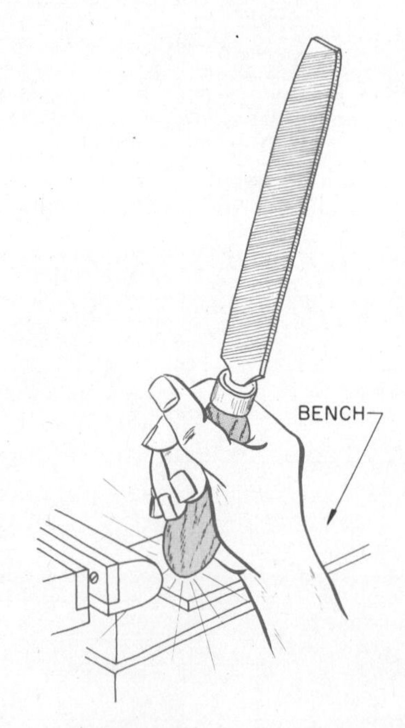

Fig. 14-3. Strike Handle on Solid Surface to Drive in the File Tang

14-4. Fitting File Handle

To fit a common handle on a file, the handle must be drilled for the tang of the file. Drill a hole equal in diameter to the average width of the **tang** to a depth equal to the length of the tang. Next, drive the handle on the tang by striking on a solid surface, Fig. 14-3. Be very careful not to split the handle. The entire tang should fit into the handle. Predrilled handles and types which screw onto the tang simplify the fitting.

14-5. Sizes of Files

Files are made in many sizes. The **length,** which is always given in inches, is the distance from the **point** to the **heel,** without the **tang.** See Fig. 14-1. The common sizes of files are the 6″, 8″, 10″, and 12″ lengths.

As the length of a file is increased, its width and thickness are also increased. The jeweler's file is available in 5″ or 6″ lengths, including the handle, and is usually sold in sets as shown in Fig. 14-4.

14-6. Shapes of Files

The shape of a file is its general outline or **cross section.** Files are made in hundreds of shapes. A **blunt file** is one which has the same width and thickness from the heel to the point. A **tapered file** is one which is thinner or narrower at the point. Table 9 shows the most commonly used shapes. These are:

1. **Warding**
2. **Mill**
3. **Flat**
4. **Hand**
5. **Pillar**
6. **Square**
7. **Round or rat-tail**
8. **Half-round**
9. **Three-square**
10. **Knife**

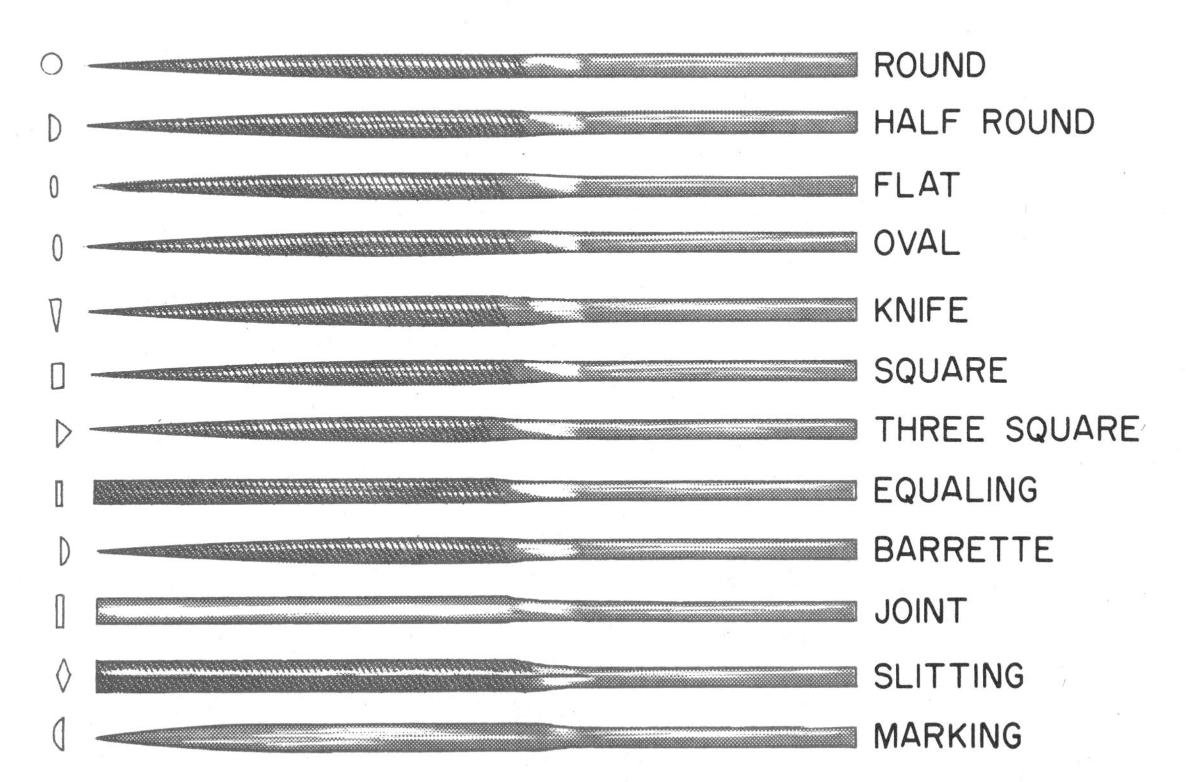

Fig. 14-4. Jeweler's Files

TABLE 9

Files and Their Uses

Name and Shape of File	Kind of Cut	Uses
WARDING FILE, parallel faces, edges taper to the point. Very thin.	Double-cut	Filing notches as in keys. Much used by **locksmiths.** Gets name from **ward,** meaning a notch in a key.
MILL FILE, tapered or blunt.	Single-cut	**Drawfiling,** finishing, and lathe work. Also used for finishing brass and bronze.
FLAT FILE, tapered in width and thickness.	Double-cut	One of the most commonly used files for general work.
HAND FILE, equal in width and tapered in thickness, one safe edge.	Double-cut	Finishing flat surfaces. Has one **safe edge** and, therefore, is useful where the flat file cannot be used.
PILLAR FILE, equal in width and tapered in thickness. One safe edge. Narrower and thicker than hand file.	Double-cut	Used for narrow work, such as **keyways** (see § 26-26), slots, and grooves.
SQUARE FILE, tapered or blunt.	Double-cut	Filing square corners. Enlarging square or **rectangular**[1] openings as **splines**[2] and **keyways.**
ROUND FILE (rat-tail), tapered.	Single-cut or Double-cut	Filing curved surfaces and enlarging round holes and forming **fillets.**[3]
HALF-ROUND FILE, tapered. Not a half circle; only about one-third of a circle.	Double-cut	Filing curved surfaces.
THREE-SQUARE FILE, tapered.	Double-cut	Filing corners and angles less than 90°, such as on **taps, cutters,**[4] etc., before they are hardened.
KNIFE FILE, tapered in width and thickness, shaped like a knife.	Double-cut	Filing narrow slots, notches, and grooves.

[1]**Rectangular** means having four 90° angles.

[2]A **spline** is a long **feather key** (see § 26-26) fastened to a shaft so that the pulley or gear may slide along the shaft lengthwise, both turning together.

[3]A **fillet** is a curve that fills the angle made by two connecting surfaces to avoid a sharp angle.

[4]A **cutter** is a sharp tool fixed in a machine for cutting metal.

14-7. Cuts of Files

Cuts of files are divided into three groups, as shown in Fig. 14-5. These groups are:

1. **Single-cut**
2. **Double-cut**
3. **Rasp-cut**

A **single-cut file** has single rows of cuts across the face of the file. The **teeth** are like the edge of a chisel.

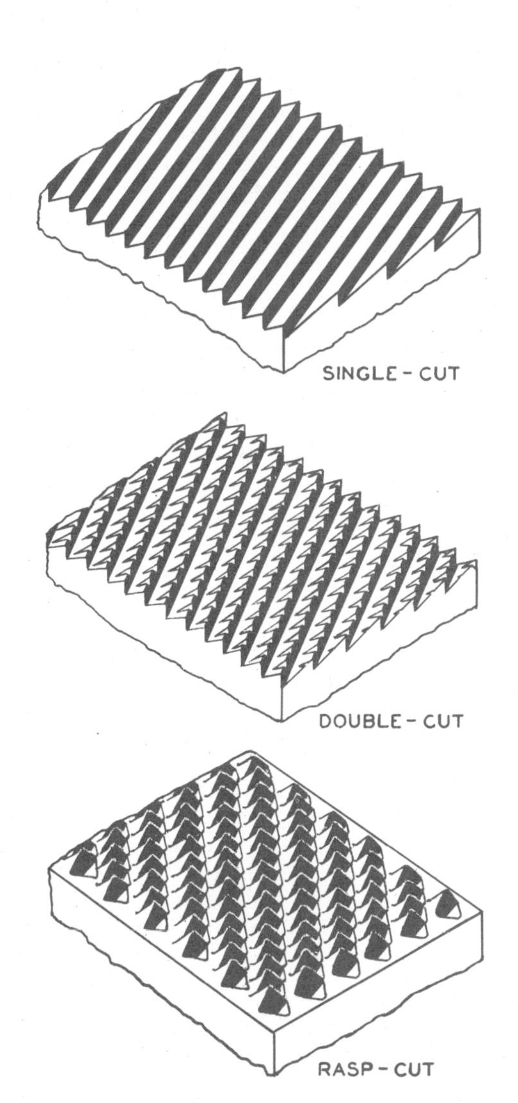

Fig. 14-5. Cuts of Files

A **double-cut file** has two sets of cuts crossing each other which give the teeth the form of sharp points. These files cut faster but not as smoothly as the single-cut file.

The teeth on a **rasp-cut file** are not connected and are formed by raising small parts of the surface with a punch. The rasp-cut file is used mostly for shaping wood.

14-8. Spacing Between Teeth

Single-cut and double-cut files are further divided according to the coarseness or spacing between the rows of teeth, Fig. 14-6. The six main kinds of spacings are:

1. **Rough**
2. **Coarse**
3. **Bastard**
4. **Second-cut**
5. **Smooth**
6. **Dead smooth**

Files with the greatest spacings between the teeth are called **rough;** those with the least spacings between the teeth are known as **dead smooth.** The teeth on the dead smooth file are very fine.

The files most often used are the bastard, second-cut, and smooth. The rough, coarse, and dead smooth files are used only on special jobs.

The coarseness of a file changes with its length. The larger the file, the coarser it is. Thus, a rough cut on a small file may be as

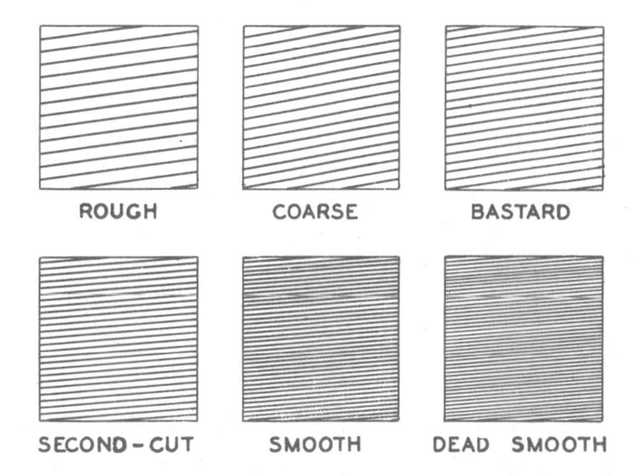

Fig. 14-6. Spacings between Teeth

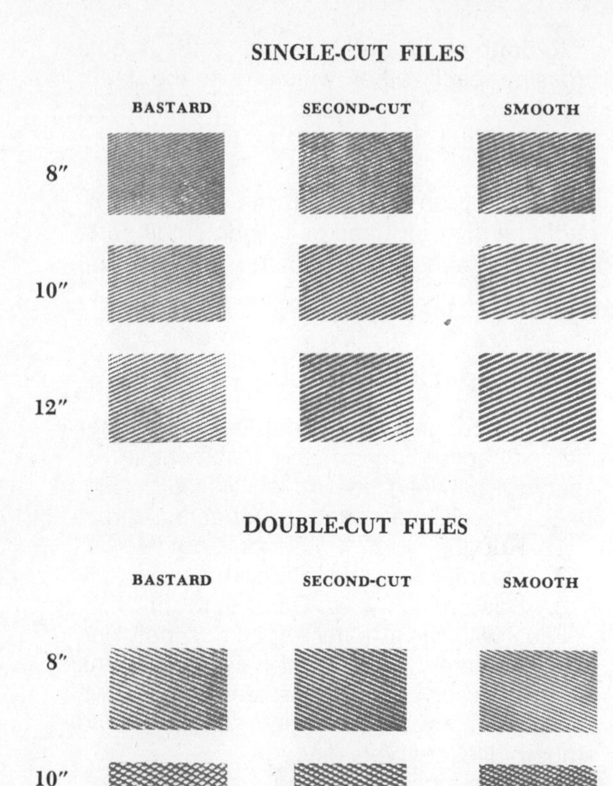

Fig. 14-7. Relation between Teeth Spacings and Lengths of Files (Actual Sizes)

fine as a second-cut on a large file. Compare the spacings between the teeth of different sizes of files in Fig. 14-7.

14-9. Ordering Files

One manufacturer advertises 5000 different sizes, shapes, and cuts of files. When ordering a file, include the following information:

1. **Length**
2. **Shape**
3. **Cut**
4. **Spacing** between teeth

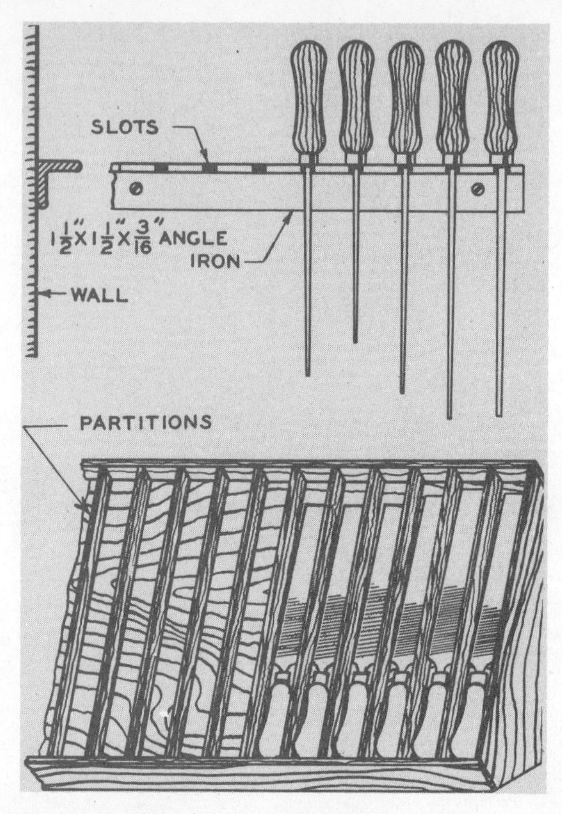

Fig. 14-8. Good Ways to Keep Files

Example:

1. 10-inch
2. Flat
3. Single-cut
4. Bastard

14-10. Storing Files

Take good care of files. They should be placed where they will not rub or strike each other. Never throw them together on the bench or into a drawer with hammers, screwdrivers, cold chisels, etc. Good care will save the cutting edges of the files and cause them to last longer.

Hanging files, as in the upper part of Fig. 14-8, is good because a file with a loose handle cannot be hung up and should be started on its way for a new handle. Another good way to keep files is to place **partitions** or walls between them, Fig. 14-8.

14-11. Choosing File for Work

Table 9 on page 122 gives information about the kinds of files and their uses. Each file is made for a certain kind of work. The **double-cut hand file** cuts metal quickly; a 10″ or 12″ **bastard file** is generally used for rough filing at the bench; a **second-cut file** is used to bring the surface of the work closer to a finish; the **single-cut mill file** is used for finish cuts. The size of the work should be considered when selecting the file size to be used.

Words to Know

file
- edge
- face
- heel
- point
- tang

file cuts
- double-cut
- rasp-cut
- single-cut

file shapes
- flat
- half-round
- hand
- knife
- mill
- pillar
- round or rat-tail
- square
- three-square
- warding

safe edge

spacing between teeth
- bastard
- coarse
- dead smooth
- rough
- second-cut
- smooth

Review Questions

1. What is the tang of a file?
2. What is the purpose of a safe edge on a file?
3. How should a handle be fitted on a file?
4. How is the length of a file measured?
5. What is a blunt file? A tapered file?
6. Make a sketch of the cross-section shape of a flat file; a half-round file; a square file; a three-square file.
7. Make a sketch of how the teeth appear on a single-cut file; a double-cut file.
8. What is a jeweler's file?
9. What information is needed when ordering a file?
10. How should files be stored when not in use?

UNIT 15

Filing

15-1. Filing Is an Art

To beginners the file seems very easy to use, but good filing takes much practice. It is one of the most difficult operations in the mechanical field. Few people, except experienced filers, know the different kinds of files and how to use them correctly. More files are worn out by abuse than by use.

15-2. Holding Work for Filing

The workpiece should be held tightly in a vise or otherwise secured so it cannot move. If in a vise, the part to be filed should be near the vise jaws to keep it from chattering, Fig. 15-1.

Polished surfaces should be protected from the rough, hardened steel vise jaws by covering the jaws with vise jaw caps.

15-3. Body Position for Filing

The body position for filing is the same as for sawing with the hand hacksaw. Face the work and stand with one foot ahead of the other, Fig. 11-6. Make long, slow, steady

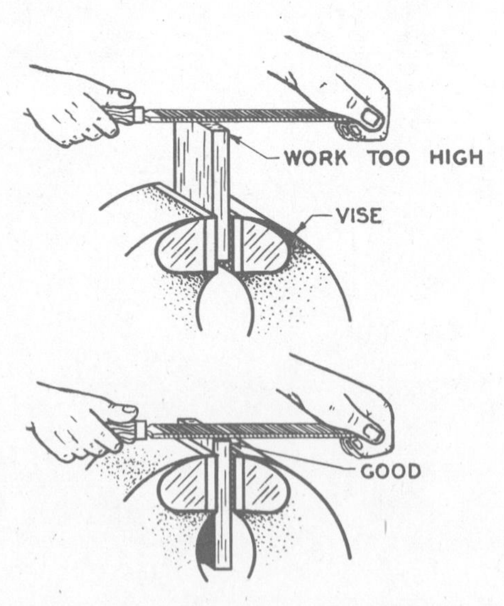

Fig. 15-1. Chattering Results When the Work Is Held Too High in the Vise

Fig. 15-2. Get Files with Handles

strokes by moving the arms while swaying the body back and forth to maintain good balance.

15-4. Using a File

The file is used for cutting all kinds of metal except hardened steel. Many files are dulled because the teeth touch the hardened jaws of a vise during the filing process. The file should be used only on metal which is not as hard as the file itself.

Insist on using a file with a handle, Fig. 15-2. Use only a file with sharp teeth. Grasp the handle of the file with the favored hand and hold the palm against the end of the handle with the thumb on top. For heavy filing, place the palm of the hand on the point of the file with the fingers pressing against the underside, Fig. 15-3. For light filing, the thumb of the left hand should be placed on the top of the file, Fig. 15-4.

Place the point of the file on the **work;** cut by pressing down on the **forward stroke,** known as the **cutting stroke** (Fig. 15-5), and lifting a little on the **return stroke** to prevent dulling the file. The file cuts only on the cutting stroke. Keep the file level on the work; otherwise the filed surface will be rounded and uneven instead of flat. Use the full length of the file and avoid jerky motions. This is called **cross-filing.**

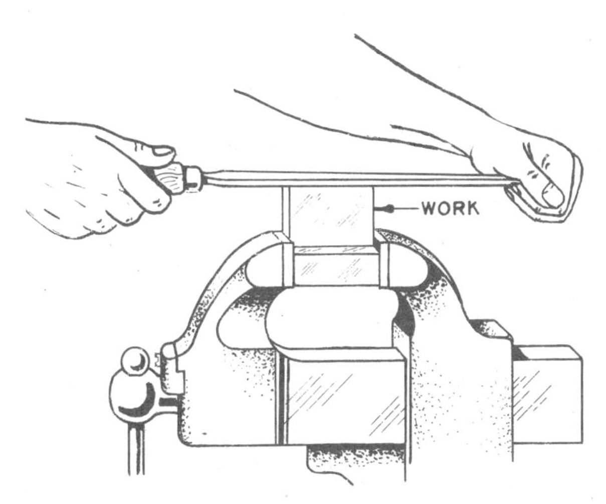

Fig. 15-3. Holding the File for Heavy Filing

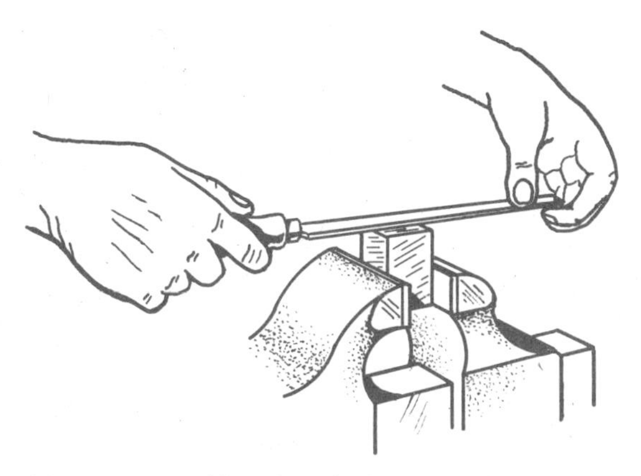

Fig. 15-4. Holding the File for Light Filing

15-5. Speed

In filing, the greatest fault with beginners is too much speed; press hard on the file, but take slow strokes. The harder the metal to be filed, the slower the stroke should be.

15-6. Pressure

Be sure that the file really cuts on the forward stroke. Rubbing, slipping, or sliding over the metal without cutting makes the cutting edges smooth, bright, and shiny like glass. It also makes the teeth blunt and dulls them. A worn file needs more pressure than a new file.

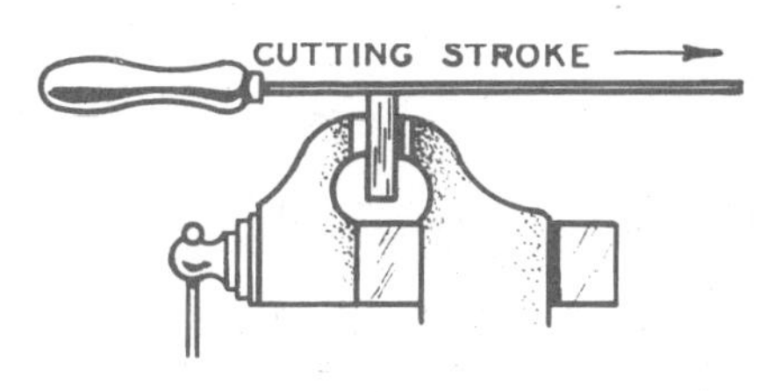

Fig. 15-5. Cutting Stroke on a File

15-7. Holding File for Small or Fine Work

Only on small or fine work, where the file is too small to hold in both hands, may it be held in one hand. In this case, the forefinger, instead of the thumb, is placed on top (Fig. 15-6), because the forefinger guides the direction of the file. Note how the left hand also helps to guide the file.

15-8. Drawfiling

Drawfiling is done to get a finely finished surface, Fig. 15-7; such a surface is flatter than one made by ordinary filing, which is known as **cross-filing.** A **mill file** is recommended for this purpose. Drawfiling is done by holding the file with one hand on each end of the file, the thumbs about ½" to ¾" from the work on each side. Hold the file flat on the work and draw it across the work. Press down only on the cutting stroke.

15-9. Filing Flat, Square, and Angular Surfaces

Flat work should be tested often with a **straightedge** held in different positions against the light. The straight edge of a steel rule or square may be used for this purpose, Fig. 15-8. The work is flat if no light can be seen between the straightedge and the surface. It may also be tested for

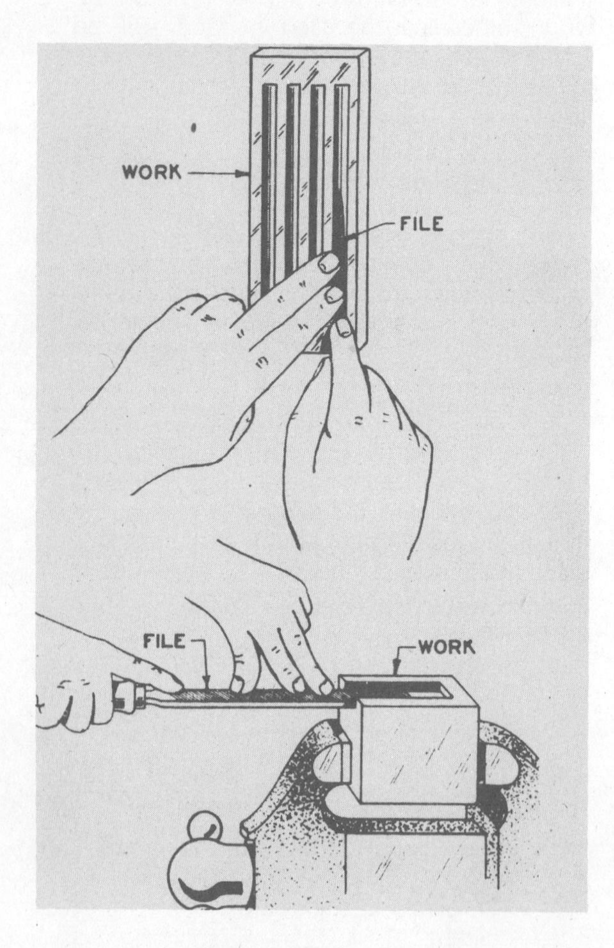

Fig. 15-6. Using the Point of the File for Small Openings

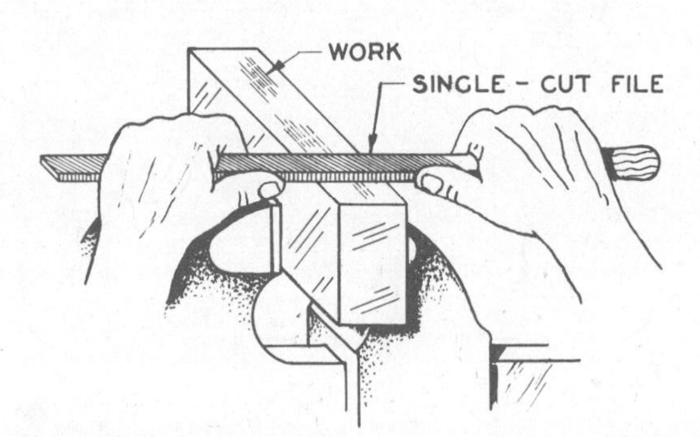

Fig. 15-7. Drawfiling

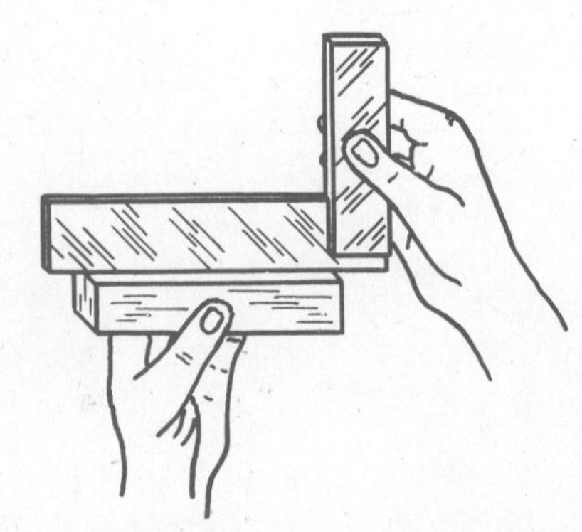

Fig. 15-8. Testing Flat Work with Straightedge

flatness on a **surface plate,** as explained in section 16-5.

Square work should be tested often with a square, Fig. 15-9. The **bevel protractor** should be used to test all angles except **right angles,** Fig. 15-10.

15-10. Filing Cast Iron

Before filing cast iron with a **scaled**[1] **surface,** remove the scale by tapping with a large **flat cold chisel.** Wear goggles when removing the scale. The file will be ruined in a few strokes if you file the scale. Sometimes it is necessary to chip the scale off with a hammer and cold chisel.

Do not touch the filed part of cast iron with the fingers or hand because no matter how dry the skin may appear there is always enough oil given off and left on the metal so that the file will slip over this touched part on the next few strokes. This dulls the file. Blow away any filings which you wish to remove.

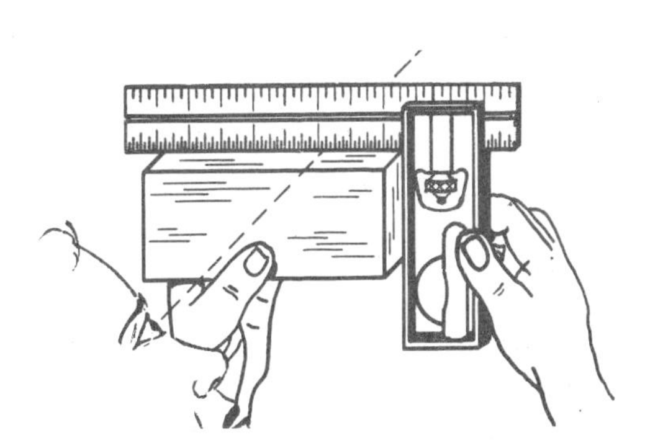

Fig. 15-9. Testing the Squareness of Work with a Double Square

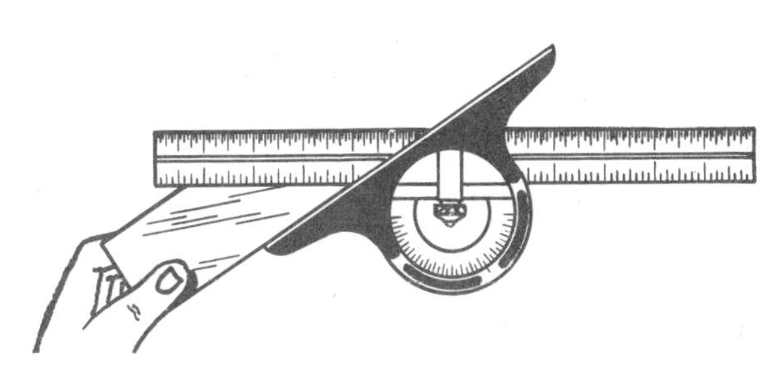

Fig. 15-10. Testing an Angle with a Bevel Protractor

[1]**Scale** is a thin, black, hard skin or crust which forms on metal, especially when heated.

15-11. Filing Soft Metals

Special files are made for cutting soft metals such as **brass, bronze, aluminum, lead, solder,** and **babbitt.** These metals are explained in Unit 20. Filing soft metals with the ordinary file causes the teeth to become quickly clogged with chips which are hard to remove.

15-12. How Should a File Be Cleaned?

File teeth often become clogged with chips and filings called **pins.** These small chips stick in front of the teeth and scratch the work. Keep the file free from chips and filings by brushing it with a wire brush called a **file card,** Fig. 15-11.

Rub the file card over the file in the direction of the cuts. Sometimes the pins stick so tightly that the file card will not remove them.

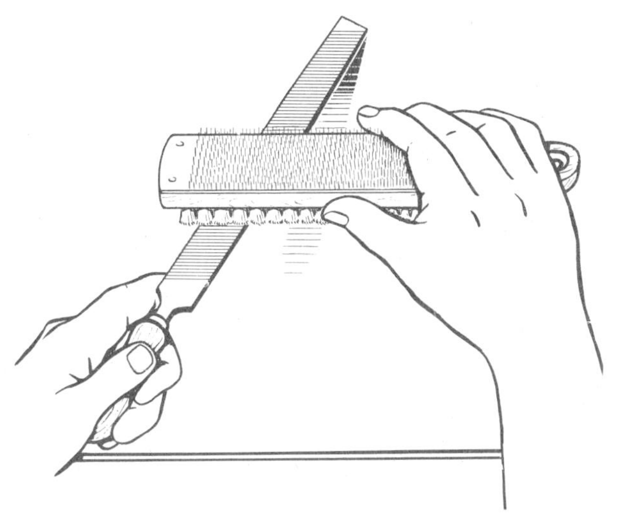

Fig. 15-11. Cleaning a File with File Card and Brush

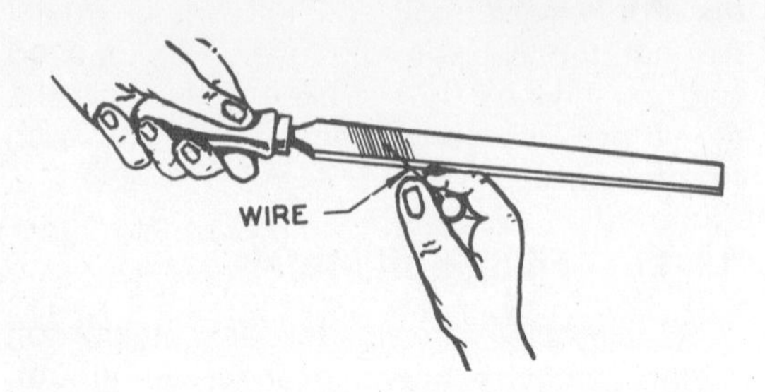

Fig. 15-12. Remove Chips with Wire to Clean File Teeth

In such cases, use a thin piece of sheet metal or a pointed wire to remove them, Fig. 15-12. Keeping the file clean will save much hard work and will help you to do better work. A file can be cleaned more easily if chalk is rubbed on the teeth before it is used. Never try to clean a file by tapping it on the bench, vise, or work. Files break easily because they are very **hard** and **brittle.** Oil may be removed from a file by rubbing chalk into the teeth and then brushing with a file brush. Clean files give the best results. A new file clogs more than a used file because the teeth on the new file are of uneven height and, therefore, the longer teeth cut off bigger chips which stick in front of the teeth.

Words to Know

cast iron	file card
cross-filing	scale
drawfiling	scaled surface

Review Questions

1. How should work be held for filing in order to keep it from chattering?
2. How may finished surfaces be protected from the rough jaws of a vise?
3. Should a file be used without a handle? Why?
4. How fast should you file?
5. What happens to the file teeth if the file slips or slides over the metal?
6. How should a file be held for draw filing?
7. Why should you not file the scale on cast iron?
8. How should the scale on cast iron be removed before filing?
9. Why should you not touch the filed part of cast iron with the fingers when filing?
10. Explain how to keep files clean and free of chips.

UNIT 16

Scrapers and Scraping

16-1. What Does Scraping Mean?

Scraping means shaving or paring off thin slices or flakes of metal to make a fine, smooth surface.

16-2. Reasons for Scraping

Scraping is an art. Much practice is needed to make a good true surface. It is slow, expensive work and is seldom done today. It has been largely replaced by surface grinding and more accurate machining and die-casting processes. However, it is still used in fitting large machine parts and in making machine repairs.

Scraping is most ofen used in fitting soft bearings[1] to a shaft,[2] in correcting minor imperfections in machining, or in decorating flat machined surfaces, such as the ways on a lathe or drill press.

Machined surfaces usually are not perfectly true for a number of reasons: (1) file scratches or tool marks need removing, (2) metal may be of unequal hardness or the metal may have been sprung while being clamped for machining, or (3) the cutting tool may wear or have been sprung during the machining. If a very true surface is needed, the high spots must be located and removed. This may be done by hand scraping or by one of several types of grinding machines. See Unit 58.

16-3. Scrapers

Scraping is done with tools called **scrapers**, Fig. 16-1, which have very hard **cutting edges.** Scrapers are made in many shapes, depending on the work to be done. Those used on flat surfaces are known as **flat scrapers** and **hook scrapers.** Those used on curved surfaces are known as **three-cornered** scrapers and **half-round** scrapers. The three-cornered scraper may also be used to remove sharp corners or burrs.[3] The half-round scraper, or bearing scraper, has

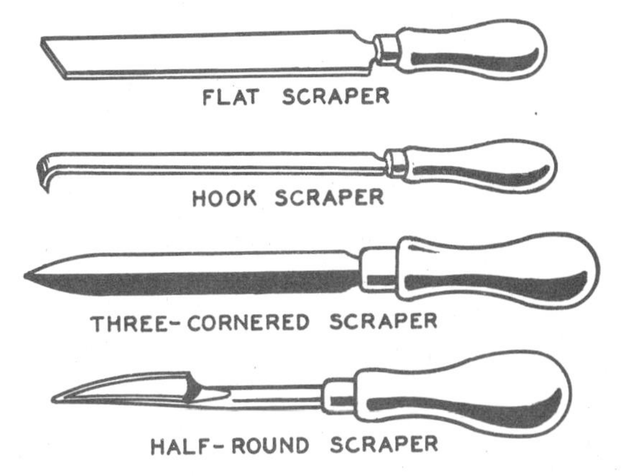

Fig. 16-1. Scrapers

[1]A **bearing** is the part which holds or supports a shaft and inside of which the shaft turns.

[2]A **shaft** is an axle or round bar of metal on which a pulley or gear is fastened or supported.

[3]A **burr** is a thin, rough edge made in cutting.

two cutting edges on the concave (hollowed) side.

Old files make excellent scrapers. First, grind off the teeth on all sides and then sharpen the cutting edge, Fig. 16-2. The cutting edge of the flat scraper is at the end. It should be slightly rounded, looking at the broad side, and should be sharpened last.

16-4. Holding a Scraper

Scrapers are held as shown in Figs. 16-3 and 16-4. The handle of the scraper for flat scraping should be held and steadied under the right arm while the blade is guided by both hands; **pressure** is thus applied to the cutting edge. As with files, all scrapers except the hook scraper cut on the pushing stroke. Lift the scraper a little on the return stroke. Experience teaches how hard one should press and at what angle the tool cuts the metal easily.

16-5. Scraping a Flat Surface

A **surface plate** is used to find the high spots on a flat surface. The surface plate should be larger than the surface to be scraped. The top of the surface plate is covered with a very thin film of **Prussian blue.** This blue paint comes in a tube, is about as thick as toothpaste, and may be applied with a small brush. The surface to be scraped is then laid on the surface plate and moved back and forth. Thus, the high spots which have to be removed will be marked with Prussian blue from the surface plate. If too thick a coat is put on the sur-

Fig. 16-3. Scraping a Flat Surface

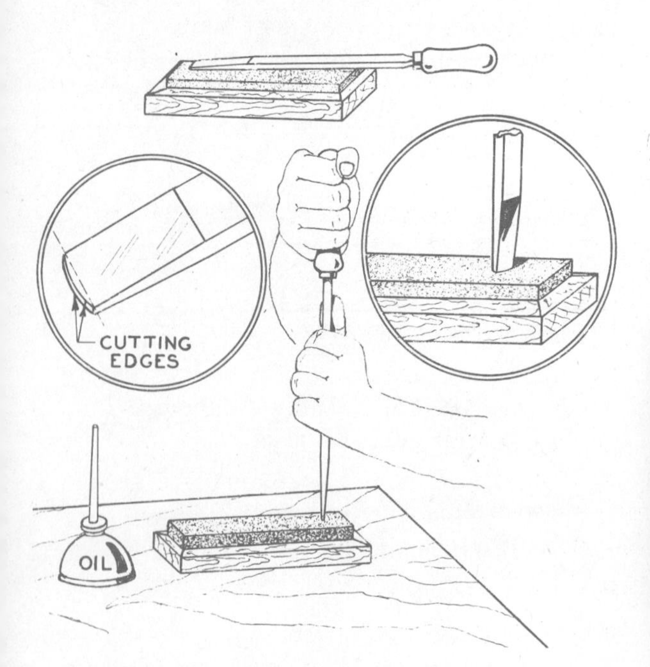

Fig. 16-2. Oilstoning a Flat Scraper

Fig. 16-4. Scraping a Round Bearing

face plate, the low spots on the work will be marked as well as the high ones.

Finding the high spots is called **spotting.** Remove the blue marks with a flat scraper. See Fig. 16-3. After scraping off the high spots, test again on the surface plate and again scrape off the high spots. Repeat these operations until the blue marks increase and are evenly spread over the surface of the work. Study where and how much to scrape off. **Red lead** may be used instead of Prussian blue.

16-6. Scraping a Round Bearing

Round bearings must fit on other sliding or running parts; any **shaft bearing** is an example of this. See Fig. 16-4. The same way of marking and scraping the high spots of flat surfaces is also used for round and curved surfaces, Fig. 16-5. For round surfaces the Prussian blue is put on the shaft or part which is fit into the hollow surface. The half-round scraper, and sometimes the three-cornered scraper, is used for scraping round and curved surfaces.

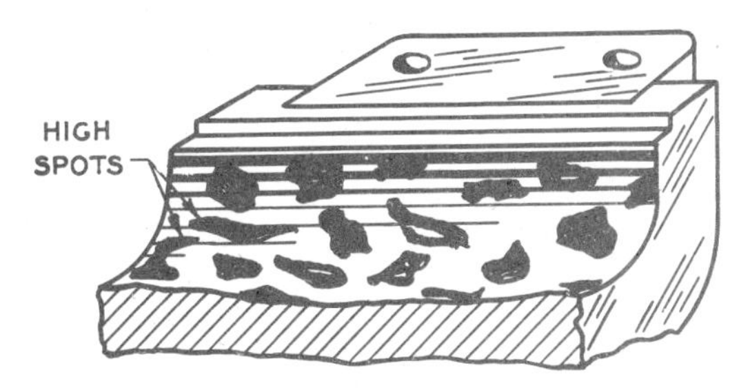

Fig. 16-5. High Spots to Be Scraped Off a Bearing

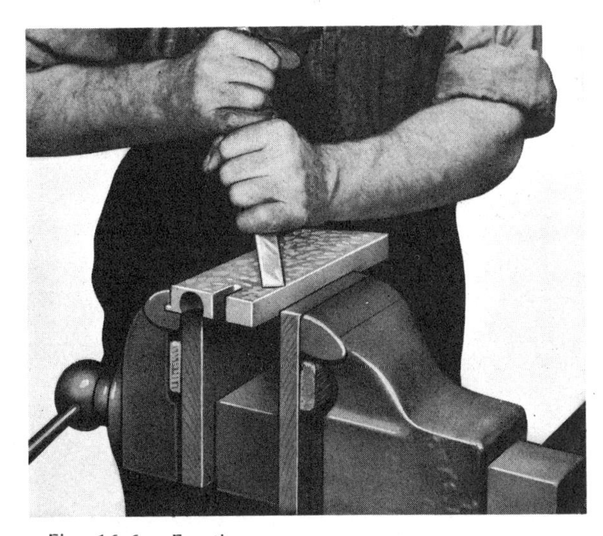

Fig. 16-6. Frosting

16-7. Frosting or Flowering

This is a finish which is an imitation of frost, patchwork, or checkerboard design, Fig. 16-6. It is also called **spotting** or **flaking** and is made by scraping off the high spots, as was explained in the scraping of flat surfaces. The strokes should be very short, about ¼″ to ½″. Change the direction of the strokes after each marking.

Words to Know

bearing scraper
burr
flaking
flat scraper
flowering
frosting
half-round scraper
hook scraper
Prussian blue
red lead
scraper
scraping
spotting
three-cornered scraper

Review Questions

1. Give several reasons for scraping.
2. How can an old file be made into a scraper?
3. What is Prussian blue?
4. Why should Prussian blue be applied in a thin layer?
5. What kind of a scraper should be used when scraping a flat surface?
6. What kind of a scraper should be used when scraping a curved surface?
7. Describe the procedure for scraping flat surfaces.
8. What is meant by frosting, flowering, or spotting?

Mathematics

1. How much would it cost to scrape a surface plate if a toolmaker scraped 26 hours, and the shop charge was $12.00 an hour?

Occupational Information

1. In what work is knowledge of and ability to do scraping necessary?

PART V

Getting Acquainted with Metals

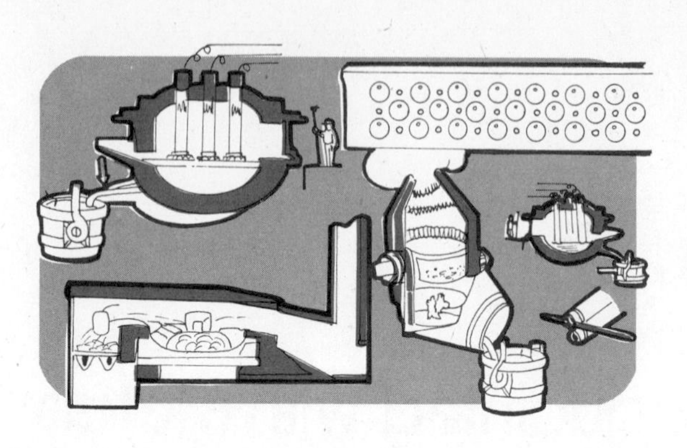

UNIT **17**

Iron

17-1. Iron

Iron is our commonest and most useful metal. We are living in an age of iron and steel. The purest iron comes from the sky in the form of "shooting stars" or **meteors** which are visible at night. Pure iron does not rust in water; however, iron rusts because it contains impurities. Pure iron is seldom used in industry because it is too soft for most work. It is so soft that it can be scratched with the fingernail.

Iron is used in three forms:

1. **Cast iron**
2. **Wrought iron**
3. **Steel**

The first two are explained in sections 17-4 and 17-6; steel is explained in Units 18 and 19.

17-2. Iron Ore

Iron ore is not pure as it comes from iron mines, Fig. 17-1. Most of the iron ore in the United States is found in the mines of Michigan and Minnesota, known as the **Lake Superior Region** or **Mesabi Range.** Deposits are also mined in Alabama. Iron ore is also imported from foreign countries.

At first, iron ores were rich enough in iron to be used just as they were mined. Now, however, the depletion of the rich deposits has forced the use of iron ores with a much lower iron content. Low-grade ores, such as **taconite,** are first enriched in processing plants near the mines by separating the excess rock material from the ore. The ores are crushed to a fine powder and magnetic separation is used to remove those particles with high iron content. The iron-rich powder thus obtained is mixed with powdered coal to form pellets of rough ball shape. The pellets are then baked to make them hard enough to keep their shape during handling and shipping.

Fig. 17-1. The Rouchelau Open-Pit Mine on the Mesabi Range in Minnesota
Note provision for both rail and truck transportation.

Iron ore is carried from the mine processing plants in railroad freight cars that operate on high-loading docks. The ore is dumped into bins holding several carloads each. From these bins the ore is dumped into large boats, Fig. 17-2. A boat is loaded in a very short time, less than 3 hours. These boats then carry the ore to Chicago, Gary, Detroit, Cleveland, and other steelmaking cities along the Great Lakes. They are called **ore boats** and carry about 12,000 tons of ore in one load. Steps in the purifying of iron and making of steel are shown in Fig. 18-1.

Fig. 17-2. Unloading an Ore Boat

Fig. 17-3. Blast Furnaces Viewed across Stockpile of Iron Ore

17-3. Pig Iron

Making pig iron is the first step in the purifying of iron and the making of steel. See § 18-1 and Fig. 18-1. Iron ore becomes pig iron when the impurities are burned out in a **blast furnace.** It is called a blast furnace because a blast of hot air is forced into it near the bottom. This looks like a tall chimney on the outside, 50′ to 100′ high, Figs. 17-3 and 17-4. It is a large, round, iron shell from 10′ to 20′ in diameter, covered on the inside with (refractory) bricks or clay which can withstand great heat, Fig. 17-5. Each blast furnace has four or five giant stoves which first heat the air.

It takes about 2 tons of iron ore, 1 ton of coke, ½ ton of limestone, and 4 tons of air[1] to make 1 ton of pig iron. The iron ore, coke, and limestone are dumped into the top of the blast furnace which holds about 1000 tons. The burning coke and the blast of very hot air melt the iron ore. The limestone mixes with the ashes of the burnt coke and

Fig. 17-4. Blast Furnace

[1]A ton of air would be a column about 18 inches square and 1 mile high.

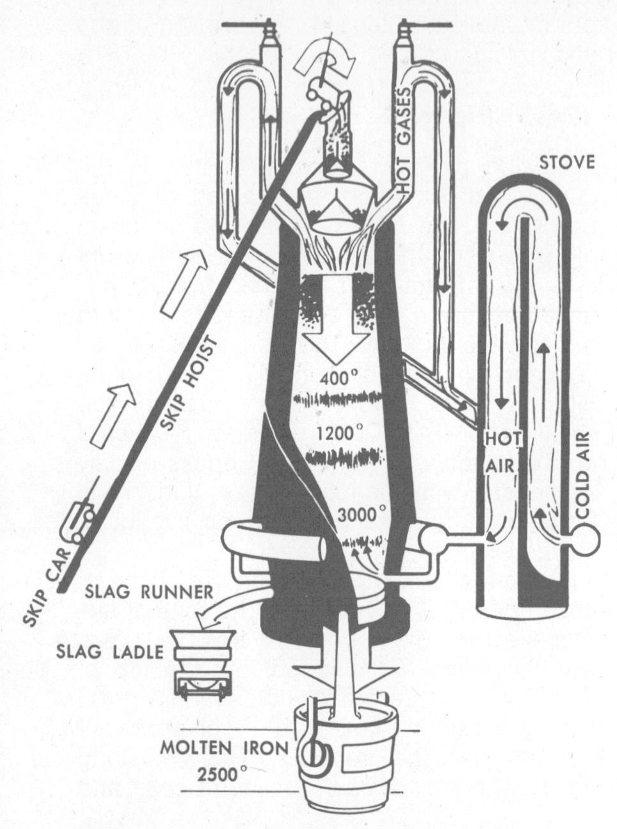

Fig. 17-5. Cross Section of a Blast Furnace

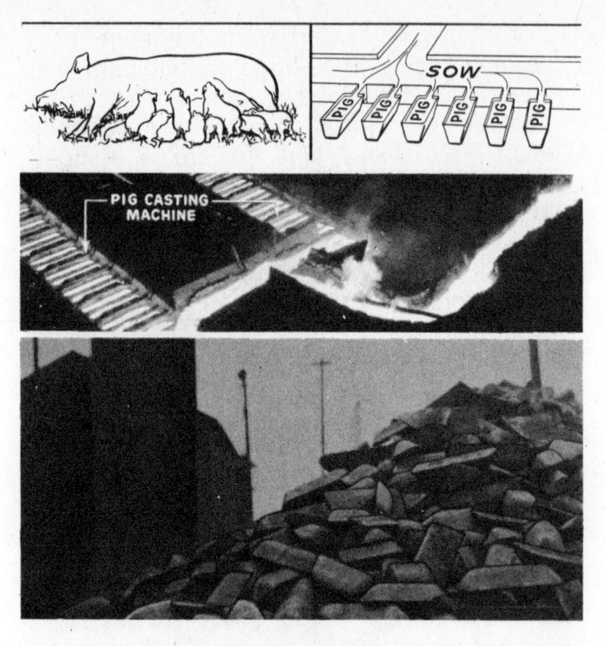

Fig. 17-6. Pigs

with the rock and earth of the iron ore. The mixture forms a waste which is called **slag.** As the iron melts, it drips to the bottom of the furnace.

Steelmakers are constantly seeking improvements in their operations. Some are experimenting with using a mixture of pulverized or powdered coal and oil in the blast furnace to speed up this operation.

When the furnace is emptied, the melted iron flows out into a ladle for transfer to a steel furnace or into a trough and then into iron or sand **molds.** See § 39-1. When it cools, **pigs**[2] are formed; hence, the name **pig iron,** Fig. 17-6. The slag which floats on top of the melted iron in the furnace is drained off through a separate hole. The blast furnace works continuously.

Pig iron is very hard and **brittle.** It has three uses:

1. **Cast iron**
2. **Wrought iron**
3. **Steel**

Pig iron contains about 93% pure iron and from 3% to 5% **carbon.** The remainder is **silicon,**[3] **sulfur,**[4] **phosphorus,**[5] and **manganese.**[6]

[2]The **molds** used in forming pig iron are a number of parallel trenches connected by a channel running at right angles to them. The iron cools in these and forms slabs or bars. These are called **pigs**; they are connected at one end to the long bar called the sow, in comparison to a **sow** with her little pigs.

[3]**Silicon** is a chemical which is in clays, sand, and rocks. It gives hardness to iron.

[4]**Sulfur** is a yellow, flammable nonmetal. Excess sulfur and phosphorus are injurious in iron and steel. Too much sulfur makes these metals weak and brittle, causing cracks. Modern steel mills attempt to remove all impurities (including carbon) and then carefully add elements as needed for various uses. A little sulfur is added to some steels for better machinability.

[5]**Phosphorus** is a poisonous, active nonmetal. It causes brittleness and coarse grain in iron and steel.

[6]**Manganese** is a grayish white metal, hard and brittle, that resembles iron but is not magnetic. It is used in making steel.

17-4. Cast Iron

Pig iron which has been remelted and poured into a product shape is **cast iron.** It is often called pig iron because some kinds of cast iron contain the same materials. Pig iron is remelted in a **cupola furnace** that is really a small blast furnace. The melted pig iron is poured into **sand molds** of various shapes which later become parts of machines or other objects. The iron is then called **cast iron** and the object is a **casting.**

17-5. Kinds of Cast Iron

There are several different kinds of cast iron. The properties of the different kinds of cast iron vary to a large degree according to amount, form, and arrangement of the carbon content in the iron. The carbon content[7] actually may range from about 1.7% to 6% in different kinds of cast iron. However, most grades of cast iron have carbon content ranging from about 2% to 4.5%. When broken, cast iron has a crystallike grain structure at the fracture. The following are several common kinds of cast iron.

Gray Cast Iron

This basic type of cast iron usually contains from 1.7% to 4.5% carbon. It melts at about 2200° F. Most of the carbon is in a **free** state, scattered in the form of **graphite** (carbon) flakes throughout the crystalline grain structure of the metal. This arrangement of carbon makes the cast iron brittle. Thus, it fractures easily from sharp blows. It has a gray crystalline color where fractured. The gray color is due to the tiny flakes of graphite (carbon) mixed in with the grains of iron.

Gray cast iron is the cheapest kind of metal. It is used to make large pipes, steam radiators, water hydrants, frames for machines, and other machine parts which must be large and heavy but in which impact strength is not very important. Gray cast iron can be machined easily. Several different grades of gray cast iron are available, each having different properties.

[7]One percent (1.00%) carbon is known as **100-point** carbon, **point 100** carbon, or **100 carbon.**

White Cast Iron

This kind of cast iron is so named because of its white, crystalline color at the fracture when broken. The carbon content usually ranges from about 2% to 3.5%. However, most of the carbon in white cast iron is in a chemically **combined** state. Thus, it forms a very hard substance called **cementite,** which is **iron carbide** (Fe_3C). White cast iron is so hard that it cannot be machined, except by grinding. Its direct use is limited to castings requiring the surfaces to withstand abrasion and wear. The major use of white cast iron, however, is in making **malleable** cast iron.

Malleable Iron

Metal is malleable if it can be hammered into different shapes without cracking. **Malleable iron** is **cast iron** which has been made soft, tough, strong, and malleable. It generally has about 2% to 2.65% carbon content. It is produced by converting white cast iron to malleable iron by heating at a high temperature for a prolonged period of time (100 to 120 hours) in a heat-treatment furnace. This **heat-treatment** process, called **malleableizing,** changes the arrangement of the carbon from its combined state as **cementite** (Fe_3C) to free carbon. The free carbon forms **aggregates** or globules of free carbon at the prolonged high temperature. The surrounding iron then becomes soft and machinable (see section 19-15 for definition of machinability). Malleable castings have many of the tough characteristics of steel. Several different grades of malleable iron are available. They are used for making tough castings for automobiles, tractors, and many kinds of machinery parts.

Ductile Cast Iron

This kind of cast iron is also known as **nodular iron** (a small rounded lump of carbon aggregate in soft iron) or **spheroidal graphite iron.** The carbon in the grain structure of ductile iron is in the free state, in ball-like forms called **nodules.** The iron surrounding the tiny balls of graphite (carbon) is soft, tough, and machinable. Ductile cast iron is produced very much like gray cast iron. Magnesium alloys and certain other elements are added to a ladle of gray iron before it is poured into molds to make castings. These additives, together with proper heat treatment, cause the carbon in the molten iron to form balls or nodules as it cools and becomes solid.

Ductile cast iron has properties similar to malleable iron. It is tough, machinable, and possesses many of the characteristics of steel. It is used for making tough castings for automobiles, farm machinery, and many other kinds of machinery. Further information concerning the different kinds and grades of cast iron is available in standard handbooks for machinists.

17-6. Wrought Iron

Wrought iron is purified **pig iron.** It contains almost no **carbon,** only about 0.04%.

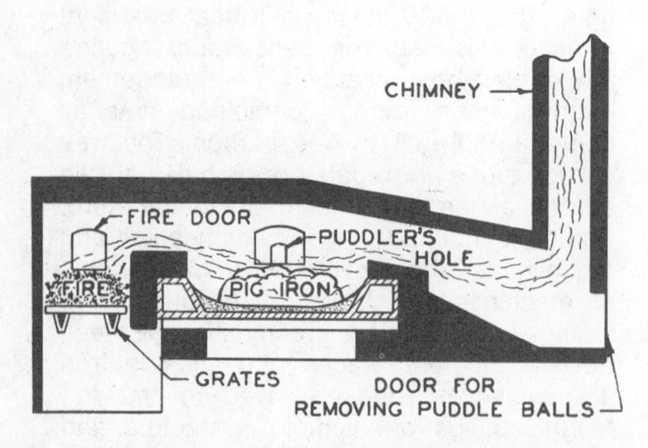

Fig. 17-7. Puddling Furnace

Wrought iron is the purest form of iron in commercial use and is therefore often called just **iron.** Wrought iron as applied to contemporary metal home furnishings is a misnomer, as today these are usually mild steel. Wrought iron now is not common, but it does have several desirable characteristics not found in steel. These are due to its peculiar grain structure caused by stringy particles of slag running throughout the metal.

An old way of making wrought iron is in a **puddling furnace,** Fig. 17-7, which looks like a large baker's oven in which the flame is over the metal. Pig iron melts at about 2100° F. while wrought iron melts at about 2700° F. After the pig iron is melted, it is stirred or **puddled.** This is done so that every part of the melted iron is touched by the flame. Most of the carbon and other impurities are thus burned out. The less carbon that iron contains, the higher is its **melting point.** The heat of the furnace is kept high enough to melt pig iron but not high enough to keep wrought iron in liquid form. Thus, as the melted iron becomes purified, it becomes pasty. This paste is worked into lumps or balls, called **puddle balls,** which are then taken out of the furnace and squeezed, hammered, and rolled into bars while hot. The bars are piled or fastened into bundles and then reheated, welded together by rolling, and rolled into finished shapes of bars, **sheets,** and **plates** ready to be sold and used.

Wrought iron is very tough, **malleable,** and bends easily, cold or hot. It looks stringy when broken. Good wrought iron can be tied into a knot while cold without breaking. It is, therefore, used for work which needs much hammering, bending, twisting, and stretching, such as ornamental iron work, rivets, bolts, wire, nails, horseshoes, chain links, pipes, etc. You can make many beautiful things for your home out of wrought iron, such as lamps, lanterns, candle holders, flower pot holders, door knockers, smoking stands, etc. Wrought iron rusts very slowly and is easy to weld.

Words to Know

blast furnace
carbon
casting
cupola
ductile cast iron
gray cast iron
iron
iron ore
malleable iron
manganese
phosphorous
pig iron
puddle ball
puddler
puddling furnace
silicon
slag
steel
sulfur
white cast iron
wrought iron

Review Questions

1. What is iron ore? Where is it found in the United States?
2. What is pig iron? Why is it so named?
3. Describe the furnace and the process for making pig iron.
4. For what is pig iron used?
5. What is cast iron? For what is it used?
6. Of what importance is carbon in making iron?
7. What percent is 200-point carbon?
8. What is malleable iron? For what is it used?
9. What is wrought iron? For what is it used?
10. What is white cast iron and what is its chief use?
11. What is ductile cast iron and for what is it used?
12. What is taconite? Describe how it is processed.

Mathematics

1. If a blast furnace holds 1000 tons of iron ore, coke, and limestone, how many carloads will it take to fill it if each car holds 50 tons?
2. How many carloads will it take to fill a 12,000-ton ore boat if a car holds 50 tons?
3. How many pounds of carbon are there in one ton of cast iron if it is 500-point carbon?

Social Science

1. Of what value is iron to civilization? Write a story.
2. Write a story telling what might happen if the United States were suddenly unable to get a supply of iron.

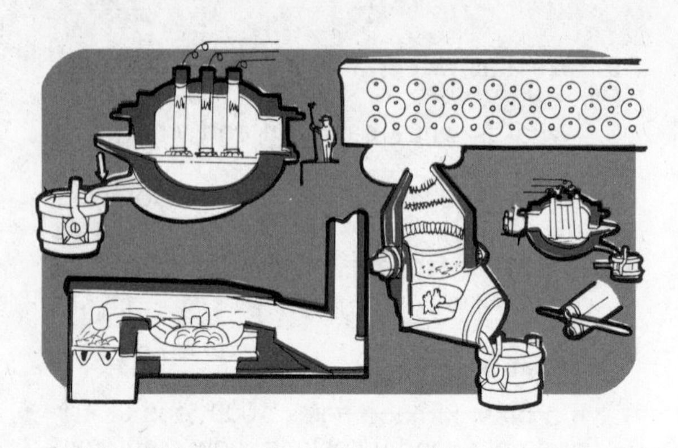

UNIT 18

Steel

18-1. Steel

Steel is iron which contains carbon in any amount up to about 1.7%. It is a bluish gray metal which, when broken, looks like crystals or like the break in a lump of sugar. The kinds and uses of steel are many. For certain uses, especially for tools, there is no metal with the abilities of steel. It is brighter and stronger than iron. Steel can be made so hard that it will cut iron.

18-2. Kinds and Grades of Steel

There are two kinds of steel: **carbon steel** and **alloy steel**. Each kind is divided into several grades.

Carbon Steel

1. **Low-carbon steel**; also known as **machine steel, machinery steel,** and **mild steel.**
2. **Medium-carbon steel.**
3. **High-carbon steel**; also known as **tool steel.**

Alloy Steel

1. **Special alloy steel**, such as **nickel steel, chromium steel**, etc.
2. **High-speed steel.**

18-3. Iron + Carbon = Steel or Cast Iron

Carbon is found in the form of coal, diamonds and **graphite**. Charcoal and coal are mostly carbon. Steel is between **wrought iron** and **cast iron** in carbon content. The dividing line between steel and cast iron is usually about 1.7% carbon.

Carbon is easily united with **oxygen**. It makes steel stiff, strong, and hard. Steel is graded by its percentage of carbon. As the carbon in iron and steel is reduced, the **melting point** is increased (see Table 16, page 169).

Figure 18-1 shows the steps in purifying iron ore and making steel. Table 10 gives the carbon content and uses of each kind of **carbon steel.**

Steel can be made from these metals:

1. Pig iron or cast iron by taking out some of the **carbon**.
2. Wrought iron by adding **carbon**.
3. Wrought iron, pig iron, and old scraps of iron and steel by melting them together, then testing and adding or burning out **carbon** until the melted metal contains the amount desired.

18-4. Five Ways of Making Steel

There are five basic ways of making steel:

1. Bessemer converter
2. Open-hearth furnace
3. Crucible furnace
4. Electric furnace
5. Basic-oxygen process

18-5. Bessemer Converter

The Bessemer way of making steel is named after Sir Henry Bessemer, who in-

HOW STEEL IS MADE

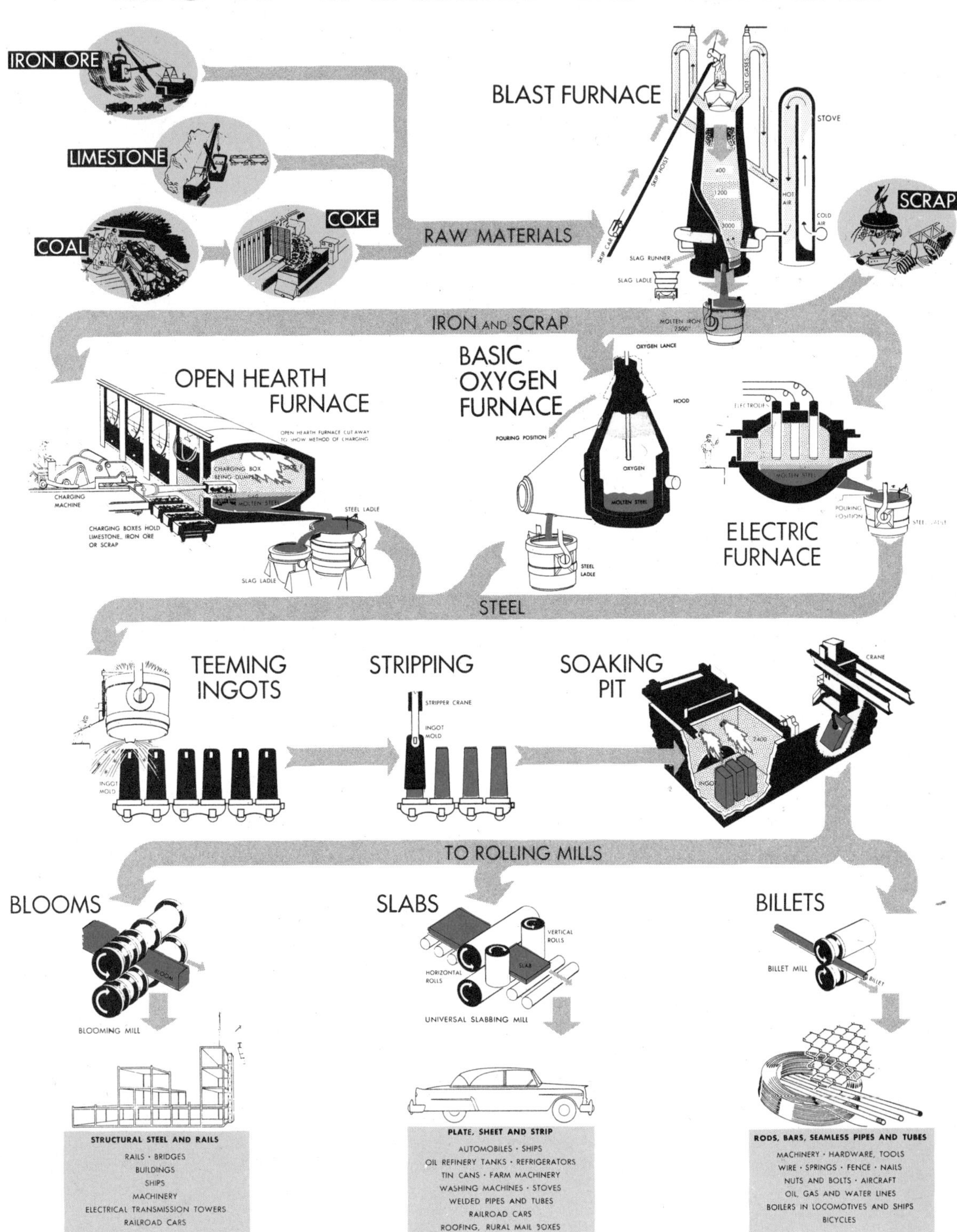

Fig. 18-1. How Steel Is Made

TABLE 10

Carbon Steels and Their Uses

Percent of Carbon in Steel	Uses
Low-Carbon:	
0.05-0.20	Automobile bodies, buildings, pipes, chains, rivets, screws, nails.
0.20-0.30	Gears, shafts, bolts, forgings, bridges, buildings.
Medium-Carbon:	
0.30-0.40	Connecting rods, crank pins, axles, drop forgings.
0.40-0.50	Car axles, crankshafts, rails, boilers, auger bits, screwdrivers.
0.50-0.60	Hammers, sledges.
High-Carbon:	
0.60-0.70	Stamping and pressing dies, drop-forging dies, drop forgings, screwdrivers, blacksmiths' hammers, table knives, setscrews.
0.70-0.80	Punches, cold chisels, hammers, sledges, shear blades, table knives, drop-forging dies, anvil faces, wrenches, vise jaws, band saws, crowbars, lathe centers, rivet sets.
0.80-0.90	Punches, rivet sets, large taps, threading dies, drop-forging dies, shear blades, table knives, saws, hammers, cold chisels, woodworking chisels, rock drills, axes, springs.
0.90-1.00	Taps, small punches, threading dies, needles, knives, springs, machinists' hammers, screwdrivers, drills, milling cutters, axes, reamers, rock drills, chisels, lathe centers, hacksaw blades.
1.00-1.10	Axes, chisels, small taps, hand reamers, lathe centers, mandrels, threading dies, milling cutters, springs, turning and planing tools, knives, drills.
1.10-1.20	Milling cutters, reamers, woodworking tools, saws, knives, ball bearings, cold cutting dies, threading dies, taps, twist drills, pipe cutters, lathe centers, hatchets, turning and planing tools.
1.20-1.30	Turning and planing tools, twist drills, scythes, files, circular cutters, engravers' tools, surgical cutlery, saws for cutting metals, tools for turning brass and wood, reamers.
1.30-1.40	Small twist drills, razors, small engravers' tools, surgical instruments, knives, boring tools, wire drawing dies, tools for turning hard metals, files, woodworking chisels.
1.40-1.50	Razors, saws for cutting steel, wire drawing dies, fine cutters.

vented it in England in 1856. Steel made this way is called **Bessemer steel**. It was first made in this country in 1864 at Wyandotte, Michigan. The Bessemer method is a quick way to make steel; a ton can be made in about one minute.

The main difference between steel and **pig iron** is the amount of **carbon** they contain. Pig iron contains from 3% to 5% carbon, while steel contains less than 2%. Thus, in making steel from pig iron, it is necessary to take out some carbon. The carbon and other impurities are burnt out of melted pig iron by blowing a current of cold air through it. It seems strange that it burns more (in fact it boils) when cold air is blown into it, but that is what happens. This is because the carbon in the pig iron unites with the **oxygen**[1] in the air. When this happens, a more forceful burning takes place. It is the same as when a cold wind makes a bonfire burn more forcefully.

The Bessemer converter is a large tank shaped like a pear with an open top, Fig. 18-2. It is made of steel and brick which can withstand great heat and is supported on two sides so that it can be tipped to pour out the melted metal. It is 15′ to 22′ high and holds 30 tons of metal. The melted metal is first poured into the converter. The air pressure is great enough to keep the melted metal from flowing into the holes. The oxygen of the air burns out nearly all of the carbon. When the carbon burns, a white flame about 25′ or 30′ high shoots up from the mouth of the converter with a mighty roar and a great shower of sparks like a volcano. This is a magnificent sight at night. The color of the flame changes as the carbon burns out. When the flame dies out, it is a sign that most of the carbon has been burnt out, and the air is then shut off. The exact amount of

[1]**Oxygen** is a colorless, odorless, tasteless gas. One-fifth of the air is oxygen. Without oxygen there can be no burning; thus, if a piece of wood or charcoal that is merely glowing is put into a jar of oxygen, it will burst into flame; likewise, the flame of a candle placed into a jar with no oxygen will burn out.

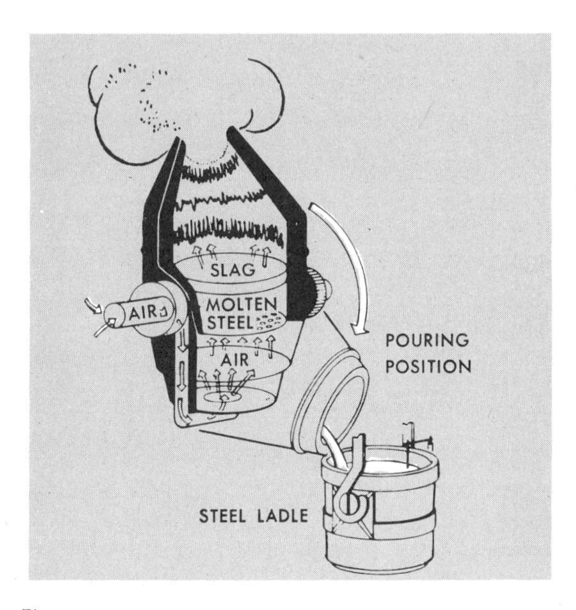

Fig. 18-2. Bessemer Converter

carbon necessary is then thrown in, thus making steel.

This melted steel is poured into large buckets called **ladles**. Through a hole at the bottom of the ladle, the melted steel is then poured into **ingot molds**, which are forms for making the steel into blocks, Fig. 18-1. These blocks of metal are about 20″ square, from 4′ to 6′ long, weigh about a ton, and are called **ingots**, Fig. 18-1. The ingots, still red-hot but cold enough to stand alone, are rolled into shapes ready for use. See Fig. 18-7.

Bessemer steel is an inexpensive type of steel used for nails, screws, wire, shafts, rails, and building materials such as beams. Because of its good machinability when .05% to .33% sulfur is added, it is well suited to making machine parts.

18-6. Basic-Oxygen Process (BOP)

The **Basic-Oxygen Process**, called the **BOP** process, is somewhat similar to the Bessemer process of making steel. A **basic-oxygen furnace**, Fig. 18-1, is used to make steel by this process.

The similarities and differences between the **BOP** and the Bessemer processes can be understood by studying Figs. 18-1 and 18-2. With the **BOP** process, pure oxygen is blown into the molten iron through a lance which enters from the top of the furnace. With the Bessemer process, air composed of oxygen and nitrogen is blown through the molten metal from the bottom. With the **BOP** process, the pure oxygen creates more heat and burns the impurities out of the molten iron much more rapidly than does the atmospheric air.

For many years it was known that oxygen could be used to improve the steel-making processes, but pure oxygen was too expensive to be used for this purpose. Since 1950, methods of producing pure oxygen at low cost have been available, making its use for steel-making possible. Most of the steel produced in the United States is now made by the basic-oxygen process.

Steel made by the **BOP** process is of a high quality. Since pure oxygen is used to burn out the impurities in the molten iron, nitrogen which makes steel brittle does not enter the process. Thus, the chemical specifications of the steel can be controlled more accurately with the **BOP** process, and it is a more rapid way of making steel. A furnace of the type shown in Fig. 18-1 produces about 80 tons of steel per hour.

The furnace is tilted on its side for charging. Molten iron and scrap are charged into its mouth. It is then tilted to an upright position, and pure oxygen is blown into the furnace under high pressure, thus burning out the impurities. Burned lime, converted from limestone, is also added to the furnace with the oxygen to increase the removal of impurities. When the impurities have been burned out of the molten iron, the necessary elements are added to meet the specifications for the steel required. The furnace is then discharged by tilting it on its side and pouring the molten steel into a large ladle.

18-7. Open-Hearth Furnace

Much of the steel in this country is made in the **open-hearth furnace**, Fig. 18-3. The open-hearth method of making steel can be controlled better than the Bessemer method because the melted metal can be tested for carbon content and more carbon added at any time during the heating.

The open-hearth furnace, which is somewhat like a baker's oven (Fig. 18-3), holds up to 200 tons of metal. It has two pairs of rooms with brickwork built like a checkerboard. Pig iron, wrought iron, and old scraps of iron and steel are placed on a saucer-shaped **hearth**. Hot air and gas are used for heating. The gas passes through the heated brickwork in one of the rooms. Air passes through a different room of heated brickwork. The gas and air then combine and make a very hot flame. One pair of rooms is heated while the other pair is being used. As the one used becomes cool, the air and gas are made to pass through the heated one. The flames touch the metal from above. Thus, a very high temperature keeps the iron in a liquid form. Samples of the white-hot metal are taken, cooled, and tested to find out if they contain the amount of carbon desired. If too much carbon has been burned out, more can be added. When the melted metal contains the right amount of carbon, it is poured into **ingot molds**. Figure 18-4 shows ingots being held at a high temperature in a soaking pit. They may be rolled, drawn, or extruded in the next forming step.

Steel made in an open-hearth furnace is called **open-hearth steel** and is used for bridges, rails, bolts, screws, shafts, etc. It is also used for making high-grade tool steel. An advantage of the open-hearth furnace is that old scrap iron and steel, as well as pig iron, can be used.

18-8. Crucible Furnace

The crucible process is the oldest method used for making **high-carbon steel** and **alloy steel**. High-carbon steel is made by melting wrought iron and scrap steel in a **crucible** — a melting pot shaped like a barrel, about 20″ high and 1′ in diameter, made of graphite or clay which can withstand great heat. The amount of **carbon** desired is then placed on top of the wrought iron and steel. A cover is placed tightly over the top, and a number of these crucibles are put in a hot furnace. The melted iron mixes with the carbon, thus making steel. The melted steel is then poured into **ingot molds.**

Alloy steel is made the same way except that additional materials, such as **chromium, tungsten,** etc., are also put in the crucible. Steel made in a crucible is called **crucible steel**. It is a higher grade, stronger, and more expensive steel than either Bessemer

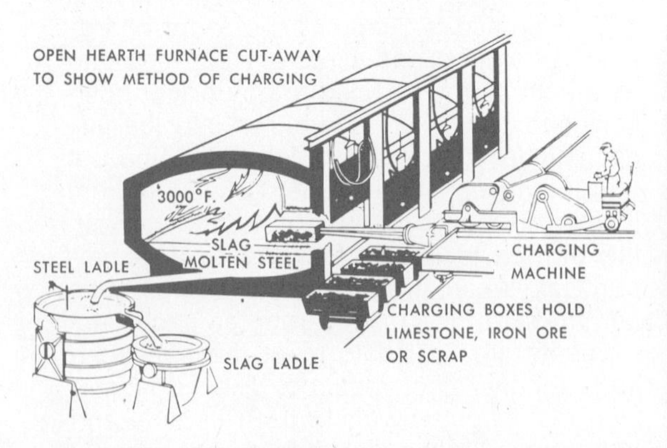

Fig. 18-3. Open-Hearth Furnace

Fig. 18-4. Ingots in Soaking Pit

steel or open-hearth steel. Crucible steel is used for razors, pens, knives, needles, dies, tools, machine spindles, gears, gear shafts, and other machine parts where stiffness and unusual strength are needed. The crucible furnace has been almost completely replaced by the electric furnace, which is a large arc-heated crucible.

18-9. Electric Furnace

The electric furnace (see Figs. 18-5 and 18-6) is used when close control of temperature and amounts of alloying elements is important. Higher temperatures can be reached with the electric furnace than are possible with other steel-making furnaces. This high-grade steel is called **electric steel.** **High-carbon steel**, special alloy steel, and high-speed steel are made this way. They are used for cutting tools, dies, and the like.

Electric **arc furnaces** give very close control of the grain structure of steel. Electric **vacuum-induction** furnaces give close control over the chemistry of steel.

18-10. Low-Carbon Steel

Low-carbon steel is also known as **machine steel, machinery steel,** and **mild steel.** It contains about .05% to .30% carbon. See Table 10. It is made in the Bessemer converter, the basic-oxygen furnace, and the open-hearth furnace. It is used for forge work, rivets, chains, and machine parts which do not need great strength. It is also used for almost every purpose that wrought iron is used. In fact, the production of wrought iron has greatly decreased for this reason.

Some of this steel is rolled while cold between highly polished rollers under great pressure. This gives it a very smooth finish and exact size. It is then called **cold-rolled steel.**

18-11. Medium-Carbon Steel

Medium-carbon steel has more carbon and is stronger than low-carbon steel. It is also more difficult to bend, weld, and cut than low-carbon steel. It contains about .30% to .60% carbon. Medium-carbon steel is used for bolts, shafts, car axles, rails, etc. See Table 10.

It is frequently hardened and tempered by heat treatment. Medium-carbon steels can be hardened to a Rockwell-C hardness of about 40 to 60, depending on the carbon content and the thickness of the material. See Unit 46.

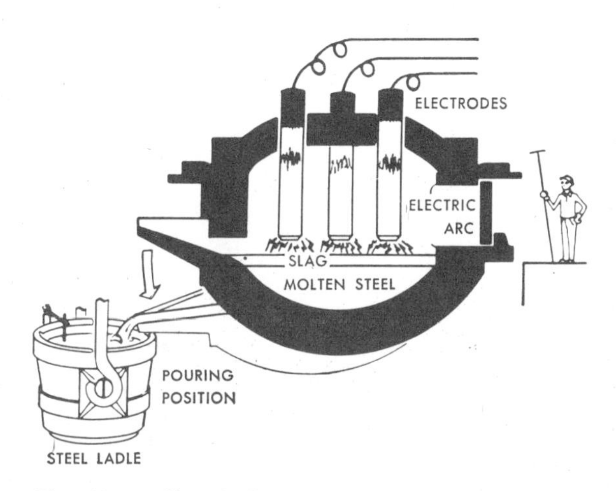

Fig. 18-5. Electric Furnace

Fig. 18-6. Charging Steel Scrap into an Electric Furnace

18-12. High-Carbon Steel

High-carbon steel, known as **tool steel** and **carbon-tool steel**, generally contains about .60% to 1.50% carbon. The best grades of this steel are made in the crucible and the electric furnace. It is called tool steel because it is used to make such tools as drills, taps, dies, reamers, files, cold chisels, crowbars, and hammers. See Table 10. It is hard to bend, weld, and cut.

When it is heated to a red heat and suddenly cooled in water or oil, high-carbon steel becomes very hard and very brittle. This is called **hardening**. The more carbon the steel contains and the more quickly it is cooled, the harder it becomes. High-carbon steel can be hardened to about Rockwell-C 60 to 66, which is hard enough for metal-cutting tools.

High-carbon steel is rolled to the desired shape and is often ground to provide a smooth finish. Round bars which are ground and polished are called **drill rod** which is used for making drills, reamers, taps, punches, and dowel pins.[2]

Fig. 18-7. Ingot Entering Blooming Mill (Top) and a General View of Rolling Beams (Bottom) in a Steel Mill

Notice the blooming mill at lower left and the steel taking finished shape at the far end.

18-13. Hot-Rolled Steel

After pig iron or wrought iron has been made into steel, some of it is made into large blocks called **ingots**. See Fig. 18-4. These ingots are rolled while hot between heavy, powerful, steel rollers each about 2′ to 3′ in diameter the same way that clothes are squeezed through a clothes wringer, Fig. 18-7. The hot steel is rolled between many sets of rollers; each set is a little closer together than the preceding set. When finished, this steel is called **hot-rolled steel** because it was rolled while it was hot. Work that must be bent or twisted may be made of hot-rolled steel.

Some ingots are passed through rollers with grooves that press the hot steel into smaller sizes and different shapes, Fig. 18-1. Soon the ton of red-hot steel gets longer and thinner; it comes from the rollers like a fiery snake about 200′ long — all in a few minutes. Round, flat, and various other shapes are thus formed.

Some ingots are rolled into wide sheets. **Sheet steel** is thinner than .250″, and the thickness is expressed by a gage number. Plate steel generally is heavier, in fractional thicknesses of 1/4″ and up.

Pipe and tubing are made by **drawing** or pulling a flat piece of steel through a bell-shaped ring called a **bell**, Fig. 18-8.

[2]A **dowel pin**, in metalwork, is a metal pin used to keep two parts in a certain relation to each other and to keep them from moving or slipping.

After cooling, hot-rolled steel has a thin, black, hard skin or crust which is called **scale**.

18-14. Cold-Rolled Steel

Cold-rolled steel is made from hot-rolled steel, that is, steel that was rolled while hot and rolled again when cold. The hot-rolled steel is a little larger than the final size of the cold-rolled steel, thus, it is reduced only a small amount by the cold rolling.

The bars or rods of hot-rolled steel, when cold, are first put in water containing **sulfuric acid**. This is known as **pickling**. It also removes the black skin or **scale** from the surface of the steel. The sulfuric acid is washed off by dipping the bars in pure water and then in **lime water**.[3] When dry, the bars are rolled while cold between highly finished rollers under great pressure. This gives them a smooth, bright finish and a very exact size. They are then called cold-rolled steel and are often used without any more **finishing** or **machining**.[4]

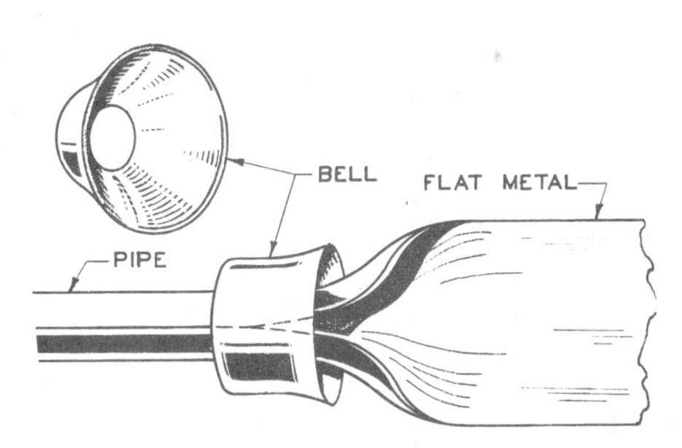

Fig. 18-8. Drawing a Flat Piece of Metal through a Bell to Form a Pipe

[3]**Lime water** is lime mixed with water. It keeps acid, which may still be on the steel, from eating further into it. After the steel is dry, the lime water which remains on the steel keeps it from rusting until it can be rolled.

[4]**To machine** means to cut, using a machine.

18-15. Cold-Drawn Steel

Cold-drawn steel is also made from hot-rolled steel which has been cooled and pickled and which is a little larger than the final size of the cold-drawn steel. The size is thus reduced only a small amount at a time by **drawing** or by pulling the bar or rod of cold steel through **straightening rolls.**

Fig. 18-9. Drawing Cold Steel through a Die, Making Cold-Drawn Steel

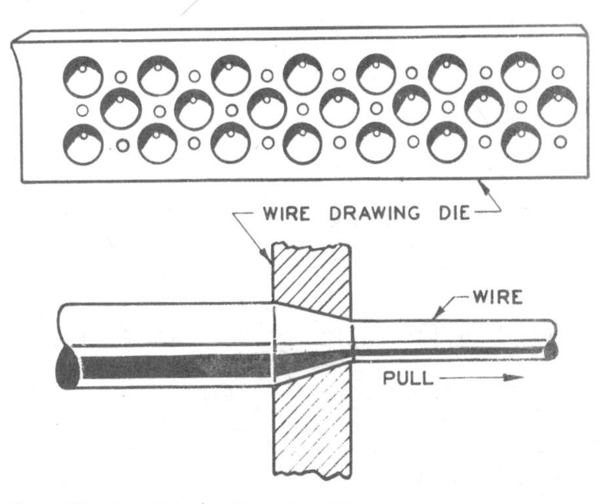

Fig. 18-10. Die for Drawing Wire

Fig. 18-11. Drawing Steel Wire

Cold-drawn steel has a smooth and bright surface and is very exact in size and shape. It is often used without any more finishing or machining, Fig. 18-9.

A **wire drawing die** (also called a drawplate) is used for drawing wire as shown in Fig. 18-10. Wire is made by drawing the steel through this die, Fig. 18-11.

18-16. Cast Steel

To **cast** means to form into a certain shape by pouring into a form. When melted steel is poured into molds in the same way as **cast iron** to make a certain shaped object, the resulting metal is called **cast steel**. The object is known as a **steel casting.**

Words to Know

alloy steel
arc furnace
basic-oxygen process
Bessemer converter
carbon steel
carbon tool steel
cast steel
chromium
cold-drawn steel
cold-rolled steel
crucible furnace
drawplate
drill rod
electric furnace
high-carbon steel
high-speed steel
hot-rolled steel
induction furnace
ingot
ingot mold
low-carbon steel
medium-carbon steel
mild steel
open-hearth furnace
pickling
pipe drawing
special alloy steel
steel casting
tool steel
tungsten
wire drawing die

Review Questions

1. From what materials is steel made?
2. What is the difference between pig iron and steel?
3. Describe the Bessemer process of making steel. How does it differ from the basic-oxygen process?
4. What is an ingot?
5. Describe the open-hearth process of making steel.
6. Describe the crucible process of making steel.
7. How is steel made in the electric furnace?
8. Name the two classes of steel.
9. Name the three grades of carbon steel and list their carbon content.
10. What kinds of steel are made in the basic-oxygen furnace? Open-hearth furnace? Electric furnace? Crucible furnace?
11. What is hot-rolled steel?
12. What is drill rod? How is it made?
13. What is cold-rolled steel?
14. What is cold-drawn steel?
15. What is cast steel?
16. What kind of steel would you use to make a cold chisel?

Mathematics

1. If one cubic inch of rolled steel weighs 0.2833 pounds, how much will a bar 4″ x 4″ x 4″ weigh?

Drafting

1. Draw a wire drawing die which can be held and used in a vise in the shop.

Social Science

1. Write a story telling of the value of steel to civilization.

Career Information

1. Make a list of jobs available in steel-making. Find out their educational requirements and their rate of pay.

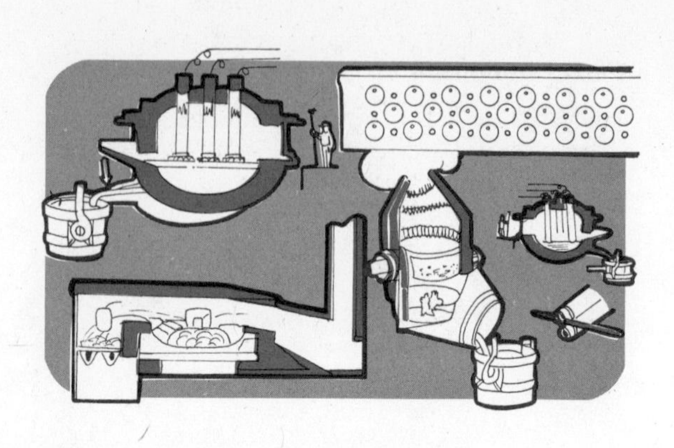

UNIT 19

Steel Alloys

19-1. What Is an Alloy?

An **alloy** is a mixture of two or more metals melted together to form a new metal different from either of the original metals. If a metal is mixed with even a small percentage of another metal, it changes; the color may be changed, it may become harder, the **melting point** may be lowered, etc. For example, **copper** and **zinc** melted together make **brass;** brass is thus an alloy. **Lead** and **tin** melted together make **solder;** solder is an alloy.

19-2. Special Alloy Steel

Alloy steel contains one or more of the following metals: **nickel, chromium, manganese, molybdenum, tungsten,** and **vanadium.** By adding these metals to steel, special steel can be made that is harder, tougher, or stronger than ordinary carbon steel. These alloy steels are described in the following sections.

19-3. Nickel Steel

Nickel steel contains **nickel,** which adds strength and toughness to steel. Nickel steel does not rust easily and is very strong and hard. It is also **elastic;** that is, it can stand vibration, shocks, jolts, and wear. It is used for wire cables, shafts, steel rails, automobile and railroad car axles, and **armor plate.**

19-4. Chromium Steel

Chromium, also known as **chrome,** gives hardness to steel, toughens it, makes the grain finer, and causes it to resist rust, stains, shocks, and scratches. Chromium steel is used for safes, rock crushers, and automobile bearings.

Chromium is the basis for **stainless steel,** also known as **high-chromium steel,** which contains from 11% to 26% chromium. It has a lasting, bright, silvery gloss and is used for sinks, tabletops, tableware, pots and pans, cutting tools and instruments, candlesticks, plates for false teeth, dental tools, ball bearings, fine measuring tools and instruments, moldings, automobile parts, and valves for airplane engines. It does not rust or **corrode.**[1] It is sometimes brightly polished and used for mirrors.

19-5. Chrome-Nickel Steel

Chrome-nickel steel, also known as **nickel chromium steel,** contains both **chromium** and **nickel.** It is hard and strong and is used for **armor plate** and automobile parts, such as gears, springs, axles, and shafts. Nickel and chromium also are alloyed in large proportions to produce the nickel-chromium type **stainless steels.**

[1]**Corrode** means to wear away gradually, as iron does by rusting. It is a chemical reaction of impurities in the metal to moisture and oxygen.

19-6. Manganese Steel

Manganese steel contains **manganese** which is a hard, brittle, grayish white metal. It purifies and adds strength and toughness to steel. Manganese steel remains hard even when cooled slowly; it is so very hard that it is difficult to cut. **Wear** makes the surface harder. It is usually **cast** into shape. Manganese steel can stand hard wear, strain, hammering, and shocks. It is used for the jaws of rock and ore crushers, steam shovels, chains, gears, railway switches and crossings, and safes.

19-7. Molybdenum Steel

Molybdenum steel contains **molybdenum;** it is called "Molly" for short in steel mills. Molybdenum, a silvery white metal which is harder than silver, adds strength and hardness to steel and causes it to stand heat and blows. Molybdenum steel is used for automobile parts, high-grade machinery, wire as fine as 0.0004″ in diameter, ball bearings, and roller bearings.

19-8. Tungsten Steel

Tungsten is a rare, heavy, white metal which has a higher melting point than any other metal. Tungsten adds hardness to steel, makes a fine grain, and causes it to withstand heat. Tungsten is used as an alloying element in tool steels, high-speed steels, and in cemented carbide. It is also used to armor plate.

Cemented-tungsten carbide, also called **tungsten carbide,** is the hardest metal made by man, being nearly as hard as diamond. It is made by molding **powder metals,** including tungsten and carbon. Tungsten-carbide metal-cutting tools retain their hardness at red-hot temperatures (as high as 1700° F.) without significant softening. The metal is expensive, so a small piece is **silver soldered** on the tip of a **cutting tool.** It is also used for **wire drawing dies.** Such tools, known as **carbide tools,** cut two to four times faster than high-speed steel. They must be sharpened on **diamond grinding wheels** or silicon-carbide grinding wheels.

19-9 Vanadium Steel

Vanadium steel contains **vanadium** which is a pale, silver gray metal. It is brittle and resists **corrosion.** Vanadium gives lightness, toughness, and strength and makes a fine grain in steel. Vanadium steel can withstand great shocks. It is used for springs, automobile axles and gears, and for other parts that vibrate when in use.

Chromium-vanadium steel, also called **chrome-vanadium steel,** is hard and has great **tensile strength.**[2] It can be bent double while cold and is easy to cut. Chromium-vanadium steel is used for automobile parts such as springs, gears, steering knuckles, frames, axles, connecting rods, and other parts which must be strong and tough but not brittle.

19-10. High-Speed Steel (HSS)

High-speed steel, known as **high-speed tool steel** or **self-hardening steel,** is an alloy steel. Its carbon content may range from about 0.70% to 1.50%. Several different grades are available. It generally contains one or more special alloys, such as **chromium, vanadium, molybdenum, tungsten,** and **cobalt.** The first four of these elements are carbide formers. They combine with carbon to form carbides such as chromium carbide, vanadium carbide, etc. These carbides are very hard and wear-resistant; therefore, they make good cutting tools. Cobalt is not a carbide former, but it increases the **red-hardness** of the cutting tool. Thus, the tool retains its hardness at higher temperatures. High-speed steel-cutting tools retain their hardness without significant softening at temperatures up to

[2]**Tensile strength** means the ability of a metal to withstand stretching without tearing apart. It is indicated by the weight in pounds per square inch (of the cross-section area of the metal being tested) which will cause it to pull apart.

about 1100° F., a temperature indicated by a dull, red heat. On the other hand, carbon-tool steel-cutting tools start to soften significantly at temperatures above 450° F. This temperature is indicated by a dark brown or purple color.

High-speed steel is made in an electric furnace. It is used for cutting tools such as drills, reamers, countersinks, lathe-tool bits, and milling cutters. It is called high-speed steel because cutting tools made of this material can be operated at speeds twice as fast as those for tools made of carbon-tool steel. High-speed steels cost about two to four times as much as carbon-tool steels.

19-11. Cast Alloys

A number of cast alloys have been developed for use in making metal-cutting tools. Some trade names[3] include Stellite, Rexalloy, Armaloy, and Tantung. The cast alloys are used as brazed tips on tool shanks, as removable tool bits in lathe toolholders, and as inserts in toolholders and milling cutters. They retain their hardness at high temperatures, up to about 1500° F. Cast alloy-cutting tools may be operated at cutting speeds 50% to 75% faster than for high-speed steel-cutting tools.

Cast alloys are metals composed of the following elements: cobalt 35% to 50%, chromium 25% to 35%, tungsten 10% to 20%, nickel .01% to 5% and carbon 1.5% to 3%. Small amounts of other elements are sometimes added. Since they do not contain iron, they are not steels. They are very hard and cannot be machined, except by grinding.

19-12. Spark Test of Iron and Steel

The **spark test** is one way of discovering what kind of iron or steel was used in an object. Different kinds of iron and steel will produce different sparks. The iron or steel is held lightly against the grinding wheel and the sparks carefully watched. Four things should be noticed about the sparks:

1. The color of the spark.
2. The shape of the spark as it leaves the grinding wheel and after it explodes.
3. The quantity of sparks.
4. The distance the sparks shoot from the grinding wheel.

The color, shape, quantity, and length of the sparks tell how much **carbon** the steel contains, Fig. 19-1. The sparks are small pieces of metal that have been cut away and heated by rubbing. Instead of the sparks of carbon steel dying out after they leave the grinding wheel, they suddenly broaden, light up, become much brighter, and then disappear. This shows that something is happening after the spark leaves the wheel. The carbon in the particle of metal unites with the **oxygen** in the air. Carbon unites readily with oxygen, causing more forceful burning. The more carbon in the steel, the brighter and greater in number are the sparks.

The sparks of steel containing much carbon are very numerous and follow around the **circumference** of the wheel. Also, the more carbon in the steel, the more the sparks look like stars. As the carbon in the steel decreases, the sparks become a darker red and appear more or less like straight lines instead of stars.

A good way to find the amount of carbon in a piece of steel is to compare its sparks with the sparks of a piece of steel of which the carbon content is known. Files contain a high percentage of carbon, usually about 1.30%. If the steel that is being tested gives more starlike sparks than the file, it contains more carbon. If fewer starlike sparks of a darker red color are produced, the steel contains less carbon than the file. Table 11 describes the different sparks of iron and steel.

The beginner should experiment with pieces of iron and steel, the carbon contents of which are known; then he may

[3]A **trade name** is a name or term used by a manufacturer or distributor for a particular product or group of products. **Trademarks** are registered with the copyright office and may not be used by any other person or company.

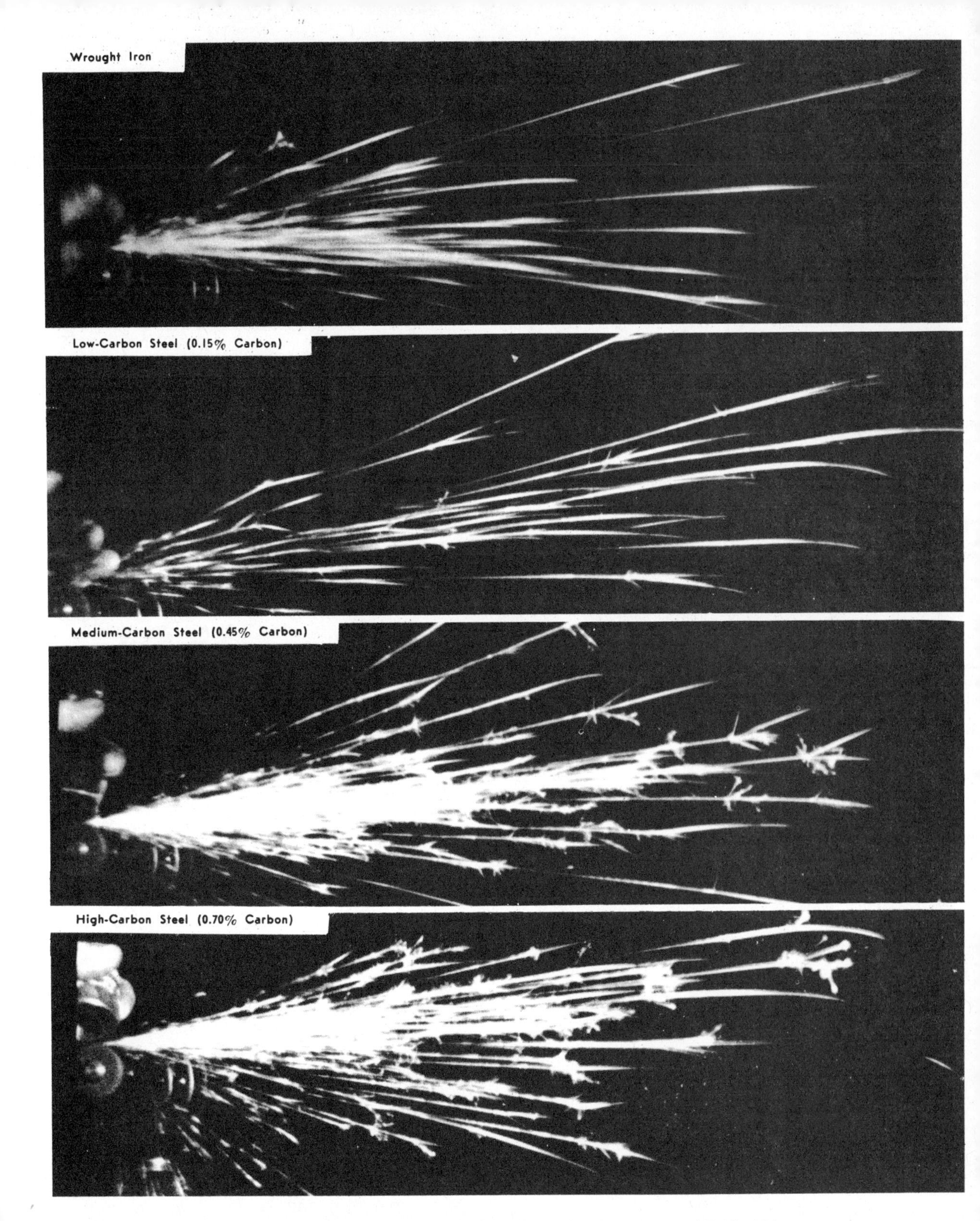

Fig. 19-1. Spark Tests of Iron and Steel (See Section 19-12 and Table 11.)

TABLE 11

Descriptions of Iron and Steel Sparks

Metal	Descriptions of Sparks
WROUGHT IRON	Long, dull red, with dark tips. Shoot off in straight streaks which widen in the middle and then disappear.
STEEL:	
Low-carbon (also called **Machine Steel** or **Mild Steel**)	Same as wrought iron sparks, except brighter and explode or branch out a little.
Medium-Carbon	Bright with starlike explosions. More and brighter as carbon is increased. The more carbon in the steel the more the sparks look like stars.
High-Carbon (also called **Tool Steel** or **Carbon Tool Steel**)	White and explode immediately into bright stars, then disappear. Many sparks follow around wheel. Thus high-carbon steel burns easier and quicker than low-carbon steel; this must be remembered when heating high-carbon steel to keep from burning it.
ALLOY STEEL: High-Speed Steel	Sparks of high-speed steel are dark, broken lines which end in dull red, pear-shaped sparks a short distance from wheel. Some sparks shoot or branch off at **right angles** to main stream of sparks. Many alloy steels give no sparks at all.

TABLE 12

SAE Steel Numbering System

Kind of Steel	Numbers
Carbon Steels	1xxx
Nonsulfurized Carbon Steels	10xx
Example	1018
Resulfurized Carbon Steels (Free Machining)	11xx
Example	1112
Example	1113
Nickel-Chromium Steels	3xxx
1.25% Nickel, 0.65% Chromium	3140
Molybdenum Steels	4xxx
Molybdenum 0.20 or 0.30%	40xx
Example	4024
Chromium Steels	5xxx
Chromium 0.25, 0.40, or 0.50%	50xx
Example (0.25% Chromium, 0.46% Carbon)	5046
Example (1.45% Chromium, 1.00% Carbon)	52100
Chromium-Vanadium Steels	6xxx
Nickel-Chromium-Molybdenum Steels (Nickel less than 1.00%)	8xxx
Silicon-Manganese Steel	92xx
Nickel-Chromium-Molybdenum (Nickel 3.25%)	93xx

gradually add different alloy steels. He will thus form pictures in his own mind of the different sparks for different metals.

19-13. SAE Steel Specifications

The **Society of Automotive Engineers, Inc.,** developed a number system which indicates the chemical composition of the different kinds of steels. This system is known as the **SAE Steel Numbering System** and also as the **SAE Steel Specifications.** Any **steel catalog** gives this numbering system. Examples of it are given in Table 12.

Each kind of steel has a number with four or five digits, usually four digits. The first digit of the number tells what kind of steel it is: 1 is a **carbon steel,** 3 is a **nickel-chromium steel,** etc. For the simple **alloy steels,** the second digit of the number generally (but not always) tells the approximate percentage of the major alloy represented by the first digit. The last two or three digits tell the average percentage of carbon the steel contains in **points** or hundredths of 1% (one percent, 1.00%). Thus, 3140 means that nickel-chromium steel is about 1% nickel and 0.40% carbon.

Some steels have numbers with five digits; the last three digits indicate the percentage of carbon, as, for example, 52100 means that the **chromium steel** contains about 2% chromium (1.45) and about 1.00% carbon.

A different kind of numbering system is used for the identification of tool steels and stainless steels. The identification system for these steels is included in standard handbooks for machinists.

19-14. AISI Steel Specifications

The **American Iron and Steel Institute** (AISI) has developed a system which also states the kind of furnace used to make the steel. Letters are used together with the same numbers used in the SAE systems.

B is acid Bessemer-carbon steels.

C is basic open-hearth or basic-electric furnace carbon steels.

E is electric furnace alloy steels.

Thus, the letter E in the number means that the steel was made in the electric furnace.

19-15. Properties of Metals

There are many different types of steel and other metals used in modern industry. Each is selected for its special properties. In fact, engineers sometimes specify the properties required in a particular situation. Then metallurgists and other scientists must develop an alloy or combination of materials having the required properties. This has been especially true for space vehicles.

Common important properties are density (relative weight), tensile strength, hardness, hardenability, ductility, malleability, brittleness, toughness, elasticity, fusibility, weldability, response to heat treatment, heat resistance, corrosion resistance, and machinability. People working with metal should be acquainted with these properties. Manufacturer's catalogs and technical handbooks give specific data for those needing it.

Tensile strength is the strength necessary to pull apart a piece of metal. It is usually expressed in pounds per square inch (psi). This is also called **ultimate** tensile strength. **Yield point** is the point (usually much less) at which the metal first begins to stretch with no increase in load.

Hardenability is that property which enables a metal to harden completely through to its center when heat treated. Some steels rank low in hardenability. This means that they **harden** to a shallow depth at the surface, and the core is softer at the center of the metal. The alloy steels generally rank higher in hardenability than the carbon steels of similar carbon content.

Hardness is the resistance to being dented or penetrated. Some steels are hard enough to cut softer steels. Hardness of some steels can be increased by heat treating. Several kinds of hardness testers are available.

Ductility means that the metal can be drawn out or stretched without breaking. Aluminum, steel, and copper are very ductile and can be easily formed into wire.

Malleability means that the metal can be hammered, rolled, or bent without cracking or breaking. The more malleable a material, the easier it is formed.

Brittleness means that a material suddenly cracks and breaks easily. Glass is brittle, as are gray cast iron and hardened steels. Usually brittleness is related to hardness.

Toughness in metal refers to its resistance to breaking, bending, stretching, or cracking. It is related to hardness and brittleness. High-carbon steel is very brittle when hardened, but **tempering** reduces hardness and increases toughness.

Elasticity is the property which enables metal to be bent or twisted and still return to its original shape without being deformed. Spring steel must be elastic.

Fusibility is the property which enables metal to liquefy easily and join with other metals while liquid.

Weldability is the degree to which a metal may be welded with good fusion and a minimum loss of other properties. Gray cast iron is more difficult to weld without special techniques because the heat required for fusion frequently causes brittleness. The weld may often crack in the process of cooling.

Heat treatment is the heating or cooling of ferrous metals in the solid state to change

TABLE 13

Physical Properties of Steels

AISI No.	Condition of Steel	Tensile Strength (psi)	Brinell Hardness	Machinability Rating (B 1112=100)
C 1018	Hot-Rolled	69,000	143	52
C 1018	Cold-Drawn	82,000	163	65
B 1112	Cold-Drawn	82,500	170	100
B 1113	Cold-Drawn	83,500	170	130
Ledloy 375	Cold-Drawn	79,000	155	220
C 1045	Cold-Drawn	103,000	217	60
C 1095	Hot-Rolled	142,000	293	...
C 1095	Water Quenched at 1450° F, Tempered at 800° F:	200,000	388	...

their mechanical, microstructural, or corrosion-resisting properties. **Hardening temperature** is the point at which the grain structure of steel becomes fine and may be hardened. **Heat-resisting metals** (high-speed steel, cast alloys, and titanium alloys) retain their strength and resist oxidation at relatively high temperatures.

Corrosion resistance is the ability of steel to resist rusting and other chemical action. Chromium, nickel, and titanium in stainless steels increase this resistance. Anodizing and cladding of aluminum also increase corrosion resistance.

Machinability is the ease with which metal can be machined while maintaining maximum tool life, cutting speed, finished appearance, or any combination of these factors. Traces of finely dispersed lead in steel greatly increase machinability with no great loss of other properties. Sulfur in steel also increases machinability but is detrimental to welding and other hot forming qualities. Resulfurized and leaded steels are widely used for machining parts in lathes and automatic screw machines. Machinability ratings have been computed on the basis of B1112 steel equaling 100. High numbers indicate comparatively better machinability.

Table 13 indicates some properties of several types of steel which are frequently used in industry and in metalworking classes.

Words to Know

AISI
alloy
carbide tool
cast alloy
chrome
cobalt
corrosion
hardenability
high-speed steel
manganese
molybdenum
nickel
powder metal
SAE
spark test
stainless steel
tensile strength
tungsten
tungsten carbide
vanadium

Review Questions

1. What is an alloy?
2. Why are metals alloyed?
3. Make a list of the alloy steels mentioned in this unit and give a typical use for each.
4. What is high-speed steel? For what is it used? Why is it called high-speed steel?
5. What are cast alloys? For what are they used?

6. What is the use of tungsten carbide?
7. Of what use is the spark test?
8. What is the composition of SAE 1018 steel? SAE 4130 steel?
9. Define machinability.
10. What is the meaning and importance of tensile strength?
11. How do hardenability and hardness differ?
12. Define toughness as it applies to steel.

Social Science

1. Write a report on how stainless steels have benefitted our society.

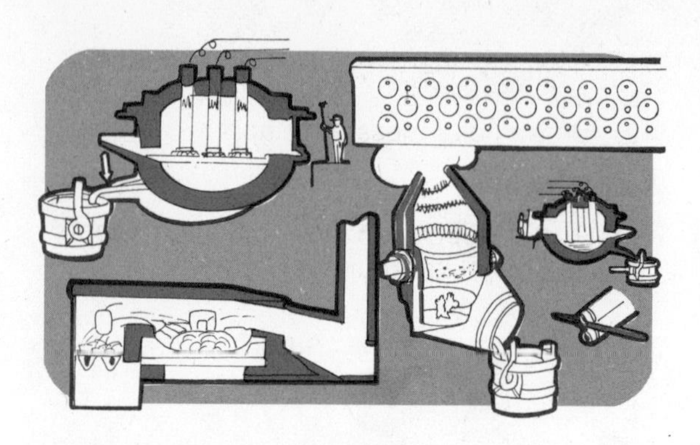

UNIT 20

Nonferrous Metals

20-1. Aluminum

Aluminum is a brilliant, silvery white metal. It is one of our most useful metals. Although it costs four or five times as much per pound as iron or steel, it weighs only about one-third as much. It also machines much faster (two to three times deeper cuts at double speeds). It costs less to transport, usually needs no finish to prevent rusting, is quite maintenance free, and has a natural surface beauty. Thus, it may be more economical than other metals when all things are considered.

Fig. 20-1. Skin-Mill Cuts Integral Ribs into Type 7075 Aluminum Plate in Making 100-Foot Rivetless Wing Structures for Supersonic Aircraft

Aluminum is a good conductor of electricity and heat, yet reflects heat when highly polished. It can be drawn into very fine wire, spun or stamped into deep forms, and hammered or rolled into thin foil sheets, some only 0.00025″ thick.

Aluminum melts at about 1200° F. and is cast in molds when 100° to 300° above that. As this is about one-half the temperature required for iron or steel, aluminum is preferred for casting in schools or other situations where initial equipment cost and safety are important.

Working with Aluminum

At one time, joining aluminum parts was largely limited to using screws, rivets, or other mechanical means, but this problem has been solved by improved welding and brazing techniques. Sometimes entire sections are machined from one large plate for greater strength, Fig. 20-1.

Working with aluminum is somewhat different than working with steel, but not necessarily difficult. Different cutting angles and speeds are recommended as well as different techniques in welding, brazing, or soldering.

Be sure to check instructions and available references for these specifications before beginning work.

Pure aluminum is too soft for many uses, but numerous modern alloys make it ideal for many jobs. Some of the common uses for aluminum are: aircraft and rocket parts; bodies for railroad cars, trucks, and trailers; pistons, blocks, and heads for engines; tubing; window frames; structural members; cooking utensils, machine tool housings; and foil and collapsible tubes for packaging.

Refining Aluminum

Aluminum is made from an ore called **bauxite**, Fig. 20-2. Important deposits are in Arkansas, Washington, Oregon, and parts of Canada. It usually is mined in open pits, then refined where cheap electrical power is available in quantity. One-sixth of the earth's crust is aluminum ore, but it is difficult to extract the pure metal.

Crushed bauxite is changed chemically to aluminum oxide, a white powder called **alumina**. This is smelted into aluminum by removing the oxygen in large electrolytic tanks called **reducing pots**, Fig. 20-3. Electricity passes from carbon anodes through a mixture of the alumina and molten **cryolite** (sodium aluminum flouride). This heats the mixture, and molten aluminum is deposited at the bottom where it can be drained off, Fig. 20-4. Later the metal is often alloyed, cast, rolled, or **extruded**[1] into many shapes, Figs. 20-5-20-10.

Fig. 20-2. Arkansas Bauxite Mine

Fig. 20-3. Breaking Top Crust of Alumina in Electrolytic Reduction Pot

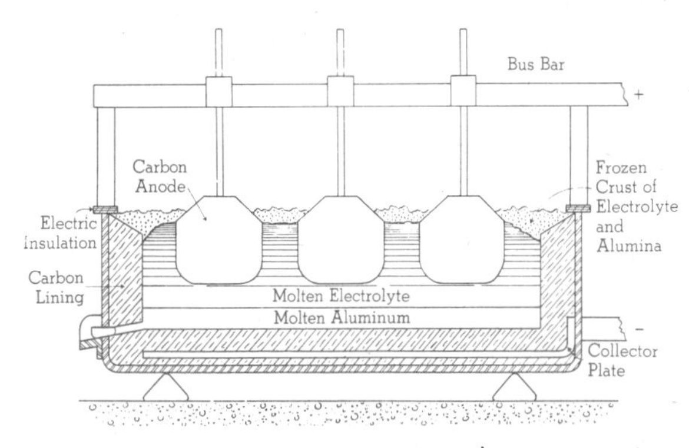

Fig. 20-4. Electricity in an Aluminum Reduction Pot Passes from Carbon Anodes through the Electrolyte Depositing Molten Aluminum on the Carbon Lining

[1]**Extrude** means to push heated material through a specially shaped opening to form a long strip in that shape. Extruding is much like squeezing toothpaste from a tube.

Fig. 20-5. Aluminum Alloy Ingots Clad with High Purity Liners Enter Hot Rolling Mill
Liners form thin corrosion-resistant coating on finished sheet.

Fig. 20-6. Aluminum Extrusion Emerging from Hydraulic Press

20-2. Aluminum Alloys

Other metals are often added to the pure aluminum to improve its physical properties. Industry uses many varieties of aluminum alloys. These are formed by using various methods of tempering or hardening the metal. The tensile strength for various aluminum alloys may range from 13,000 to 81,000 pounds per square inch. In Table 14 four typical alloys are compared with soft and hardened steel.

TABLE 14
Physical Properties[1] of Aluminums

A.A. No.[2]	Old No.	Hardness (Br)	Tensile Strength (Psi)	Cold Work-ability	Machin-ability	Weldability Gas	Weldability Arc	Weldability Spot	Corrosion Resistance
1100-0[3]	2S	23	13,000	A+	D	A	A	B	A
2024-T36[4]	24S	130	73,000	E	A	D	B	A	B
6061-T6[5]	61S	95	41,000	C	B	(Good after heat treating)			
7075-T6[6]	75S	150	76,000	D	A	D	D	B	B
C1018 Hot-Rolled Steel		143	69,000						
C1095 Tempered Steel		388	200,000						

[1]Code for working properties: A=excellent, B=superior, C=good, D=poor, E=not recommended.
[2]That part of the Aluminum Association Number following the hyphen denotes the amount and kind of temper treatment used in manufacture.
[3]No. 1100-0 is 99% pure soft aluminum.
[4]No. 2024 contains 4.5% copper, 1.5% magnesium, and 0.6% manganese.
[5]No. 6061 contains 1.0% magnesium, 0.6% silicon, 0.25% copper, 0.25% chromium.
[6]No. 7075 contains 5.6% zinc, 2.5% magnesium, 1.6% copper, and 0.3% chromium.

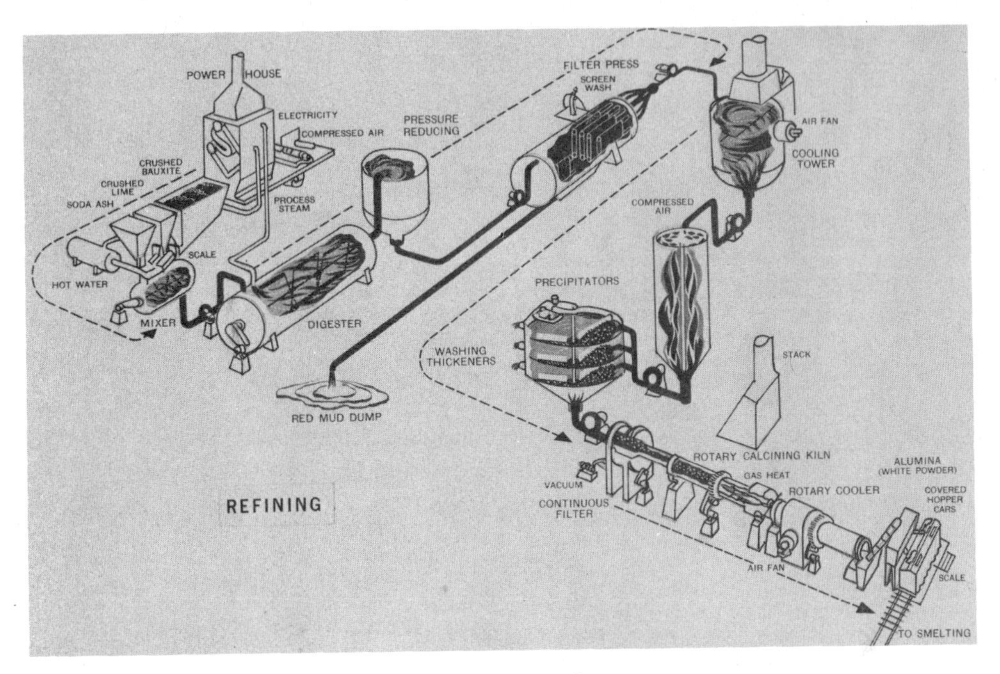

Fig. 20-7. Refining Aluminum

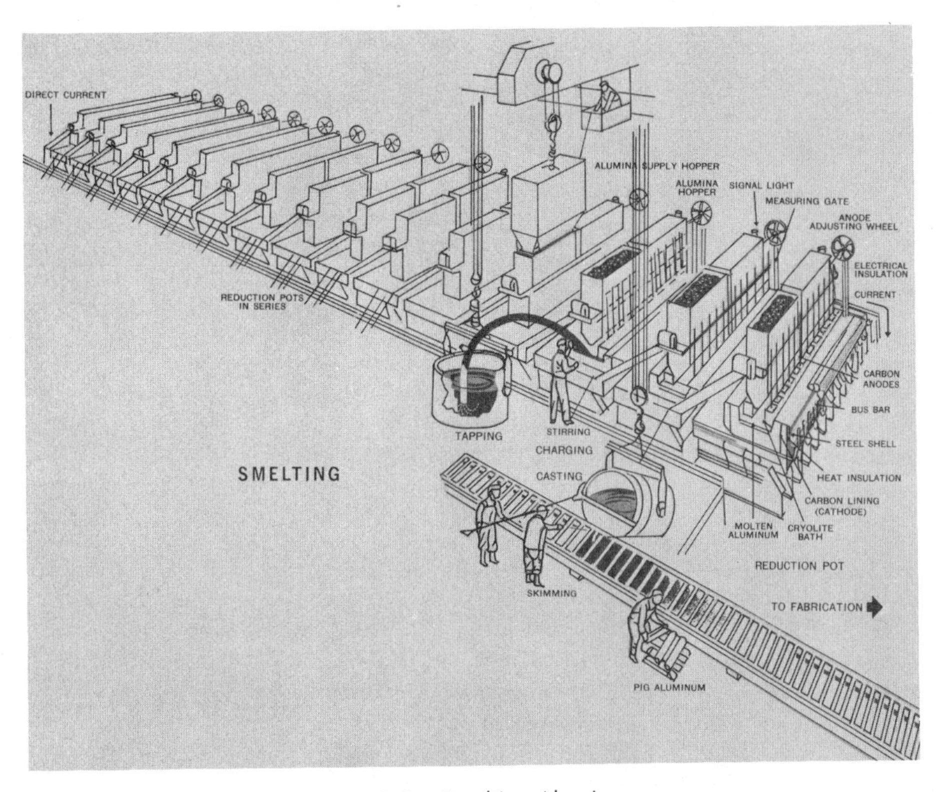

Fig. 20-8. Smelting Aluminum

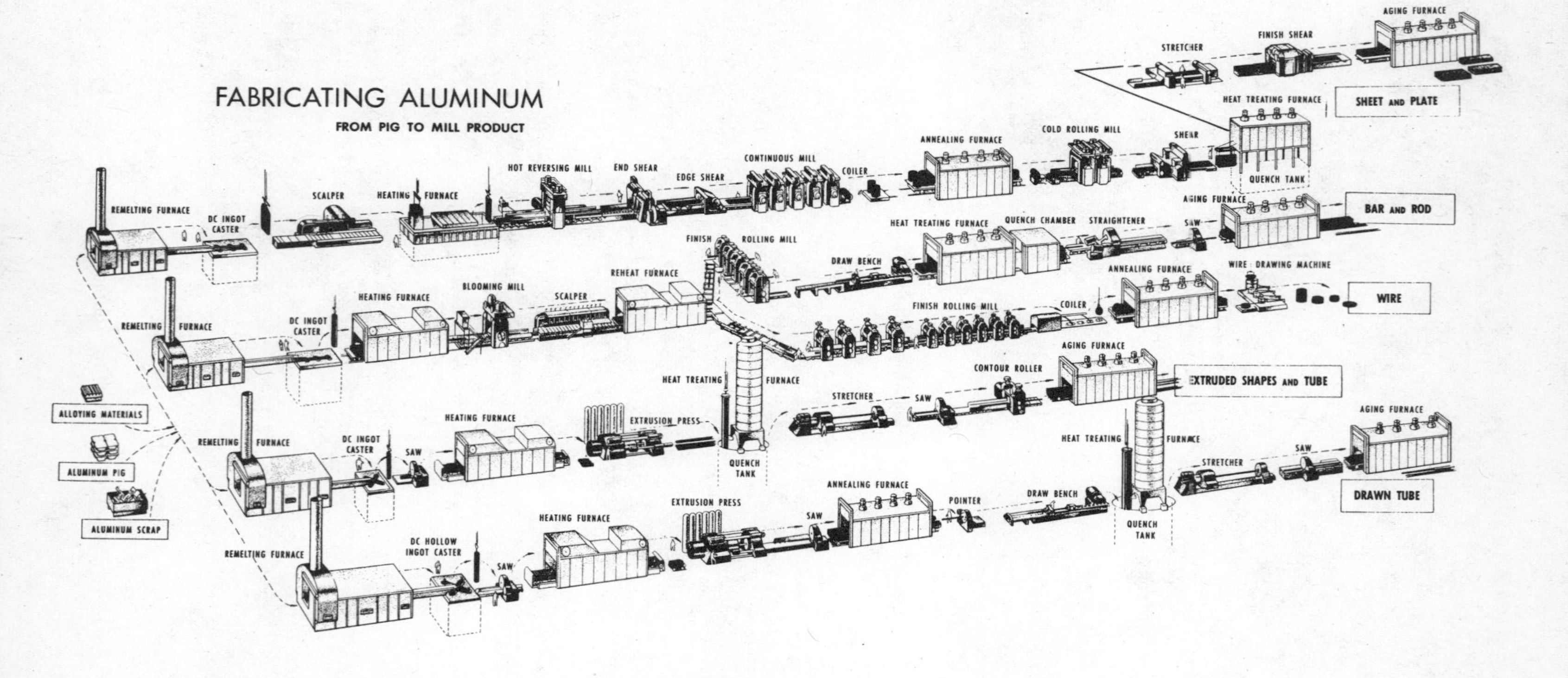

Fig. 20-9. Fabricating Aluminum Stock

Fig. 20-10. Copper Wire Will be Drawn from These Cast "Wire Bars"

Aluminum Association Alloy Designations

In 1954 the Aluminum Association adopted a standard alloy designation system. In the A.A. number, the first digit identifies the major alloying element as follows: 1—99% pure aluminum or better, 2—copper, 3—manganese, 4—silicon, 5—magnesium, 6—magnesium and silicon, 7—zinc, 8—other elements, 9 is special. These are the main alloying metals. Refer to the footnote on Table 14 showing the composition of each alloy.

The other three digits have technical meanings which will be omitted here. The part of the number following the hyphen (-T36) is the temper designation, and a "-0" (1100-0) indicates the alloy to be soft and workable.

Typical Aluminums

Number 1100-0 pure aluminum is naturally soft, ductile, and more resistant to chemical attack than any of the alloys. It is the form most commonly used for hammering and shaping when maximum strength is not needed. However, it gradually becomes hardened as it is worked. When this happens, it must be **annealed**. This is done by heating the aluminum to 650° F. (indicated by the heat at which blue carpenters' chalk turns whitish) and then allowing the heated metal to air cool slowly.

Number 2024 is often used for structural or machining applications. Number 6061 has a number of applications which would require high tensile strength and good welding properties, such as railings and protective guards. Number 7075 is used for aircraft and other work where the highest strength is required. These three are strong heat-treatable alloys.

Metal suppliers furnish data books and charts which list the alloys, meanings of number designations, properties of the alloys, and recommended applications. Refer to these for more complete data. Two broad classes of alloys are the wrought alloys (for cold working) and the casting alloys.

20-3. Copper

Copper is the oldest metal known to man. It is a tough, reddish brown metal. As found in copper mines, located in Arizona, Michigan, Montana, and Utah, it is known as **copper ore**. It is sold in the form of wire, bars, plates, and sheets.

Copper is the second-best carrier of electricity; silver is the best. Copper is used for electric, telephone, and telegraph wires and cables. It is also used for water heaters, wash boilers, pipes, kettles, window and door screens, roofing, etc.

Beryllium copper is copper to which a small amount of beryllium has been added. See section 20-4. This very tough alloy is used for corrosion-resistant springs; because it is non sparking, it is used for wrenches and other metal objects in plants which manufacture explosives.

In art metalwork, copper is used for bowls, vases, ashtrays, etc. Copper is used in **brass, bronze, monel metal,** and **German silver** (which contains about 50% copper, 30% **zinc**, and 20% **nickel**). German silver is used as a substitute for silver in making inexpensive jewelry. Many beautiful copper articles for the home can be made in the school shop or at home by **hammering** or **spinning**, as explained in Unit 34.

The surface of copper becomes green in moist air. This may be seen on copper which has not been cleaned for some time. See § 35-9. Copper hardens when hammered but can easily be softened or **annealed**.

20-4. Beryllium

Beryllium is a gray metal about the same color as steel. It is expensive; light in weight; and more heat resistant, harder and more brittle than the other light metals (aluminum and magnesium). A small percentage of beryllium, usually less than 2%, makes a very strong alloy when it is added to copper or nickel. Its light weight and heat resistance are valued in aerospace applications. The first of our space capsules used a heat shield of beryllium. Pure beryllium is also used in nuclear reactors as a moderator and reflector. Since beryllium dust is toxic, workers are required to wear respirators while machining it, and machines are equipped with vacuum dust collectors.

Fig. 20-11. Galvanized Steel

20-5. Brass

Brass is a yellow **copper-base** alloy in which the principal alloying element is zinc. There are many varieties of brass, ranging from about 60% copper and 40% zinc to as high as 90% copper with only 10% zinc. It can be made harder by adding **tin** or certain other alloying elements. Brass neither discolors nor corrodes as fast as copper. It is used for ornamental work, musical instruments, inexpensive jewelry, screws, door hinges, window locks, small gears, and other parts for watches and clocks.

Many lovely brass articles for the home can be made in the school shop or the home workshop by **hammering** or **spinning**. Like copper, brass hardens when hammered, but by heating and cooling (annealing), it can be softened.

20-6. Bronze

Bronze is a **copper-base alloy** in which the principal alloying element is tin. Sometimes it also contains **zinc** or other alloying elements. It is harder and lasts longer than **brass**. It is also more expensive because of the high cost of tin. Bronze is used for bells, statues, medals, propellers, **bushings**,[2] bearings for machines, etc. The one-cent coin, or penny, is bronze. "Bronze" welding rods are more like brass, and today there is no clear defining line between these two alloys of copper. See Table 15.

20-7. Zinc

Zinc is a brittle, bluish white metal. It is used as a **coating** for iron and steel for protection against rust. This coating with zinc is called **galvanizing** and is done by dipping the metal into melted zinc. As the zinc cools, it forms into **crystals** which make the spotted color on **galvanized steel**, Fig. 20-11. **Galvannealed** metal is heated after being gal-

[2]A **bushing** is somewhat like a bearing; it is a removable sleeve or collar, as on the handlebar post on a bicycle.

vanized, producing a coating of alloyed zinc and iron or steel. These coated metals are used for wire fences, eave troughs, metal roofing, water tanks, water pipes, buckets, automobile frames, signs, etc.

Zinc is also used in **German silver, brass, bronze,** and **dry cell batteries**. Like copper and brass, zinc hardens when hammered. It can be softened by heating and slow cooling. Zinc-based alloys are also widely used in producing die castings for items such as engine blocks for small gas engines, housings for small engines, carburetors, parts of typewriters, car door handles, and parts of portable electric tools.

20-8. Silver

Silver is a beautiful, shiny, white metal. It is found in the form of **silver ore**. Pure silver is soft. It is used for ornamental work, jewelry, tableware, mirrors, and coins. United States silver coins formerly contained nine parts silver and one part copper. Silver is the best carrier of electricity.

Sterling silver is silver with only a little copper added to make it harder. It is used for the best tableware and jewelry.

German silver is made of copper, zinc, and nickel; it does not contain silver. See Table 15.

20-9. Gold

Gold is a precious, heavy, beautiful, bright yellow metal. Grains and **nuggets** of gold are found in river gravels and sands of the seashore. Gold is also found in rocks in the form of **gold ore.** Pure gold is too soft for articles of general use and is therefore mixed with copper, silver, or other metals.

Gold can be **hammered** into very thin leaves, called **gold leaf**, much thinner than the thinnest tissue paper. The art of covering something with gold leaf or gold powder is called **gilding**. One pound of gold can be made into a wire one mile long.

Gold is used for ornamental work, jewelry, coins, and fillings in teeth. United States gold coins were made of 90% gold and 10% copper.

The purity of gold is measured in karats. Pure gold is 24 karats. Thus, an 18-karat gold ring is made of 18 parts by weight of gold and 6 parts by weight of some other metal. Jewelers abbreviate karat as **k**; for example, 14k gold. They use the spelling **carat** for the weight of precious stones.

White gold, a silvery metal used for jewelry, is 15% to 20% nickel added to gold, thus changing the color from gold to white.

TABLE 15

Composition and Uses of Alloy Metals that Contain No Iron

Metal	Percent								Uses
	Antimony	Copper	Gold	Lead	Nickel	Silver	Tin	Zinc	
Babbitt	7	4					89		Bearings and bushings for machines and engines.
Brass (Average)		75						25	Ornamental work, bearings, bushings, musical instruments, inexpensive jewelry, screws, door hinges, window locks, parts for watches and clocks.
Bronze		90					10		Bells, statues, medals, coins, propellets, machine bearings, bushings.
German silver		50			20			30	Inexpensive jewelry.
Monel metal		33			67				Chemical equipment, motor boat propellers, cooking utensils.
Pewter	5	3					92		Tableware, ornamental work.
Solder				50			50		For fastening metals together.
Sterling silver		7.5				92.5			Tableware and jewelry.
White gold			80		20				Jewelry.

Green gold has a greenish cast and is used for jewelry. The 15-karat green gold, for example, is 15 parts gold, 8 parts silver, and 1 part copper.

20-10. Magnesium

Magnesium is a silvery white, light, malleable metal that is much lighter than aluminum. It is abundant in nature but is always alloyed with another metal because of its high cost and low tensile strength. It is usually alloyed with aluminum, contributing to the strength and heat resistance of aluminum. In the pure form it burns easily, giving off an intense white light, so it must be handled with care. It is found in sea water, in minerals, in the chlorophyll of green plants, in seeds, and in animal bones.

Magnesium is used chiefly where light weight is important. For this reason it has many applications in transportation vehicles, especially in aircraft, missile, and space vehicles.

20-11. Lead

Lead is a very heavy, bluish gray, poisonous metal. Men working with lead must guard against a disease called **lead poisoning**. Lead is found in the form of **lead ore**. It is the softest metal in general use. Lead is so soft that it can be scratched with the fingernail and can easily be cut with a knife. When freshly cut, it is very bright; this brightness soon disappears when exposed to the air. Water and air, however, have less effect upon lead than upon any other metal. It is used for **pipes** and in **storage batteries**.

White lead, a pigment used in making some paints, is made from lead. Lead is also used in pewter, solder, and other metals. A small amount of lead added to other metals improves their machinability.

20-12. Tin

Tin is a shiny, silvery metal. It is found in the form of **tin ore**. After the tin is removed from the ore, it is poured in the form of blocks which are called **blocked tin**. "**Tin cans**" are made of steel and then coated with tin. The tin is less than 1% of the weight of the can. Tin does not rust. Very few articles are made of pure tin. It is used in making **bronze, babbitt metal, pewter, solder**, and other metals; when used, it always increases the hardness and whitens the articles.

Tin is soft and can be **hammered** or rolled into very thin sheets. **Tinfoil**, which is tin in sheets, is made as fine as .0002″ thick. It was long used for wrapping tea, tobacco, drugs, cheese, candy, etc., to keep away air and moisture. Because tin is expensive, **aluminum foil** and plastic are now used. Tubes, as for toothpaste, once were made of tin.

Tin plate is "**sheet iron**" or **sheet steel** coated with tin and used for pots, pans, cans, pails, metal roofing, etc. It is often incorrectly called "tin." Copper kettles, used for cooking, are coated with tin to keep the poison from copper out of the cooked food.

20-13. Babbitt

Babbitt, also called **babbitt metal**, was invented by Isaac Babbitt. There are two kinds of babbitt. When the base metal or principal metal is lead, it is called lead-base babbitt. When the base metal is tin, it is called tin-base babbitt.

It is an **alloy** made of **lead, tin, copper,** and **antimony**.[3] It does not rust and is used for bearings in machines and engines because it is strong, tough, and enduring. See Tables 15 and 16.

20-14. Pewter

Pewter is a silvery white metal. It is made of 92 parts **tin**, 5 parts **antimony**, and 3 parts **copper**. Other grades of pewter contain a little more or less tin. The lower grades

[3]**Antimony** is a bright, silvery white, hard, and brittle metal. It is melted into other metals to give them hardness. See Section 25-9.

TABLE 16
Melting Points of Metals

Metal	Degrees Fahrenheit
Solder, 50-50 (See Section 28-5.)	400
Pewter	420
Tin	449
Babbitt	462
Lead	621
Zinc	787
Magnesium	1204
Aluminum	1218
Bronze	1675
Brass	1700
Silver	1761
Gold	1945
Copper	1981
Iron, Cast	2200
Steel	2500
Nickel	2646
Iron, Wrought	2700
Tungsten	6150

contain some **lead**. The best pewter contains the least lead because it gives pewter a dull appearance.

Pewter is also called **Britannia metal** because it was first made in Britain. It is used for tableware and ornamental work. Pewter can be made bright and cheerful by polishing as explained in section 35-4. See Tables 15 and 16.

20-15. Nickel

Nickel — a hard, tough, shiny, silvery metal — is found in the form of ore. It does not rust and can be polished to a very bright, silvery finish. It is therefore used for **plating** iron and brass to improve appearance. Nickel plating may be used under chromium plating on trim for automobiles and appliances. It is also used to toughen steel. Such steel is called **nickel steel**. The five-cent coin (a nickel) is made of one part nickel and three parts **copper**.

Monel metal is a white metal containing about two-thirds **nickel** and one-third **copper**, along with small amounts of other elements. It is strong, tough, does not rust, and shines like silver. Monel metal is used for chemical and cooking equipment, motor boat propellers, and the like.

White gold, a silvery metal used for jewelry, is 15% to 20% **nickel** added to **gold**. **German silver** contains about 20% **nickel**. See Tables 15 and 16.

20-16. Titanium

Titanium is a silvery white metal with high strength and heat resistance. It weighs about 44% less than steel alloys, yet its tensile strength is equal to or greater than common structural alloys. Temperatures up to 800° F. do not weaken the metal and it will tolerate up to 2000° F. for short periods of time. Because of these properties, it is used for many supersonic aircraft parts.

Titanium is relatively inert and is used to replace bone and cartilage in surgery. It is also used as a liner for pipes and tanks in the food-processing industries. Titanium dioxide is a bright white pigment used in the ceramics and rubber industries, and it is rapidly replacing the poisonous white lead pigment used in paints.

Words to Know

alumina
aluminum
antimony
babbitt metal
bauxite
beryllium
carat
cryolite
extrude
galvanized steel
galvannealing
German silver
gilding
gold
gold leaf
karat
magnesium
monel metal
nickel plating
nugget
pewter
silver
sterling silver
tinplate
titanium
white gold
white lead

Review Questions

1. Describe copper. In what states is copper mined?
2. Name the important properties of copper.

3. What is brass? For what is it used?
4. What is bronze? For what is it used?
5. Why is the Statue of Liberty, which is molded sheet copper, green in color?
6. Describe zinc. What are its main uses?
7. How is the purity of gold measured?
8. What is aluminum? What are its important properties?
9. What is lead? How is it used?
10. Describe tin. For what is it used?
11. What is babbitt metal? What is its major use?
12. For what is pewter used?
13. Describe nickel. Name several uses for it.
14. What is beryllium? For what is it used?
15. What is titanium? Name its major properties and uses.

Mathematics

1. How many ounces of copper, zinc, and nickel are there in a pound of German silver if it contains 50% copper, 30% zinc, and 20% nickel?

Social Science

1. Write a report on how the discovery of the use of copper changed life during the Bronze Age.

Career Information

1. What is lead poisoning? Write about it and tell in which occupations it must be guarded against.

PART VI

Threads, Dies, and Taps

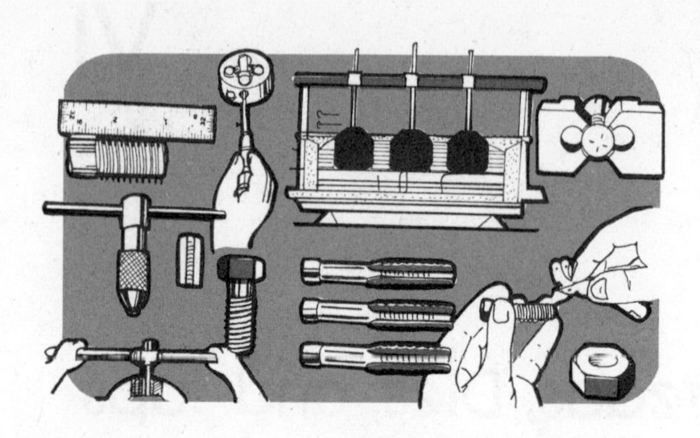

UNIT 21

Screw Threads

21-1. What Is a Screw Thread?

The winding groove around a bolt, screw, or in the hole of a nut forms a thread, also called a **screw thread,** Fig. 21-1. The thread on a rod or screw is an **external thread,** Fig. 21-1. The thread on the inside of a hole or nut is an **internal thread,** Fig. 21-1. A thread is an **inclined plane** that spirals around the surface of a bolt or nut. The screw is also one of the six basic machines.

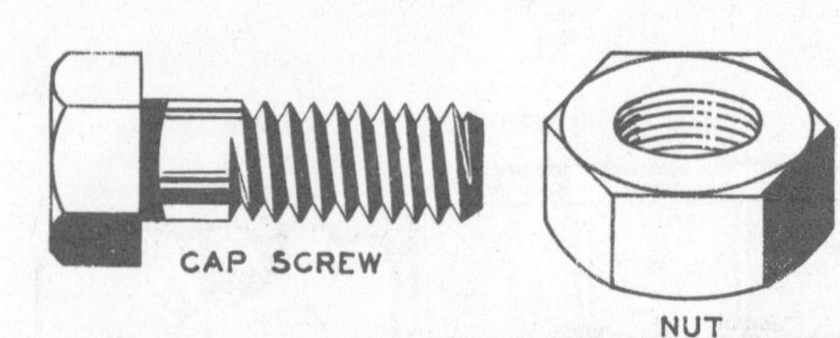

Fig. 21-1. Threads on Screw and Nut

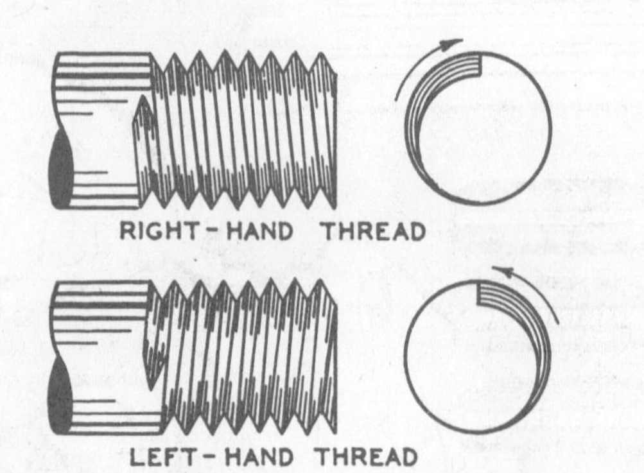

Fig. 21-2. Right-Hand and Left-Hand Threads

21-2. Right-Hand and Left-Hand Threads

A nut which is turned **clockwise** to screw it onto a bolt has right-hand threads (RH), Fig. 21-2. Most threads are right-hand. A nut which is turned **counterclockwise**[1] to screw it onto a bolt has **left-hand threads** (LH). For example, the **shaft** on a grinder, which has two grinding wheels (see Fig. 43-2), has left-hand threads on one end and right-hand threads on the other end. The threads on the bolt in each case are also left-hand or right-hand as are the threads on the nut.

21-3. Single, Double, Triple, and Quadruple Threads

A **single thread** is formed by cutting one groove, Fig. 21-3. Most screws and bolts are single threaded. A **double thread** has two grooves, a **triple thread** has three grooves, and a **quadruple thread** has four grooves alongside each other. Double, triple, and quadruple threads are also known as **multiple threads,** which means more than one thread.

In one turn on a single thread the nut moves forward the distance of one thread, on a double thread it moves twice as far, on a triple thread it moves three times as far, and on a quadruple thread, four times as far.

[1]**Counterclockwise** (CCW) means opposite to the direction in which the hands of a clock turn.

Whether a piece is single threaded or multiple threaded may be determined by looking at the end of the bolt or screw and counting the grooves that have been started.

21-4. Major Diameter of a Thread

The **major diameter** was formerly known as the **outside diameter,** abbreviated OD. It is the largest diameter of a straight external or internal thread, Fig. 21-4. On an external thread, the major diameter is the actual diameter measured across the outside diameter of the thread, Fig. 21-4. On an internal thread, the major diameter is the diameter at the bottom or root of the thread, Fig. 21-5.

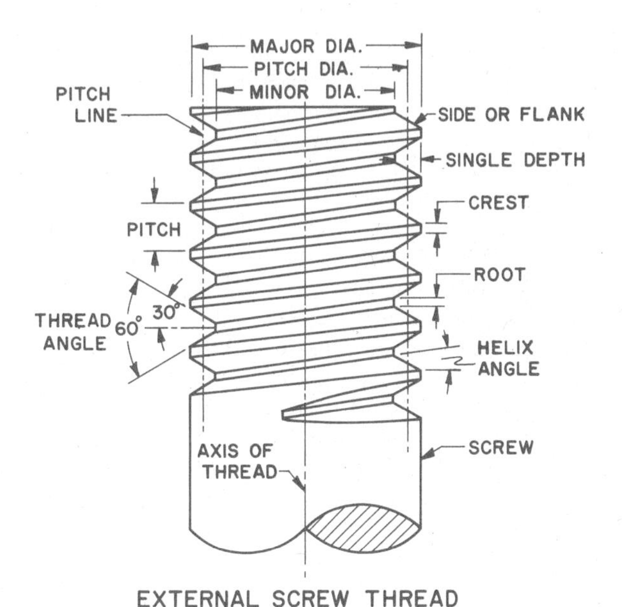

Fig. 21-4. Principal Parts of a Screw Thread

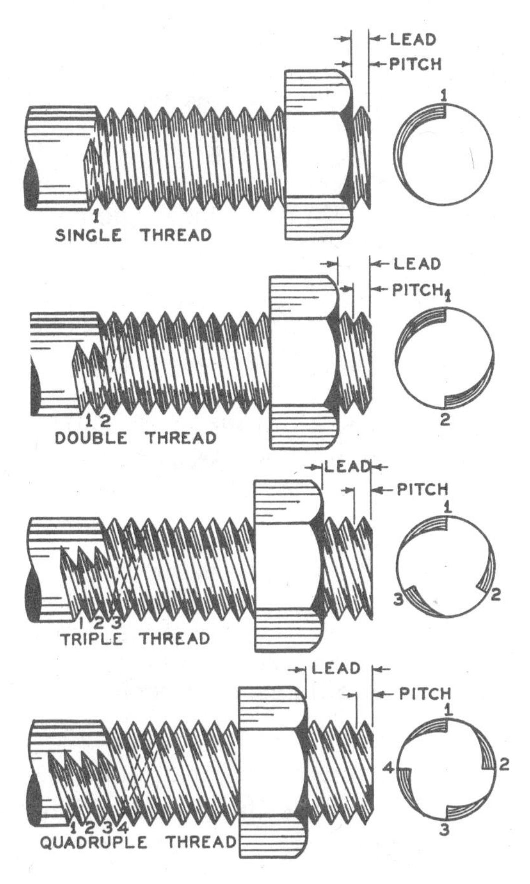

Fig. 21-3. Single, Double, Triple, and Quadruple Threads

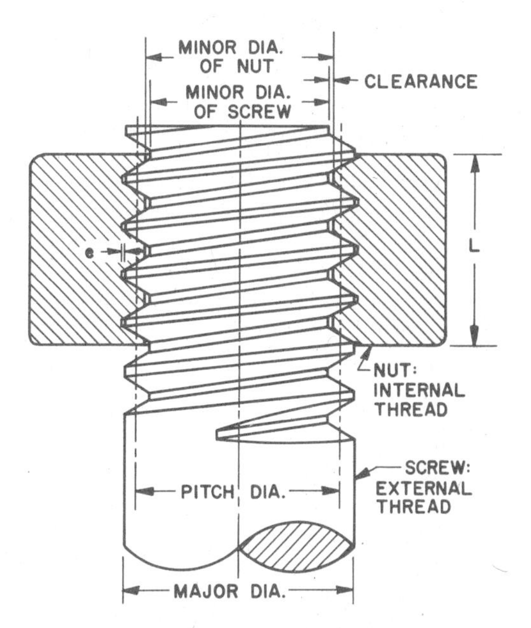

Fig. 21-5. Comparison between the Minor Diameters of a Screw and a Nut, Showing Clearance
External threads and internal threads have the same basic pitch diameters.

The major diameter is similar to the nominal (basic) size. The **nominal size** is used for general identification, such as identifying the thread size as ¼″ or ½″ diameter. The **basic size** is the size from which size limits (see § 46-6) are derived by applying **tolerances** (see § 46-7) and **allowances** (see § 46-8). Thus, the basic size of the thread on a ½″ diameter rod or bolt is ½″, and this is also the nominal size.

In modern production, tolerances and allowances are applied to the major diameter and other parts of screw threads. Therefore, the actual major diameter is usually a few thousandths of an inch smaller or larger than the basic size. The major diameter of an external thread is usually smaller. However, on an internal thread, the major diameter is usually larger than the basic size. Tolerances and allowances also apply to the minor diameter and the pitch diameter of screw threads. The tolerances and allowances for screw threads of various sizes are available in standard handbooks for machinists.

21-5. Minor Diameter of a Thread

The **minor diameter** was formerly known as the **root diameter (RD).** It is the smallest diameter of a straight external or internal thread, Figs. 21-4 and 21-5.

The bottom surface which joins the two sides of the thread groove is called the **root,** Fig. 21-4. The root of an external thread is at its minor diameter, Fig. 21-4. The root of an internal thread is at its major diameter, Fig. 21-5.

The top surface which joins the sides of a thread is called the **crest,** Fig. 21-4. The crest of an internal thread is at its minor diameter.

The minor diameter (**RD**) of a screw thread is equal to the **major diameter** (**OD**) minus the **double depth** (**DD**).

Thus: RD = OD − DD

21-6. Thread Depth

The **depth of thread** (**D**), which is also called **height,** is the vertical distance between the top (crest) and the bottom (root) of the thread groove. This is the **single depth of thread.**

21-7. Double Depth of Thread

The **depth of thread** multiplied by two equals the **double depth of thread** (**DD**).

Thus: D × 2 = DD

The double depth of thread must be known to find the **minor diameter** of the thread.

21-8. Pitch Diameter (PD)

On a perfect thread, the **pitch diameter** (**PD**) is an imaginary line which passes through the thread at a point where the width of the thread and the width of the groove are equal, Figs. 21-4 and 21-5. Through careful control of the pitch diameter when cutting threads, the proper fit or class of thread can be produced. The pitch diameter may be measured with a screw thread micrometer, Fig. 9-5.

The pitch diameter (**PD**) equals the **outside diameter** (**OD**) minus the **single depth** (**D**) of an external thread. See Fig. 21-4.

Thus: PD = OD − D

Thread **clearance** (also called **crest clearance**) is provided between internal and external mating threads, as shown in Fig. 21-5.

21-9. Finding Number of Threads Per Inch with a Steel Rule

The **number of threads per inch** on a bolt may be found by placing the edge of a steel rule on the threads as shown in Fig. 21-6. The 1″ mark should be directly above one of the threads. Then count the number of grooves or spaces in 1″; this number will be the **number of threads per inch.**

21-10. Finding Number of Threads Per Inch with a Screw-Pitch Gage

The quickest and most exact way to find the number of threads per inch on a bolt or

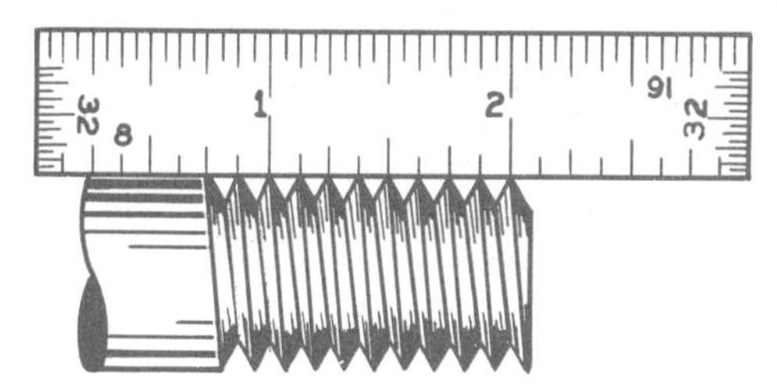

Fig. 21-6. Finding Number of Threads per Inch with a Steel Rule

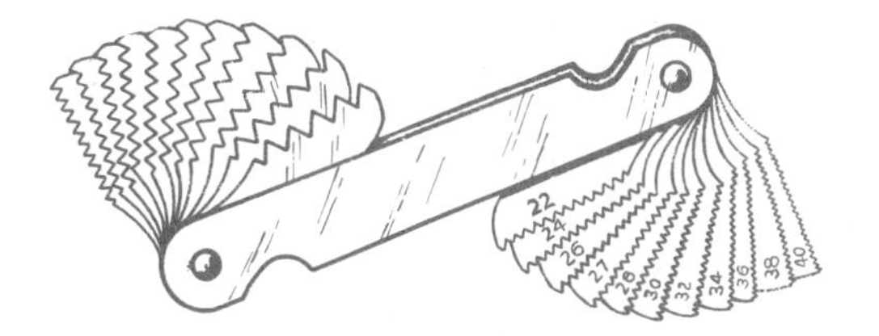

Fig. 21-7. Screw Pitch Gage

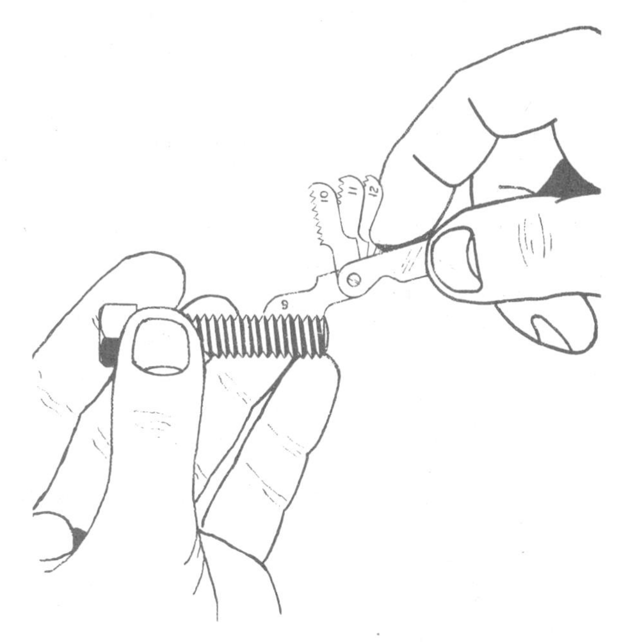

Fig. 21-8. Using a Screw Pitch Gage to Find the Number of Threads per Inch

a nut is with a **screw pitch gage,** Fig. 21-7. Try different **blades** of the gage on the threads until you find one that fits, Fig. 21-8. The number stamped on the blade is the **number of threads per inch.**

21-11. Pitch of Thread

The pitch (**P**) of the thread is the distance from a point on one thread to a corresponding point on the next thread. See Figs. 21-4 and 21-10. The **pitch** equals 1″ divided by the number of threads per inch. For example, a screw having 8 threads per inch has a pitch of ⅛″; that is, 1″ ÷ 8 = ⅛″. This means that the screw has threads that are ⅛″ apart from center to center.

21-12. Lead of Thread

The **lead** of a thread (pronounced to rhyme with bead) is the distance the screw moves into the nut in one complete turn, Fig. 21-3. Thus, on a **single thread** the lead is the same as the **pitch;** on a **double thread** the lead equals twice the pitch; and on a **triple thread** the lead equals three times the pitch, etc.

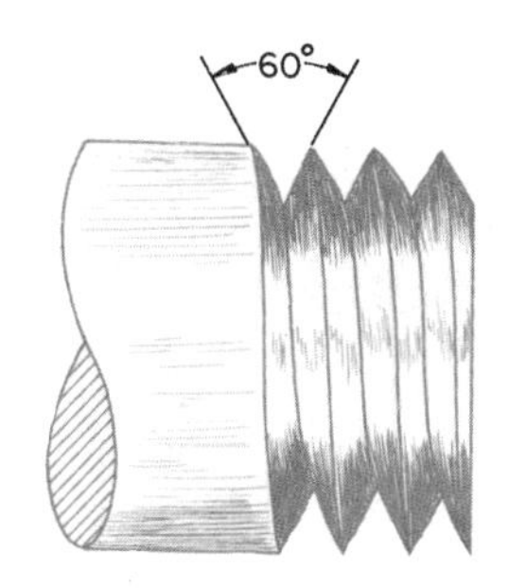

Fig. 21-9. Sharp V-Thread

21-13. Sharp V-Thread

The simplest form of thread is the sharp V-thread. See Fig. 21-9. It is necessary to know about this thread because other threads are formed from it. Its form is like an **equilateral triangle;** that is, a triangle with three equal sides and three equal angles. The sides of the V-thread are V-shaped and form a 60° angle.

The sharp edges of this thread are easily nicked and wear off quickly; for this reason it is not commonly used. It is, however, used where tight fits are necessary, such as the threads on a pipe.

21-14. Development of Screw Threads in the United States

The cross-sectional shape of a screw thread is called its profile or form. To overcome the objections of the sharp V-thread form, Fig. 21-9, the United States Standard Form Thread was developed, Fig. 21-10. It was developed during the 1860's, was first adopted by the United States Navy in 1868, and later became widely adopted by American industries. Some manufacturers, however, continued to make threads according to their own systems. Often, bolts made by one manufacturer would not fit nuts made by another.

During World War I the need arose for more standardization of screw threads. Congress, in 1918, established the National Screw Thread Commission to study the problem and adopt new screw thread standards for American industries and government services. Its aim was to eliminate unnecessary thread sizes and to use existing sizes as much as possible.

The efforts of the committee and other cooperating agencies produced the American National Screw Thread Standard in 1924. The thread profile was named the American National Form Thread and was substantially the same as the United States Standard Form Thread, Fig. 21-10. Another early thread standard, the Society of Automotive Engineers (SAE) Thread, adopted about 1911, was incorporated (with modifications) into the new thread standard and became known as the National Fine Thread. The SAE Extra-Fine Thread Series was added (with modifications) to the American National Screw Thread Standard approved in 1933.

In 1948, the United States, Canada, and Great Britain agreed to adopt the Unified Screw Thread to provide interchangeability of threaded parts in these countries. American industry has now virtually completed the transition from use of the American National threads to use of Unified threads.

21-15. Unified Screw Threads

The Unified Screw Thread is essentially the same as the American National Form Thread except that it generally has rounded roots and may have either rounded or flat crests, as in Fig. 21-11. American industries use flat crests, while the English prefer rounded crests. The rounded root is optional on both the external and the internal threads for most applications; however, for some applications the root must be rounded.

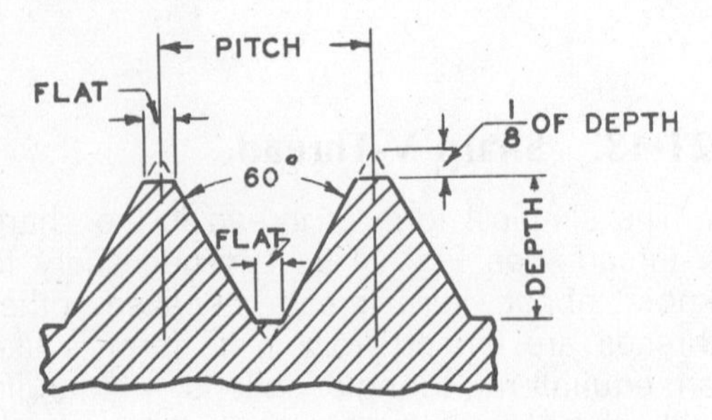

Fig. 21-10. External American National Form Thread (Formerly United States Standard Form Thread)

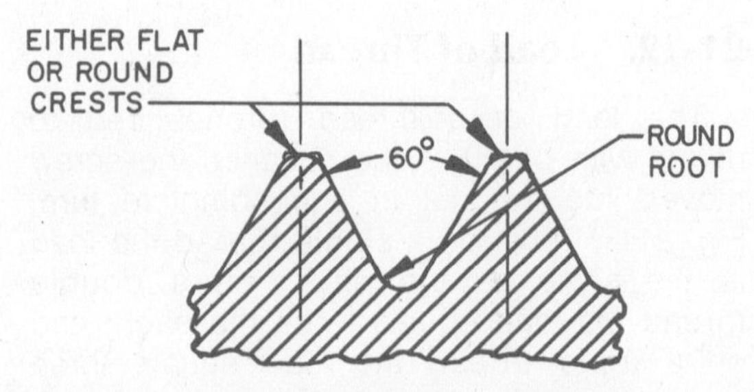

Fig. 21-11. Unified Screw Thread Form

The calculations for an external **Unified Form** Thread may be made with the following formulas (See Figs. 21-10 and 21-11.):

$$\text{Pitch} = P = \frac{1}{\text{number of threads per inch}}$$

$$\text{Depth} = D = 0.61343 \times P$$

$$D = \frac{0.61343}{\text{number of threads per inch}}$$

$$\text{Flat at crest} = F = \frac{P}{8}$$

The depth of the external Unified Form Thread is slightly less than the depth for the American National Form Thread. However, these two kinds of threads are essentially the same, and permissible production tolerances permit them to be used interchangeably in most instances.

The depth of the internal Unified Form Thread is equal to 0.51427 × P, or 0.51427 divided by the number of threads per inch.

The Unified Screw Thread System includes several series of threads. The following common screw thread series are included in the American Standard, **Unified Screw Threads:**[2]

1. Unified National Coarse Thread (UNC): adopted from the NC thread. (See Table 17.)
2. Unified National Fine Thread (UNF): adopted from the NF thread. (See Table 17.)
3. Unified National Extra-Fine Thread (UNEF): adopted from the NEF thread. (See Table 17.)
4. Other less-common series, such as the 8-thread series (8UN), the 12-thread series (12UN), and the 16-thread series (16UN) are also included.

The **UNC, UNF,** and **UNEF** thread series are interchangeable with the **NC, NF,** and **NEF** series on most of the common sizes.

[2]**Unified and American Screw Threads** (ASA B1.1-1949) and later edition, **Unified Screw Threads** (ASA B1.1-1960), published by the American Society of Mechanical Engineers, New York.

TABLE 17

Screw Threads

Diameter			Threads Per Inch		
No.	Inch	Decimal Equivalent	UNC (NC) (USS)	UNF (NF) (SAE)	UNEF (NEF) (EF)
0		.0600		80	
1		.0730	64	72	
2		.0860	56	64	
3		.0990	48	56	
4		.1120	40	48	
5	1/8	.1250	40	44	
6		.1380	32	40	
8		.1640	32	36	
10		.1900	24	32	
12		.2160	24	28	32
	1/4	.2500	20	28	32
	5/16	.3125	18	24	32
	3/8	.3750	16	24	32
	7/16	.4375	14	20	28
	1/2	.5000	13	20	28
	9/16	.5625	12	18	24
	5/8	.6250	11	18	24
	11/16				24
	3/4	.7500	10	16	20
	13/16				20
	7/8	.8750	9	14	20
	15/16				20
	1	1.0000	8	14	20

21-16. Unified Thread Classes (Fits)

Six **classes of threads,** formerly called **fits,** are used with the standard Unified screw threads. The term **fits** has been dropped. The six classes include three external classes and three internal classes. The external thread classes are designated 1A, 2A, and 3A. The internal classes are designated 1B, 2B, and 3B. Thread fit means the degree of tightness between mating threads. Normally, class 2A and class 2B threads are mated. However, any unified class of external thread may be mated with any internal class, as long as the product meets the specified fit requirements. The following classes of mated threads have tolerances which permit the indicated type of fit.

Classes 1A and 1B: loose fit
Classes 2A and 2B: free fit
Classes 3A and 3B: close fit

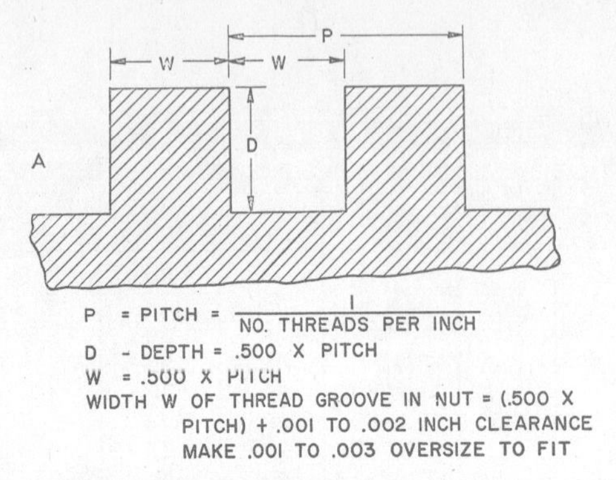

Fig. 21-12. Square Thread

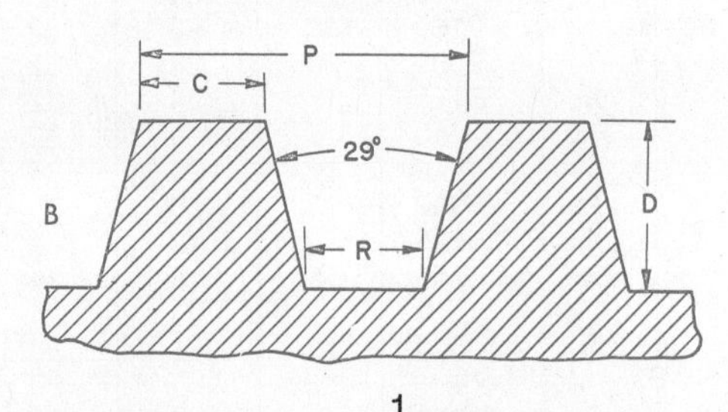

$$P = \text{Pitch} = \frac{1}{\text{No. Threads Per Inch}}$$

D = Depth = 1/2 P

C = Flat on top of thread = .3707P − .259 × (pitch diameter allowance on external thread)

R = Flat on bottom = .3707P − .259 × (minor diameter allowance on external thread − pitch diameter allowance on external thread)

Fig. 21-13. Acme Thread

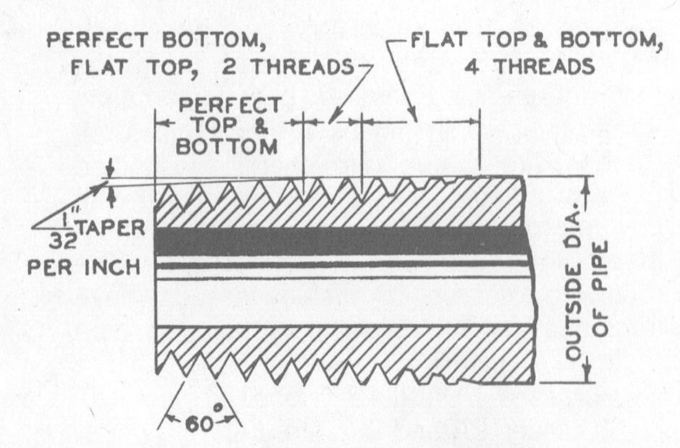

Fig. 21-14. Taper Pipe Threads (NPT)

When classes 2A and 3B are mated, the tolerances permit an intermediate fit ranking between the free and close fits.

21-17. Square Thread

A square thread is formed like a square, Fig. 21-12. The depth and width of the groove are equal. The height and width of each ridge between the grooves are also equal. Thus, the groove and ridge form two squares. The screw on a **machinist's vise** sometimes has square threads. There is a trend, at present, to use the acme thread instead of the square thread.

21-18. Acme Thread

Acme threads are usually used on the **lead screw** of a **lathe.** They are also used on many other kinds of machine tools. The angle of the thread is 29°, Fig. 21-13. Information concerning pitches, fits, tolerances, and allowances for acme threads is available in handbooks for machinists.

21-19. Pipe Threads

American National Standard pipe threads are used for assembling pipes and pipe fittings. Four types of pipe threads are used:

American National Standard Taper Pipe Threads (NPT)

American National Standard Railing Joint Taper Pipe Threads (NPTR)

American National Standard Straight Pipe Threads (NPS)

American National Standard Dryseal Taper Pipe Threads (NPTF)

The NPT thread is commonly used for general purpose pipe-fitting jobs requiring a low-pressure seal against liquid or gas leakage. Use of a pipe compound or sealant is normally required for this type thread. The NPT threads are tapered ¾″ per foot on the diameter, which equals ¹⁄₁₆″ per inch, Fig. 21-14. The angle between the sides of the thread is 60°. Several threads on the end are perfect; the next two threads have per-

TABLE 18

Pipe Dimensions

Pipe Diameters				
Nominal Size	Actual Inside	Actual Outside	Threads Per Inch	Tap Drill Size
1/8	0.270	0.405	27	11/32
1/4	0.364	0.540	18	7/16
3/8	0.494	0.675	18	19/32
1/2	0.623	0.840	14	23/32
3/4	0.824	1.050	14	15/16
1	1.048	1.315	11 1/2	1 5/32
1 1/4	1.380	1.660	11 1/2	1 1/2
1 1/2	1.610	1.900	11 1/2	1 23/32
2	2.067	2.375	11 1/2	2 3/16
2 1/2	2.468	2.875	8	2 5/8

fect bottoms but flat tops; the last four threads have flat tops and bottoms. The farther the tapered threads are screwed together, the tighter the joint becomes. Table 18 provides basic data for different sizes of pipe.

The NPTR thread provides a rigid, mechanical, tapered thread joint for use in railing construction.

There are several types of NPS threads:

1. The NPSC thread is used in pipe couplings.
2. The NPSM thread is designed for use in free-fitting mechanical joints.
3. The NPSL is for use in loose-fitting mechanical joints with locknuts.
4. The NPSH is for loose-fitting mechanical joints in hose couplings.

Dryseal threads are used in applications which require a gas or liquid tight thread joint without the use of pipe compound or sealant. The four forms of Dryseal threads are listed below:

NPTF — Dryseal Standard Pipe Thread

PTF-SAE Short — Dryseal SAE Short Taper Pipe Thread

NPSF — Dryseal Standard Fuel Internal Straight Pipe Thread

NPSI — Dryseal Standard Intermediate Internal Straight Pipe Thread

Further information concerning the different types of pipe threads is available in handbooks for machinists.

Words to Know

acme thread
depth of thread
double depth of thread
double thread
Dryseal thread
extra-fine thread
lead of thread
left-hand thread
major diameter
minor diameter
multiple thread
nominal size
pipe compound
pitch diameter
pitch of thread
quadruple thread
right-hand thread
root diameter
screw-pitch gage
sharp V-thread
single thread
square thread
tapered
triple thread
Unified Form thread

Review Questions

1. Are most screw threads right-hand or left-hand?
2. How can you tell whether the thread on a bolt is right-hand or left-hand?
3. What is a double thread? Triple thread? Quadruple thread? Multiple thread?
4. What is meant by major diameter?
5. What is meant by minor diameter?
6. What is the pitch diameter?
7. What is the pitch of a thread?
8. What is the lead of a thread? What is the difference between lead and pitch?
9. Describe two ways to measure the number of threads per inch.
10. Describe a V-thread.
11. Describe the Unified Screw Thread System.
12. Describe the six classes of threads used with the Unified screw thread.
13. Describe a square thread and give an example of where it is used.
14. Describe an acme thread and give an example of where it is used.
15. List the four types of pipe threads. Which is used for general purpose plumbing?
16. How do NPT threads differ from Unified screw threads?
17. Where can additional information be obtained on screw threads?

Mathematics

1. What is the minor diameter of a 1″ — 8 UNC thread?
2. What is the minor diameter of a 1″ — 14 UNF thread?
3. What is the pitch diameter of a ¾″ — 10 UNC thread?

Drafting

1. How are threads represented on working drawings?
2. Make a drawing of a standard form of screw thread and label its parts.

UNIT 22

Threading Dies

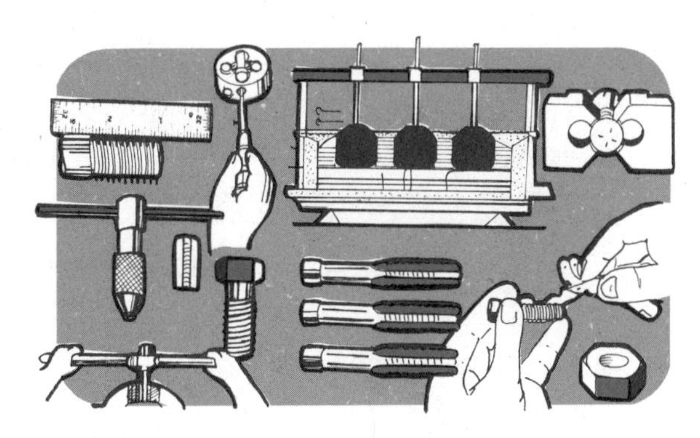

22-1. What Is a Threading Die?

A threading die is a round or square block of **hardened steel** with a hole containing **threads** and **flutes** which form **cutting edges.** It is used to cut threads on a round bar of metal, such as the threads on a bolt.

One kind of threading die is shown in Fig. 22-1. It is called a **round, adjustable, split die. Adjustable** means that it can be set to cut larger or smaller. Another kind is the **two-piece die,** Fig. 22-2. It is made in two halves which match each other. **Dies and taps** (see § 23-1) may be bought in a set called a **screw plate,** Fig. 22-3.

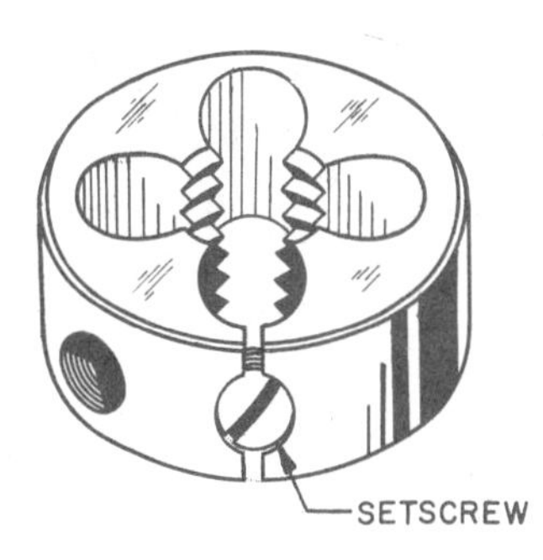

Fig. 22-1. Round-Adjustable Split-Threading Die

22-2. Left-Hand and Right-Hand Dies

A left-hand die cuts **left-hand threads.** It has the letter L stamped on it. If the L is not on the die, it is a **right-hand die.**

22-3. Sizes of Threading Dies

The size of the threading die and the number of threads per inch are stamped on the die, as ¼-20 which means that the die will cut a thread with a **major diameter** of ¼" and has 20 threads per inch. The sizes of screw threads as given in Table 17, page 177, are also the sizes of threading dies.

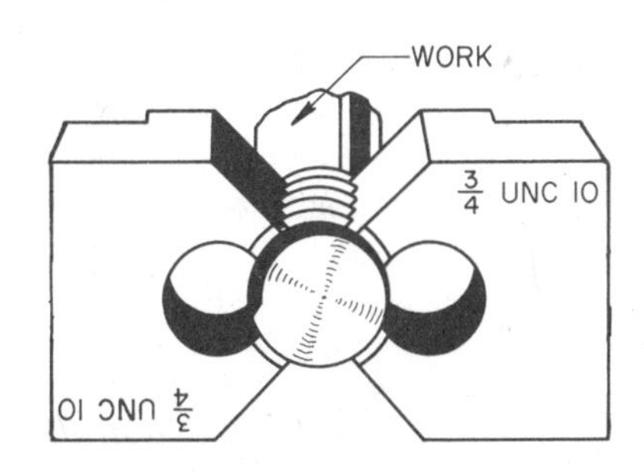

Fig. 22-2. Two-Piece Threading Die

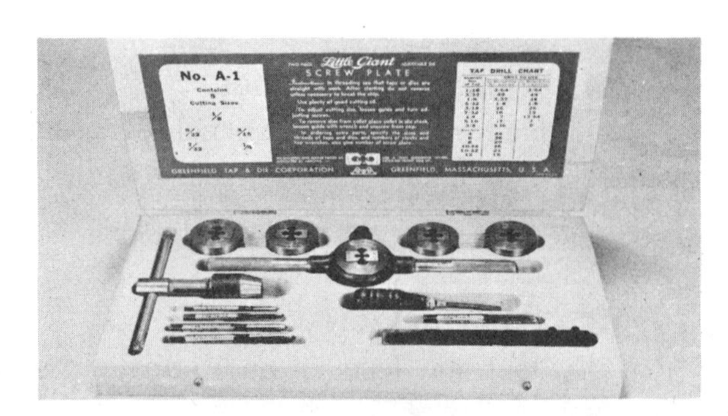

Fig. 22-3. Screw Plate (Set of Taps and Dies)

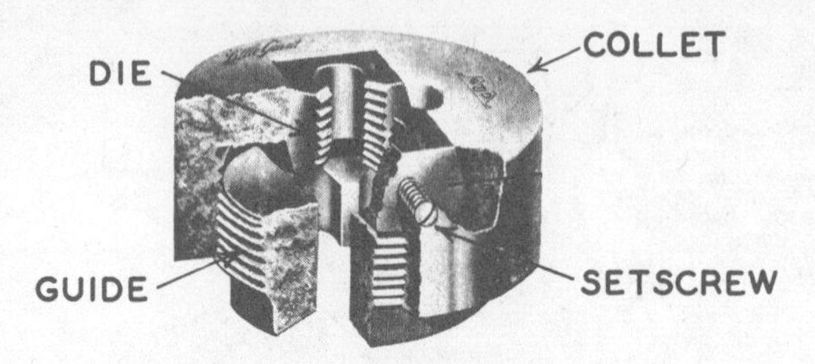

Fig. 22-4. Die, Collet, and Guide

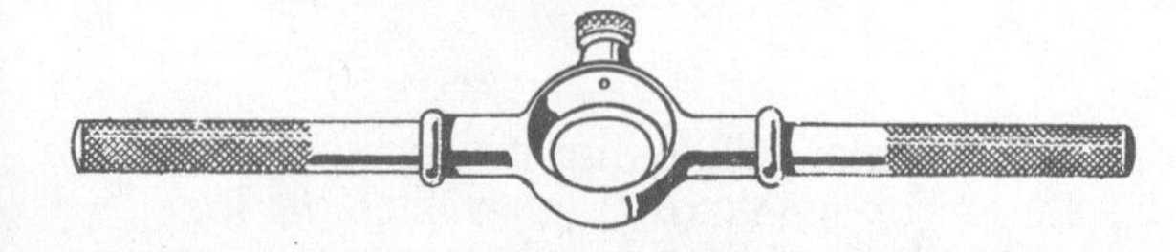

Fig. 22-5. Diestock

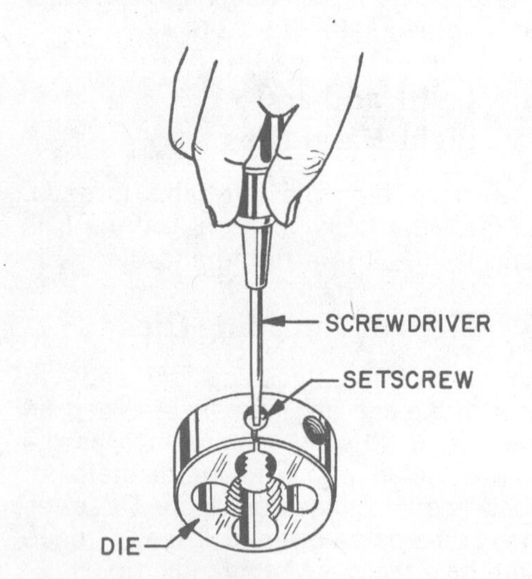

Fig. 22-6. Setting a Die

22-4. Collets and Guides

The threading dies fit into a holder called a **collet**, Fig. 22-4. The **guide** is a ring which is held in place under the die and fits around the bar or bolt to be threaded. It guides the die so that it will go on the work squarely.

22-5. Diestock

The tool for holding and turning the threading die and collet is called a **diestock.** It is often just called a **stock,** Fig. 22-5.

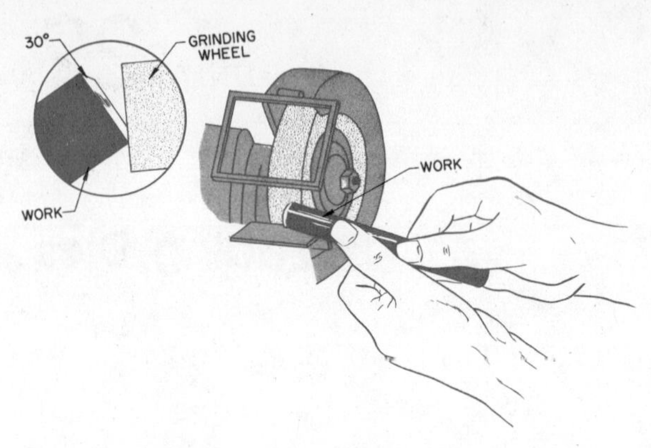

Fig. 22-7. Beveling the End of Work Before Threading

22-6. Setting the Threading Die

Most dies can be set to cut the thread a little oversize or undersize. This is done by turning the **setscrew** with a screwdriver, Fig. 22-6; the setscrew is sometimes on the top instead of at the side of the die.

22-7. Threading with Stock and Die

Cutting threads on a round rod or bolt with a die and stock is called **threading.** The end of the work should be beveled[1] to make starting easier; this may be quickly done on a grinder, Fig. 22-7. Hold the work to be threaded upright in the vise. Fasten the die in the diestock. The threads are beveled a little on one side of the die (Fig. 22-8) to make starting easier and to form the thread gradually. Always start cutting a thread with this beveled side.

Place the die over the end of the work. Grasp the diestock with both hands near the die (Fig. 22-9), press down firmly upon the work, and at the same time, slowly screw it on the work **clockwise.** The die cuts the thread as it goes. Be sure that the die goes on squarely; much skill is needed to do this. After the thread is started, grasp the two diestock handles and with a steady movement, continue screwing the die on the work, Fig. 22-10. Then it is no longer necessary to press down because the die will draw itself on the work when turned.

[1]**Beveled** means tapered, sloped, slanted, inclined.

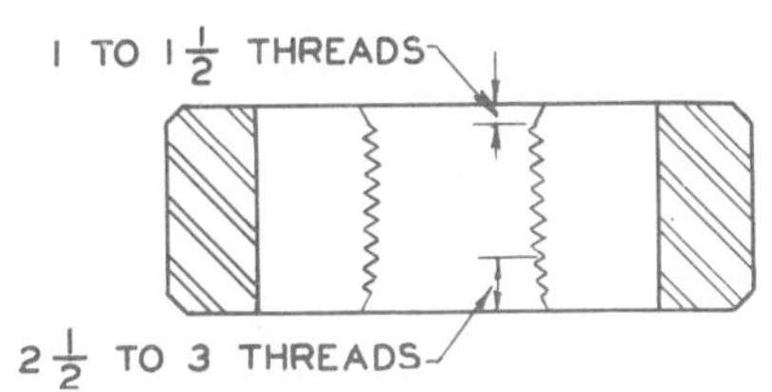

Fig. 22-8. Bevel on Threading Die

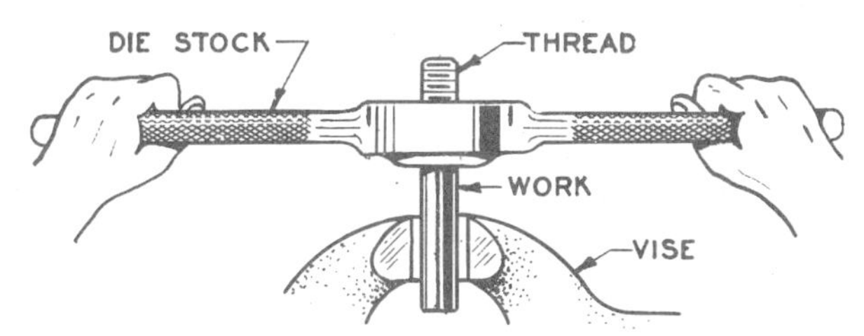

Fig. 22-10. Cutting the Thread with the Stock and Die

Back up the die every now and then to break and clean away the chips and to make the threads smooth. "Back up one step and go ahead two." Continue in this way until you have the length of thread you want. Clean the thread, die, and other parts when the threading is done.

22-8. Cutting Fluids for Threading

Use **lard oil** or **mineral-lard oil** when threading steel and only enough to keep the work moist. Oil which runs all over the work, the bench, and the floor is wasted. A cutting fluid is not necessary when threading cast iron. Cutting fluid recommendations for other metals are given in Table 33, page 452.

22-9. External Thread Measurements

External threads must be cut to the correct depth to produce the proper **fit** or **class of thread.** If the pitch diameter is too large, the thread fits too tightly. If it is too small, it fits too loosely. The exact pitch diameter may be measured with a thread micrometer to determine whether it is within specified size limits for the fit or class of thread desired. Tables which show the maximum and minimum pitch diameter dimensions for various fits and classes of screw threads are included in standard handbooks for machinists

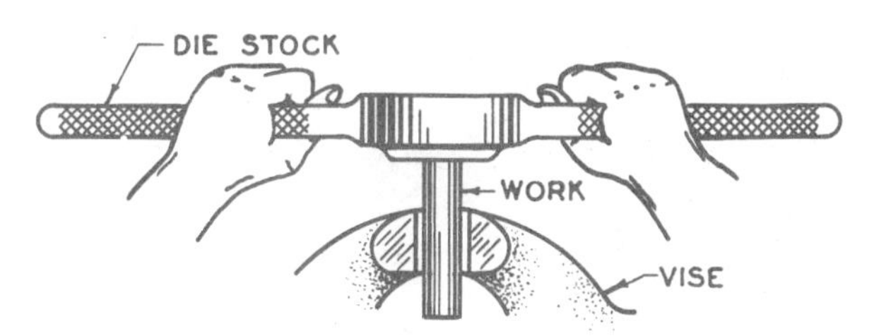

Fig. 22-9. Starting to Cut a Thread with a Stock and Die

The correct fit or class of screw threads can also be determined in the following ways:

1. With a thread roll snap gage, Fig. 46-20. (See § 46-17.)
2. With a thread ring gage, Figs. 46-18 and 46-19. (See § 46-17.)
3. By testing how it fits in a standard mating nut or threaded hole.

Words to Know

bevel	round, adjustable, split die
collet	screw plate
cutting fluid	thread micrometer
diestock	thread ring gage
left-hand die	thread roll snap gage
right-hand die	two-piece die

Review Questions

1. What is a threading die?
2. Describe a round, adjustable, split die.
3. What is a two-piece die?
4. What is a screw plate?
5. What does ¼-20 on a die mean?
6. What is a diestock?
7. Why should the end of a rod be beveled before threading?
8. Why is it necessary to back up the die when threading?
9. What kind of cutting fluid should be used when threading steel?
10. List several ways in which the fit or class of thread can be determined.

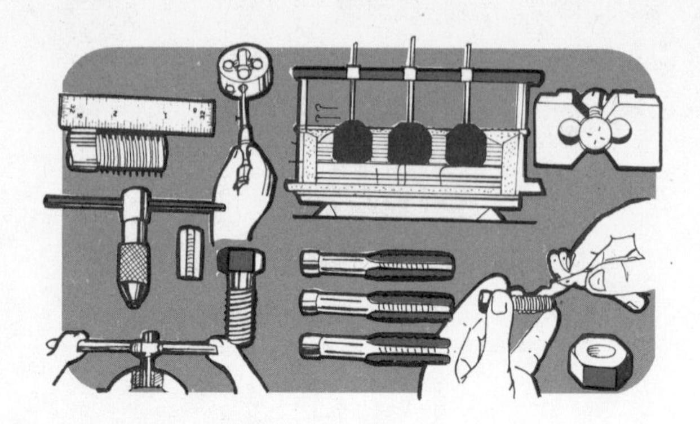

UNIT 23

Threading with Taps

23-1. Taps

A tap is a screwlike tool which has threads like a bolt and three or four **flutes** cut across the threads, Fig. 23-1. It is used to cut threads on the inside of a hole, as in a nut. The edges of the thread formed by the flutes are the **cutting edges,** Fig. 23-2. One end of the tap is square so that it can be turned with a wrench.

Taps are made from **carbon steel** or **high-speed steel** and are **hardened** and **tempered.** A set of taps includes a taper tap, a plug tap, and a bottoming tap, Fig. 23-1.

The end of the **taper tap** has about six threads **tapered** so that it will start easily. The threads are cut gradually as the tap is turned into the hole. The taper makes it easier to keep the tap straight.

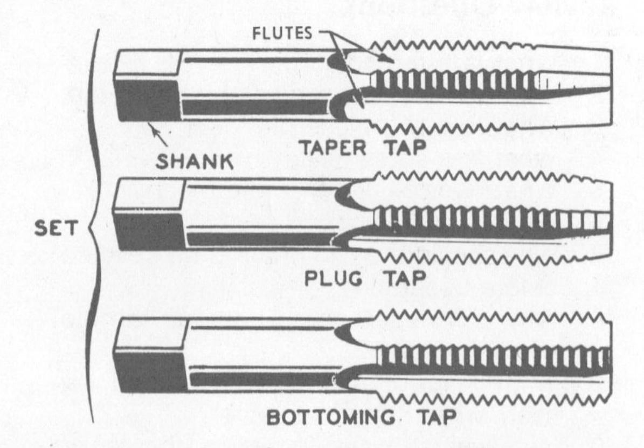

Fig. 23-1. Set of Taps

The **plug tap** has three or four threads tapered at the end and is used after the taper tap.

The **bottoming tap** has a full thread to the end and is used to cut a full thread to the bottom of a hole.

23-2. Tap Styles

Taps are made in a variety of styles in order to suit various hand- and machine-tapping operations. **Hand taps** are straight fluted, have three or four flutes, and have a cutting edge parallel to the tap centerline, Fig. 23-1. Chips tend to collect in the flutes of these taps. Unless the tap is backed out to clear it of chips, the tap may bind in the hole and break. Several types of machine taps are designed for efficient chip removal, thereby permitting rapid thread cutting with a minimum of tap breakage.

Gun taps are straight fluted, with two, three, or four flutes depending on the size of

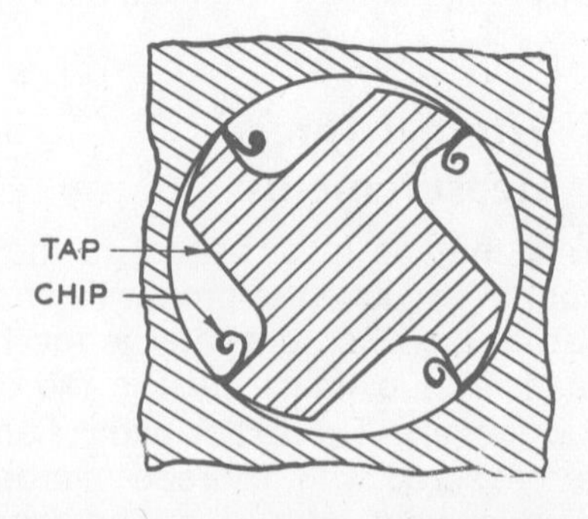

Fig. 23-2. How the Cutting Edges of a Tap Cut

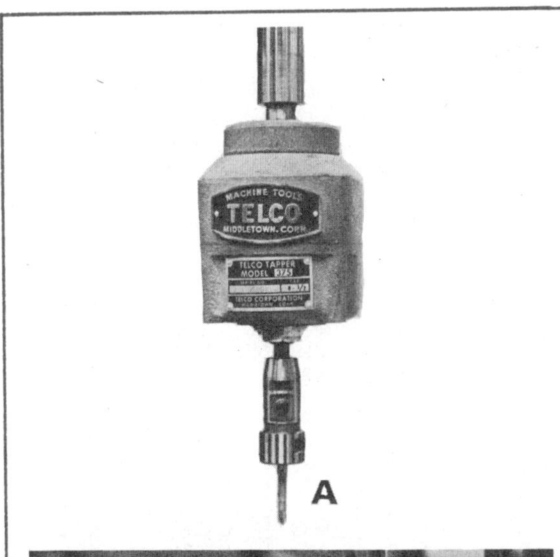

Fig. 23-3. (A) Plug Gun Tap; (B) The Tap in Action

the tap. The cutting edges are ground at an angle to the tap centerline. The angular cutting edges cause the chips to shoot ahead of the tap, Fig. 23-3, A and B. Plug-type gun taps are designed for tapping open or

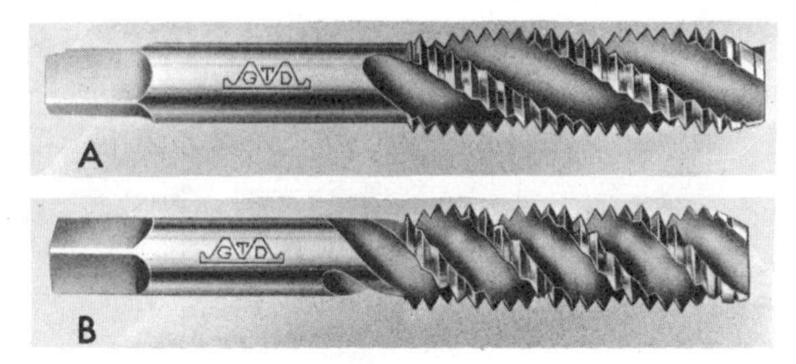

Fig. 23-4. Spiral Fluted Taps
(A) Low-Angle; (B) High-Angle

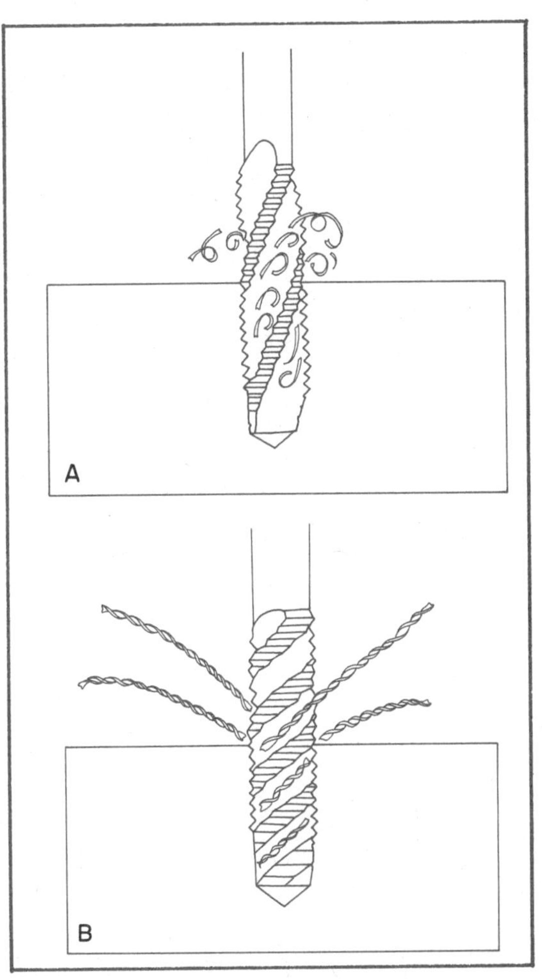

Fig. 23-5. Cutting Action of Taps (A) Low-Angle Spiral Fluted Tap; (B) High-Angle Spiral Fluted Tap

TABLE 19

Sizes of Taps, Tap Drills[1], and Clearance Drills[2]

Size of Tap		Outside Diameter (Inches)	Root Diameter (Inches)	Size of Tap Drill (In Inches) For 75% Thread Depth			Clearance Drill (Inches)		Clearance (Inches)
UNC NC (USS)	UNF NF (SAE)			Number and Letter Drills	Fractional Drills	Decimal Equivalent	Size	Decimal Equivalent	
	#0-80	0.0600	0.0438	. . .	3/64	0.0469	#51	0.0670	0.0070
#1-64	. . .	0.0730	0.0527	53	. . .	0.0595	#47	0.0785	0.0055
	#1-72	0.0730	0.0550	53	. . .	0.0595	#47	0.0785	0.0055
#2-56	. . .	0.0860	0.0628	50	. . .	0.0700	#42	0.0935	0.0075
	#2-64	0.0860	0.0657	50	. . .	0.0700	#42	0.0935	0.0075
#3-48	. . .	0.0990	0.0719	47	. . .	0.0785	#36	0.1065	0.0075
	#3-56	0.0990	0.0758	45	. . .	0.0820	#36	0.1065	0.0075
#4-40	. . .	0.1120	0.0795	43	. . .	0.0890	#31	0.1200	0.0080
	#4-48	0.1120	0.0849	42	. . .	0.0935	#31	0.1200	0.0080
#5-40	. . .	0.1250	0.0925	38	. . .	0.1015	#29	0.1360	0.0110
	#5-44	0.1250	0.0955	37	. . .	0.1040	#29	0.1360	0.0110
#6-32	. . .	0.1380	0.0974	36	. . .	0.1065	#25	0.1495	0.0115
	#6-40	0.1380	0.1055	33	. . .	0.1130	#25	0.1495	0.0115
#8-32	. . .	0.1640	0.1234	29	. . .	0.1360	#16	0.1770	0.0130
	#8-36	0.1640	0.1279	29	. . .	0.1360	#16	0.1770	0.0130
#10-24	. . .	0.1900	0.1359	25	. . .	0.1495	13/64	0.2031	0.0131
	#10-32	0.1900	0.1494	21	. . .	0.1590	13/64	0.2031	0.0131
#12-24	. . .	0.2160	0.1619	16	. . .	0.1770	7/32	0.2187	0.0027
	#12-28	0.2160	0.1696	14	. . .	0.1820	7/32	0.2187	0.0027
1/4"-20	. . .	0.2500	0.1850	7	. . .	0.2010	17/64	0.2656	0.0156
	1/4"-28	0.2500	0.2036	3	. . .	0.2130	17/64	0.2656	0.0156
5/16"-18	. . .	0.3125	0.2403	F	. . .	0.2570	21/64	0.3281	0.0156
	5/16"-24	0.3125	0.2584	I	. . .	0.2720	21/64	0.3281	0.0156
3/8"-16	. . .	0.3750	0.2938	. . .	5/16	0.3125	25/64	0.3906	0.0156
	3/8"-24	0.3750	0.3209	Q	. . .	0.3320	25/64	0.3906	0.0156
7/16"-14	. . .	0.4375	0.3447	U	. . .	0.3680	29/64	0.4531	0.0156
	7/16"-20	0.4375	0.3725	. . .	25/64	0.3906	29/64	0.4531	0.0156
1/2"-13	. . .	0.5000	0.4001	. . .	27/64	0.4219	33/64	0.5156	0.0156
	1/2"-20	0.5000	0.4350	. . .	29/64	0.4531	33/64	0.5156	0.0156
9/16"-12	. . .	0.5625	0.4542	. . .	31/64	0.4844	37/64	0.5781	0.0156
	9/16"-18	0.5625	0.4903	. . .	33/64	0.5156	37/64	0.5781	0.0156
5/8"-11	. . .	0.6250	0.5069	. . .	17/32	0.5312	41/64	0.6406	0.0156
	5/8"-18	0.6250	0.5528	. . .	37/64	0.5781	41/64	0.6406	0.0156
3/4"-10	. . .	0.7500	0.6201	. . .	21/32	0.6562	49/64	0.7656	0.0156
	3/4"-16	0.7500	0.6688	. . .	11/16	0.6875	49/64	0.7656	0.0156
7/8"- 9	. . .	0.8750	0.7307	. . .	49/64	0.7656	57/64	0.8906	0.0156
	7/8"-14	0.8750	0.7822	. . .	13/16	0.8125	57/64	0.8906	0.0156
1"- 8	. . .	1.0000	0.8376	. . .	7/8	0.8750	1 1/64	1.0156	0.0156
	1"-14	1.0000	0.9072	. . .	15/16	0.9375	1 1/64	1.0156	0.0156

[1]If you cannot get the size of tap drill given here, see Table 34, "Drill Sizes," on page xxx to find the size of drill nearest to it; be sure to get a drill a little larger than the **root diameter** of the thread. (See Section 21-7.)

[2]The drill that makes a hole so that a bolt or screw may pass through it is called a **clearance drill.** This drill makes a hole with a **clearance** for the **nominal diameter of thread** (see Section 21-4). The **clearance** equals the difference between the clearance drill size and the nominal diameter of the bolt or screw:

Clearance drill = Diameter of bolt or screw + Clearance

Example: The clearance for a 1/4" bolt or screw is 1/64"; the size of the clearance drill should, therefore, be 1/4" + 1/64" or 17/64".

through holes, while bottom-type gun taps are designed for tapping blind holes, producing fine chips which can readily escape.

Helical fluted taps, commonly known as **spiral fluted taps,** are designed to efficiently lift the chips out of the hole being tapped. For this reason, they are well suited for tapping blind holes, Fig. 23-4. Low-angle spiral fluted taps are best for tapping ductile materials like aluminum, copper, or die cast metals. High-angle spiral fluted taps work best on tough metals, such as carbon and alloy steels, Fig. 23-5. They are made with two, three, or four flutes, depending on the tap diameter and are available in plug and bottoming types.

Serial taps are usually made in sets of three. They have one, two, or three identifying rings cut at the end of the shank, Fig. 23-6. These taps are designed for cutting threads in very tough metals. The taps are similar in appearance to the taper, plug, and bottoming taps, but they differ in pitch diameter and major diameter. Each tap is designed to remove part of the metal which must be cut away to produce the thread. The No. 1 tap is used first to make a shallow thread, then the No. 2 tap is used, and the No. 3 tap cuts the thread to final size.

Cam ground taps, Fig. 23-7, have no cutting edges and therefore produce no chips. Threads are formed by forcing the metal to flow around the threads on the tap. This produces a strong thread, because the grain of the metal is forced to follow the thread profile and the surface is somewhat work-hardened. Because of the high pressures involved, thread depths are less than those produced by conventional taps. Tap drills must be larger than comparable conventional taps. The manufacturer's tap drill size recommendations should be carefully followed.

23-3. Left-Hand and Right-Hand Taps

A left-hand tap cuts **left-hand threads.** It has the letter L stamped on the shank. If there is no L, it is a **right-hand tap.**

23-4. Sizes of Taps

Taps are made the same sizes as bolts and screws. The size (**outside diameter**) of the tap and the **number of threads per inch** are stamped on the **shank** of the tap. For example, ¼-20 means that the outside diameter of the tap is ¼" and that there are 20 threads per inch. Table 19 gives the sizes of taps.

23-5. Tap Drills

The hole to be tapped must be the right size. If the hole is too large, a **full thread** will not be formed; if too small, the tap may break. The drill which is used to make a hole before tapping is called the **tap drill.** Note that this hole must be smaller than the **major diameter** of the tap so as to leave

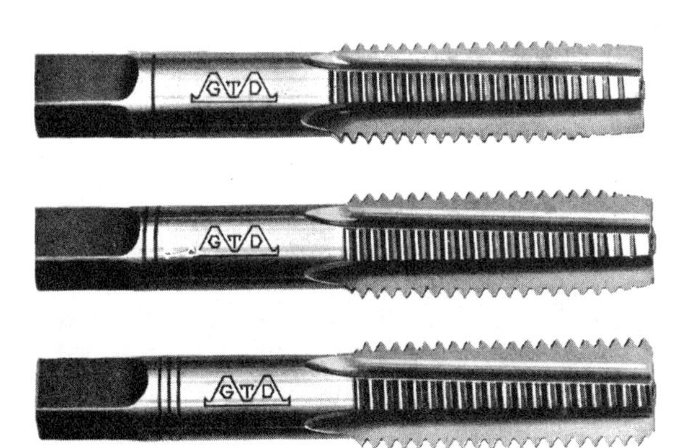

Fig. 23-6. Serial Taps

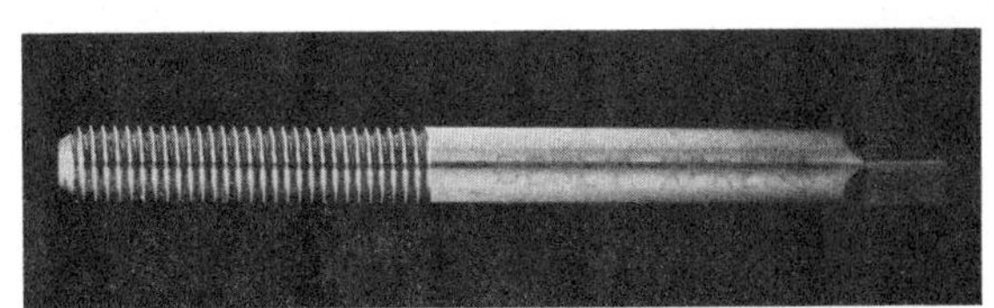

Fig. 23-7. Cam Ground Tap

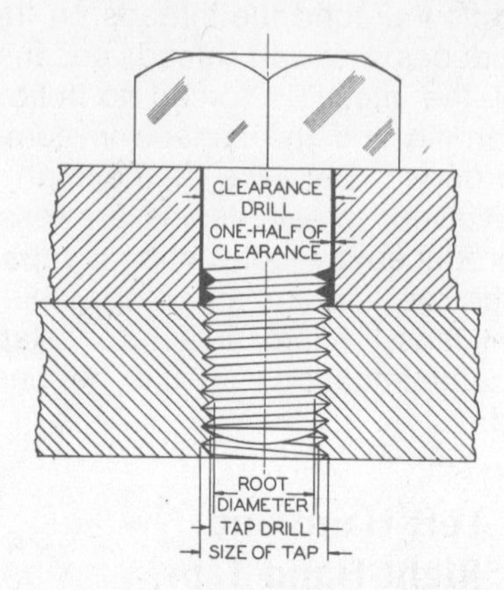

Fig. 23-8. Comparing Sizes of Tap, Tap Drill, and Root (Minor) Diameter of Thread

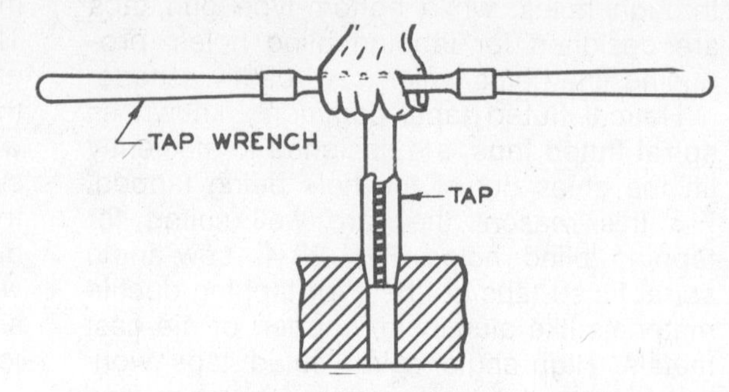

Fig. 23-10. Start the Threads with a Taper Tap

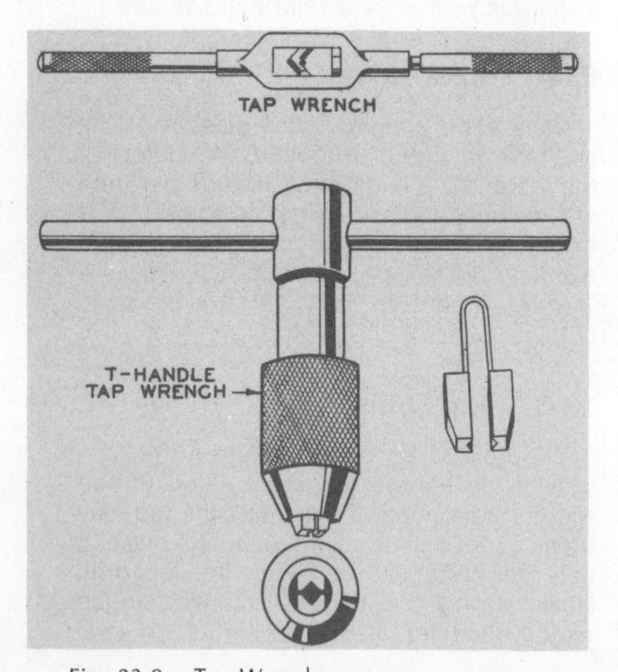

Fig. 23-9. Tap Wrenches

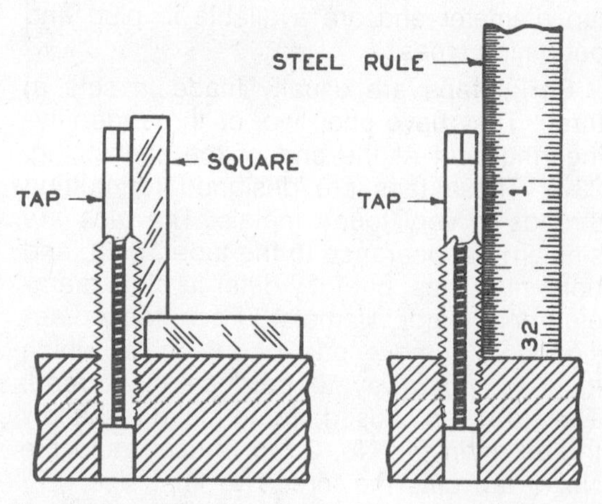

Fig. 23-11. The Tap Must Be Square with the Work

enough metal from which to form the threads, Fig. 23-8.

The right size tap drill must be used. Its size should be a little larger than the diameter at the bottom or **root** of the thread, known as the **minor diameter.** To save the time of figuring, **tap drill sizes** are put in the form of a table such as Table 19. The tap drill, recommended for average work, produces threads which are about 75% of the depth of external threads.

23-6. Tapping

Cutting **inside threads** as in a **nut** is called **tapping.** After the hole has been drilled with the **tap drill,** it is ready for tapping — that is, the cutting of the threads by the tap.

Clamp the work in the vise with the hole in an upright position. First use a **taper tap.**

If a taper tap is not available, use a plug tap. Clamp its square end in the **tap wrench,** Fig. 23-9. The **T-handle tap wrench** is used for holding small taps. Grasp the tap wrench with the right hand directly over the tap and place the tap in the hole. Press down and start to screw the tap **clockwise** into the hole, Fig. 23-10. A steady, downward pressure on the wrench is needed to get the thread started.

Skill is needed to keep the tap square with the work. Place a square or a wide steel rule several places against the tap, as in Fig. 23-11, to make sure that the tap is square with the work. If it is not square, back it out of the hole a little, straighten it, and with some side pressure screw it into the hole again. This must be repeated until the tap is well started in the hole and cannot get out of square. Use the correct cutting fluid for the material being tapped.

Continue to turn the tap with one hand until the thread has a good start. Then it is no longer necessary to press down because the tap will draw itself into the work when turned. Grasp the tap wrench by the two handles; with a slow, firm, steady movement, continue screwing the tap into the hole, Fig. 23-12. Back up the tap now and then to break and clear away the **chips** and to make the threads smooth. "Back up one step and go ahead two."

Tapping may also be done on the drill press, as explained in section 54-14. Taps should be cleaned after being used.

23-7. Tapping Blind Holes

A **blind hole** is one which goes only part way through the work. If the hole does not go through the work, the **taper tap** (see Fig. 23-1) should be followed by a **plug tap.** If full threads to the bottom of the hole are wanted, a **bottoming tap** must also be used. Clean out the hole now and then so that the **chips** collecting in the bottom of the hole will not keep the tap from going to the bottom.

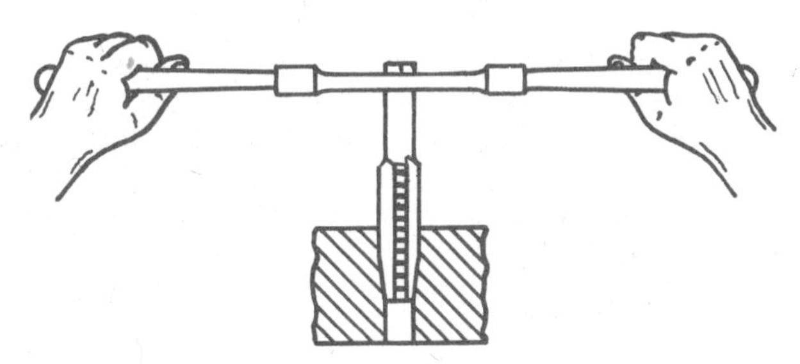

Fig. 23-12. Hold the Tap Wrench in This Manner After Thread Has Been Started

23-8. Cutting Fluids for Tapping

Cutting fluids for tapping are the same as for **threading,** explained in section 22-8.

23-9. Causes of Broken Taps

Taps break for the following reasons:

1. A hole that has been drilled too small needs more pressure and strain than the tap can stand and, therefore, it breaks.
2. A tap that is lopsided in the hole will stick tight and then break.
3. The tap wrench acts like a **lever** and with it a great twisting force can be put upon the tap. Misuse of this force breaks many taps.
4. Lack of cutting fluid where required will cause a tap to stick tight and break.
5. Failure to back up the tap will cause the **chips** to crowd in front of the **cutting edges.** More force is then needed to push these larger chips, and this extra force is often more than the tap can stand.
6. Turning a tap after the bottom of the hole is reached will cause it to break.

23-10. Removing a Broken Tap from a Hole

A broken tap stuck in a hole may cause much trouble because it is difficult to remove. It is often necessary to make repeated efforts to remove a broken tap; much of the worker's time is lost and the job is slowed up. Thus, it is expensive. A broken tap may be removed from a hole with a **tap extractor,** as shown in Fig. 23-15.

A tap broken near the top of the hole may sometimes be removed by placing a dull **cape chisel** in the **flute** of the tap and striking light blows with the hammer, as shown in Fig. 23-13. Penetrating oil may help to free the broken tap. Another way is to drop a little **nitric acid**[2] in the hole. The acid will eat the steel and loosen the tap so it can be removed with the cape chisel, as just explained. Be sure to wash the acid out of the hole afterward to keep the acid from further eating the threads.

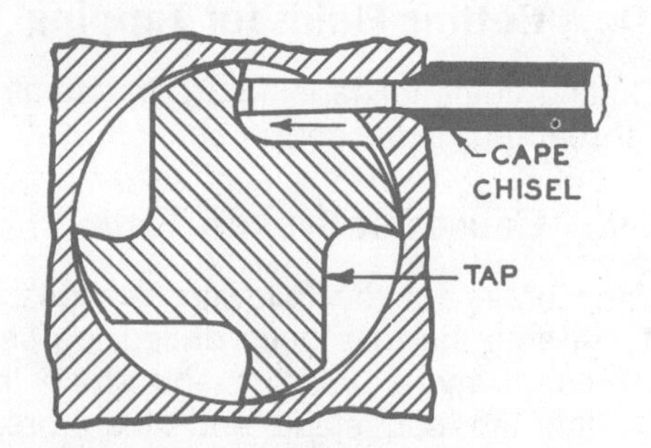

Fig. 23-13. Removing a Broken Tap with a Cape Chisel and Hammer

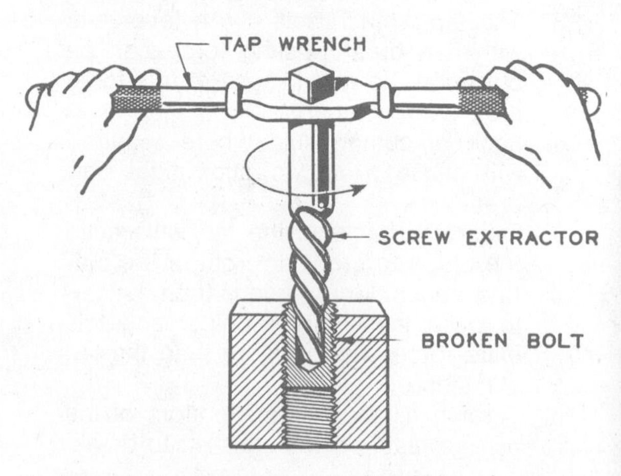

Fig. 23-14. Removing a Broken Bolt with a Screw Extractor

[2]**Nitric acid** is a powerful acid made up of nitrogen, hydrogen, and oxygen. It is poisonous and eats most metals.

Safety Note

Working with acid is dangerous. Be very careful. Be sure your instructor approves.

23-11. Removing a Broken Bolt from a Hole

A broken bolt or screw that is not hardened may be removed from a hole with a **screw extractor,** Fig. 23-14. First drill a hole in the broken bolt; then put the correct size screw extractor in the hole and with a **tap wrench,** turn it **counterclockwise.** The screw extractor acts like a corkscrew. It grips into the sides of the hole; when the right force is used, the bolt begins to turn and come out.

If a screw extractor is not handy, drive a **diamond-point chisel,** shown in Fig. 13-1, into the drilled hole and remove as shown in Fig. 23-14.

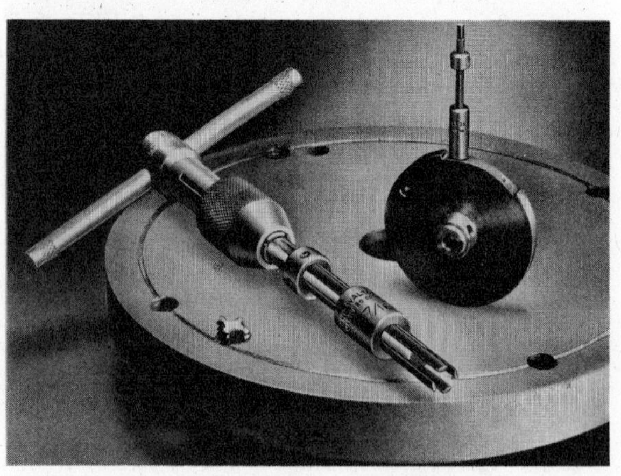

Fig. 23-15. Tap Extractor Is Used to Remove Broken Taps

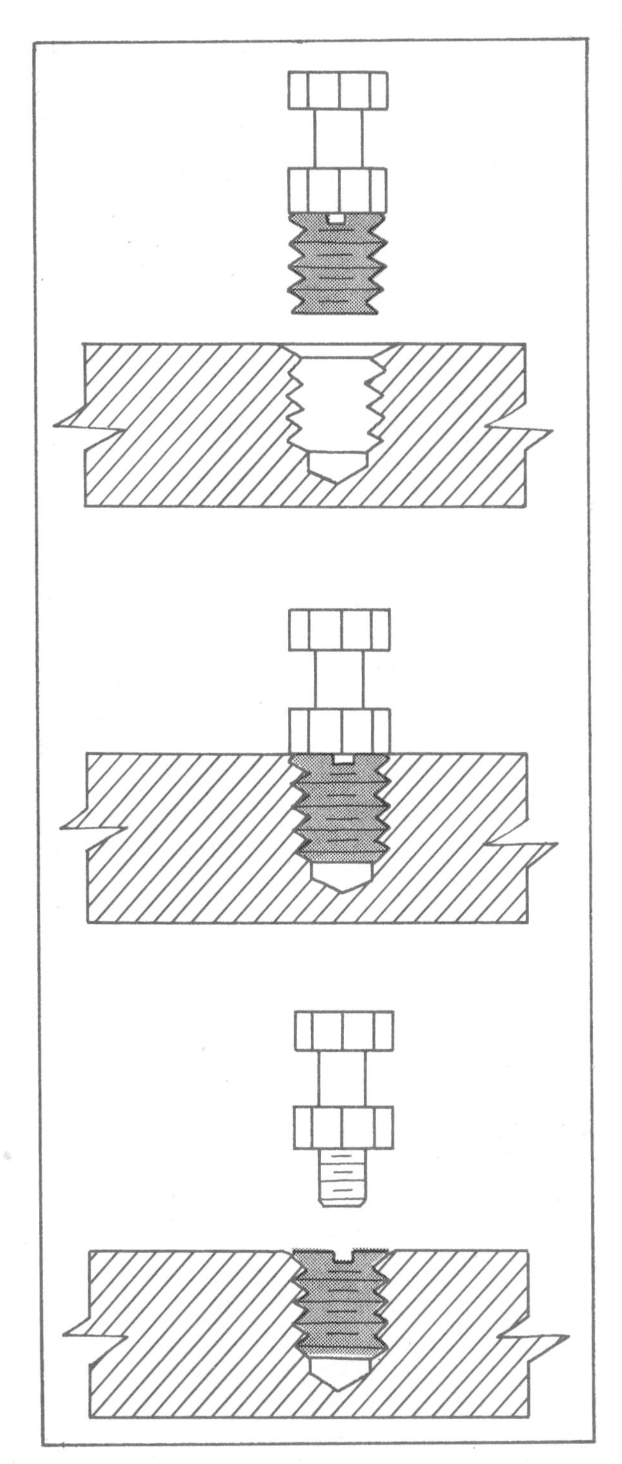

Fig. 23-16. Installation of Thread Insert

23-12. Tap Size Limits

Taps are available with either **cut** threads or **ground** threads. The ground threads have a precision-ground finish, and they produce tapped threads to a very accurate size. They are also more expensive than taps with cut threads. Both carbon tool steel taps and high-speed steel taps are available with cut threads. However, high-speed steel taps are also available with ground threads.

The size of the pitch diameter of a tap determines the depth and the fit of a tapped thread. Internal threads may be tapped for various classes of threads, such as classes 1B, 2B, and 3B. The thread may be either a loose or tight fit, depending on the pitch diameter of the tap.

Taps with ground threads are available with standard, oversize, or undersize pitch diameters. The size limits of the pitch diameter are indicated by a **pitch diameter limit number,** such as L1, H1, H2, or H6, on the shank of ground thread taps. When purchasing taps of this type, the **limits code number** and the fit or class of thread to be tapped should be specified. If they are not specified, the supplier generally sends taps with pitch diameter size limits which produce tapped threads for a Class 2 fit for National Form threads or Class 2B for Unified Form threads. These fits are generally used on most commercially available bolts, screws, and nuts.

The recommended taps, including limits code numbers for various thread fits and classes of threads, are given in standard handbooks for machinists. For example, a ⅜-16 UNC ground thread tap for a Class 2B thread would be identified with the letter G and with the additional number H5; for a Class 3B thread, the code number would be GH3.

23-13. Internal Thread Measurement

Internal threads must be cut to the correct depth in order to produce the proper fit or class of thread. If the pitch diameter of the internal thread is too large, the bolt or screw will fit too loosely. If too small, it will fit too tightly. Internal threads may be checked for the correct fit or class of thread with thread plug gages, as shown in Fig. 46-17. The procedure is explained in section 46-17.

23-14. Thread Inserts

Thread inserts are threaded steel bushings. They are used for replacing worn or damaged threads or for providing wear resistant threads in low tensile strength metals, such as cast aluminum and magnesium. Special tools are required for installing some types of thread inserts. However, some types of thread inserts are installed without special tooling. See Fig. 23-16.

Words to Know

blind hole
bottoming tap
cam ground tap
gun tap
hand tap
left-hand tap
machine tap
plug tap
screw extractor
serial taps
tap extractor
taper tap
tapping lubricant
T-handle tap wrench
thread insert

Review Questions

1. What is a tap and for what is it used?
2. Describe a taper tap. Plug tap. Bottoming tap. For what is each used?
3. How does a machine tap differ from a hand tap?
4. How do cam ground taps differ from conventional taps?
5. What does $\frac{5}{16}$-18 stamped on the shank of a tap mean?
6. What is a tap drill?
7. What size tap drill is used for a $\frac{1}{4}$-20 thread? 10-32 thread?
8. What is a tap wrench? Name two types.
9. When hand tapping, how can the tap be kept square with the work?
10. What kind of lubricant should be used for tapping steel? Aluminum?
11. Why is it necessary to back up hand taps while tapping?
12. How can a broken tap be removed from a hole?
13. Name six causes of broken taps.
14. How can a broken bolt be removed from a hole?
15. What are thread inserts? For what are they used?

PART VII

Fitting and Assembling

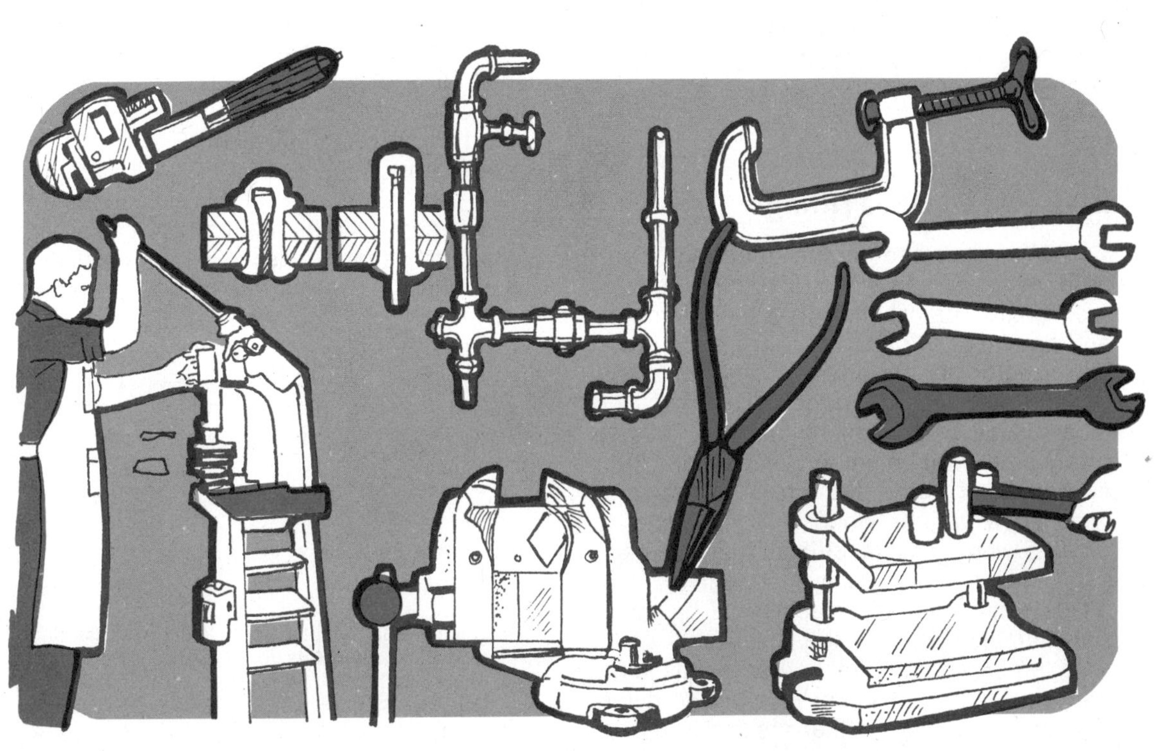

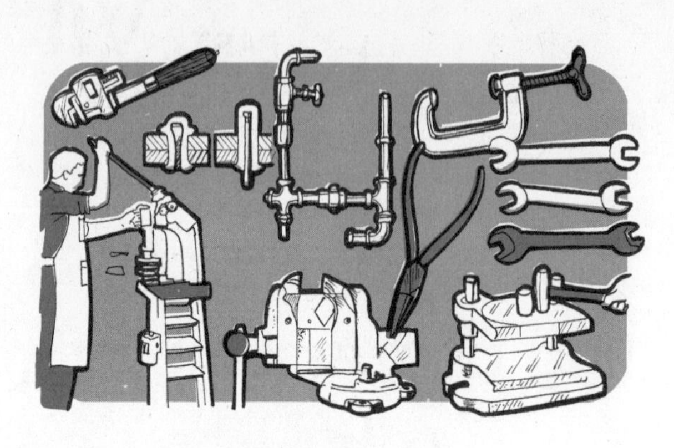

UNIT 24

Fits and Fitting

24-1. Meaning and Importance of Fitting

In metalwork, **fitting** means preparing mating parts to touch or join each other in such a way that one will turn inside another, one will slide upon another, or the parts will hold tightly together so that they cannot move upon each other.

Sawing, chipping, filing, scraping, grinding, or cutting by machine may be necessary to make the parts fit. Judgment and great skill are needed in fitting; these qualities are especially necessary to perform the **operations** just mentioned.

24-2. Who Does Fitting?

The **aircraft-and-engine mechanic** and **auto mechanic** fit wheels, pistons, valves, crankshafts, camshafts, etc. The **diemaker** and **diesinker** fit parts of dies to each other. The **erector** fits the different parts of a machine together. The **gagemaker** must do much fitting when making gages. The **gas fitter, pipe fitter, plumber,** and **steam fitter** fit pipes and **pipe fittings.** An **inspector** must know various kinds of fits in order to inspect work. The **machinist** must fit parts such as wheels, pulleys, bearings, and shafts when repairing machines. The **metal patternmaker** must fit the different parts of a pattern. A **sheet metalworker** makes joints that must fit. A **toolmaker** must fit the different parts of the tools that are made.

24-3. Kinds of Standard Fits

The term **fit** is used to signify the range of tightness which exists between two mating parts. The kind of fit which exists is a result of the application of tolerances (explained in § 46-7) and allowances (explained in § 46-8). When parts are produced within the maximum and minimum size **limits** (see § 46-6) specified on a drawing, they may be assembled with the desired kind of fit. Some parts have to fit together tightly while others have to fit loosely.

A system of **standard fits** has been established for the design and assembly of mating parts. It is the American Standard (now USAS) **Preferred Limits and Fits for Cylindrical Parts** (ASA B4.1-1955).[1] It includes three general groups of classifications for fits between plain (nonthreaded) cylindrical parts. The general classifications are designated with symbols for educational purposes only. The symbols are not to be shown on drawings. Instead, the dimensional size limits for each part are specified on the drawing for the specific kind of fit desired. Three general groups of classifications of fits, including their symbols, are as follows:

[1]Extracted and adapted from American Standard, **Preferred Limits and Fits for Cylindrical Parts** (ASA B4.1-1955), with the permission of the publisher, The American Society of Mechanical Engineers, New York.

Running and Sliding Fits (RC)
Locational Fits (LC, LT, and LN)
locational clearance fit (**LC**)
locational transition fit (**LT**)
locational interference fit (**LN**)
Force Fits (FN)

These letter symbols are used with additional numbers to designate specific classes of fits within each general group classification. Examples of running or sliding fits include **RC** 1 for close-sliding fits, **RC** 4 for close-running fits, **RC** 7 for free-running fits, and so on. The many specific classes of fits (indicated by two letters and a number) are included in tables in handbooks for machinists. Each symbol represents a complete fit, including minimum and maximum clearance or interference. The minimum and maximum size limits for mating parts up to 20″ diameter also are given in the tables.

Fits of threads and **classes of threads** are explained in section 21-16.

24-4. Running and Sliding Fits (RC)

The **sliding fit** is a snug or close fit with some clearance between two parts so that one will move upon or against the other. There should be no wobbling. Examples of sliding fits are a **piston** sliding up and down in a **cylinder** of an automobile or the parts sliding on the **bed** of a **lathe.** See Fig. 55-1.

The **running fit** is used where one part runs or turns inside another; for example, the **shaft** running in a **bearing,** Fig. 24-1. The fit must not be so tight that it keeps the shaft from turning.

24-5. Locational Fits

The **locational fits** are intended for the purpose of accurately locating the position of mating parts. Some mating parts must be located rigidly and accurately, while others may be located with some looseness for ease in assembling them. Hence, three general groups of locational fits were established in the American Standard, now USAS (ASA B4.1-1955).[2]

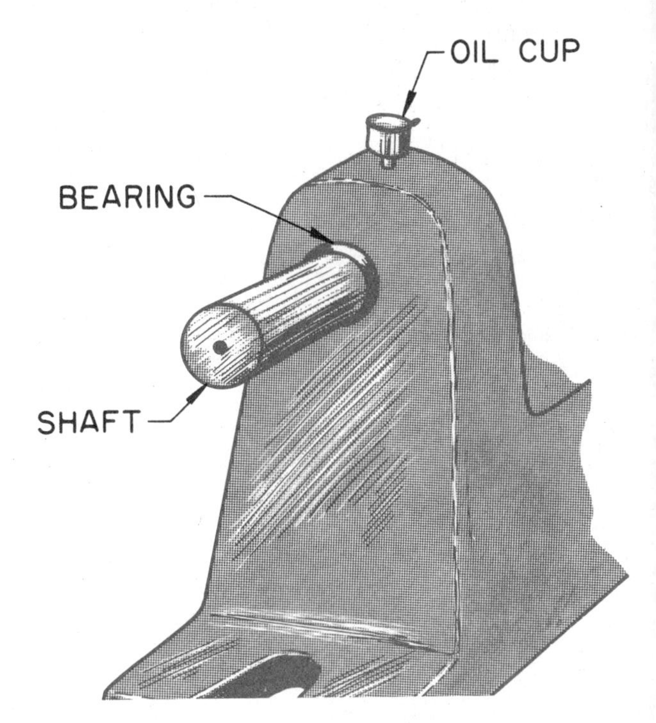

Fig. 24-1. Shaft Running in a Bearing

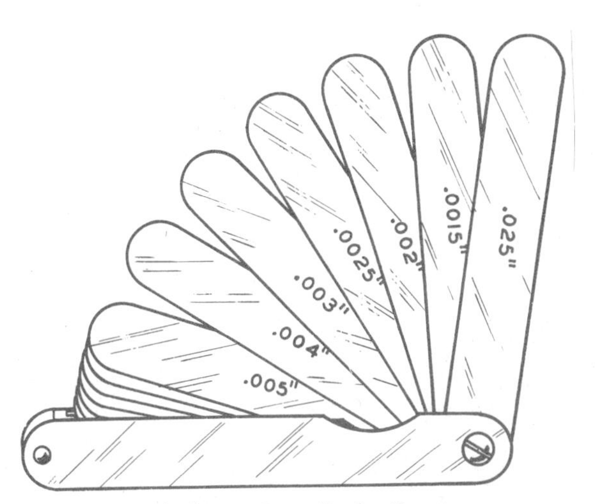

Fig. 24-2. Thickness Gage (Feeler Gage)

[2]See footnote 1, page 194.

Locational clearance fits (LC) are intended for use in the assembly of stationary parts where some clearance is permissible between the mating parts. Fits within this group may range from snug fits to fits with a medium amount of clearance.

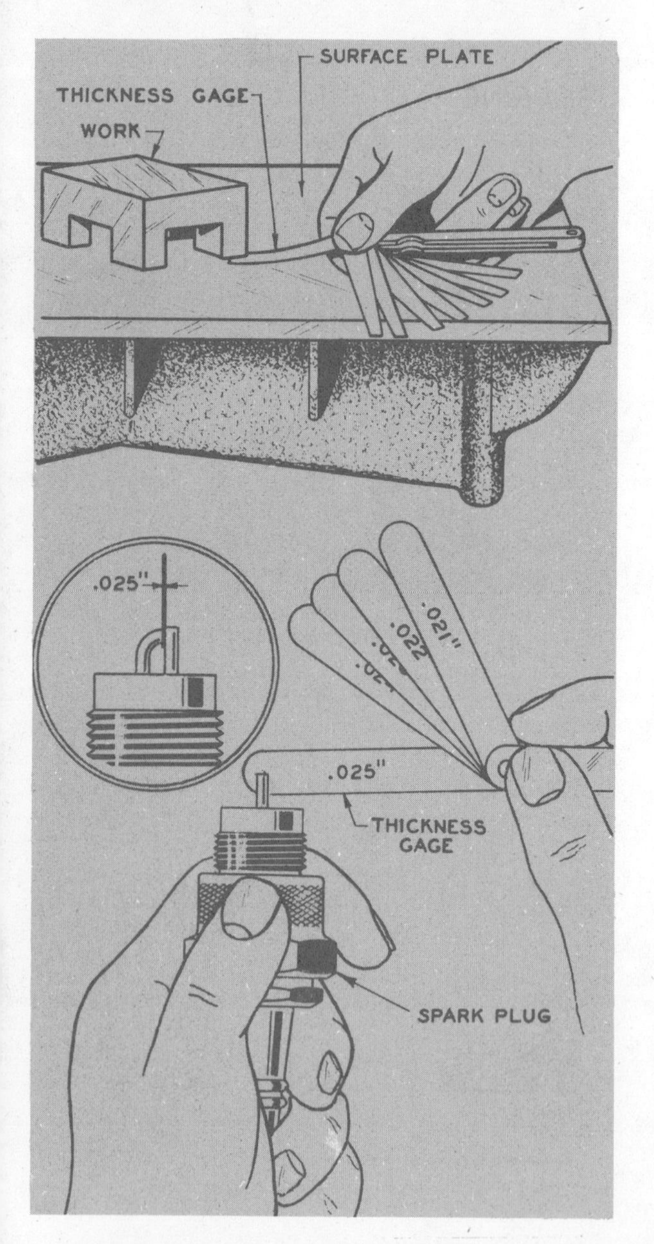

Fig. 24-3. Using a Thickness Gage

Locational interference fits (LN) are used where accuracy of location and rigidity of mating parts are most important. These fits are used to transmit a frictional load from one part to another because of their tight fit. To meet these conditions, mating parts must be assembled with force fits.

Locational transitional fits (LT) are ranked between clearance fits and interference fits. They are used where accuracy of location is important and where a small amount of either clearance or interference is permissible between the mating parts.

24-6. Force Fits (FN)

Force fits or **shrink fits** include several classes of fits which involve interference between mating parts. Generally, the hole size is a standard or basic size. The shaft is slightly larger, thus causing definite interference. Parts assembled with a force fit may be assembled in three ways: (1) they may be driven together with a hammer, thus forming a **drive fit;** (2) they may be pressed together with an arbor press, Fig. 25-35, or other kind of large press, thus forming a regular **force fit;** (3) they may be assembled with a **shrink fit,** which will be explained shortly.

Force fits are used where parts must be held together very tightly. With these fits, parts can be fastened together almost as tightly as though they were made from one piece. Force fits are used to assemble gears, pulleys, bearings, collars, and similar parts on shafts. Railroad car wheels are put on their axles in this way under pressures of 100 to 150 tons with large presses. The hole diameter of the wheel is a little smaller than the shaft diameter of the axle.

Shrink Fits

These fits are classified under force fits, since they involve interference between mating parts. However, where heavy driving or pressing forces are not practical or possible, the parts are sometimes installed by **shrinking** one part on the other. A pulley, gear, or collar may be fastened to a shaft by shrinking it on the shaft. The hole diameter is slightly smaller than the shaft. The pulley or part with the hole is made larger by heating it (to cause the hole to expand) and slipping it over the shaft. The pulley is then allowed to cool and shrink on the shaft. Thus, the pulley is held on the shaft with much greater pressure than if it were pressed or driven on. The pulley and shaft are nearly as tight as though they were one piece.

24-7. Allowances for Different Fits

When a **shaft** is fitted to a **bearing** (see Fig. 24-1), the diameter of the shaft should be a little smaller than the diameter of the bearing to allow space for a **film of oil** between the surfaces and to allow the shaft to get larger from the heat caused by rubbing. The differences in the diameters is called the **allowance** for the fit. For example, if the shaft diameter is 0.500″ and the hole diameter is 0.501″, then the allowance is 0.001″. The amount of allowance depends upon:

1. Size of work
2. Kind of metal
3. Amount of metal around hole
4. Smoothness of hole and shaft

24-8. Thickness Gage

The thickness gage, also known as **feeler gage,** is made up of a number of thin, steel **blades** which fold into a handle like the blades of a pocket knife, Fig. 24-2. The thickness is marked on each blade. The blades are used to measure small spaces between surfaces, as in Fig. 24-3.

Words to Know

allowance
arbor press
drive fit
feeler gage
force fit
locational fit
press fit
running fit
shrink fit
sliding fit
thickness gage
tolerance

Review Questions

1. What does fitting mean?
2. List the metal-removal operations used in fitting parts together.
3. Name three kinds of standard fits which are used in the assembly of non-threaded cylindrical parts.
4. List three general groups of locational fits and their symbols.
5. Give an example of where a sliding fit is used.
6. Give an example of where a running fit is used.
7. Describe how a pulley can be assembled with a shaft using a shrink fit.
8. How does a locational fit differ from a running fit?
9. List three general kinds of force fits.
10. Give several examples of where force fits would be used.
11. For what purpose is an arbor press used?
12. What is meant by allowance?
13. What is meant by tolerance?
14. What is a thickness gage? For what is it used?

Mathematics

1. If the shaft diameter is 2″ and the hole diameter is 2.002″, what is the allowance for the running fit?
2. If steel expands .0000065″ per inch for each degree of increase in temperature above 68° F., how much would a 2″ diameter shaft expand if it were heated to 180° F.?

UNIT 25

Assembly Tools

Fig. 25-1. Assembling Transmission and Engine

25-1. What Does Assembling Mean?

Assembling means putting the parts of something together. For example, automobile parts are put together into complete automobiles, Fig. 25-1. Tools for holding, setting, and fastening are needed.

This unit describes tools used in assembling which may also be used in other

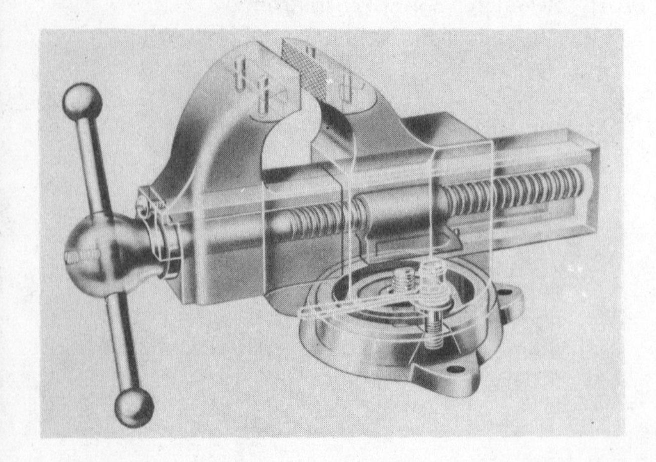

Fig. 25-2. Machinist's Vise

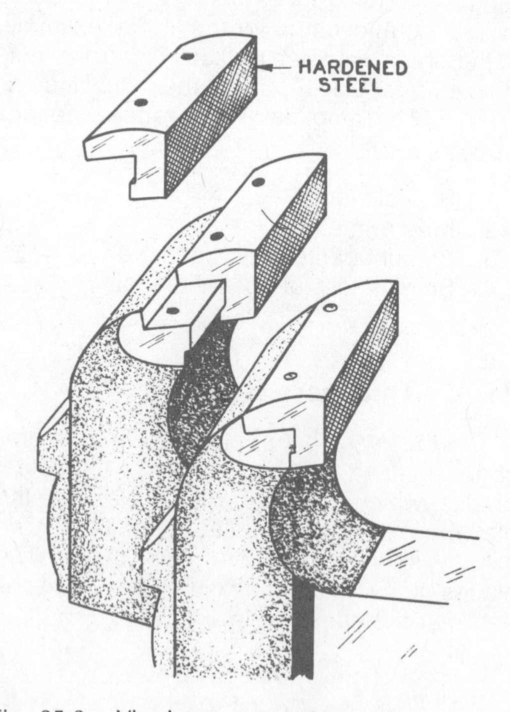

Fig. 25-3. Vise Jaws

work. The opposite of **assemble** is **disassemble,** which means to take apart.

25-2. Machinist's Vise

The machinist's vise, Fig. 25-2, is fastened near the edge of the bench with bolts. It is often used to clamp parts together while they are being assembled. The size of a vise is determined by the width of the jaws, which are made of hardened steel, Fig. 25-3. Thus, a 3″ vise is one whose jaws measure 3″ in width.

The handle of the vise acts as a lever; thus, the screw and lever help to make a powerful clamp. The way to tighten a vise is shown in Fig. 25-4.

25-3. Soft Jaws

Soft jaws, Fig. 25-5, are made of copper, brass, wood, leather, or other soft material. They are slipped over the steel jaws of the vise so as not to scratch or **nick** the **finished surfaces** on the work.

25-4. C-Clamp

The C-clamp is shaped like the letter C, Fig. 25-6. It is made in many sizes and is used to clamp parts together while they are being assembled, Fig. 25-7.

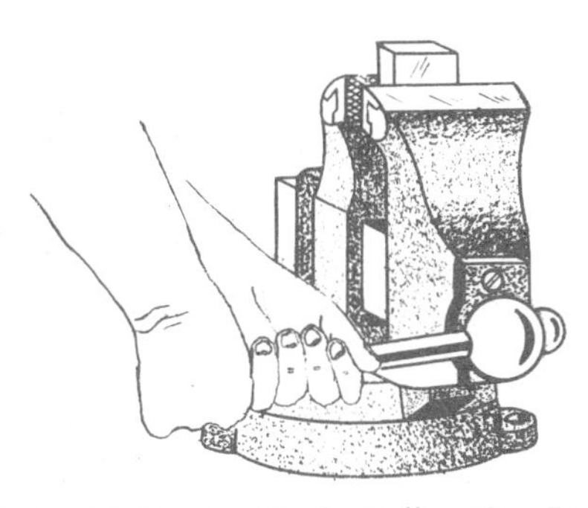

Fig. 25-4. Tightening Vise by Pulling One End of the Handle

Fig. 25-5. Soft Jaws

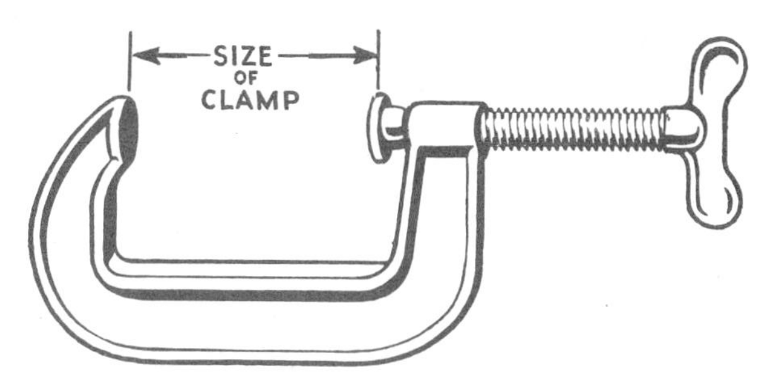

Fig. 25-6. C-Clamps

Fig. 25-7. Using C-Clamps

25-5. Parallel Clamp

The parallel clamp has two steel jaws which are opened or closed by turning two screws. See Fig. 6-25. It is used to hold small work. The jaws should always be **parallel;** this is why the clamp is called a **parallel clamp.**

25-6. Pliers

There are many kinds of pliers, some of which are shown in Fig. 25-8. They are handy tools and are used for cutting small wire and for holding, gripping, twisting, turning, pulling, and pushing.

The **slip-joint plier,** also known as **combination plier,** is used for gripping; it can also cut small-size wire. The **slip joint** makes it possible to grip large parts.

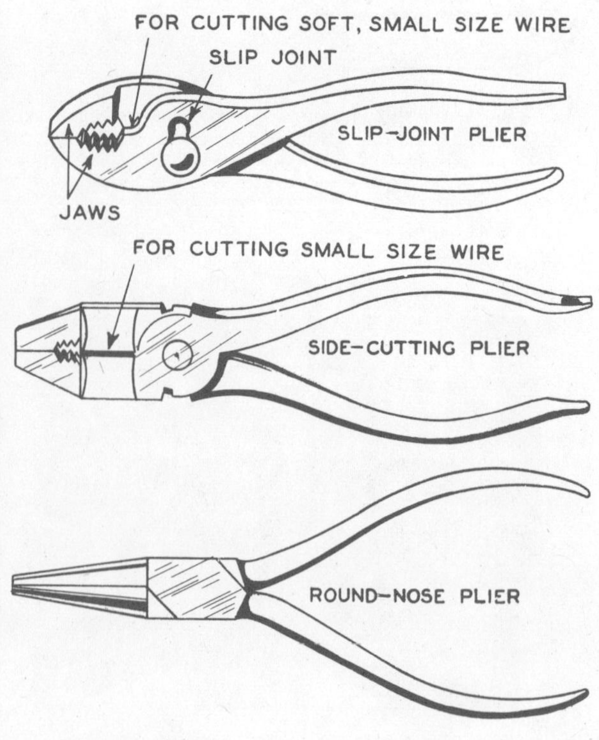

Fig. 25-8. Pliers

The **side-cutting plier** is especially useful for cutting wire and nails. It is used by **electricians** for cutting electric wires. The flat square **jaws** are useful for bending corners on thin metal.

The **flat-nose plier** is shown in Fig. 31-13.

The **round-nose plier** is used to bend small wire and thin metal and to hold small parts.

Some pliers have long jaws and are called **long-nose pliers.**

25-7. Machinist's Hammer

The machinist's hammer, also called **ball peen hammer,** has been described in sections 6-5 and 13-12. In assembly work it is used for striking and driving. Figures 26-31 — 26-33 show how it is used for **riveting.** Use a light hammer for light work and a heavy hammer for heavy work.

25-8. Soft Hammers and Mallets

Striking two **hardened steel** parts against each other is dangerous, because hardened steel is **brittle;** it breaks easily when struck by another piece of hardened steel. In section 15-12 it was mentioned how easily a file made of hardened steel can be broken. Thus, it is also dangerous to strike two hardened-steel hammers together or to strike any other hardened-steel object with the hardened-steel **face** of a **machinist's hammer.** A **chip** of hardened steel (which may break off under these conditions) has very sharp edges, flies as fast as a bullet, and can cause great injury if it strikes someone.

Soft hammers—with **heads** made of lead, copper, brass, or other soft materials — are used to strike hardened-steel surfaces. They are also used to strike thin or soft metals or **finished surfaces** so that nicks will not be made.

25-9. Lead Hammer

A lead hammer, Figs. 25-9 and 25-10, is used for striking **finished surfaces** where the steel hammer would dent or nick the surface. The **head** of a lead hammer is made of **lead.** The **handle** is usually a piece of pipe around which the melted lead is poured.

Lead hammers can be made in the shop with a **combination hammer mold,** Fig. 25-10. One end of the pipe handle is clamped in the **mold.** Put pieces of lead in the **ladle,** which is a dipper fastened to the mold; then heat until the lead is melted. Tip the ladle, and the melted lead will flow through a hole into the mold and around the end of the pipe.

A lead hammer made of pure lead may be too soft and **mushroom** too easily. It will last longer if a little **antimony** is melted into the lead when making the lead hammer. Lead hammers with battered or **mushroomed faces** should be remelted and made over into new ones. See section 40-1.

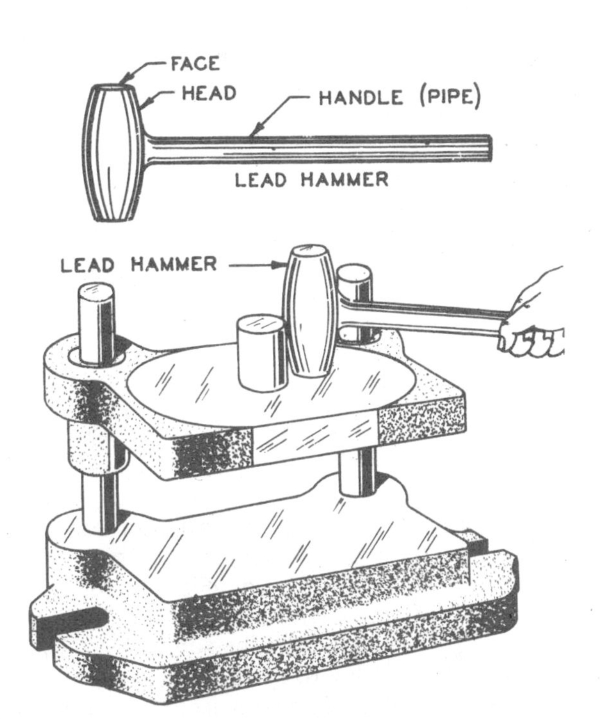

Fig. 25-9. Hammering with a Lead Hammer

25-10. Mallet

A mallet is usually made of wood, leather, or rubber, Fig. 25-11; it is used for the same purposes as a soft hammer.

25-11. Screwdrivers

Screwdrivers are used to turn or **drive** screws with **slotted** heads. They are made in many sizes and several shapes, Fig.

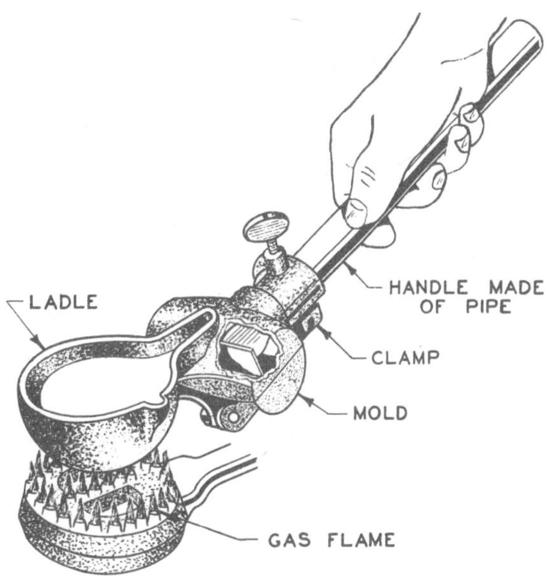

Fig. 25-10. Combination Hammer Mold for Making a Lead Hammer

Fig. 25-11. Using a Mallet

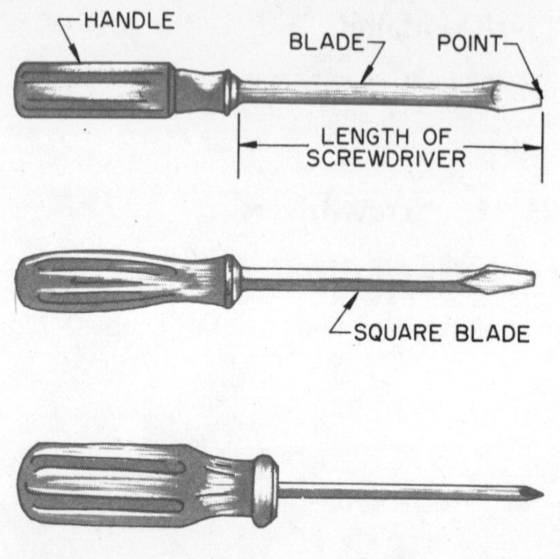

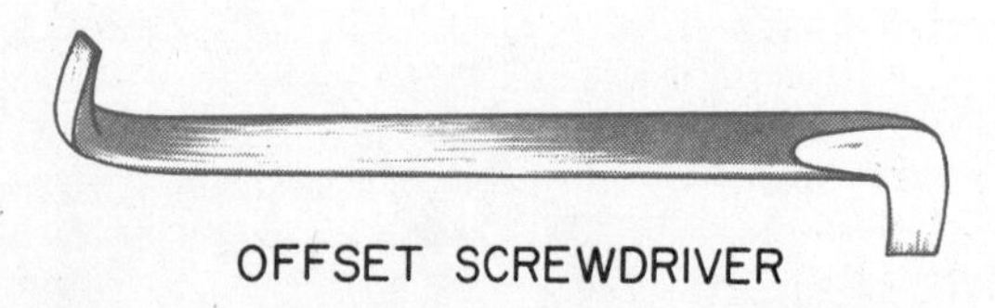

Fig. 25-12. Screwdrivers

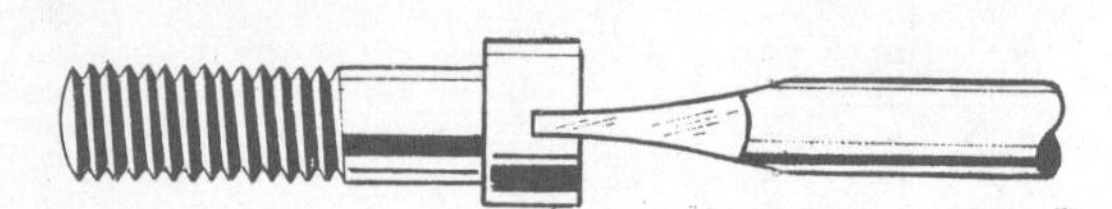

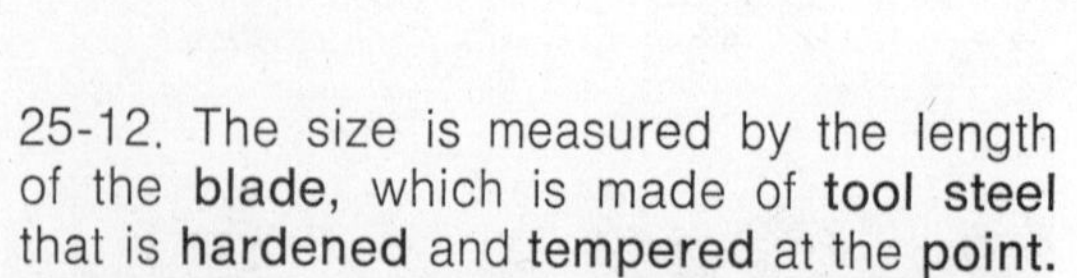

Fig. 25-13. Shape of a Screwdriver Point

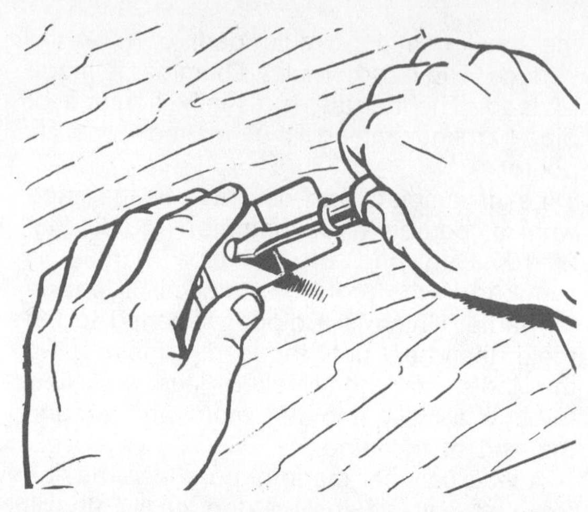

Fig. 25-14. Hold Small Work in the Bench When Using a Screwdriver

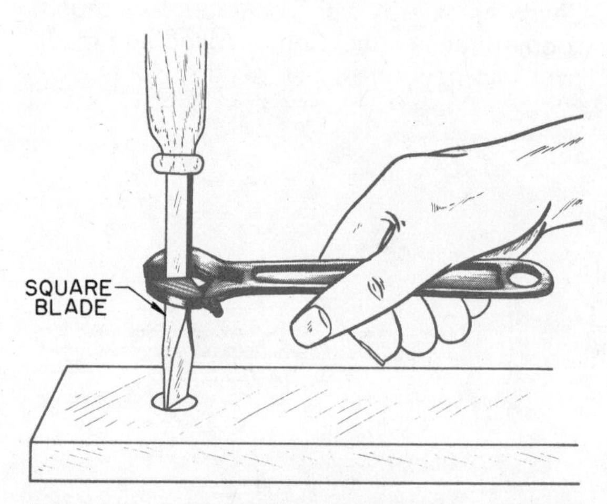

Fig. 25-15. Turning a Screwdriver with a Wrench

25-12. The size is measured by the length of the **blade,** which is made of **tool steel** that is **hardened** and **tempered** at the **point.**

The screwdriver point should be correctly shaped; it must fit the slot in the screw, Fig. 25-13. The **hollow-ground** sides of the point must press against the sides of the slot.

To avoid injuring the hand if the screwdriver slips, lay small work on the bench when using a screwdriver, Fig. 25-14, instead of holding it in the hand. **Burrs** made by a screwdriver slipping out of a screw slot should be filed off to prevent cutting the hands.

The blades on some of the larger screwdrivers are square; a wrench may be used to turn such a screwdriver, Fig. 25-15.

An **offset screwdriver** has a bent handle. It is used where a straight screwdriver will not reach, Fig. 25-16.

The use of a screwdriver for **recessed screws** is shown in Fig. 25-17.

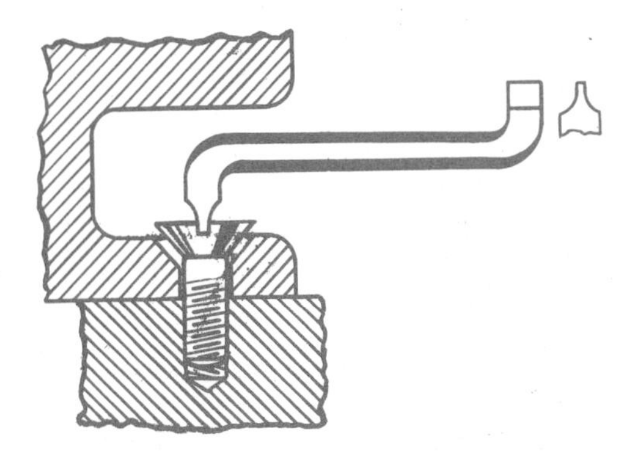

Fig. 25-16. Using an Offset Screwdriver

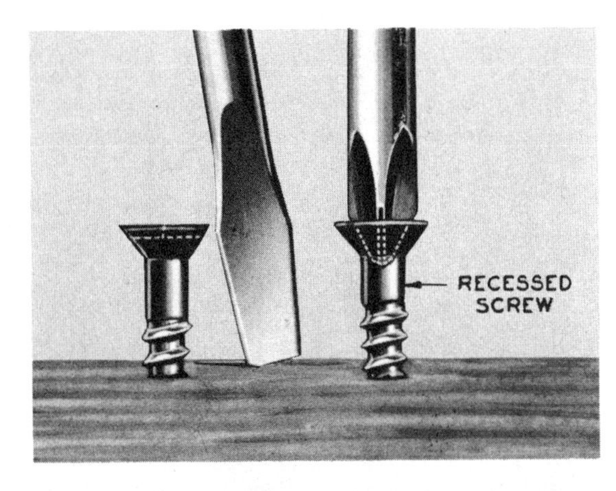

Fig. 25-17. Using a Screwdriver for Recessed Screws
Regular screwdriver may slip and damage work or screw.

25-12. Wrenches

There are many kinds of wrenches. Some are **adjustable,** which means that they can be made larger or smaller to fit different

Fig. 25-18. Monkey Wrench

sizes of bolts and nuts. Others are **non-adjustable;** that is, they fit only one size bolt or nut. The basic types of wrenches fit these classifications.

Adjustable Wrenches

Monkey wrench
Adjustable-end wrench
Adjustable S-wrench
Vise-grip wrench
Pipe wrench

Nonadjustable Wrenches

Open-end wrench
Box wrench
Socket wrench
Spanner wrench

25-13. Monkey Wrench

The monkey wrench, Fig. 25-18, is named after its inventor, Charles Moncky. It is used for tightening or loosening bolts and nuts and can be set to fit many sizes.

When a monkey wrench is used, the **jaws** should be tight on the nut and should be pointed in the same direction that one intends to pull. In other words, the wrench should be turned in the direction shown by the arrows in Fig. 25-19 in order to avoid spreading the jaws. The fixed **jaw** is the stronger of the two and can, therefore, stand more strain.

25-14. Adjustable Wrenches

The **adjustable-end wrench,** Fig. 25-20, is a strong tool which is used for general work in the shop. Its jaws are pointed at

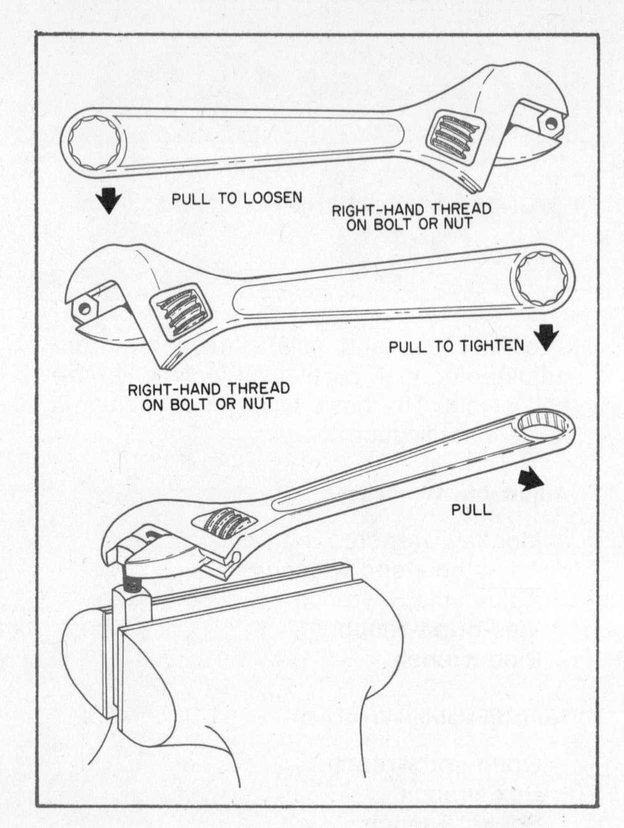

Fig. 25-19. Point Jaws in Direction of Pull

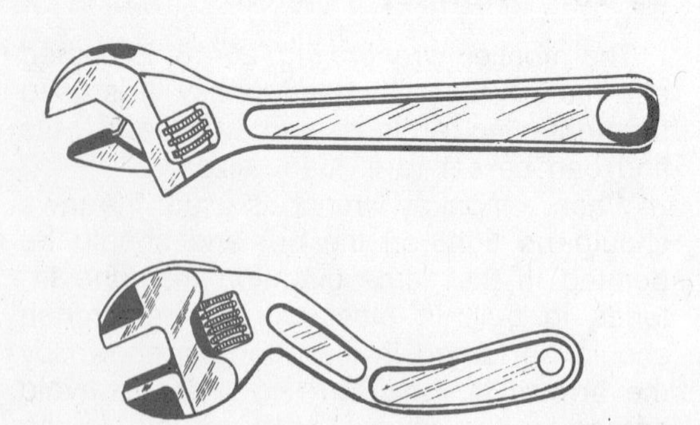

Fig. 25-20. Adjustable Wrenches

such an angle that it can be used in close corners and unhandy places.

The **adjustable S-wrench,** Fig. 25-20, is for general use in the shop. Its S-shape makes it useful in many places where a straight-handled wrench cannot be used.

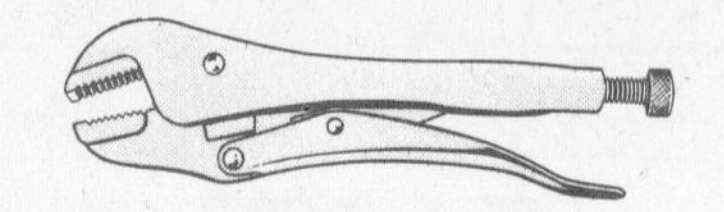

Fig. 25-21. Vise-Grip Wrench

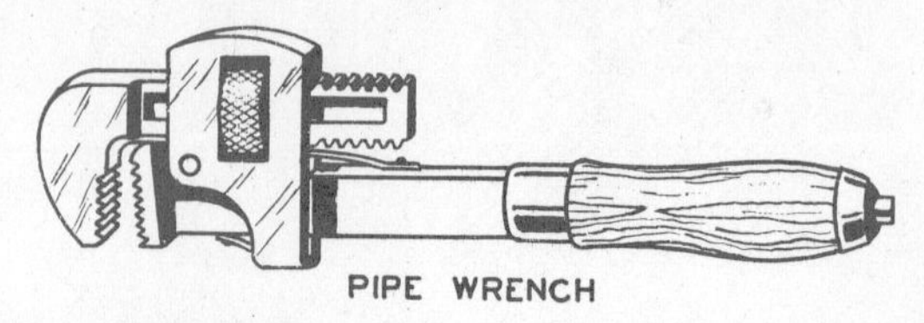

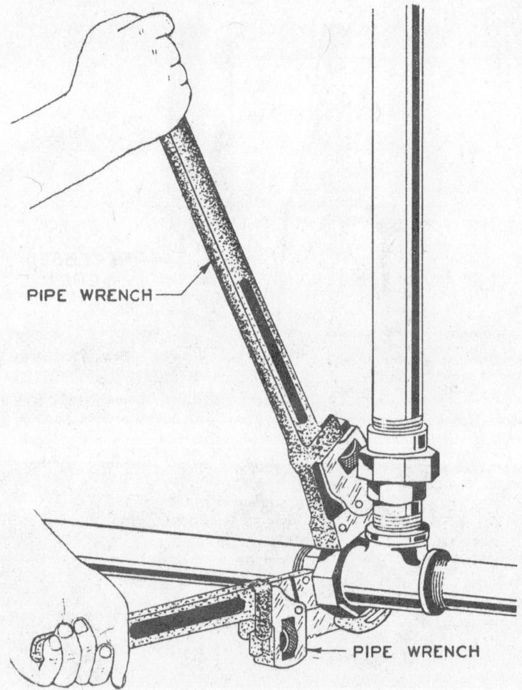

Fig. 25-22. Using Pipe Wrenches

25-15. Vise-Grip Wrench

The **vise-grip wrench** (Fig. 25-21) also called the vise-grip plier, is a handy tool

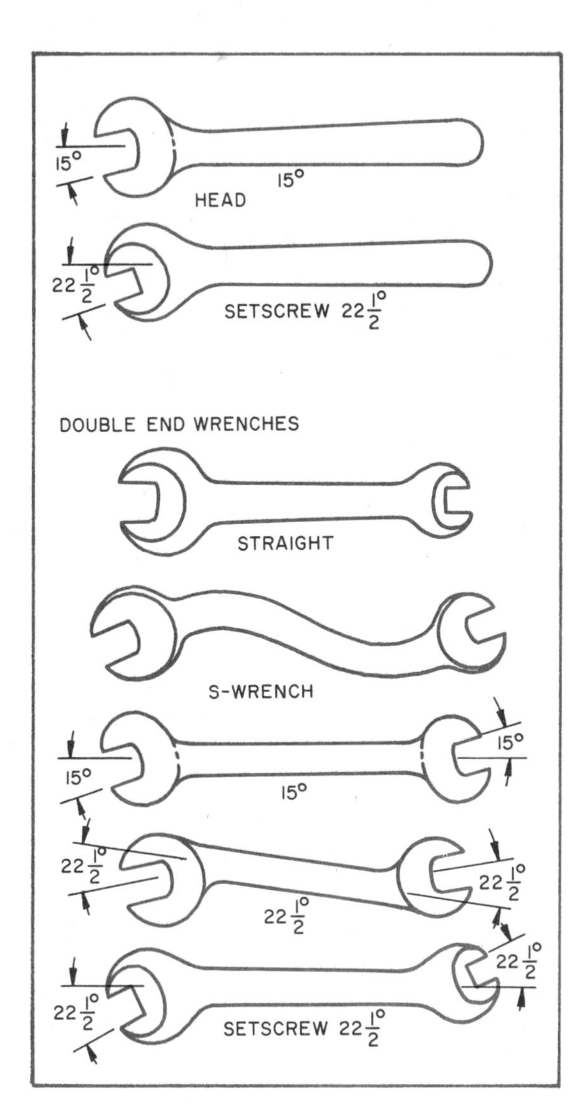

Fig. 25-23. Open-End Wrenches

and is used by many mechanics. It does many things faster and easier than any other tool; it acts like a vise, clamp, plier, pipe wrench, open-end wrench, or locking tool. The vise-grip wrench holds round, square, or other shaped objects. It works in close places and the strong steel jaws lock to the work and will not slip.

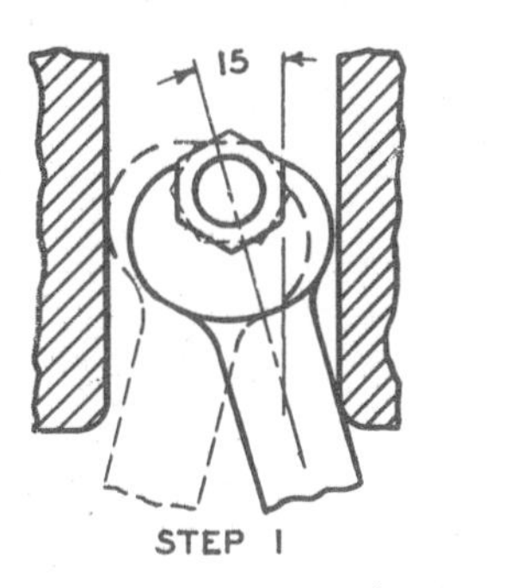

Fig. 25-24. Use of 15° Wrench

25-16. Pipe Wrench

The pipe wrench is used to hold or turn pipes or other round pieces of metal, Fig. 25-22. The **teeth** should be kept clean to keep the wrench from slipping. When using the pipe wrench, point the **jaws** in the direction you intend to pull, as in Fig. 25-19.

25-17. Open-End Wrenches

A number of open-end wrenches are shown in Fig. 25-23. Some have a **single end,** others have a **double end,** some are **straight,** some are **S-shaped,** while on others the **head** makes a 15° or 22½° angle with the center line of the handle.

The S-shaped, the 15°, and the 22½° wrenches are used in narrow spaces where a straight wrench cannot make a quarter of a turn (90°) for a **square nut** or a sixth of a turn (60°) for a **hexagon nut** (which has six sides). The use of the 15° wrench is shown in Fig. 25-24. The uses of the S-shaped and 22½° wrench are about the same.

25-18. Sizes of Open-End Wrenches

Table 20 on page 206 tells the size of the open-end wrench needed for a certain-sized **bolt, nut,** or **cap screw.** The size of an open-end wrench is measured by the size of the opening. Each wrench is stamped with its size.

TABLE 20

Open-End Wrench Sizes for American Standard Bolts, Nuts, and Cap Screws

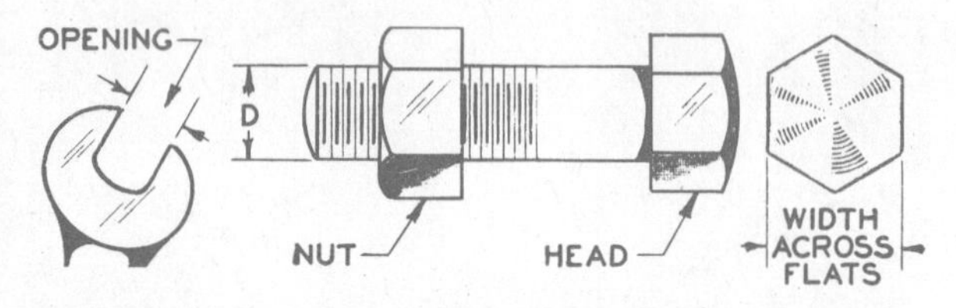

All dimensions given in inches.

Wrench Opening / Width Across Flats of Bolt Heads and Nuts	Unfinished and Semi-finished Bolts (D)	Finished Bolts / Nuts and Jam Nuts (D)	Cap-Screws (D)	Machine Screw Nuts and Stove Bolt Nuts (D)
5/32	...	...	...	#0, #1
3/16	...	...	...	#2, #3
1/4	...	...	...	#4
5/16	...	...	...	#5, #6
11/32	...	...	...	#8
3/8	1/4	...	...	#10
7/16	...	1/4	1/4	#12, 1/4
1/2	5/16	...	5/16	...
9/16	3/8	5/16	3/8	5/16
5/8	7/16	3/8	7/16	3/8
3/4	1/2	7/16	1/2	
13/16	...	1/2	9/16	
7/8	9/16	9/16	5/8	
15/16	5/8	...	...	
1	...	5/8	3/4	
1 1/8	3/4	3/4	7/8	
1 5/16	7/8	7/8	1	
1 1/2	1	1	1 1/8	
1 11/16	1 1/8	1 1/8	1 1/4	
1 7/8	1 1/4	1 1/4		

The size of a **bolt** is measured by the diameter of the body, including the threads. A **nut** is measured by the diameter of the bolt which it fits. Thus, a wrench with a 3/8″ opening will fit a 1/4″ bolt, a No. 10 machine screw nut, or a No. 10 stove bolt nut.

For an explanation of the meaning of **American Standard** (now United States American Standards) in Table 20, see section 21-15.

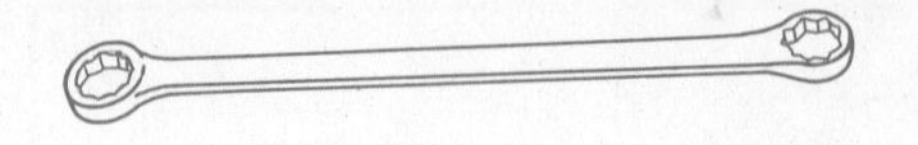

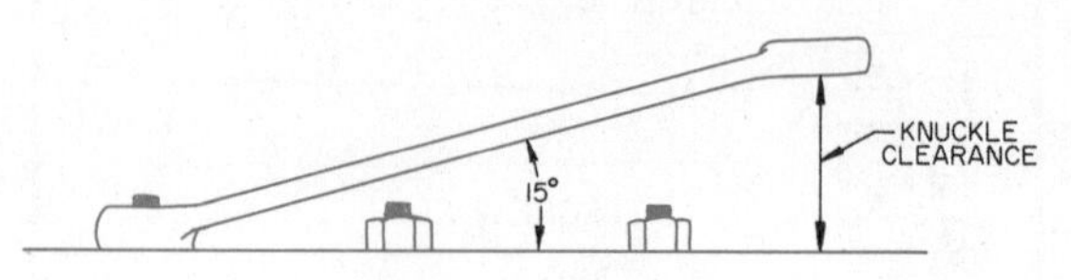

Fig. 25-25. Box Wrench

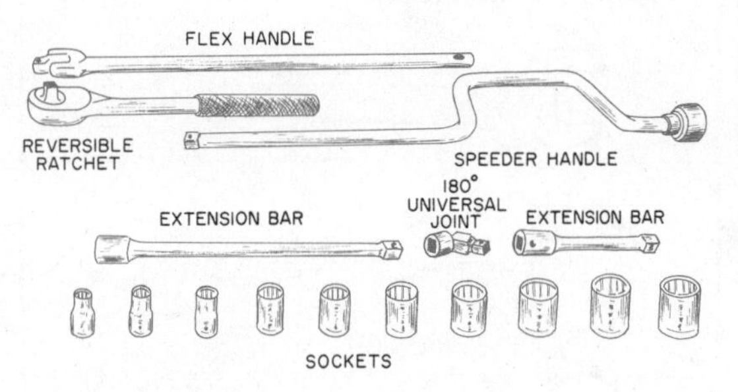

Fig. 25-26. Socket Wrench Set

25-19. Box Wrenches

Box wrenches have closed ends, Fig. 25-25; that is, the **head** of the wrench goes completely around the nut or bolt head. Their handles are usually offset 15°, which provides wrench clearance for nearby obstacles and the user's knuckles as well.

25-20. Socket Wrenches

Socket wrenches, Fig. 25-26, are used to turn nuts and bolts which are in hard-to-reach places or are below the surface of the work. A set of socket wrenches makes a very versatile tool kit. It usually includes an assortment of sockets which engage the nut or the bolt head; one or more extension bars; a hinge handle (commonly called a breaker bar) for loosening difficult nuts and bolts; a reversible ratchet for normal loosening and tightening; a speed handle for rapid assembly or disassembly of threaded parts;

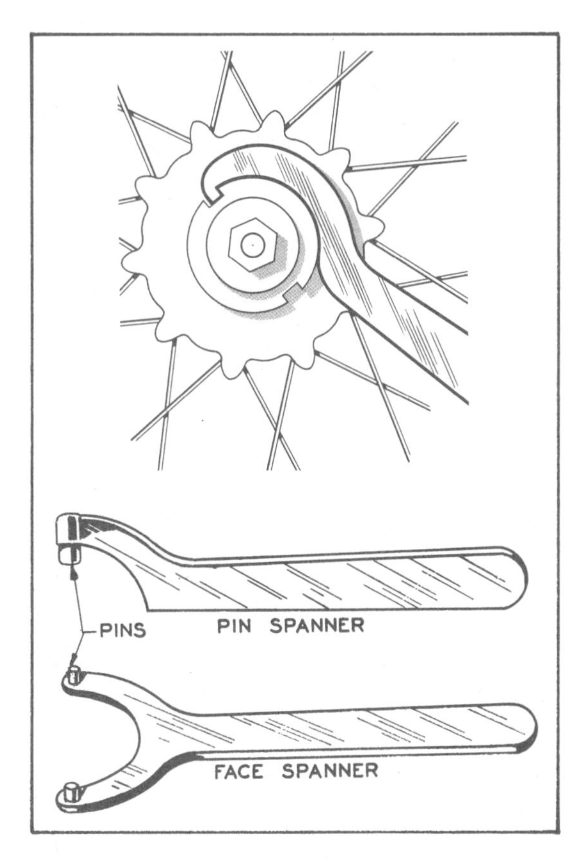

Fig. 25-27. Spanner Wrenches

and a universal joint, which is used when the wrench handles cannot be positioned at right angles to the nut or bolt centerline.

25-21. Spanner Wrenches

Spanner wrenches have one or two **pins**, Fig. 25-27. These pins fit into holes or slots in round nuts or threaded collars to loosen or tighten them.

25-22. Hexagonal Wrenches

Hexagonal wrenches are also called hex keys and Allen wrenches. They are used with socket cap screws and socket set screws, Fig. 25-28. They are available in a wide range of sizes, either singly or in sets.

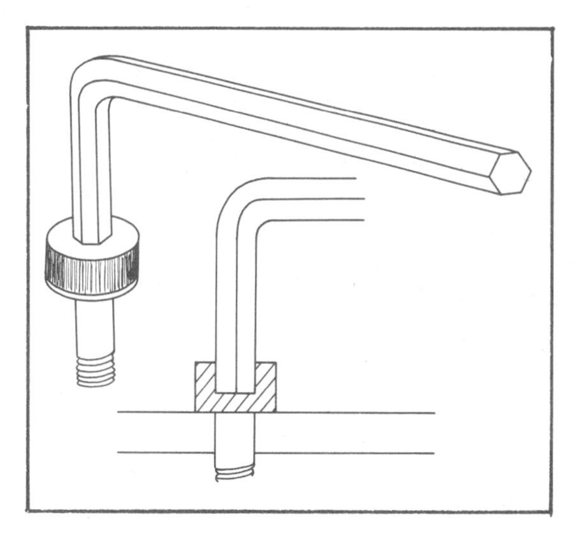

Fig. 25-28. Hexagonal (Socket Head) Wrench

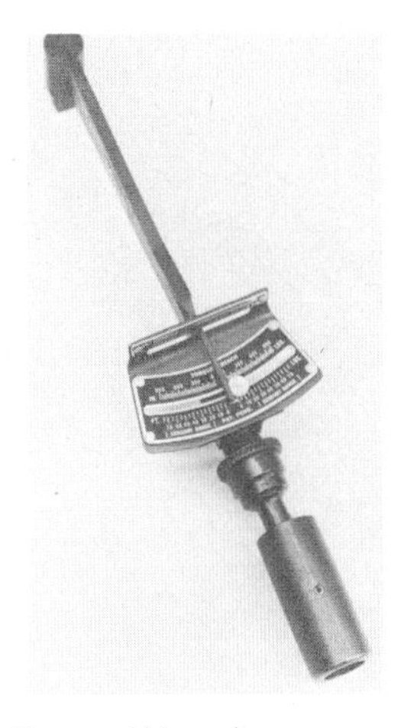

Fig. 25-29. Torque Wrench

Handle length is proportional to the size of the wrench, but wrenches can be obtained with extra-long handles for reaching screws in confined locations.

25-23. Torque Wrenches

Torque means a turning or twisting force applied to a bolt or shaft, and it is measured in foot-pounds. The torque wrench, Fig. 25-29, is used when several bolts or nuts must all be tightened a uniform amount. It is also used to prevent overtightening a single fastener. Uneven or excess stresses and strains may cause the work to warp or be pulled out of shape. Thread stripping and bolt breakage are avoided by intelligent use of the torque wrench.

25-24. Powered Wrenches

Electric and air-powered wrenches, Fig. 25-30, enable workers to rapidly assemble and disassemble parts with a minimum of physical effort. The wrenches can be preset for a given amount of torque. Other powered units are available for driving screws, tapping threads, chipping, and riveting.

25-25. Using Wrenches

Always use a wrench that fits the bolt or nut snugly. When using an **adjustable wrench**, set the **movable jaw** until it fits

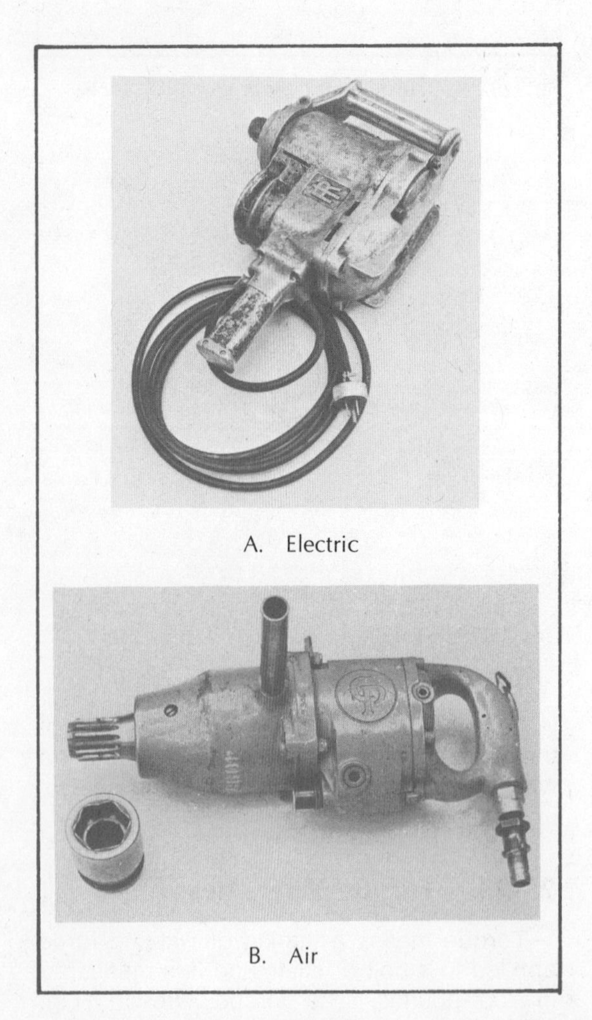

A. Electric

B. Air

Fig. 25-30. Powered Wrenches

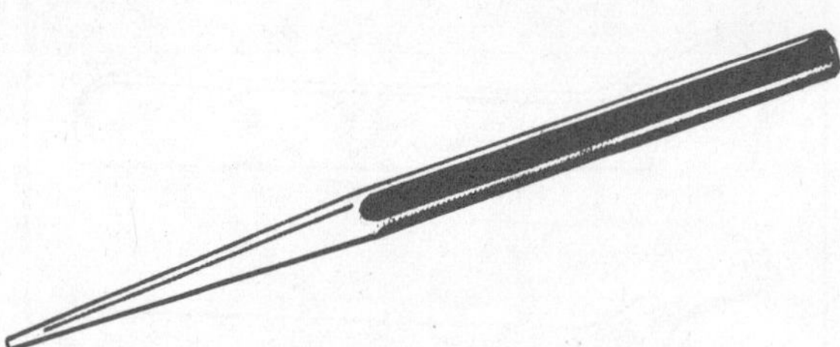

Fig. 25-31. Drift Punch

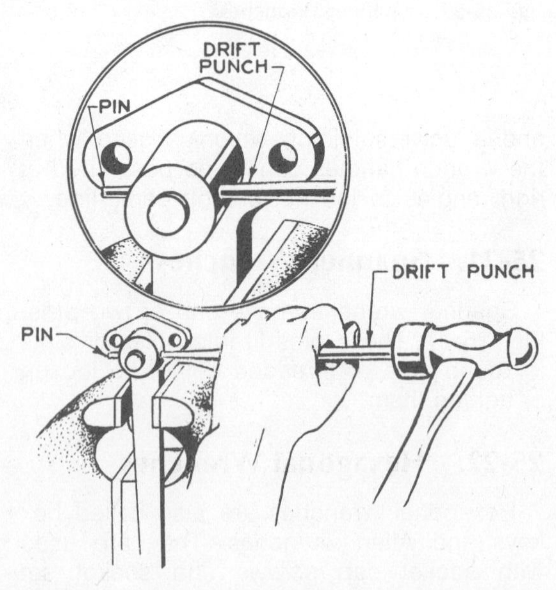

Fig. 25-32. Aligning Holes and Loosening a Pin with a Drift Punch

tightly on the bolt or nut. Use small wrenches for small bolts and nuts and large wrenches on large bolts and nuts.

A wrench is a **lever.** The longer the handle, the greater is the multiplication of force of the lever. It is very important that you study how much a bolt or nut should be tightened with a wrench. You can tell by the **feel** or sense of touch whether a bolt or nut is about to twist off or whether the threads are beginning to **strip.**

25-26. Drift Punch

A drift punch, sometimes called a **tapered punch** or **aligning punch,** Fig. 25-31, is smooth and **tapered.** It may be used to arrange holes in a straight line, to drive out **pins** as shown in Fig. 25-32, or to drive out **rivets.**

Fig. 25-33. Pin Punch

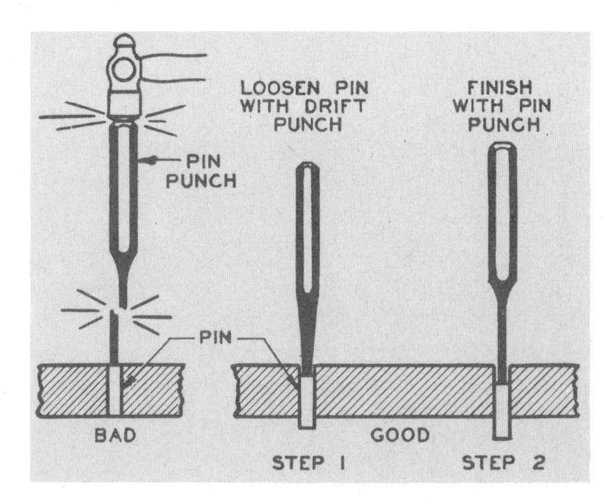

Fig. 25-34. Use of the Pin Punch

25-27. Pin Punch

A pin punch, Fig. 25-33, has a straight end and is used to drive out **cotter pins** and **tapered pins.** The pins should first be loosened with a **drift punch** and then driven out with the pin punch, Fig. 25-34.

25-28. Arbor Press

An arbor press is a machine for pressing parts of machinery together or forcing them apart, such as pressing a shaft in or out of a pulley or gear, Fig. 25-35.

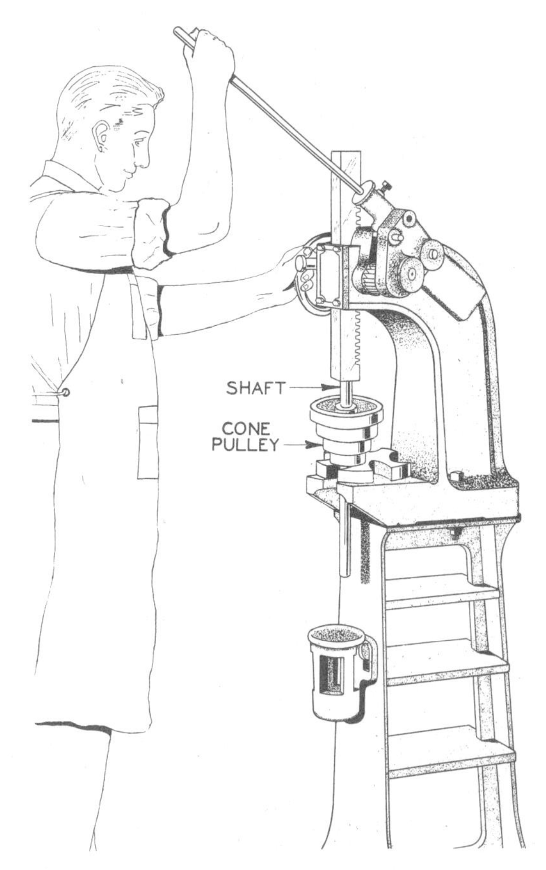

Fig. 25-35. Pressing a Shaft into a Pulley with an Arbor Press

Words to Know

adjustable wrench
box wrench
combination hammer mold
combination plier
double-end wrench
drift punch
face spanner
15° wrench
hexagonal wrench
long-nose plier
machinist's vise
mallet
monkey wrench
offset screwdriver
open-end wrench
pin punch
pin spanner
pipe wrench
powered wrench
recessed screw
round-nose plier
side-cutting plier
slip-joint plier
socket wrench
soft hammer
spanner wrench
S-wrench
tapered pin
tapered punch
torque wrench
22½° wrench
vise-grip wrench

Review Questions

1. How is the size of a machinist's vise determined?
2. How may finished surfaces be protected from becoming marred when clamped in a machinist's vise?
3. Name three kinds of pliers and give several uses for each.
4. Why are soft hammers used to aid assembly of hardened steel parts instead of hard hammers?
5. Describe how lead hammers are made.
6. How is a mallet different from a soft hammer?
7. What is the purpose of a square blade on a screwdriver?
8. Describe an offset screwdriver. For what is it used?
9. In what direction should the jaws point when using an adjustable wrench or a pipe wrench?
10. How does an open-end wrench differ from a box wrench?
11. Name two types of spanner wrenches.
12. What are socket wrenches? What special advantages do they have over box and open-end wrenches?
13. For what are hexagonal wrenches used?
14. Describe a torque wrench and tell how it is used.
15. How are powered wrenches powered?
16. For what is a drift punch used?
17. For what is a pin punch used?
18. For what is an arbor press used?

Mathematics

1. How many foot-pounds of torque would be applied to a bolt by a 10″ long wrench if 60 pounds of force were used?

Career Information

1. Investigate what working on an assembly line is like. Find out the level of skill required, pay, working conditions, etc.

UNIT 26

Fasteners

Bolts, Screws, Nuts, Washers, Shims, Pins, Keys, and Rivets

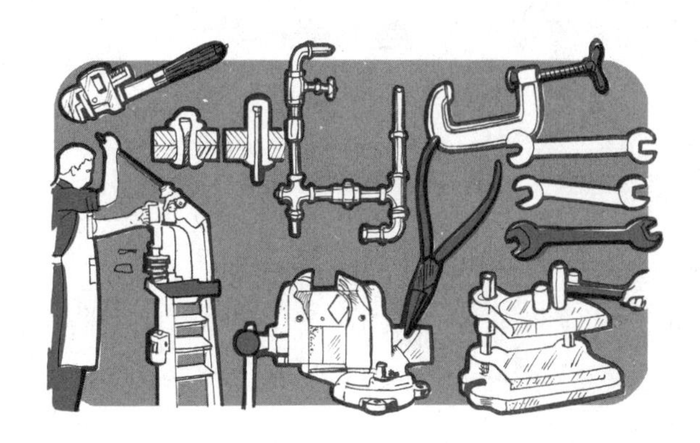

26-1. Metal Fasteners

Metal parts can be held together with any of several kinds of **metal fasteners,** such as rivets, bolts, screws, pins, and numerous special fastening devices. The worker must use good judgment in deciding upon the best kind of fastener.

26-2. Bolts and Screws

Bolts and screws are made in many shapes and sizes. The sizes of bolts and screws are measured by the diameter and length of the **body**; the **head** is not included in the length except on flat-head bolts and screws, Fig. 26-1. The kinds most used are shown in Fig. 26-2.

Rough and **semifinished** bolts and screws are rolled, pressed, hammered, or punched out of cold or hot metal. **Finished** bolts and screws are cut out of a bar of steel by a screw machine, which is a special automatic lathe.

26-3. Use of Bolts and Screws

Bolts and screws are usually used to fasten together parts which have to be separated later. A **bolt** is used where one can get at both sides of the work with wrenches. A **screw** is used where only one side can be reached with a wrench or screwdriver.

26-4. Carriage Bolts

A carriage bolt has a **round head,** Fig. 26-2. The part of the body under the head is square. It has a black **rough finish** and has the **Unified National Coarse thread.**

A carriage bolt is usually used to fasten a wooden part to metal. The square part under the head sinks into the wood, and thus the bolt cannot turn while the nut is being **screwed** on.

26-5. Machine Bolts

A machine bolt (see Fig. 26-2) has either a square or **hexagonal** head. It is made with a black, **rough finish** or **finished** all over. A machine bolt has a **Unified National Coarse thread (UNC)** or a **Unified National Fine thread (UNF).**

26-6. Tap Bolts

A tap bolt is like a **machine bolt** except that the whole body is threaded. It may be used with or without a nut, Fig. 26-2.

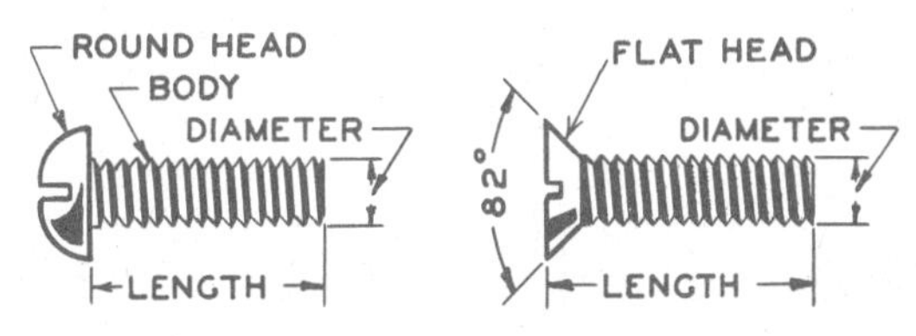

Fig. 26-1. Measurements of Bolts and Screws

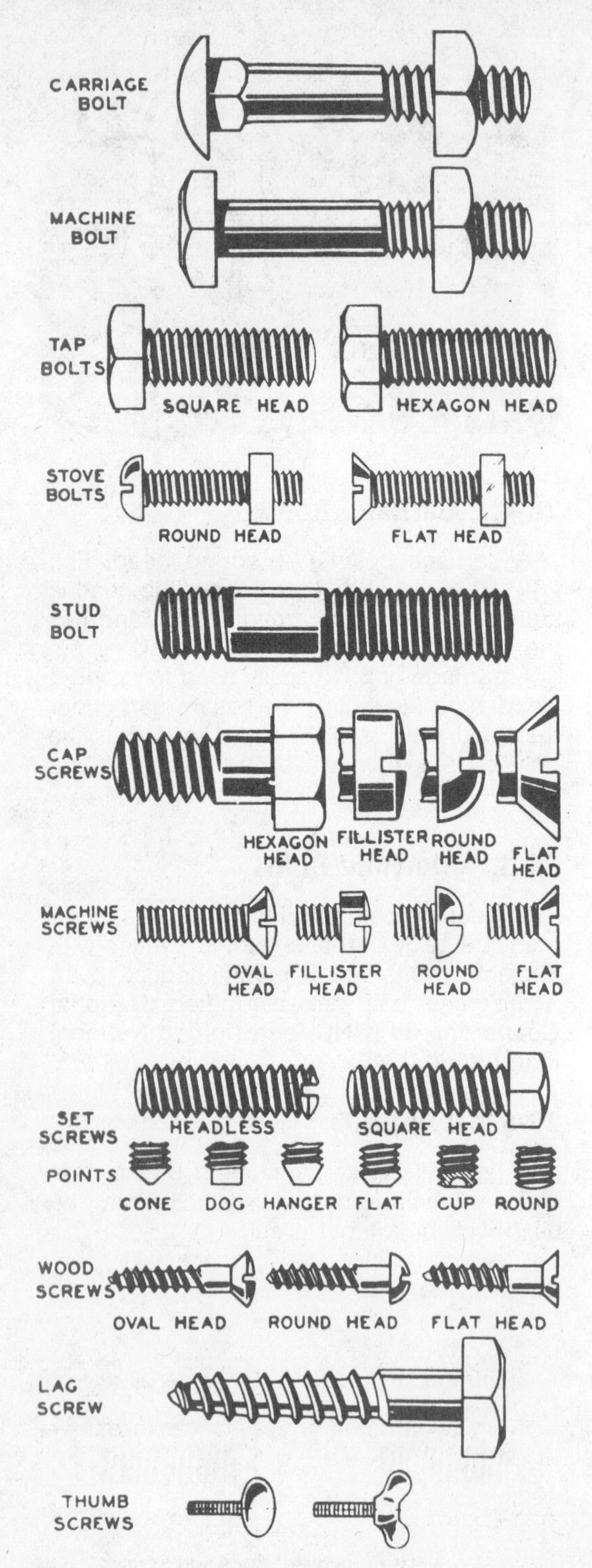

Fig. 26-2. Types of Bolts and Screws

26-7. Stove Bolts

A stove bolt has either a round or flat head which is slotted so that it can be turned with a screwdriver, Fig. 26-2. It is made with the **UNC thread.**

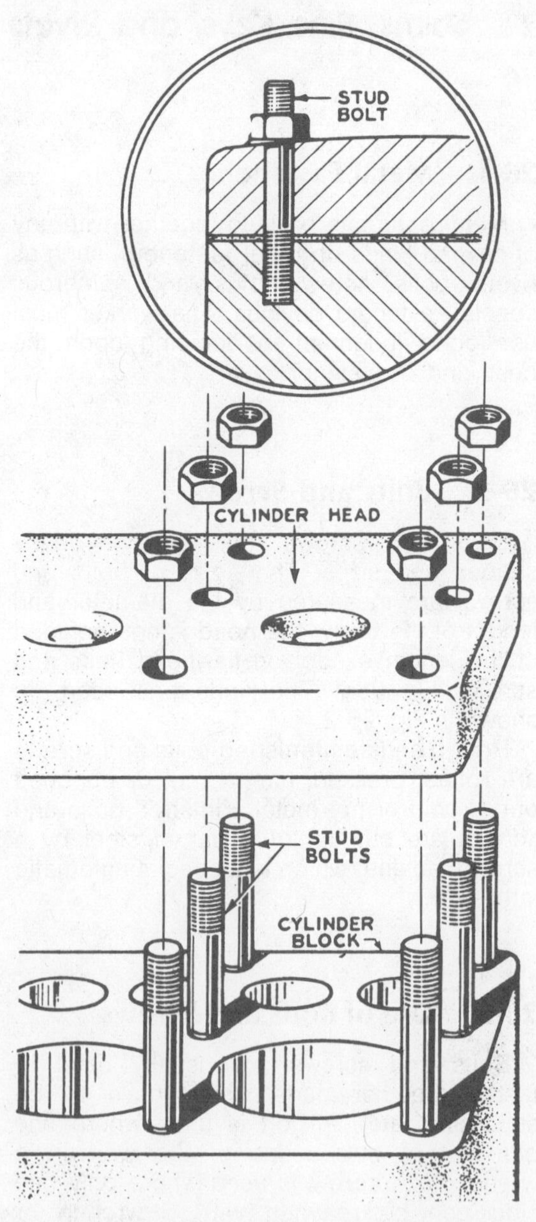

Fig. 26-3. Use of Stud Bolts on an Automobile Engine

26-8. Stud Bolts

A stud bolt has no head and is threaded on both ends, Fig. 26-2. One end of the stud bolt has more threads than the other. Its use is shown in Fig. 26-3. The **cylinder head** of an automobile engine is fastened to the **cylinder block** by screwing the nuts on the **stud bolts.**

26-9. Cap Screws

Cap screws are made with heads of several different shapes, Fig. 26-2. They are usually **finished** all over and are made with **UNC** or **UNF threads.** Cap screws are used when it is not convenient to get at both sides of the work with wrenches. The head of the cap screw presses against the top piece and holds the parts together, as shown in Fig. 26-4.

26-10. Machine Screws

Machine screws are made with heads of several different shapes (see Fig. 26-2) and are made with either the **UNC** or **UNF thread.** They are made of steel, aluminum, or **brass.** The smaller diameters are measured by **gage numbers.** The sizes range from number 0 (.060″) to 3⁄8″ in diameter. Note in Table 29, page 423, that the gage numbers are the same for both machine screws and wood screws.

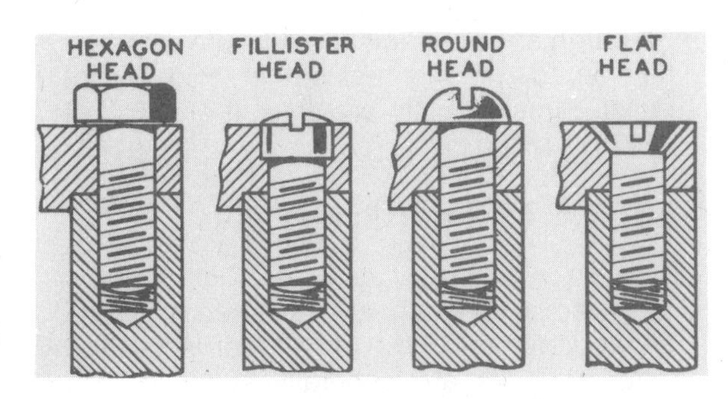

Fig. 26-4. Position of Cap-Screw Heads

26-11. Setscrews

Setscrews are made with **square heads** or are **headless,** Fig. 26-2. Both kinds are made with different **points.** Setscrews are **case-hardened** and are used to fasten pulleys and collars on shafts, as shown in Fig. 26-5. **Headless setscrews** are described in the next section.

26-12. Headless Setscrews

The headless setscrew is made for safety, Fig. 26-5. Screws with heads are dangerous on moving parts because the worker may be caught and injured.

There are two kinds of headless setscrews. One kind has a slot for a screw-

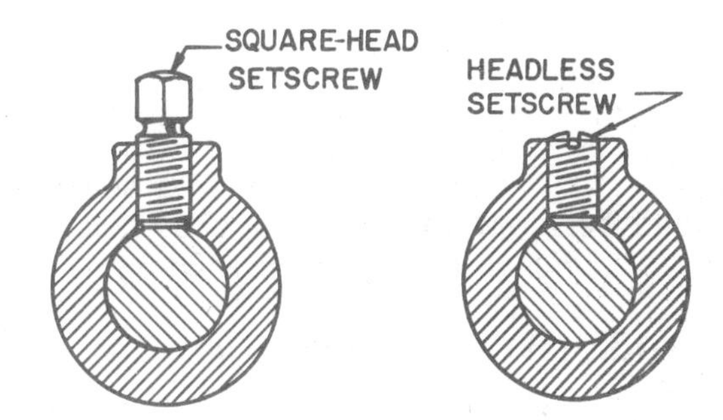

Fig. 26-5. Setscrews

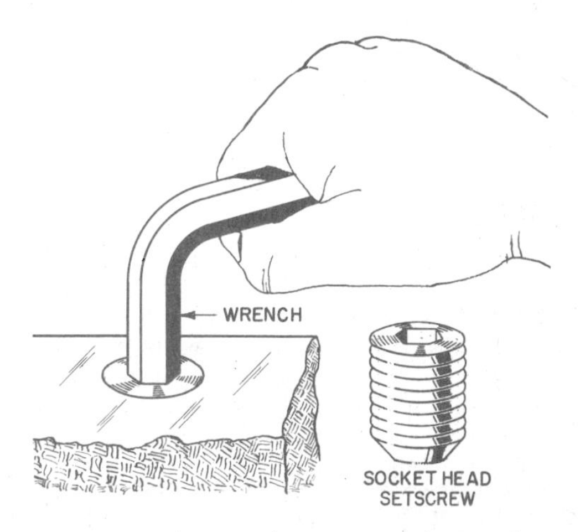

Fig. 26-6. Socket-Head Setscrew and Wrench

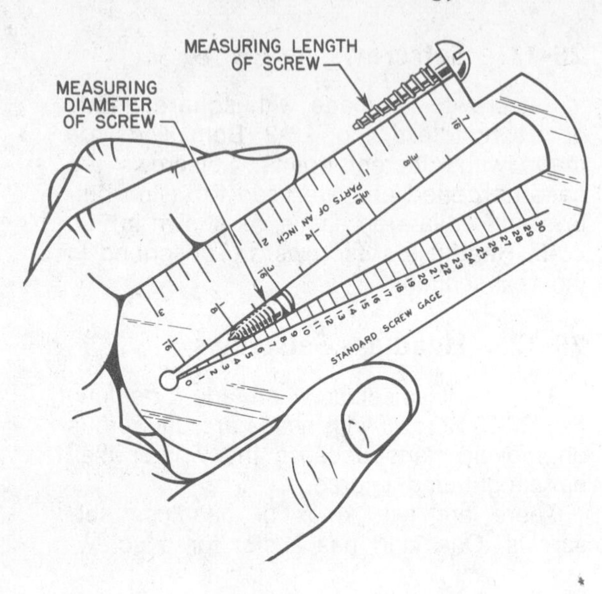

Fig. 26-7. Gaging Wood Screw with Screw Gage

driver. The other kind, known as a **socket-head setscrew,** has a **hexagonal** hole. A special wrench is needed, Fig. 26-6.

26-13. Wood Screws

Wood screws are often used to fasten metal parts to wood. They are made with flat, round, or oval heads, Fig. 26-2. The heads are slotted or recessed so they can be turned with screwdrivers. The angle of the flat head is 82°. Wood screws are made of steel, brass, and aluminum. Steel wood screws come in either bright[1] or blued finish, or they are plated with cadmium, nickel, or chromium.

The diameter of a wood screw is measured on the body under the head by the **American Standard Screw Gage.** Figure 26-7 shows how a wood screw is measured by placing it in the opening of the **screw gage** until it touches on both sides; the

[1]**Bright finish** is the natural color of steel made shiny by the finishing cut on a machine or by polishing.

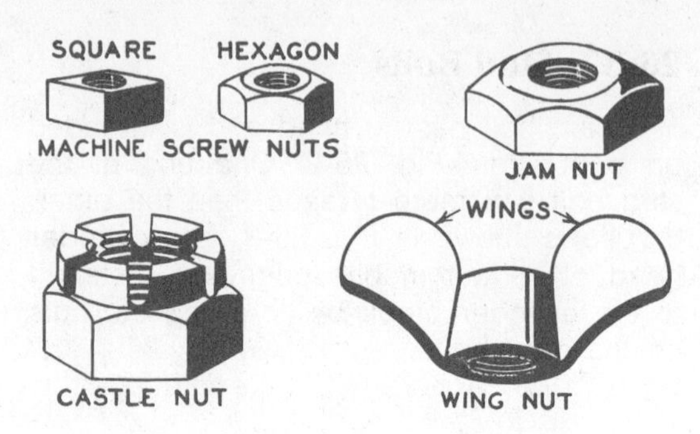

Fig. 26-8. Types of Nuts

number where it touches is the **gage number.**

26-14. Lag Screws

A lag screw has a **square head** like a bolt and is threaded like a **wood screw,** Fig. 26-2. It is used for heavy work such as fastening a machine to a wooden floor.

26-15. Thumbscrews

A thumbscrew is a screw with one or two **wings** or with a **knurled head.** It is used where a screw must be turned by the thumb and finger, Fig. 26-2.

26-16. Nuts

There are many different shapes and sizes of nuts; samples of these are shown in Fig. 26-8. The size of a nut is measured by the diameter of the bolt it fits. In other words, a ½" nut fits a ½" bolt.

Rough and **semifinished** nuts are pressed, hammered, or punched out of cold or hot metal. **Finished** nuts are cut out of a bar of steel by machine.

26-17. Machine Screw Nuts

A machine screw nut (see Fig. 26-8) is six-sided, or hexagonal. The thread may be either UNC or UNF. Stove, carriage, and machine bolts are generally supplied with square nuts.

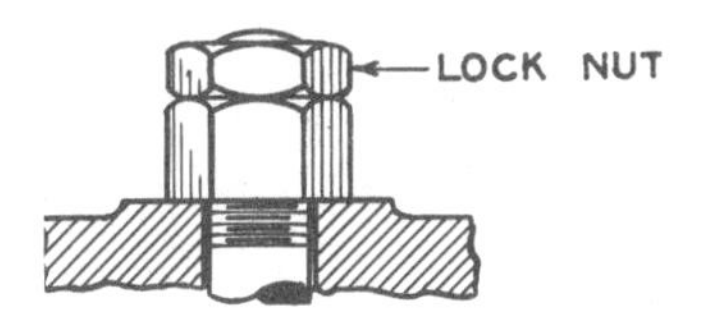

Fig. 26-9. Lock Nut

Fig. 26-10. Washers

26-18. Jam Nuts and Lock Nuts

A jam nut (see Fig. 26-8) is sometimes called a **lock nut** or **check nut.** It is thinner than an ordinary nut and is used as a lock to keep another nut from loosening by vibration, Fig. 26-9. Although the jam nut is usually put on last, the thick nut may be put on last to make use of the greater strength. Another type of lock nut, a preassembled washer and nut, is also available.

26-19. Castle Nuts

A castle nut (see Fig. 26-8) has slots across the top. The parts which extend upward make it look like a castle, hence the name. A **cotter pin** is slipped in a slot and through a hole in the bolt to lock the nut to the bolt and thus keep the nut from jarring off. Castle nuts are usually used to hold wheel bearings and wheels in place.

26-20. Wing Nuts

A wing nut (see Fig. 26-8) has two thin, flat **wings** and is used where a nut has to be turned with the thumb and finger.

26-21. Washers

Washers serve several purposes in fastener assemblies. They are used primarily as a bearing surface for bolts, nuts, and screws. They also serve to distribute the load over a greater area, protect the surface, and prevent movement of parts. They are sometimes used with rivets when fastening leather, fiber, canvas, and similar soft materials.

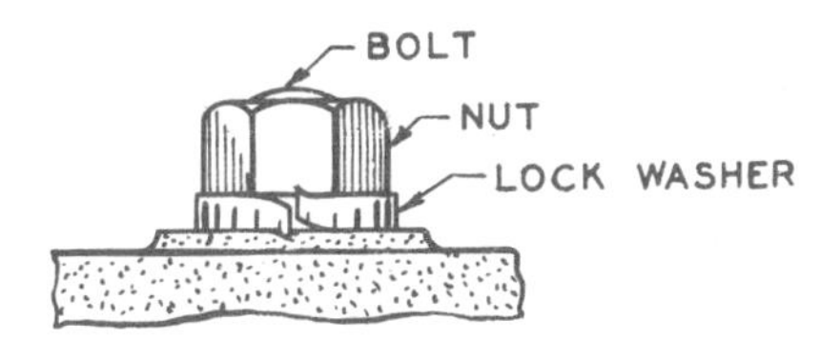

Fig. 26-11. Use of Lock Washer

The common **flat washer** is a thin, round, metal disk with a hole in the middle, Fig. 26-10. It is used as a **bearing surface** under a nut or under the head of a bolt or screw.

The size of a washer is measured by the diameter of the bolt that it fits; thus, a ½″ washer is for a ½″ bolt. Flat washers are sold by the pound.

26-22. Lock Washers

Lock washers serve as a spring takeup between bolts or screws and the workpiece. They also serve to lock the nut or screw in place, thus preventing movement or loosening due to vibration. The **helical spring-type** lock washer, Fig. 26-10, looks like a coil from a spring. Lock washers of this type are available in light, medium (regular), heavy, and extra-heavy types for screws and bolts from size No. 2 (.086″) to 3″ diameter. They are hardened and tempered and are used under a screw or nut to lock it in place so it will not jar loose, Fig. 26-11.

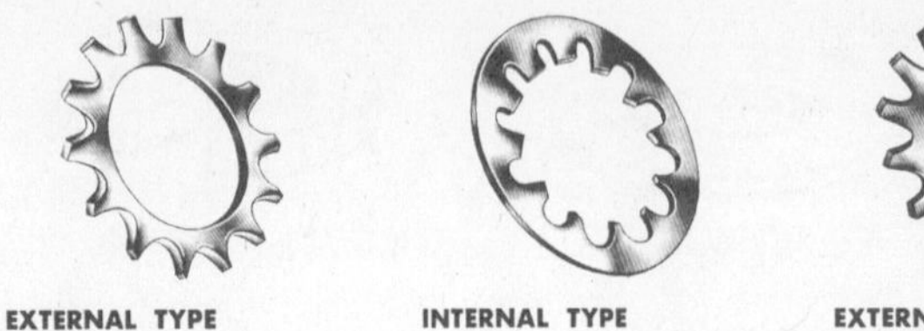

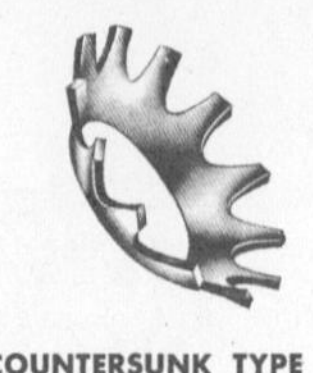

Fig. 26-12. Common Tooth-Type Lock Washers

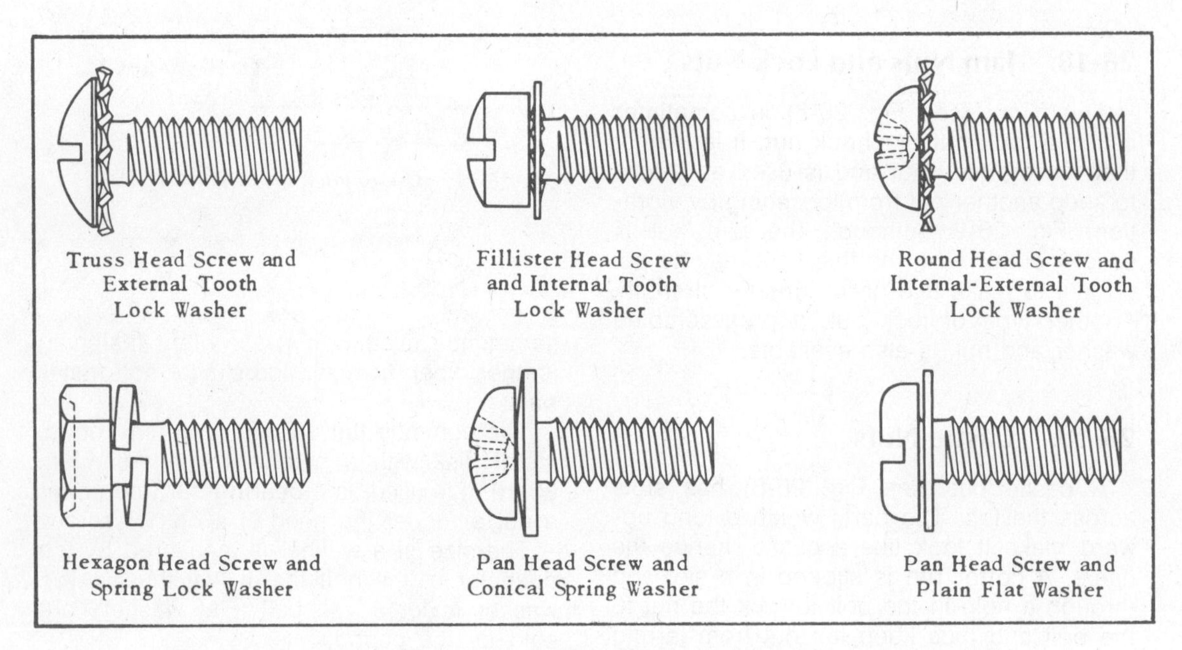

Fig. 26-13. Preassembled Screw and Washer Assemblies (ASA. B18.12—1962)

Tooth-type lock washers, Fig. 26-12, of hardened steel will wedge into the bearing surfaces to prevent bolts, nuts, or screws from turning or loosening due to vibration. Several standard types are shown in Fig. 26-12.

Preassembled **screw and washer assemblies,** called **sems** (Fig. 26-13) have a lock washer fitting loosely below the screw head. The expanded rolled thread diameter prevents the washer from falling off. They are used for more rapid assembly on modern assembly lines. Preassembled **lock washer and nut units** are also available. These also speed up assembly work.

26-23. Shims

A shim is a thin sheet of metal, wood, or paper placed between two surfaces to keep them a certain distance apart so that the shim is a support. The two halves of a **bearing** around a **shaft** may be separated a little by placing shims between them as shown in Fig. 26-14. This lessens the tightness on the shaft. As the bearing wears down and

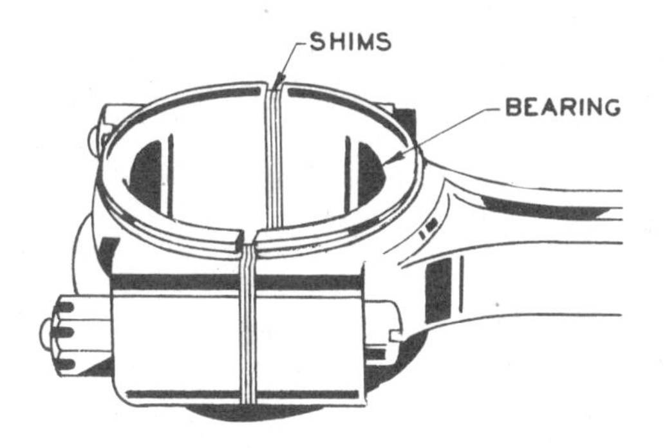

Fig. 26-14. Use of Shims between the Halves of a Bearing

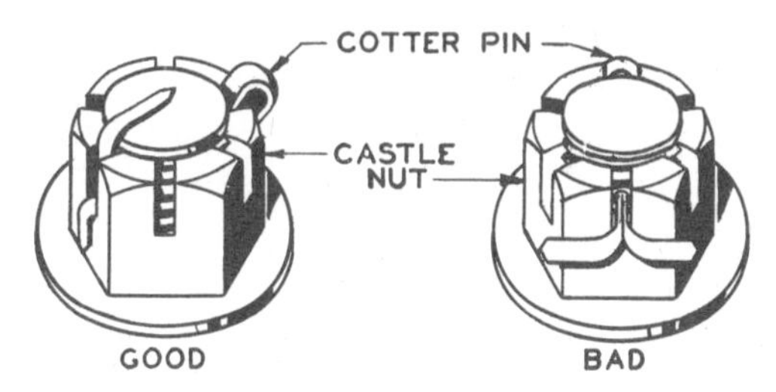

Fig. 26-15. Use of Castle Nut and Cotter Pin

gets loose, a shim may be removed to get a closer fit. A shim is used as a support when, for example, it is placed under a leg of a machine so that it will be level.

26-24. Cotter Pins

A cotter pin, also called a **cotter key,** is made of wire. It is slipped through a hole in a bolt behind a nut to keep the nut from turning. Note the right way to lock the nut, Fig. 26-15. The head of the cotter pin should fit into the slot of the nut; one leg should be bent over the end of the bolt and the other leg should be bent over the side of the nut.

26-25. Tapered Pins

A tapered pin is often used on a job such as fastening a pulley or collar to a shaft, Fig. 26-16. The **taper** equals ¼″ per foot. Tapered pins are made in lengths from ⅜″ to 6″. The hole into which the pin fits is first drilled and then **reamed** with a **taper-pin reamer,** which has the same taper as the pin. Taper pins are made in 17 standard sizes which are designated by numbers. The sizes are included in handbooks for machinists.

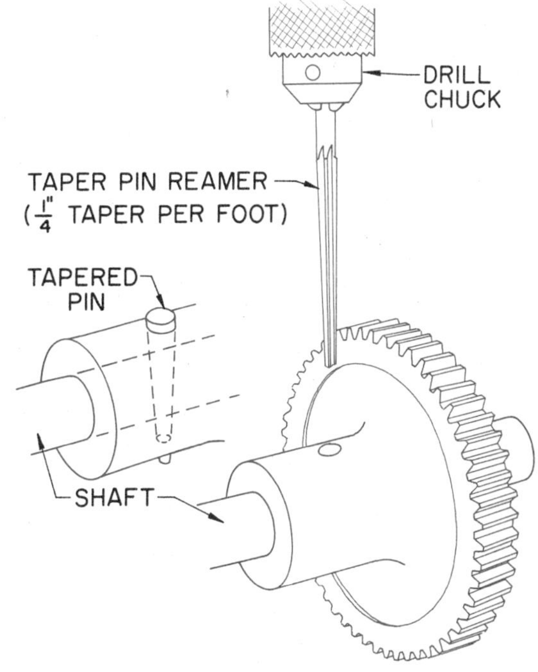

Fig. 26-16. Reaming a Hole with a Taper Pin Reamer for a Tapered Pin

26-26. Keys

Keys are made in several shapes. They are used to keep pulleys and gears from moving on **shafts,** Fig. 26-17. Half of the key fits in a **keyway,** which is a slot in the shaft, and the other half fits into a slot in the pulley or gear; the pulley or gear is thus fastened to the shaft.

The **square key,** also known as a **feather key,** is the one that is most used.

The **gib-head key** is useful where it is necessary to remove the key from one side of the pulley or gear. A **wedge** may be used to back the key out of the hole.

26-27. Retaining Rings

Retaining rings, Fig. 26-18, are a relatively new type of fastener used in assembling parts of modern metal products. They generally are inserted and seated in either internal or external grooves with a special plier-like tool, Fig. 26-19. Self-locking types do not require a seating groove. The principal function of retaining rings is to provide a shoulder for holding, locking, or positioning parts of assemblies. Internal-type retaining rings are used in bored holes. External types are used on shafts or studs.

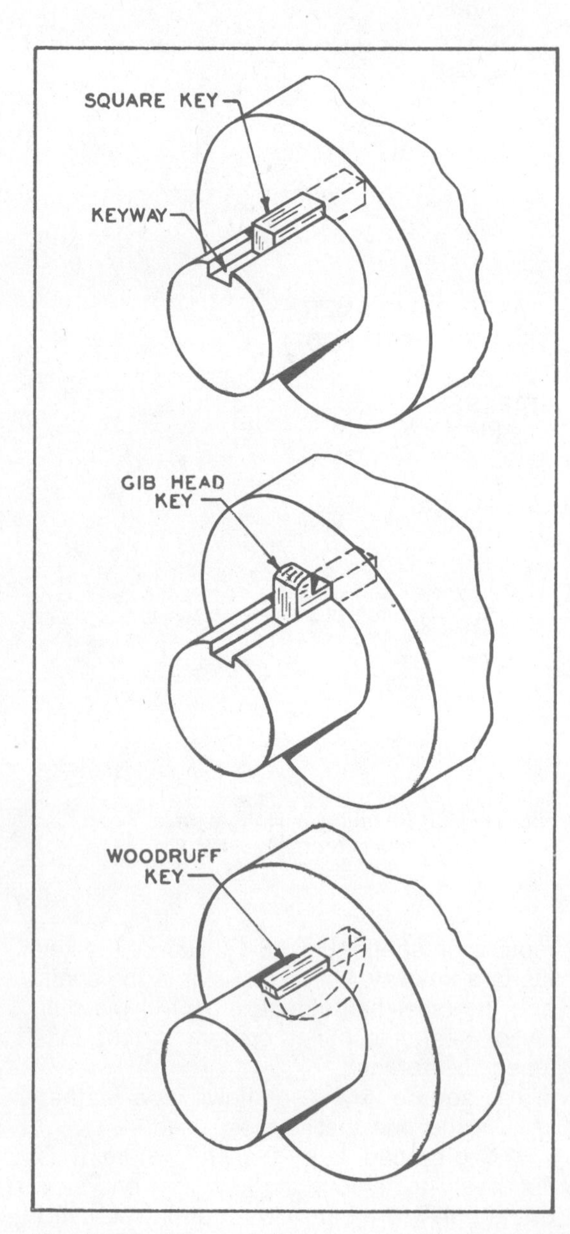

Fig. 26-17. Types of Keys

26-28. Self-Tapping (Sheet Metal) Screws

Self-tapping screws (Figs. 26-20, 26-23, 26-24, and 26-25) cut their own threads in soft steel, aluminum, and other soft metals. They are driven into punched or drilled

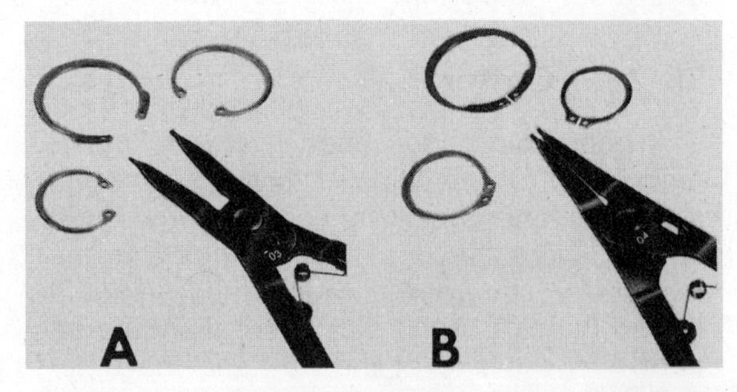

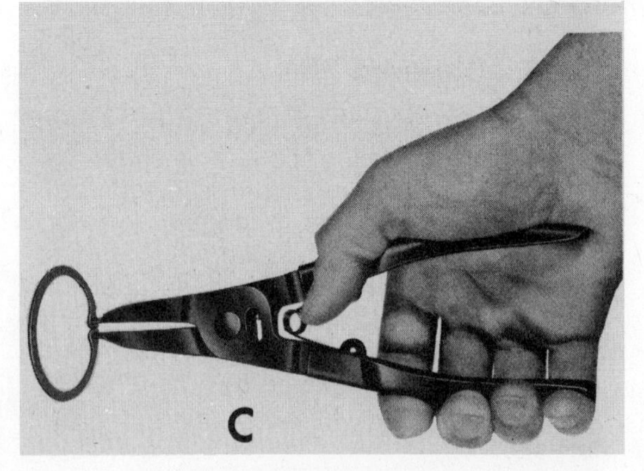

Fig. 26-19. Retaining Ring Pliers Used to Install Retaining Rings
A. Internal Ring
B. External Ring Pliers
C. Insert Pliers into the Ring

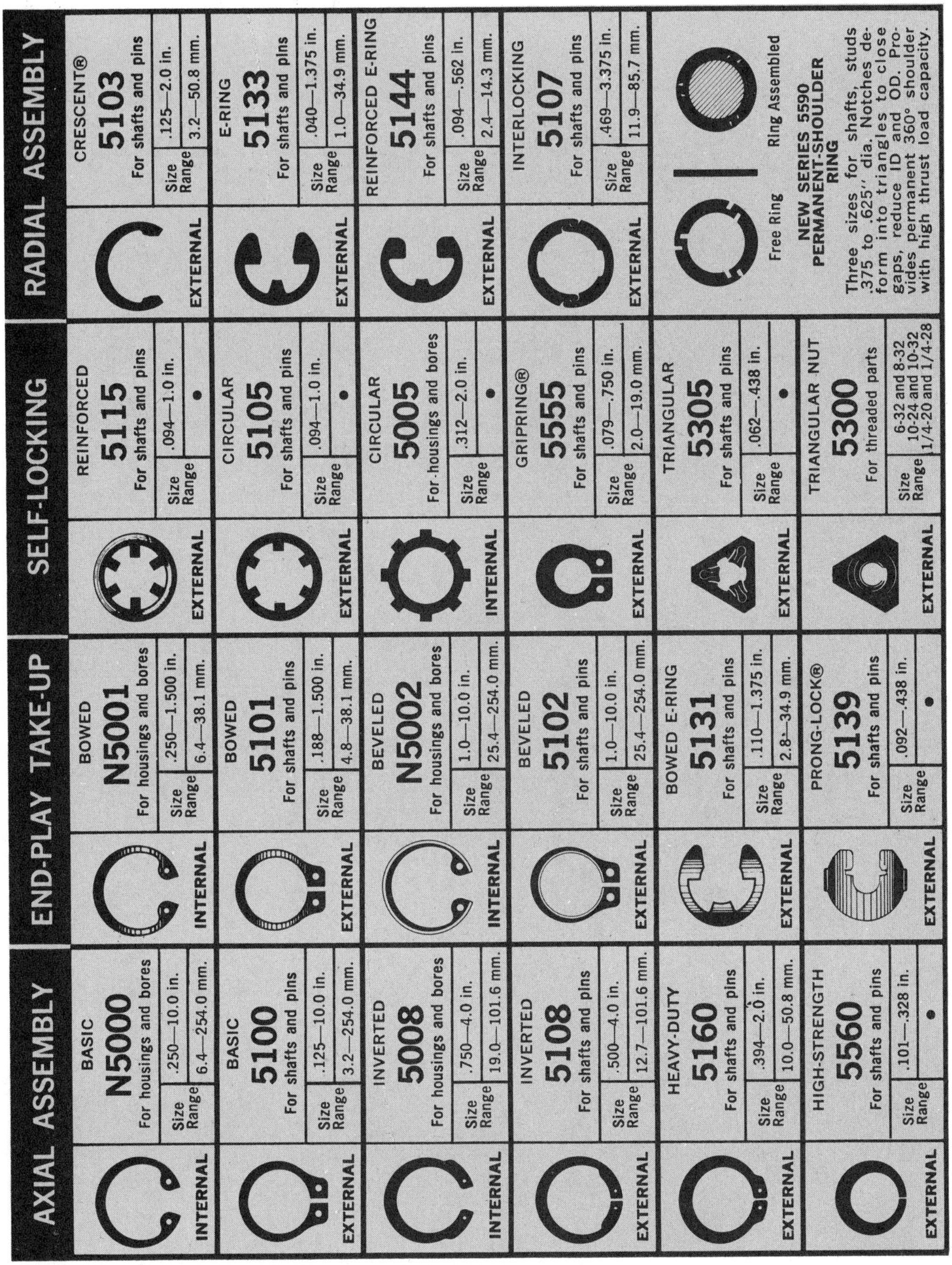

AXIAL ASSEMBLY

Type	Style	No.	Use	Size Range (in.)	Size Range (mm.)
INTERNAL	BASIC	N5000	For housings and bores	.250—10.0 in.	6.4—254.0 mm.
EXTERNAL	BASIC	5100	For shafts and pins	.125—10.0 in.	3.2—254.0 mm.
INTERNAL	INVERTED	5008	For housings and bores	.750—4.0 in.	19.0—101.6 mm.
EXTERNAL	INVERTED	5108	For shafts and pins	.500—4.0 in.	12.7—101.6 mm.
EXTERNAL	HEAVY-DUTY	5160	For shafts and pins	.394—2.0 in.	10.0—50.8 mm.
EXTERNAL	HIGH-STRENGTH	5560	For shafts and pins	.101—.328 in.	•

END-PLAY TAKE-UP

Type	Style	No.	Use	Size Range (in.)	Size Range (mm.)
INTERNAL	BOWED	N5001	For housings and bores	.250—1.500 in.	6.4—38.1 mm.
EXTERNAL	BOWED	5101	For shafts and pins	.188—1.500 in.	4.8—38.1 mm.
INTERNAL	BEVELED	N5002	For housings and bores	1.0—10.0 in.	25.4—254.0 mm.
EXTERNAL	BEVELED	5102	For shafts and pins	1.0—10.0 in.	25.4—254.0 mm.
EXTERNAL	BOWED E-RING	5131	For shafts and pins	.110—1.375 in.	2.8—34.9 mm.
EXTERNAL	PRONG-LOCK®	5139	For shafts and pins	.092—.438 in.	•

SELF-LOCKING

Type	Style	No.	Use	Size Range (in.)	Size Range (mm.)
EXTERNAL	REINFORCED	5115	For shafts and pins	.094—1.0 in.	•
EXTERNAL	CIRCULAR	5105	For shafts and pins	.094—1.0 in.	•
INTERNAL	CIRCULAR	5005	For housings and bores	.312—2.0 in.	•
EXTERNAL	GRIPRING®	5555	For shafts and pins	.079—.750 in.	2.0—19.0 mm.
EXTERNAL	TRIANGULAR	5305	For shafts and pins	.062—.438 in.	•
EXTERNAL	TRIANGULAR NUT	5300	For threaded parts	6-32 and 8-32 10-24 and 10-32 1/4-20 and 1/4-28	

RADIAL ASSEMBLY

Type	Style	No.	Use	Size Range (in.)	Size Range (mm.)
EXTERNAL	CRESCENT®	5103	For shafts and pins	.125—2.0 in.	3.2—50.8 mm.
EXTERNAL	E-RING	5133	For shafts and pins	.040—1.375 in.	1.0—34.9 mm.
EXTERNAL	REINFORCED E-RING	5144	For shafts and pins	.094—.562 in.	2.4—14.3 mm.
EXTERNAL	INTERLOCKING	5107	For shafts and pins	.469—3.375 in.	11.9—85.7 mm.

Free Ring | Ring Assembled

NEW SERIES 5590 PERMANENT-SHOULDER RING

Three sizes for shafts, studs .375 to .625'' dia. Notches deform into triangles to close gaps, reduce ID and OD. Provides permanent 360° shoulder with high thrust load capacity.

Fig. 26-18. Truarc (Trade-Mark) Retaining Rings

	Type	USA Standard	Manufacturer
THREAD FORMING TAPPING SCREWS		AB*	AB*
	NOT RECOMMENDED – USE TYPE AB**	A	A
		B	B
		BP	BP
		C	C
THREAD CUTTING TAPPING SCREWS		D	1
		F	F
		G	G
		T	23
		BF	BF
		BT	25
		U	U

Fig. 26-20. Type Designation of Tapping Screws and Metallic Drive Screws
See Figs. 26-23 and 26-24

holes which are slightly larger than the minor diameter of the screw head. As they are driven into the hole, they produce threads in either one or both parts being fastened.

Self-tapping screws eliminate the need for a tapping operation. They are used for economical assembly of sheet metal and other sheet materials such as plastic, plywood, asbestos, and fiber materials. They are used by sheet metal-workers for installing heating, ventilation, and air-conditioning ducts and equipment. They are used for assembling many parts on automobile bodies, radio and television chassis, stoves, refrigerators, and other appliances used in the home.

There are many kinds of self-tapping screws available. They are available with a wide variety of head styles, Fig. 26-21. A variety of driving recesses, including clutch heads, slotted heads, phillips recessed heads, and hexagonal heads are available. See Fig. 26-22. Some of the common head styles used on self-tapping screws are shown in Figs. 26-23 and 26-24. Self-tapping screws are made in diameters according to screw gage numbers, which are the same as those used for wood screws and machine screws. See Table 29, page 423. Most of the many types of self-tapping

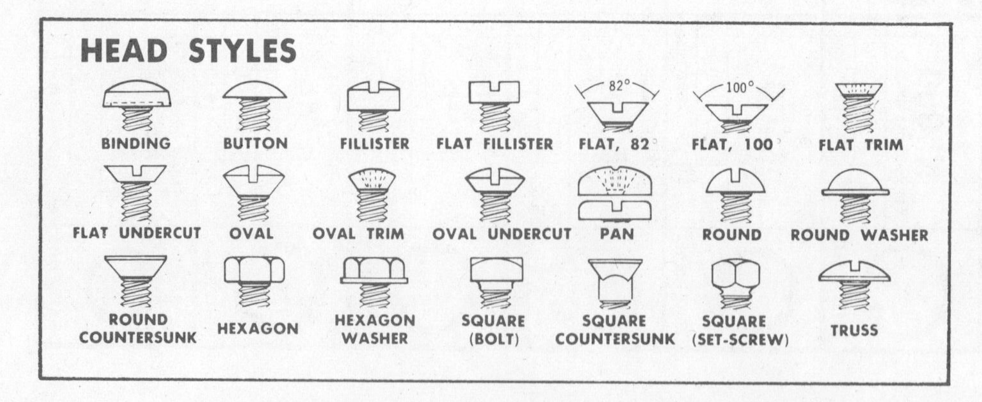

Fig. 26-21. Kinds of Head Styles Used on Threaded Fasteners

screws may be classified under the following three headings:

1. Thread-Forming (sheet metal) Screws, Fig. 26-23.
2. Thread-Cutting Screws, Fig. 26-24.
3. Metallic Drive Screws, Fig. 26-25.

Thread-Forming Screws

Self-tapping screws within this classification include types A, B, C, AB, and BP, Figs. 26-20 and 26-23. These thread-forming screws form **chip-free** mating threads by squeezing and displacing metal as they are driven into punched or drilled holes. Types A and B, Fig. 26-20, are also known as **sheet metal** screws. They are available with a variety of heads, Fig. 26-21.

Type-A (sheet metal) screws have a relatively wide-spaced, coarse-pitch thread with a **gimlet** point. They are used for fastening thin sheet metals with thicknesses up to and including 20 gage or 1/32 inch. They are also used for fastening resin-impregnated plywood and asbestos compositions. Because of the gimlet point, exact alignment of the workpieces is not always necessary. This type of thread point starts into the workpiece rapidly. It also exerts very high pressures on the material as it forms the thread. Hence, screws of this type are not recommended for use with brittle plastics or other brittle materials. The more recently designed type-AB screw is rapidly replacing the type-A screws.

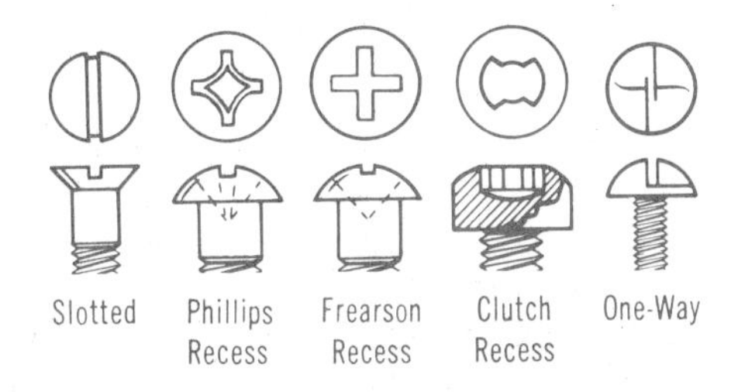

Fig. 26-22. Kinds of Driving Recesses Used on Screws

Type-B (sheet metal) screws have a narrower spaced, finer pitch thread than the type-A screws, and they have a blunt point, Fig. 26-20. The blunt point requires good alignment of the workpieces and holes. Type-B screws are recommended for both thin and thicker sheet metals, ranging from 0.015″ to 0.200″ thickness. They are also recommended for use on aluminum and other nonferrous metal castings, asbestos compositions, and fiber materials.

Type-BP (sheet metal) screws, Fig. 26-20, have essentially the same threads as the type-B screws, except that the type-BP has a sharp point to aid in correcting misalignment of the holes.

Type-AB (sheet metal) screws, Fig. 26-20, have the same thread as the type-B screws, but they also have a gimlet point as on the type-A screws. The type-AB screws are

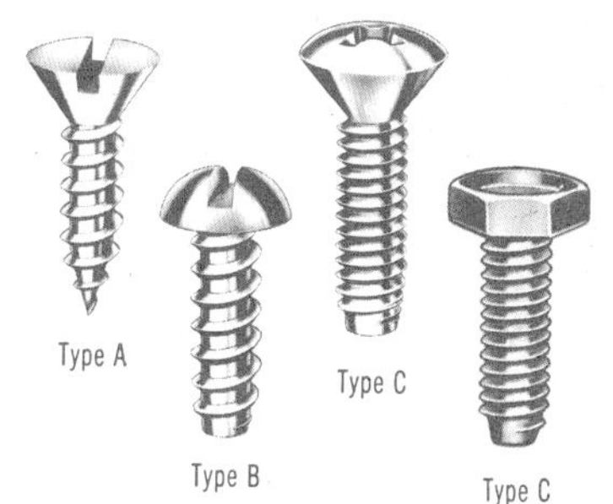

Fig. 26-23. Thread-Forming Tapping Screws

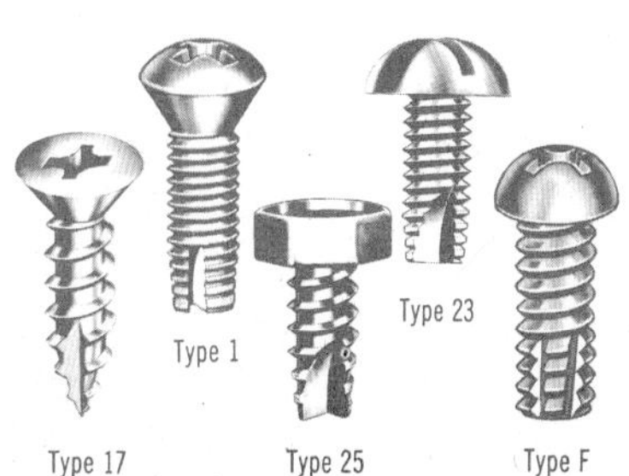

Fig. 26-24. Thread-Cutting Tapping Screws

rapidly replacing type-A screws and also type-BP screws. They are recommended and preferred for most applications for which both type-A and type-BP screws are used. They are used for thin sheet metals, resin-impregnated plywood, wood, and asbestos compositions where a sharp point is preferred. Use No. 6 screws for thin sheet metals up to 20 gage. Use larger screws for thicknesses up to 18 gage.

Type-C self-tapping screws have finer threads which are essentially the same as the threads on standard machine screws. They also have a blunt point which requires good hole alignment. Type-C screws are recommended for thicker sheet metals, ranging from 0.030 to 0.100″ thickness. They are recommended where the chips from **thread-cutting** screws are objectionable.

Sizes

Thread-forming screws range from No. 0 to ½″ diameter and from 3/16″ to 2″ length. Common gages used for sheet metal products in the school shop include No. 6 diameter by ⅜″ length, No. 6 diameter by ½″ length, and No. 8 diameter by ½″ length.

Procedure for Installing Self-tapping Screws

1. Locate and prick punch the location for the hole.
2. Punch or drill the holes. The drill or punch should be slightly larger (several thousandths of an inch) than the minor diameter of the screw. For example, a 7/64″ drill is used for a No. 6 type-A screw. A ⅛″ drill is used for No. 8 type-A screw. The proper size can be estimated by holding the drill behind a screw. The minor diameters or hole sizes for the various kinds and sizes of self-tapping screws are included in standard handbooks for machinists.
3. Align the holes with a punch. Clamp the two pieces together firmly and insert the screw with a screwdriver.

Thread-Cutting Screws

Screws in this classification include several types of self-tapping screws, Figs. 26-20 and 26-24, which form threads by actual removal of metal chips with a cutting action. The screw is hardened steel, and it cuts the mating thread in a manner similar to a tap cutting a standard machine screw thread.

Several types of self-tapping, thread-cutting screws are shown in Fig. 26-20. Type-BF and BT screws have blunt points and spaced threads as on type-B threads. They are used with thin materials, plastics, soft nonferrous metals, and die castings.

Type-D, F, G, and T thread-cutting screws, Fig. 26-20, have fine threads which are similar to standard machine screw threads. These screws are hardened and have blunt ends. The ends have tapered entering threads with one or more chip cavities for cutting mating threads.

Thread-cutting screws are used for fastening sheet metal, structural steel, cast iron, nonferrous forgings, plastics, aluminum, and zinc. They are available in sizes ranging from No. 0 to ½″ diameter and in lengths from ⅛″ to 2½″. They are also available with a wide variety of head shapes.

26-29. Metallic Drive Screws

Metallic drive screws, type-U, Figs. 26-20 and 26-25, are made of hardened steel and have multiple threads and a pilot-type nose. They are a thread-forming type screw and form chip-free threads by displacing metal. The screw is driven or forced into drilled or punched holes in workpieces to be held together, thus forming threads. The holes should be slightly larger in diameter than the pilot end of the screw. They are intended for permanent assembly. Metallic drive screws are used on ferrous and nonferrous castings, plastics, and sheet metals from 0.060″ to ½″ thickness. They should not be driven into materials of thickness less than the diameter of the screw.

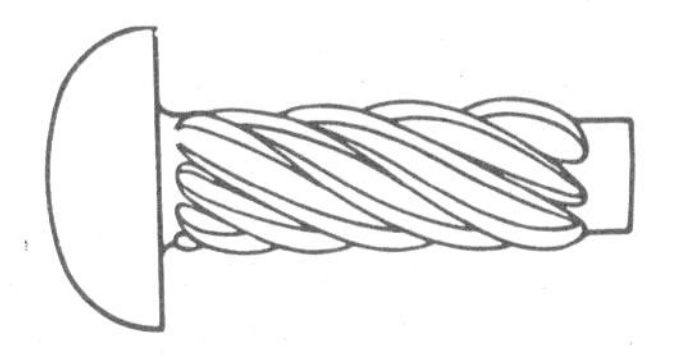

Fig. 26-25. Metallic Drive Screws (Type U)

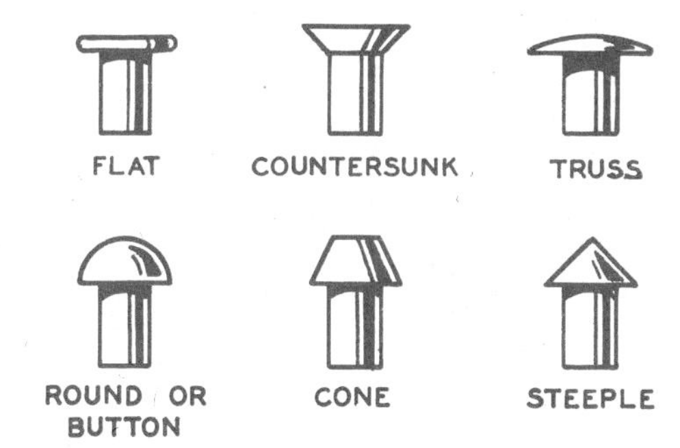

Fig. 26-26. Rivets

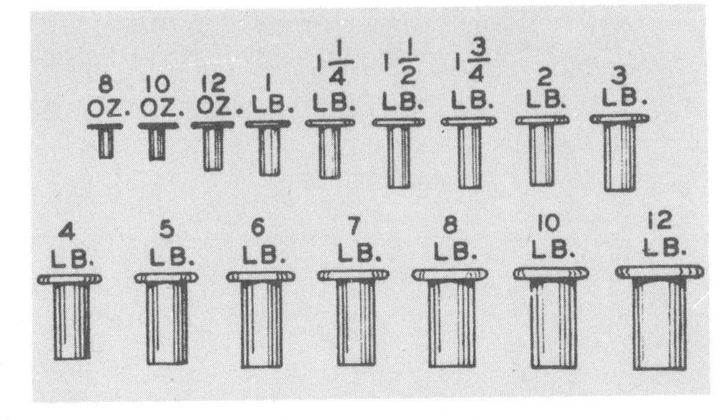

Fig. 26-27. Tinners' Rivets (Actual Size)

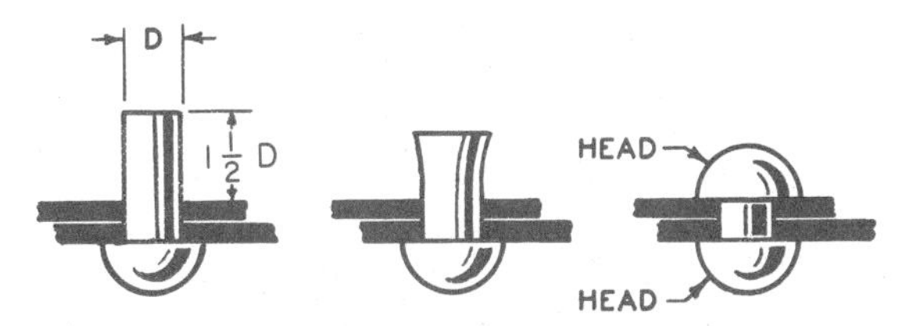

Fig. 26-28. Amount of Rivet Needed to Form Head

26-30. Rivets and Riveting

Rivets are metal pins that look like bolts without threads. They are made of different metals, such as soft iron or steel, aluminum, copper, and brass. They are available in many different sizes and shapes of heads. The most common kinds of heads are shown in Fig. 26-26. They may be either **solid,** as shown in Figs. 26-26 and 26-27, or they may be of **tubular** or special form, as shown in Fig. 26-34.

Riveting is the fastening of pieces of metal or other material together with **rivets,** Figs. 26-26 and 26-27. Rivets are used to hold pieces together permanently. The rivet is put through holes in the pieces to be fastened together, Fig. 26-28. The rivets may be solid or they may be hollow with a tubular form, Fig. 26-34. The small end of solid rivets is hammered into the form of a head, as in Figs. 26-30 and 26-31. Hollow rivets are **clinched** at the small end with a special riveting tool. See Figs. 26-35 and 26-36.

Many kinds of rivets are used for fastening pieces of metal together. Nonmetallic materials such as leather, plastic, fiber, and canvas are also fastened with rivets. Special kinds of rivets are used for fastening these materials. In this unit, we are principally concerned with the kinds of rivets and riveting procedures used for riveting metals.

Many metal-fastening jobs which were formerly riveted are now welded. This is true of structural-steel products such as bridges and steel frames of large buildings. However, riveting is still widely used for fastening metals and other nonmetallic materials.

Rivets are used for fastening metals which are not easily welded, or where welding is not practical. Rivets are used instead of welding in cases where the heat required for welding would reduce the strength of the metal or cause it to warp severely. They are often used to fasten aluminum sheet metal in the construction of aircraft, small boats, and other aluminum products. In modern manufacturing plants, riveting is done rapidly and economically with special riveting tools and machines.

The heads of rivets are sometimes used to decorate and add beauty to an object.

Size

The size of a rivet is measured by the diameter and length of the body. The head is not included in the length except on those designed to be countersunk. They generally are available in diameters ranging from 1/8″ to 3/8″, and in lengths from 1/4″ to 3″. The kinds most commonly used for hand riveting in school shops and maintenance shops are solid rivets 1/8″, 5/32″, and 3/16″ diameter with flat or round heads. Rivets 1/8″ in diameter with round heads are commonly used in making art metal projects.

Solid rivets generally are sold by the pound and in boxes of 100 or 1000.

Tinner's rivets, Fig. 26-27, are used for riveting sheet metal. They are made of soft steel, either plain or coated with tin. The tin coating makes them easier to solder and makes them rust resistant. The sizes of tinner's rivets are given in ounces or pounds per 1000; a 6-oz. rivet means that 1000 rivets weigh 6 ounces; a 2-lb. rivet means that 1000 of these rivets weigh 2 pounds. As the weight increases, so does the diameter and length. See Fig. 26-27.

26-31. Choosing a Rivet

Choose a rivet that is .003″ to 1/64″ smaller in diameter than the holes in the pieces to be riveted. It should be long enough to extend through the pieces to be riveted, plus enough metal from which to form a **head,** which is about 1½ times the diameter of the rivet, Fig. 26-28.

If the rivet is part of the design of the project, select a round-head rivet. If the rivet is not to be noticed, use a flat-head rivet. If the back of the work must be flush when using round-head rivets, countersink the hole and rivet as shown in Fig. 26-33. Allow the rivet to extend a small amount above the surface, just enough to fill the countersunk hole when headed. The rivet selected generally should be made of the same material as the metal being riveted.

Fig. 26-29. Rivet Set

26-32. Rivet Spacing

As a general rule, rivets should not be spaced closer together than three times the diameter of the rivet. Generally, for adequate strength, they should not be spaced farther apart than 24 times the diameter. Thus, the minimum space between two 1/8″ diameter rivets is 3 × 1/8″, or 3/8″ apart. The maximum distance recommended between two 1/8″ diameter rivets is 24 × 1/8″, or 3″. See Fig. 26-38.

26-33. Rivet Set

A rivet set is a **hardened-steel** tool with a hollow in one end, Fig. 26-29. It is used to shape the end of a rivet into a round, smooth head. Rivet sets are made in various sizes, designated by the following numbers: 00, 0, 1, 2, through 8. A No. 8 rivet set generally is used for 10- or 12-ounce tinner's rivets. The No. 8 is the smallest size, and the No. 00 is the largest.

26-34. How to Rivet

Holes are first punched or drilled through the metal pieces. On thin sheet metals, the holes are punched in one of three ways: (1) with a pin punch, Fig. 31-8; (2) with a solid punch, Fig. 31-8, and (3) with a hand punch, Fig. 31-8. The procedure for punching holes is explained in section 31-9. On thick sheet metals or on metal bars, the holes generally are drilled.

The holes in the two pieces should be carefully laid out. Put the rivet through the holes, press the pieces together, and place the head of the rivet on something solid, such as a steel block. The head of the rivet can be kept from flattening by resting it on a **riveting block,** which is made of steel and has a hollow like the shape of the rivet head, Fig. 26-30. Strike blows in the center of the end of the rivet with either the **face** or the **peen** of a **ball peen hammer** until the end of the rivet is spread out a little. Then strike it with the peen until it is quite round on top like a mushroom, Fig. 26-31. Too much hammering will bend the metal out of shape. A **riveting hammer,** Fig. 31-5, may be used instead of a ball peen hammer for riveting if desired. The flat face of the hammer may be used to flatten the rivet slightly and fill in the hole. The head is then rounded off with lighter blows. Next, place the **rivet set** on the hammered end of the rivet and strike the rivet set, Fig. 26-32. A round, smooth head is thus formed.

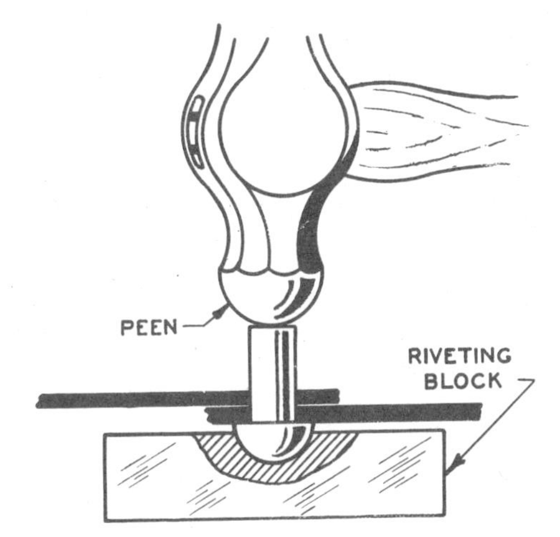

Fig. 26-30. Strike First in the Center of Rivet

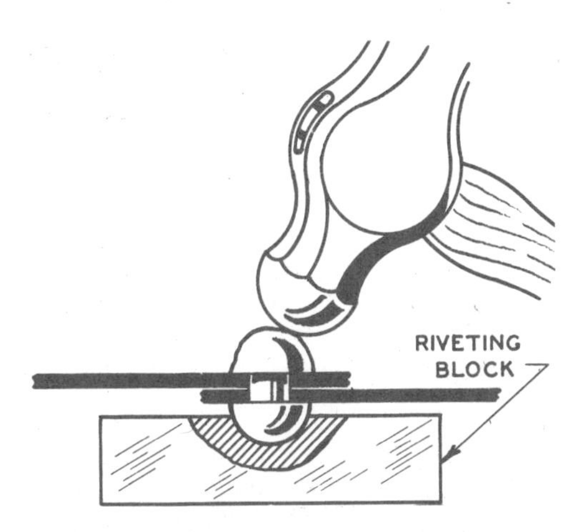

Fig. 26-31. Rounding the End of the Rivet by Peening

26-35. Hot and Cold Rivets

Iron and steel get larger when heated; that is, the metal **expands.** Iron and steel expand 1/8″ to 3/16″ per foot when red hot. A piece of steel 1′ long when cold will measure 12 1/8″ to 12 3/16″ when red hot. Thus, an iron rivet gives more strength when heated before riveting because it **shrinks** as it cools, holding the pieces more tightly together. Large rivets, such as those used to rivet structural steel beams together, are hammered when hot. Small rivets are hammered when cold.

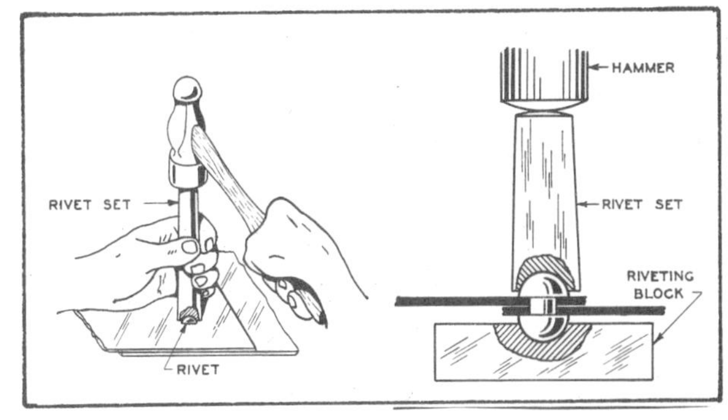

Fig. 26-32. Forming the Rivet Head with the Rivet Set

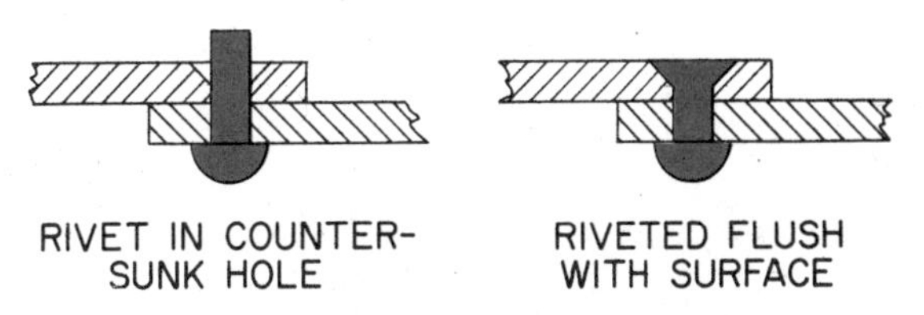

Fig. 26-33. Form Rivet Head Flush with Surface in Countersunk Hole

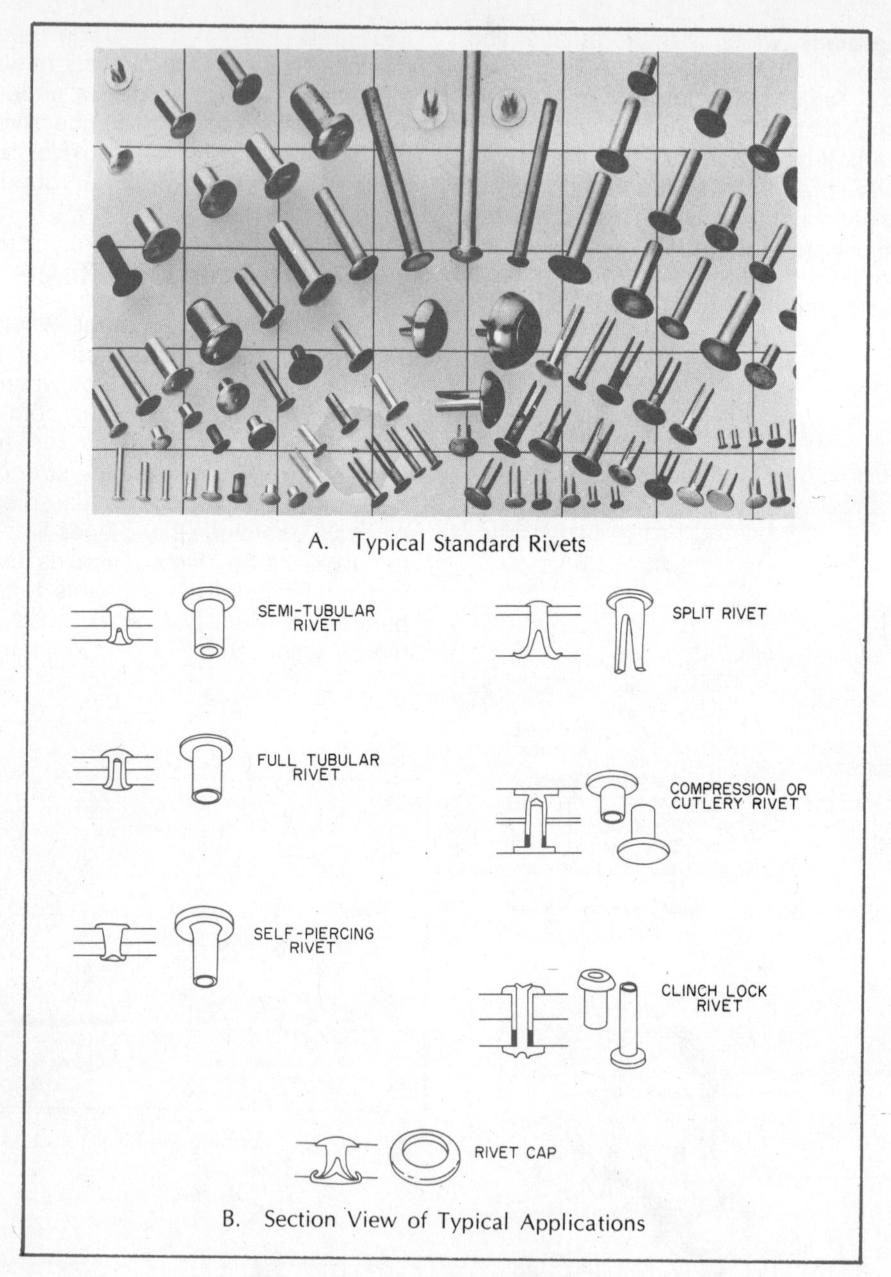

Fig. 26-34. Standard and Special Tubular Rivets

26-36. Removing Rivets

Rivets are used to fasten pieces together permanently, but it is sometimes necessary to remove them. Figure 13-9 shows how the head of a rivet is cut off with a **cold chisel** and hammer. After the head is cut off, the rest of the rivet may be driven out with a **drift punch** and **hammer.** Blind rivets may be removed by drilling.

26-37. Tubular and Special Rivets

A wide variety of tubular rivets of standard and special design are used in producing many appliances and hardware items. Several common types are shown in Fig. 26-34. In modern production procedures, these rivets generally are inserted in the stock and clinched with special riveting machines or tools.

26-38. Blind Rivets

Blind rivets are so named because they can be inserted and set from the same side of the workpiece. Solid rivets and other kinds of rivets generally require access to both sides of the work being riveted. Blind rivets, often called **Pop**[1] rivets, clinch inside with a pull from the tool from the outside, as shown in Fig. 26-35. The stem, or mandrel, fractures and breaks off when the head

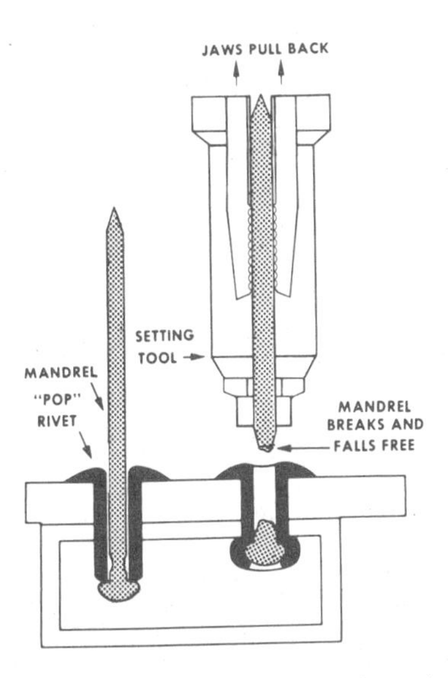

Fig. 26-35. Blind Rivets Clinch Inside with a Pull from the Tool Outside

[1]**Pop Rivet** is the trademark for blind rivets manufactured by the USM Fastener Company, Division of the United Shoe Machinery Corporation.

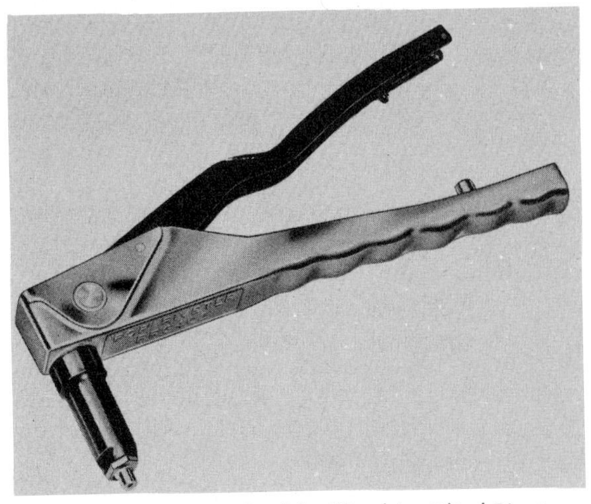

Fig. 26-36. Riveting Tool for Clinching Blind Rivets

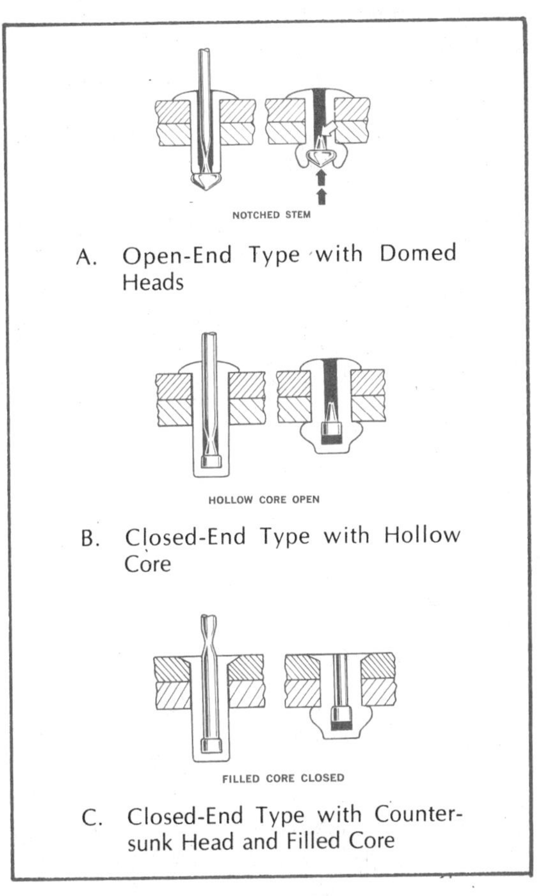

Fig. 26-37. Common Pull-Stem Types of Blind Rivets

is clinched. In school shops, maintenance shops, and home workshops, a plierlike tool, Fig. 26-36, is used for clinching blind rivets. Power-operated setting tools are often used in industrial plants where many rivets are used.

Blind rivets may be used for fastening sheet metal, thin flat metal bars, and non-metallic materials, such as fiber or plastic. They may be used for fastening metal to metal, plastic to metal, and fiberglass to metal. They are used on items such as automobiles, aircraft, appliances, furniture, sheet metal duct work, toys, and for numerous other purposes.

Types

Blind rivets are available in a wide variety of different designs, types, sizes, and kinds of materials. Three common **pull-stem** types, shown in Fig. 26-37, include the following:

1. Open-end type with domed head. (Also available with countersunk head.)
2. Closed-end type with hollow core and domed head. (Also available with countersunk head.)
3. Closed-end type with countersunk head and filled core. (Also available with domed head.)

An example of the use of open-end type rivets with domed heads is shown in Fig. 26-38. The closed-end type[2] seal is liquid and pressure-tight when set. It may be used for containers and items which must be pressure tight.

Other types of blind rivets include the **drive-pin** type and the **explosive-**type, Fig. 26-39. The former is clinched with a drive pin, while the latter is clinched by chemical expansion with an explosive charge. The charge is activated by a hot iron or similar tool.

Blind rivets are available with domed heads or countersunk heads (120°), as shown in Fig. 26-37, or with larger flanged heads. They are made of the following materials: aluminum, steel, copper, monel, and stainless steel. They generally are made in diameters ranging from 3/32" to 1/4" and in various lengths up to 3/4".

Rivet Selection

For best results, the length of the rivet selected should be such that it will clinch with a short head, as shown in Figs. 26-37 and 26-38. However, with careful riveting procedures, rivets of the same length may

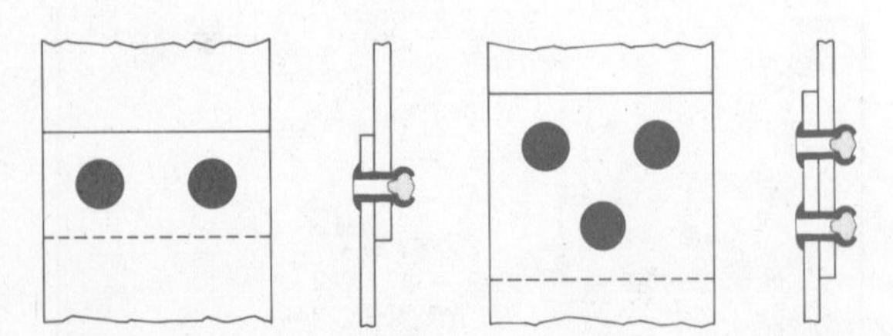

Fig. 26-38. Lap Joints Riveted with Blind Rivets (Open-End Type Rivets)

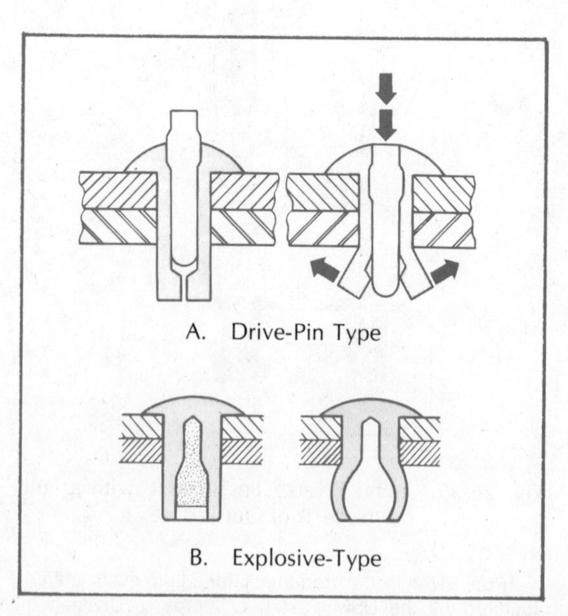

Fig. 26-39. Blind Rivets

[2]Patent held by the USM Fastener Company, Division of United Shoe Machinery Corporation.

Fig. 26-40. Blind Rivets in this Assembly are the Same Length

be used for materials of various thickness, as shown in Fig. 26-40. The rivet generally should be made of metal which is similar to the metal being riveted. For added strength, the rivets may be located closer together or in double rows, as shown in Fig. 26-38.

26-39. How to Order Metal Fasteners

In order to assure getting the correct metal fasteners, they must be completely described on a bill of materials or order blank as follows:

1. Name of fastener
2. Quantity needed
3. Kind
   a. Kind of material (steel, brass, etc.)
   b. Kind of finish (plain, blued, etc.)
4. Size
   a. Diameter (in inches or gage number) × length
   b. Thread information, if threaded
5. Shape
   a. Shape of head
   b. Shape of point (setscrews only)

Supplier's catalogs should be consulted for listings of sizes and types of fasteners available.

Words to Know

American Standard Screw Gage
blind rivet
blued finish
bright finish
carriage bolt
castle nut
cotter pin
drive screw
explosive rivet
gage number
gib-head key
jam nut
keyway
lag screw
oval head
Pop rivet
retaining rings
rivet set
self-tapping screws
setscrew
socket-head setscrew
spring washer
square key
stove bolt
stud bolt
tap bolt
taper-pin reamer
thread-cutting screws
thread-forming screws
thumbscrew
tinner's rivet
tubular rivet
wing nut
Woodruff key

Review Questions

1. How are bolts and screws measured?
2. For what are carriage bolts used? Stove bolts? Stud bolts?
3. Name several kinds of cap screws.
4. Why do set screws have different shape points?
5. Name two types of set screws and give an example of how each is used.
6. Name three kinds of heads on set screws.
7. Describe a lag screw. For what is it used?
8. Describe a thumbscrew. For what is it used?
9. How is the size of a nut measured?
10. For what is a jam nut used?
11. For what is a castle nut used?
12. Describe a wing nut.
13. Describe two types of lock washers.
14. For what purpose are shims used?
15. How is a cotter pin used?
16. How are taper pins used?
17. What is the difference between a square key, a gib-head key, and a Woodruff key?
18. What are retaining rings, and how are they used?
19. List three types of self-tapping screws. Describe how each type is used.
20. Make a bill of materials for five different kinds of screws.

21. Why are metal pieces sometimes riveted together instead of being bolted or welded?
22. List several kinds of products which use rivets in their construction.
23. Of what metals are rivets commonly made?
24. Describe the four shapes of rivet heads that are most commonly used.
25. How is the size of a rivet measured?
26. How much larger in diameter should the hole be than the rivet?
27. What happens if the rivet diameter is too small?
28. List a general rule which may be used regarding the minimum and maximum spacing between rivets.
29. What is a rivet set?
30. How can you keep the head of a round-headed rivet from being flattened when it is being set?
31. Why do hot rivets hold tighter than cold rivets?
32. Describe blind rivets and explain how they are used.

Mathematics

1. A tapered pin 3″ long is 3/16″ in diameter at the small end. What is the diameter of the long end?
2. How many 8-oz. tinner's rivets are there in a pound?
3. How long should a 1/8″ diameter rivet be to go through two pieces of 3/16″ thick metal if it is to be set with a round head?

Career Information

1. What are some of the dangers of structural iron work?
2. Make a list of occupations in which a knowledge of metal fasteners is required.

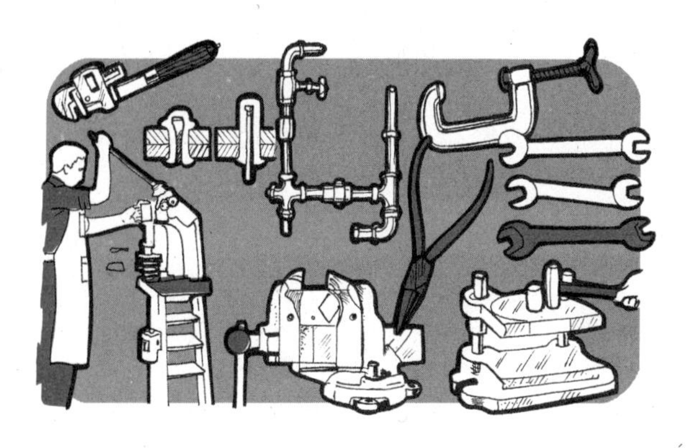

UNIT 27

Pipe, Pipe-Fitting Tools, and Tubing

27-1. Uses of Pipe and Tubing

Pipe and tube systems are used in industry to deliver and distribute fluids of many kinds. Common fluids are water, gas, steam, and compressed air. Food processors use piping to move such fluids as ketchup, soup, beverages, and chocolate. Chemical and petroleum processing requires complex systems of pipes.

Electric wire and cable is often installed in **conduit,** a kind of tubing. Motor vehicles use tubing to carry gasoline, water, oil, and hydraulic fluid. Piping and tubing are also used for various structural applications such as stair railings, aircraft and race car frames, flag poles, and television antenna towers.

Common pipe is made of low-carbon steel or malleable iron. Galvanized pipe is low-carbon steel coated with zinc. Tubing is readily available in various alloys of steel, stainless steel, copper, brass, and aluminum. Seamless pipe and tubing are made from a solid piece of metal; there is no joint or seam.

27-2. Who Uses Pipe-Fitting Tools?

Pipe-fitting means measuring, cutting, fitting, and putting pipes and **pipe fittings** together. A person who fits pipe is called a **pipe fitter**. A **plumber** installs and repairs water pipes, sinks, bathtubs, etc. A **steam fitter** installs pipes and fittings for steam heating, etc. A **gas fitter** installs gas stoves, gas heaters, gas meters, etc. The **electrician** installs pipes, called **conduit,** inside of which electric wires are placed. The **auto mechanic** repairs gasoline and oil tubing on automobiles, buses, trucks, and tractors. The **airplane mechanic** repairs gasoline and oil tubing on airplanes.

Pipe-fitting tools can also be very useful in making repairs on the automobile and around the home or farm. Every **mechanic** must at times do pipe-fitting, either in the home, on the car, in the shop, or at work.

27-3. Pipe Sizes

Many years ago Robert Briggs **standardized** the sizes of pipe. Thus, we have the **Briggs Pipe Standard,** also called the **American Standard,** which is used in the United States.

Pipe comes in lengths of 12′ to 20′. It is always measured by the diameter of the hole. Thus, a ½″ pipe has a ½″ hole. Hence, ½″ is called the **nominal size** or **nominal diameter** of the pipe. The size of the hole, however, is usually a little larger or smaller than the nominal size, depending upon the wall thickness of the pipe. The actual size of the hole is called the **actual diameter.** The **outside diameter** must be kept a certain size so that the **threads** will be a **standard size.**

Table 18 gives the dimensions of pipes up to 2½″. When the diameter of the pipe is doubled, the **area** increases four times.

27-4. Pipe Threads

For information on types of pipe threads and pipe sizes, refer to section 21-16.

27-5. Pipe Fittings

Pipe fittings, called **fittings** for short, are shown in Fig. 27-1. They are usually made of **cast iron** or **malleable iron** and are used to make turns and to change from one size of pipe to another. They are screwed on the threads on the ends of pipes.

A short piece of pipe with threads on both ends on the outside is called a **nipple.** A piece of pipe that is so short that the threads cover it entirely is called a **close nipple.**

A **coupling** has threads on the inside and is used to connect two pieces of pipe. A **reducing coupling** is screwed on the outside of a pipe to change to a smaller or larger size.

When the threaded ends of two pipes meet, the best way to connect them or draw them together is with a **union;** it can be taken apart easily to make repairs without taking all of the pipes apart.

A **cap** is screwed over the end of a pipe to close the hole. A **plug** is screwed into a pipe fitting to close the hole.

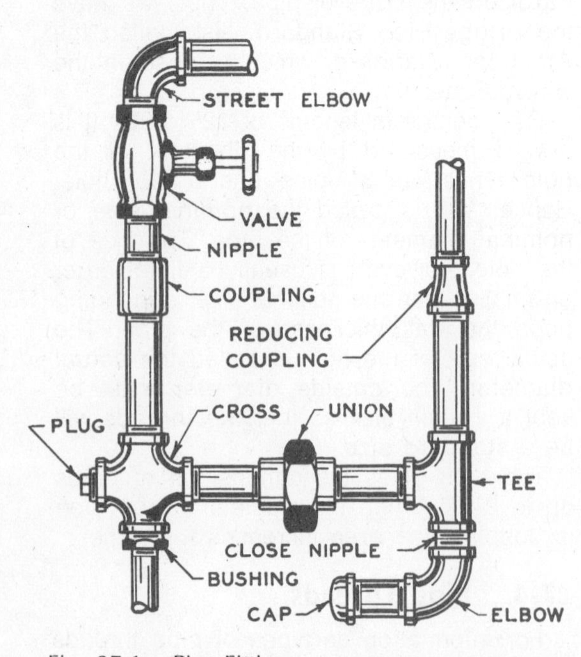

Fig. 27-1. Pipe Fittings

A **bushing** is used to connect two pipes or pipe fittings of different sizes.

An **elbow,** or **ell,** is used to make a turn. A **street ell** has threads on the outside on one end; the other end has threads on the inside.

A **tee** is shaped like a T and is used to connect a pipe from the side of another. A **cross** is shaped like a cross and is used where four pipes come together.

A **valve** is used to turn the flow on or off.

The size of a fitting is measured by the diameter of the hole in the pipe on which it is to be screwed. Thus, a ½″ elbow fits on a ½″ pipe, although the hole in the elbow is large enough to fit the threads on the outside of the pipe.

27-6. Pipe Vise

A vise which is specially made to hold pipe is called a **pipe vise,** Fig. 27-2. The **jaws** of a **machinist's vise** can grip a piece of pipe only in two places, while the pipe vise grips it in four places. In order to hold pipe in the machinist's vise so that it would not turn between the jaws, it would be necessary to tighten the jaws so much that they would crush or flatten the pipe. A pipe vise should be fastened to the end of a bench so that long pipe can be cut and **threaded.**

27-7. Pipe Wrench and Pipe Tong

The **pipe wrench** has been described in section 25-16. Large pipe and **pipe fittings** may be screwed together by using a **pipe tong,** also called **chain tong,** Fig. 27-3.

27-8. Pipe Cutter

A pipe cutter is a tool for cutting pipe, Fig. 27-4. It has three small, round cutters made of **hardened steel.**

The pipe cutter is slipped over the pipe so that the three cutters touch the pipe where the cut is to be made. By turning the handle, the cutters are pressed against the pipe, Fig. 27-5. At the same time, the whole

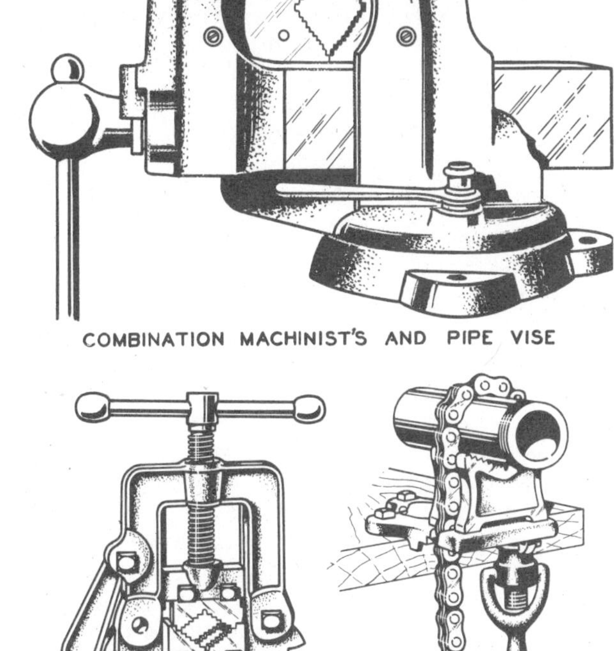

Fig. 27-2. Pipe Vises

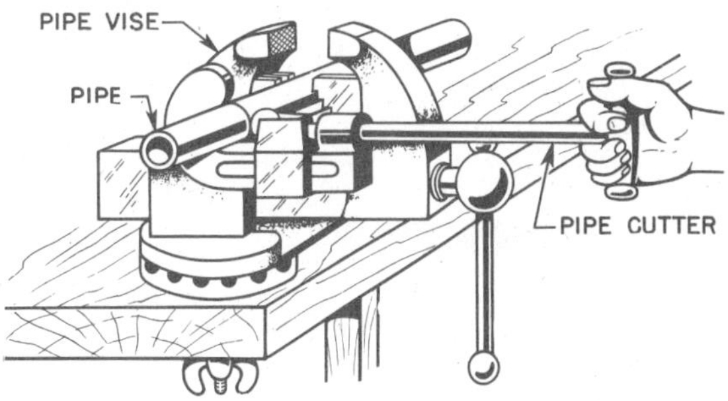

PIPE VISE
PIPE
PIPE CUTTER

Fig. 27-5. Cutting Pipe with Pipe Cutter

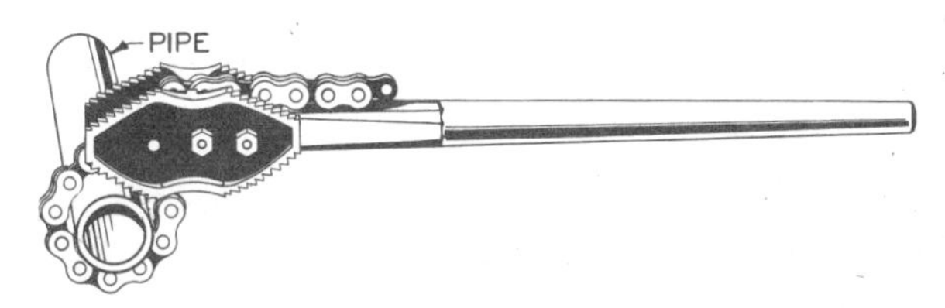

Fig. 27-3. Pipe Tong

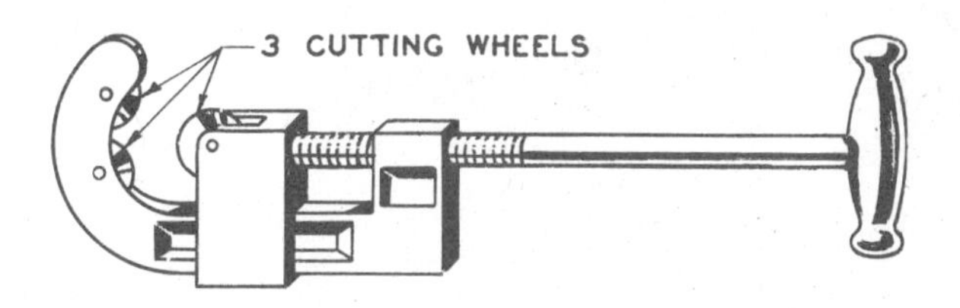

Fig. 27-4. Pipe Cutter

tool should be swung around the pipe which causes the cutters to cut a groove around the pipe. The handle is then turned again to press the cutters further into the pipe, and the groove is cut deeper. This is repeated until the pipe is cut off.

A **hacksaw** may be used instead of a pipe cutter.

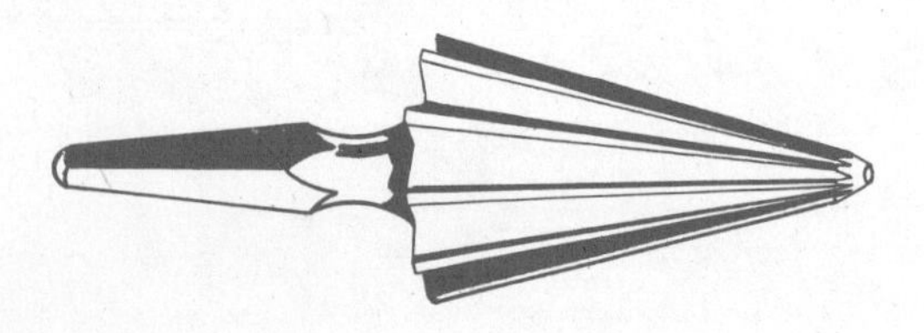

Fig. 27-6. Pipe-Burring Reamer

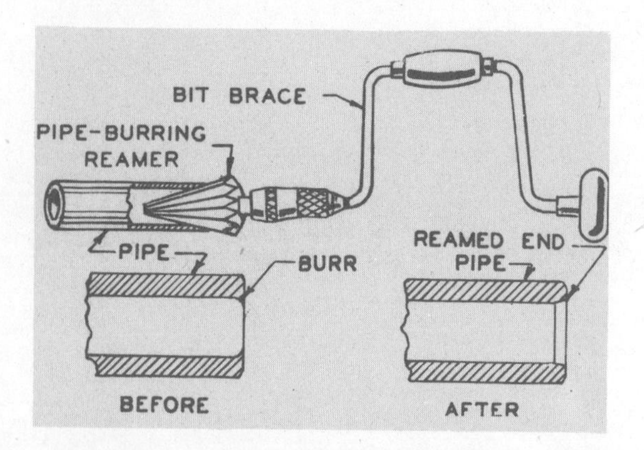

Fig. 27-7. Using Pipe-Burring Reamer

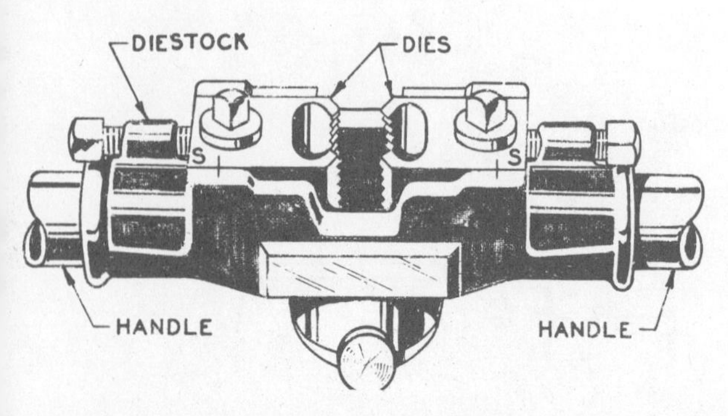

Fig. 27-8. Pipe Dies and Diestock

27-9. Pipe-Burring Reamer

The pipe-burring reamer, Fig. 27-6, is used to cut away the rough edges or **burrs** inside the end of the pipe after it has been cut with a **pipe cutter,** Fig. 27-7. These rough edges make the hole in the pipe smaller and slow up the flow of water or other fluid through the pipe.

27-10. Pipe Dies and Diestock

Threads on pipes are cut with **pipe dies,** Fig. 27-8. These look something like the dies described in Unit 22. The dies are held in a **diestock.** The threads inside NPT pipe dies, however, are tapered ¾″ per foot, just the same as the NPT pipe threads.

Pipe dies are used as shown in Fig. 27-9. They are marked with the size of pipe on which they are to cut threads. Thus, ½″ pipe dies will cut threads on a ½″ pipe.

Some pipe dies are made in two parts so that they can be made to cut smaller or larger by setting them closer or farther apart. Such dies are called **adjustable pipe dies.**

Most pipe dies and diestocks have an S and short lines stamped on them which mean that when the S on the die and the S on the diestock come together the dies will cut a **standard** size thread.

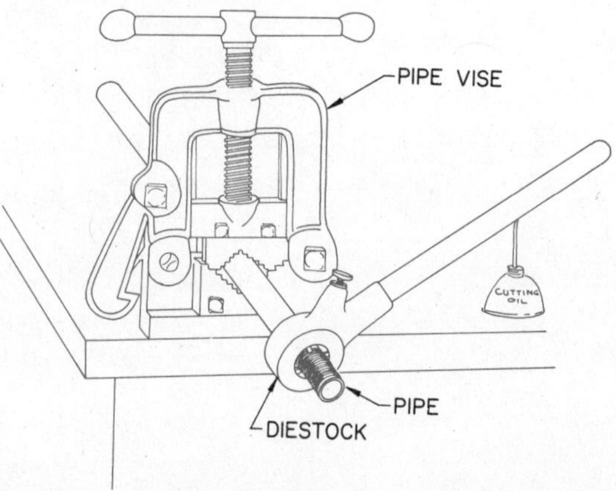

Fig. 27-9. Threading a Pipe

27-11. Tap Drill Sizes for Pipe

A **tap drill** is the drill used to drill a hole before it can be tapped. Thus, a tap drill for a ½″ pipe tap is much larger than the beginner would suspect; the tap drill for a ½″ pipe tap is 23⁄32″. See Table 18.

27-12. Pipe Reamer

A drilled hole that is to be tapped should be tapered with a **pipe reamer**, Fig. 27-10. This makes the hole more like the shape of the **pipe tap,** Fig. 27-11; therefore, the tapping is easier and the wear on the tap is reduced.

A pipe reamer is used when good work is desired and especially when large holes are to be tapped. Oftentimes the hole is tapped right after drilling without the use of the pipe reamer.

27-13. Pipe Tap

An NPT pipe tap, Fig. 27-11, is used to cut threads on the inside of a pipe or pipe fitting. Such threads are **tapered** and, therefore, the tap is also tapered. Thus, when the pipe and its fittings are screwed together, they make a tight joint and the water, steam, or gas will not leak out.

A pipe tap measures larger than the size marked on it. Note how much larger in diameter a ½″ pipe tap is than a ½″ tap for tapping nuts, Fig. 27-11. This is because a ½″ pipe tap, for example, is made to cut threads in the hole of a fitting which will fit over ½″ pipe.

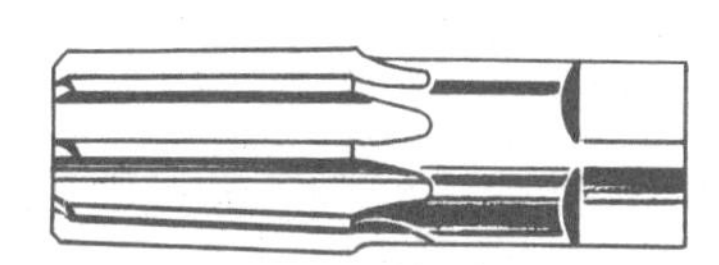

Fig. 27-10. Pipe Reamer

27-14. Put Pipe Compound on Joints

Before screwing pipe and pipe fittings together, a little **red lead** or pipe compound should be smeared on the threads. This makes a tight joint so that the water, steam, or gas will not leak out. It also keeps the metal from rusting.

27-15. Pipe Welding

Joints on pipes are often made by **welding** the pipes together, using the **oxyacetylene flame,** Fig. 27-12, or the **electric arc.**

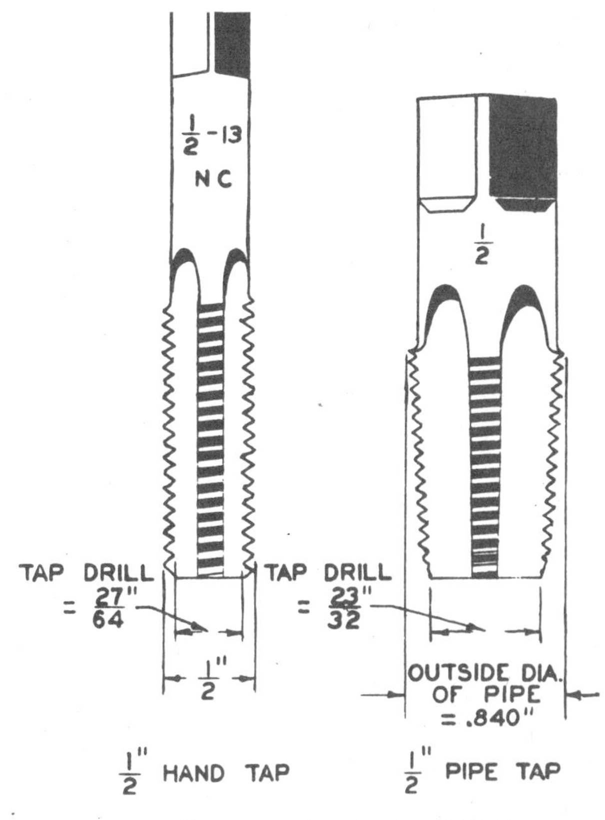

Fig. 27-11. Difference in Sizes between NPT Pipe Tap and Tap for Tapping Nuts with Unified National Form Threads (Actual Size)

Fig. 27-12. Welding Pipe

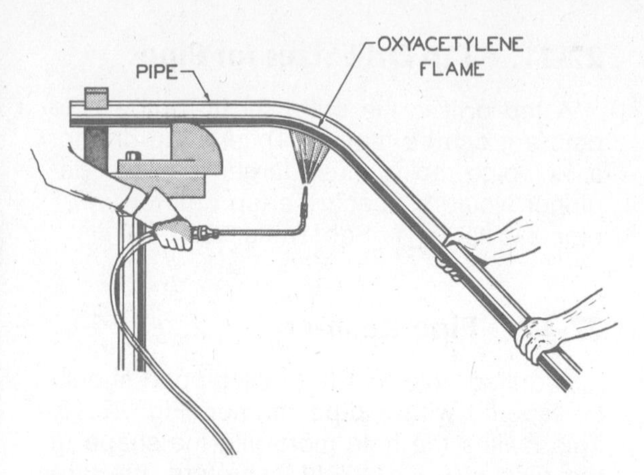

Fig. 27-14. Heating and Bending a Pipe

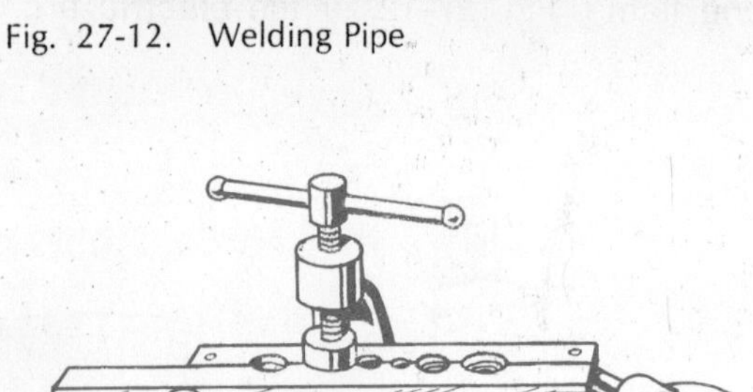

Fig. 27-13. Flaring Tool

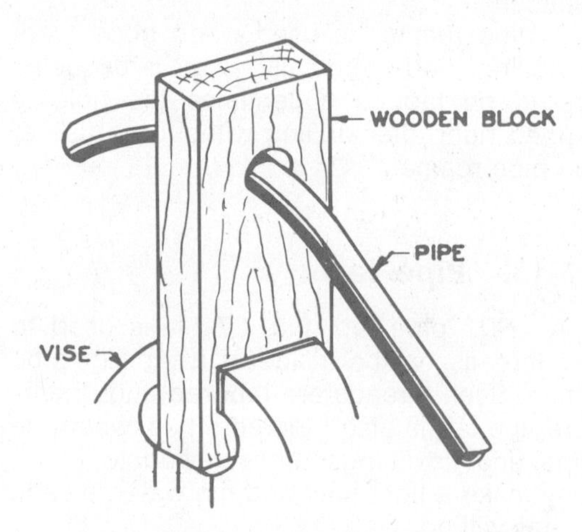

Fig. 27-15. Bending a Pipe through a Hole in a Wood Block

27-16. Tubing

Thin-walled pipe is called tubing. Tubing is usually measured and ordered by the outside diameter and the thickness of the tube wall. It is made in sizes from ⅛″ to 10″. The wall thickness of tubing is specified in decimal parts of an inch; for example 5/16″ diameter x .020″ wall.

27-17. Tube Fittings

Tube fittings are used to connect tubing in the same way as **pipe fittings** are used to connect pipes. They also have the same names as pipe fittings. See Fig. 27-1.

27-18. Flaring Tool

The end of a tube must be spread out or **flared** before a **tube fitting** can be connected to it. This is done with a **flaring tool,** Fig. 27-13.

27-19. Bending Pipe or Tubing

When a pipe or tube is bent, the outer part of the bend is stretched, while the inner part is squeezed together. Because of this, the pipe or tube flattens and often breaks when bent. To prevent this, pack the pipe or tube tightly with fine, dry sand and plug the ends so that the sand will not drop

Fig. 27-16. Bending Pipe with a Hickey

out. The sand must be dry or an explosion will result from heating the pipe to bend it. Keep the **seam** of the pipe on the side of the bend. Next, heat the place where the bend is to be made and then bend it. Figure 27-14 shows a pipe being heated with an **oxyacetylene flame** while being bent.

A long, curved bend may be made by slipping the pipe into a hole in a post, Fig. 27-15, and by bending a little at a time. Or, a **hickey** may be used to make the bend, Fig. 27-16. A hickey may be made by screwing a large **tee** (see Fig. 27-1) on a pipe three or four feet long. Again, the bending must be done a little at a time.

27-20. Drilling with a Star Drill or Masonry Drill

It is sometimes necessary to drill a hole into a brick, stone, or concrete wall to let a pipe pass through or to fasten something to the wall. This kind of drilling is done with a **star drill** or a masonry drill, Fig. 27-17. The point of a star drill is shaped like a cross or star. The drill is struck lightly with a hammer and turned a little between blows.

A masonry drill is a twist drill with a **carbide tip.** It is used like any other twist drill.

Fig. 27-17. Drilling a Hole in a Brick Wall with a Star Drill

Words to Know

adjustable pipe die
chain tong
close nipple
coupling
cross
elbow or ell
flaring tool
galvanized iron pipe
hickey
malleable iron pipe
masonry drill
nipple
pipe-burring reamer
pipe bushing
pipe cutter
pipe fittings
pipe plug
pipe reamer
pipe tap
pipe tee
pipe tong
pipe vise
reducing coupling
seamless tubing
star drill
street ell
tube fitting
union
valve

Review Questions

1. Name several common uses for pipe and tube.
2. What is meant by the nominal diameter of pipe? Actual diameter? Outside diameter?
3. How many times does the area of a pipe increase when the diameter is doubled?
4. Why is a pipe vise better for holding pipe than a machinist's vise?
5. For what are chain tongs used?
6. For what is a burring reamer used?
7. Why should the burrs be removed from the inside on the ends of the pipe?
8. What is the tap drill size for a 1/8" pipe tap? For a 1/2" pipe tap?
9. To what diameter of the pipe does the size of the tap refer?
10. For what is a pipe reamer used?
11. How is tubing measured?
12. Describe how a pipe or tube can be bent without flattening or breaking it.
13. What is seamless pipe or tube?
14. For what is a flaring tool used?
15. Why are pipe dies adjustable?
16. What cutting tools are available for drilling holes in brick or concrete block walls?
17. List the three kinds of American National Standard pipe threads.

Mathematics

1. What is the area of the inside of a 1/2" pipe?
2. What is the outside diameter of a 1/2" pipe?
3. If a 1" pipe delivers 2.3 gallons of water per minute at a velocity of 50 feet per minute, how much will a 2" pipe deliver at the same velocity?

Social Science

1. Discuss the advantages of having properly installed sanitary plumbing.

Career Information

1. List the advantages and disadvantages of being a plumber or another kind of pipe fitter.

PART VIII

Adhesion and Fusion Bonding

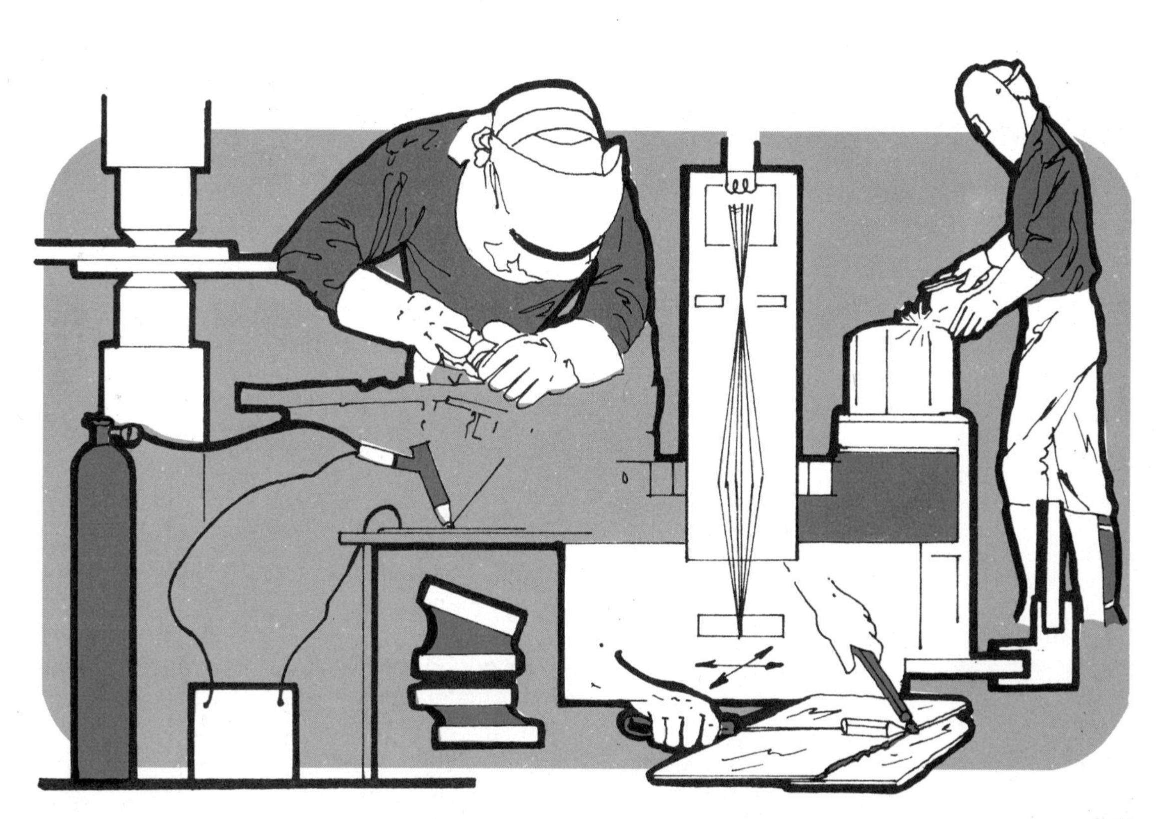

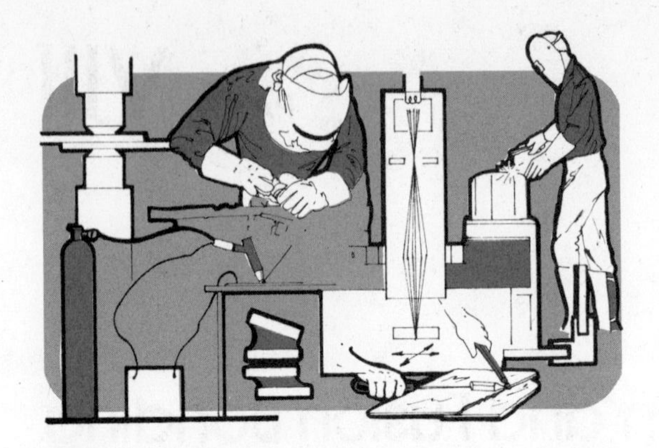

UNIT 28

Soldering and Brazing

28-1. What Is Soldering?

Soldering is a process of fastening metals together with a nonferrous metal of low melting point which adheres to the surfaces being joined. If the solder melts below 800° F., it is called **soft soldering**; if the solder melts above 800° F., it is called **hard soldering** or **brazing.** Soldering and brazing are different from welding. In welding, the pieces being joined are melted and fused together. The filler rod for welding is basically the same as the pieces being joined; this is not true of soldering or brazing.

28-2. Who Solders and Brazes?

Electricians and electronic technicians solder wire joints for electric circuits. Sheet metal workers assemble and repair gutters, spouting, ductwork, and sheet metal roofing by soldering. Plumbers must solder copper plumbing pipe and pipe fittings together. Jewelers repair ornaments, rings, and other jewelry by hard soldering. Auto body service workers often use soldering and brazing in their work.

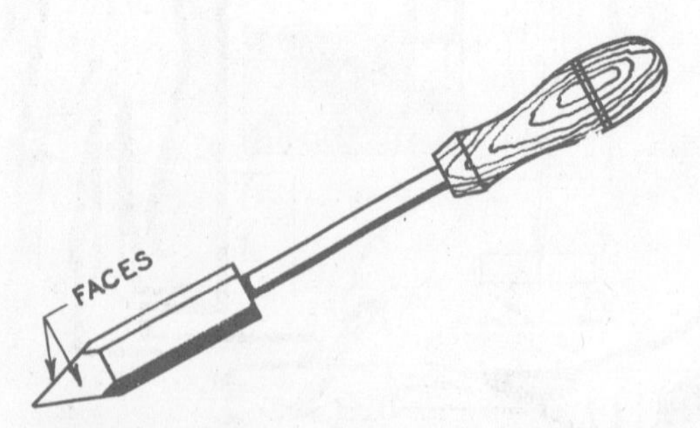

Fig. 28-1. Soldering Copper

28-3. Tools and Materials Needed for Soft Soldering

A heat source, such as a soldering copper or a bottled-gas torch, soft solder, and flux are needed for soft soldering. These are explained in the following sections.

28-4. Soldering Copper

A soldering copper is a bar of copper, usually octagonal in cross section, pointed at one end and fastened to a steel bar with a wooden handle on the other end. See Fig. 28-1.

Soldering coppers are made in different weights, from 1 ounce to 3 pounds. The weight is usually given per pair of coppers. For example, a 2-pound copper means that a pair weighs 2 pounds and that each copper weighs 1 pound. The size of the copper to be used depends upon the work; use a heavy copper on heavy work and a light copper on light work.

An **electric soldering copper** and instant-heating soldering guns are heated by electricity; they connect to any 115-volt electric outlet.

Soldering coppers are generally used for sheet metal work; electric soldering coppers and guns are used chiefly for soldering electric wiring.

TABLE 21

Fluxes for Soft Soldering

Metal To Be Soldered	Flux	Chemical Name of Flux
Brass Copper	Cut acid Rosin Sal ammoniac[1]	Zinc chloride Colophony Ammonium chloride
Zinc Galvanized iron (zinc coated)	Cut acid Raw acid (muriatic acid)	Zinc chloride Hydrochloric acid
Iron Steel	Cut acid Sal ammoniac	Zinc chloride Ammonium chloride
Tin Tin plate	Rosin Cut acid	Colophony Zinc chloride
Pewter	Rosin Tallow	Colophony
Nickel Silver	Cut acid	Zinc chloride
Aluminum	Special fluxes by different manufacturers	

[1]**Sal ammoniac is ammonium chloride.** It is a white solid substance that looks like rock salt or rock candy; it is used as a soldering flux. Sal ammoniac changes directly from a solid to a gas.

28-5. Solder

The best **soft solder** is an **alloy** made of **lead** and **tin.** For general work, a solder made of one-half lead and one-half tin is used; it is called **half-and-half** or **fifty-fifty** solder and melts at about 400° F. The solder must have a lower **melting point** than the metals to be joined.

Solder comes in bars 12″ long and also in the form of wire on spools. A handy way to buy solder is in the form of hollow wire, the center of which is filled with **rosin** or **acid flux.**

Special solders are made for soldering **aluminum.** It takes more heat to melt **aluminum solder** than solder made of lead and tin.

Many commercial **paste-type solders** are available for use on various metals. Plumbers use them to sweat solder joints and fittings on copper tubing.

Other special kinds of soft solder are available. One is a silver-tin alloy with a low melting point for use on stainless steel. It will make solder joints which are up to ten times stronger than lead-tin solder.

28-6. Fluxes for Soldering

Flux cleans the metal and helps the solder to flow. Table 21 gives the kinds of fluxes that are used for soldering different metals. Ready-made fluxes in liquid, crystal, and paste forms are commercially available and can be used for ordinary soldering. Some of these fluxes are for general-purpose use on all common metals, and they produce good results. Flux makes the metal surface chemically clean and stops corrosion, permitting the solder to stick tightly and make a good joint.

Soldering salt makes water slightly acid when the commercial crystals are added to water. For bright tinned parts, mix 1 ounce in 8 ounces of water. For bright copper and brass, mix 1 ounce in 4 ounces of water. For galvanized iron, mix 1 ounce in 2 ounces of water. Various kinds of soldering salts are available. It is best to follow the manufacturer's recommendations when preparing soldering flux solutions with soldering salts. The flux must not be stronger than necessary.

Powdered **rosin**[1] may be used as a flux for soldering tin or copper; it is sprinkled in the joint and when heated flows into the seam. Rosin should be used for soldering **electrical joints.**

A mixture of 1 ounce of glycerin and 5 drops of hydrochloric acid can be used as flux for soldering **pewter.** The **tallow** from

[1]**Rosin,** also called resin, is a sticky hard substance. It is made from turpentine obtained from pine trees. Rosin is used in making varnish. Musicians use rosin on the bows of violins, cellos, etc. Baseball pitchers rub a finely ground rosin on their hands.

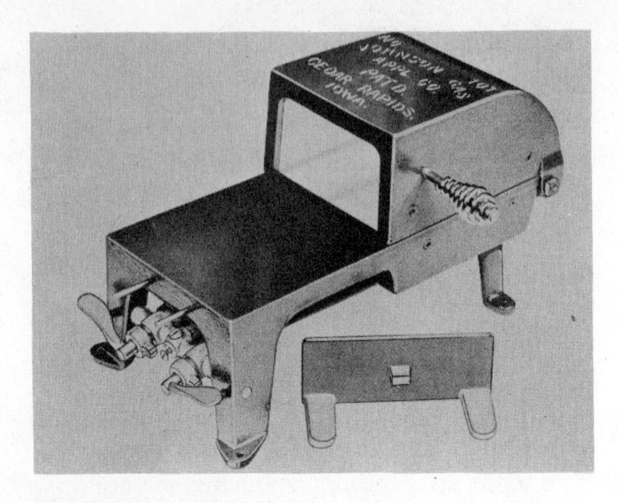

Fig. 28-2. Bench-Type Gas Furnace

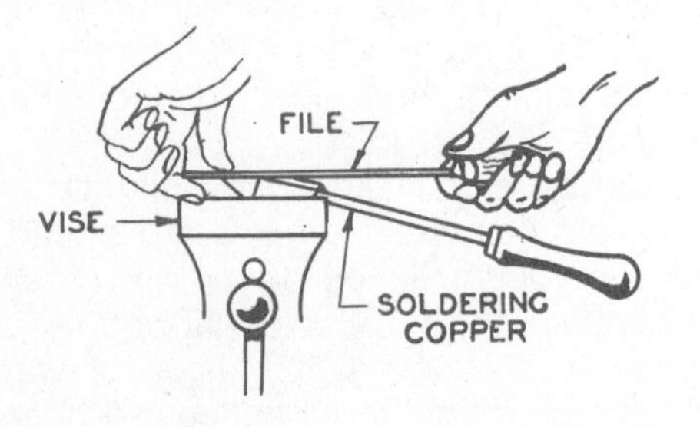

Fig. 28-3. Cleaning a Soldering Copper with a File

a tallow candle may also be used for soldering pewter. Special fluxes are made for soldering aluminum.

28-7. Heat for Soldering

Gas furnaces are often used in shops to heat **soldering coppers,** Fig. 28-2. The gas and air must be mixed to give a blue flame. Burning gas alone gives a yellow flame. The blue flame is hotter than the yellow flame. A portable gasoline torch or small, portable bottled gas torch is used for outside work. An **electric soldering copper** may be used; it can be connected to any electric light socket.

28-8. Tinning the Soldering Copper

A well-tinned soldering copper will hold a drop of solder on the point and make soldering easier. The **faces** on the point of the soldering copper (see Fig. 28-1) must be cleaned and coated with solder before the copper can be used for soldering. This is called **tinning.**

To tin the copper, clean the four faces of the point with a file, Fig. 28-3. If the copper is too hard to file easily, it may be softened by heating it red-hot and quenching it in water. Keep the point bright. Always remember that dirt and rust are enemies of good soldering. Heat the copper until it melts solder. It does not need to be red hot. Rainbow colors show up when the copper begins to overheat; this is the danger sign. Continue with one of the following three ways, depending upon the materials that are handy:

1. Dip point quickly into and out of a jar of **zinc chloride;** then melt solder on the faces and wipe with a dry cloth.
2. Rub points in drops of solder on a block of **sal ammoniac.** The solder will melt, run on the sal ammoniac, and then stick to the copper, Fig. 28-4.
3. Rub the point in some solder and powdered **rosin.**

28-9. Joints for Soldering

Sheet metal joints to be soldered are usually **lock joints, lap joints,** or **butt joints,** Fig. 28-5. The parts to be joined should fit each other exactly; the better the fitting, the stronger will be the soldered joint.

28-10. Cleaning Surfaces

The surfaces to be soldered must be physically and chemically clean. All visible rust, dirt, and grease must be removed. New metal usually needs no physical cleaning but old metal may have to be polished with steel wool or abrasive cloth until bright

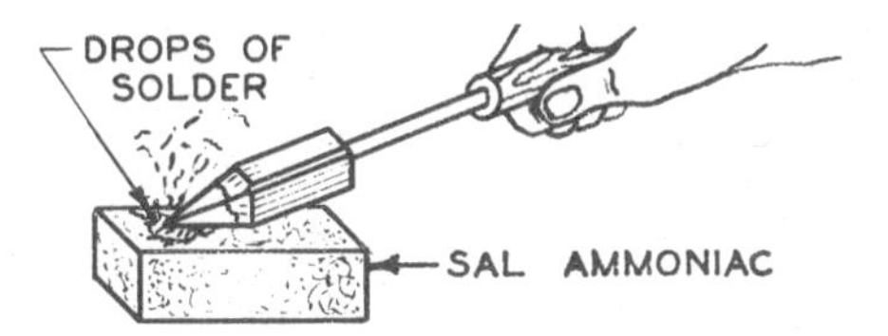

Fig. 28-4. Tinning a Soldering Copper on a Block of Sal Ammoniac

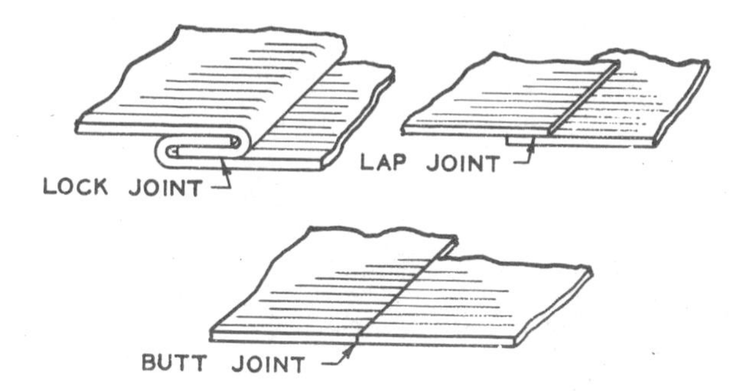

Fig. 28-5. Joints

metal appears. Any metal surface which has just been cleaned immediately begins to combine with oxygen in the air to form a layer of **oxide.** No matter how thin the oxide, it makes soldering impossible. The surfaces must be made chemically clean by applying a flux which thoroughly removes the oxide film.

A soft-lead (pencil) coating or a thin film of grease or shoe polish on the surface where the solder is **not wanted** will allow the solder to flow only where it is needed. This produces neater work.

28-11. Holding Work

Place the parts ready for soldering together, and fasten them so that they cannot move while being soldered. If possible, clamp them together in a vise or C-clamp. The joint to be soldered should not touch the vise or other metal because the heat will be carried away from the joint and make soldering impossible.

28-12. Heating Soldering Copper

The soldering copper must be **tinned** and heated. Heat is used to melt the solder and to heat the pieces which are to be joined. If the point of the soldering copper is red hot, the tinning will burn off, and then the copper must be retinned. It is best to place the large part of the copper in the flame with the point out of the flame. It will remain hot longer and the point will not become **pitted;** that is, full of small holes. The copper should be heated until it melts solder but should not be hot enough to burn off the tinning.

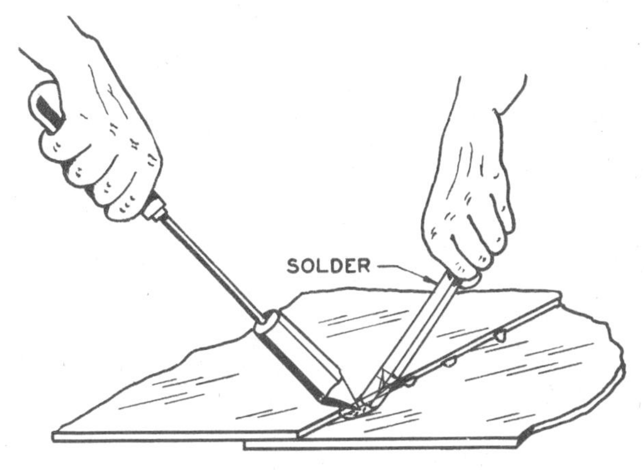

Fig. 28-6. Tacking a Seam with Drops of Solder While Holding the Lap Down Firmly

28-13. Soldering a Seam

The surfaces to be soldered must be clean and must be held together. Hold the soldering copper in the right hand and the solder in the left hand. **Tack** the seam together; that is, connect the parts in several places with the hot copper and a few drops of solder, Fig. 28-6, so that the parts will not buckle. Note that a well-tinned copper will hold a drop of solder on the point and make soldering easier.

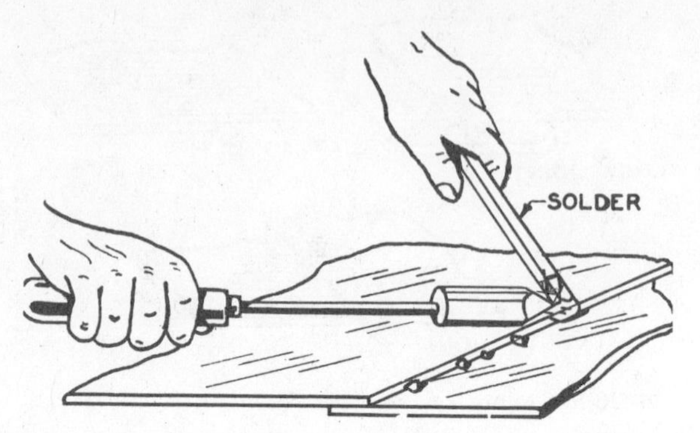

Fig. 28-7. Holding Soldering Copper on Edge of Seam

Next, hold the hot tinned copper on the edge of the seam and touch the copper with the solder, Fig. 28-7. Move the copper slowly along the seam, a short distance at a time. As the copper melts the solder, put a little along the edge of the seam.

For neat work, remember to use only a little solder; too much solder on a joint is hard to remove. Also, a thin film of solder makes a stronger joint. The solder must be between the surfaces to be joined, not around them. Then move the copper back to the starting point and hold one of its flat **faces** on the seam until the metal gets hot enough to cause the solder that was put along the edge to flow into the seam. Give it time to flow into place before going ahead.

28-14. Sweat Soldering

In sweat soldering, the heat is applied to the pieces to be joined and the solder is melted by its heat. As it melts, the pieces of solder form balls, then flow along the joint. First, clean the surfaces or edges to be joined, put on a little flux, add small pieces of solder, and then press the parts together. Heat until the solder melts and joins the parts together.

28-15. Cleaning Joint After Soldering

When zinc chloride, sal ammoniac, an acid, or any corrosive-type paste flux has been used as a flux for soldering, it is necessary to wash the joint in cold, running water. This is done to wash away all traces of the flux. If this is not done, the joint and the metal touched by the flux will turn to a black, dirty color.

28-16. Reasons for Hard Soldering

Hard soldering (silver soldering or brazing) is used where a strong joint is needed or where the parts will afterwards have to be in greater heat than the **melting point of soft solder.**

The most widely used hard solders are the **silver** and **silver-alloy solders.** Most of them melt at temperatures from 1100° to 1300° F. Hard soldering is used for joining such metals as copper, silver, and gold in jewelry and art metal-work. It can also be used on all carbon and alloy steels. It is often used to join band saw blades.

28-17. How to Do Hard Soldering

The parts to be joined must be carefully cleaned, fitted snugly together, and then coated with the proper flux. The parts must be fitted much more closely than for soft soldering, because the silver solder flows out very thinly and cannot be used to fill small gaps. For best results, use a **commercial flux** recommended for silver soldering. If a commercial flux is not available, powdered **borax** can be substituted. Place pieces of silver solder on the joint; then heat the parts until the solder melts, runs into the joint, and fastens the parts together. See Fig. 28-8.

Heat for hard soldering is applied directly from a flame or torch. Gas-air torches usually have only one hose furnishing a gas. Natural, propane, or acetylene gas may be used, but each requires a different tip. Air

Fig. 28-8. Silver Soldering

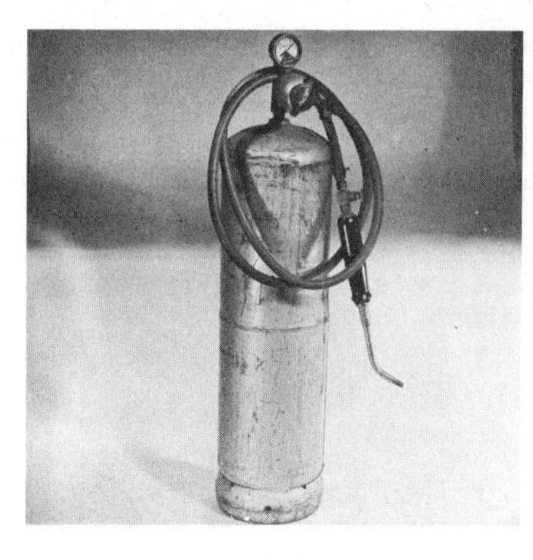

Fig. 28-9. Gas-Air Torch with Small Acetylene Tank

is drawn from the room. Some may use a small pump to furnish low-pressure air through a second hose. This raises the melting temperature above the 1300° F., which is not possible without air pressure. Figure 28-9 shows a gas-air torch with an acetylene tank as commonly used by plumbers and jewelers. The oxyacetylene torch (with heat to 6000° F.) can be used with care.

28-18. Brazing and Its Applications

Brazing is defined by the American Welding Society as a group of welding processes which use a filler rod of a nonferrous metal or alloy with a melting point above 1000° F., but lower than the melting point of the metals being joined. Hence, brazing and hard soldering really mean the same thing, because they are performed with nonferrous filler rods above a red heat. See Table 26. There is no fusion or melting together of the metals being joined in brazing, but there is a very strong bond between them. Bronze welding rods are used in brazing and **bronze welding.** Silver solder and silver alloys are used in silver brazing or hard soldering.

Brazing is done when a strong joint is needed or where the parts will afterwards be in a greater heat than the melting point of soft solder. Brazing is widely used to repair gray iron and malleable iron castings. The high heat of fusion welding destroys the heat-treatment properties of malleable iron castings, and it makes gray iron castings very brittle. Brazing can also be used to join thin sheet metal and thin pipe with less skill than is usually required to weld them.

28-19. Brazing and Welding Fluxes

Many different fluxes have been scientifically developed for specific types of brazing and for welding of various metals. They are available commercially from welding supply dealers. Because of these fluxes, many metals which were formerly difficult to braze or weld may now be brazed or welded with comparative ease.

28-20. Rods for Brazing and Bronze Welding

Bronze rods are used for **brazing** and **bronze welding** because they make strong joints. These rods are about 60% copper and 40% zinc. In addition, other elements such as iron, tin, manganese, and silicon are included in small amounts to give them better brazing and bronze welding characteristics. Most of the modern bronze welding rods melt at approximately 1600° F. Some of these rods produce a bond with a tensile strength of more than 50,000 pounds per square inch on steel and cast iron. Bronze welding rods are available with or without flux coatings. Coated rods are widely used because they simplify brazing and produce better joints.

28-21. How to Braze

The parts to be brazed must be carefully cleaned, fitted snugly together, and then sprinkled with the proper flux. For best results, use a flux-coated rod or a commercial flux recommended for the type of metal being brazed. A different flux is usually recommended for brazing cast iron than for brazing steel. The joint is heated to a dark red color with an oxyacetylene torch. The bronze brazing rod is melted and flowed over the joint in a thin layer until it is thoroughly **tinned** or coated with bronze. Add additional flux by dipping the hot end of the brazing rod into the flux. The flux will adhere to the rod. Continue applying heat to the joint and the rod, forming a puddle of molten bronze, until the joint is built up.

Words to Know

acid flux
aluminum solder
borax
brazing
50-50 solder
flux
hard soldering
oxide
paste-type solder
rosin flux
sal ammoniac
soft soldering
soldering copper
soldering furnace
soldering salt
sweat soldering
tinning
zinc chloride

Review Questions

1. How does soft soldering differ from hard soldering?
2. What tools and materials are needed for soft soldering?
3. Of what metals is soft solder made? Hard solder? A brazing rod?
4. Why is flux necessary for soldering and brazing?
5. Name some of the fluxes used for soft soldering. For hard soldering. For electrical connections.
6. Describe how a soldering copper is tinned.
7. Tell how to solder two pieces of metal together.
8. What is meant by sweat soldering?
9. Why and how should a soldered joint be cleaned after a corrosive-type flux is used?
10. Name several applications for hard soldering and brazing.
11. What kinds of filler rods are used for brazing? What is their approximate melting temperature?

Career Information

1. What occupations require an ability to solder or braze?
2. What dangers are involved in soldering?

UNIT 29

Welding

29-1. What Is Welding?

A weld occurs when sufficient temperature, pressure, or a combination of both causes some of the metal from each of the pieces being joined to flow (melt) and blend together. When heat alone is used, the weld is called a **fusion weld.** Welding heat may be produced by the burning of gases, with high-amperage electric current, by friction, or by other means.

Pressure welding usually involves heating the areas to be welded together to a plastic state and then forcing them together with mechanical, hydraulic, or air pressure. Under certain conditions, however, welds can be made using pressure alone.

29-2. Why Is Welding Important?

Welding is widely used in industry as a joining technique both for fabrication in production and for repairs. It has almost completely replaced riveting in assembling the structural steel members of bridges and buildings. Barges and ships, boilers and large storage tanks, pipe lines, railroad cars, and automobiles are assembled largely by welding. In many cases welded steel construction has replaced the use of iron castings in machine bases, frames, and bodies. Large frames are welded on the site. See Fig. 29-2. Appliances such as toasters, dishwashers, washing machines, and clothes driers rely heavily on **spot welding** for their assembly.

29-3. Welding Processes

Figure 29-1 shows how the **American Welding Society** has classified the welding processes in use today. Each of the following terms is basic to an understanding of welding processes.

Brazing is the joining of metals by melting a nonferrous filler rod at a temperature above 800° F. but below the melting point of the metals being joined. Brazing is explained in sections 28-18 to 28-21.

Forge welding is pressure welding by heating and hammering pieces together (see § 37-7).

Oxyacetylene welding, the most common form of **gas welding,** is explained in sections 29-5 to 29-16.

Shielded Metal-Arc Welding is the most common form of **arc welding** and is explained in sections 29-17 to 29-33.

Resistance welding uses the heat generated by electric current passing through a small area of the metals being joined. Pressure forces the heated areas together until they have fused. **Spot welding,** a common form of resistance welding, is covered in section 29-35.

Thermit welding processes use the intense heat of a chemical reaction to fuse metals.

Induction welding uses high-frequency current as the source of heat.

Inert gas, such as helium or argon, is used in several processes to prevent the weld from becoming contaminated by scale, oxidation, or other impurities. The gas keeps

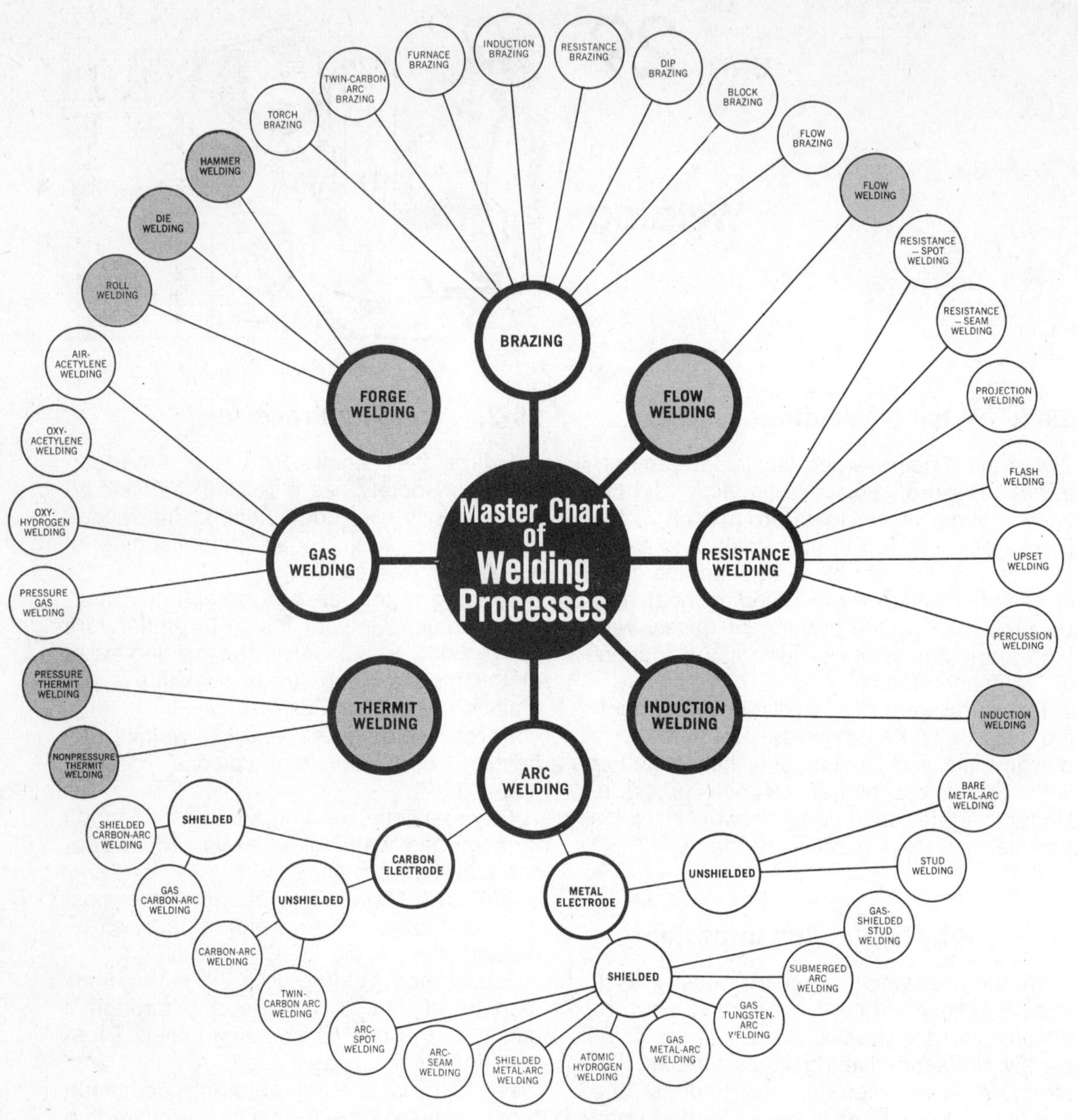

Fig. 29-1. Chart of Welding Processes

air away from the weld much like a CO_2 fire extinguisher puts out a fire by keeping air away. Most arc welding is shielded by a gas formed when the coating on the electrode is heated.

The processes listed in the outer circle make interesting topics to read about in reference books in the library. Most of the processes have important industrial uses.

29-4. Who Welds?

A person that earns his living by welding is a welder and usually a brazer, also.

Fig. 29-2. Welding Cross Members on a Frame Assembly

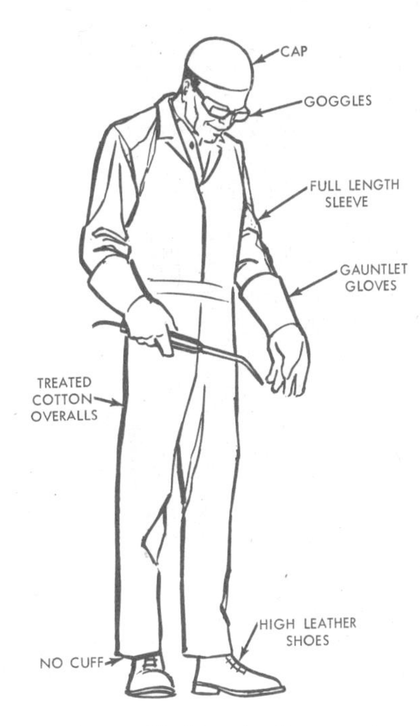

Fig. 29-3. Proper Clothing for Oxyacetylene Welding

Welders are found in most manufacturing plants, foundries, construction sites, boiler and tank shops, welding shops, railroad shops, and shipyards. Welding processes are probably used in every metal industry in some manner.

29-5. Oxyacetylene Welding

The **oxyacetylene weld** is a **fusion weld.** The heat is produced by burning **acetylene gas** with **oxygen gas,** giving a temperature of about 6000° F. The edges of metal to be welded melt and fuse together. When the metal becomes solid, the pieces are welded together. The weld may be made so strong that the two pieces are like one piece.

Often, metal is added to the joint by being melted from a **filler rod.** This rod is usually made of the same metal as the pieces being welded. The pieces of metal to be joined are called the **base metal.** All of the melted metal from both the base metal and filler rod is called the **weld metal.**

29-6. Proper Clothing

Protective clothing should be worn by the welder using oxyacetylene equipment. See Fig. 29-3. This includes the following items:

1. A cloth or leather cap.
2. Goggles with colored lenses.
3. Coveralls of treated cotton or a full-length apron.
4. Trousers without cuffs and a shirt with full-length sleeves.
5. Leather shoes.
6. Gloves with cuffs that overlap the sleeves.

29-7. Oxyacetylene Welding Equipment

Oxyacetylene welding requires some special equipment, Fig. 29-4. The pieces of equipment are:

1. A cylinder of **oxygen.**
2. A cylinder of **acetylene.**
3. Special valves called **regulators** for the cylinders.
4. A **welding torch,** sometimes called a **blowpipe,** and various sizes of **torch tips.**

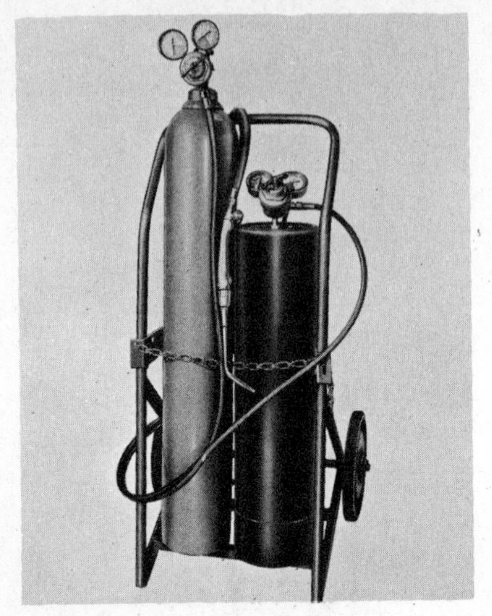

Fig. 29-4. Equipment for Oxyacetylene Welding

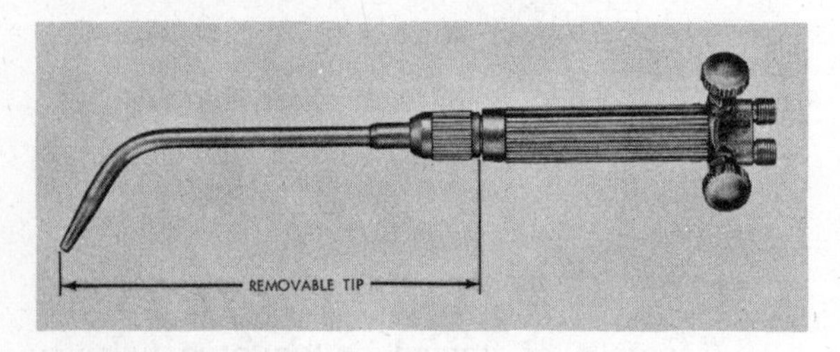

Fig. 29-5. Removable Tip of Welding Torch

5. **Hoses** to carry the gases from the cylinders to the torch.
6. A **lighter** that can make a spark to ignite the mixture of gases.
7. A pair of **tongs** or **pliers** to handle pieces of hot metal.

29-8. The Welding Torch

The welding torch mixes the gases from the oxygen and acetylene cylinders. It is made so that it can control the flow of each gas to produce the size and type of flame needed. With the welding torch, the welder can direct a maximum amount of heat to a small spot or area.

TABLE 22

Tip Sizes, Regulator Pressures, and Rod Sizes for Welding

Tip Size No.	Gas Pressure Psi[1] Oxygen	Gas Pressure Psi[1] Acetylene	Metal Thickness, Inches	Filler Rod Dia., Inches
1	1	1	. . .	. . .
2	2	2	1/32 (22 gage)	1/16
3	3	3	1/16 (16 gage)	1/16
4	4	4	3/32 (13 gage)	1/16, 3/32, or 1/8
5	5	5	1/8 (11 gage)	3/32 or 1/8
6	6	6	3/16	1/8
7	7	7	1/4	3/16
8	8	8	5/16	3/16
9	9	9	3/8	3/16
10	10	10	7/16	3/16 or 1/4

[1]psi means pounds per square inch.

29-9. Torch Tips

There is a certain size and kind of flame that is best for welding each thickness of metal. Since the **tip of the torch** controls the flame, welding torches are made so that the tips can be easily changed. See Fig. 29-5.

Be sure to use the correct size wrench when changing tips. Some tips with rubber gaskets must be screwed only finger tight. Table 22 gives information that will help in selecting tips, rods, and gas pressures. Follow the manufacturer's recommendations. Each tip has the size number marked on it.

Sometimes the hole in the torch tip (called the **orifice**) becomes partly clogged with a small piece of carbon, dust, dirt, etc. The gases passing through the tip may make a different kind of hissing sound, and the flame may have an odd shape. When this happens, it is important that a special tip cleaner be used to clean the tip. See Fig. 29-6. Do not try to use a piece of wire, wood, or a wire brush to open the hole, because the tip may be damaged. Some tip cleaners carry the same number as the tip. Clean the tip by inserting a tip cleaner of proper size in the hole in the torch tip.

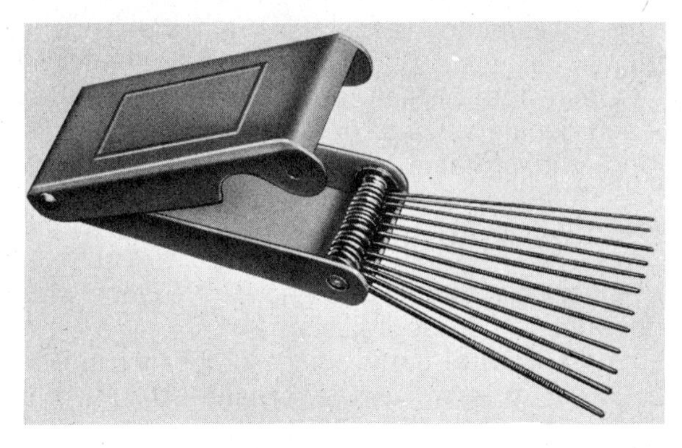

Fig. 29-6. Tip Cleaners

Fig. 29-7. Oxygen Cylinder Pressure Regulator

29-10. Gas Pressure Regulators

In the acetylene cylinder the pressure of the acetylene gas can vary from about 15 pounds per square inch (when it is almost used up) to 250 psi (when it is full). The pressure in the oxygen cylinder can vary from about 15 psi (almost used up) to 2200 psi (full).

The welder works with only from 1 to 10 psi of each gas supplied to the welding torch. Therefore, he needs a device on each cylinder that will deliver a constant supply of gas at a reduced pressure. The device that does this is a **gas pressure regulator.**

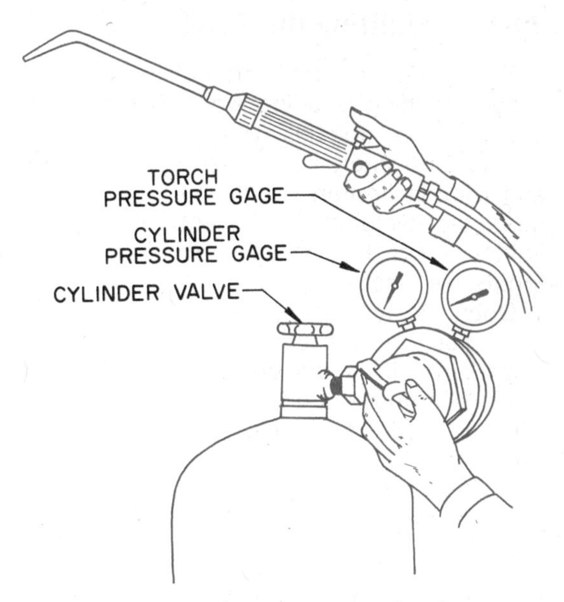

Fig. 29-8. Setting the Gas Pressure Regulators

See Fig. 29-7. Many welders shorten the name to **regulator.**

29-11. Setting the Regulators

Open the oxygen cylinder valve — slowly at first — until the pressure gage stops increasing, then more quickly until the valve is fully opened. See Fig. 29-8. Open the acetylene cylinder valve 1/4 to 3/4 of a turn. Keep the wrench on the valve so that in an emergency the valve can be closed quickly. If either cylinder gage shows less than 15 psi, the cylinder should be replaced.

Open the oxygen **needle valve** on the torch about one turn. The oxygen hose is the green hose. Turn the oxygen regulator handle clockwise until the gage shows the proper working pressure (see Table 22); then close the needle valve on the torch.

Open the acetylene needle valve on the torch about one turn. The acetylene hose is the red hose. Turn the acetylene regulator handle clockwise until the gage shows the proper working pressure. Close the needle valve on the torch.

29-12. Lighting the Torch

Before lighting the torch, be sure the proper clothing is being worn. See § 29-6. This includes goggles with colored lenses.

A **torch lighter**, also called **spark lighter** and **friction lighter**, should be kept with or near the welding equipment. The torch lighter is used to make a spark that will **ignite** the acetylene gas.

Open the acetylene needle valve on the torch about a half turn. Point the tip of the torch away from the body and away from the gas cylinders. Ignite the acetylene gas with the torch lighter, Fig. 29-9. Do not use matches or a cigarette lighter to light the torch. The acetylene will burn a dark orange color and may give off a black soot. The acetylene valve is adjusted properly when the flame almost separates from the torch tip.

Open the oxygen needle valve. Adjust each needle valve to obtain the desired size and kind of flame.

29-13. Adjusting the Flame

Once the torch is lighted, the oxyacetylene flame can be adjusted to:

1. A **neutral flame** — the proper mixture of oxygen and acetylene for most welds. See Fig. 29-10.
2. A **carburizing flame** — a low-temperature flame for torch brazing, this flame has too much acetylene (or too little oxygen) for welding. See Fig. 29-11.

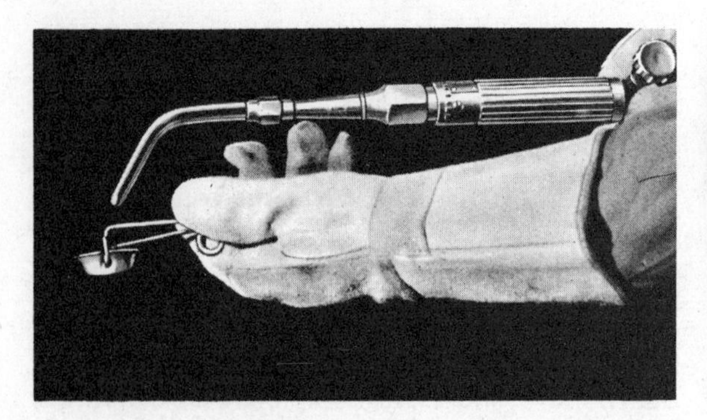

Fig. 29-9. Lighting the Torch

Fig. 29-11. Carburizing Flame — Excess Acetylene

Fig. 29-10. Neutral Flame — Equal Volumes

Fig. 29-12. Oxidizing Flame — Excess Oxygen

3. An **oxidizing flame** — this flame has an excess of oxygen (or insufficient acetylene). It will not make a strong weld — always harmful. See Fig. 29-12.

29-14. Tacking

In welding, a **tack** is a weld at one spot, Fig. 29-13. Form a puddle of melted metal from the edges of both pieces to be welded. Put the end of the filler rod into the puddle and melt it to fill the gap between the pieces. Remove the rod and flame. Tack two ends to hold the pieces in position for welding.

29-15. Welding with an Oxyacetylene Flame

Adjust the colored goggles over the eyes to fit comfortably. Put on the welding gloves; light the torch and adjust flame to neutral. Set and clamp the pieces to be welded. Tack each end of the weld to be made.

The procedure now is to trace a path with the flame that will (1) melt the edges of the pieces to be welded, and (2) melt just enough metal from the filler rod to fill the gap between the pieces. Starting at the right, hold the torch so that its tip forms an angle of about 45° with the work. See Fig. 29-14. The tip of the inner cone of the flame should almost touch the work.

Play the flame over both pieces to be welded, moving it in a small circle until each edge just begins to melt and a puddle forms. Put the end of the filler rod into the puddle. Now start moving the flame around three sides of the rod so that a half circle or U-shape is traced with the flame. At the same time, begin to move the rod and the flame slowly to the left along the gap. Continue the movement until the weld is finished. If the torch and rod are moved too rapidly, the gap will not be filled. If the movement is too slow, the flame will melt a hole through the metal. It takes a lot of practice and experimenting to become good at this movement.

Welding from right to left or front to back as in Fig. 29-14 is called the **forehand technique**. It is commonly used on thin metals. On thick metals the **backhand technique** is used. In this, the rod is moved from side to side and the flame held steadily, welding from the opposite end — left to right or back to front.

If the welding must be stopped for only a few minutes, it is all right to shut off the flow of gases at the torch. When the welding is to be stopped for a longer time, as for

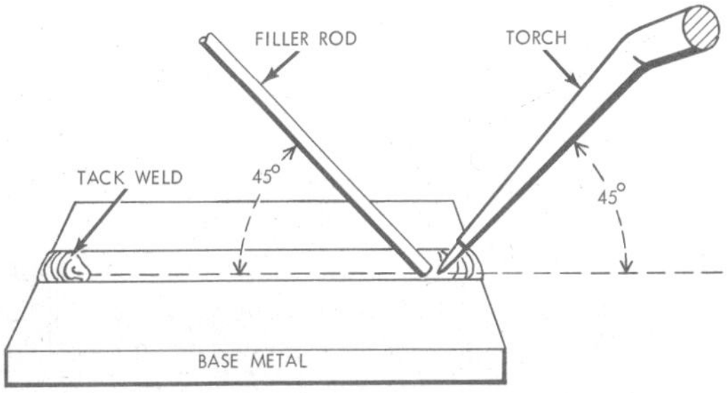

FILLER ROD
TORCH
TACK WELD
45°
45°
BASE METAL

Fig. 29-13. Tack Welding a Butt Joint

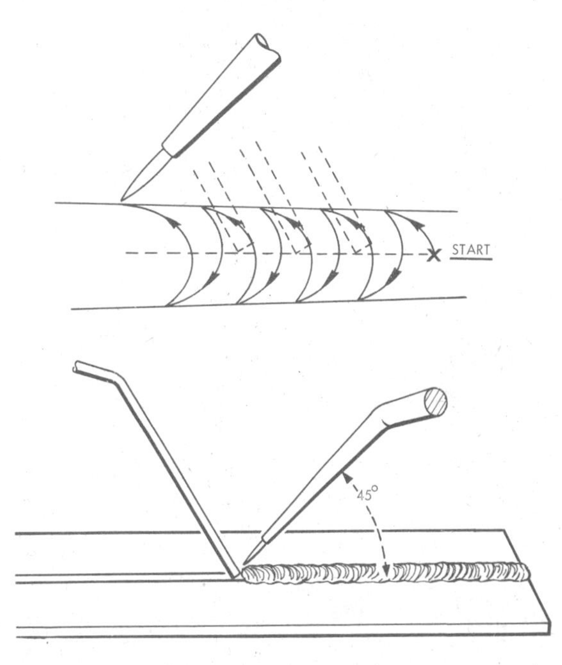

Fig. 29-14. Forehand Oxyacetylene Welding

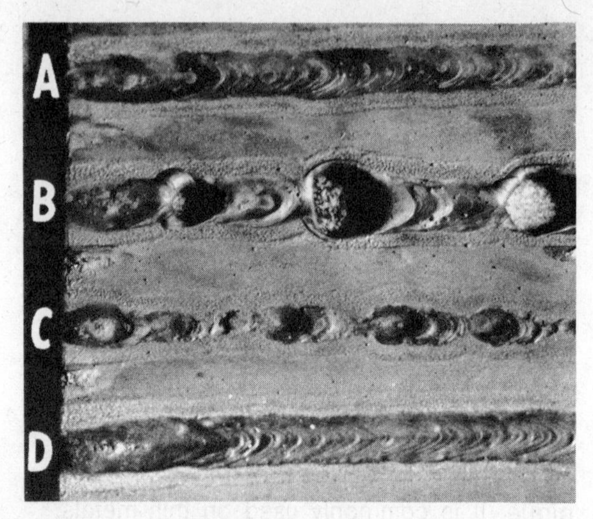

Fig. 29-15. Appearance of Oxyacetylene Welds
A and D — Satisfactory Welds.
B — Excessive Heat. Torch Moved Too Slowly.
C — Insufficient Heat. Torch Moved Too Fast.

Fig. 29-16. Motor-Driven Generator

lunch, end of the class period, or overnight, the acetylene and oxygen tank valves must be closed.

Fig. 29-17. Engine-Driven Generator

29-16. Inspecting the Oxyacetylene Weld

When the weld is finished, remove the scale with a wire brush or tool. Look for errors that can be corrected when making the next weld. Study Fig. 29-15 for examples of good and bad welds.

29-17. Welding with Electric Current

There are many welding processes that use electricity. The names of some are **shielded metal-arc, carbon-arc, atomic-hydrogen arc, tungsten inert gas-arc** (TIG), and **spot weld.** See Fig. 29-1. Each process has advantages over the others for a particular type of work.

Only the shielded metal-arc weld will be described in detail in this book. Some of the techniques, however, do apply to other kinds of arc welding. The shield is a vapor formed when the flux coating on the electrode is heated. Manual shielded-arc welding is widely used in constructing machinery of all kinds, structural steel work, and all tvpes of maintenance and repair welding.

29-18. Current for Arc Welding

Either of two kinds of electrical current may be used for electric welding: direct current or alternating current. Direct current means that the current always flows in one direction. Alternating current means that the current rapidly changes its direction again and again.

Direct current is usually shortened and called DC. Alternating current is abbreviated AC.

29-19. Types of Machines for Electric Arc Welding

There are three types of arc-welding machines in common use:

1. The direct current generator, driven by an electric motor or gasoline engine. See Figs. 29-16 and 29-17.
2. The alternating current transformer-type welder, which operates from a powerline with a built-in transformer.
3. The direct current transformer-type welder, which operates from a power line with built-in transformer and rectifier.

To produce the heat necessary to melt metal, the electric current must be changed from the 110 or 220 volts provided by power companies. The voltage must be reduced or stepped-down, allowing the amperage to be increased or stepped-up. A typical small arc-welding machine will produce 150-250 amperes at 20 to 40 volts.

Each type of welding machine has controls for adjusting the amperage, which is determined by the size of electrode being used and the thickness of the metal being welded. When the current setting is too high, the melted puddle is too large, the electrode melts too rapidly, and there is considerable spatter. When the current setting is too low, the puddle is too small, it is difficult to maintain the arc, and the electrode is likely to stick to the base metal. In general, an amperage setting equal to the decimal equivalent of the diameter of the rod being used will be found to be on the high side of an acceptable range. For example, with 1/8" rod this would mean 125 amperes; with 3/16" rod, 185 to 190 amperes. These should be considered as trial settings.

Some welders have a low and a high welding range with overlapping current ranges. For example, a 180-ampere welder may have a low-current range of 20 to 115 amperes, and a high-current range of 60 to 180 amperes. One could weld with a 3/32" diameter electrode on either current range using 60 to 80 amperes. The low-current range provides a higher open-circuit voltage and a more stable arc, thus making it easier to weld thin metals. The high-current range provides a lower open-circuit voltage which produces better quality welds with thicker rods on thicker metals.

29-20. The Welding Arc

In some ways the welding arc is like the sun. The arc gives off heat, a brilliant white light, infrared rays, and ultraviolet rays. If the body is not properly protected from the arc, it can be burned in much the same manner as sunburn. The eyes, arms, legs, and feet should be protected.

29-21. Helmet and Face Shield

The **helmet** and **face shield** are made from lightweight, pressed-fiber material, Fig. 29-18. Their black color reduces reflection. They have a window made of special dark-colored glass, called a lens, through which the welder looks at the welding arc. The special lens absorbs infrared and ultraviolet rays. It should be a number 10 lens for electric welding. This is much darker than the number 5 or 6 lens used in oxyacetylene goggles. To protect the more expensive, colored lens from the spatter of the weld, a clear, easily replaceable cover lens is used over the dark lens. See Fig. 29-19.

Looking at the arc without a shield will burn the eyes. This will depend on the length of time the eyes are exposed to the

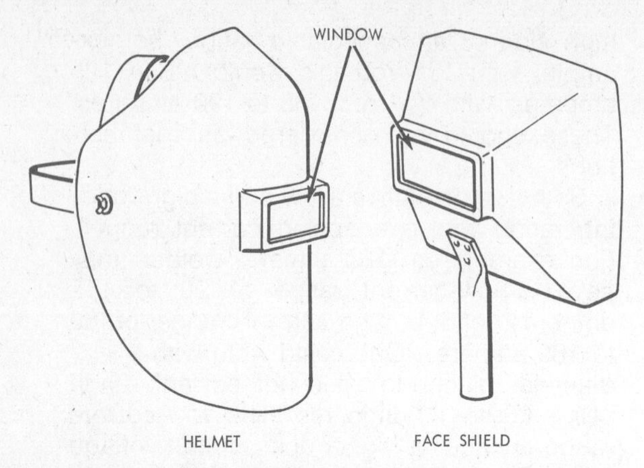

Fig. 29-18. Helmet and Face Shield for Arc Welding

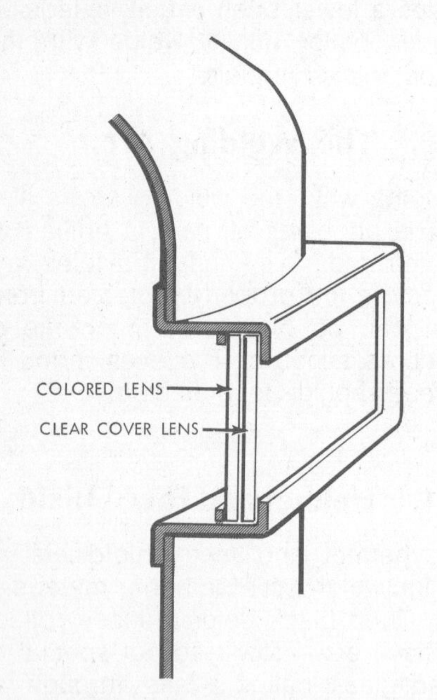

Fig. 29-19. Cutaway Showing Typical Construction of Helmet Window

arc light. Spatter from an arc can also cause serious damage.

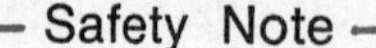

Safety Note

Always wear clear safety goggles underneath the welding helmet. This protects the eyes when chipping and brushing the weld flux or **slag.**

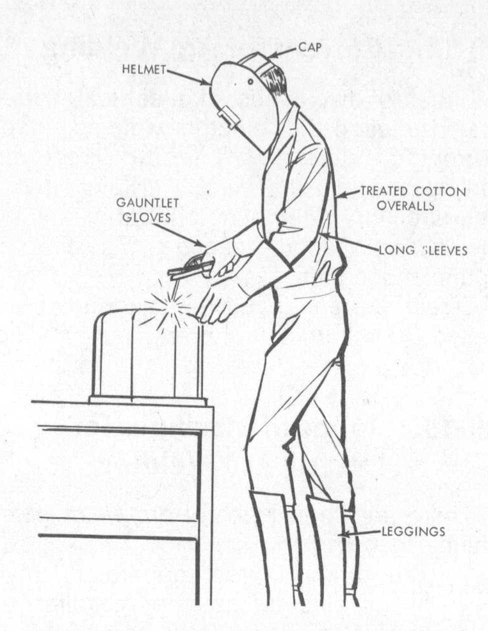

Fig. 29-20. Proper Clothing for Arc Welding

29-22. Clothing for Electric Welding

Heavy, long-sleeved, fire-resistant cotton coveralls should be worn. Pay special attention that the sleeves and legs are not turned up to form cuffs. Clothing having no front pockets, or pockets with button cover flaps, is recommended. Pieces of hot metal from the weld often catch in cuffs and sometimes cause painful burns before they can cool or be removed. Some welders like to wear leather aprons, sleevelets, leggings, and spats. Most welders use a glove that overlaps the sleeve for protection from spatter of melted metal and the arc. Gloves also help to handle hot metal. See Fig. 29-20.

29-23. Welding Shop Tools

The commonly used welding shop tools are shown in Fig. 29-21 and are listed below:

1. Wire brush — for cleaning the work and the weld.
2. Chipping hammer — to remove burrs and slag.

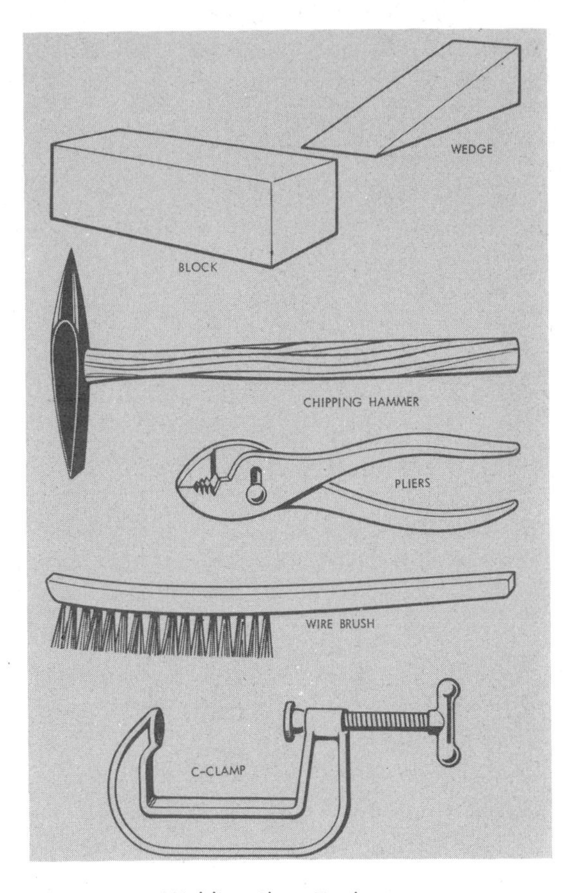

Fig. 29-21. Welding Shop Tools

3. Hammer — for bending and shaping metal.
4. Wedges — to position the work for welding.
5. Clamps — to hold the work for welding.
6. Pliers or tongs — for handling pieces of hot metal.

29-24. The Electric Circuit for Arc Welding

An electric circuit is a path over which an electric current can flow. If the path is interrupted or incomplete at any point, the circuit is said to be open. Current will not flow in an **open circuit.** Current will only flow in a completed, or **closed circuit.** The diagram in Fig. 29-22 is an electric arc welding circuit. The circle with a positive (+) on one side and a negative (−) on the other is a symbol for the welding machine. Starting from the — side, trace the path of the current. From the machine, it leads to the electrode holder, the **arc, the material being welded,** and then back to the + side of the welder. If the arc is broken, the path is interrupted and current will not flow. One cable (−) is connected to the electrode; the other (+) to the work. The cable connected to the work is called the ground cable.

29-25. Polarity

Look at Fig. 29-22 again. The path from the + side of the welder leads to the work. The path from the electrode leads to the − side of the welder. When the electrode leads to the − side of the welder, the circuit is said to have **straight polarity.** Straight polarity is sometimes called **electrode negative.**

If the + and the − are changed around, Fig. 29-22, the polarity is reversed and the circuit is said to have **reverse polarity**. Reverse polarity is sometimes called **electrode positive** because the electrode leads to the positive side of the welder.

Straight polarity is the normal setup for DC welding, but some types of welds can be made more easily with reverse polarity.

Straight polarity and reverse polarity refer only to direct current circuits (DC). If an AC machine is being used, the polarity reverses with each cycle of the current.

29-26. Welding Arc Temperatures

When electricity flows continuously through the air gap between a welding electrode and metal to be welded, the flow is called a **welding arc**, Fig. 29-23. Such an arc gives off a brilliant white light and extreme amounts of heat. Temperatures of

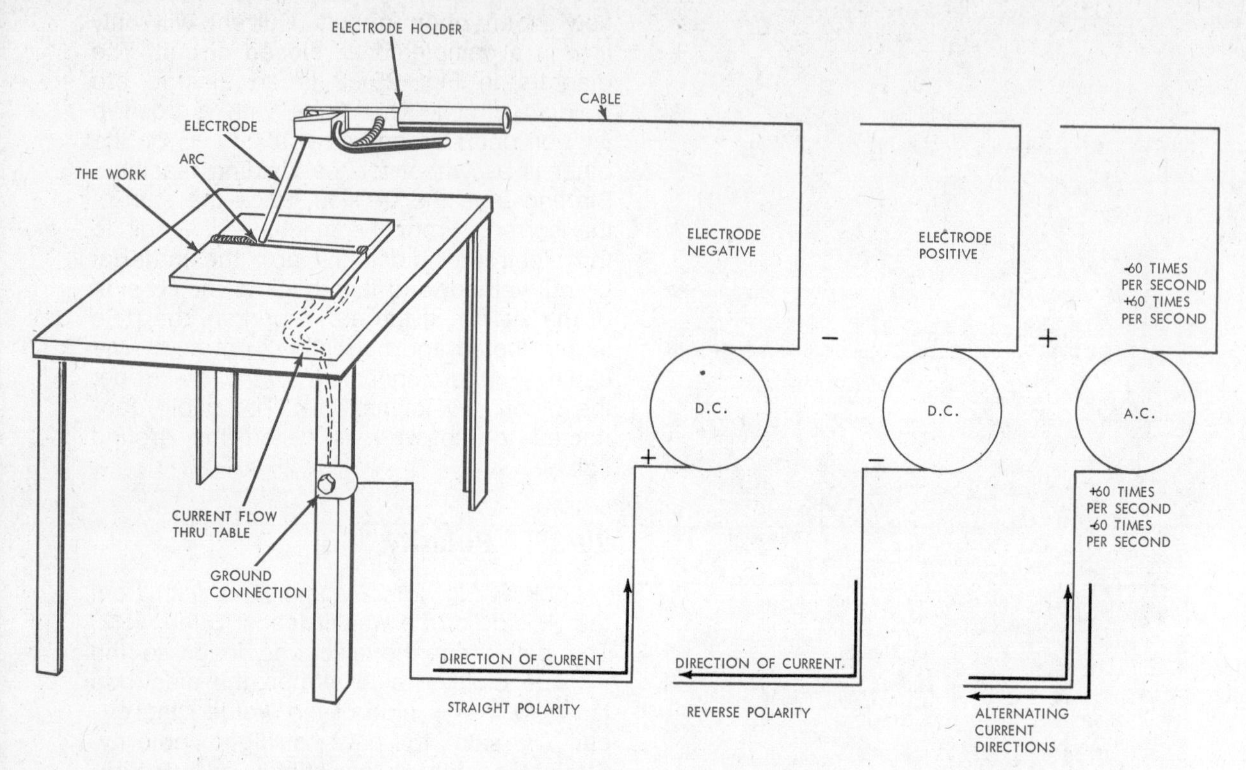

Fig. 29-22. Electric Arc Welding Circuits

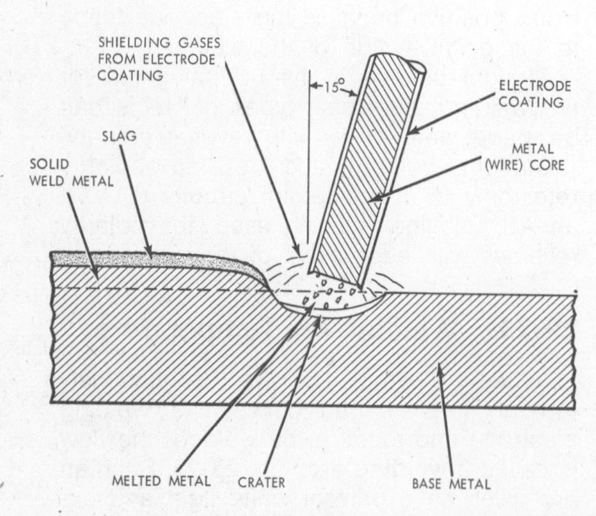

Fig. 29-23. The Welding Arc

10,000° F. have been measured in the welding arc.

29-27. The Electrode

As shown in Fig. 29-23, the **electrode** is a rod that carries current between the electrode holder and the welding arc. The welder can make the welding arc long or short by moving the electrode up or down. Electrodes are widely used in 14″ lengths and in a range of diameters from 1/16″ to 3/8″.

The electrode is also used to supply additional metal to the weld. Because it supplies metal to the weld, the electrode is used up (consumed) and must be replaced frequently.

Electrodes of carbon or tungsten are used for metal cutting and for welds which do not need additional metal.

29-28. The Coated Electrode

When melted metal is exposed to the air, it has a tendency to combine with nitrogen and oxygen from the air. These **nitrides** and **oxides**, called impurities, make the weld metal **brittle**. Brittle metal is not strong. For most welds it is best to have the weld as strong or stronger than the metal being welded. **Shielded-arc welding** is very popular for making strong welds that are not brittle.

To shield the arc means to protect it (and the melted metal) from the surrounding air. This is done by using a special electrode, called a **coated electrode**, in the electrode holder. As the coating burns off in the arc, it forms gases that surround the arc and protect the arc from the air. The electrode coating is also a **flux**. A flux is a material that will quickly combine with the nitrides, oxides, and other undesirable impurities in the melted metal. These impurities are lighter than the melted metal and therefore float on top of the weld metal. When the metal cools, the impurities form a crust on top of the weld. This crust is called **slag**, Fig. 29-23, and can be easily chipped away with a chipping hammer, Fig. 29-21.

29-29. Electrode Numbering System

Examples of numbers that identify welding electrodes are E6013, E7025, and E9030. They belong to a system developed by the **American Welding Society**. The system uses a letter followed by **four digits**.[1] See Table 23. The first two digits denote tensile strength. Numbers in which the first digit is 6 (as E6013) are mild steel; other numbers (as E7025, E9030, etc.) are alloy steels.

Features and traits of welding electrodes vary from one manufacturer to the next,

[1]Digit means any of the ten figures 0, 1, 2, 3, 4, 5, 6, 7, 8, 9, by which all numbers may be expressed.

TABLE 23

American Welding Society (AWS) Numbering System for Coated Electrodes

E • • • •

Letter — 1st Digit — 2nd Digit — 3rd Digit — 4th Digit

Letter or Digit	Description
E	The electrode is made for electric arc welding.
1st and 2nd Digits	These indicate the minimum tensile strength of the wire used to make the electrode. (60 means 60,000 pounds per square inch.)
3rd Digit	This shows the welding position for which the electrode can be used. If the 3rd digit is: 1—the electrode may be used for any welding position: for flat welds (See Fig. 29.24), for vertical welds (See Fig. 29.25), for horizontal welds (See Fig. 29.26), for overhead welds (See Fig. 29.27). 2—the electrode may be used for flat or horizontal welds. 3—the electrode may be used for flat welds only.
4th Digit	This indicates the kind of current and type of coating.

Fig. 29-24. Flat Welding

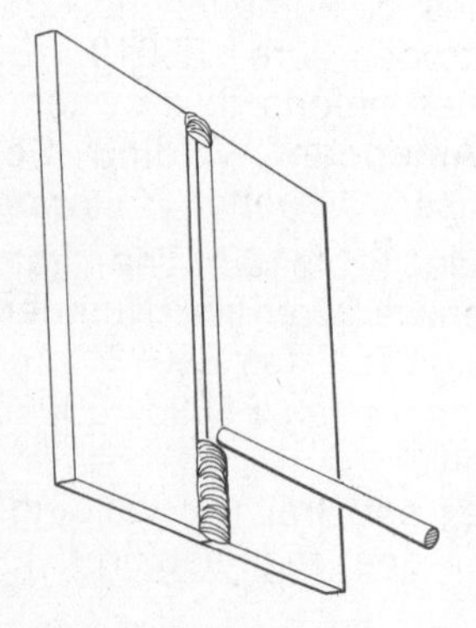

Fig. 29-25. Vertical Welding

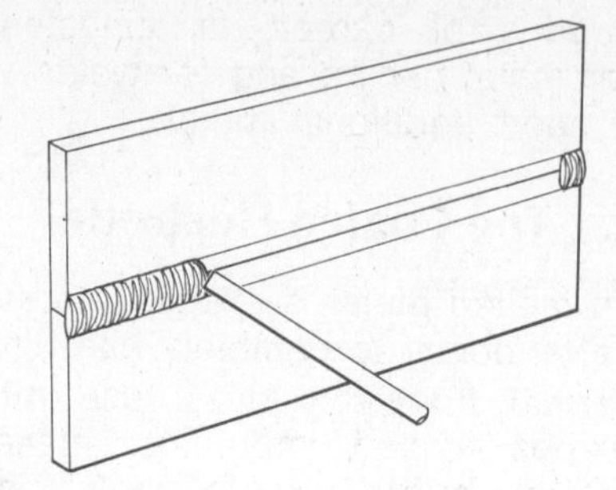

Fig. 29-26. Horizontal Welding

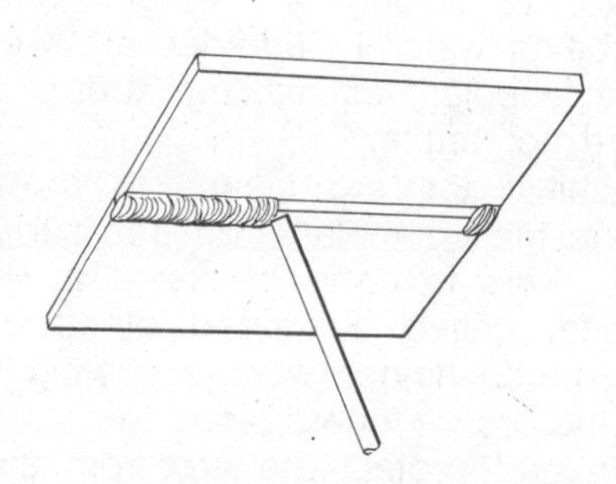

Fig. 29-27. Overhead Welding

even though they have identical numbers under the AWS numbering system. Therefore, it is always best to use the current and usage recommendations of the individual manufacturer. Most dealers can supply electrode selection charts and handbooks which list the characteristics and uses of their electrodes.

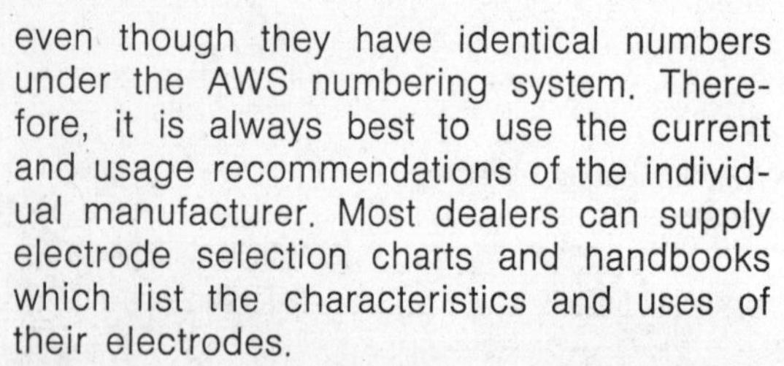

A good electrode for the beginner who is learning to weld is the ⅛″ diameter E6013 rod. It is a mild steel electrode which can be used to weld in all positions, on AC or DC, and on light- or heavy-gage mild steels. It has a steady arc, a smooth bead, high tensile strength, and it starts and restarts easily. The same electrode in a 3/32″ diameter can be used on sheet metal and other thin steels.

29-30. Starting the Arc

Select the proper electrode for the job and clamp it in the electrode holder. If using a DC welder set the polarity to straight or reverse. See section 29-25. Make an approximate current setting — low for welding thin metal, higher for thicker metal. Take notice that you are wearing the proper clothing. See section 29-22. Put on the welding gloves and have a helmet or face shield ready. Turn on the welding machine.

Either of two methods can be used to start the arc. With a relaxed but firm grip on the electrode holder, put the end of the electrode near the spot where the weld will start. Before the arc is started, protect your eyes with the helmet or face shield. Experienced welders learn to put the shield over their faces and start the arc at the same time.

Method 1 — Down-Up Method

See Fig. 29-28. Lower the electrode straight down and touch the plate lightly. Quickly pull the electrode up and away —a

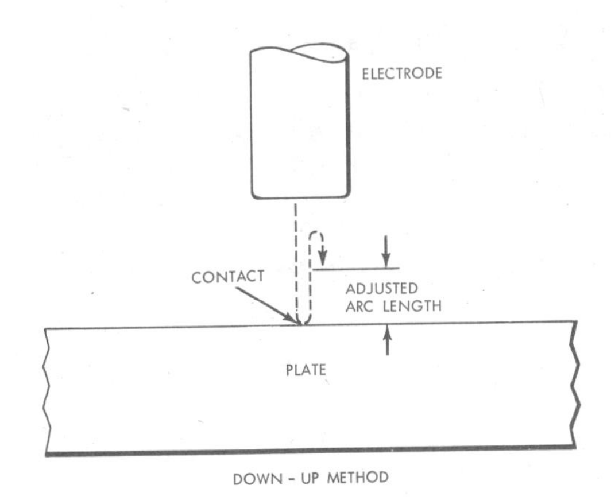

Fig. 29-28. Starting the Arc — Down-Up Method

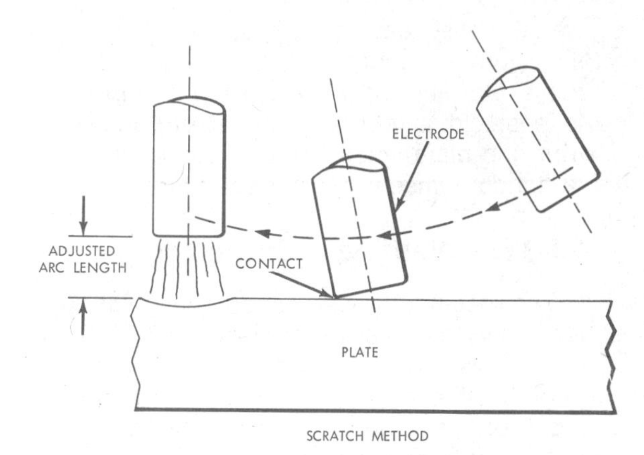

Fig. 29-29. Starting the Arc — Scratch Method

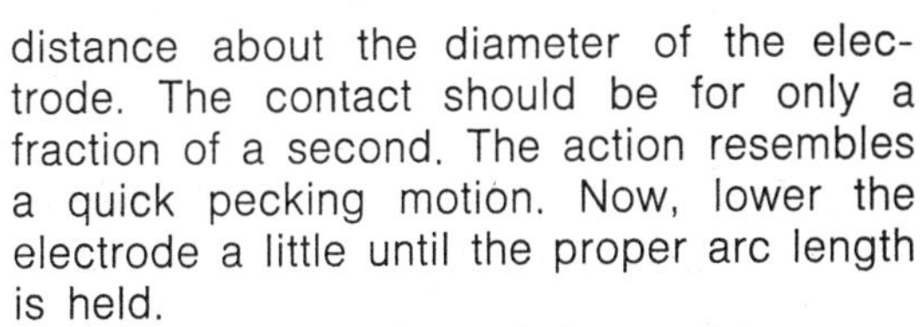

distance about the diameter of the electrode. The contact should be for only a fraction of a second. The action resembles a quick pecking motion. Now, lower the electrode a little until the proper arc length is held.

This method is best. It is used by many experienced welders. At first, it is more difficult to start the arc by this method on a cold metal workpiece. With some practice it becomes easy.

Method 2 — Scratch Method

See Fig. 29-29. This method of starting the arc is something like striking a match. Hold the end of the electrode near the plate. With a sweeping motion, scratch the plate lightly with the electrode. Follow through by moving the electrode slightly away from the plate, stopping at the spot where the weld is to start. Adjust the length of the arc.

This method is usually a little easier to learn, particularly when striking an arc on a cold metal workpiece. It is not as good as the down-up method because it is harder to find the spot where the weld is to start. It also puts arcing scars on the plate.

Need For Practice

Starting the arc (sometimes called **striking the arc**) requires much practice. The

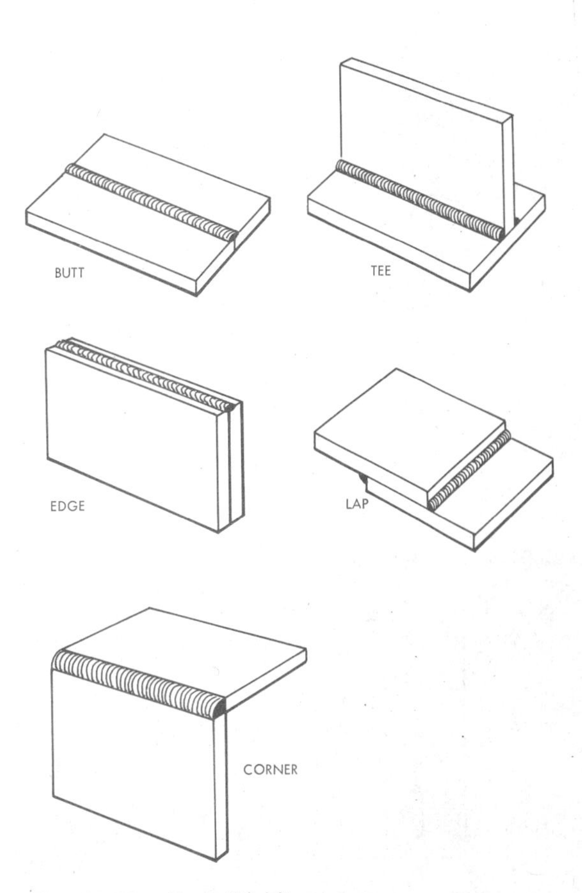

Fig. 29-30. Basic Welding Joints

welder must learn to adjust the setting on the current generator, test the polarity, move the electrode, handle the helmet and face shield, and break the electrode away from the plate when it sticks. He must learn to handle himself and the equipment safely.

29-31. Welding with the Arc

The beginner must experiment with different kinds and sizes of electrodes, various current settings, straight and reversed polarity, the length of the arc, and movement of the electrode before he can expect to make good welds. A weld should not be made over another weld until the slag has been thoroughly chipped away.

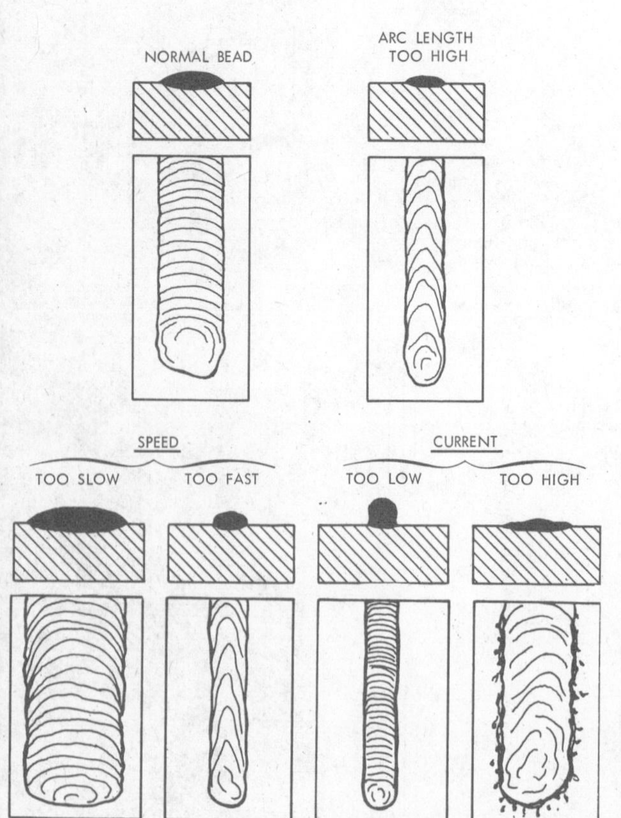

Fig. 29-31. Appearance of Shielded Arc Welds

29-32. Joints Used in Welding

The basic joints used in gas and arc welding are the **butt joint, tee joint, corner joint, lap joint,** and **edge joint**. (See Fig. 29-30. There are many variations of these.

29-33. Inspecting the Arc Weld

After making the weld, chip the slag away with a chipping hammer. Be sure to wear safety glasses with clear lenses while chipping. Carefully inspect the weld. Figure 29-31 shows an example of a good weld. It also shows some bad welds and tells why they are bad.

Always inspect each weld before going on to the next. Try to determine the reasons for defects and improve the next weld.

29-34. Cutting Metal with Welding Equipment

Metal may be cut by the oxyacetylene process and with the electric arc process. Plates of steel are often shaped in this way, since it is faster than sawing. With specialized equipment, elaborate contours are formed easily and the edge produced

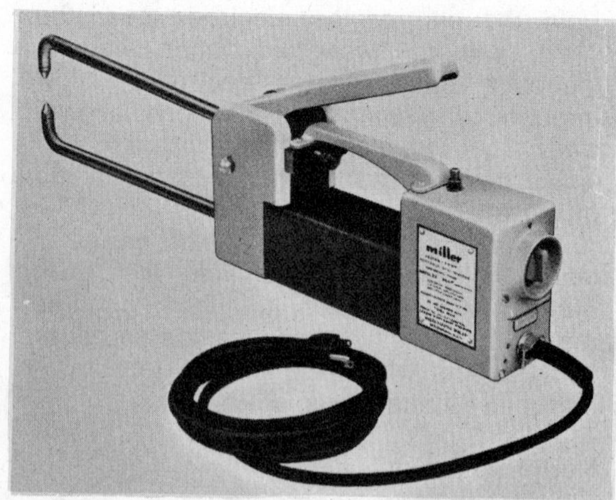

Fig. 29-32. Portable Spot Welder

requires little more finishing than if it were made by sawing.

When cutting with the electric arc, a carbon or tungsten electrode is used. Much more current is used for cutting than for welding.

29-35. Electric Spot Welding

Electric **spot welding** is a form of **resistance welding**. Spot welding is done by passing high current at a low voltage through a small spot on two pieces of metal, usually sheet metal, for a short period of time. It is done with a spot welder, Figs. 29-32 and 29-33. Resistance to the flow of current through the metal at the spot causes heat which makes the spot weld. The pieces of metal must be held together under moderate pressure during the weld and for a few seconds after, while the weld cools. See Fig. 29-34.

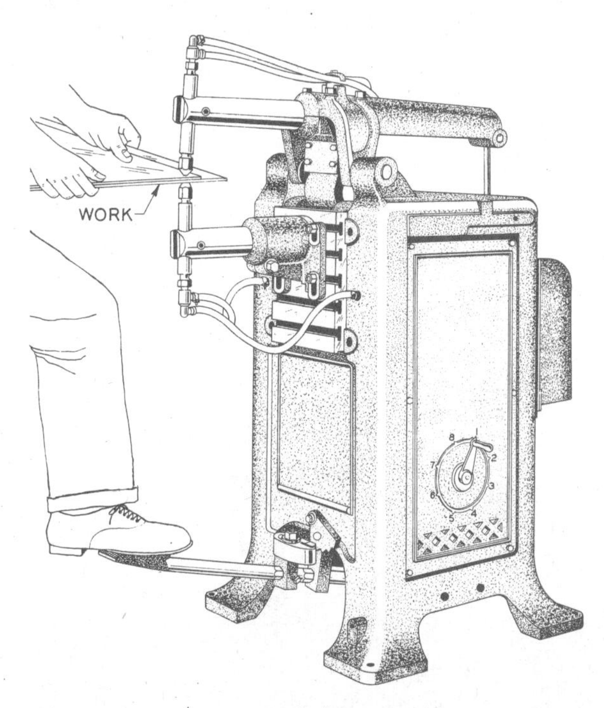

Fig. 29-33. Water-Cooled Spot Welder

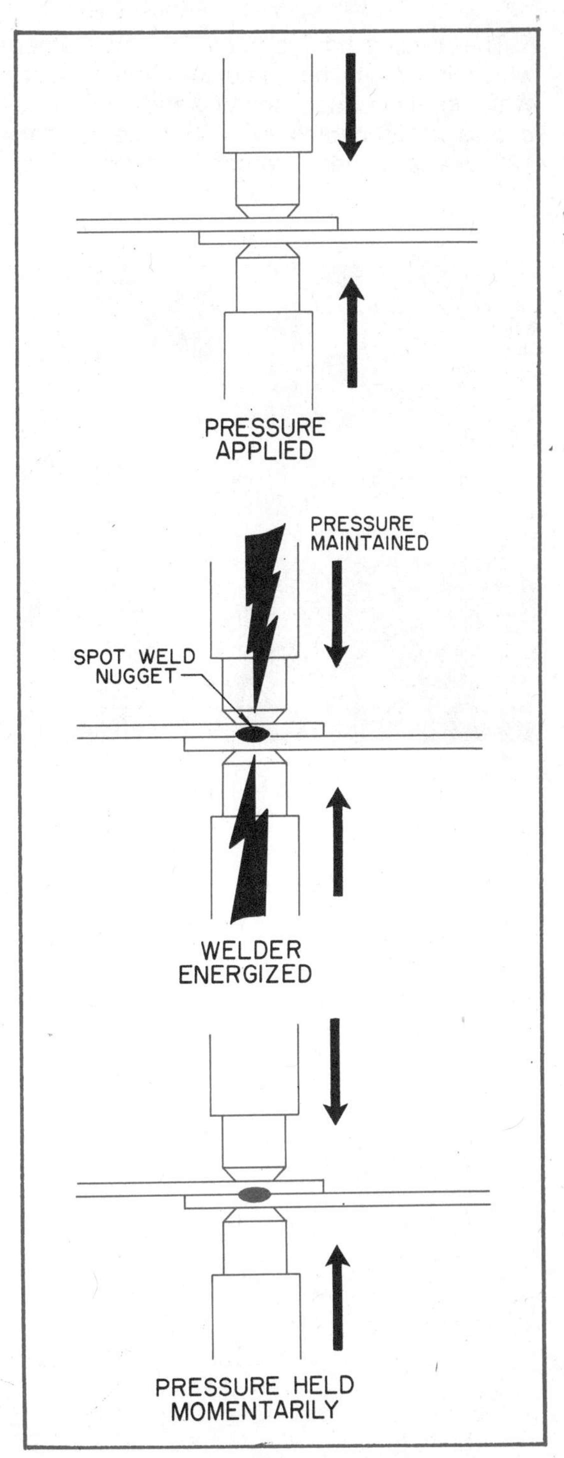

Fig. 29-34. Spot Welding Sequence

The welding time is controlled by a timer which is usually built into the spot welder. Welding time commonly varies from **3 cycles** to **120 cycles**. With 60-cycle current, 120 cycles means 2 seconds of time. Two pieces of 20-gage sheet steel may be spot welded in approximately a 15- to 20-cycle period. If the welding time is too long, the weld will be pitted from excess heat. If the time is too short, the weld will come apart. The points on the welding tongs should be properly dressed to shape with a file, and they should be replaced when badly worn or burned. The tongs should also be adjusted for the correct pressure for the thickness of the weld desired. Too much pressure will cause pitting of the weld.

Fig. 29-35. TIG Welding Equipment

Although spot welding is most frequently used to weld sheet metal joints, it may also be used to spot weld sheet metal to small diameter rods or flat bars.

29-36. Advanced Welding Processes

The increasing use of difficult-to-weld metals in aircraft and aerospace vehicles, together with the search for more efficient methods of welding, has resulted in development of several advanced welding techniques. **Tungsten inert gas (TIG),**

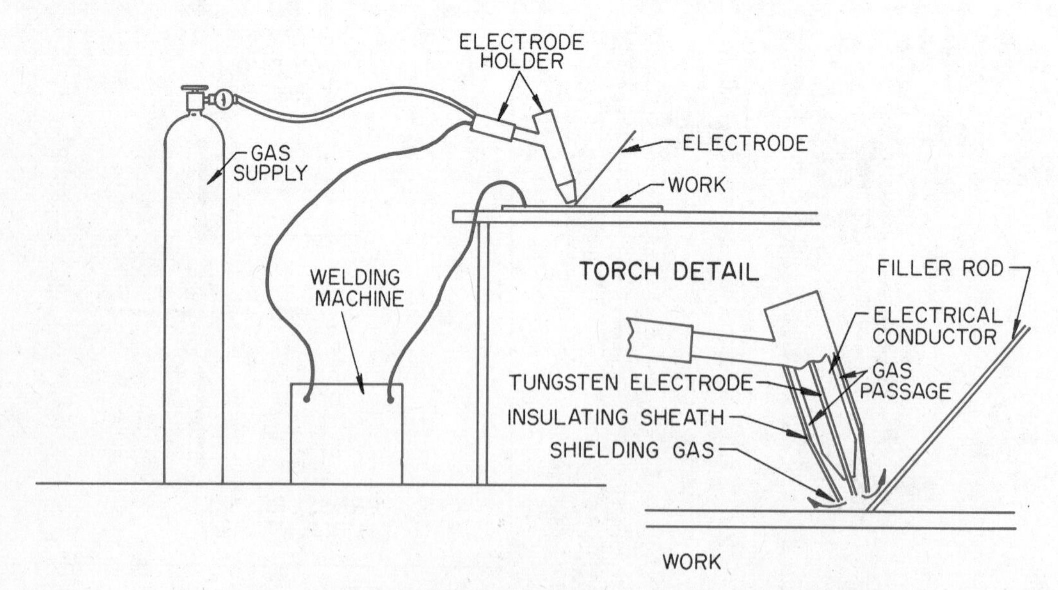

Fig. 29-36. TIG Welding System with Cutaway of TIG Welding Torch

metal inert gas (MIG), **electron beam (EB)**, **laser**, and **inertia welding** are rapidly becoming commonplace.

TIG is an arc-welding process which uses a nonconsumable electrode of tungsten and an inert gas shield of **argon** or **helium**, although **carbon dioxide** may be used for welding steel. See Figs. 29-35 and 29-36. Originally developed for welding magnesium, it also finds wide use for welding aluminum, copper, stainless steel, and other difficult-to-weld metals. Due to the inert gas shielding, TIG welds are very clean and strong. Skilled operators can frequently make welds that are almost invisible.

MIG is also an arc-welding process using inert gas shielding, but it differs from TIG in that it uses a consumable electrode in wire form. See Fig. 29-37. The wire electrode contributes filler metal to the weld and is fed through the welding gun at a rate controlled by the welding machine operator. The combination of gas shielding and automatic electrode feeding makes possible clean, sound welds at a rate far greater than can be obtained with conventional equipment. For this reason, MIG welders are rapidly replacing conventional welders for production work. See Fig. 29-38.

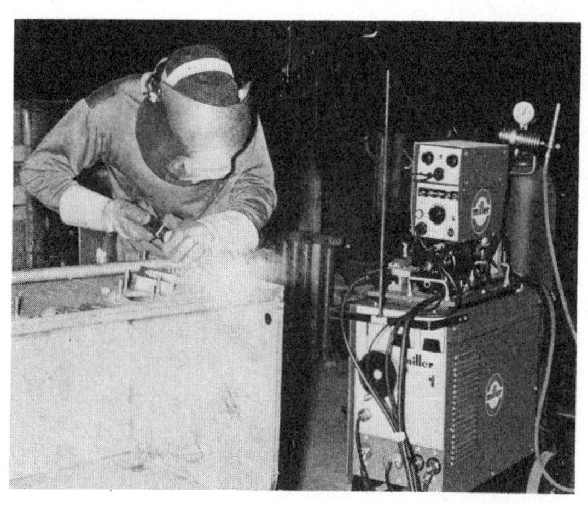

Fig. 29-38. MIG Welding Fittings on a Transformer Case

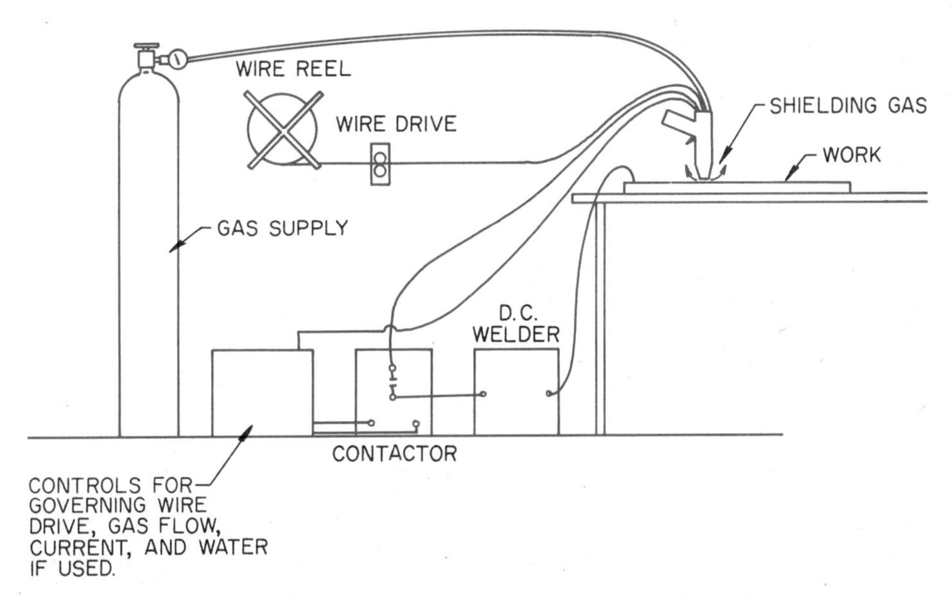

Fig. 29-37. MIG Welding System

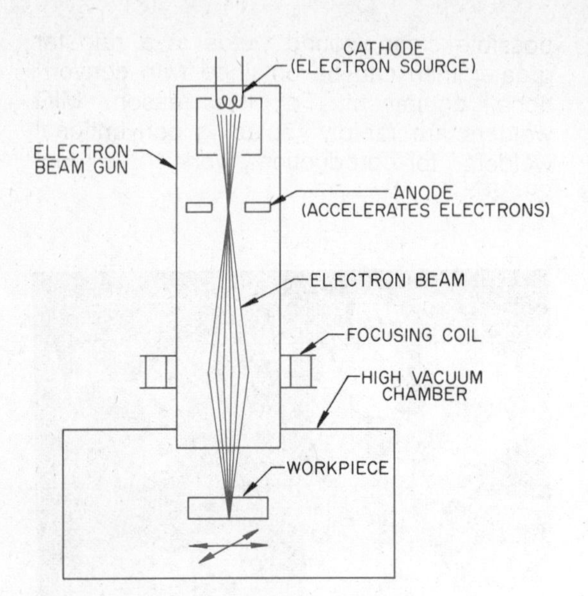

Fig. 29-39. Electron Beam Welding System

Fig. 29-40. Hard Vacuum Electron Beam Welding Equipment

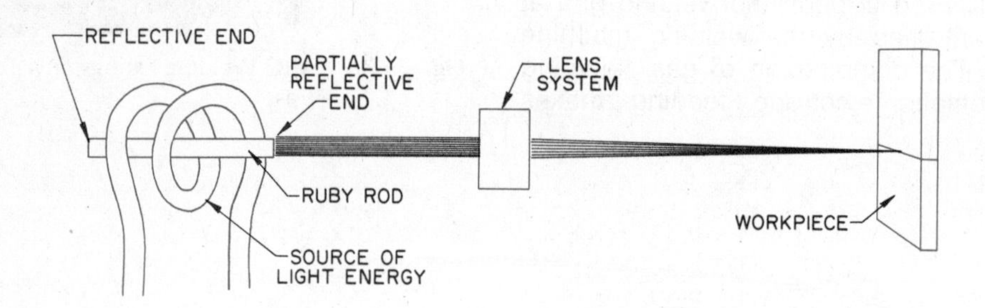

Fig. 29-41. Laser Welding System

In **electron beam welding**, a concentrated beam of electrons bombards the base metal, causing it to melt and fuse together. See Figs. 29-39 and 29-40. The process is most efficient when done in a vacuum, therefore, the size of the vacuum chamber limits the size of the work-pieces which can be welded. Advantages include the ability to produce welds of extremely high purity, ability to melt any known material, ability to weld dissimilar metals, and ability to make welds with depths as great as 6 inches.

Electron beam welding is costly due to the high initial cost of the equipment, and especially so when done in a high vacuum due to the time lost in pumping down the vacuum chamber between welds. When welds are not made in a vacuum, many of the advantages of the process are sharply reduced.

The energy source for **laser welding** is a concentrated beam of light. (Laser is an acronym for Light Amplification by Stimulated Emission of Radiation.) See Fig. 29-41. Instead of providing a continuous source of welding heat, lasers release their energy in bursts or pulses at a rate of 6 to 10 a minute. Since each pulse lasts only a few millionths of a second, the metal is liquid too short a time for chemical reaction to occur, and no protection is needed to obtain sound welds.

Laser systems can be precisely controlled and have sufficient power to weld and even vaporize any known material. Other advantages include the ability to make welds through transparent coverings and to make welds in locations impossible to reach with conventional welding gear. Depth of penetration is, however, presently limited to a few hundredths of an inch.

Inertia welding is a process which uses friction to generate the welding heat. The inertia welding machine rotates one of the parts being welded while applying pressure to force the two parts together. The friction thus obtained generates enough heat to soften the metal to a plastic state. Hydraulic pressure forces the softened metal together to form the weld. Full-strength welds are made in only a few seconds. No special preparation of the surfaces being joined is required, and there is no need for fluxes or filler metals. The localized heating makes possible distortion-free welded assemblies of parts with widely different thicknesses. Dissimilar metals, such as steel and titanium, may also be welded by this process.

Words to Know

arc weld
argon
backhand technique
base metal
butt joint
carbon dioxide
carburizing flame
coated electrode
corner joint
edge joint
electrode
electron beam weld
filler rod
forehand technique
forge welding
fusion weld
gas pressure regulator
helium
inert
inertia weld
lap joint
laser weld
MIG
neutral flame
nitrides
oxides
oxidizing flame
oxyacetylene weld
pressure weld
resistance weld
reverse polarity
shielded-arc welding
slag
spot welding
straight polarity
tack weld
tee joint
TIG
weld nugget
weld metal

Review Questions

1. Describe briefly how to make each of the following types of welds: fusion weld, pressure weld, and inertia weld.
2. What temperature can the oxyacetylene flame produce? Electric arc?
3. What is the difference between "base metal" and "weld metal"?
4. Describe how a person doing oxy-acetylene welding should be dressed for maximum safety.
5. What tip size, regulator settings, and filler rod diameter should be used to weld 11-gage steel?
6. Why is it necessary to use a gas pressure regulator?
7. Why should the wrench be kept on the acetylene cylinder?
8. Describe the correct procedure for setting up an oxyacetylene system for welding with a No. 5 tip.
9. What kind of flame should be used to make most welds?
10. What effect does an oxidizing flame have on the weld quality? Carburizing flame?
11. What is the purpose of a "tack"?
12. What can happen if the weld is made too slowly?
13. In what ways is the electric welding arc like the sun?
14. Why should the welder wear clothing without pockets and cuffs?
15. What is the purpose of the coating on arc-welding electrodes?

16. Are all E6013 electrodes the same? Why?
17. Name the joints commonly used in welding.
18. Why should each weld be inspected?
19. During the electric welding process, why should regular safety goggles be worn as well as the welding helmet?
20. What is the difference between straight polarity and reverse polarity? Which is most commonly used?
21. What is the purpose of having an electrode numbering system?
22. Describe the spot welding process. For what is it used?
23. How does TIG welding differ from MIG welding?
24. How do electron beam and laser welding differ? Name several advantages of each.
25. List some advantages of inertia welding.

Drafting

1. Design an all-welded work bench.
2. Make isometric drawings of the basic joints used in welding.

Career Information

1. What abilities are needed to learn to weld?
2. What hazards must the welder guard against in the normal course of his work?
3. How much are welders paid in your area?
4. Are welders required to be licensed in your state?

UNIT 30

Adhesive Bonding

30-1. Meaning of Adhesive Bonding

Adhesive bonding is the process of joining product components with **nonmetallic** glues or adhesives to make permanent fastenings. Soldering and brazing employ metals as adhesives and are not generally classified as adhesive-bonding methods. They are discussed in Unit 28.

30-2. Structural and Nonstructural Bonding

Adhesives have been used to fasten materials together for thousands of years. Until recently, however, the adhesives available were only suited for use in non-structural metalworking applications. Adhesives were not used where failure of the adhesive would affect the structural soundness of the assembly. Nonstructural adhesive bonding is widespread. For example, nonstructural adhesive bonding in automobile construction includes attachment of rubber seals to hoods, trunks, and doors; fastening upholstery materials to metal body parts; attaching metal trim; and as sealing material between metal parts fastened together by conventional means.

Structural (load-bearing) use of adhesive-bonded metals was pioneered by the aircraft industry during World War II. Since then, the obvious advantage of adhesive bonding for many purposes has stimulated the development of new adhesives suitable for a wide range of structural requirements.

Structural adhesive bonding is now widely used in the aerospace, automotive, appliance, instrument, and other metalworking industries, Fig. 30-1. This unit is concerned primarily with structural adhesive bonding of metals.

30-3. Advantages of Adhesive Bonding

Adhesive bonding has distinct advantages which can only be realized with intelligent selection of adhesives, proper joint design, care in surface preparation, adhesive application, and assembly.

1. Loads are distributed more evenly in bonded joints. There are no high-stress concentrations such as occur in spot welded, riveted, or bolted joints. In some cases this allows the use of thinner materials without loss of joint strength, resulting in cost and weight savings.
2. **Fatigue resistance** is improved due to the ability of adhesives to stretch or compress under various load conditions. For example, adhesive-bonded helicopter rotor blades have a service life of over 1,000 hours, while riveted blades have a service life of only 80-100 hours.
3. There is no loss of part strength or shape distortion with adhesive bonding. Strong assemblies are made without the

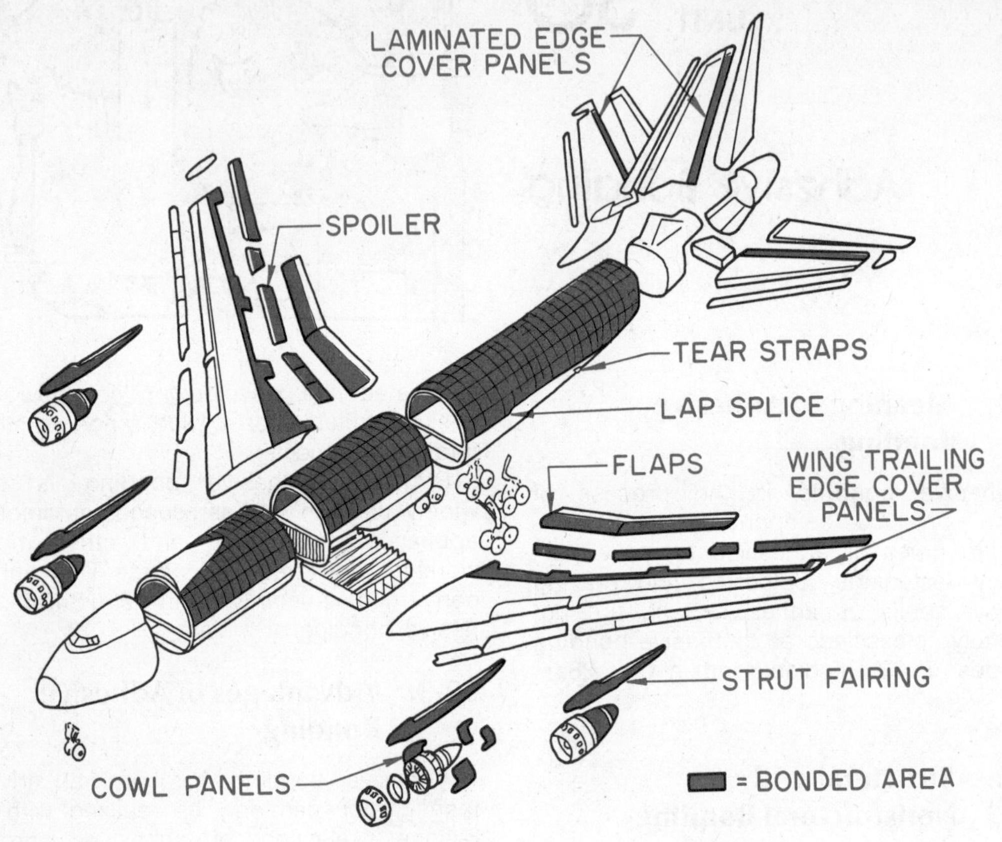

Fig. 30-1. Structural Bonding Application on Boeing 747

holes required for mechanical fastening and without the distortion resulting from high temperatures during welding.

4. Adhesive bonding permits greater design flexibility. Almost any combination of materials can be adhesive bonded, regardless of their shape, thickness, chemical, or mechanical properties. This is of particular advantage in joining delicate assemblies which would be distorted or damaged when joined by conventional methods. For example, adhesive bonding has made possible the development and use of **honeycomb** construction in aircraft and space vehicles. Honeycomb cores of fragile aluminum foil, stainless steel, or

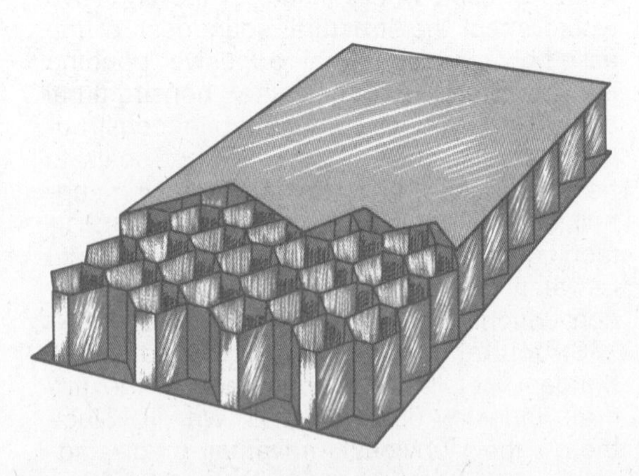

Fig. 30-2. Honeycomb Panel Construction

plastic are successfully bonded to thin sheets of various aircraft metals, resulting in extremely rigid but lightweight parts, Fig. 30-2.

5. Adhesive-bonded joints are automatically sealed against transmission of fluids and most gases, eliminating the need and expense of separate sealing operations. The adhesive also helps reduce sound transmission. Between different metals, the adhesive can insulate against **electrolytic corrosion**. Adhesives, however, may be formulated to be electrically conductive, semiconductive, or nonconductive.
6. Cost reductions are possible because more expensive conventional fastening operations such as drilling, riveting, or welding are eliminated. Savings may also come from use of thinner or lower-cost materials and from the reduced number of finishing operations required.

30-4. Disadvantages of Adhesive Bonding

1. **Performance limitations**. A critical disadvantage is that most adhesives fail rapidly above 500° F. However, service temperatures are slowly being raised with the addition of heat-resistant fillers. Also, research with **ceramic adhesives** indicates that service temperatures of 1000° F. are possible.

   Some adhesives lose strength in situations where temperatures are alternately high and low. Other adhesives are subject to attack by bacteria, mold, moisture, solvents, and even vermin. Adhesives must be selected carefully to meet performance requirements to assure a trouble-free service life.
2. **Application problems**. Surface cleanliness is very critical and parts must be closely fitted to provide a uniform gap of correct size. Adhesives must be handled with care so as to prevent contamination. Special fixtures are often needed to assure correct alignment of parts during assembly. Worker protection must be provided when using toxic adhesives and solvents, and flammable solvents. Some adhesives require **curing** by heating under pressure for long periods of time.
3. **Inspection problems**. Quality of bonds is difficult to assess. Precise bond strength can only be determined by destructive testing.

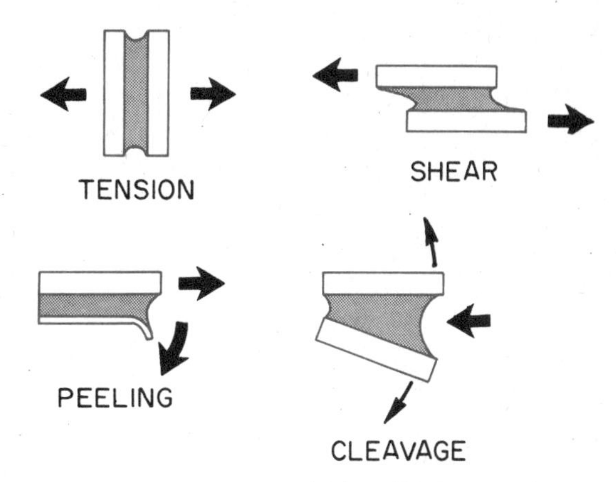

Fig. 30-3. Forces to Which Adhesive Bonds Are Subjected

30-5. Joint Design

Adhesive-bonded joints are subject to any one or a combination of **shearing, tension, cleavage,** or **peeling** forces, Fig. 30-3. Joints should be designed to take advantage of the higher resistance of adhesives to shear and tension forces. Care should be taken to provide sufficiently large bonding areas to withstand calculated loads. Joints with **mechanical interlocking** are stronger than plain bonded joints and also speed assembly because they are **self-aligning.** Figure 30-4 illustrates several preferred joint designs.

30-6. Types of Adhesives

Structural adhesives may be either **thermoplastic** materials, **thermosetting** ma-

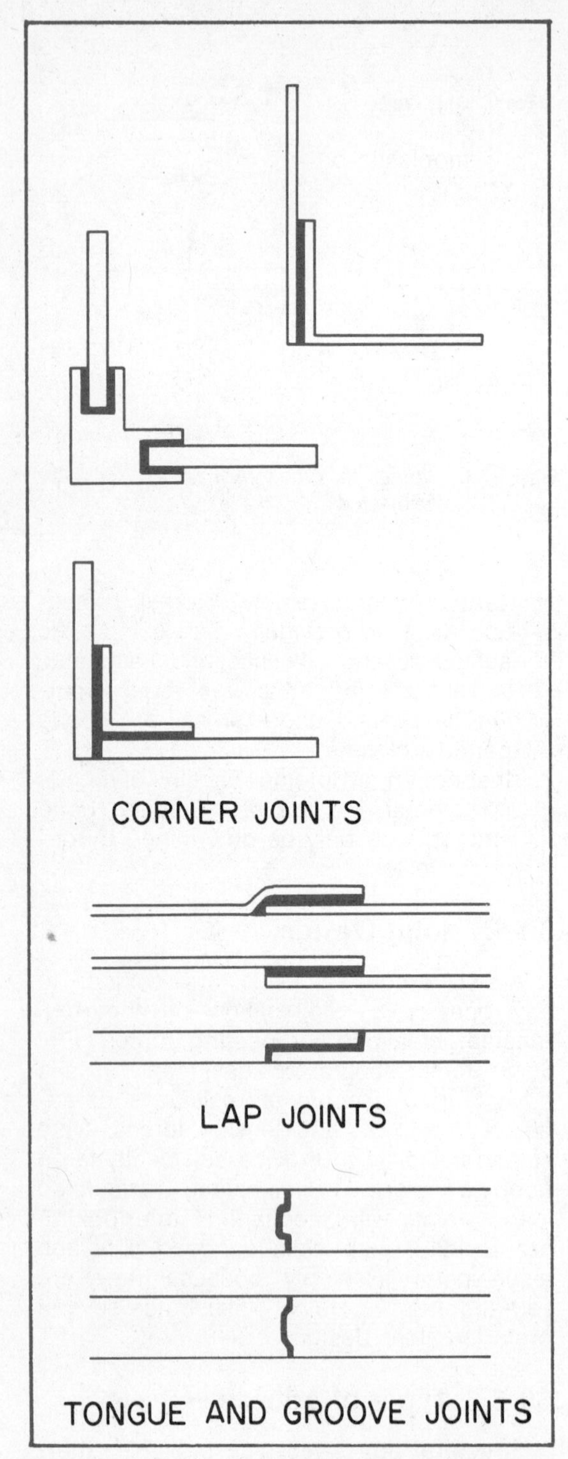

Fig. 30-4. Joints Suitable for Adhesive Bonding

terials, or a combination of both. Thermoplastic adhesives soften when heated and harden on cooling, a cycle that can be repeated over and over. This property permits removal and replacement of parts damaged in service. Thermoplastic adhesives are preferred for assembly line operations because of their ease of application and fast setting times. However, they are not suitable for applications subject to high loads over long periods of time because they tend to **creep** or slip. Thermoplastic adhesives are based on **acrylic, vinyl, nylon,** and **cellulosic** plastics.

Thermosetting adhesives permanently set with heat and pressure and cannot be resoftened. They are harder, less flexible, have higher strength, are more expensive, and more difficult to use than thermoplastic adhesives. **Phenolic, epoxy, polyurethane, polyester,** and other thermosetting plastics are used in making thermosetting adhesives. **Silicones** and **rubbers** provide flexible thermosetting adhesives.

In order to obtain desired properties, thermoplastic materials are sometimes mixed with thermosetting materials to produce a semirigid adhesive.

Most structural adhesives for bonding metals are either **phenolics**, which have been modified with epoxy, vinyl, or rubber, or **epoxies**, which are either used plain or modified with phenolics, **nylon, polyamides**, or **polysulfide** rubbers.

Shear strengths of these adhesives range from a low (at 75° F.) of about 200 psi for one rubber-modified phenolic adhesive to a high of 6,500 psi for an unmodified epoxy adhesive. Most, however, have a shear strength between 2000 and 4000 psi.

30-7. Application Methods and Techniques

Surface Preparation

Porous materials such as wood, paper, and leather are much more easily bonded with adhesives than are metals, for metals are essentially **non-porous**. Porous materials readily absorb adhesive which effectively increases the contact area between parts, thereby improving joint strength. In bonding metals, adhesives are not absorbed but must stick or cling to relatively smooth surfaces. Surface preparation, therefore, is of critical importance since every bit of joint surface must be **wetted** with adhesive in order to obtain maximum joint strength.

Dirt should be washed off with **detergents** or commercial **alkaline cleaners.** Remove oil or grease by washing with solvents such as **acetone, carbon tetrachloride, toluene,** or **xylene. Vapor degreasing** in closed tanks with such solvents as **trichloroethylene** or **perchloroethylene** is sometimes done. If solvent cleaning is done by wiping, the solvent should not be allowed to dry on the surface but should be wiped off with clean rags or paper towels.

Scale, rust, or **oxide films** which cannot be removed with solvents are removed mechanically with abrasives or chemically by **pickling** in acid solutions. After pickling, thorough rinsing in clean water is necessary. Parts should either be bonded immediately after drying or wrapped in a moisture-proof barrier to prevent oxidation or other contamination until ready for bonding.

A good test of surface cleanliness is the **water break test**. On clean surfaces water will flow out to form a continuous sheet or film, but "breaks" in the film will occur at unclean spots. Smooth surfaces are preferred, since rough surfaces require thicker layers of adhesive, wasting material and weakening the bond.

Application Methods

Adhesives are available as liquids, syrups, powders, and pastes; and in sheet, film, and tape forms. Spray systems are often used for applying liquid adhesives. Fast and inexpensive application of liquids, syrups, and pastes is accomplished by brushing, troweling, or roller coating. An efficient way of applying powders is by first heating the parts to be joined above the melting point of the powder, then applying the powder by dipping, spraying, or dusting.

Thermoplastic or **hot-melt** adhesives use electrically heated dispensers to melt the adhesive and air pressure for application. Sheet, film, and tape forms of adhesive are simply cut to size and shape and placed on one of the surfaces to be bonded. They are held in place prior to bonding by tacking with a hot air or soldering gun.

Stronger bonds are obtained with thin layers of adhesive than with thick layers. A thickness of .003″ to .006″ is recommended for most structural bonding with unreinforced adhesives. Care should be taken, however, to assure that enough adhesive is applied to thoroughly wet both surfaces and to prevent voids in the bond.

Hardening or Curing of Adhesives

Adhesives become hardened or **set** by cooling, evaporation of solvents (drying), or by chemical reaction (curing). Thermoplastic adhesives are applied hot and only require cooling to set or harden. Solvent adhesives usually require predrying to remove most of the solvent before assembly. Solvents trapped in the joint after assembly retard the drying process and may prevent the development of full bond strength. Drying of solvents may be speeded by drying with hot air either in or out of an oven.

Some adhesives are applied and allowed to dry completely, at which point they become inactive. Then, just before assembly, they are reactivated with a special solvent.

Thermosetting adhesives are cured by chemical reaction. Epoxies only require sufficient pressure to keep the joints closed until curing is complete. Some cure at room temperature while others require heating to between 250° and 350° F.

Phenolic adhesives require heating in the 325-350° F. temperature range while under pressure. This may be accomplished in the following ways:

1. By clamping and oven heating;

2. By use of **autoclaves** (pressure vessels) which can provide both heat and pressure; or
3. By using presses with heated tooling. Use of presses is preferred because of higher production capabilities. Another method of curing involves passing an electric current through heating elements placed in the bond line with the adhesive. The heating elements are fine wires or graphite-impregnated cloth which remain in the joint but do not adversely affect bond strength. This technique enables large objects to be bonded without the use of costly ovens or autoclaves.

Quality Control

The best assurance of obtaining sound adhesive bonds is in establishing and following rigid process controls. This requires the selection of an appropriate adhesive and then **in-process inspection** to assure correct surface preparation, adhesive application, and drying or curing. A sound quality control program, however, also requires inspection of completed assemblies.

Destructive testing of standard specimens provides a reliable gage of product quality. Specimens are either prepared and processed along with the production assemblies or are made to duplicate the same conditions of manufacturing. Properties of the production assemblies are inferred from subjecting the specimens to standard tests. Another method of obtaining samples for testing involves cutting a small circle or "button" from the object at a location which will not affect its strength.

Visual inspection may detect gaps at the edges of joints indicating lack of adhesive or poor fit. Waviness and bulging also indicate joint separation. Tapping with the edge of a coin produces a hollow sound over unbonded areas. Small vacuum cups may be used to find any separations between thin-face panels and cores in laminated assemblies.

Several techniques of testing with ultrasonic sound waves provide a means of inspecting for voids. Voids are detected because they reflect or retard passage of sound waves rather than passing them as freely as bonded areas do. A nondestructive method of testing bond strength is by measuring the sound generated at the bond line while it is subjected to **dynamic loading**. A sharp increase in noise occurs at about 90% of bond strength, enabling failure strength to be closely estimated.

Words to Know

acrylic
adhesive bonding
alkaline cleaner
autoclave
carbon tetrachloride
cellulose
ceramic adhesives
chemical reaction
cleavage
creep
curing
destructive testing
detergents
dynamic loading
electrolytic corrosion
epoxy
fatigue resistance
honeycomb core
hot-melt adhesive
in-process inspection
mechanical interlocking
nondestructive testing
nonmetallic
nonstructural
nylon
oxide film
peel
perchloroethylene
phenolic
pickling
polyamide
polyester
polysulfide
polyurethane
scale
self-aligning
shear
silicones
structural
tension
thermoplastic
thermosetting
toluene
toxic
trichloroethylene
ultrasonic sound waves
vapor degreasing
vinyl
water break test
wetted
xylene

Review Questions

1. What is meant by adhesive bonding?

2. How does structural bonding differ from nonstructural bonding? Give examples of each.
3. Name several industries which are leaders in the use of structural adhesive bonding.
4. List six advantages of adhesive bonding.
5. List six disadvantages of adhesive bonding.
6. Name the forces which act on adhesive bonds in such a manner as to pull them apart.
7. Which of the above forces are adhesive bonds best able to resist?
8. Why are adhesive-bonded joints designed for mechanical interlocking preferred?
9. Of what materials are thermoplastic adhesives made?
10. How are thermoplastic adhesives applied and set?
11. Of what materials are thermosetting adhesives made?
12. How are thermosetting adhesives cured or set?
13. What type of adhesives are used in most structural bonding applications?
14. List the forms in which adhesives are available.
15. What cleaning methods and materials are used in preparing metal surfaces for adhesive bonding?
16. Why is it recommended that adhesive bonding occur immediately after parts have been cleaned and dried?
17. Describe the water break test for surface cleanliness.
18. List several methods of applying adhesives in liquid, syrup, or paste form.
19. Describe how powdered adhesives are applied.
20. Why do solvent-type adhesives require predrying before assembly?
21. List several methods of curing thermosetting adhesives.
22. What quality control procedure is the best guarantee of obtaining consistently sound adhesive bonds?
23. Name two methods used to obtain samples for destructive testing of adhesive bonds?
24. What defects in adhesive bonding can be detected visually?
25. What is the purpose of tapping an adhesive-bonded joint with the edge of a coin?
26. How are sound waves used in inspecting adhesive bonds for voids? For strength?

PART IX

Fabricating Sheet Metal

UNIT 31

Sheet Metalwork Hand Tools and Cutting Tools

31-1. Meaning of Sheet Metal and Importance of Sheet Metalwork

Metal sheets less than ¼″ thick (.250″) are generally termed **sheet metal**, and thickness is designated by gage numbers or by decimal parts of an inch. Metal sheet ¼″ and more in thickness is termed **plate** and its thickness is given in fractional parts of an inch.

Sheet metalwork means work involved in making and installing objects of sheet metal. Sheet metalworkers build and install ducts for heating and air-conditioning systems, exhaust hoods and ventilator systems, metal roofs and ceilings, and structural steel buildings. They also construct sheet metal parts for boats and ships, buses, trailers, aircraft, and space vehicles.

Sheet metal is a versatile material and is often used in place of wood. Objects of sheet metal are to be found everywhere. Street and road signs, tool parts, toys, washing and drying machine cabinets, television and file cabinets, desks, and mailboxes are common sheet metal products. The sheet metals most commonly used are steel, galvanized steel, tinplated steel, stainless steel, aluminum, copper, and brass.

31-2. Who Does Sheet Metalwork?

A person who works primarily with sheet metal is called a **sheet metalworker**. The heating, ventilating, air-conditioning, transportation, construction, and appliance industries all have need for skilled sheet metalworkers. Auto body repair requires workers with ability to fashion and reshape sheet metal parts. Other metalworkers such as machinists, tool-makers, and diemakers occasionally make gages, templates, and other parts of sheet metal. Carpenters use sheet metal "flashing" around chimneys and in "valleys" where roof lines intersect.

31-3. Useful Sheet Metal Products

The beginner will learn a great deal about sheet metalwork by making tool trays, tool racks, dust pans, flower boxes or planters, and funnels. Larger articles such as mailboxes, bird feeders, outdoor lamps, toolboxes, minnow buckets, wastebaskets, and watering troughs are often made in schools. Careful attention to the product design, choice of materials, quality of layout, and care in construction and finishing is necessary if your product is to be of commercial quality.

31-4. Patterns and Patternmaking

Objects made from sheet materials, whether cloth shirts, leather pocketbooks, plastic lampshades, or metal wastebaskets, require the use of a **pattern**. The pattern is made to the exact size and shape of the flat sheet material needed to form the object. A pattern is often called a **stretchout**, because it shows what the object looks like

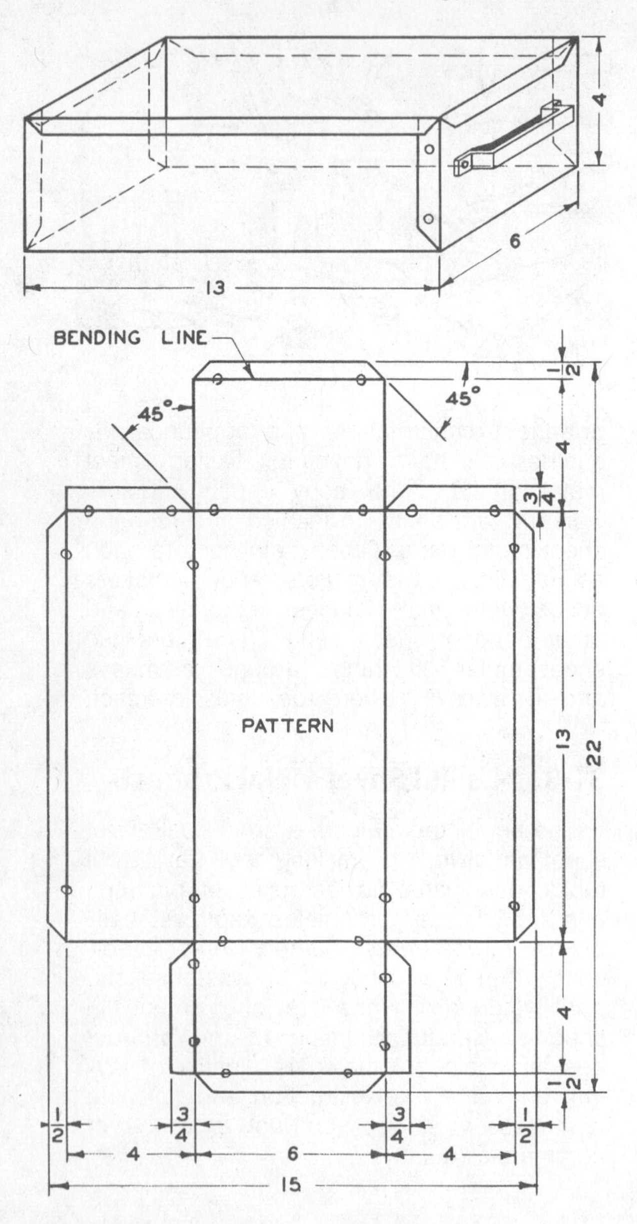

Fig. 31-1. Pattern for a Metal Box

when stretched flat. A metal pattern is called a **template**.

The pattern for a box, for example, may be a flat piece of paper, cardboard, or sheet metal cut to the outline of the unfolded shape of the box, Fig. 31-1. Note how bending lines are shown; a small freehand circle is drawn at each end of the line.

The marking of lines on sheet metal is called **laying out**. To lay out the patterns for the different sheet metal jobs, a knowledge of geometry is required. The more geometry you know, the more different types of patterns you can lay out. There are three ways to lay out a pattern:

1. By drawing lines on paper and then transferring them to the sheet metal by using carbon paper.
2. By drawing lines on paper as before, then taping the paper to the sheet metal to keep it from slipping while making small prick punch marks through the paper. Make punchmarks at all corners, intersections of lines, ends of lines, and centers of arcs and circles. Curves are made by putting the punch marks close together. After punching, the paper is removed and the punch marks are used as guides for scribing the lines, circles, and curves. Use a steel rule or square and a **scriber** or **scratch awl** (see Fig. 31-3) to scribe the straight lines and a **divider** or **trammel** to make the circles and arcs. See Figs. 6-5, 6-6, and 31-4.
3. By measuring and scribing the lines, circles, arcs, and curves directly on the sheet metal. If two or more pieces are to be cut alike, use the first piece of metal that has been cut as the template for the others.

31-5. Sheet Metal and Wire Sizes

Sheet metal thicknesses and wire diameters are specified either by **gage numbers** or by decimal fractions of an inch. Sheet metal and wire gages have slots which correspond to the different gage sizes, Fig. 31-2. On one side of the gage is stamped the gage number; on the other side is stamped the decimal equivalent of the gage number. Table 29, page 423, gives the names of the different gages, what they

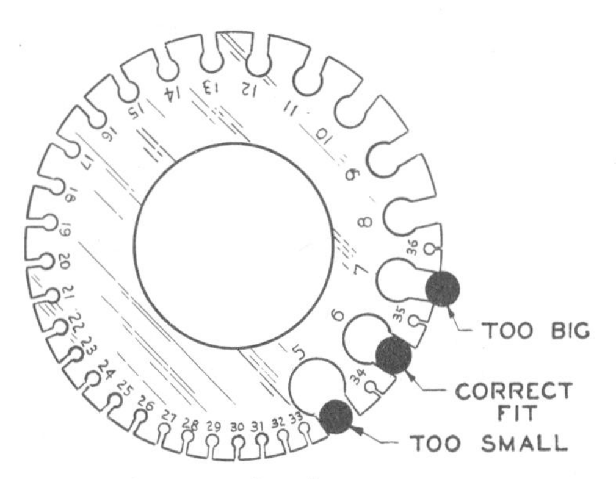

Fig. 31-2. Sheet Metal and Wire Gage

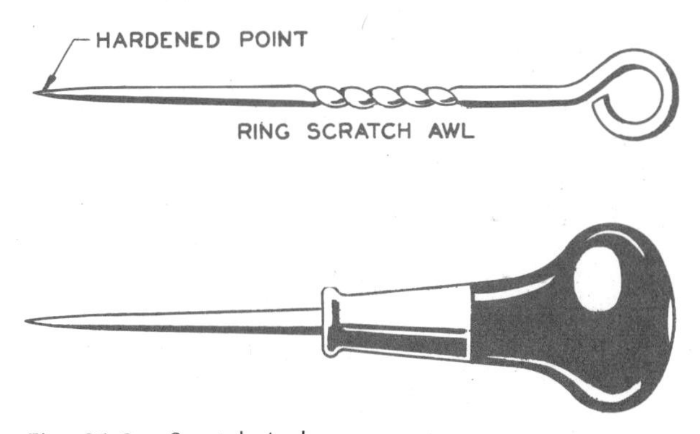

Fig. 31-3. Scratch Awls

measure, and the decimal equivalent of the gage number. Note that the **Manufacturer's Standard Gage for Steel Sheets** is used to measure the thickness of iron and steel sheets. The thickness of most nonferrous metals, such as aluminum, copper, and copper alloys, is now specified in decimal parts of an inch. These were once made to the Brown and Sharpe or American Standard Gage.

Be sure you use the gage which is stamped according to the material you wish to measure, as "U.S. Standard Gage," "American Steel & Wire Gage," "American Steel & Music Wire Gage," etc.

When ordering sheet metal or wire, one should always specify the decimal thickness of the material in inches, and give the name of the gage and the gage number, if any.

31-6. Scratch Awl

A scratch awl is a steel tool with a sharp point on one end, Fig. 31-3. It is used to scratch layout lines on sheet metal. Some scratch awls have wooden handles. A scriber (see Fig. 6-3) may be used instead.

31-7. Wing Divider

The divider used most often by sheet metalworkers is the **wing divider**, shown in Fig. 31-4. It is used to scribe arcs and circles.

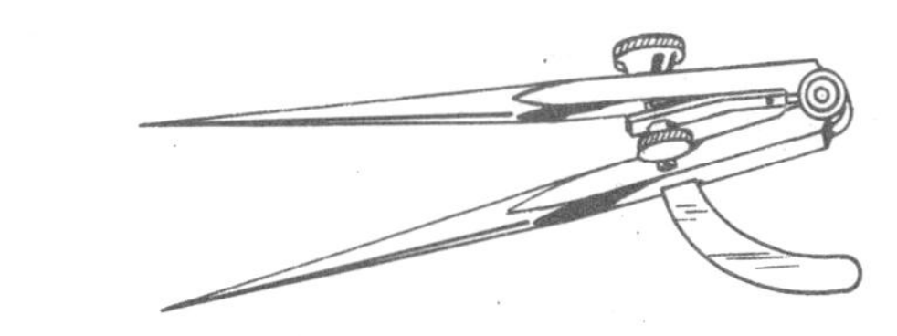

Fig. 31-4. Wing Divider

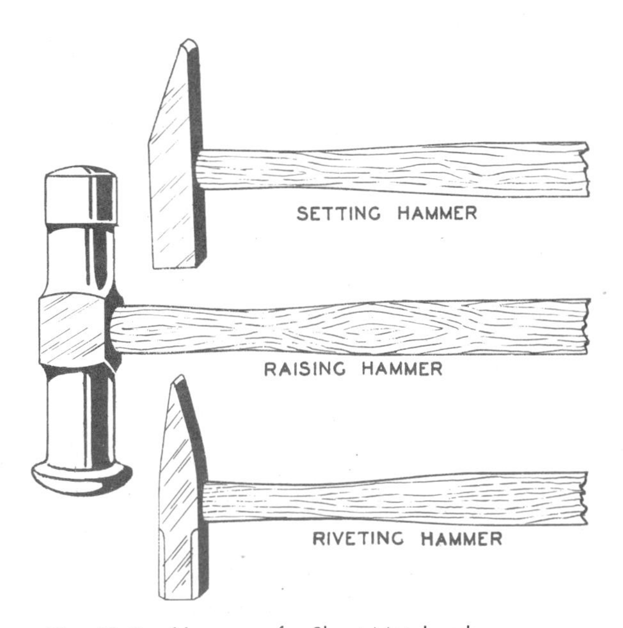

Fig. 31-5. Hammers for Sheet Metalwork

31-8. Hammers

Several kinds of hammers are used in sheet metalwork, Fig. 31-5. The **riveting**

hammer is used for setting solid rivets. The setting hammer is used for bending or tucking in the edges of sheet metal; it is especially useful for **setting down** the edges when making a **double seam**, Fig. 31-6.

Raising hammers are used for producing curved sheet metal surfaces which cannot be made with forming machines. There are many shapes of raising hammers; some of their uses are explained in section 34-3. See also Fig. 34-9. The polished faces on raising hammers should be carefully protected from nicks, because the nicks show on the work and spoil its appearance.

To keep from stretching and nicking sheet metal, it should be struck with a mallet, Fig. 31-7. Mallets are available in different sizes and shapes and are made of wood, rawhide, or plastic.

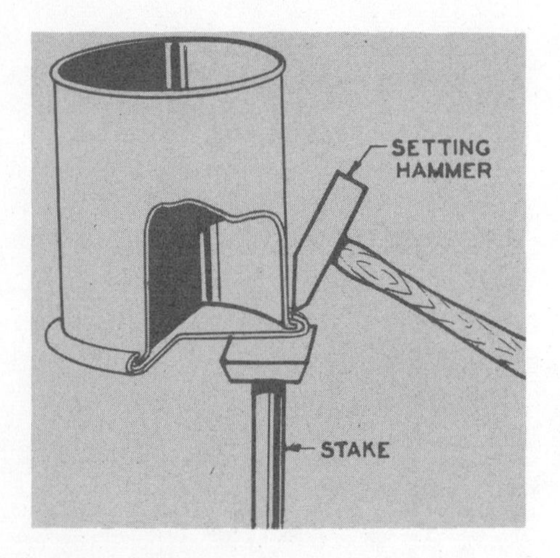

Fig. 31-6. Setting Down with a Setting Hammer

Fig. 31-7. Forming a Corner with a Mallet

31-9. Punches

The hand punches in Fig. 31-8 may be used to punch holes in sheet metal. A pin punch, Fig. 26-15, also may be used. The **hollow punch** makes large holes to allow passage of bolts, cables, and pipes, while the other punches make small holes for rivets, nails, screws, etc.

To use the **solid punch** or hollow punch, lay the metal on a block of lead or the end grain of hard wood, place the punch on the sheet metal, and strike a heavy blow with a hammer, Fig. 31-9.

The **hand punch** is used as you would use a paper punch. It can only punch holes near the edge of the metal.

A hand-operated turret punch, Fig. 31-10, can punch accurate, burr-free holes of a variety of sizes and shapes very efficiently.

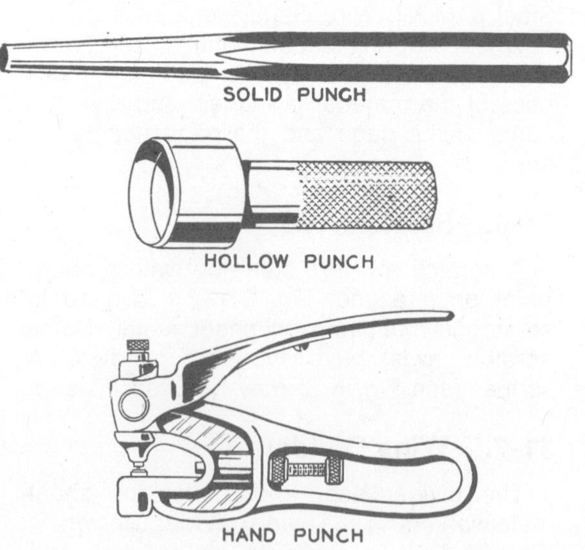

Fig. 31-8. Types of Punches

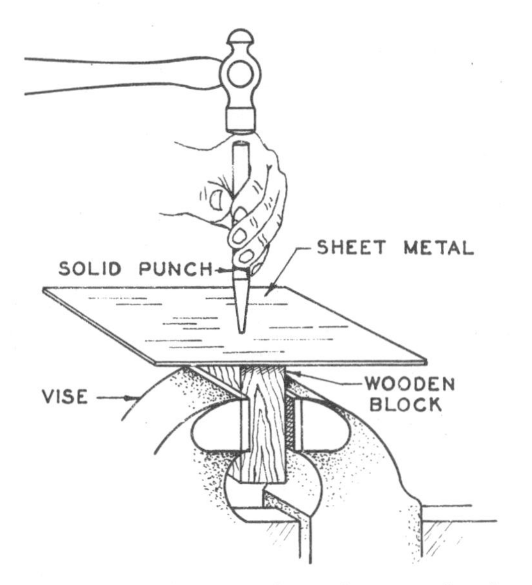

Fig. 31-9. Punching a Hole in Sheet Metal with a Solid Punch

Fig. 31-10. Turret Punch Press

31-10. Pliers

The square jaws of the **flat-nose plier**, Fig. 31-13, are used to bend square corners in sheet metalwork. The **side-cutting plier** and the **round-nose plier** are described in section 25-6.

31-11. Hand Seamer

The **hand seamer**, Fig. 31-11, is a portable tool for making hems, seams, and other straight line bends in thin sheet metal where bending machines are not available. In the shop it is used in places inaccessible to bending machines. A vise-grip version is also useful as a convenient clamp for temporarily holding parts together, Fig. 31-12.

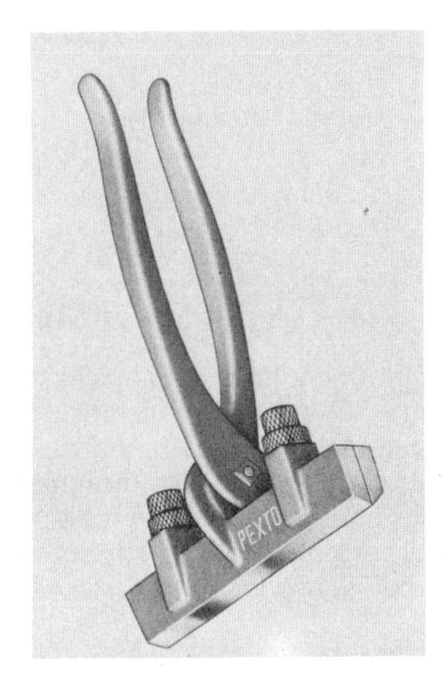

Fig. 31-11. Hand Seamer

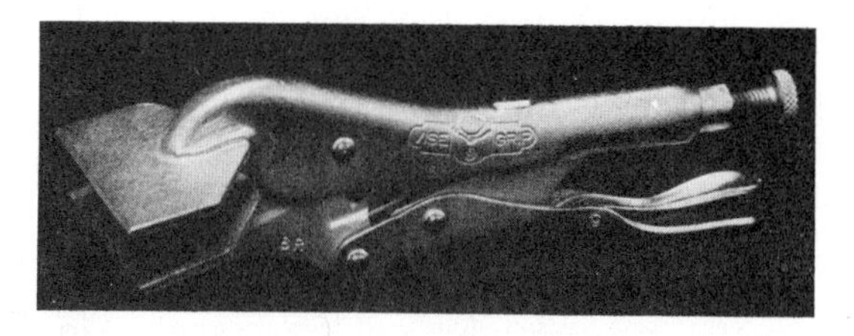

Fig. 31-12. Vise Grip Style Hand Seamer

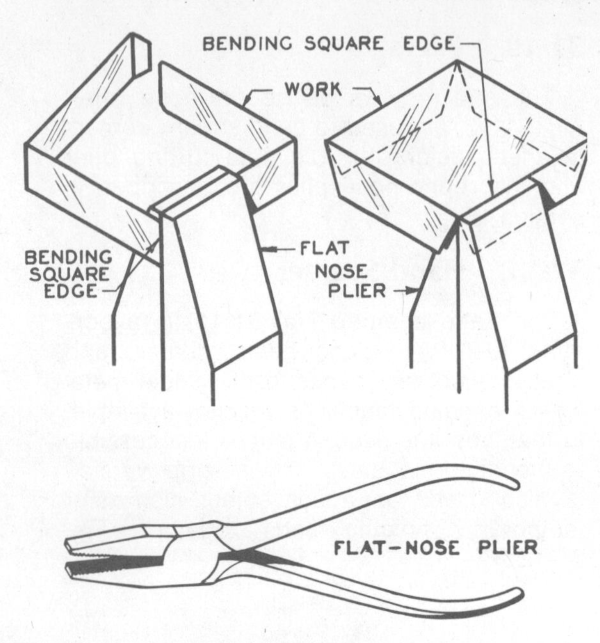

Fig. 31-13. Using a Flat-Nose Plier

31-12. Ways to Cut Sheet Metal

There are many tools available for cutting sheet metal. There are **tin snips, aviation snips, double-cutting shears, bench shears, squaring shears, notcher, ring and circle shears,** and **lever shears. Nibblers** and **portable electric shears** are also used to cut thin-gage sheet metal.

31-13. Tin Snips

Tin snips, Fig. 31-14, are used like scissors to cut thin soft metal. They should be used to cut 20-gage or thinner metal. To cut to a corner, the snips should be set so that the point will finish in the corner. Keep the bolt tight; the blades must fit closely against each other.

A left-hand snip should be used by a left-handed person. The scroll, or hawk's bill snip, is used for cutting curves, Fig. 31-15.

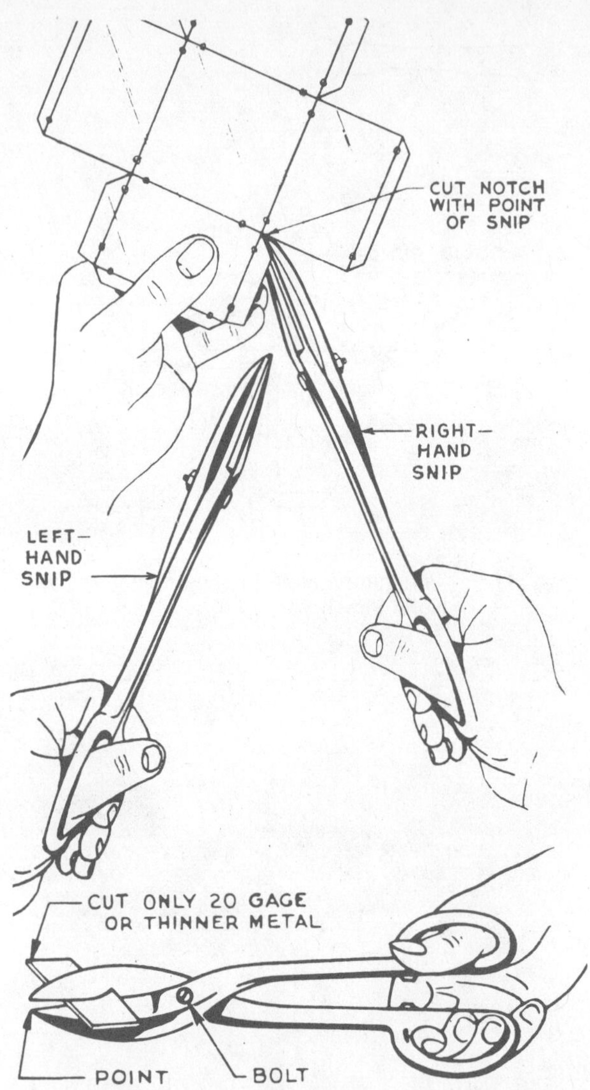

Fig. 31-14. Using Snips

31-14. Aviation Snips

Aviation snips are also used for cutting thin-gage metal. They are made in left, right, and straight versions, Fig. 31-16. The blades of aviation snips are **serrated** which makes them grip the metal better than tin snips. In addition, the aviation snips are compound levered, requiring less manual

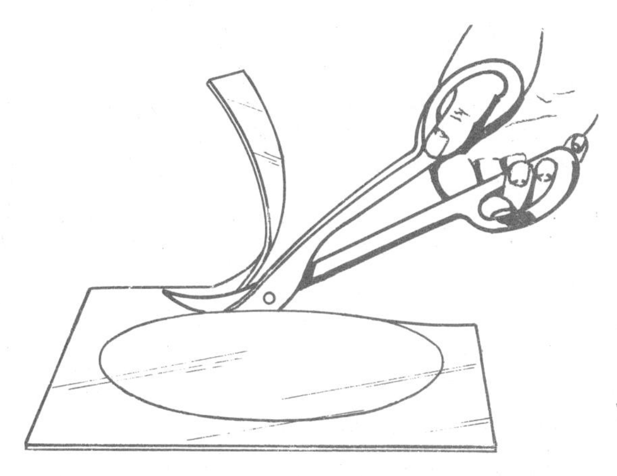

Fig. 31-15. Using Hawk's Bill Scroll Snips

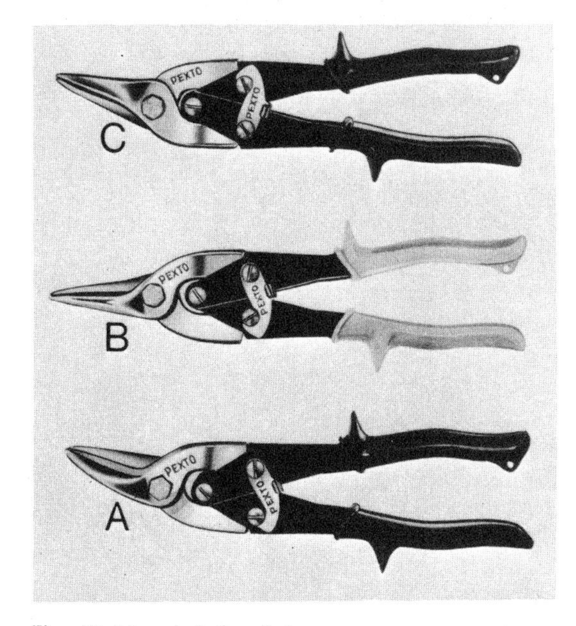

Fig. 31-16. Aviation Snips
A. Cuts Left
B. Cuts Straight
C. Cuts Right

force than tin snips to cut metal of the same thickness.

31-15. Double-Cutting Shear

The double-cutting shear has three blades and is used to cut around cans, stove and furnace pipes, etc., Fig. 31-17. The pointed lower blade is pushed through the metal to start the cut.

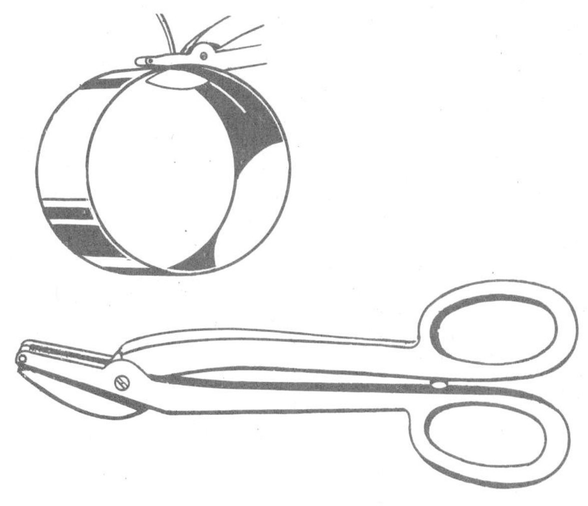

Fig. 31-17. Using Double-Cutting Shear to Cut a Cylinder

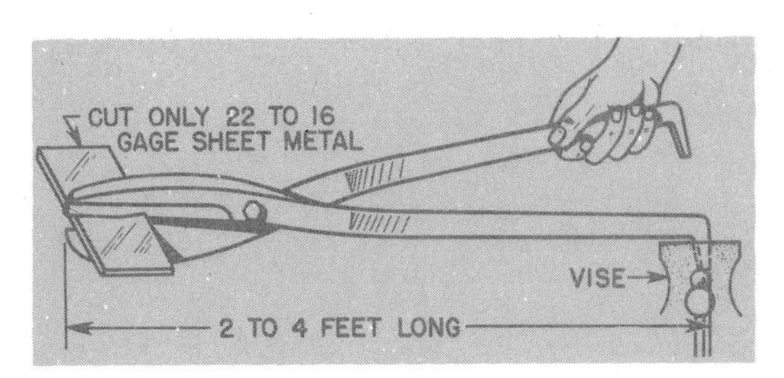

Fig. 31-18. Bench Shear

31-16. Bench Shear

The bench shear, Fig. 31-18, is a large pair of scissors from 2′ to 4′ long. One handle may be held in a vise or **bench plate** (see Fig. 32-4) while the other handle is moved up and down to do the cutting. The bench shear can cut metal as thick as 16-gage mild steel.

31-17. Squaring Shear

The squaring shear, Fig. 31-19, is operated by foot and is used to cut metal no thicker than 16-gage mild steel. It is especially useful for cutting strips of sheet metal and for cutting and trimming the edges square to each other; hence, the name. **It should never be used for cutting wire or nails since this causes the blade to be nicked and ruins it for use with sheet metal.**

The **side guides** on the table help to keep the metal square with the cutting blades. The squaring shear is useful for cutting many pieces of the same size because the **back gage** can be locked at any desired setting. The fingers should be kept away from the blade, which should be guarded. The **treadle** should be fitted with a stop to prevent it from going all the way to the floor and possibly injuring the operator's foot.

31-18. Notcher

The notcher is a hand-operated machine which makes a 90° cut or notch in a workpiece, Fig. 31-20. It greatly speeds the work of cutting out the corners of a workpiece which will become a box, pan, or tray. The **tab notcher** speeds the cutting of patterns which require a tab for corner fastening, Fig. 31-21.

31-19. Ring and Circle Shear

The ring and circle shear, Fig. 31-22, is used to cut circular pieces of metal such as those to be made into covers or bottoms of buckets.

31-20. Lever Shear

A lever shear is used for cutting small rods and cutting notches and corners in heavier-gage metals, Fig. 31-23.

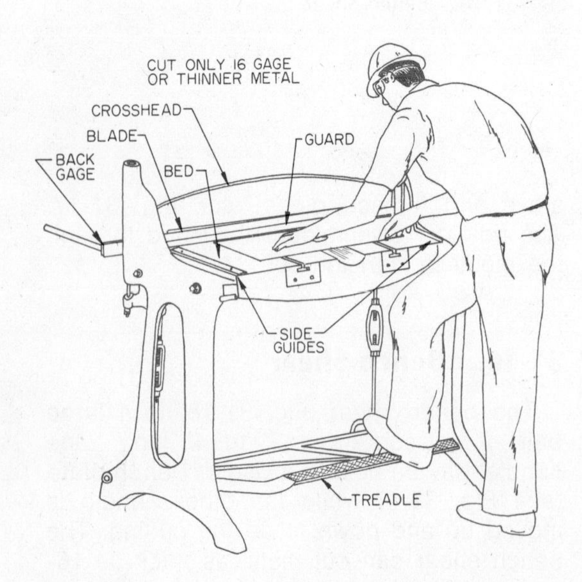

Fig. 31-19. Squaring Shear

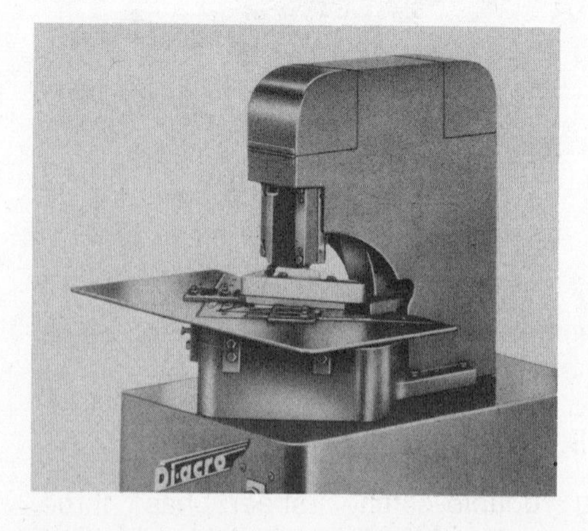

Fig. 31-20. Notcher

31-21. Nibblers and Portable Electric Shears

Electrically powered nibblers, Fig. 31-24, and portable electric shears, Fig. 31-25, cut thin-gage sheet metal rapidly. They are useful in production work where the quantity of cutting would make the use of hand-operated shears tiresome and inefficient.

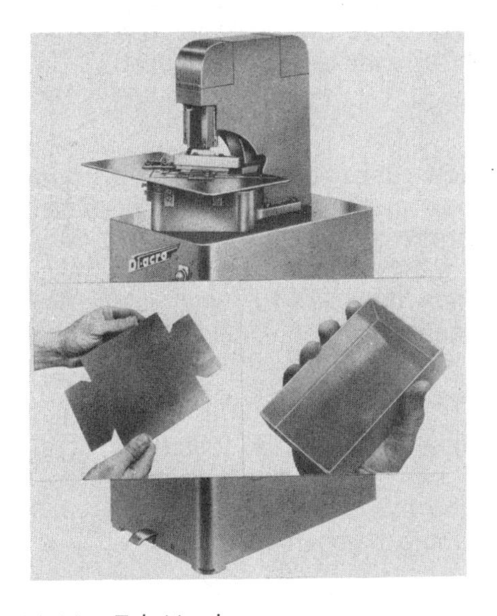

Fig. 31-21. Tab Notcher

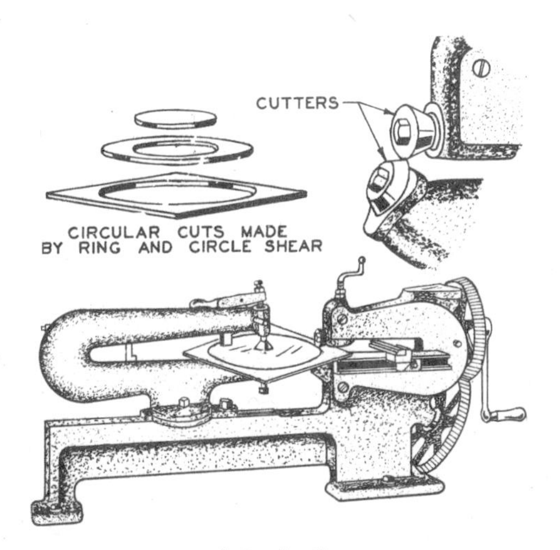

Fig. 31-22. Ring and Circle Shear

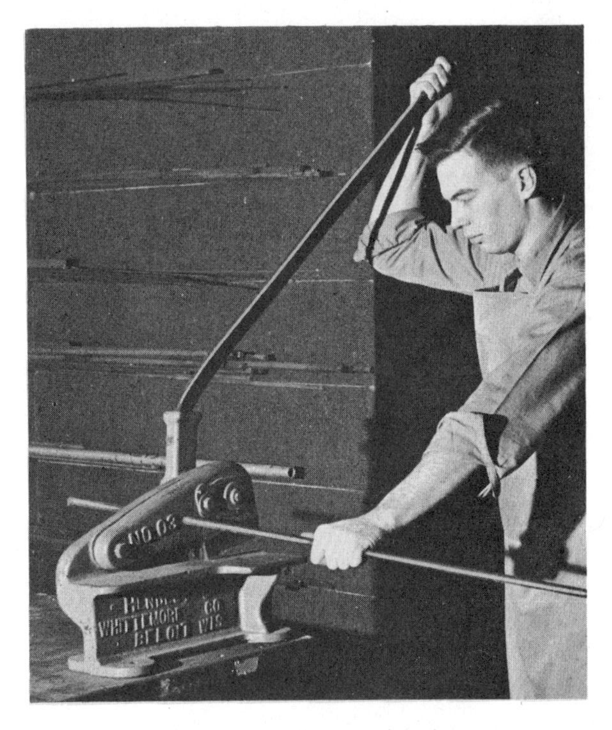

Fig. 31-23. Using Lever Shear and Rod Parter

Fig. 31-24. Bench-Mounted Nibbler

Fig. 31-25. Portable Electric Sheet Metal Shear

Words to Know

aviation snips
bench shear
double-cutting shear
flat-nose plier
hand punch
hand seamer
hollow punch
left-hand snip
lever shear
nibbler
pattern
raising hammer
ring and circle shear
riveting hammer
scroll snips
setting hammer
sheet metal gage
solid punch
squaring shear
stretchout
template
tin snips
turret punch
wing divider
wire gage

Review Questions

1. What metal thickness represents the division between sheet metal and plate?
2. For what purpose are patterns used?
3. How does a pattern differ from a template?
4. What tools are used to scribe arcs or circles on sheet metal?
5. What is the name of the gage used to specify sheet steel thickness?
6. What information should be given when ordering sheet metal or wire?
7. For what is a setting hammer used? Raising hammer?
8. Why are mallets sometimes used instead of hammers?
9. Describe how to use solid or hollow hand punches.
10. Of what advantage is a turret punch?
11. What is the maximum thickness of metal which can safely be cut with tin snips?
12. Why are aviation snips able to cut metal more easily than tin snips?
13. What is a squaring shear? What thickness of metal can it safely cut?
14. For what is a notcher used? Tab notcher?
15. What cutting machine is designed for making circular cuts?
16. Name several uses for a lever shear.
17. For what is a nibbler used?

Mathematics

1. How does a knowledge of geometry help in the layout of sheet metal patterns?
2. If a 2′ x 8′ sheet of 26-gage galvanized steel costs $4.10, what is its cost per square foot?

Drafting

1. Make a pattern or stretchout for a one-piece cookie tray which measures 10″ x 15″, has sides ⅝″ high, and a 3/16″ double hem around the top.

Career Information

1. In what occupations is a knowledge of sheet metalwork important?
2. What education is necessary in order to become a skilled sheet metalworker?
3. What are pay rates for sheet metalworkers in your area?

UNIT 32

Bending Sheet Metal

32-1. Hems and Seams

Several kinds of hems and seams are shown in Fig. 32-1. A **hem** is an edge or border made by folding. It stiffens the sheet of metal and does away with the sharp edge. A **seam** is a joint made by fastening two edges together.

The **single hem** is made by folding the edge of the sheet metal over to make it smooth and stiff. It is normally made in the bar folder. See Fig. 32-8. It may also be made by bending over the **hatchet stake** (see Fig. 32-5) and then finishing with a **mallet.** Small pieces may be pinched and bent in a vise and then finished with a mallet.

The **double hem** is made by folding the edge over twice to make it stiff and smooth.

The **wired edge** is smooth and very strong. It is made by first making an **open fold.** An **open fold** is a rounded bend. It is made on the bar folder or on the **turning machine,** Fig. 32-14. To make an open fold, set the stop adjustment indicator of the bar folder at 1½ times the diameter of the wire, and set the gap between the folding edge and the bending wire equal to the diameter of the wire.

The wire is then placed in the open fold, and the fold is finished with a **setting hammer,** Fig. 32-2. Or, after placing the wire in the fold, the metal may be bent over a little with a mallet and finished on the **wiring machine,** Fig. 32-15. The making of a wired edge on a circular object, as on the top edge of a wastebasket is explained in sections 32-12 and 32-13.

The **grooved seam** is made by hooking two single hems together and then locking them with a **hand groover** (see Fig. 32-6) or with a **grooving machine** (see Fig. 32-7).

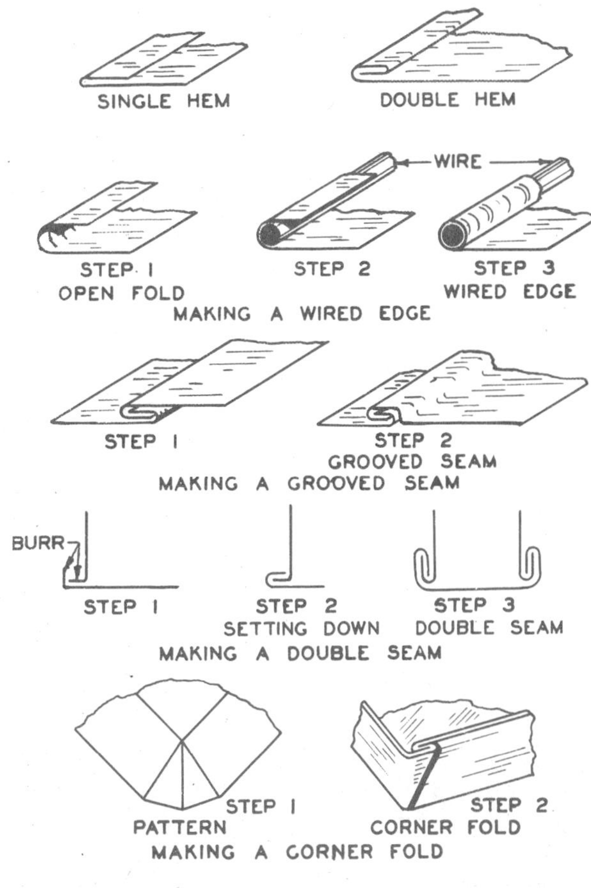

Fig. 32-1. Hems and Seams

The **double seam** fastens the bottom to a pail or can. See §§ 31-8, 32-4, and 32-14 to 32-16.

The **corner fold** is useful on corners of pans such as baking pans, which are made to hold liquids.

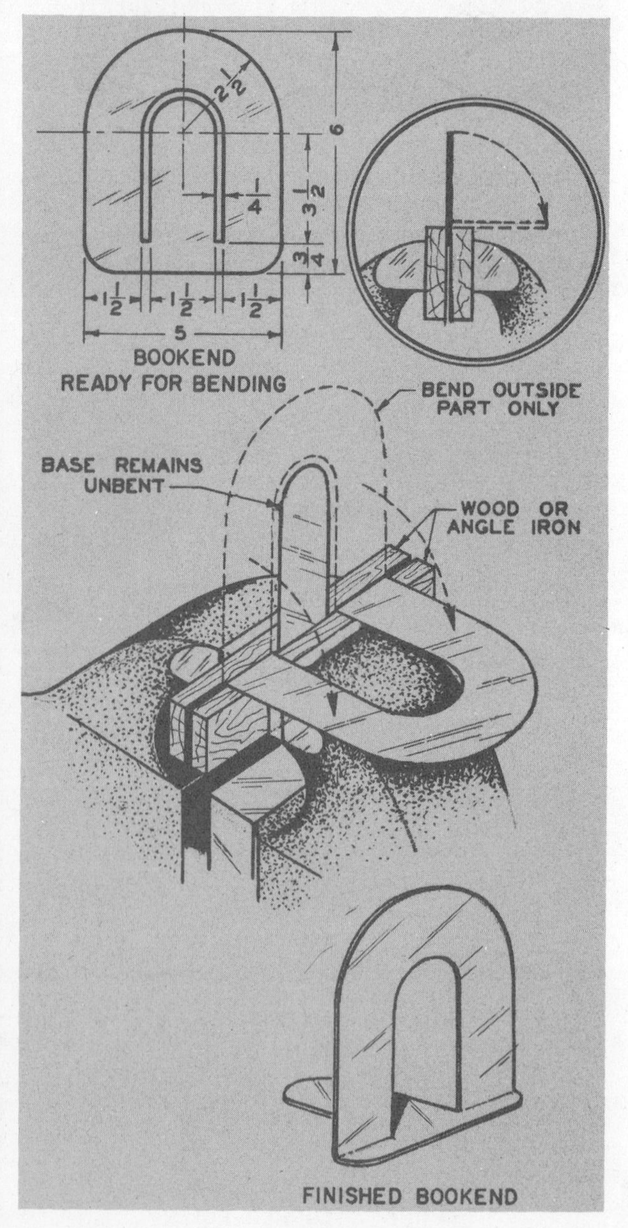

Fig. 32-3. Bending a Metal Bookend in a Vise

32-2. Bending Sheet Metal in a Vise

It is sometimes convenient to bend sheet metal in a vise. The bookend in Fig. 32-3 is an example. Note that only the outside is bent; the inside must remain unbent because it is the base upon which the bookend rests.

32-3. Bench Plate

The **bench plate,** or **stake plate,** Fig. 32-4, is fastened to the bench with bolts.

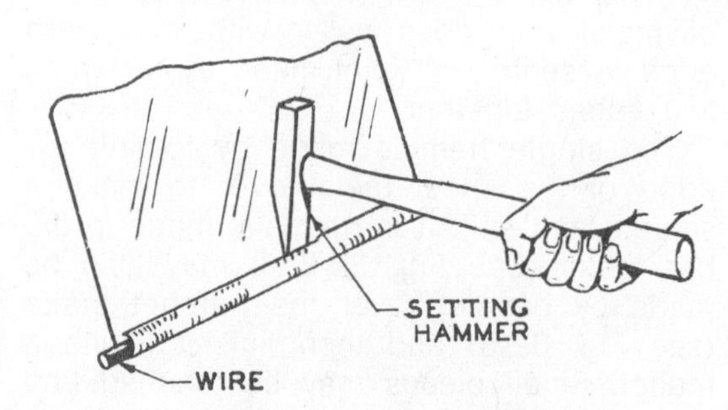

Fig. 32-2. Finishing a Wired Edge with Setting Hammer

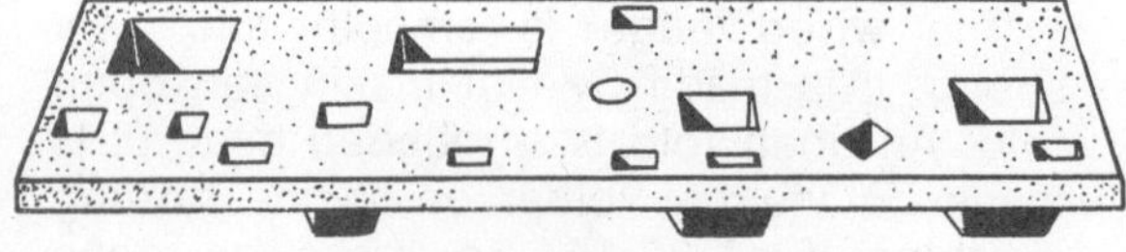

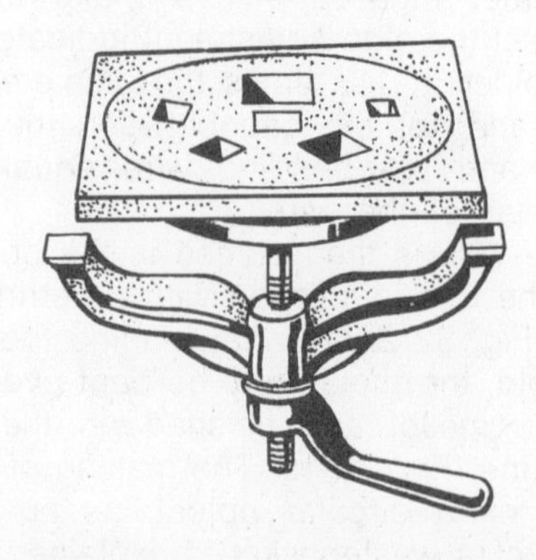

Fig. 32-4. Bench Plates

The holes in the bench plate are used to hold the bench shear, stakes, etc. See Figs. 32-5 and 31-18.

32-4. Stakes

Many bends and forms in sheet metalwork must be made on **stakes,** Fig. 32-5. The stake is supported by inserting the tapered square end in a matching hole in the bench plate.

The **double-seaming stake** is used to make the **double seam.**

The **beakhorn stake** is used for riveting, forming round and square surfaces, bending straight edges, and making corners.

The **bevel-edged square stake** is used to form corners and edges.

The **hatchet stake** is used to make straight, sharp bends and for folding and bending edges.

Small tubes and pipes may be formed on the **needle-case stake.**

Cone-shaped articles may be formed on the **blowhorn stake.** The use of the **hollow mandrel stake** is shown in Fig. 32-6.

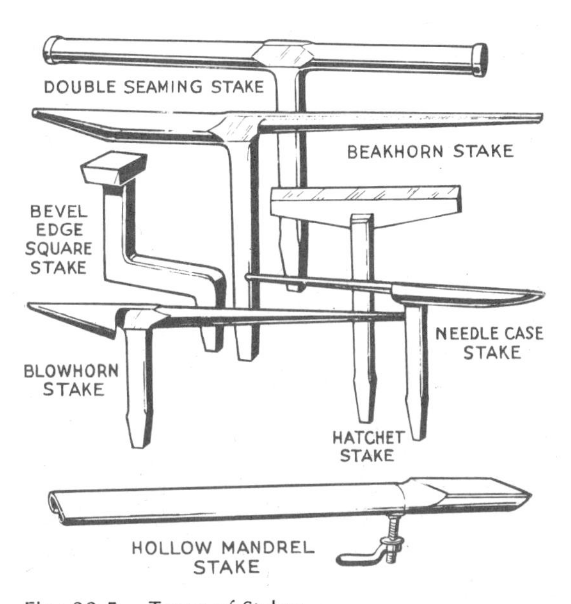

Fig. 32-5. Types of Stakes

Sheet metal should be hammered with a **mallet** to keep from stretching and nicking it.

32-5. Hand Groover

The hand groover is used to **groove** and flatten a seam as shown in Fig. 32-6. Groove the ends first, then finish the rest of the seam.

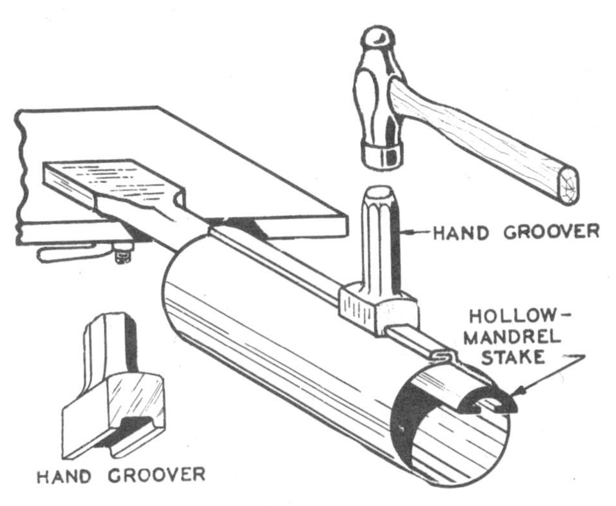

Fig. 32-6. Grooving a Seam with Hand Groover

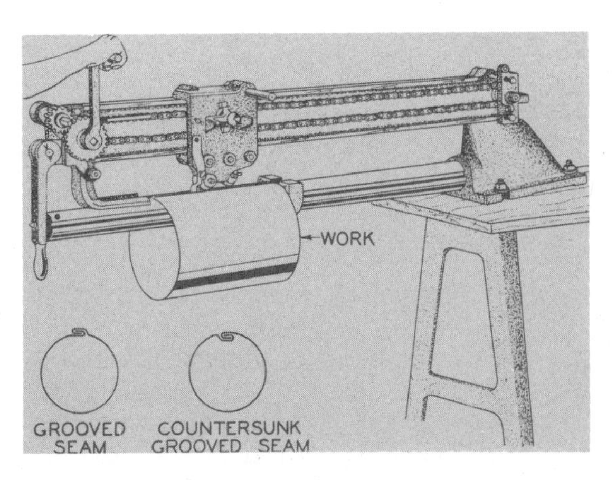

Fig. 32-7. Grooving Machine

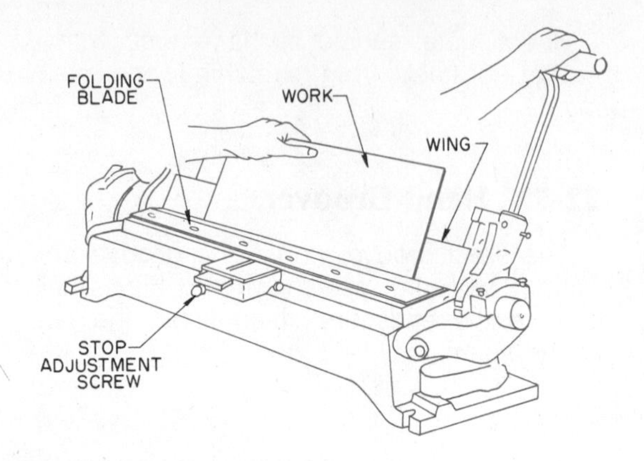

Fig. 32-8. Bar Folder

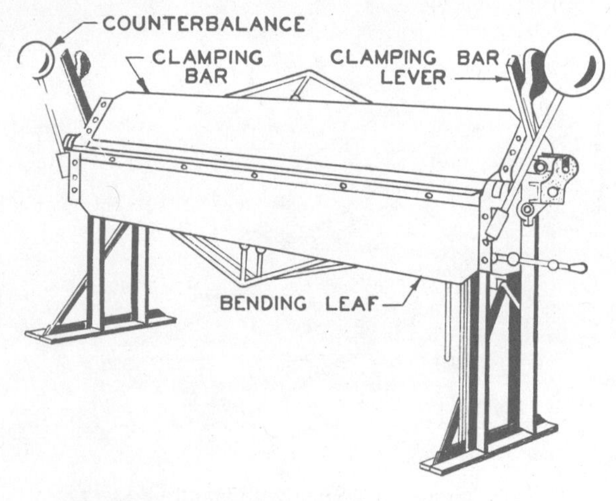

Fig. 32-9. Brake

32-6. Grooving Machine

The grooving machine, Fig. 32-7, is a machine which makes **grooved seams** and **countersunk seams.** It **grooves** and flattens the seams which have been started on the **bar folder.**

32-7. Bar Folder

The bar folder, Fig. 32-8, is a machine for folding or bending sheet metal edges such as are used for seams and hems. Folds are limited to a width of 1″ or 1¼″, depending on the size of the bar folder.

Fig. 32-10. Press Brake for Forming Sheet Metal

32-8. Brake

A brake, Fig. 32-9, is a machine for bending and folding sheet metal. Unlike the bar folder, the brake can bend or fold the metal any distance from the edge. **Moldings** can be made on the brake by using a **mold,** Fig. 32-12.

32-9. Press Brake

The hand-operated **press brake,** Fig. 32-10, is similar to the power-operated press brakes used for bending heavy-gage sheet metal and plate. The model pictured however, has a capacity of 16-gage mild steel. Dies (jaws between which the metal is formed) are available to form straight line bends, offset bends, and bends of different radii.

32-10. Box and Pan Brake

The box and pan brake was designed to allow boxes, pans, or trays to be folded from one piece of metal. The upper jaw is made of a number of blocks of different width which can be put together in any combination so as to make a bend of any width desired, Fig. 32-11. This permits the sides

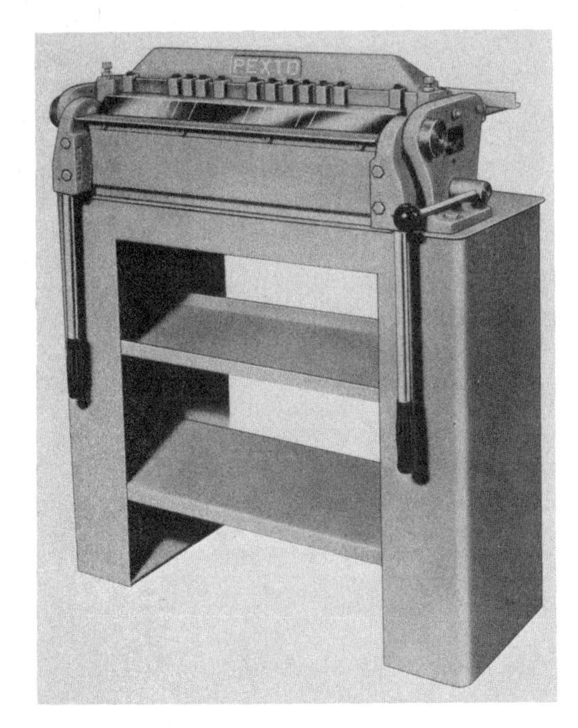

Fig. 32-11. Box and Pan Brake

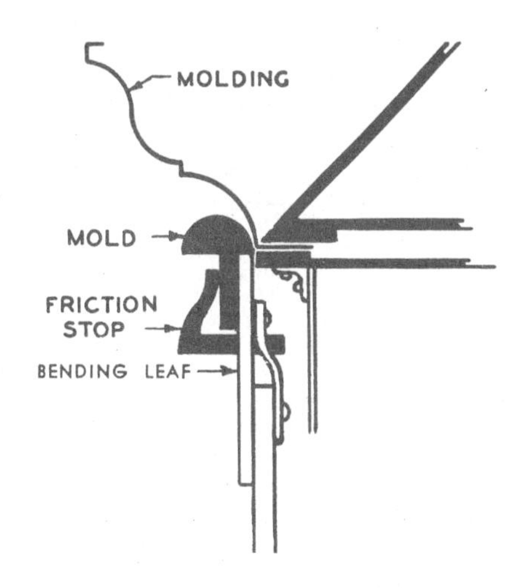

Fig. 32-12. Making Molding with a Mold on a Brake

to be bent between the opposite sides which have already been bent.

32-11. Slip-Roll Forming Machine

Stove pipes, cans, etc., are formed out of flat sheet metal on the **slip-roll forming machine,** Fig. 32-13. It has three **rolls** which can be set different distances apart; the curves are formed between the rolls. If the metal to be formed has a **wired edge,** the wired edge may be slipped into one of the **grooves** at one end of the rolls.

32-12. Turning Machine

Sometimes a **wired edge** (see Fig. 32-1) is made, as, for instance, on the top edge of a wastebasket. A **turning machine,** Fig. 32-14, is used to make the rounded edge into which the wire is placed.

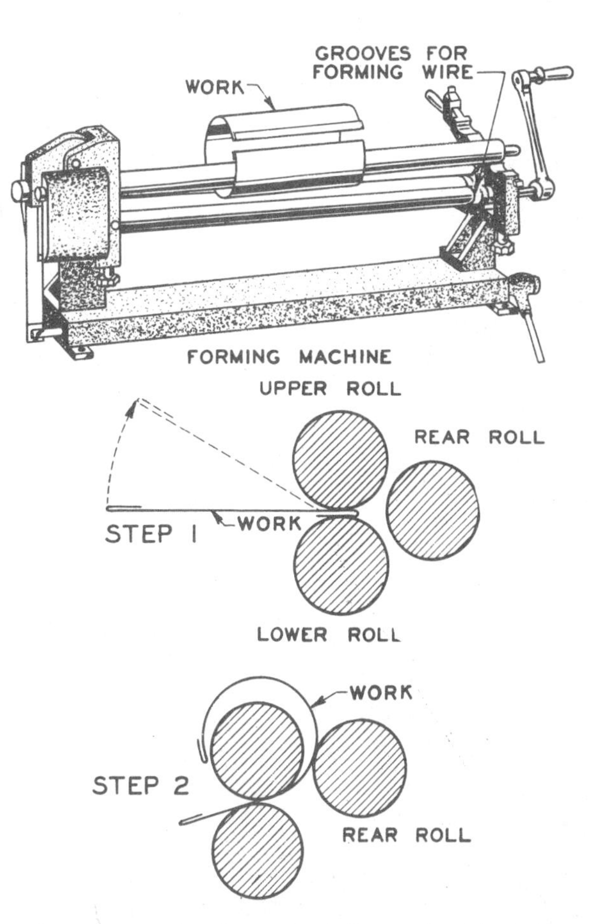

Fig. 32-13. Forming with Rolls on the Slip-Roll Forming Machine

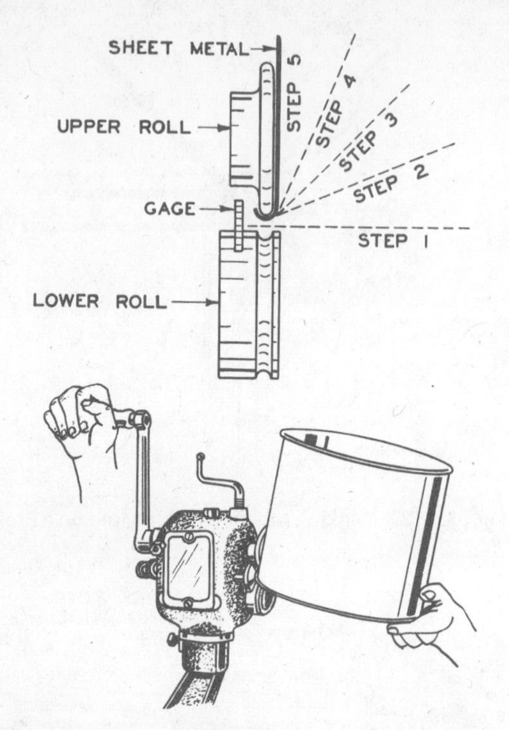

Fig. 32-14. Turning Machine

To make this rounded edge, set the **gage** about 2½ times the diameter of the wire away from the center of the groove in the **roll.** Be sure the rolls are set to fit each other. Then place the work on the lower roll and against the gage. Screw down the top roll so that it grooves the work a little when turning the crank. Next, screw the upper roll down a little more, raise the work a little with the left hand, and turn the crank again. The groove is thus made deeper.

Repeat these steps several times, each time screwing the top roll down a little more and raising the work with the left hand until the work held with the left hand is straight up and down and touches the side of the top roll. The groove, or rounded edge, is now ready for the wire and may be taken out of the machine.

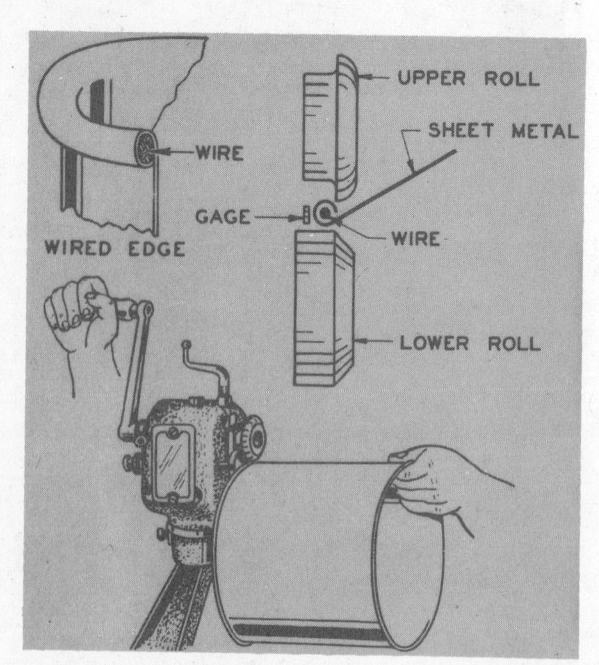

Fig. 32-15. Wiring Machine

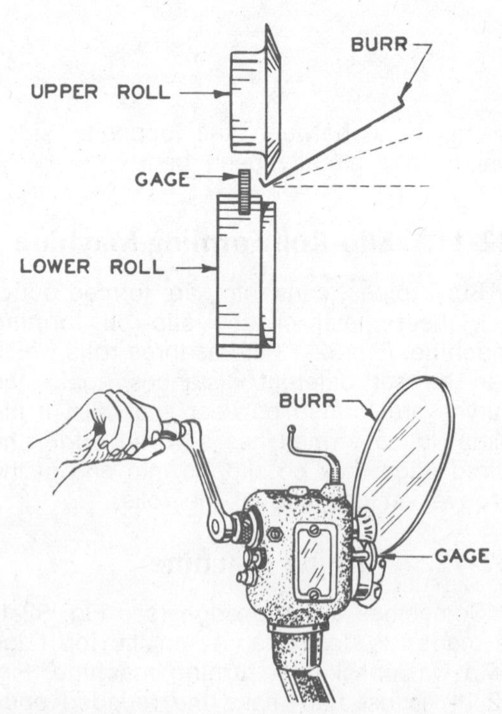

Fig. 32-16. Burring Machine

32-13. Wiring Machine

After the edge of the work has been made round in the **turning machine,** the wire is placed into the rounded edge which is then hammered over a little with a mallet. The edge of the metal may then be completely pressed over the wire with a **wiring machine,** Fig. 32-15. The **wired edge** may be used on flat work as well as round work.

32-14. Burring Machine

The burring machine, Fig. 32-16, is used to make a **burr** on the edge of the bottom for a can and on the end of a **cylinder.** The making of such burrs is the first step in making a **double seam** with the **double-seaming machine.** See **double seam** in Figs. 32-1 and 32-18. The handling of the metal with the left hand is the same as when turning the edge on the **turning machine.** The beginner should practice on scraps of metal.

32-15. Setting-Down Machine

After the **burrs** on the end of a cylinder and on the edge of the bottom for a can have been made with the **burring machine,** the **seams** are closed or **set down** on a **setting-down machine,** Fig. 32-17. If a setting-down machine is not available, seams may be set down with the use of a setting hammer and a stake.

32-16. Double-Seaming Machine

After the burrs have been made on the **burring machine** and set down on the **setting-down machine,** the seam or edge can be turned up against the sides of the can with a **double-seaming machine,** Fig. 32-18, thus making a **double seam.** If a double-seaming machine is not available, the seam or edge can be turned up against the sides of the can with a hammer and stake.

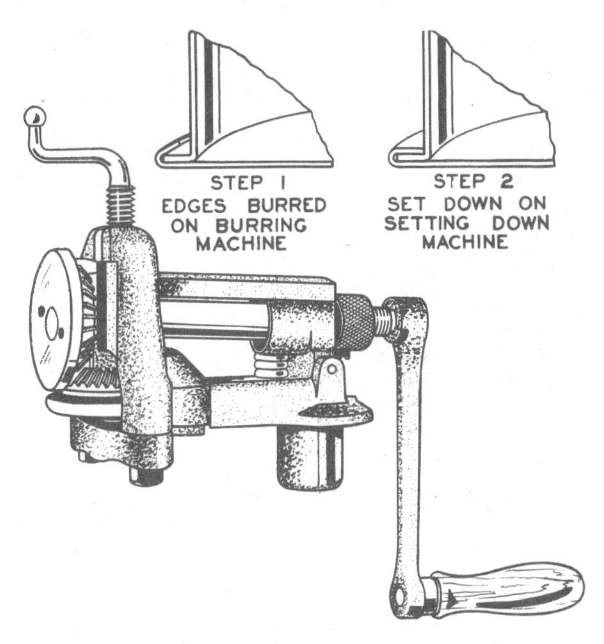

Fig. 32-17. Setting Down Machine

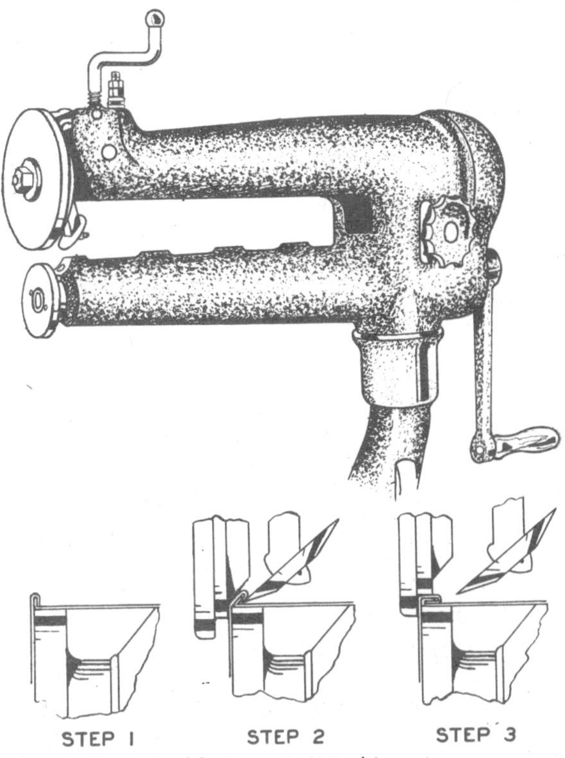

Fig. 32-18. Double-Seaming Machine

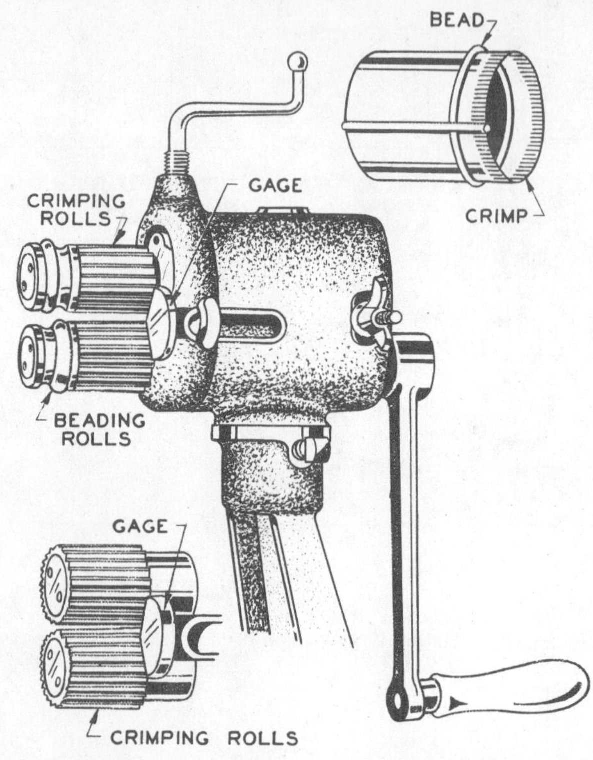

Fig. 32-19. Crimping and Beading Machine

32-17. Crimping and Beading Machine

The making of the wavy end on a stove or furnace pipe is called **crimping.** It makes the end of the pipe smaller so that it will fit into another pipe.

A **bead** is made as an ornament and in order to stiffen the object. Stove and furnace pipes are beaded on one end.

Crimping and beading are done on a **crimping and beading machine**, Fig. 32-19, or crimping may be done in a **crimping machine,** which does only the crimping; beading may be done on a **beading machine,** which only does beading.

Words to Know

bar folder
beakhorn stake
bench plate
blowhorn stake
box and pan brake
brake
burring machine
countersunk grooved seam
crimping and beading machine
double hem
double-seaming machine
double-seaming stake
grooved seam
grooving machine
hand groover
hatchet stake
hollow mandrel stake
needle-case stake
open fold
press brake
setting-down machine
single hem
slip-roll forming machine
turning machine
wired edge
wiring machine

Review Questions

1. What is the purpose of single and double hems?
2. Describe a wire edge.
3. Describe how to make a grooved seam.
4. For what is a bench plate used?
5. For what are stakes used?
6. What is the difference between a bar folder and a brake?
7. What is the unique feature of the box and pan brake?
8. Describe how a press brake differs from a conventional brake.
9. For what is the slip-roll forming machine used?
10. For what is the turning machine used?
11. For what is the wiring machine used?
12. For what is the burring machine used?
13. For what is the setting-down machine used?
14. For what is the double-seaming machine used?
15. For what is the crimping and beading machine used?

UNIT 33

Sheet Metal Manufacturing Processes

33-1. Importance of Sheet Metal Manufacturing Processes

Literally millions of sheet metal parts are produced daily by the thousands of metal-cutting and forming machines in our factories. **Punch press** and **press brake** processes account for the greatest part of the production of sheet metal parts. Other important sheet metal manufacturing processes include **roll forming, metal spinning, shear spinning,** and **chemical milling. High-energy-rate** processes for forming sheet metal include **explosive, electrohydraulic,** and **electromagnetic** forming.

33-2. Presswoking Processes

Sheet metal pressworking operations may be classified under the four headings of **drawing, shearing, bending,** and **squeezing.**

Drawing operations include the following:
1. **Shell drawing**
2. **Embossing**
3. **Stretch forming**

Shearing operations include the following:
1. **Blanking**
2. **Punching**
3. **Perforating**
4. **Lancing**
5. **Shaving**

Bending operations include the following:
1. **Angle bending**
2. **Curvilinear bending**
3. **Beading**

Squeezing operations include the following:
1. **Coining**
2. **Burnishing**

33-3. Drawing Operations

Shell drawing is the forming of one-piece cylindrical or rectangular containers from sheet metal. **Matched metal dies** are primarily used for this process, Fig. 33-1. Other shell-drawing processes substitute rubber pads for the expensive female die. The **Guerin process** is depicted in Fig. 33-2, and the **Marform process** in Fig. 33-3. The addition of the pressure pad makes the Marform process capable of deeper drawing than the Guerin process.

Embossing is a very shallow drawing operation usually done for decorative purposes, such as raised lettering or other designs, Fig. 33-4.

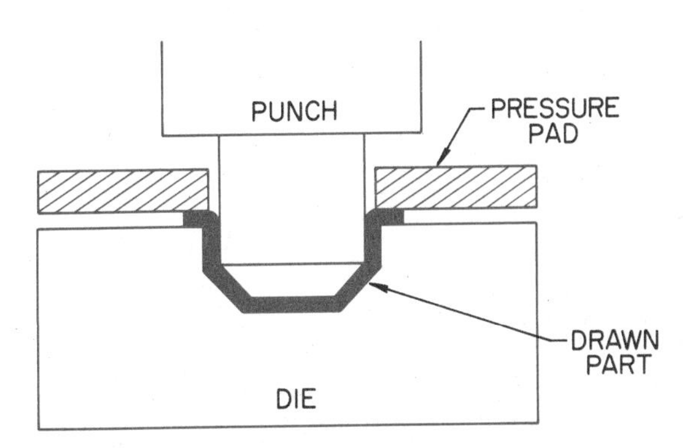

Fig. 33-1. Shell Drawing with Matched Metal Dies

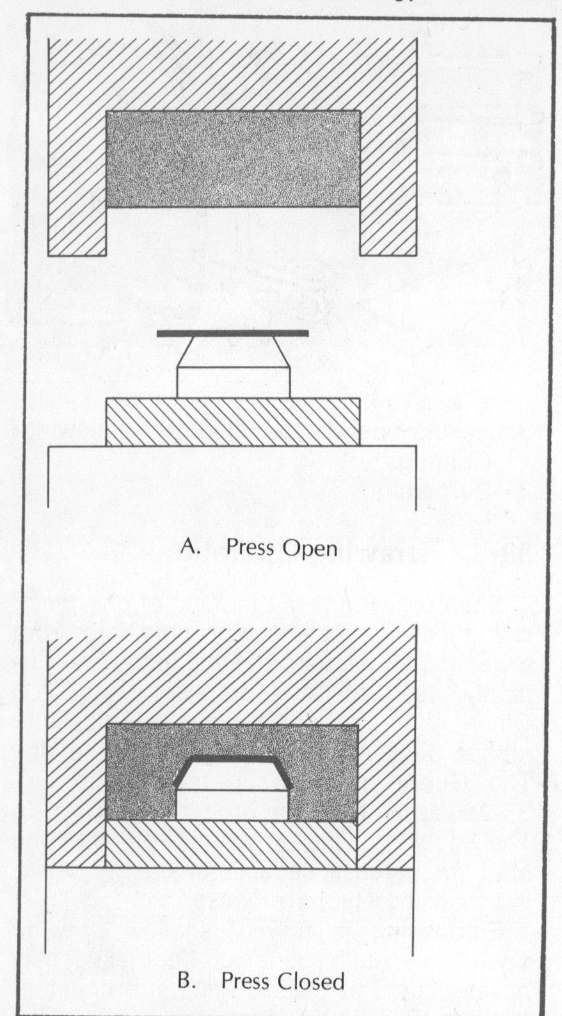

Fig. 33-2. The Guerin Process for Sheet Metal Forming

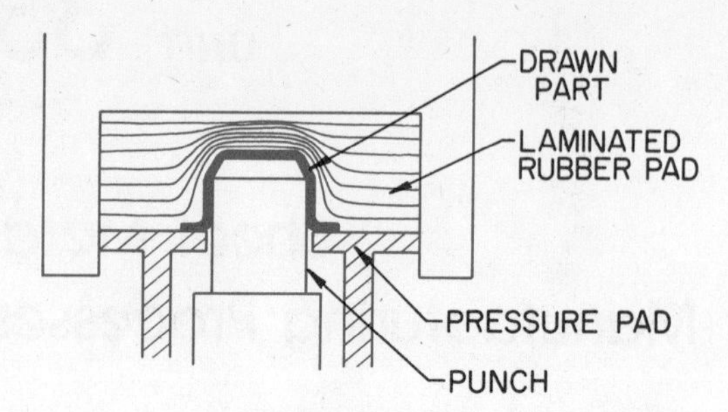

Fig. 33-3. The Marform Process for Sheet Metal Forming

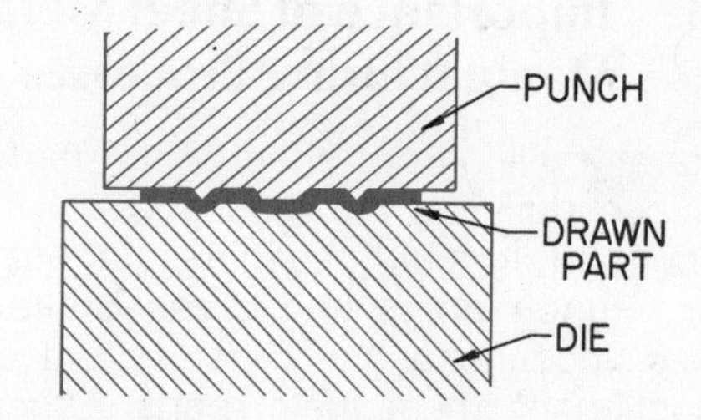

Fig. 33-4. Embossing with Matched Metal Dies

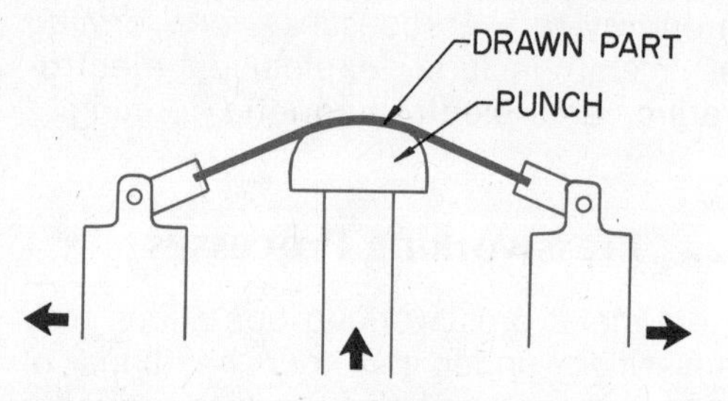

Fig. 33-5. Stretch Forming Process

Stretch forming was developed for the aircraft industry for economical shallow drawing of large areas, Fig. 33-5. The metal is first stretched tight then is formed over a single punch.

33-4. Shearing Operations

Blanking and **punching** are both shearing operations employing a punch and a die as in Fig. 33-6. Their difference is one of

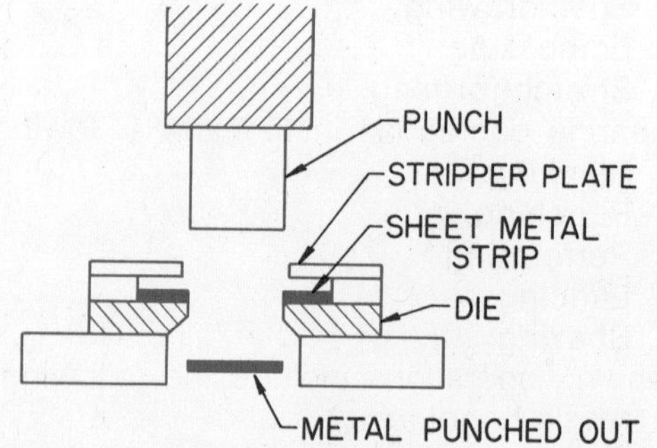

Fig. 33-6. Basic Arrangement for Blanking or Punching

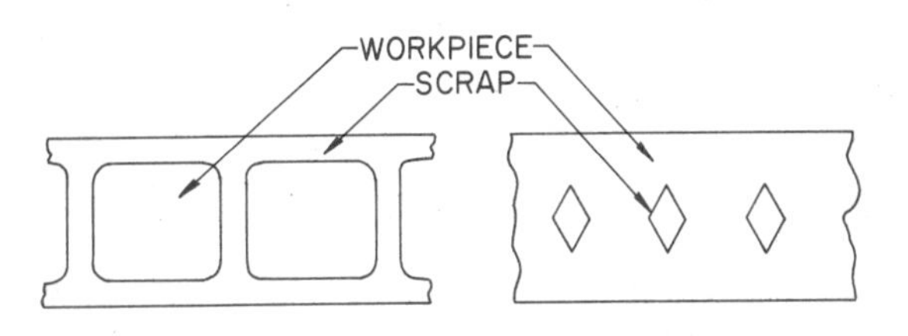

Fig. 33-7. Difference Between Punching and Blanking

Fig. 33-8. Decorative Patterns in Sheet Metal Made by Perforating

definition. In punching, the metal removed becomes scrap, while in blanking, the metal remaining is scrap, Fig. 33-7.

Perforating is simply punching a large number of holes close together. It is often done for the purpose of forming a decorative pattern, as in Fig. 33-8.

Lancing is a piercing operation which can vary from providing a simple slit to creating expanded metal. No metal is lost in this process.

Shaving, as the name implies, removes only a very small amount of metal. It is a punch- and die-cutting operation which is

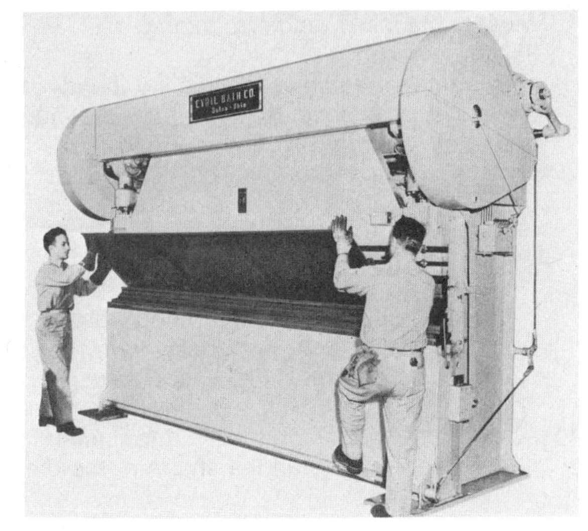

Fig. 33-9. Industrial Press Brake

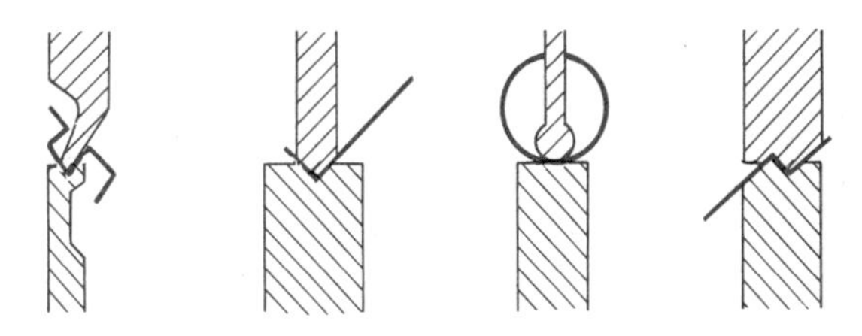

Fig. 33-10. Various Bends Made on Press Brakes

done on parts already punched. Its purpose is to produce smoother edges, edges of more accurate shape, or more accurately sized pieces than can be obtained from the initial punching operation.

33-5. Bending Operations

Angle bending is chiefly a press brake operation but it can be done in punch presses as well. Figure 33-9 shows a typical press brake and Fig. 33-10 shows types of bends made in a press brake by using various dies.

Curvilinear shapes can be bent in a press brake by making small bends close together either with a round-nose die or with a standard angle bending die, Fig. 33-11.

Beading is forming a tubelike bend on the edge of a sheet primarily for the purpose of adding strength. Figure 33-12 shows how a bead can be formed in a press brake.

33-6. Squeezing Operations

Coining is the pressworking process used to make coins, medals, and parts calling for fine detail or exact size. It is done in close-fitting closed dies with pressures high enough to cause the metal to flow, thereby taking on the shape of the die, Fig. 33-13. Pressures as high as 100 tons per square inch are sometimes required.

Burnishing is done to improve the finish on drawn parts. The part is forced through a slightly tapered die, having its small end slightly smaller than the workpiece. The rubbing against the highly polished sides of the die smooths the workpiece.

33-7. Roll Forming

Roll forming of sheet metal is done in a machine which progressively bends the metal to its desired shape by passing it through successive sets of forming rolls, Fig. 33-14. Highly complex shapes may be rapidly formed. Typical products are moldings, downspouts and gutters for buildings, and welded tubing and pipe.

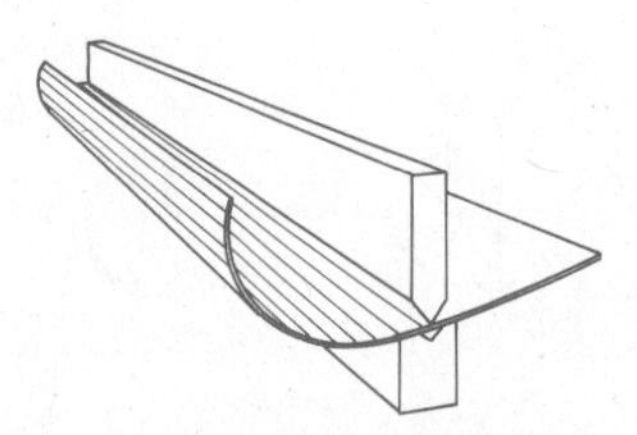

Fig. 33-11. Curvilinear Shape Bent in Press Brake

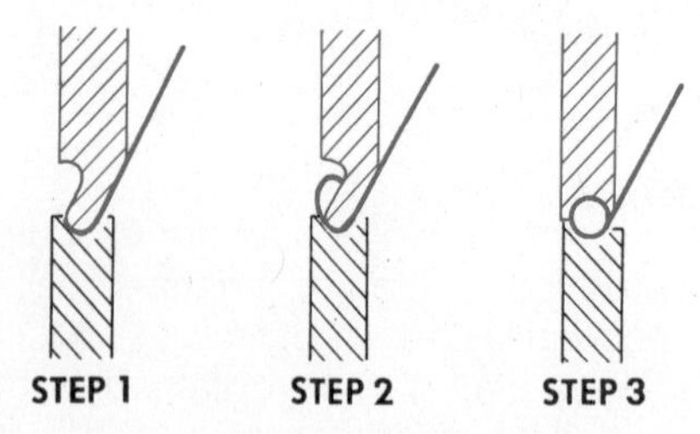

Fig. 33-12. Sequence of Bends and Dies Used for Forming a Bead on a Press Brake

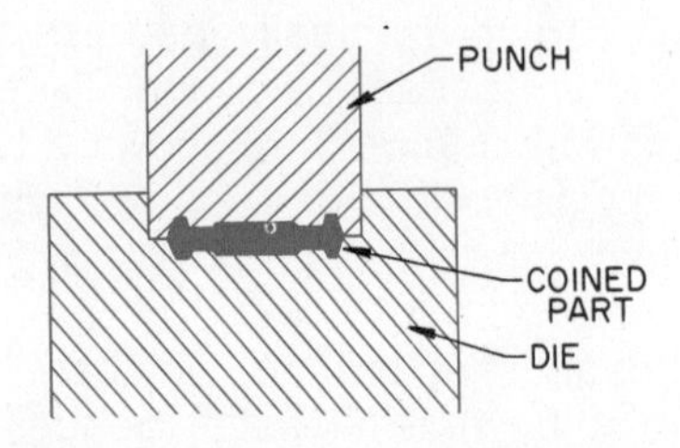

Fig. 33-13. The Process of Coining

Fig. 33-14. Roll Forming Machine

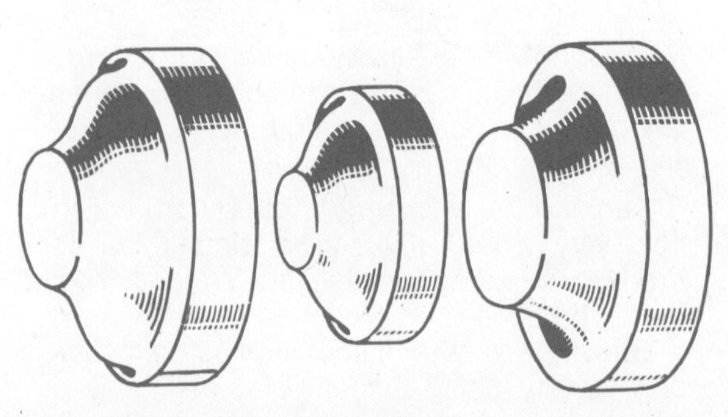

Fig. 33-15. Spinning Chucks

33-8. Metal Spinning

Metal spinning is a process by which a rotating disc of sheet metal is formed into a hollow shape by gradually forcing the metal to conform to the shape of a male form called a **spinning chuck,** Fig. 33-15. Spinning chucks for short production runs are commonly made of close-grained hardwoods, such as maple or birch. Steel chucks are used when the quantity of pieces to be produced is high enough to offset the cost of their construction. Spinning tools are hand-held and are made of hardened and polished steel, Fig. 33-16. Soft metals, however, such as pewter and the softer aluminum alloys, may be spun with a simple hardwood tool, Fig. 33-17.

A part is spun by first mounting the spinning chuck on the **headstock spindle** of the spinning lathe. The disc of metal is then centered on the small end of the chuck and is held in place by pressure from a **follow block** which is attached to a ball bearing center in the tailstock, Fig. 33-18. The metal disc should be lubricated with grease, tallow, soap, or paraffin. This allows the spinning tool to slide easily over the metal and helps to produce a smooth, polished surface.

Discs up to 7″ in diameter may be revolved as fast as 1800 rpm. Larger discs should be revolved at slower rpm's. Thin metal should also be revolved slowly, sometimes as slow as 300 rpm. The disc should first be rotated at slow speed while the metal is quickly formed around the base of

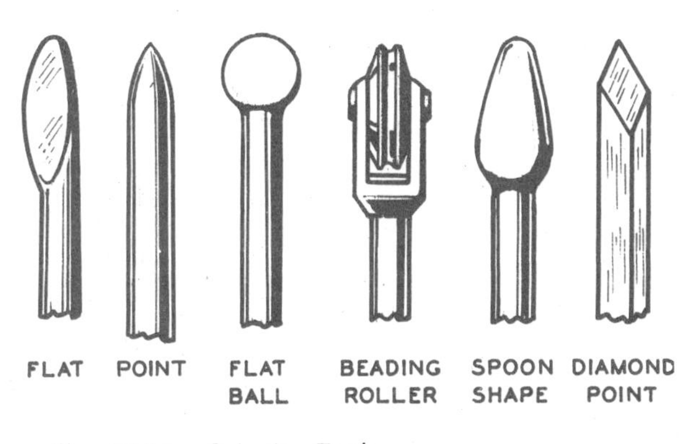

Fig. 33-16. Spinning Tools

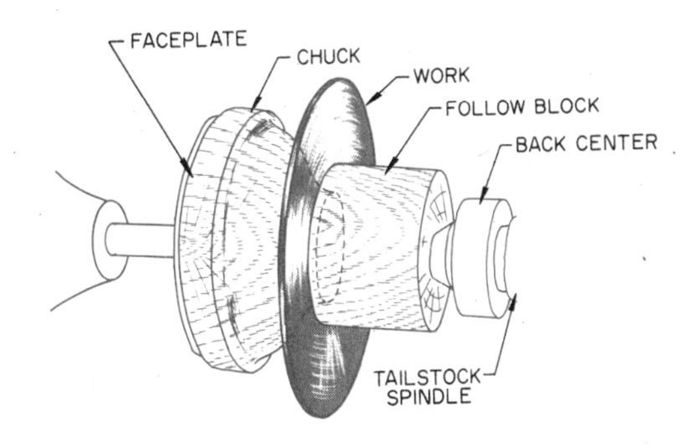

Fig. 33-18. Lathe Setup for Metal Spinning

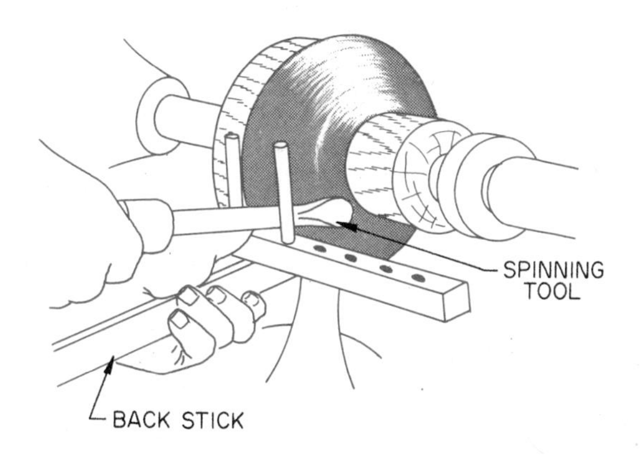

Fig. 33-17. Spinning a Large Piece with Wood Tool and Backing Stick

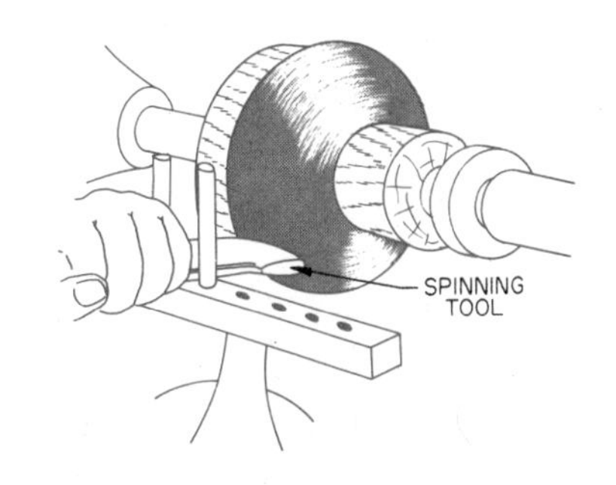

Fig. 33-19. Spinning

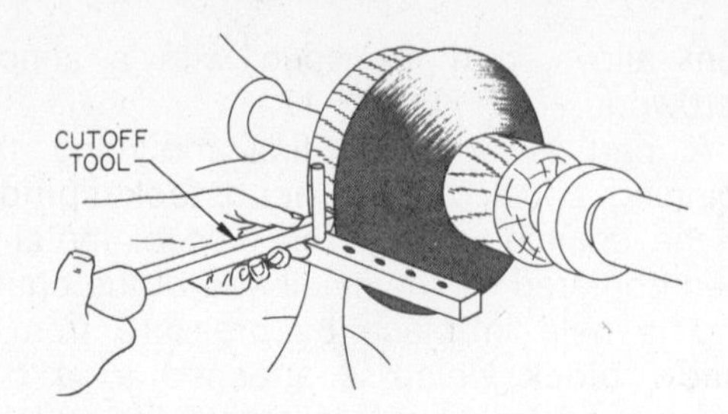

Fig. 33-20. Trimming the Blank Round with the Cutoff Tool

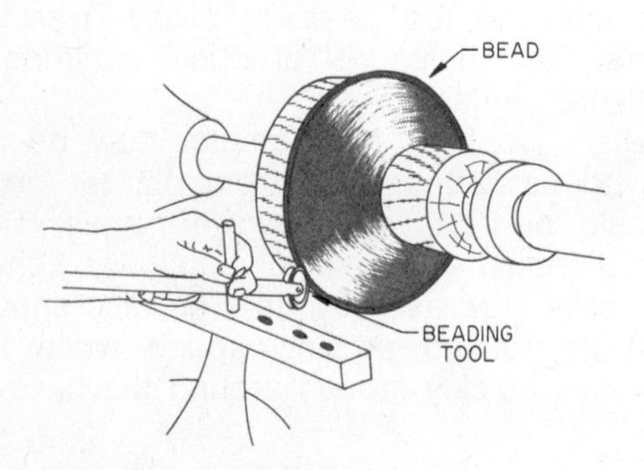

Fig. 33-21. Forming a Bead

the chuck. The disc can then be brought to the desired spinning rpm without danger of its being thrown from the lathe.

The spinning tool is then moved slowly across the face of the disc, making more strokes toward than away from the chuck, Fig. 33-19. This is necessary in order to shrink the disc of metal down to the diameter of the chuck. Care must be taken not to work the tool too much in one direction since this can cause thinning of the metal to the point of rupture. The disc should be trimmed periodically in order to keep it round, Fig. 33-20.

Any wrinkles which develop should be removed immediately. This may be done by pressing a **back stick** against the back of the disc while at the same time pressing the front of the disc with the spinning tool, Fig. 33-17. Parts spun of thin metal are often strengthened by rolling a **bead** on their edge, Fig. 33-21.

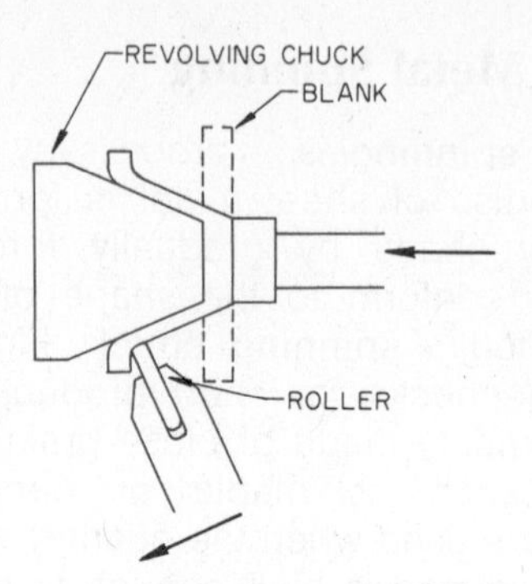

Fig. 33-22. Shear Spinning Process

Metals such as copper and brass **work harden** during spinning and are **annealed** as often as necessary to allow the metal to be spun down to the chuck.

33-9. Shear Spinning

Shear spinning is a variation of metal spinning which uses powerful machines with power-operated tools. Instead of using thin sheet metal blanks which must be shrunk down to the chuck diameter, blanks for shear spinning are much thicker and only slightly larger in diameter than the largest diameter of the chuck. Tool forces are sufficient to cause the metal to become plastic and flow ahead of the tool, thus taking on the shape of the chuck, Fig. 33-22. **Power spinning, floturning,** and **hydrospinning** are other names for shear spinning.

Parts made by this process benefit from the cold working of the metal. Improved tensile strength, resistance to fatigue, and good surface finish result. Spinning is not limited to relatively soft metals, such as aluminum, copper, brass, and mild steel. Shear spinning makes possible the spinning of difficult metals such as high-strength copper alloys, tool steels, stainless steels, and titanium.

33-10. Chemical Milling

Chemical milling is a process of shaping metal by using strong **acid** or **alkaline** solutions to dissolve away unwanted metal. The process is very simple and requires neither highly skilled labor nor expensive equipment. It has these advantages:

1. It does not cold work the metal as conventional machining does.
2. Very large parts can be machined.
3. Any number of the workpiece surfaces can be machined at the same time.
4. Removal of metal from complex surfaces is easily accomplished.
5. Since there are no mechanical cutting pressures involved, thin, delicate workpieces such as metal honeycomb can be safely machined.

The procedure for chemical milling begins with the cleaning of the metal. Masking of neoprene rubber or vinyl plastic is then sprayed or flowed on and cured by baking. Areas to be chemically milled are then scribed with the aid of templates, and the masking is removed from these areas. The part is then submerged in the chemical milling solution until the unwanted metal is dissolved away. Rinsing with clean water and demasking complete the process.

Chemical milling is most widely used in the aircraft and aerospace industries to remove unwanted weight from complex airframe parts without sacrificing strength. **Chemical blanking** is a variation of chemical milling which produces sheet metal shapes by cutting all the way through the sheet.

Fig. 33-23. Explosive Formed Propellent Case for a Space Vehicle

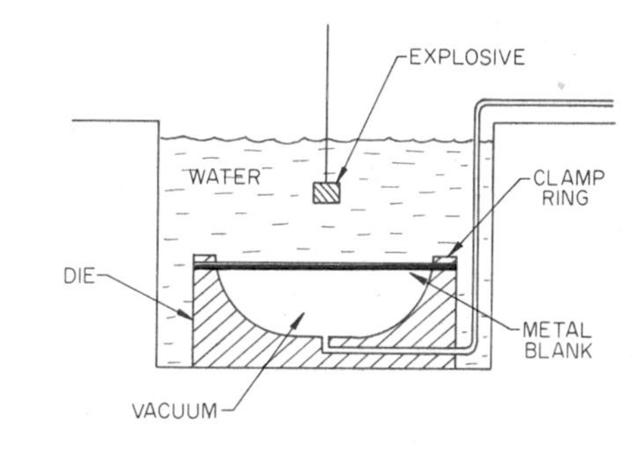

Fig. 33-24. Explosive Forming Process

33-11. High-Energy-Rate-Forming (HERF) Processes

Explosive forming is a HERF process which uses the energy released by detonating powerful explosives. The process is used in the aircraft and aerospace industries to make low-volume quantities of large sheet metal parts with complex shapes, Fig. 33-23. High-strength materials which are difficult to shape by conventional methods are readily processed by explosive forming.

There are two types of explosive forming. **Pressure forming** uses the gas pressure produced by relatively slow burning explosives, such as gunpowders, and must be done in a closed container. **Shock forming** is done in open or partially open containers, the energy from rapidly burning explosives being transmitted to the workpiece through water, oil, plastic, or other liquid medium, Fig. 33-24.

Electrohydraulic forming, sometimes called **spark forming,** is quite similar to shock forming with explosives, but energy rates are lower. The shock waves are created by one or more spark discharges in a liquid medium, as shown in Fig. 33-25. The process has most of the advantages of explosive forming.

Electromagnetic forming uses the force of a sudden and intense magnetic field generated by an electric coil placed inside, around, or next to the workpiece. The workpiece is strongly repelled by the magnetic field, forcing it into the shape of the non-magnetic die, Fig. 33-26. The process is widely used for sizing, bulging, and assembling tubing, but flat pieces can also be formed.

Words to Know

acid solution
alkaline solution
angle bending
annealed
back stick
beading
blanking
burnishing
chemical milling
coining
curvilinear bending
dies
electrohydraulic forming
electromagnetic forming
embossing
explosive forming
follow block
Guerin process
headstock spindle
HERF
lancing
Marform process
metal spinning
perforating
press brake
punching
punch press
roll forming
shaving
shear spinning
shell drawing
spinning chuck
stretch forming
work harden

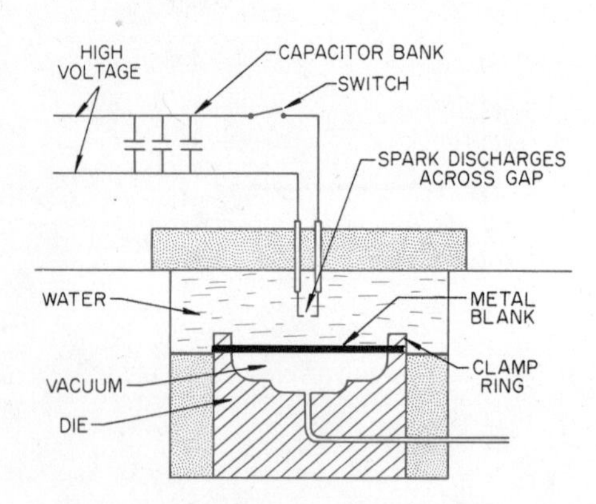

Fig. 33-25. Electrohydraulic Forming Process

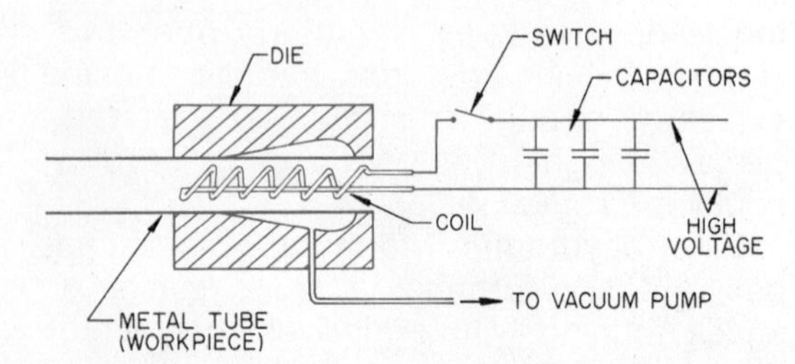

Fig. 33-26. Electromagnetic Forming Process

Review Questions

1. List the four basic types of presswork-ing operations.
2. Name three kinds of drawing operations.
3. Name five kinds of shearing operations.
4. Explain how the Guerin and Marform processes differ from the matched metal die-drawing process.
5. How does embossing differ from coining?
6. Explain the difference between punching and blanking.
7. What is the main difference between perforating and lancing?
8. How does a press brake differ from a punch press?
9. Explain the roll-forming process and name several products made that way.
10. Name several advantages shear spinning has over metal spinning.
11. How are wrinkles removed during metal spinning?
12. How many rpm's should metal spinning discs be revolved?
13. What is meant by work hardening? Why is it a problem in metal spinning? How can the hardness be removed?
14. Describe chemical milling and name several advantages of the process.
15. How do explosive forming and electrohydraulic forming differ?
16. How is electromagnetic forming done? What are its main uses?

PART X

Shaping and Decorating Art Metal Objects

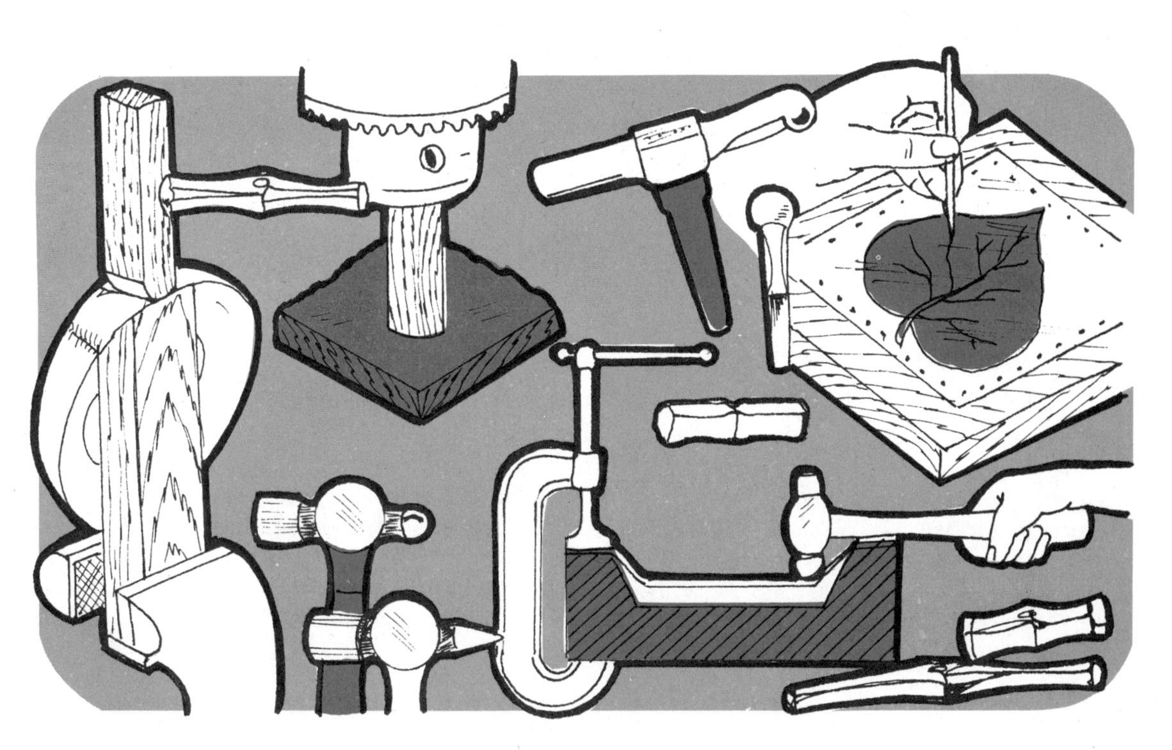

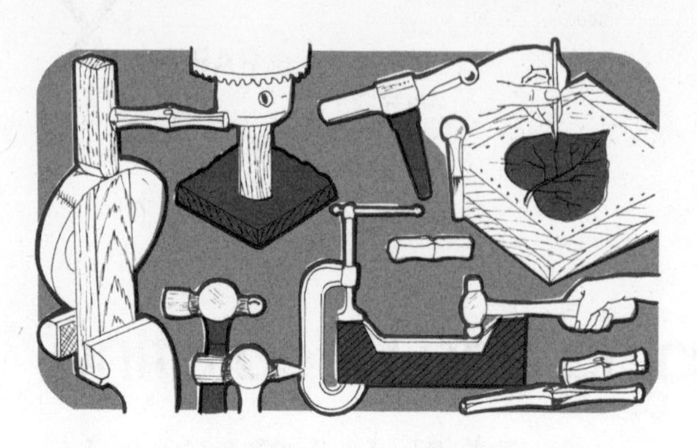

UNIT 34

Art Metalwork

34-1. What Is Art Metalwork?

Art metalwork is the name given to the process of making either decorative or utilitarian objects of metal by hand. Skilled artists and craftsmen are capable of designing and making beautiful art metal objects. These can be tableware, such as eating utensils, bowls, trays and pitchers; candlesticks; trophies; and jewelry, such as bracelets, rings, brooches, and pins. See Fig. 34-1. Many such objects are hammered into shape from metal in sheet form. Platinum, gold, silver, pewter, copper, brass, aluminum, nickel silver, and stainless steel are the most commonly used art metals.

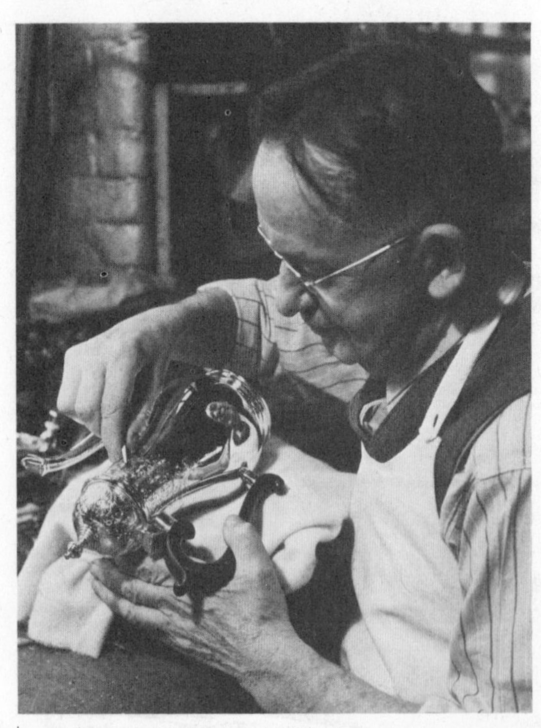

Fig. 34-1. Engraving a Hand Crafted Sterling Teapot

34-2. Who Does Art Metalwork?

Art metalwork is done by the **jeweler, goldsmith, silversmith, coppersmith,** and **artist.** Other kinds of workers who must know how to shape metal by working it with hand tools are the **sheet metalworker, ornamental ironworker, blacksmith, auto body service worker,** and **mechanics who repair aircraft bodies.**

34-3. Shaping Metal by Raising

Hammering a soft metal thins it and causes it to stretch. Repeated hammering in the same spot with a rounded hammerhead will cause the metal to develop a bulge. By careful hammering and using overlapping blows, the bulging can convert a flat disc of metal into a curved or hollowed shape such as a spoon, saucer, or bowl. This process is called **raising,** Fig. 34-2.

Raising may be done in several ways. A block of hardwood, such as maple or birch, into which a shallow depression is hammered into the end grain, works well for shallow objects, Fig. 34-2. If the object is to be circular, the diameter of the starting blank can be determined as shown in Fig. 34-3.

Cut a disc to the required size and lay out circles about ¼" apart. The circles should be drawn in pencil with a compass, not scratched in with dividers, Fig. 34-4. Use metal without scratches, since scratches are very difficult to remove, and they will not be hammered out in the raising process.

Hold the metal over the raising block and hammer with a **raising hammer** or the **peen** of a ball peen hammer. See Figs. 34-5 and 34-2. Begin by hammering lightly and evenly on the outside circle. Lift the hammer only about 2". The marks should overlap one another. Striking harder in some places than in others will make the object lopsided. Evenly spaced hammer marks add beauty to the object. If a wrinkle forms at the edge, hammer it out at once, using great care. After the first circle has been hammered,

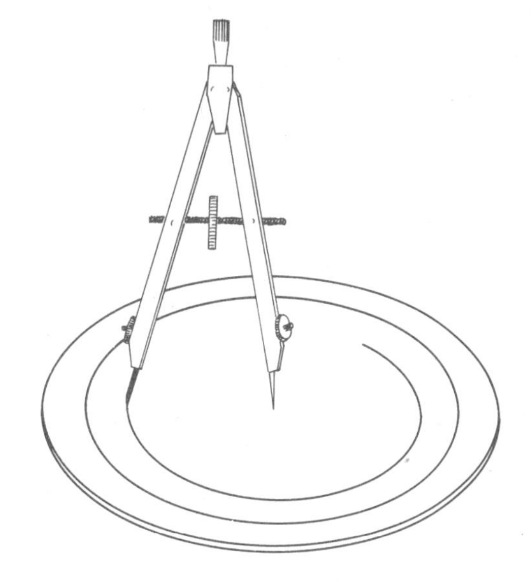

Fig. 34-4. Layout Circles on Blank with a Compass

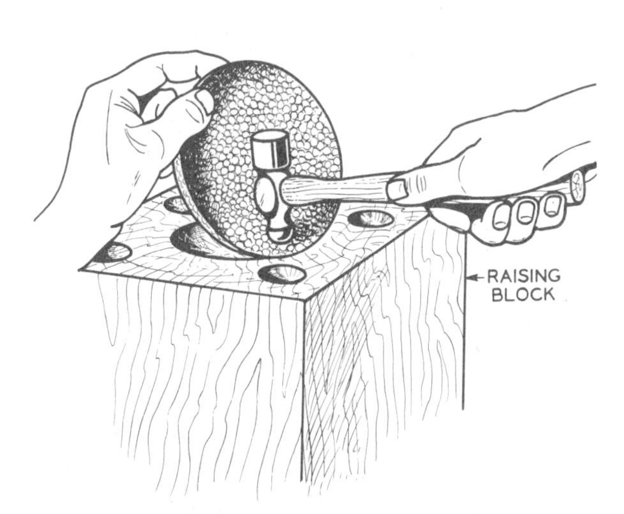

Fig. 34-2. Raising a Bowl

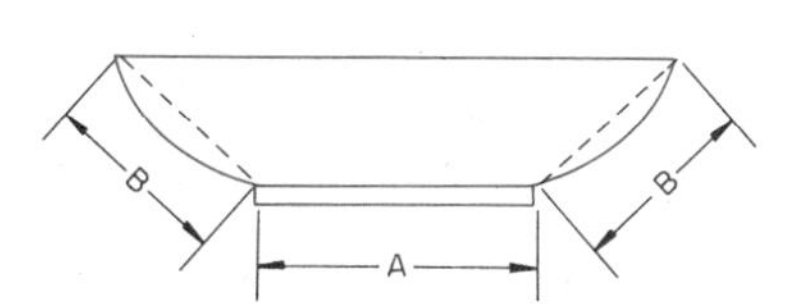

Fig. 34-3. Figuring Blank Diameter for a Bowl

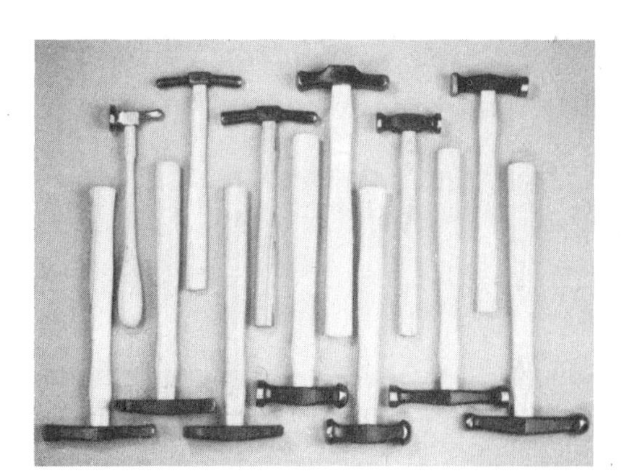

Fig. 34-5. Raising Hammers

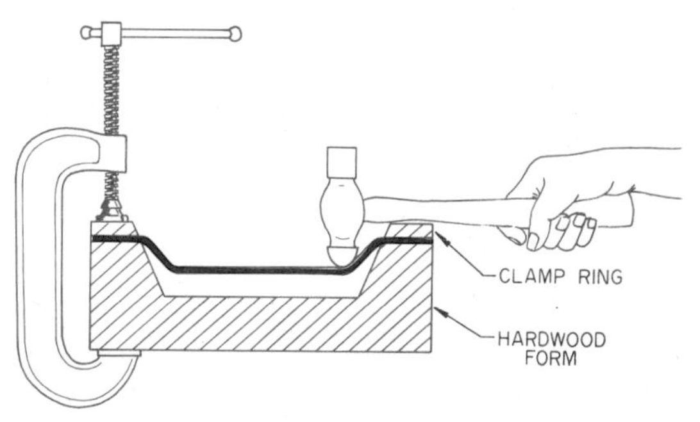

Fig. 34-6. Hammering Metal into Hardwood Form

Fig. 34-7. Raising a Deep Form on a Stake

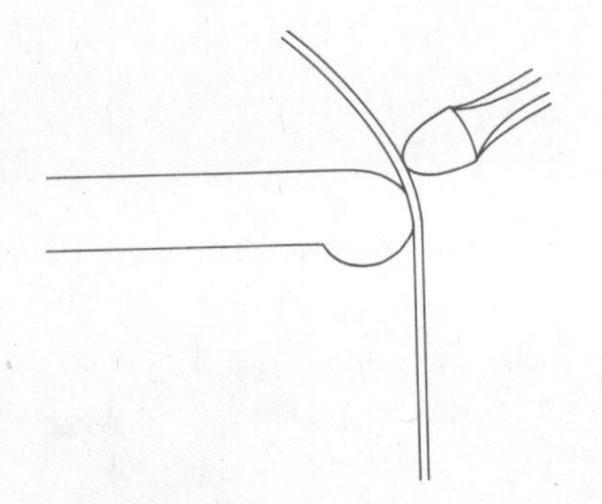

Fig. 34-8. Strike Metal Just Ahead of Where It Touches the Stake

continue by hammering the second circle, then the third, and so on until the object is the desired shape.

A sandbag and a mallet can also be used for raising. Beginners will find it easier to hammer the metal into a hardwood form from which the desired shape of the bowl has been cut, Fig. 34-6.

Deep forms can be made by raising the metal over a stake, Fig. 34-7. The raising operation is easiest to start on the hardwood block or sandbag. After the form has been developed somewhat, change to the stake and continue to hammer in a circular path with overlapping blows. Care should be taken that the hammer blows strike the metal just ahead of the point where the metal touches the stake, Fig. 34-8.

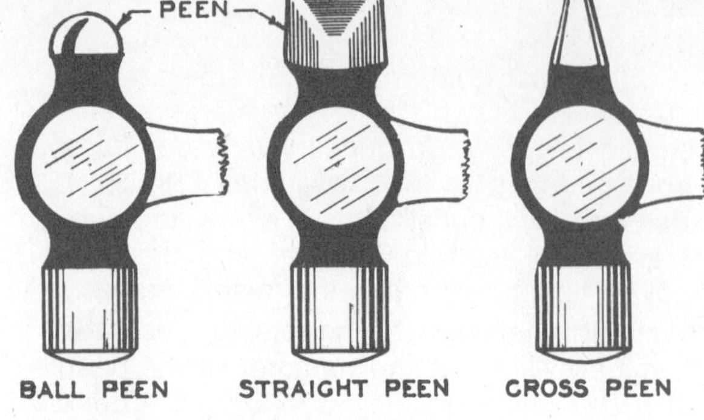

Fig. 34-9. Peening Hammers

Hammering makes copper and brass hard, stiff, and so stubborn that it is hard to work; it gets so hard it will even crack. It must be softened periodically by **annealing,** as explained in the next section. Aluminum hardens a little; pewter does not harden.

34-4. Annealing Copper or Brass

Copper and brass become hard from hammering, stretching, pressing, and bending. They may be **annealed** to make them soft again. This is done by heating the metal until rainbow colors begin to show and then cooling it quickly in water. Or, before heating, the copper or brass may be wiped with an oily cloth, heated until the oil burns off the metal, and then quickly cooled in water. Annealed copper and brass is dark and dirty and should be cleaned by dipping in a pickling solution of five parts water to one part sulfuric acid.

34-5. Hammered Finishes

Sometimes a hammered finish is desired on art metal objects. Hammered finishes are obtained by **peening** with the **peen end** of hammers, Fig. 34-9. The round marks in Fig. 34-10 are made with a **ball peen hammer** which has a ball-shaped peen. The

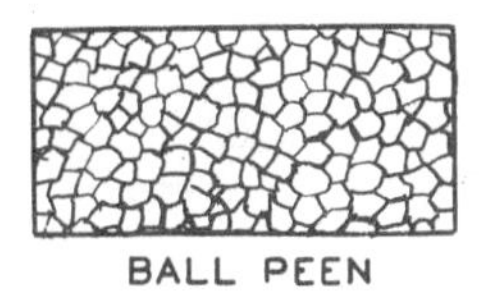

Fig. 34-10. Hammered Finishes

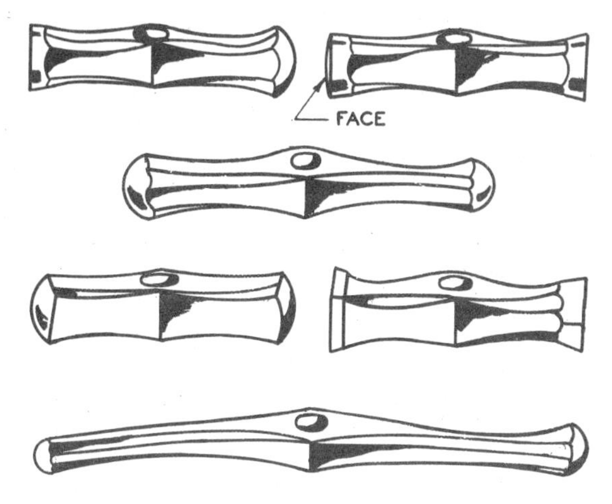

Fig. 34-11. Planishing Hammers

long narrow marks are made with a **cross peen hammer** or with a **straight peen hammer.**

If hammer marks are wanted on both sides of the metal, fasten one hammer with the peen side up in a vise, hold the metal on the peen, and strike it with the peen of the second hammer. Other marks of interest may be made with blunt punches and chisels.

34-6. Planishing

Planishing means to make smooth. This may be done with a hammer which has a smooth, flat (or almost flat) face; it is called a **planishing hammer,** Fig. 34-11. It levels the uneven surface made by raising the metal, stiffens it, hardens it, and, when skillfully done, makes a beautiful finish. The

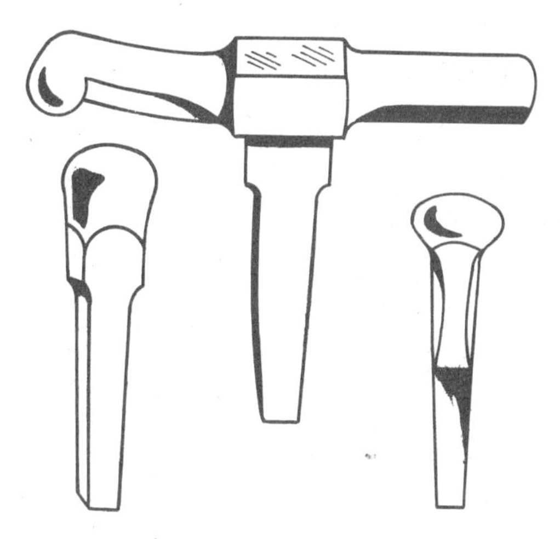

Fig. 34-12. Planishing Stakes

finely polished surfaces on planishing hammers and stakes should be carefully protected from nicks and scratches because they are transferred to the work and spoil its appearance.

To planish a raised bowl, place it over a planishing stake, Fig. 34-12, and hammer lightly with a planishing hammer, Fig. 34-11. The hammering should proceed in the same manner as for raising, except that the hammer blows should strike the metal at the same point that it touches the stake.

34-7. Fluting

Making grooves, as on the sides of plates, trays, and bowls, is called **fluting,** Fig. 34-13.

First, divide the circle into the desired number of equal parts and draw lines with a pencil to show where the **flutes** are to be made. Then make a **fluting block.** Cut the shape of the flute into the end of a block of hardwood with a **coarse** file. A **rasp-cut file** is best. All edges on the fluting block must be rounded. Next, hold the block in a vise with the fluted end up and hammer the metal into the flute, using a tool with a round end which is very well polished. Fluting

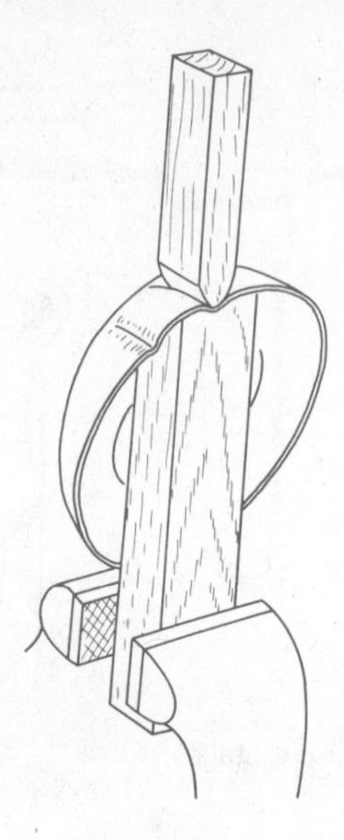

Fig. 34-13. Fluting

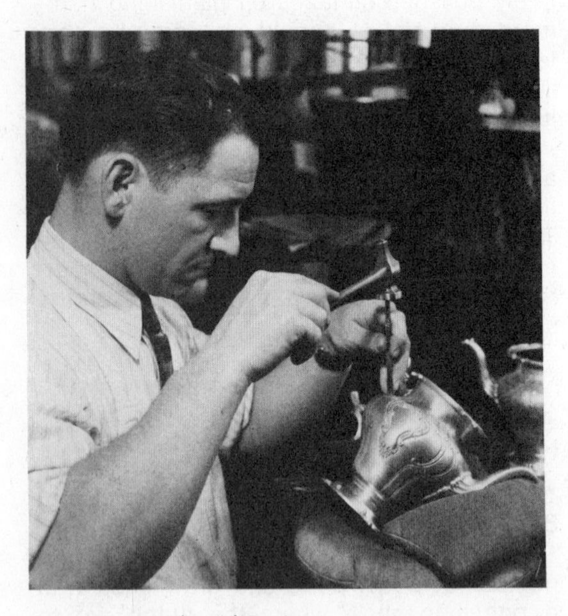

Fig. 34-14. Chased Design

helps to curve the top of the dish inward, thus making it smaller in diameter at the top and adding beauty to the design.

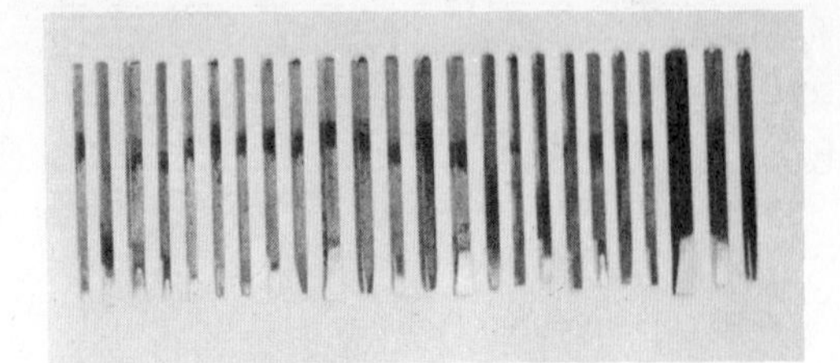

Fig. 34-15. Chasing Tools

34-8. Chasing

Deepening the outline of a design on metal is called **chasing,** Fig. 34-14; it is also called **repousse work,** which is a French word. It is done with **chasing tools,** also called **tracers,** which look like small cold chisels and punches with blunt, rounded, highly polished ends and edges, Fig. 34-15. Simple chasing tools may be made of large nails by rounding off and polishing the points.

The design to be chased is first drawn on paper and then transferred to the sheet metal by using **carbon paper.** Another way is to paste the paper with the design on the metal with **rubber cement** or **shellac.** The metal with the design drawn on it is then fastened to a softwood board. Small screws placed about 1″ apart are better, because tacks and small nails pull out when the metal is hammered. Another way to hold the metal is with **pitch,** as explained in section 34-9.

It is important to note in Fig. 34-16 how the chasing tool is held. Chasing is fine work and great care must be taken to see that the marks are made in the right places on the metal. The tool is held firmly between the thumb and the first and middle fingers. The ring finger and the little finger are pressed on the metal. This manner of holding the tool helps to guide it.

When a line is made, the chasing tool is held on the metal all the time, even between hammer blows. It is important that you have good light; there must be no shadow on the

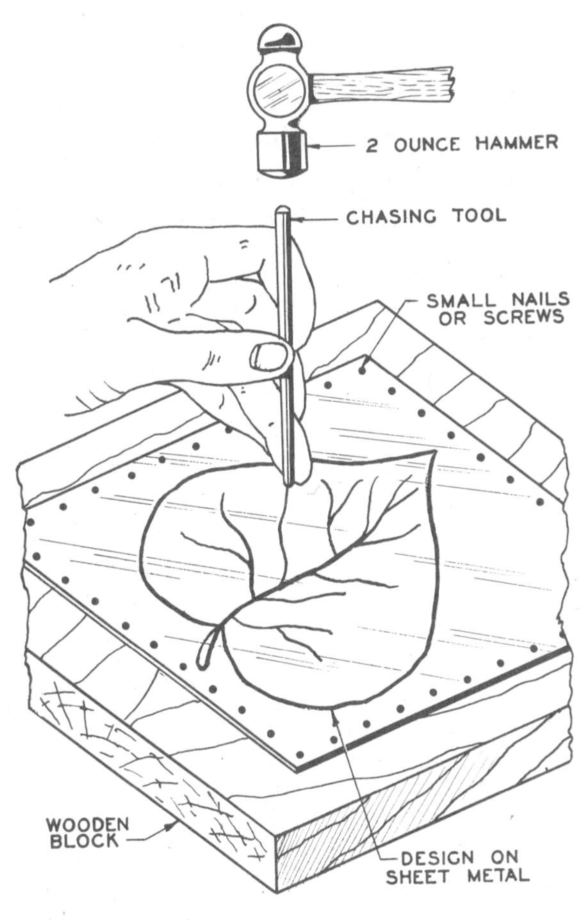

Fig. 34-16. Chasing

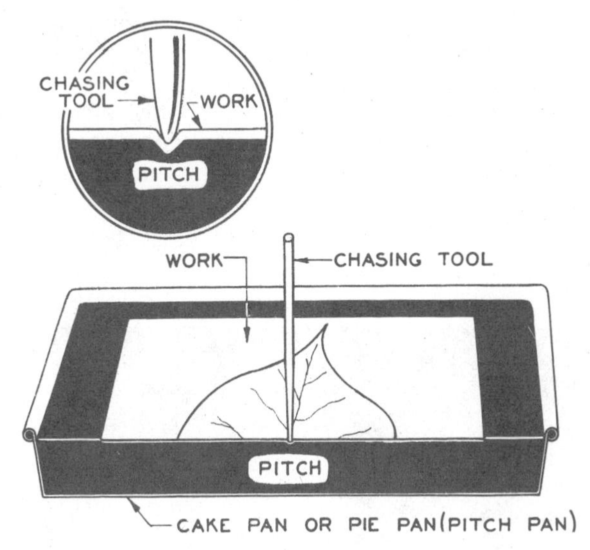

Fig. 34-17. Using Pitch for Chasing

part to be chased. Light should come from over the right shoulder. The line to be chased should be between you and the chasing tool. Move the tool a very small distance toward you on the line between blows without lifting it. First, chase the entire outline lightly, using a light hammer weighing about 2 ounces. Deepen the grooves a little at a time, going over and over again.

34-9. Pitch

Another way to hold the metal for **chasing** is with a **pitch** made of three ingredients:

Rosin — 1 lb.
Plaster of paris[1] — 1 lb.
Beef tallow — 2 oz.

Mix as follows:

1. Using a low flame, melt a little rosin slowly in an old coffeepot (easy to handle, pour, and store away).
2. Add plaster of paris, a little at a time, and stir until well mixed.
3. Add melted beef tallow and stir thoroughly.

Rosin, when cold, is brittle. Tallow softens it; plaster of paris hardens it. The whole mixture should be like rubber when cold. If it is crumbly, add tallow; if sticky, add rosin and plaster of paris. In warm weather, more plaster is needed; cold weather calls for more tallow. **Linseed oil** or **turpentine** may be used instead of tallow.

Now, suppose that the pitch is to be used to hold the metal and to chase the design in Fig. 34-16. First, heat the pitch and pour

[1]**Plaster of paris** is soft, white powder like flour. It makes a paste when mixed with water. When allowed to stand in the air, the paste sets (dries) quickly and becomes hard. It is called plaster of paris because it was first mined near Paris, France.

it into a pie pan or a cake pan, then called a **pitch pan.** Next, rub oil on the back of the metal; this will cause it to stick to the pitch better and to be more cleanly removed later. Warm the metal and place face upward on the pitch. Press the metal down evenly to remove air pockets. The work is now ready for chasing. Figure 34-17 shows how the pitch forms a perfect support for different shapes when chasing.

34-10. Metal Tooling

Metal tooling is similar to chasing but is done on **metal foil** with hand-held **modeling tools.** Hammering is not necessary. The making of metal foil pictures, such as in Fig. 34-18, is a popular craft activity but is also done by professional artists. Metal of 30 to 36 gage (Brown and Sharpe) is generally considered to be foil and is .010″ or less in thickness. Copper, brass, and aluminum foils are the most popular metals used for tooling.

Tools made of hardwood dowels are very satisfactory. However, modeling tools for leatherworking are also excellent for tooling metal foil, Fig. 34-19.

Designs are transferred to the metal in the same manner as for chasing. After the design is established, set it immediately by going over it with a tracing tool. Sharp detail is obtained by working on a hard surface. Areas which call for the surface to be raised are worked from the back of the foil with ball- or spoon-shaped tools while it is on a soft resilient surface. Use corru-

Fig. 34-18. Tooled Articles

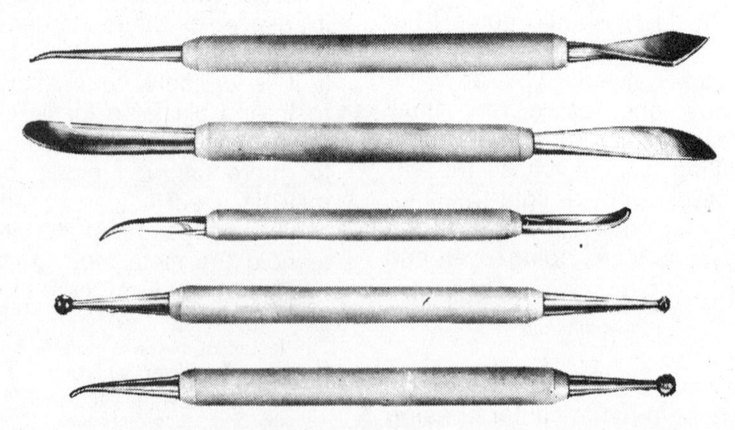

Fig. 34-19. Leather Modeling Tools May be Used for Metal Tooling

gated cardboard, 1/8″ sponge rubber, or a pad of newspapers for this purpose.

Backgrounds of various patterns are often provided to add interest. Tapping lightly with a tool which has a small, rounded point makes a **stippled** background. Window screen and sandpaper background patterns can be obtained by placing the pattern material on a hard surface, covering with the foil, and then rubbing the foil with a blunt tool. Before mounting the tooled foil, protect the raised areas by filling the cavities on the back with plastic wood or plaster of paris.

Words to Know

annealing
art metalwork
chasing tool
cross peen hammer
fluting
hammered finish
metal foil
metal tooling
modeling tool
pitch
planishing
plaster of paris
peening
raising
repousse
straight peen hammer

Review Questions

1. What is meant by art metalwork?
2. What is raising?
3. Describe three methods of raising.
4. What effect does hammering have on metal?
5. What is meant by annealing?
6. How are copper and brass annealed?
7. What is peening and how is it used in art metalwork?
8. What is meant by planishing?
9. What kind of tools are used for planishing?
10. What is fluting? How is it done?
11. What is meant by chasing?
12. Describe two ways metal may be held for chasing.
13. Why should oil be put on the back of the metal before placing it on the pitch?
14. How does metal tooling differ from chasing?

Mathematics

1. What is the correct blank size for raising a popcorn bowl whose dimensions are: 5″ top diameter; 2½″ bottom diameter; 2¼″ deep?

Career Information

1. Write a paper describing how a knowledge of and skill in art metalworking are used in an occupation of your choice.

UNIT 35

Metal Finishing

35-1. What Is Meant by Metal Finishing?

Metal finishing is the final treatment given to a metal surface in order to improve its appearance, make it wear longer, or protect it from rusting. It is also done to improve electrical conduction, neutralize the reaction of a metal to certain chemicals, and to improve the value of an object, as when **electroplating** with precious metals.

This unit will deal mostly with finishes that are used for decorative treatment of metal objects. Some metal finishes are described in other units. **Filing** and **drawfiling** are described in Unit 15; **scraped, frosted, spotted,** and **flaked** finishes in Unit 16; **abrasive polishing** in Unit 20; **hammered** finishes in Unit 34; and **grinding** in Unit 58.

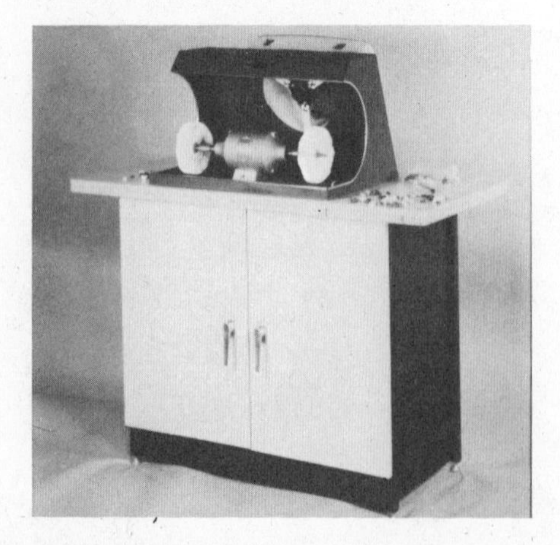

Fig. 35-1. Buffing Machine

35-2. Buffing

Buffing means to polish to a smooth, bright finish by rubbing the metal surface with a **buffing wheel** to which a **polishing compound** has been applied. On most metals, this process can produce a mirror finish, which is often the desired finishing treatment. Buffing is also done to smooth a metal surface in preparation for electroplating.

Buffing wheels are made of cloth, felt, or leather. Leather and felt make hard wheels which are used with coarse polishing compounds for fast initial polishing of rough surfaces. Cloth wheels of **cotton muslin** are used for intermediate buffing, and soft **cotton flannel** wheels and fine polishing compounds are used to obtain smooth, bright finishes. **Goblet** buffing wheels are ball shaped so they can polish the inside of bowls and goblets. Some buffing wheels are set with fiber or wire bristles and are, therefore, revolving brushes. Wire brushes are used for deburring, cleaning, and some finishing operations. A **satin finish** can be obtained on aluminum by wire brushing, and light scratch brushing is recommended just prior to chemical coloring treatment as an aid in obtaining uniform coloring.

Buffing machines, also called **buffers** and **buffing heads,** look very much like ordinary grinding machines, Fig. 35-1. Some buffing wheels have arbor holes and are attached to the arbor ends with a washer and nut, Fig. 35-2. Jewelers' buffing wheels have only a pinhole at center and are screwed directly on a tapered buffing spindle, Fig. 35-3.

35-3. Polishing Compounds

Polishing compounds, or **buffing compounds,** are the abrasive materials applied to the buffing wheels which enable the worker to cut and polish the metal. Some of the cutting materials used are **lime, tripoli,**[1] **crocus** and **rouge,**[2] **emery** and **aluminum oxide flour,** etc. These are mixed with **tallow** or some other heavy grease and pressed into bars or cakes. Coarse compounds are used for roughing, while fine compounds are used for final polishing, see Table 24.

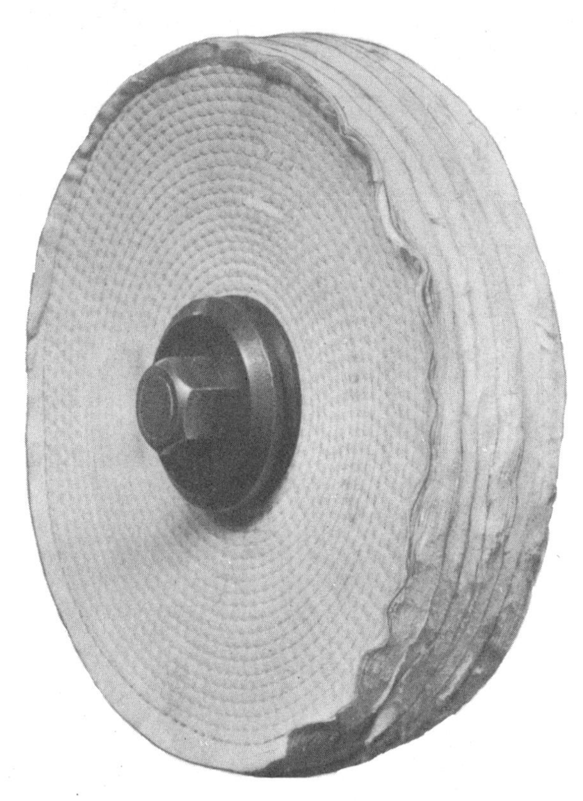

Fig. 35-2. Washer and Nut Method of Attaching Buffing Wheel to Motor Arbor

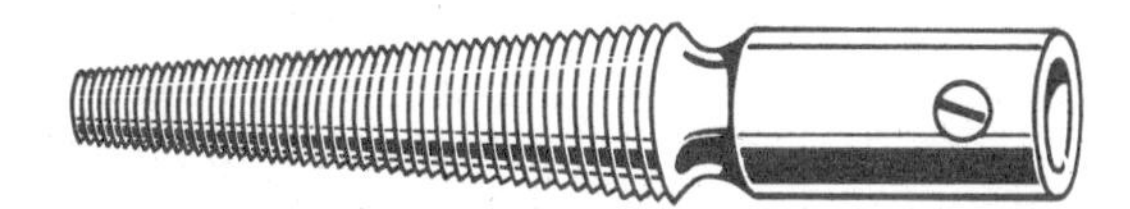

Fig. 35-3. Tapered Spindle used with Jeweler's Buffing Wheels

35-4. How to Buff

Choose a polishing compound according to the kind of metal to be buffed. Put it on the buffing wheel by holding it against the edge of the revolving wheel. It is best to put only a little polishing compound on the wheel at a time. When the surface of the wheel is coated with the compound, it is then ready for buffing.

The work should be held on the underside of the front of the wheel so that if it is pulled out of the hands, it will fly **away from the operator,** not toward him, Fig. 35-4. Again, when the work is held on the underside of the wheel, the dust will also fly away from the operator. If considerable buffing is to be done, a **respirator** should be worn to filter out the dust and grit created by the

TABLE 24

Buffing Compounds and Uses
(See Section 35-3.)

Metal	Compound	
	Roughing	Finishing
Aluminum	Tripoli	Rouge
Brass	Tripoli	Lime
Copper	Tripoli	Lime
Pewter	Tripoli	Rouge
Steel	400 Silicon Carbide	Rouge

[1]**Tripoli** is a weathered, decomposed limestone; also called rottenstone.

[2]**Rouge** is a soft iron oxide. It comes in different shades of red; the darker the color, the harder the rouge. The lighter product is called **rouge** and the darker **crocus.**

Fig. 35-4. Buffing

Fig. 35-5. Respirator for Protection Against Dust Inhalation

Safety Note

Leaded primers and paints should never be used indoors. There are many cases of infant poisoning due to teething or chewing objects painted with leaded paints.

polishing operation, Fig. 35-5. The work should be moved and turned as it is held against the wheel. In this way, the wheel rubs every corner and curve of the work. Let the buffing compound do the work of polishing. Pushing the workpiece hard against the wheel only creates friction and makes the workpiece hot. If the polishing is proceeding too slowly, try using a coarser compound first.

Change from a wheel with a coarse polishing compound to a wheel with a finer compound when all scratches and blemishes have been removed. Polish with the fine compound until you are satisfied with the finish. If the work comes from the wheel looking greasy and dirty, too much polishing compound has been put on the wheel. The grease can be removed by washing with hot water and mild soap or washing soda.

35-5. Painted, Enameled, and Lacquered Finishes

Paint is often used to decorate and protect metal finishes. It is available in many colors and dries with a **flat,** or dull, finish. **Flat black paint** is a popular interior finish for wrought iron products. **Aluminum paint** provides good protection for iron and steel fences, signposts, flagpoles, and other metal products exposed to the weather. A **primer** is a paint which will **adhere** well to a metal surface and is therefore used as a first coat. Some metals require a special primer. Galvanized steel, for example, should be prime-coated with a **zinc chromate** primer. A good primer for iron and steel surfaces exposed to the weather is **red lead.**

Enamel is a type of paint that is available in a wide range of colors, dries hard, and provides a choice of either a **high gloss** or a **semigloss** finish. Quick-drying enamels will dry in about four hours. Some enamels are baked onto the metal, as is the enamel on many appliances, machines, and tools.

Lacquer is a very quick-drying finishing material. It is widely used as a metal finish and is harder and tougher than enamel or paint. **Clear lacquer** is colorless; it allows the color of the metal to show while protecting it from tarnishing or rusting. **Flat lacquer** has no gloss. Most lacquers are made of synthetic materials and require special solvents for thinning and cleaning. While some lacquers are designed for application by brushing, others are compounded for spraying. Lacquers are also available in many colors.

Lacquer is quite **flammable** and must be kept away from open flames. Containers should be closed immediately after use, and, to protect against **noxious** and **explosive fumes,** lacquer should be applied only in a well-ventilated room.

Bronzing is giving an article a metallic bronze color. Powdered brass or bronze is often used for bronzing. If the entire surface is to be a bronze color, the bronze powder may be made into a bronze paint by mixing it with **banana oil.** The bronze paint can then be put on with a brush.

If the article is to be only partly bronzed, or **spotted,** it can first be coated with paint, enamel, or lacquer. While it is still sticky, bronzing powder may be dusted on with a pepper shaker or it may be blown on with a **powder blower,** Fig. 35-6. Colored bronzing powders are also available.

35-6. Surface Preparation for Painting

Painted, enameled, or lacquered finishes will not stick to the metal unless the metal has been properly cleaned. **Scale** and **rust** should be removed from old metal by **wire brushing, polishing with abrasive cloth,** or

Fig. 35-6. Powder Blower

sand blasting. All dirt and dust should be removed by brushing or wiping. This is followed by washing or wiping with a solvent in order to remove all traces of oil or grease.

For surfaces to be painted or enameled, **mineral spirits** is a recommended solvent. **Lacquer thinner** should be used to degrease surfaces to be lacquered. Solvents such as **benzene** and **carbon tetrachloride** are **toxic** (hazardous to health) and should be avoided. **Gasoline** should not be used because it is so **highly flammable. Kerosene** and **diesel fuel** leave oily residues which will interfere with paint **adhesion.**

Surfaces which have been degreased should not be handled with bare hands, because body oil from fingerprints will contaminate the surface. Clean surfaces should have the finishing material applied as soon as possible. The need for cleanliness does not stop with the application of the first coat. Any dust, dirt, or grease that collects between coats must be removed before applying the next coat.

Paint will not adhere to newly galvanized steel surfaces without **etching.** Etching provides a dull, slightly roughened surface to which the paint will adhere. For etching, use a dilute solution, about 5% **hydrochloric, phosphoric,** or **acetic acid.** Wet the surface thoroughly and allow it to dry. Rinse with clear water and allow to dry before painting. Galvanized steel which has weathered to a dull appearance can be painted without etching.

35-7. Methods of Applying Paint, Enamel, and Lacquer

Finishes may be applied by **dipping**, **brushing**, or **spraying**. Automated finishing systems in factories use dipping and spraying methods. Custom finishing and touch-up is often done with a hand-held spray gun, Fig. 35-7. Excellent finishes can be obtained by hand spraying. Before spraying a valuable finished piece, practice adjusting and manipulating the spray gun and thinning the finishing material to proper consistency. If possible, the surface to be sprayed should be in a vertical position. The nozzle of the spray gun should be held a constant distance from the surface to be sprayed, usually 8-12″. Spraying strokes should run the length of the surface and strokes should overlap each other by about 2″. Sags and runs should be removed immediately by wiping. A rule in applying finishes is that several thin coats always produce a better finish than one thick coat. This is especially true for spraying.

Fig. 35-7. Spraying Wheel on Truck Assembly Line

Brushing is convenient for finishing relatively small objects and for touch-up. For brushing, finishing materials can usually be used directly from the can without thinning, and a brush is simpler and quicker to clean than a spray gun. Good brushing technique calls for working the finishing material onto the surface by manipulating the brush in any convenient direction, wiping the excess material from the brush, and lightly stroking the surface to make sure all the brush marks flow in the direction of the longest dimension of each surface.

Fig. 35-8. Paper Towel Wrapping Helps Hold the Shape of a Clean Brush

Fig. 35-9. Disposable Paint Brush of Plastic Foam

35-8. Cleaning Spray Guns and Brushes

Cleaning of spray guns and brushes should be done immediately after use. Spray guns should be emptied, wiped, and rinsed with an appropriate solvent. Clean solvent should be sprayed through the gun until all the passages are clean. The gun nozzle is usually disassembled for cleaning.

Reusable brushes should first be wiped on the edge of the container. Then, as much of the remaining finishing material as possible is removed by brushing on newspaper or paper towels. The brush is then rinsed in a suitable solvent two or three times. It is important that the heel of the brush (where the bristles join the handle) be flushed out. This is best done by holding the solvent-laden brush upside down and squeezing the bristles to aid the flow of the solvent through the heel. After most of the finishing material has been washed out with solvent, the brush is washed with a liquid detergent. Rinse the brush very thoroughly to remove all the detergent. Squeeze as much water as possible from the brush, reshape the brush, and then carefully wrap it in one or two layers of paper towelling to help it hold its shape, Fig. 35-8. Inexpensive disposable paintbrushes made of plastic foam are sometimes a satisfactory substitute for conventional brushes, Fig. 35-9.

35-9. Coloring Metal by Heating or with Chemicals

Many attractive **oxide colors** can be obtained on some metals by treating them with various chemical solutions or simply by heating them. This treatment causes **oxygen** to combine with the metal, forming a thin layer of colored **metallic oxide.** Oxide finishes should always be protected with a coat of lacquer. There are dozens of formulas for chemical coloring solutions, but only a few can be given here.

Copper is easier to color than any other metal. It can be colored yellow, brown, red, blue, purple, or black. The beginner can get good results with the following solution:

Ammonium sulfide .... 1 ounce
Cold water .... 1 gallon

The work is dipped into this solution. The color produced depends on the length of time the work is left in the solution. When the desired color has been obtained, dry the work in sawdust and give it a coat of lacquer.

Ammonium sulfide must be handled with great care because it stains the fingers and has a bad odor. It should be kept in a dark bottle with a glass cover. Ammonium sulfide is only good for coloring copper; it is not good for brass. **Potassium sulfide** (liver of sulfur) may be used instead of ammonium sulfide.

Brass can be given an antique green finish that will make it look as if it were very old. Make a solution of:

Household ammonia .... 4 ounces
Sal ammoniac .... 2 ounces
Common salt .... 2 ounces
Water .... 1 gallon

Large work can be brushed with the solution. Small work may be dipped into the solution. It may have to be painted or dipped several times to get the desired color. It should then be rinsed with clear water, dried, and lacquered.

One of the many formulas for chemically coloring steel blue follows:

Lead nitrate .... 1 ounce
Ferric nitrate .... ½ ounce
Sodium thiosulfate .... 4 ounces
Water .... 1 gallon

The solution should be mixed and stored in a glass, earthenware, or enameled container. The solution is used hot, 190°-210° F. Soak the clean steel workpiece in the solution until the desired color has been obtained. After careful drying, protect the finish with oil or lacquer.

Oxide colors can be obtained on polished steel by heating to between 380° and 590° F. See Table 26. The colors in order of appearance are yellow, brown, purple, violet, blue, and gray. Brighter colors are obtained on highly polished surfaces than

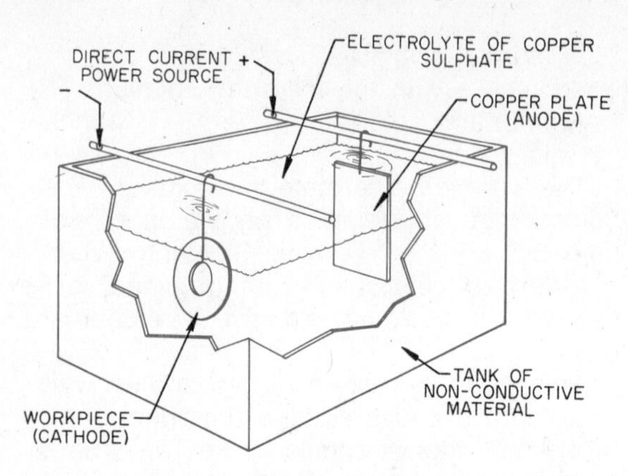

Fig. 35-10. Typical Copperplating Setup

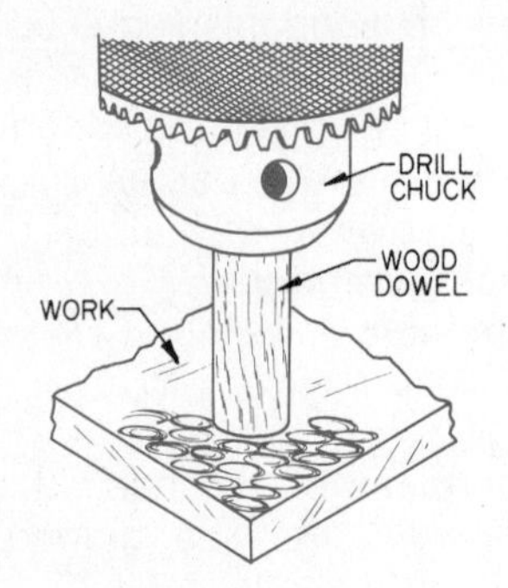

Fig. 35-11. Spot Finishing

on dull surfaces. Surfaces freshly polished with abrasive cloth are clean enough for coloring; however, buffed surfaces must be degreased with an appropriate solvent.

Since as little as 10° F. can cause a change in color, the piece must be uniformly heated if a uniform color is desired. Uniform heating may be done in a furnace or in a pot of molten lead. When using molten lead, goggles should be worn, and both the workpiece and the tools used for handling it must be dry. Any moisture introduced into the liquid lead will vaporize so fast that it will cause an explosion.

During heating, inspect the workpiece frequently. As soon as the desired color appears, withdraw it from the heat source. If overheated, the workpiece will have to be cooled, repolished, and reheated. Colored surfaces should be protected with a coat of oil or sealed with a coat of lacquer. Surface preparation for chemical coloring is the same as for electroplating. See Section 35-11.

35-10. Electrochemical Finishing

Electroplating, or plating for short, is coating an object with a thin layer of metal by **electro-deposition.** Metals which are commonly deposited by electroplating are copper, nickel, chromium, tin, zinc, brass, gold, and silver. Figure 35-10 shows a typical electroplating arrangement for electroplating copper. A **direct current** is passed from a pure copper plate (**anode**) to the workpiece (**cathode**) through a **copper sulphate** solution (**electrolyte**). This removes copper from the anode and deposits copper on the cathode. The thickness of the plating depends on how strong a current is used and how long the workpiece is left in the plating bath.

Electroforming is the name given to the process of making an object entirely by electroplating. Equipment and materials needed for electroplating are inexpensive and readily available. The following is a list of equipment needed:

- 1 earthenware or polyethylene container of 2-5 gallon capacity
- 2 dry cells or one auto battery, battery charger, or eliminator
- 1 small sheet of copper
- 1 gallon copper sulphate solution
- 1 **ammeter**
- 1 **rheostat**

The sheet of copper should be about half again as large in surface area as the article to be plated. The **positive** (+) side of the dry cells or battery should be connected to the sheet of copper. The **negative** (−) side of the dry cells or battery should be connected to the object to be plated. Connect an ammeter and a rheostat in the circuit to measure and control the current.

The current should be approximately 15 amperes per square foot of area being plated. Thus, a test strip 1″ x 6″, having 6 square inches per side or a total of 12 square inches of surface, would have 12/144ths or 1/12th of 15 amperes or 1¼ amperes as correct plating current. The voltage may vary from ¼ to 2 volts, depending on the resistance of the plating circuit, but in this problem it should not exceed 2 volts.

Anodizing is an electrochemical finish commonly applied to aluminum and magnesium. Anodizing provides improved corrosion resistance and surface hardness and is electrically insulating. Some anodized finishes are porous and can be dyed any color. Colored finishes are used on containers such as pitchers and tumblers, sports equipment, appliance trim, hardware, and novelties.

35-11. Surface Preparation for Electroplating

Electroplating will exaggerate defects in the surface finish. Therefore, if a smooth, bright plating is desired, surfaces must first be polished to a mirror finish by buffing. This is followed by solvent cleaning to remove grease, wax, buffing compound, and other **organic contamination.** A thin film of organic soil always remains after solvent degreasing and is removed by soaking in a **hot alkaline cleaner**. After a clean water rinse, all traces of oxidation are removed by dipping in an acid solution, also known as **pickling.** After another clean water rinse, the metal should be ready for plating. Parts should not be allowed to dry between steps in the plating cycle. Plating defects such as blisters, pitting, discoloration, peeling, spotting, and skip plating can result from careless and improper cleaning.

35-12. Spot Finishing

An ornamental finish called **spot finishing** can be produced on flat metal surfaces as follows: cut a piece of wood dowel about 2″ long and of the same diameter as the desired spots. A ⅜″ diameter is probably the most commonly used dowel size. Insert the dowel in a drill press chuck. Put oil and **abrasive flour** on the surface to be spotted. Run the drill press at highest speed and press lightly on the surface. Round, polished spots are thus obtained, Fig. 35-11.

35-13. Burnishing

Burnishing means to make smooth and bright by rubbing with something hard and smooth without removing any metal. Instead, the pressure flattens the points or roughness on the surface. **Roller burnishing** is sometimes used for final sizing and finishing of machined cylindrical and conical parts. **Barrel burnishing** is a finishing process which tumbles the parts in a barrel along with balls, shot, or pins with rounded ends. The peening and rubbing action thus created can produce a finish almost as good as buffing.

Words to Know

abrasive flour
acetic acid
ammonium hydroxide
ammonium sulphate
anode
anodizing
banana oil
bronzing
buffing compound
buffing wheel
burnishing
cathode
cotton flannel
cotton muslin
electroforming
electroplating
enamel
flat lacquer
flat paint
gloss enamel
goblet buffing wheel
household ammonia
hydrochloric acid
lacquer
lime
mineral spirits
negative
oxidizing
phosphoric acid
pickling
positive
potassium sulphate or liver of sulfur
primer
red lead paint
respirator
rheostat
rouge
semigloss enamel
solvent
spot finishing
tallow
thinner
tripoli
zinc chromate

Review Questions

1. What is meant by buffing?
2. Name the different kinds of buffing wheels, and tell when each is used.
3. What buffing compounds are used for polishing aluminum? Brass? Steel?
4. What is the cause of dirty, greasy-looking work that has been buffed? How should it be cleaned?
5. When buffing is done for a long period of time, why should a respirator be used?
6. Tell where the workpiece should contact the buffing wheel for safe buffing.
7. What is the difference between a paint and a primer?
8. Why are leaded paints hazardous?
9. How do enamels differ from paints?
10. Name two advantages lacquers have over paints and enamels.
11. What is bronzing? How is it done?
12. How should surfaces be prepared for coating with paints, enamels, and lacquers?
13. Name three methods of applying paints, and tell where and when each is most likely to be used.
14. Describe how to clean a conventional paintbrush.
15. How are oxide colors produced on metals?
16. What is meant by electroplating? How does it differ from electroforming?
17. What is meant by anodizing?
18. Tell how electroplating is done.
19. Why do metal surfaces have to be so clean for electroplating?
20. Tell how spot finishing is done.
21. What does burnishing mean? How is it done?

Career Information

1. What are some of the hazards involved in electroplating?
2. What precautions should be taken when painting lead-based paints?
3. What precautions should be taken when spraying lacquers?
4. List the steps involved in finishing auto bodies.

UNIT 36

Metal Marking Systems

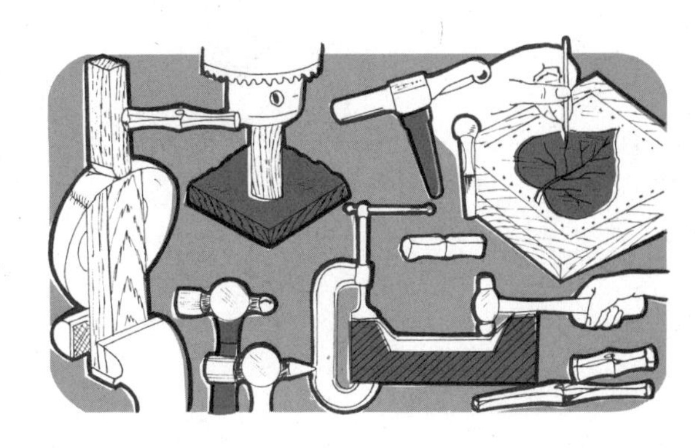

36-1. Need for Metal Marking Systems

The marking of industrial products and materials is essential to many businesses. Machines, cars, appliances, watches, and other products are marked with **serial numbers.** Many of the individual parts of these products are marked with a **part number** and sometimes a **trademark.** Serial numbers and part numbers are recorded by the manufacturer. When replacement parts are ordered, use of serial numbers and part numbers helps assure that the correct parts are supplied. Tools such as **drills, reamers, taps, dies, wrenches,** and **gages** require size and other markings for correct identification and use, Fig. 36-1. Materials made of metal are often marked to tell what kind, grade, or size they are, Fig. 36-2. Each product a student makes should be marked with his name or initials and date. This will make the product more valuable in future years. Indeed, many such articles are valued in families and are handed down from generation to generation.

36-2. Stamping

Stamping is a fast and inexpensive method of marking metal labels and parts. Equipment for stamping may be fully automatic, as when punch presses are used.

Many manually operated machines are also used for stamping. **Roll-type** marking machines use a cylindrical die for marking and are capable of very high production rates.

Steel letters and figures, Fig. 36-3, are made in sets in sizes from 1/64″ to 1″. They are used for custom marking of tools,

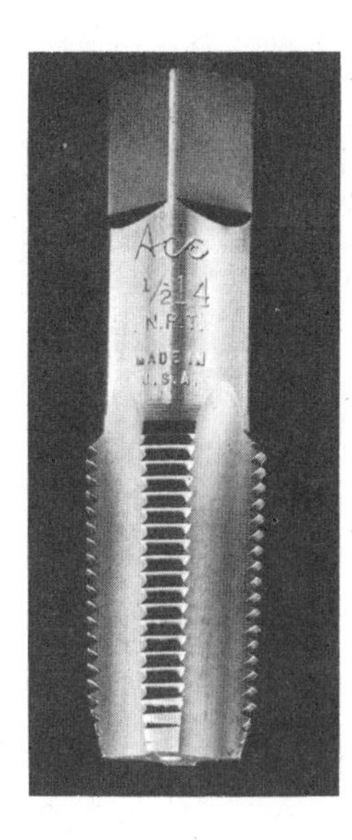

Fig. 36-1. Typical Tool Marking Showing Tool Size and Type, and Manufacturer's Trademark

Fig. 36-2. Markings on Aluminum Bar Stock Tell Size, Alloy Number, and Manufacturer

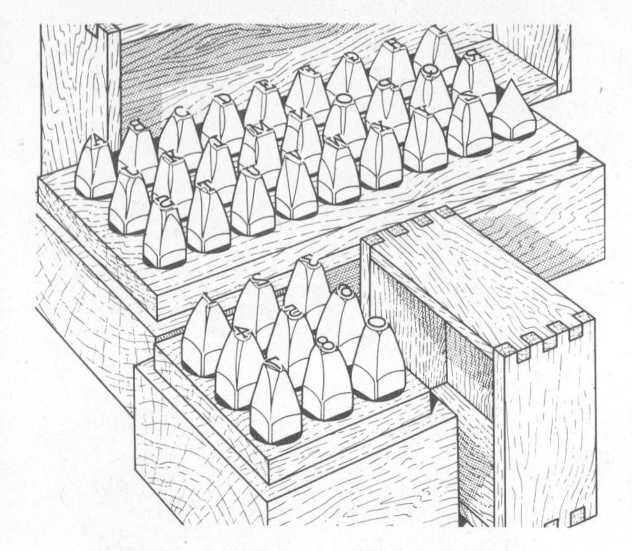

Fig. 36-3. Steel Letters and Figures

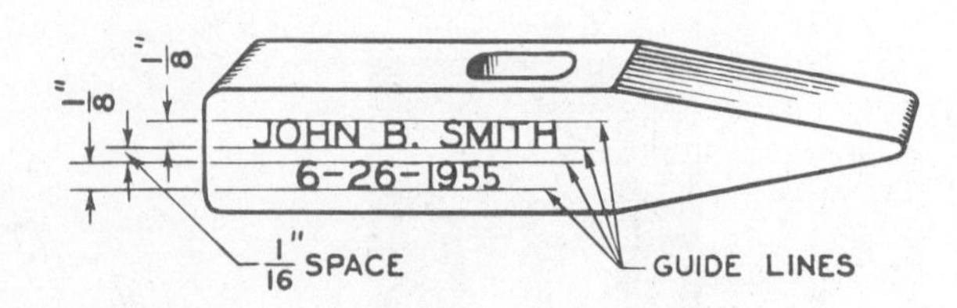

Fig. 36-4. Use Guide Lines When Stamping

labels, or parts. They may be used singly or several at a time in holders, either hand held or in a press. When using them by hand, use guidelines to help do a neat job. Figure 36-4 shows a layout for 1/8″ letters and figures. Draw the guidelines in pencil so as to avoid scratches. Try out the stamp on a metal scrap before stamping the product; unless this is done, the letter or figure is often stamped upside down.

After striking the stamp, one part of the impression is often lighter than the other parts. To correct this, carefully reset the stamp, lean it toward the direction where the stamping is lighter, and strike again. Repeat until the marking is evenly stamped. A center punch may be used for making

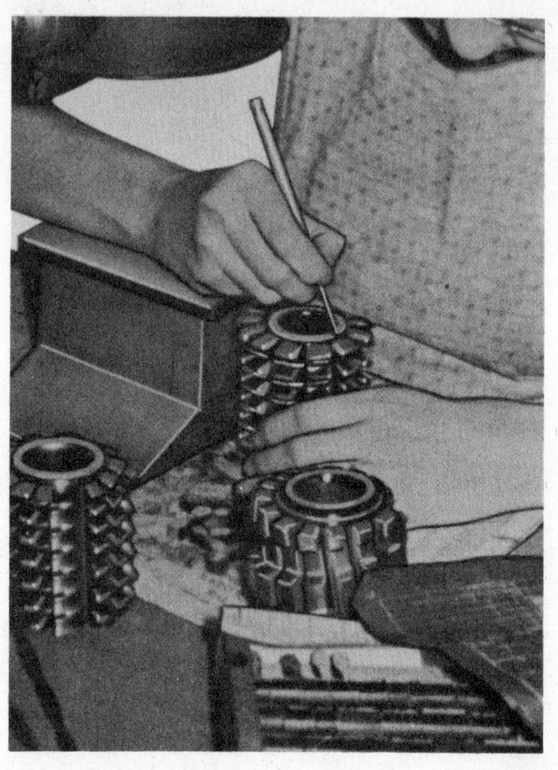

Fig. 36-5. Scratching Lines for Etching

periods, and dashes may be made with the letter "I."

Stamping tools are made of hardened and tempered steel. Therefore, any stamping of steel parts which are to be hardened should be done while the steel is soft.

36-3. Etching

Chemical etching is a way of marking or decorating by using acid to eat into the metal. See **chemical milling,** section 33-10. It is one of the few ways of marking hardened steel. Etching is sometimes used to make attractive designs on metal products. The procedure for etching is as follows:

1. Clean the metal surface.
2. Cover the surface with melted wax or **asphaltum**[1] and allow to dry.

[1]**Asphaltum** is a mineral pitch that is black or brown in color. It is another name for **asphalt.**

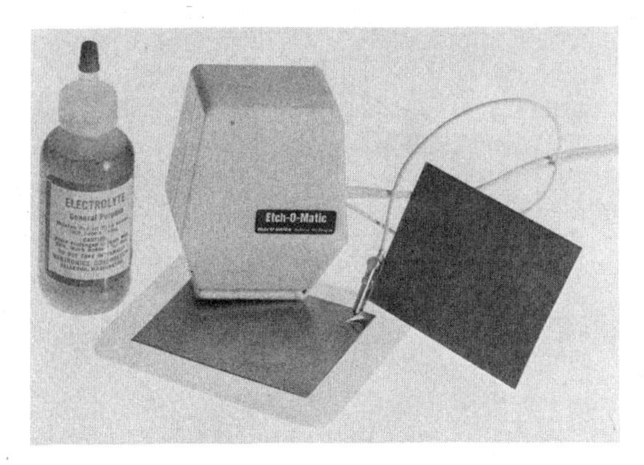

Fig. 36-6. Electro-chemical Marking System

3. Scratch the lines to be etched with a scriber or other sharp pointed tool, Fig. 36-5.
4. Place drops of **hydrochloric acid** or **nitric acid** on the scratched lines with a stick or a glass dropper.

Or, if preferred, add one part of nitric acid to two parts of water in a crock, and dip the whole article into this solution. Be sure to put the water in the crock first, and then the acid; otherwise it will explode. If the solution is too strong, add more water; it is too strong if it bubbles a lot and gives off heavy greenish yellow fumes. These acids are poisonous. They eat holes in cloth and therefore should not be spilled on clothing. The acid eats the metal only where the wax or asphaltum has been removed. After the acid has eaten deep enough, it should be washed off with water. All the remaining wax or asphaltum should then be cleaned off with mineral spirits or kerosene.

Figure 36-6 shows a tool which makes permanent marks on metal of any hardness by **electrochemical etching.** The desired design or message is drawn or typed on a **stencil** which is attached to the working end of the tool. The stencil is directly in contact with a felt pad which is soaked with an **electrolyte.** The lines typed or drawn on the stencil permit the electrolyte to pass through and contact the workpiece. The electrical circuit passing through the tool,

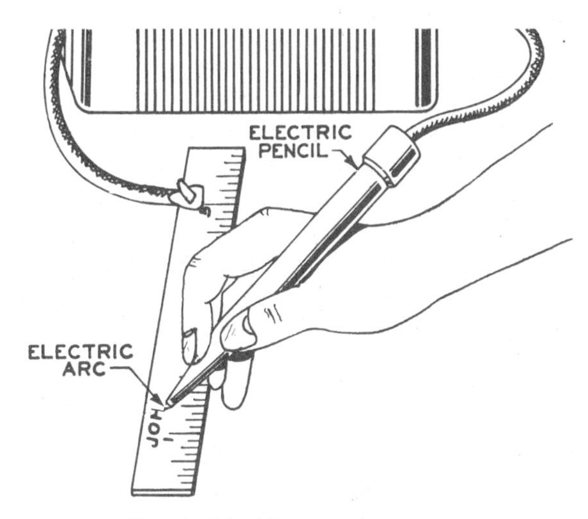

Fig. 36-7. Electrical Etching

electrolyte, and workpiece causes metal to be removed from the workpiece somewhat as in **electroplating;** see section 35-10. When alternating current is used, a very thin marking is obtained; deeper marks are made by using a direct current attachment.

Electrical etching is done with an **electric pencil** that has a **carbon point.** It is powered either by battery or with alternating current, Fig. 36-7. This system is especially useful for marking hardened steel.

36-4. Engraving

Hand-held electrically powered **engraving tools** have a pointed **carbide** tip which can cut the hardest of metals, Fig. 36-8. Cutting is accomplished by causing the tip to vibrate rapidly. The rate of vibration can be controlled by the operator to suit the hardness of the metal being engraved.

Precision engraving employs **pantograph arm engraving machines** and revolving cutters. The pantograph arrangement permits several sizes of engraved letters, figures, etc., to be made from one size of **master type.** Precision engraving is used for making signs and nameplates, graduation marks on measuring tools and

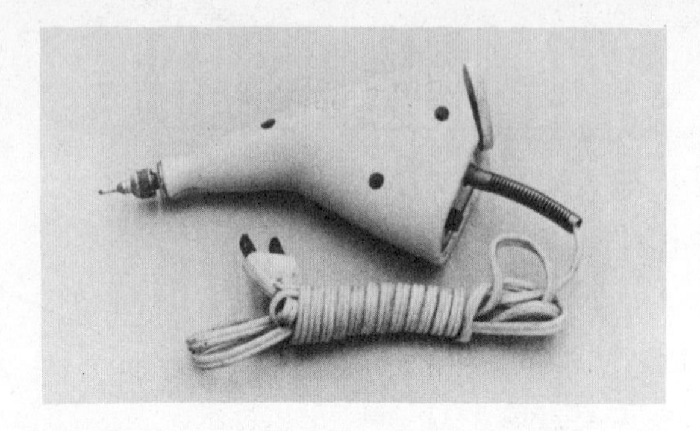

Fig. 36-8. Electrically Powered Engraving Tool

dials, in making molds for injection and compression molding, and die casting. Hardened steels are engraved with carbide or diamond cutters.

36-5. Printing

Temporary, removable markings are made on metal by printing with **rubber plates** or **type** or by **stenciling.** Rubber plates are made to customer order and are either made into hand stamps, or are used on printing presses. Rubber type is assembled into strips and is also used in hand stamping or on printing presses.

Stenciling calls for cutting the design or message through a thin piece of metal or stiff cardboard. The stencil is laid over the surface to be marked, and a **stencil brush** containing the ink is rubbed over the stencil, transferring the message through the stencil to the workpiece.

Stenciling is also done with hand stamps. As pressure is applied to the hand imprinter, ink is forced through the stencil from a pad mounted behind it, thus printing the message. This same principle is used in printing machines employing stencils.

Words to Know

asphaltum
automatic numbering machine
carbide
carbon point
chemical etching
electric pencil
electrical etching
electrochemical etching
engraving
hydrochloric acid
master type
nitric acid
part number
precision engraving
roll-type marking machine
rubber plate
rubber type
serial number
stamping
steel letters and figures
stencil
stencil brush
trademark

Review Questions

1. Why are metal marking systems necessary?
2. Describe several methods of marking metal with metal dies or stamps.
3. What is an automatic numbering machine?
4. Why can't hardened steel be marked with metal dies or stamps?
5. To assure a neat job, what precautions should be taken when hand stamping with steel letters and figures?
6. Describe how chemical etching is done, naming the materials and procedure used.
7. What hazards are associated with chemical etching?
8. How does electrochemical etching differ from chemical etching? What other electrochemical process is similar to electrochemical etching?
9. How is electrical etching done?
10. Describe how precision engraving is done and tell the kind of work for which it is used.
11. How are rubber plates and rubber type used in marking metal?
12. What is stenciling? Name three methods of using stencils for marking.

Career Information

1. List several occupations calling for a knowledge of metal marking systems. Select one, and make a detailed written or oral report about it.

PART XI

Hot Metal Forming Processes

UNIT 37

Forging and Bending

37-1. What Is Forging?

Forging is the oldest of the metalworking processes. The process consists of hammering or pressing the metal into the desired shape, either with or without the use of dies. It may be done hot or cold, but the term "forging" is usually understood to mean hot forging. Production of forged parts is now done almost entirely with machines, hand forging being limited largely to repair work and the production of custom parts.

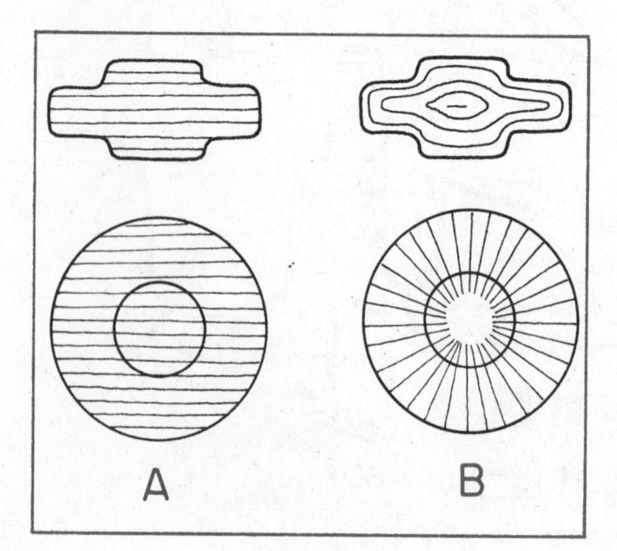

Fig. 37-1. Comparison of Grain Pattern in
A. Machined
B. Forged Part

37-2. Advantages and Disadvantages of Forging

When metal is hot, it is in a soft, **plastic** (pliable) state and is easily formed under pressure without breaking. Other advantages include the following:

1. Forged parts are stronger than machined parts of the same material. Machining cuts through the grain, whereas forging causes the grain to follow the shape of the workpiece, Fig. 37-1. Metal is strongest in the direction of grain flow.
2. Strong parts of complex shape can be produced much more economically than by machining.
3. Since shape is produced by hammering, not cutting, much less metal is lost in the process.

As for disadvantages:

1. The high forging temperatures cause rapid **oxidation,** producing a **surface scale** which results in a poor finish.
2. Because of scaling, close tolerances cannot be obtained.
3. Care must be taken to prevent contact with the metal and to avoid flying scale.

37-3. Industrial Forging Processes

Several forms of forging have been developed, making it economical to forge one piece or thousands of interchangeable parts. They are as follows:

1. Drop forging
2. Press forging
3. Hammer or smith forging
4. Upset forging
5. Roll forging
6. Swaging

Drop Forging. Drop forging is a mass production technique which hammers the metal between closed dies, Fig. 37-2. **Steam hammers** and **board hammers** are used for drop forging.

Half of the die is attached to the hammer and half to the anvil. The hot metal is placed in the lower half of the die and struck one or more times with the upper die. This forces the metal to flow in all directions, filling the die cavity. Excess metal squeezed out between the die faces is called **flash.** After the forging is completed, the flash is cut off in another press with a trimming die. Drop forging is only suitable for producing small and medium size objects, such as pliers and wrenches, gear blanks, machine parts, some fasteners, and engine parts such as connecting rods.

Press Forging. In press forging, a slow squeezing action is used to form the metal. The slow squeezing action penetrates the

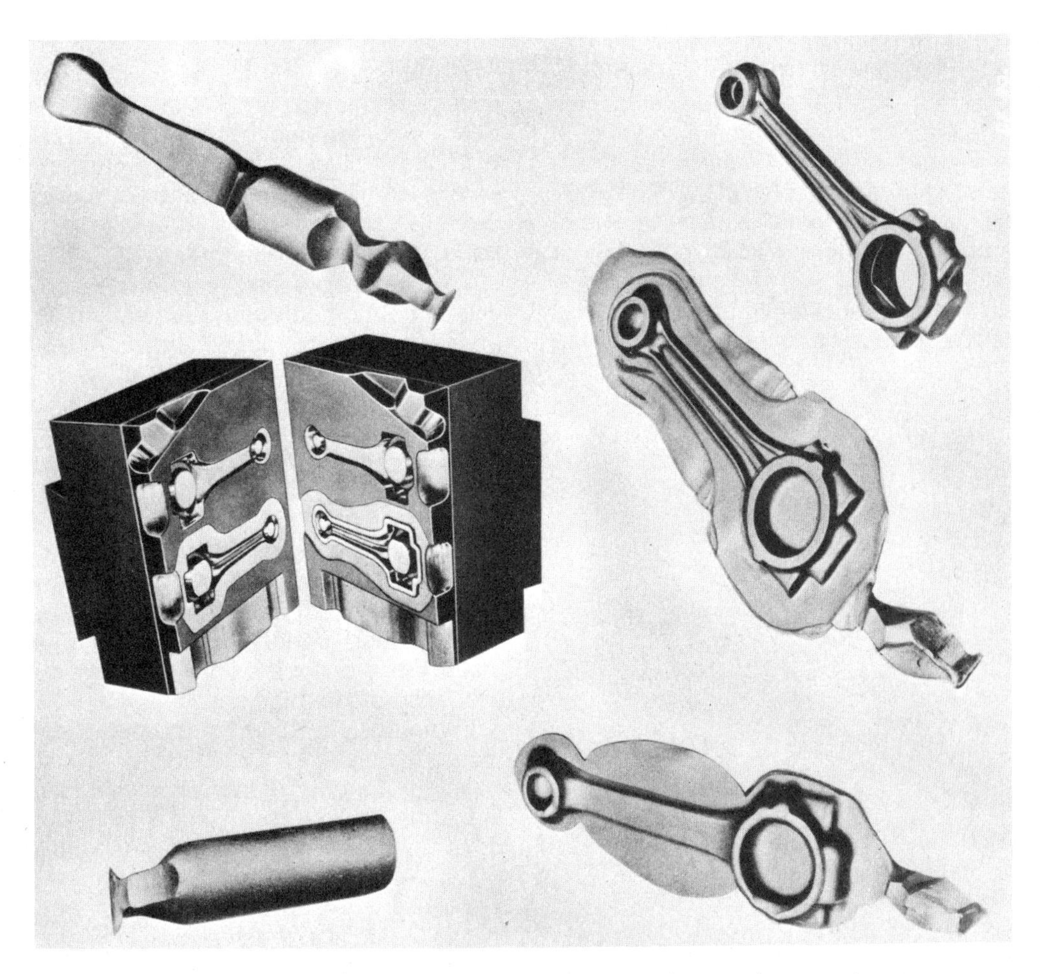

Fig. 37-2. Forging Dies and Product Resulting from Each Impression

entire workpiece, making possible the forging of large objects. Press forges are made in sizes exceeding 50,000 tons capacity. Those up to 10,000 tons capacity may be either mechanically or hydraulically operated, but the larger presses are hydraulically powered.

Press forging may use either open or closed dies. **Open dies** are often simply flat surfaces between which the metal is squeezed, but they may also be V-shaped or slightly convex. Small parts are often forged in closed dies in one stroke. Larger parts may require one or more additional strokes. Large objects such as railroad wheels, main spars for aircraft wings, and aircraft landing gear parts are press forged.

Hammer or Smith Forging. This type of forging is the same as is done by the **blacksmith,** except that the forging power comes from a steam or air hammer instead of a hammer swung by hand. Open frame steam hammers, with a capacity of up to 5,000 pounds, or double frame steam hammers, with a capacity of up to 25,000 pounds, are used.

Both the anvil and hammer are flat, and the desired shape is obtained by turning the workpiece between hammer blows. Accessories may be used for punching holes, cutting off, and for producing cylindrical shapes. Close accuracy cannot be obtained and only parts of relatively simple shape can be made. Forgings of up to 200,000 pounds are made in this manner.

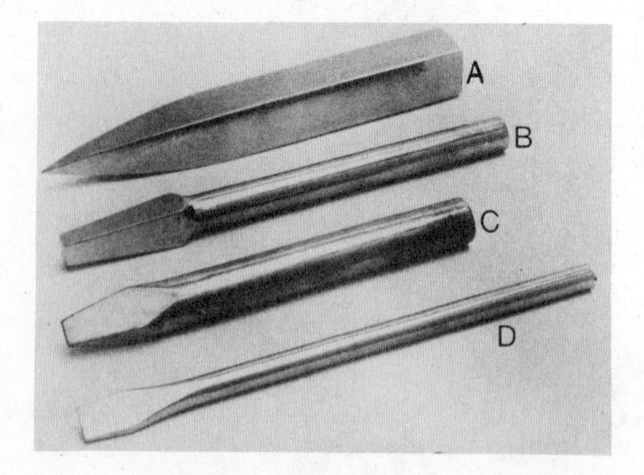

Fig. 37-3. Shapes Produced by Swaging
A. Harrow Spike
B. Auger Bit Shank
C. Soldering Iron Tip
D. Screwdriver Blade

Upset Forging. Upset forging, also called hot heading, is a process by which the cross-sectional size of a bar is increased, either at an end or at some point along its length. It is done on specially designed upsetting machines, using closed dies to control size and shape.

Typically, dies have several stations, and the part is formed progressively by moving the part from one die station or cavity to another until forging is complete.

Upset forging machines are made in several sizes, the largest capable of handling bars ten inches in diameter. Heads of bolts, valves, single and cluster gear blanks, artillery shells, and cylinders for aircraft radial engines are examples of parts made by upset forging.

This same process, when done cold, is called **cold heading.** Cold heading makes possible the economical mass production of fasteners such as nails of all types, machine and wood screws, bolts, rivets, hinge pins, etc.

Roll Forging. The purpose of roll forging is to reduce or taper the diameter or thickness of a bar, thereby increasing its length. It is done on a special machine which uses cylindrical dies with grooves of desired shape and size. The process is used for making such parts as leaf springs, axles, shafts, levers, tapered tubing, and aircraft propeller blade blanks.

Swaging. Swaging is a unique forging method which is widely used for sizing, pointing, tapering, and otherwise shaping the ends of rod or tubing, Fig. 37-3. It is done in machines like the one shown in Fig. 37-4.

In **rotary die swaging,** the forming dies in the center of the machine spindle, Fig. 37-5, are backed by blocks of metal which function as hammers. As the spindle carry-

Fig. 37-4. Swaging Machine

ing the dies and hammers revolves, the hammers are actuated by rollers surrounding the spindle. This results in a series of rapid blows to the dies, quickly shaping the workpiece.

Stationary die swaging operates in a similar manner, except that the spindle carrying the dies and hammers remains stationary while the rollers are rotated. This enables parts of other-than-round cross sections to be shaped.

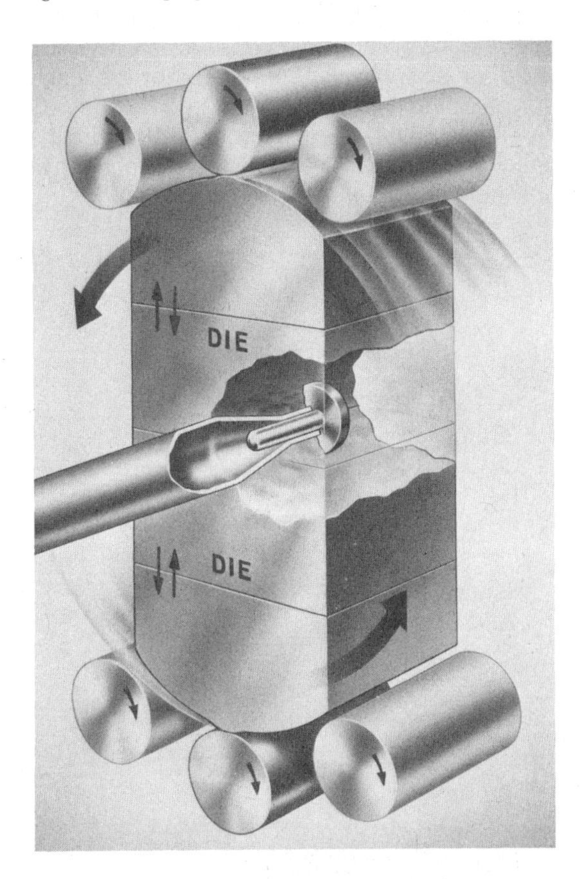

Fig. 37-5. Cutaway of Swaging Machine Die Area, Showing Dies, Hammer Blocks, and Rolls

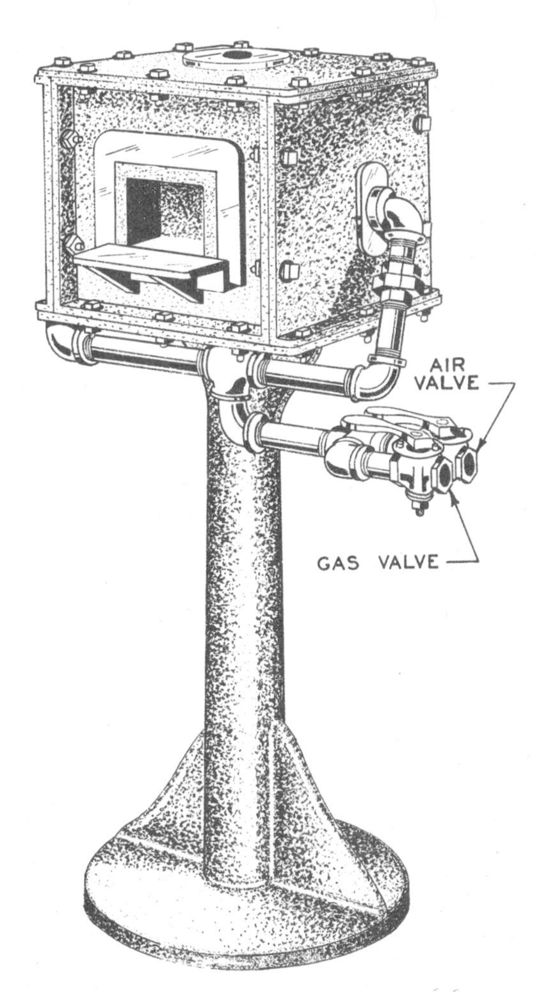

Fig. 37-6. Gas Furnace

37-4. Hand Forging

Hand forging is now done chiefly by service workers who maintain production equipment, tools, and machines. It is also done occasionally in the construction of product **pilot models** or **prototypes.**

37-5. Equipment for Hand Forging

Metal to be forged is usually heated either in a gas furnace, Fig. 37-6, or in a gas forge. Care must be taken when lighting this equipment. Automatic lighting systems with electrical ignition are safest.

Lighting procedure generally calls for the following steps:

1. Switch on the blower motor (but keep the air valve closed).
2. Open the gas valve partially.
3. Immediately ignite the gas either electrically or with a torch made of a rolled-up paper towel placed in the combustion chamber.
4. Open the air and gas valves as far as necessary to obtain a clean-burning flame of the desired size.

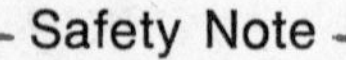

Safety Note

Under no circumstances should gas be allowed to accumulate in a furnace chamber before it is ignited, since an explosion will result.

5. When turning the forge off, always shut off the gas first, then the air.

In hand forging, metal is hammered to shape on an anvil, Fig. 37-7. The anvil body is made of soft steel, but the face is made of hardened steel and welded to the body.

The **horn** is shaped like a cone and is tough and unhardened. Rings, hooks, and other curved parts are formed on it.

The **cutting block** is between the face and the horn. Its surface is not hardened. Metal may be cut or chipped upon it with a cold chisel; the face should not be used for chipping.

The **hardy hole** is a square hole in the face of the anvil. Various tools are held in the hole for different kinds of work.

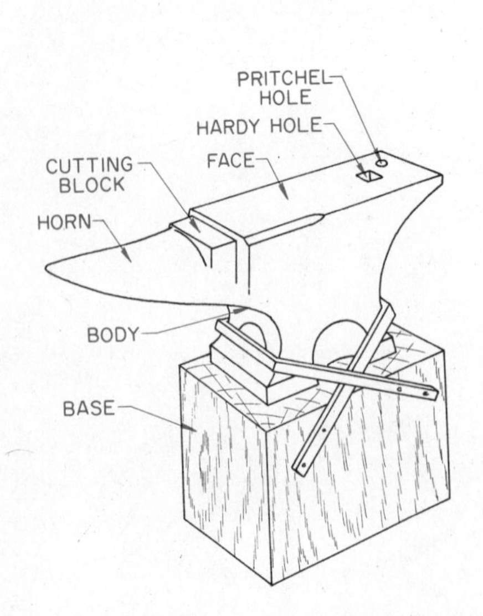

Fig. 37-7. Anvil

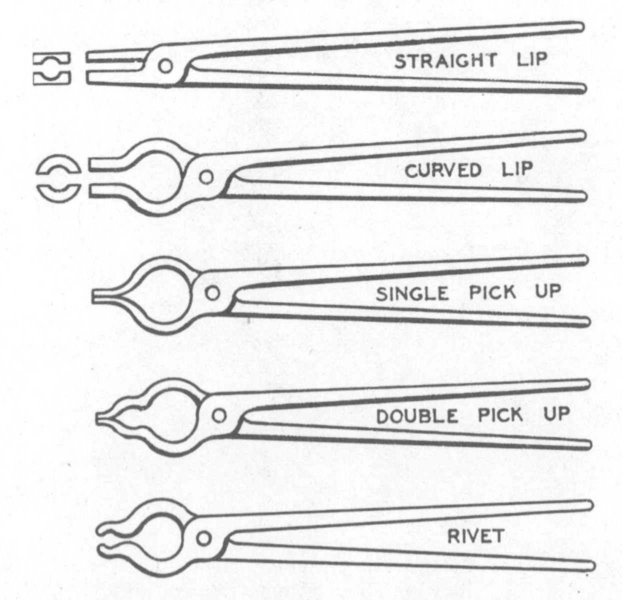

Fig. 37-8. Tongs

The **pritchel hole** is the small, round hole in the face of the anvil. It is used for bending small rods and punching holes in metal.

Anvils weigh from a few pounds to as much as 300 pounds. One weighing between 100 and 200 pounds is suitable for school use. It should be fastened securely to a stable wood or metal base which will position the face of the anvil about 30″ from the floor.

Tongs are used to hold and handle the hot metal, Fig. 37-8. Some of the types of tongs are listed below:

1. **Straight-lip tong,** also called **flat-jawed tong** — used to hold flat work.
2. **Curved-lip tong,** sometimes called **bolt tong** — used to hold round work, such as bolts or rivets. The opening behind the jaws allows space for the head of a bolt.
3. **Single-pickup tong** — used to pick up either flat work or round work.
4. **Double-pickup tong** — used to pick up either flat work or round work.
5. **Rivet tong** — used to hold square or round work, such as rivets or bolts.

Always use tongs that will grip the work firmly, Fig. 37-9. A ring, or **link,** may be slipped over the handles to hold them together and thus hold the work firmly and relieve the hand of the strain. The job is slowed up if the tongs do not fit the work. There is also danger of the hot work slipping out of the tongs, thereby injuring workers. Never leave the tongs in the fire with the work.

Hammers and sledges with many types of heads are used in hand forging. The hammers used for light forge work are shown in Fig. 34-9.

A **blacksmith hand hammer** is shown in Fig. 37-10. A **sledge** is a large, heavy hammer with a long handle. It is swung with both hands. Sledges weigh from 8 to 20 pounds.

The **set hammer** has a smooth, flat face about 1¼″ square, Fig. 37-11. It is used to make square corners and shoulders by placing it on the work and then striking the other end with a hammer or sledge.

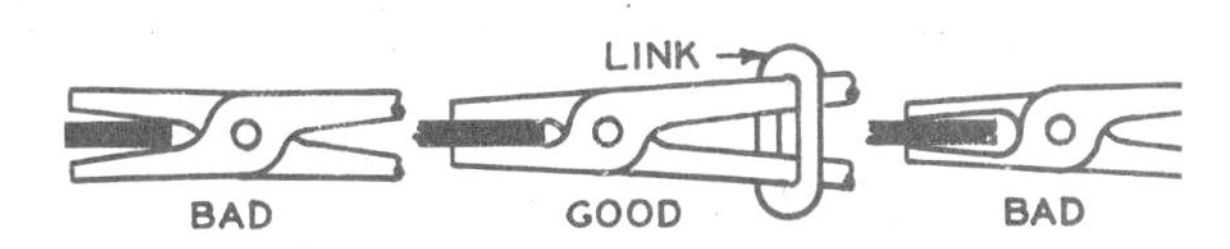

Fig. 37-9. Holding Work with Tongs

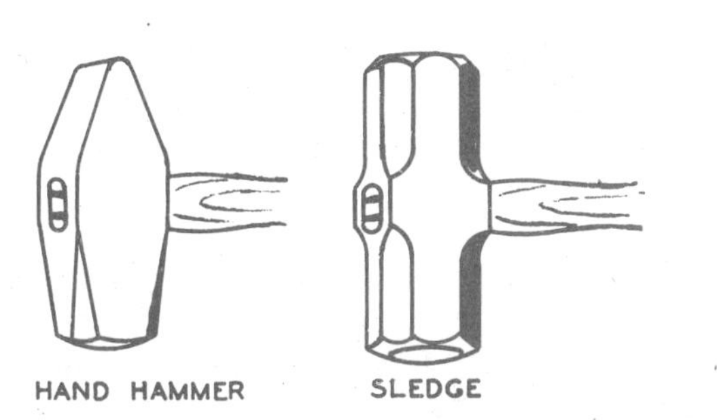

Fig. 37-10. Blacksmith Hand Hammer and Sledge

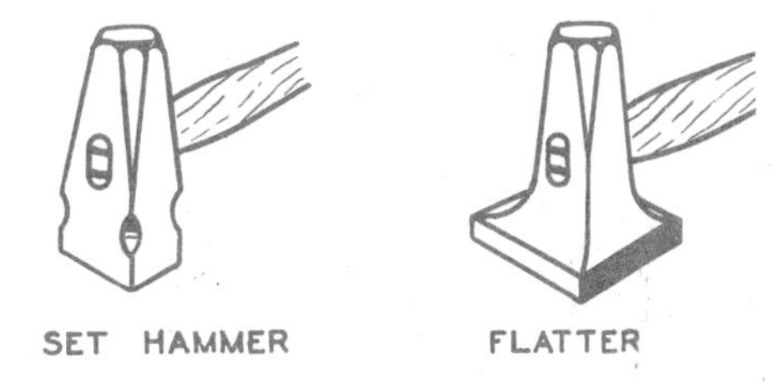

Fig. 37-11. Set Hammer and Flatter

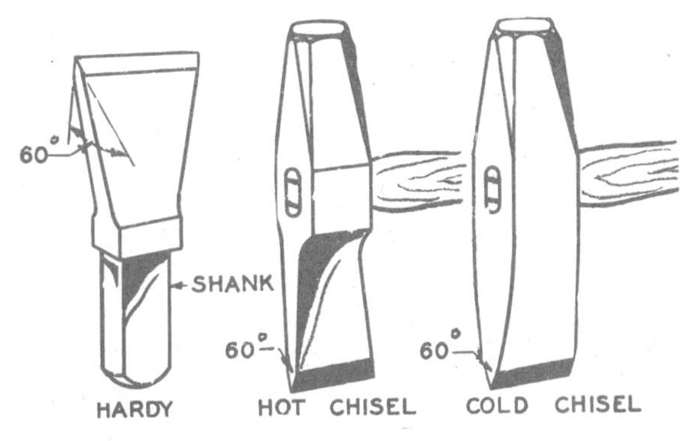

Fig. 37-12. Hardy and Blacksmith Chisels

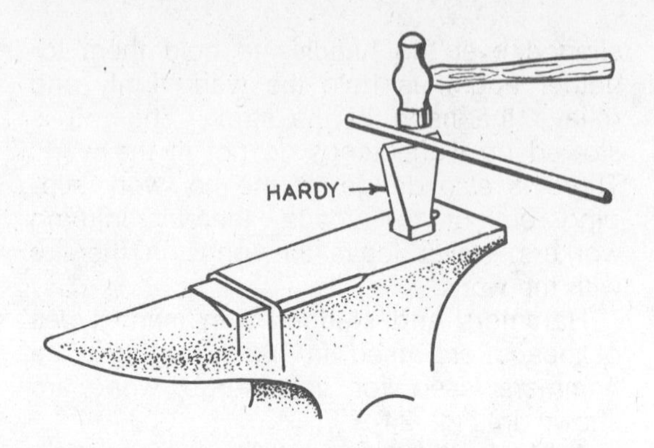

Fig. 37-13. Cutting Metal with a Hardy

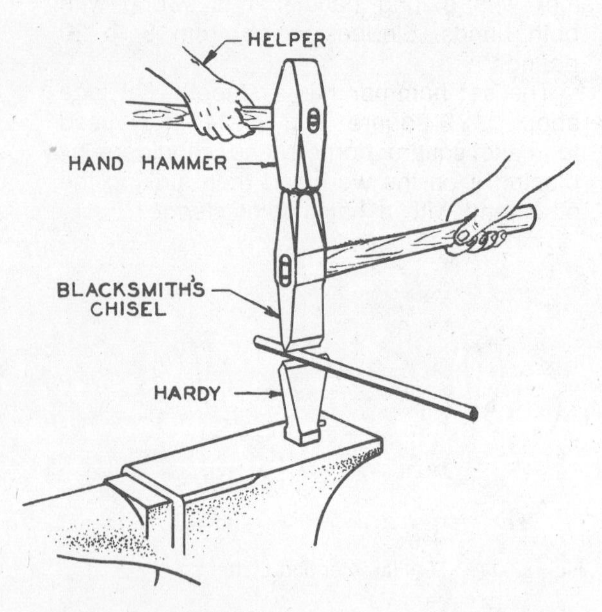

Fig. 37-14. Cutting Metal with a Blacksmith Chisel and Hardy

The **flatter** has a flat, smooth face about 2½″ square with rounded edges. See Fig. 37-11. It is used for the same kind of work as the **set hammer** except that the flatter has a larger face.

The **hardy** is a tool similar to a chisel, Figs. 37-12 and 37-13. It has a square shank and is used to cut hot and cold metal. The square shank is placed in the **hardy hole** of the anvil. The metal to be cut is then laid

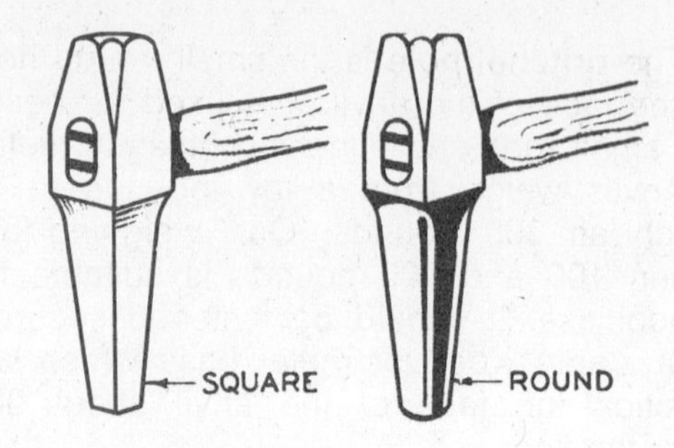

Fig. 37-15. Blacksmith Punches

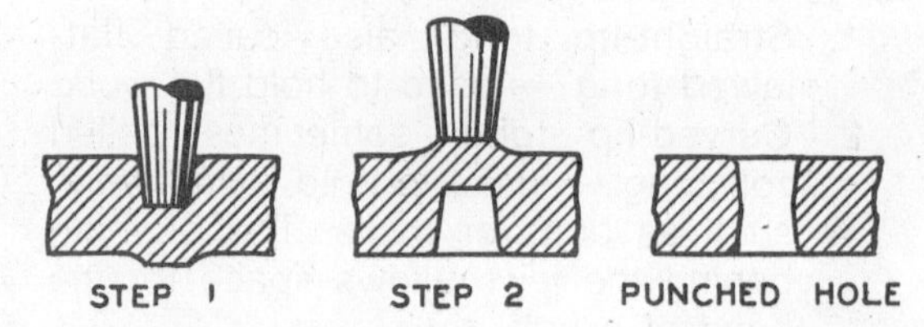

Fig. 37-16. Punching

on the **cutting edge** and struck with a hammer.

Blacksmith chisels (see Fig. 37-12), often called **cutters,** are fitted with handles. There are two kinds: one is used to cut cold metal and is called a **cold chisel,** or **cold cutter;** the other is used to cut hot metal and is called a **hot chisel,** or **hot cutter.** Note that the one used to cut hot metal is much thinner than the one used to cut cold metal; one should not be used in place of the other. Both sides of the metal should be cut or nicked only part-way through with a blacksmith chisel and hardy, Fig. 37-14, and then broken. Thin metal may be cut on the **cutting block** of the anvil. See Fig. 37-7.

Blacksmith punches (Fig. 37-15) are used to punch holes in hot metal. They are made in different sizes and shapes and are tapered.

To use a punch, heat the metal to a bright red color, lay it flat on the anvil, place the punch on the spot where the hole is to be made, and strike a heavy blow with the hammer. Continue striking the punch until

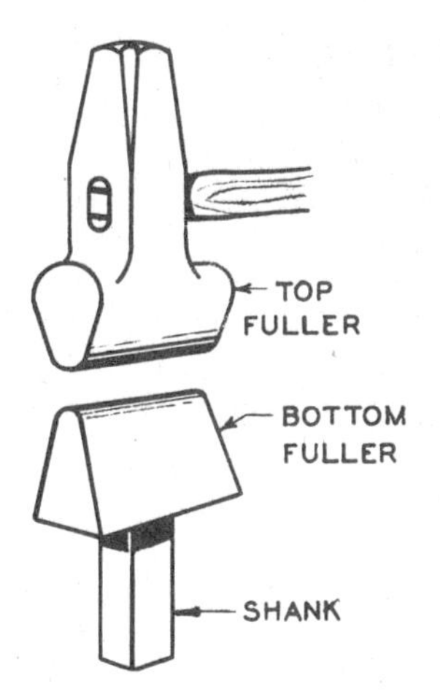

Fig. 37-17. Fullers

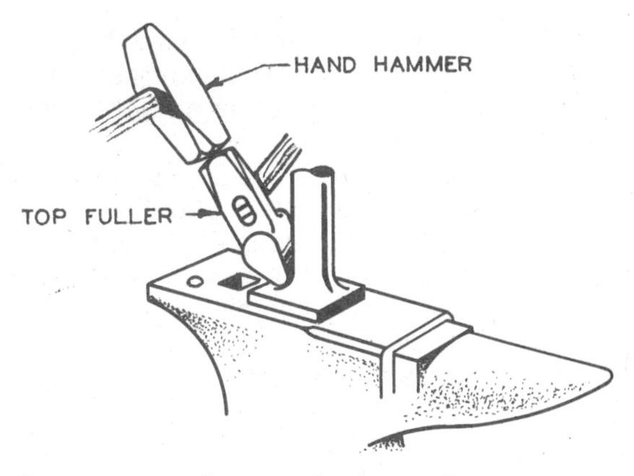

Fig. 37-19. Fullering with a Top Fuller

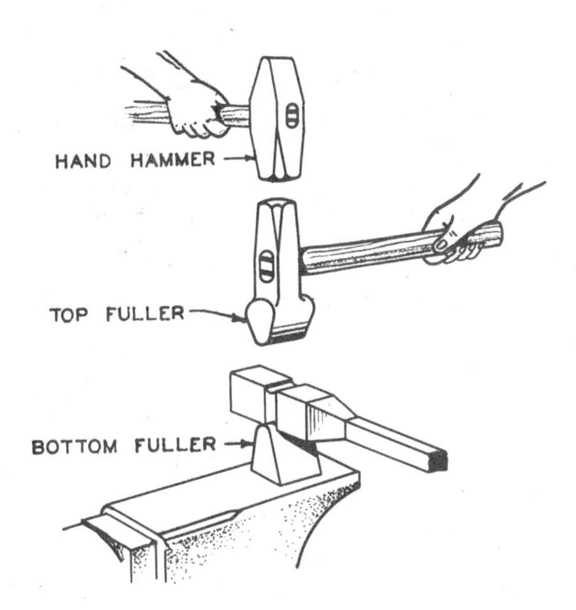

Fig. 37-18. Fullering with Top and Bottom Fullers

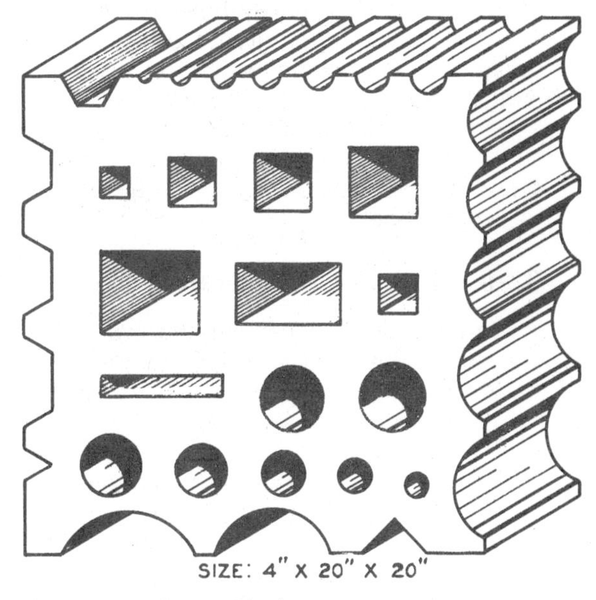

Fig. 37-20. Swage Block

it goes into the metal with difficulty, Fig. 37-16. Quickly remove the punch, cool it, and turn the metal over. Lay it so that the punched part will be exactly over the **pritchel hole.** See Fig. 37-7. Place the punch on the bulge made by punching from the other side and drive it through.

Forming tools of different shapes, used to make grooves or hollows, are called **fullers,** Fig. 37-17. They are often used in pairs. The **bottom fuller** has a square **shank** that fits into the hardy hole in the anvil. The **top fuller** has a handle.

Fullering is the using of fullers. The work is placed on the bottom fuller; then the top fuller is placed on the work and struck with a hammer, Fig. 37-18. The top fuller is also used as in Fig. 37-19 and for stretching or spreading metal just the same as when **peening** with a **cross peen** or **straight peen** hammer.

A swage block, Fig. 37-20, is a heavy block of cast iron or steel about 4″ thick and from 16″ to 20″ square. It has many different grooves and holes which are used to form metal into different shapes. The swage block can be set up in any position and often takes the place of **bottom swages.**

Swages are grooved tools used to smooth or finish round bars or surfaces, Fig. 37-21. They are often used in pairs. The **bottom swage** fits into the **hardy hole** in the anvil. The work is laid in the groove of the bottom swage. The **top swage,** which has a handle, is next placed over the work and struck with a hammer, thus making a smooth, round surface. Each swage is made for a round bar of a certain size. Swages are also made for other shapes.

37-6. Forging Temperature

Forging calls for heating the workpiece well above the **upper critical temperature** but short of a temperature which would produce extreme **grain coarsening** — which, in turn, would weaken the metal. See Fig. 38-7. An experienced worker can tell the approximate temperature of steel by its color. As steel is heated, the first **incandescent** color to appear in daylight is a dark red at about 1,050° F. This is followed by **cherry** or **bright red,** then **orange, yellow,** and **white** at about 2,200° F. At this point it begins to throw off sparks — **forge welding heat.** Table 25 gives incandescent colors of steel and their corresponding temperatures.

Forging should occur with the metal heated between bright red and yellow. If hammered at a lower temperature, **work hardening** can occur and the metal may crack or split. The metal should be heated no longer than necessary so as to minimize the production of surface scale and loss of surface carbon.

37-7. Hand Forging Techniques

Forge welding is the joining of two pieces of metal by making them soft and pasty with heat and then pressing, hammering, or melting them together. Wrought iron and steel containing up to .50% carbon can easily be welded.

Scale must be kept from forming if a good weld is to be made. **Flux** is used for this purpose. There are many kinds of fluxes. Clean **sharp sand** is a good flux for wrought iron; powdered **borax** is good for steel. The flux may be sprinkled with a long-handled

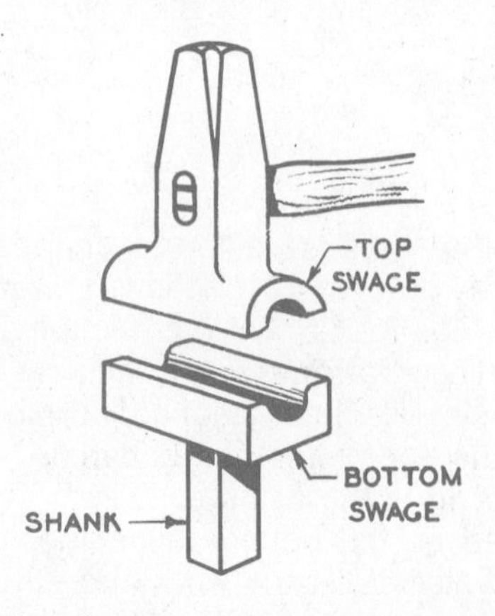

Fig. 37-21. Swages

TABLE 25

Incandescent Colors and Corresponding Temperatures of Steel

Color	Temp. F.
Faint Red	900
Blood Red	1,050
Dark Cherry Red	1,175
Medium Cherry Red	1,250
Cherry or Full Red	1,375
Bright Red	1,550
Salmon	1,650
Orange	1,725
Lemon Yellow	1,825
Light Yellow	1,975
White	2,200

spoon on the place where the weld is to be made. In heating, it melts and keeps the air, which combines with the metal to form the scale, away from the hot metal.

There are different kinds of forged welds. Only the **fagot**[1] **weld** is described here. A fagot weld is made by laying pieces of iron on top of each other and welding them into one piece as follows:

1. Heat two pieces to a bright red and put on the flux.
2. Heat to a **welding heat.** The pieces must be heated evenly so that the inside will be as hot as the outside.
3. Lay them on top of each other on the anvil and quickly strike a few light blows in the center to make them stick. Continue hammering until they are welded together.

The more often the metal is heated, the harder it is to weld. More pieces may be welded on, one at a time. Another way to make a fagot weld is to bend or fold the end of a piece of metal once or twice and weld it into a solid lump, Fig. 37-22.

Upsetting means to thicken or bulge. A bar may be upset by heating the end to a **welding heat** then placing it, hot end down, on the top of the anvil and striking the other end with a hammer, Fig. 37-23. If the bar is long, it may be grasped with both hands (if cool enough) and the end bounced or **rammed** upon the anvil, Fig. 37-24.

Heading means to form a **head** on something, as on a rivet or bolt. It is done with a **heading tool,** Fig. 37-25. This has a hole

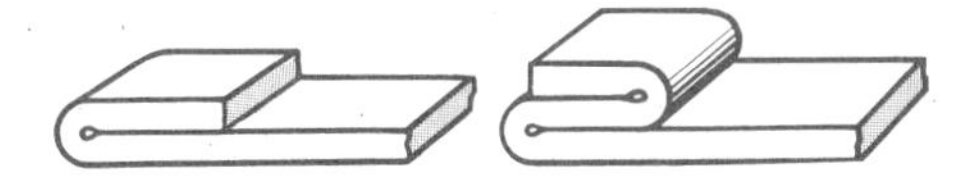

Fig. 37-22. Forged Welds (Fagot Welds)

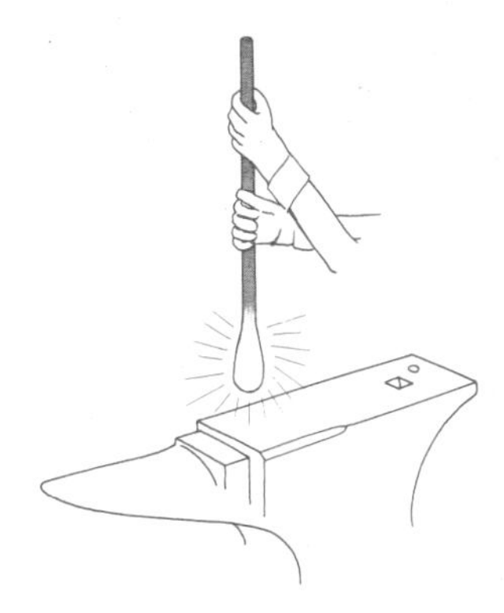

Fig. 37-24. Upsetting by Ramming

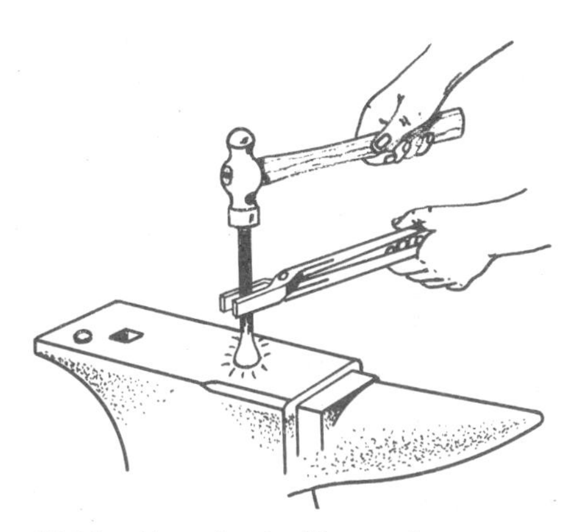

Fig. 37-23. Upsetting by Hammering

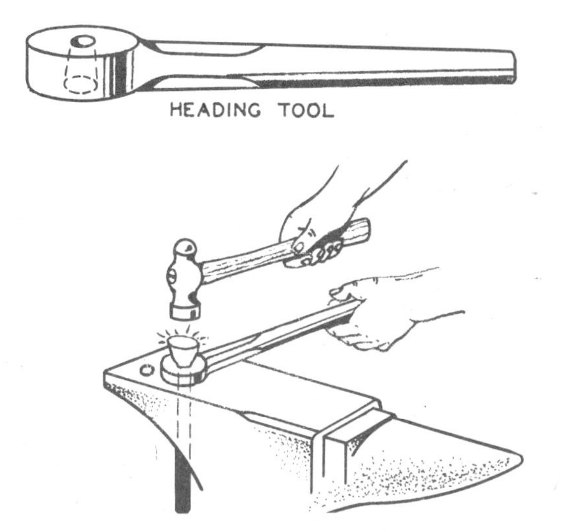

Fig. 37-25. Heading

[1]**Fagot** means to make a bundle; to tie together in a bundle, as a bundle of sticks.

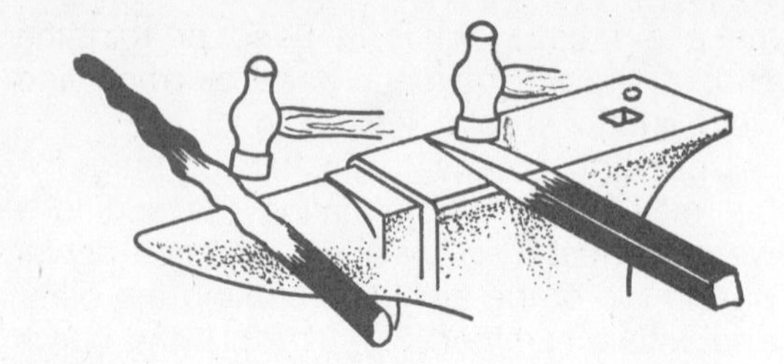

Fig. 37-26. Drawing Out Metal

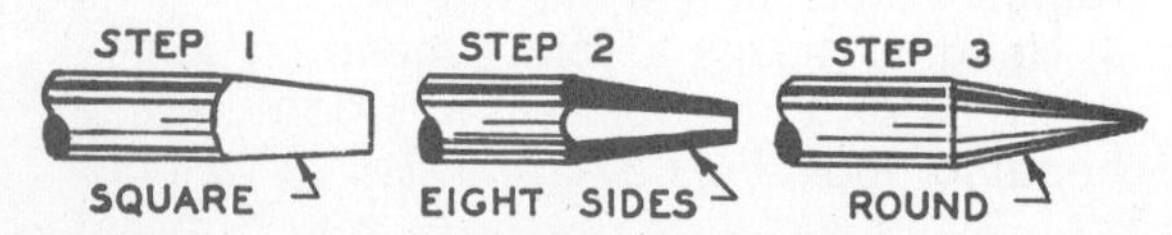

Fig. 37-27. Steps in Drawing a Round Bar to a Point

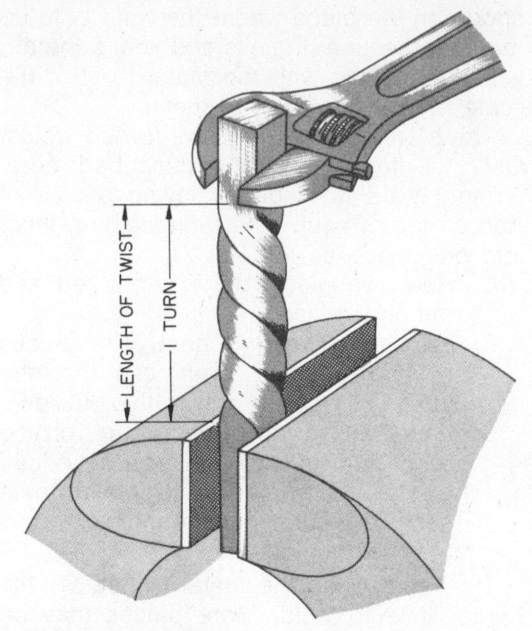

Fig. 37-28. Twisting

which is slightly **tapered.** There should be a heading tool for each size of rod. The hole should be about 1/32″ larger than the rod for rods up to ½″ in diameter and a little larger as the diameter is increased.

The steps for heading are listed below:

1. **Upset** the end of the bar.
2. Slip the bar into the small end of the hole in the heading tool and into the **hardy hole.**
3. Strike heavy blows with the hammer until the head is the right thickness.
4. Finish the sides of the head on the face of the anvil with a hammer.

Drawing out metal means stretching or lengthening it by hammering. The **tapered** part of a **flat cold chisel** is an example of drawing out metal.

First, heat the metal until it is bright red; otherwise, it will tear. To draw or stretch the metal quickly, lay it on the horn of the anvil and strike it with the hammer, Fig. 37-26. This makes a number of notches and makes the piece longer without making it much wider. The notches can then be flattened on the face of the anvil.

When drawing out metal to a round point, as on a **center punch,** it is best to make a small point first and then lengthen it. The point must be hammered only while red hot, otherwise it will tear. It should be drawn out as in Fig. 37-27.

Bending techniques include **twisting** of flat or square bars. This may be done in a vise with an **adjustable wrench,** Fig. 37-28, or a tap wrench. Metal up to ½″ square may be twisted cold, but thicker metal must be heated.

Mark the beginning and end of the twist with a chalkmark or center punch mark. Clamp the bar in the vise with one mark even with the top of the vise jaws. Place the wrench at the other mark. The length of the twist will equal the distance from the wrench to the vise jaws.

To make a long twist, slip a pipe of the length the twist is to be over the metal; then twist. It is thus kept from bending out of shape while twisting.

Offset bends, Fig. 37-29, may be made over the edge of an anvil or in a vise, Fig. 37-30.

Small rods can be bent in the **hardy hole** or **pritchel hole** of the anvil, as shown in

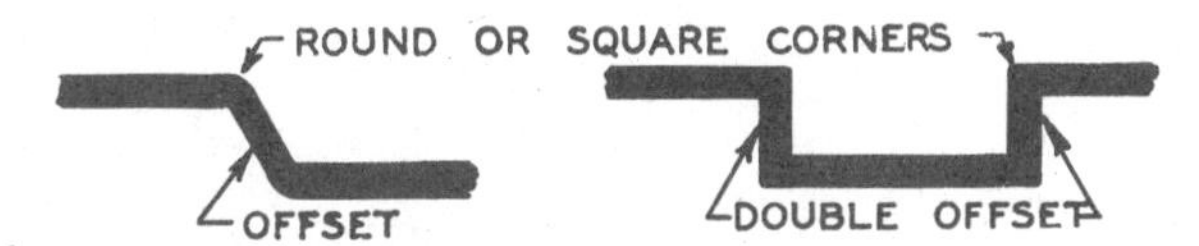

Fig. 37-29. Offsets

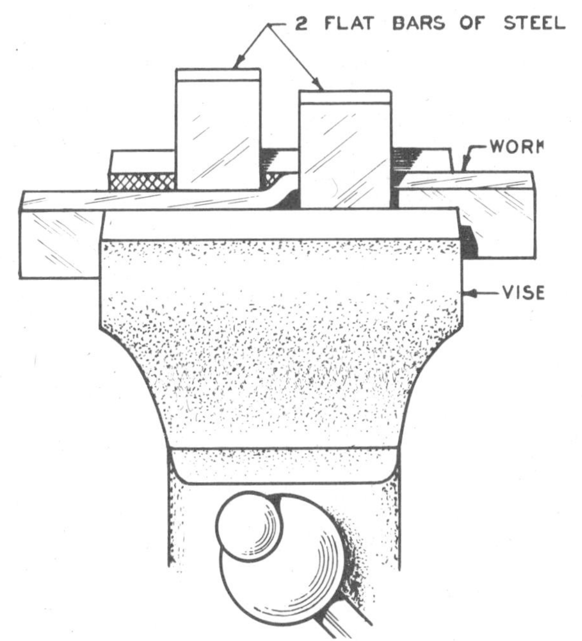

Fig. 37-30. Making an Offset in a Vise

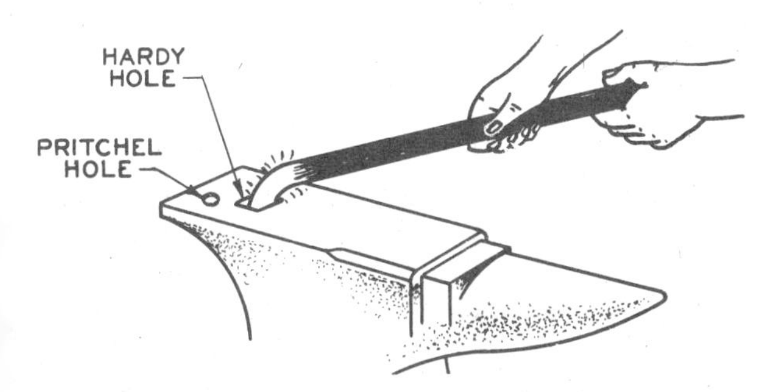

Fig. 37-31. Bending a Rod in the Hardy Hole

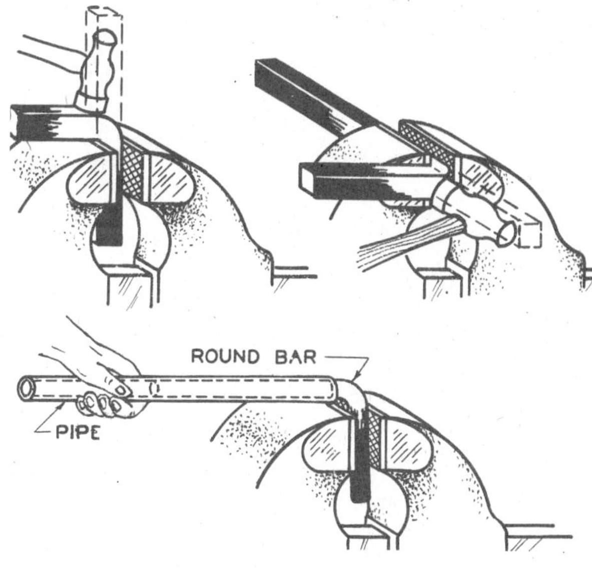

Fig. 37-32. Bending Bars or Rods in a Vise

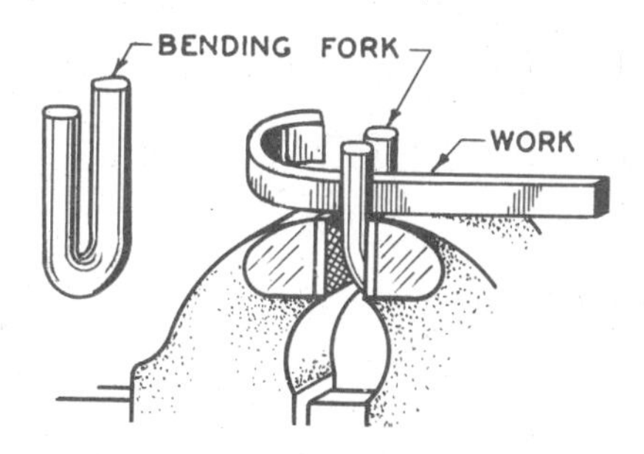

Fig. 37-33. Bending with a Bending Fork Held in a Vise

Fig. 37-31. Bars and rods can often be bent in a vise, Fig. 37-32. A rod may be bent quickly by slipping a piece of pipe over it while the rod is held in a vise and then bending it. See Fig. 37-32.

A good way to make curved bends in a vise is to first bend a 3/4″ or 1″ round rod in the shape of a U. This is called a **bending fork.** Clamp it tightly in the vise and use it for bending as shown in Fig. 37-33.

Universal bending jigs are designed to be held in a vise and are used in the same manner as a bending fork. See Fig. 37-34.

Scrolls may be made by bending the metal strip a little at a time in a bending

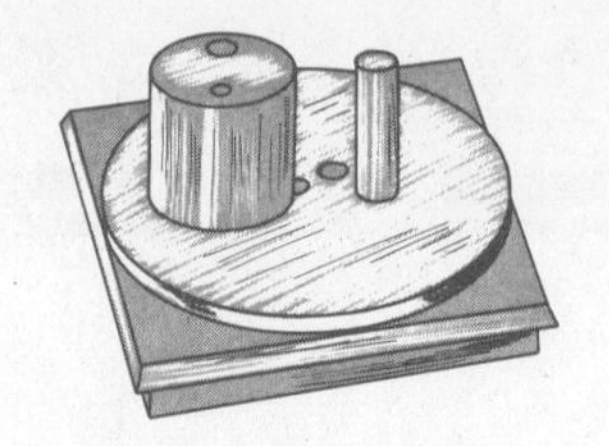

Fig. 37-34. Universal Bending Jig

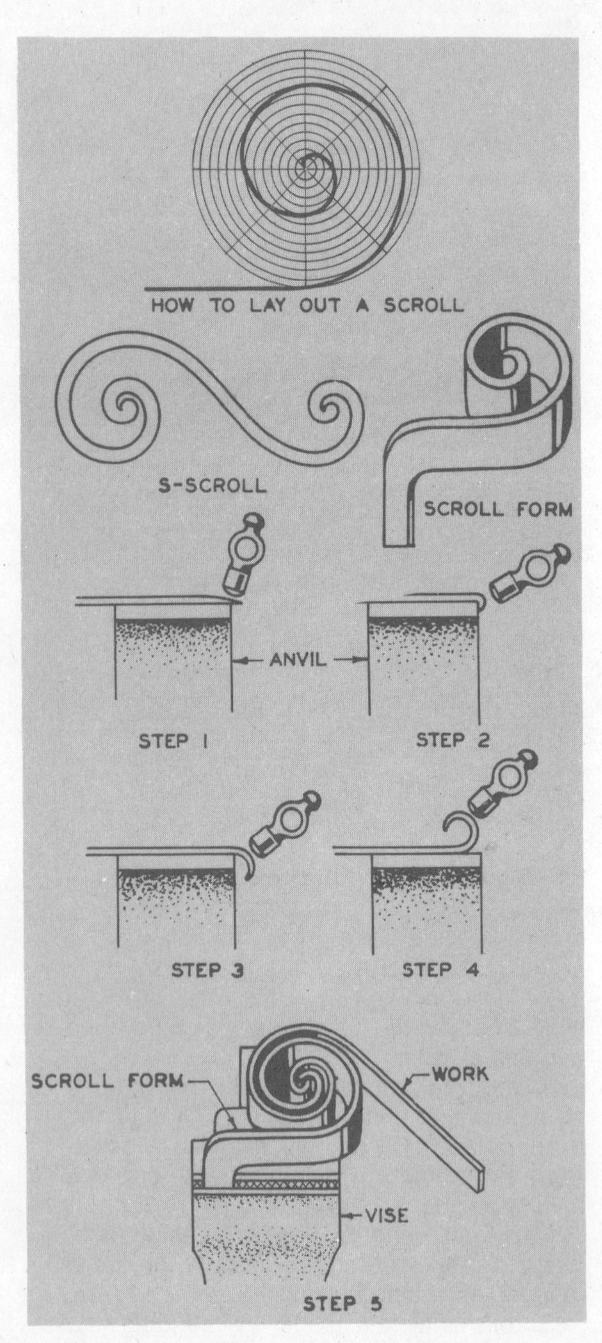

Fig. 37-35. Forming a Scroll

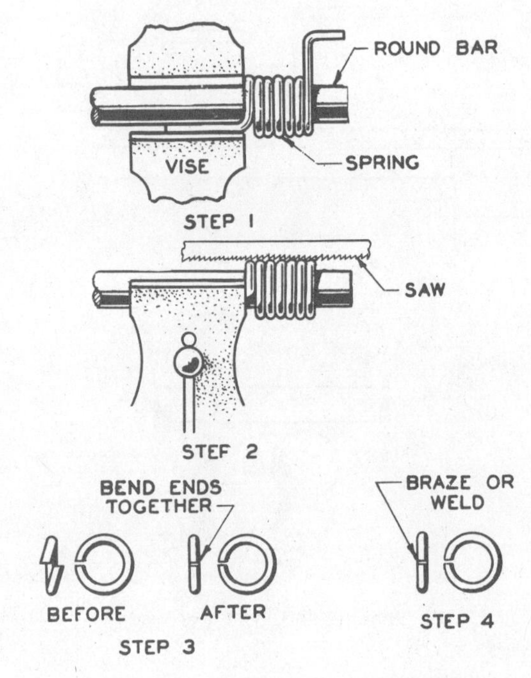

Fig. 37-36. Making Rings from a Coil or Spring

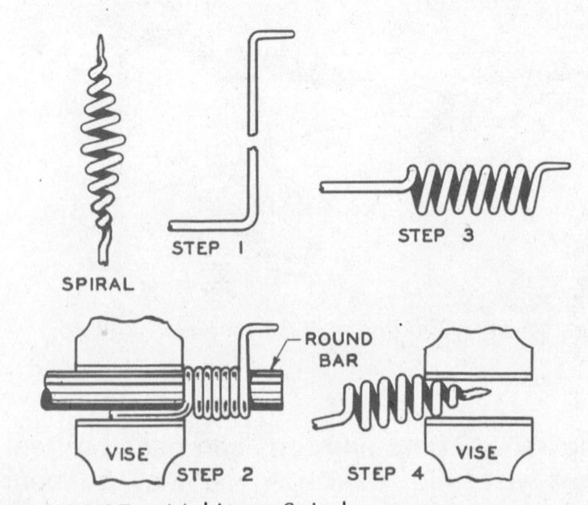

Fig. 37-37. Making a Spiral

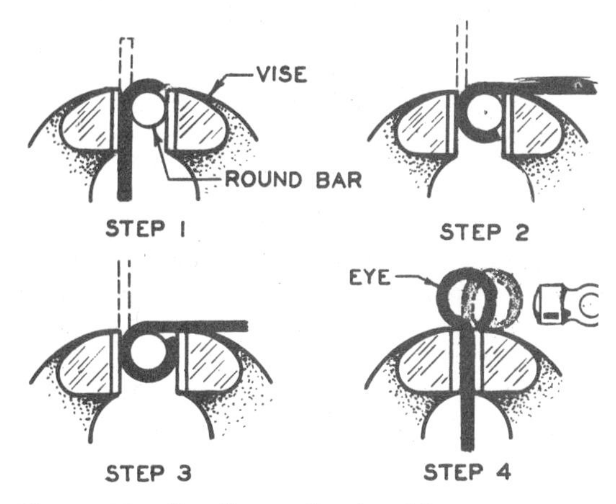

Fig. 37-38. Bending an Eye in a Vise

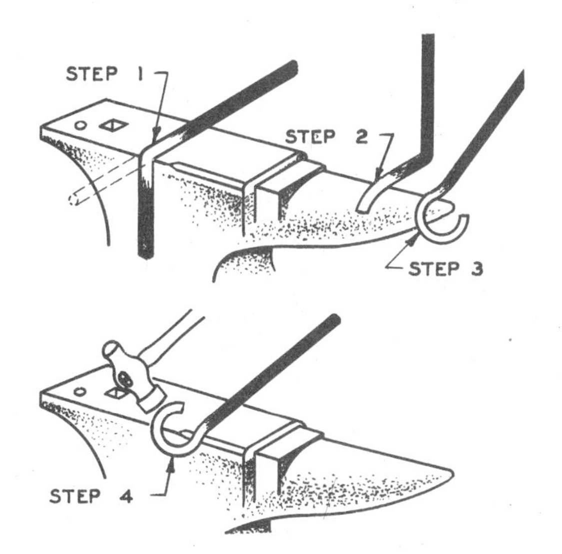

Fig. 37-39. Bending an Eye on an Anvil

Fig. 37-40. Universal Bending Machine

Fig. 37-41. Universal Bending Machine

fork or universal bending jig and comparing the shape with a scroll layout, Fig. 37-35. Use of a **scroll form,** however, is much faster, and when several scrolls of the same size are desired, results are much more uniform, Fig. 37-35.

Rings are quickly made from a coil, as shown in Fig. 37-36. A **spiral** is also made from a coil, Fig. 37-37. Forming the loop for an eye bolt may be done in a vise, Fig. 37-38, or on an anvil, Fig. 37-39.

Universal bending machines (Figs. 37-40 and 37-41), are designed to cold form rod, bar, strip, angle, and tube to any desired shape. The machines are set up quickly and are capable of producing duplicate parts. Benders are available in several hand-operated sizes, and a powered model is also available.

Words to Know

anvil
anvil face
anvil horn
bending fork
blacksmith's chisel
blacksmith's hand hammer
blacksmith's punches
board drop hammer
cold heading
drawing out
drop forging
fagot weld
flatter
forge weld
forging
fullering
gas forge
hammer or smith forging
hand forging
hardy
hardy hole
hot heading
incandescent
press forging
pritchel hole
roll forging
scroll
scroll form
set hammer
sledge
steam hammer
swage block
swaging
tong
universal bending jig
universal bending machine
upset forging
upsetting
work hardening

Review Questions

1. What is forging?
2. List three advantages and three disadvantages of forging.
3. Describe the drop forging process. Name some familiar objects that are made by drop forging.
4. How does press forging differ from drop forging? What kinds of parts are made by press forging?
5. What is meant by hammer or smith forging? What kinds of parts are made this way?
6. Describe how upset forging is done. Name some objects that are made by upset forging.
7. What kind of shaping is done by roll-forming machines?
8. What is meant by swaging and how is it done?
9. Name several common objects made by swaging.
10. How is metal usually heated for forging?
11. Describe a safe lighting and shutoff procedure for a gas furnace or forge.
12. What part of an anvil is made of hardened steel? Why?
13. On what part of an anvil is cutting done?
14. Name the two holes in an anvil and tell how they are used.
15. Which tongs are made especially for holding flat stock? Round stock?
16. For what is a hardy used?
17. For what are set hammers and flatters used?
18. How are blacksmith's chisels different from cold chisels?
19. Describe how to punch a hole with blacksmith's punches.
20. For what purposes are fullering tools used?
21. How does swaging differ from fullering?
22. What is the safe temperature range for forging?
23. What may happen to metal which is forged too cold?
24. Tell how forge welding is done.
25. Describe how to upset and make a square head on the end of a round bar.
26. What is meant by drawing out metal? How is it done?
27. Explain how to twist a flat or square bar.
28. Describe two ways of making a scroll.
29. For what kinds of work are universal bending machines well suited?

Career Information

1. Make a list of industries which use forging in the production of their products.
2. What kinds of careers can you identify that call for forging knowledge and/or skill?
3. What education and training are required to become an operator of a drop forging or other kind of forging press?

UNIT 38

Heat Treatment of Steel

38-1. What Does Heat Treatment of Steel Mean?

Heat-treatment processes involve heating and cooling of metals, in their solid states, for the purpose of changing their **properties.** The common properties of metals are explained in section 19-15. The principal properties of steel which can be changed by heat-treatment processes include hardness, brittleness, toughness, tensile strength, ductility, malleability, machinability, and elasticity.

Steel may be made harder, tougher, stronger, or softer through various kinds of heat-treatment processes. Every tool must have a certain hardness, strength, toughness, brittleness, or **grain** to do its work. All metals have a **crystalline** grain structure while in the solid state. The **grain structure** in steel may be changed in several ways with heat-treatment processes. To bring about the proper grain structure, and thus develop the desired properties, the steel is heated and then cooled in different ways. This is called **heat treating.** The following are the principal kinds of heat-treatment processes:

1. Hardening
2. Tempering
3. Annealing
4. Normalizing
5. Case Hardening
6. Flame Hardening
7. Induction Hardening

All heat-treatment processes involve heating and cooling metal according to a **time-temperature cycle** which includes the following three steps:

1. Heating the metal to a certain temperature.
2. Holding the metal at an elevated temperature for a certain period of time. (This is called soaking.)
3. Cooling the metal at a certain rate.

The procedure used in the above three steps varies for each kind of heat-treatment process.

38-2. Furnaces and Temperature Control

The various heat-treatment processes require that steel parts be heated to certain temperatures. These temperatures must be determined and controlled accurately for best results. The parts generally are heated in furnaces. However, they may also be heated with other sources of heat, such as a portable gas torch, an oxyacetylene welding torch, or even a small bunsen-type burner.

Several kinds of furnaces may be used and are heated by gas, oil, or electricity. Large industrial-type furnaces are shown in Figs. 38-11 and 38-12. Smaller furnaces are used in metallurgical laboratories, small heat-treating shops, and in school metalwork laboratories.

Kinds of Furnaces

An **electric heat-treating furnace,** which heats to temperatures within a range from

300° to 2300° F., is shown in Fig. 38-1. A gas-fired heat-treatment furnace, which heats to temperatures up to 2300° F., is shown in Fig. 38-2. The gas-fired **hardening furnace,** Fig. 38-3, is used for hardening carbon steels and high-speed steels in the temperature range from 1300° to 2350° F. The gas-fired furnace, Fig. 38-3, is a **drawing (tempering) furnace** which is used for tempering in the range from about 400° to 1150° F. A gas-fired, pot-type, liquid-hardening furnace is shown in Fig. 38-4. This kind of furnace heats salt, lead, and cyanide baths for liquid casehardening processes and for other special heat-treating processes. An ordinary **bench-type** gas furnace, such as the kind which is often used for heating soldering coppers (Fig. 28-2) may be used to heat steel for either hardening or tempering. A gas-fired forging furnace, Fig. 37-6, may also be used for heat-treatment processes.

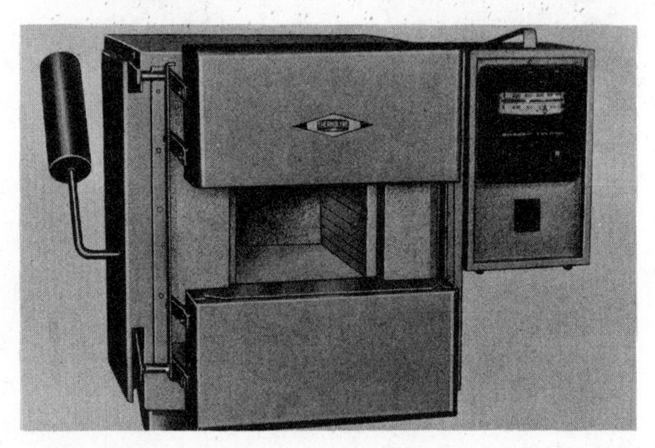

Fig. 38-1. Electric Heat-Treatment Furnace with Temperature-Indicating Control Unit

Fig. 38-2. Gas-Fired Heat-Treatment Furnace Equipped with Indicating Temperature Controls

Temperature Control

For best results, heat-treatment furnaces should be equipped with temperature-indicating and control devices, Fig. 38-5. With controls of this type and similar types, the

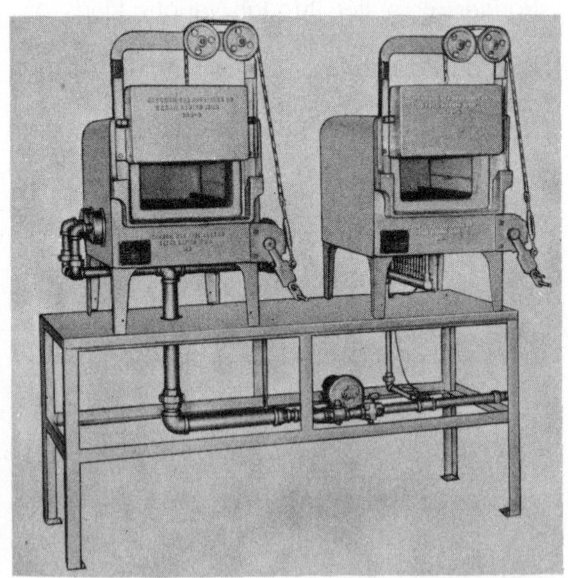

Fig. 38-3. Duo Furnace Unit: (Left) Hardening Furnace; (Right) Drawing or Tempering Furnace

desired temperature is set with a control knob. The furnace then heats to the desired temperature and maintains this temperature within a few degrees. It turns the gas or electric current on or off, intermittently, as necessary to hold a steady temperature. The furnaces in Figs. 38-1 and 38-2 are equipped with heat-indicating and temperature controls.

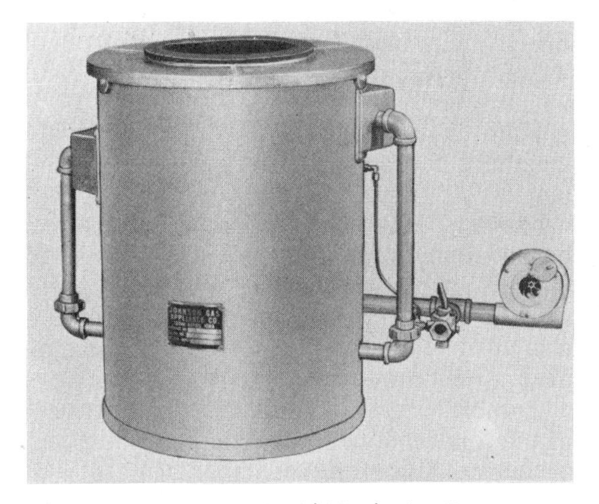

Fig. 38-4. Pot-Type Liquid Hardening Furnace

Fig. 38-5. Temperature-Indicating and Control Equipment for Gas Heat-Treatment Furnace

Temperature Colors

When clean, bright steel is heated, various colors appear at different temperatures, as shown in Table 26 below. If the furnace used for heat treating is not equipped with a temperature-indicating and control device, the temperature can be estimated by observing the color of the steel as it is heated. This is the way old-time blacksmiths determined the temperature of steel for heat-treatment purposes. This method, however, is not very accurate. Without skill and experience, the temperature can easily be off 20° to 30° in the range from 375° to 600° F. In the red-heat range, the temperature can easily be in error several hundred degrees.

Temperature-Indicating Material

An inexpensive way to determine temperatures of heated steel is through the use of temperature-indicating pellets, crayons, or paints, Fig. 38-6. These materials are made to melt at various temperatures from 100° to 2500° F. Simply select the crayon or other material which is designed to melt at the desired temperature. Rub the crayon or other identifying material on the workpiece.

TABLE 26

Typical Tempering Temperatures for Various Tools

Degrees Fahrenheit	Temper Color	Tools
380	Very light yellow	Tools which require maximum hardness: lathe centers and cutting tools for lathes and shapers
425	Light straw	Milling cutters, drills, and reamers
465	Dark straw	Taps, threading dies, punches, dies, and hacksaw blades
490	Yellowish brown	Hammer faces, shear blades, rivet sets, and wood chisels
525	Purple	Center punches and scratch awls
545	Violet	Cold chisels, knives, and axes
590	Pale blue	Screwdrivers, wrenches, and hammers

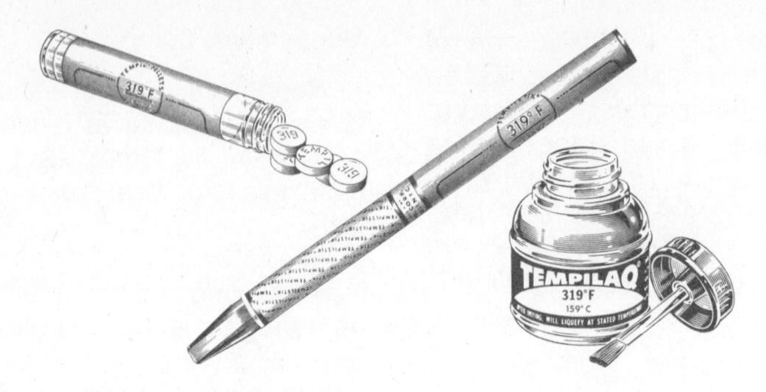

Fig. 38-6. Temperature-Indicating Products: Pellets, Crayons, and Liquid

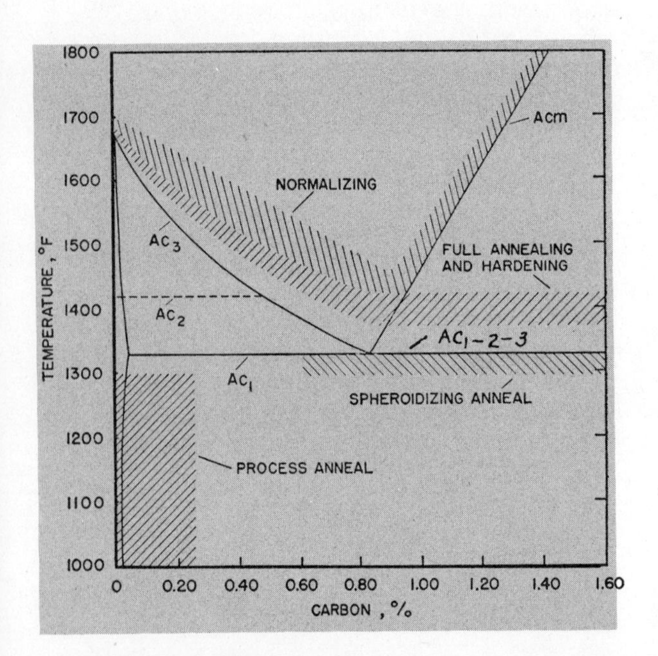

Fig. 38-7. Recommended Temperature Ranges for Heat-Treating Plain Carbon Steels (Adapted from Thomas G. Digges, Samuel J. Rosenberg, and Glenn W. Geil, *Heat-Treatment Properties of Iron and Steel*, National Bureau of Standards Monograph 88. Washington, D.C.: U.S. Government Printing Office, 1966.)

When it is heated to the desired temperature, the identifying material will melt, thus indicating the temperature of the workpiece.

A **magnet** may also be used as an aid in identifying the hardening temperature of steel. See § 38-5.

38-3. Carbon Content Affects Hardening

Plain carbon steel is composed principally of iron and carbon. The carbon content in steel enables the steel to become hardened. Hence, pure iron cannot be hardened by heat treatment. Whether steel is plain-carbon steel or alloy steel, it is the amount of carbon content which largely determines the maximum hardness obtainable by heat treatment.

The plain carbon steels may be classified in the following three general classifications according to carbon content:

Low-carbon steel, 0.05 to 0.30% carbon
Medium-carbon steel, 0.30 to 0.60% carbon
High-carbon steel, 0.60 to 1.50% carbon

The high-carbon steels can be made very hard, brittle, or tough by heat treatment. Some metal-cutting tools, such as drills, milling cutters, taps, and dies, are made of high-carbon steel with carbon content ranging from approximately 0.90 to 1.10%. See Table 10, page 144.

The medium-carbon steels can be made relatively hard by heat treatment. However, they cannot be hardened sufficiently to make drills, taps, threading dies, or similar metal-cutting dies. Some uses of medium-carbon steels are shown in Table 10, page 144.

The hardness of metals can be measured with instruments, as explained in Unit 47.

The Rockwell-C scale may be used to indicate the hardness of hardened steel. The Rockwell-B scale may be used to indicate the hardness of soft steels and nonferrous metals. These hardness values can be converted to equivalent values on other kinds of hardness scales. See Table 30, page 442 and 443.

A comparison of the maximum obtainable hardness for one kind of medium-carbon steel (1045 steel) and one kind of high-carbon steel (1095 steel) is shown in Fig. 38.8. Before tempering, the 1095 steel was hardened to a Rockwell hardness of C-66. The 1045 steel was hardened to Rockwell C-59. A hardness of Rockwell C-60 or higher, after tempering, is generally required for metal-cutting tools such as drills and files.

Low-carbon steels can be hardened only a small amount by direct hardening. However, the skin layer — the thin outside case on these steels — can be hardened by the heat-treatment process called case hardening.

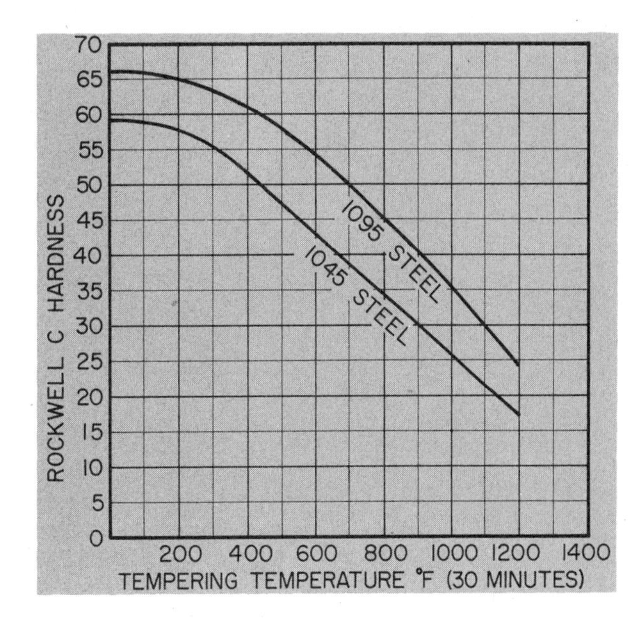

Fig. 38-8. Effect of Various Tempering Temperatures on the Hardness of Carbon Steel

Note: Surface hardness of carbon-steel bars, SAE 1045 steel ¾" square, and SAE 1095 steel ½" diameter, after tempering at various temperatures. Both steels were hardened in water quench, the 1045 steel at 1500° F., and the 1095 steel at 1450° F.

38-4. Hardening

Hardening is a heat-treatment process which makes steel harder. Medium-carbon steels and high-carbon steels are hardened by heating slowly to the proper **hardening temperature** (see Fig. 38-7) and then cooling them rapidly. They are cooled rapidly by **quenching** in water, brine, or oil. Some alloy steels are hardened by using special procedures.

The maximum hardness obtainable by heat treatment depends upon these factors:

1. The amount of carbon in steel.
2. The speed of heating.
3. The temperature at which the steel is quenched for hardening, called the hardening temperature.
4. The speed of cooling.

When high-carbon steel has been hardened, it becomes very brittle due to internal stresses which result from rapid cooling. In fact, it is frequently so brittle that if struck by a hammer it would crack or shatter. Hence, a hardened piece of high-carbon steel requires an additional heat-treatment process, called **tempering,** before it may be used. Hardened steel should be tempered immediately or as soon as possible after hardening. Occasionally, hardened steel will crack if cooled improperly or if allowed to remain hardened but not tempered for a prolonged period.

38-5. Hardening Temperature

The **hardening temperature** is the temperature at which a piece of steel should be quenched for **hardening.** The hardening temperatures for various plain carbon steels are shown in the shaded area on the chart in Fig. 38-7. The hardening temperature is about 50° to 100° F. above the AC_3 line. This represents the **upper-transformation temperature,** also called the **upper-critical temperature.**

The AC_1 line on the chart represents the **lower-transformation temperature,** also called the **lower-critical temperature.** The AC_2 line on the chart represents the **magnetic point.** When heated above the magnetic point, steel loses its magnetic properties and is no longer attracted to a magnet. For steels with more than 0.80% carbon, **note** that lines AC_3 and AC_2 join with line AC_1 on the right-hand portion of the chart, as far as hardening and annealing temperatures are concerned.

When steel is heated through the upper-transformation temperature, the grain becomes very fine; that is, the crystals get smaller. If suddenly cooled, this fine grain is trapped and the steel becomes very hard. Fine-grained steel is very strong after tempering.

At temperatures of more than 100° F. above the upper-transformation temperature, the grain again starts to coarsen, and it becomes very coarse at higher temperatures. If the steel is cooled rapidly in this condition, it will be very hard and brittle and may crack. It also will lack the desired toughness after tempering. Hence, it is important to select a hardening temperature which is within the proper hardening temperature range.

The proper hardening temperature depends upon the amount of carbon in the steel. The more carbon the steel contains, the lower its hardening temperature. Of course, you must know the approximate carbon content of the steel in order to determine the hardening temperature.

The hardening temperature can be tested and estimated with a magnet. Medium-carbon and high-carbon steels are not attracted by a magnet at the hardening temperature. Thus, a piece of steel to be hardened may be heated until a magnet no longer attracts it; it should be cooled rapidly by quenching.

38-6. Quenching Solutions

Steel is hardened by heating it slowly and uniformly to the hardening temperature (see Fig. 38-7), followed by rapid cooling in a **quenching solution.** Steel may be quenched in water, brine, or oil. Special kinds of tool steel may be quenched merely by allowing them to cool in still air.

Some quenching solutions cool more rapidly than others. The rate of cooling is most rapid with brine, less rapid with water, slow with oil, and slowest in air. If steel is quenched and cooled too rapidly, it will crack. Brine or water should be at a temperature of about 60° F. for quenching purposes. Oil, unlike water, cools best when it is at a temperature of about 100° to 140° F. Several kinds of oils are used as quenching solutions. A light grade of straight mineral oil is often recommended.

Plain-carbon steels are usually quenched in water or brine. Brine cools about twice as rapidly as water and tends to **throw** the scale away from the steel during quenching. This causes the steel to cool more uniformly. If small carbon steel parts vary in thickness, such as screwdriver blades or cold chisels, they often crack when quenched in water. This is due to the uneven rate of cooling for thick and thin sections. Parts of this type may be quenched in oil. However, remember that the parts will not be as hard after quenching in oil, which cools slower than water.

It is best to follow the steel manufacturers' recommendations when selecting quenching solutions for alloy steels or expensive tool steels. Most alloy steels must be quenched in oil to prevent cracking.

Vaporized gas bubbles will form on the surface of the hot metal when it is immersed in the quenching solution. These bubbles form a temporary insulation on the surface of the metal and cause the metal to cool slower in that area. When quenching steel, the parts should be **agitated** (moved about) in the solution. Either an **up-and-down** or a **figure-eight** movement should be used. The agitation causes the steel to cool evenly, thus preventing cracks because the gas bubbles do not have time to form on the metal's surface. Very rapid agitation can double the cooling speed. Excessive agitation of small parts in a water or brine quench may cause cracking.

38-7. Tempering

Tempering is often called **drawing** or **drawing the temper.** It is a heat-treatment process which relieves internal strain in hardened steel and thus increases its toughness. The internal stress is produced when the surface of the metal cools more quickly than its core.

Tempering makes hardened steel tougher and also softer. The purpose of tempering, however, is to make the steel tougher, not softer. If it were possible to make steel tougher without softening it, this would be ideal; however, this is not possible. Therefore, the ideal heat treatment of steel tools or parts is a combination of the right amount of hardness and toughness to do the job for which they were designed.

Hardened steel is tempered to increase its toughness so that it will not crack or fracture under heavy stress, vibration, or impact. The toughness of steel can be measured with special impact-testing machines in research laboratories. These machines measure the amount of energy, in terms of foot-pounds, required to fracture a standard-size metal specimen.

The hardness and toughness of steel are related to each other, but in opposite order. The harder the steel, the more brittle it is; the softer the steel, the tougher it is. Hence, the toughness of **hardened and tempered** steel parts may be estimated, indirectly, by determining the hardness of the steel. Hardness testing is explained in Unit 47.

The problem in tempering is to determine the correct tempering temperature. The temperature to which steel is heated for tempering depends on the following factors:

1. The type of steel (carbon steel or special alloy steel)
2. The carbon content
3. The hardness required
4. The toughness required

Recommended tempering temperatures and the color of polished steel at these temperatures are shown on Table 26. Tempering temperatures range from about 300° to 1100° F. Most carbon-steel tools, however, are tempered in the range from 380° to 600° F. Temperatures above 800° F. are used for tempering items which require extreme toughness and little hardness, such as medium-carbon steel parts for the steering mechanism on automobiles.

Tempering Procedure

Tempering should follow as soon as possible after hardening. The steel should be heated slowly and uniformly to the tempering temperature selected, see Table 26. The parts should be held at the tempering temperature (this is called soaking) for a period of one hour per inch of part thickness. This provides time for necessary atomic rearrangement (internal changes) within the grain structure of the steel. Parts ¼″ or less in thickness **do not require soaking** to obtain the required temperature. After the required soaking period, the parts may be allowed to cool in the atmosphere, or they may be quenched in water.

38-8. Annealing

Annealing is a heat-treatment process which is used to soften and improve machinability of hard or hardened steel. It relieves internal stress and strain which may be caused by machining, previous heat treatment, or by cold-working operations; some cold-working operations include rolling, stamping, and spinning. Annealing, therefore, is the opposite of hardening. Three kinds of annealing may be done:

1. Full annealing
2. Process annealing
3. Spheroidizing anneal

Full Annealing

This process relieves internal stress, produces **maximum softness** in steel, and improves machinability. It is used to soften hardened steel for remachining. If a file is fully annealed, a hole can be drilled through it. It could then be rehardened if desired. The following procedure is used for full annealing:

1. Heat the steel uniformly to the full-annealing temperature. This temperature is 50° to 100° F. above the upper-transformation temperature represented by line AC_3 in Fig. 38-7. The full-annealing temperature range is within the same range of temperature as the hardening-temperature range. It varies according to the carbon content of the steel.
2. Allow the steel to soak at the full-annealing temperature for about one hour per inch of part thickness.
3. Allow the steel to cool **very slowly.** It may be removed from the furnace and packed in ashes or lime for slow cooling, or the furnace may be shut off and the part allowed to cool in the slowly cooling furnace.

Process Annealing

This process is often called **stress-relief annealing.** It is used for relieving stresses in steel due to cold working processes such as machining, punching, or rolling. It is most frequently used with low-carbon steels. The following procedure is used for process annealing:

1. Heat uniformly to a temperature ranging from 1000° to 1300° F. (see Fig. 38-7).
2. Allow the part to soak at the desired temperature for a period of about one hour per inch of part thickness.
3. Remove the part from the furnace and allow it to cool in air.

Spheroidizing Anneal

This process involves heating steel to relatively high temperatures, usually from 1300° to 1330° F., Fig. 38-7. The steel is soaked at this temperature for several hours in order to develop a special kind of grain structure. The grain thus formed is very soft and machinable. This process generally is applied to high-carbon steels. Steel annealed by this process should cool slowly to about 1000° F. Below that temperature it may cool at any rate of speed.

38-9. Normalizing

Normalizing is a heat-treatment process which involves heating steel to the **normalizing temperature,** soaking it at this temperature for a period of time, and allowing it to cool in air. Normalizing relieves internal stresses in steel due to forging, machining, or cold working. It also removes the effects of other heat-treatment processes. It softens hardened steel and improves its machinability. It is somewhat similar to annealing, except that steel which is normalized is not as soft as when fully annealed. The following procedure is used for normalizing:

1. Heat the steel uniformly to the normalizing temperature. This temperature varies for steels of different carbon content and is shown in Fig. 38-7.
2. Allow the steel to soak at the normalizing temperature for a period of about one hour per inch of thickness.
3. Remove the steel from the furnace and allow it to cool in air.

38-10. Case Hardening

Case hardening is a **surface hardening** process which involves hardening a thin surface layer on steel, while the inner core remains quite soft, Fig. 38-9. The surface layer forms a hardened **case** over the softer steel; hence the term **case hardening.** This process generally is applied to low-carbon

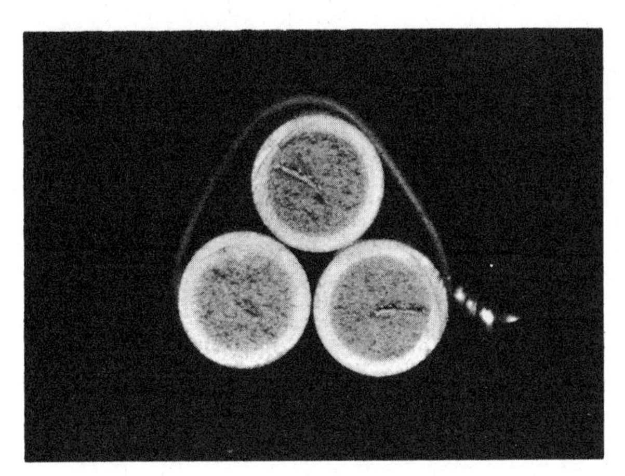

Fig. 38-9. Casehardened Steel

Fig. 38-10. Packing Parts to Be Case Hardened in Carburizing Compound

steels. Occasionally, it is applied to medium-carbon steels.

The casehardening process actually involves two important heat-treatment phases or steps: first **carburizing,** then **hardening.** It will be recalled that low-carbon steel hardens very little because it does not possess enough carbon content. Suppose that more carbon could be absorbed into the surface layer of low-carbon steel. This would cause the surface layer to become high-carbon steel. This is what carburizing does. It causes carbon to be absorbed into the surface layer of steel, thus transforming it into high-carbon steel.

During the second important phase in the casehardening process, the carburized steel is actually hardened. It is hardened by using the same procedure used for hardening high-carbon steel. Thus the casehardening process produces a hardened case or layer of steel over a softer inner core. The surface hardness value, before tempering, is usually from Rockwell C-60 to C-66. The core hardness for casehardened low-carbon steels usually ranges from about Rockwell C-20 to C-30. Since the inside core of the steel possesses some carbon content, it does harden to the degree made possible by the amount of carbon content.

Tools and parts which need increased strength or a hard-wearing surface are case hardened. Wrenches, pliers, and hammers are often case hardened. Gears, screws, bolts, and other parts which wear only on the surface are case hardened.

Carburizing is a relatively slow process. Steel must be soaked at the **carburizing temperature** for about 8 hours to carburize to a depth of 1/16″ and about 24 hours for a depth of 1/8″. Casehardened products generally are carburized to a surface depth of from 0.020″ to 0.030″ during a period of about 2 to 4 hours. The length of time required for penetration of carbon during the carburizing process varies with (1) the carburizing temperature, (2) the carburizing substance used, and (3) the depth of penetration desired.

Several kinds of **carbonaceous** substances may be used to introduce carbon into steel at the carburizing temperature. These materials include solid materials, liquids, and gases. Hence, three kinds of carburizing processes are named according to the kind of carburizing material used. Solid materials are used for **pack carburizing,** Fig. 38-10. Gases such as liquid gas

Fig. 38-11. Gas-Carburized Roller Races Ready to Be Discharged from Carburizing Furnace

Fig. 38-12. Liquid Carburizing of Gears in a Salt Bath

or natural gas are used for **gas carburizing** in special furnaces, Fig. 38-11. Special kinds of salt are heated to form a molten salt bath for **liquid carburizing** in liquid heat-treatment furnaces of the types shown in Figs. 38-12 and 38-4.

Carburized parts may be hardened by quenching directly from the carburizing temperature, or they may be allowed to cool first. When the latter procedure is used, the parts are reheated to the hardening temperature and quenched. When quenched directly from the carburizing furnace, less scale due to oxidation is built up on the parts. Hence, there is less difficulty in surface cleaning the parts.

Casehardened parts may be tempered or not, as desired. Since the inner core is relatively soft, and the hardened surface layer is very thin, there is little danger of cracking or fracturing. When deeply case hardened, the parts should be tempered as explained in § 38-7. Low tempering temperatures, 300° to 400° F., generally are used.

38-11. How to Case Harden Steel

One of the safest methods for case hardening in the school metalworking laboratory involves the use of a special type of carburizing compound. A nonpoisonous, noncombustible carbonaceous substance, such as **Kasenit** (trade name), is recommended as the carburizing material. Pack carburizing with this type of material is done in an open or well-vented container. The container may be made of heavy-gage sheet steel or plate steel. The following procedures are generally used:

Pack Method

For case depths up to 0.015″:

1. Place the steel part in an open, well-vented, shallow container. Cover the part with Kasenit or another equivalent carburizing compound. A vented cover may be placed on the container, if desired.

2. Place the container in a heat-treatment furnace and heat to 1650° F. Soak the part at this temperature for 15 to 60 minutes, depending on the depth of case desired. With this procedure, a case depth of 0.005″ to 0.020″ can be obtained.
3. Remove the part from the molten compound with dry tongs and quench in clean water immediately.
4. Temper the part, if desired. A tempering temperature of 300° to 400° F. is satisfactory for most applications.

Dip Method

For a shallow case, several thousandths of an inch in depth:

1. Heat the part uniformly to 1650° F. This will be a bright red color.
2. Dip or roll the part in Kasenit or a similar casehardening compound, and continue to heat the part for several minutes. The coating of compound will bubble and form a crust as it is absorbed into the steel.
3. With the part heated to 1650° F., quench in clean cold water.
4. To increase the depth of the carburizing, repeat Step 2 one or more times, as desired.
5. Temper, if desired.

Fig. 38-13. Flame Hardening the Ways on a Lathe Bed

38-12. Other Methods of Heat Treatment

Flame hardening is a surface-hardening process which is used on steels which have

Fig. 38-14. High-Frequency Induction Hardening of Four Track Rollers Simultaneously
A. Assembly Setup
B. Parts Being Heated
C. Parts Being Water Quenched

hardening properties. It hardens the surface to depths ranging from 1/32″ to 1/4″ deep. It is done by heating the surface layer of steel very rapidly with an oxyacetylene flame to the hardening temperature, Fig. 38-13. The surface is immediately flush quenched with a spray of water or other coolant. It is then tempered. Flame hardening is used on gear teeth, lathe parts, and other surfaces which must wear longer. They have the advantage of exterior hardness without being brittle because of the softer tough core.

Induction hardening also is a surface-hardening process which is used on steels which have hardening properties. It hardens to depths ranging up to 1/4″. It is similar to flame hardening except that high-frequency electric current is used as a source of heat for hardening. The steel is rapidly heated to the hardening temperature with high-frequency current which passes through a coil surrounding the object being treated. It is immediately flush quenched with water or another coolant, Fig. 38-14.

Words to Know

annealing
carburizing
case hardening
crystalline grain
drawing temperature
flame hardening
full annealing
gas carburizing
grain structure
hardening
heat treatment
induction hardening
Kasenit
liquid carburizing
lower-critical temperature
magnetic point
normalizing
pack carburizing
process annealing
quenching
Rockwell hardness scale
spheroidizing anneal
tempering
upper-critical temperature

Review Questions

1. What is meant by heat treatment?
2. List the principal properties of metal that can be changed by heat treatment.
3. List the principal kinds of heat treatment.
4. Explain how medium- or high-carbon tool steel is hardened.
5. What is meant by soaking during heat treatment?
6. Why is it desirable to have a temperature-indicating and controlling device on a heat-treatment furnace?
7. Name three ways the hardening temperature of steel can be determined when a temperature-controlled furnace is not available.
8. How does the carbon content of steel affect its ability to harden?
9. List three general classifications for plain carbon steels.
10. To what extent can medium-carbon steels be hardened?
11. List four factors which help determine the maximum hardness obtainable for a piece of steel.
12. Can high-carbon steel parts be used without further heat treatment after hardening? Why?
13. What is meant by the hardening temperature?
14. Is the hardening temperature the same for all steels? Why?
15. What kind of grain structure is in steel which is quenched at the proper hardening temperature?
16. List four kinds of quenching solutions. Which cools most rapidly? Which cools most slowly?
17. Why should steel parts be agitated during quenching?
18. What is the purpose of tempering? How is it done?
19. What factors must be considered in deciding what tempering temperature to use on a steel part?
20. In what range of temperatures are most carbon-steel tools tempered?
21. Why is it important that hardened steel be tempered as soon as possible?
22. List three kinds of annealing processes and explain the purpose of each kind.
23. Explain how full annealing is done.
24. Explain how normalizing is done.
25. What is meant by case hardening? For what purpose is it used?

26. List three kinds of casehardening processes.
27. What two important phases or steps are involved in casehardening?
28. What is meant by carburizing? How is it done?
29. On what kinds of steel are casehardening processes used?
30. How long does it generally take to carburize to a depth of 0.020″ to 0.030″?
31. What is the Rockwell hardness range for casehardened surfaces and for the core of casehardened parts?
32. What factors determine whether casehardened parts should be tempered?
33. Explain how to caseharden parts with Kasenit or a similar carburizing compound, using the pack method.
34. Explain how to caseharden parts with Kasenit or a similar carburizing compound using the dip method.
35. Explain the flame-hardening process and its use.
36. Explain the induction-hardening process and its use.
37. Explain several ways metals can be tested for hardness.

Career Information

1. What are the dangers involved in heat treatment?
2. In what trades is a knowledge of heat treatment of steel important?

UNIT 39

Metal Casting Processes

39-1. What Is Metal Casting?

Metal casting is a process whereby liquid metal is poured or otherwise forced into a mold. The metal cools and solidifies in the mold, taking on its shape in every detail. Casting processes make it possible to rapidly and economically produce parts of almost any complexity, of almost any size, and of any metal that can be melted. Parts made by casting are called **castings.** Because of its many advantages, casting is one of the most important methods of manufacturing metal parts. Castings are most commonly made of various alloys of iron, steel, aluminum, brass and bronze, magnesium, and zinc. A factory which specializes in making castings is called a foundry.

There are seven main types of casting processes: (1) **sand casting,** (2) **shell-mold casting,** (3) **die casting,** (4) **permanent-mold casting,** (5) **investment casting,** (6) **plaster-mold casting,** and (7) **centrifugal casting.** Table 27 compares the main features of these casting processes.

39-2. Sand Casting

Patterns

A mold for making a sand casting is made by packing a specially prepared sand around a **pattern** that has the shape of the desired casting. Patterns for sand casting are usually made of fine-grained hardwood or metals such as aluminum or magnesium. A person who makes patterns is called a **patternmaker.**

Types of Patterns

The simplest patterns are **one-piece** or **solid** patterns, Fig. 39-11. One-piece pat-

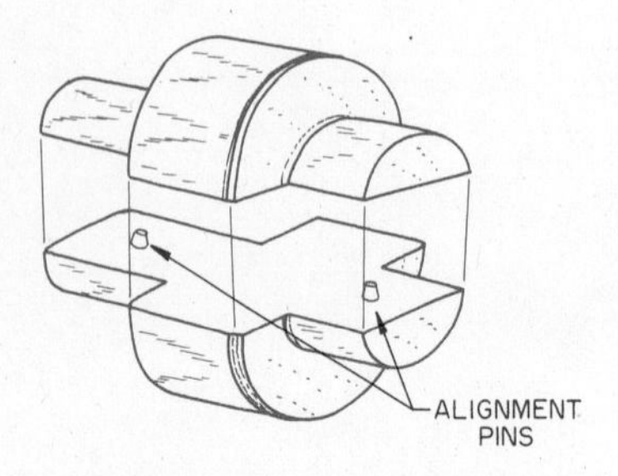

Fig. 39-1. Split Pattern

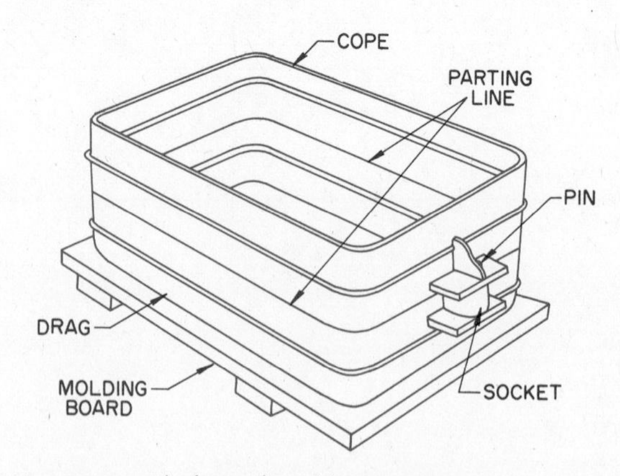

Fig. 39-2. Flasks and Molding Board

TABLE 27

Comparison of Molds Used To Make Metal Castings

	Sand Molds			Metal Molds		Special Molds	
	Green Sand Mold	Dry Sand Mold	Shell Mold	Permanent Mold	Die Mold	Investment Mold	Plaster Mold
Chief Materials for Mold	Sand + Clay Binder + Moisture	Sand + Oil Binder + (Oven Bake)	Sand + Resin Binder + (Oven Cure)	Usually Steel	Hardened Steel	Silica Sand + Special Binder + (Air & Oven Cure)	Gypsum + Water
Usual Metals Cast	Most Metals	Most Metals	Most Metals	Aluminum, Brass, Bronze, Some Iron	Alloys of Aluminum, Zinc, Magnesium	Special Alloys	Aluminum Brass Bronze Zinc (Metals that Melt Under 2000° F.)
Surface Finish of Casting	Rough	Rough	Smooth	Smooth	Very Smooth	Very Smooth	Very Smooth
Accuracy of Casting	Not Very Accurate	Not Very Accurate	Fairly Accurate	Fairly Accurate	Very Accurate	Very Accurate	Fairly Accurate
Usual Weight of Castings	Less than 1 Lb. to Several Tons	Less than 1 Lb. to Several Tons	½ Lb. to 30 Lb.	Less than 1 Lb. to 15 Lb.	Less than 1 Lb. to 20 Lb.	Less than 1 Ounce to 5 Lb.	Less than 1 Lb. to 20 Lb.
Cost of Mold	Low	Low	Medium	High	High	High	Medium

terns are used for parts of simple shape and when only a few castings are desired. **Split patterns,** Fig. 39-1, are used for parts of greater complexity; parts which cannot conveniently be molded from a one-piece pattern. A split pattern is made in two parts, being "split" along the same plane as the plane which separates the two mold halves, Fig. 39-2. One-half of the pattern forms the mold cavity in the **drag,** or lower part of the mold, and the other half of the pattern forms the mold cavity in the **cope,** or upper part of the mold.

When large quantities of castings are desired, the patterns are attached to wood or metal plates called **match plates.** They are then known as **match-plate patterns,** Fig. 39-3. The match plates eliminate the handling of loose patterns, thereby speeding production.

Cope and drag patterns work like match-plate patterns except that the cope part of the pattern is attached to one plate, and the drag part of the pattern is attached to

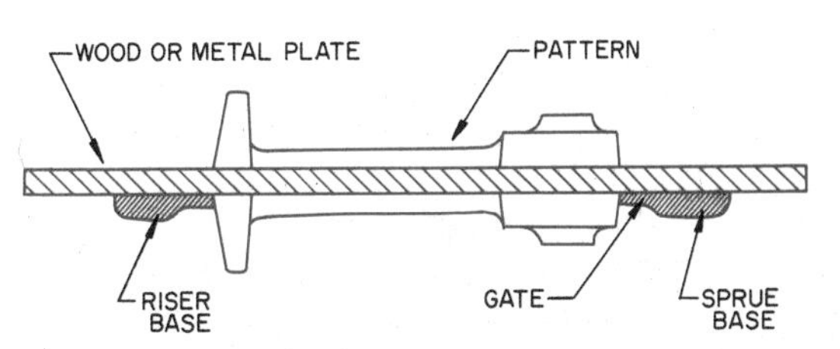

Fig. 39-3. Match-Plate Pattern

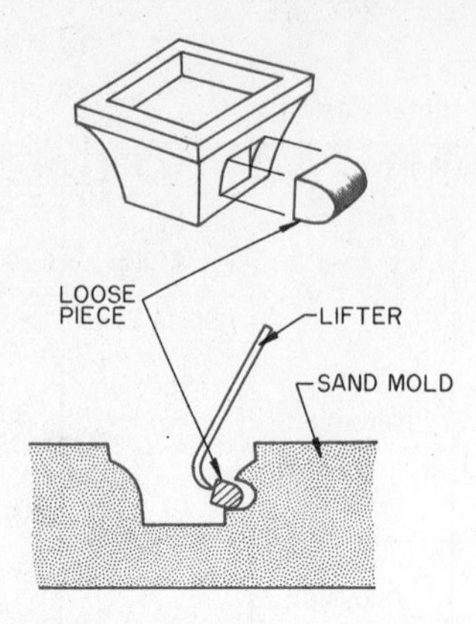

Fig. 39-4. Loose-Piece Pattern

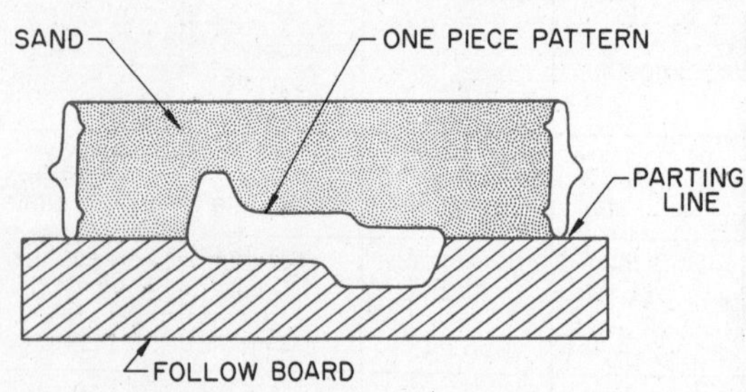

Fig. 39-5. Follow-Board Pattern and Follow-Board Molding Arrangement

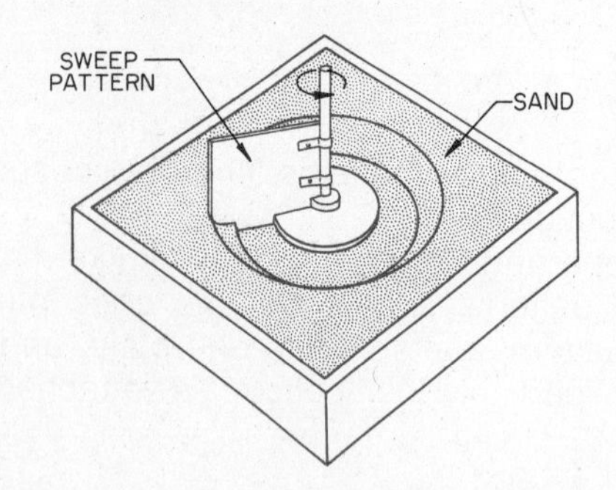

Fig. 39-6. Sweep Method of Pattern Making

another plate. The two plates make it possible for two workers to make parts of the mold at the same time.

Loose-piece patterns are required when the shape of a pattern would make it impossible for it to be removed without damaging the mold, Fig. 39-4. The pattern is constructed in such a way that the main part can be removed first, after which the loose piece can be safely withdrawn.

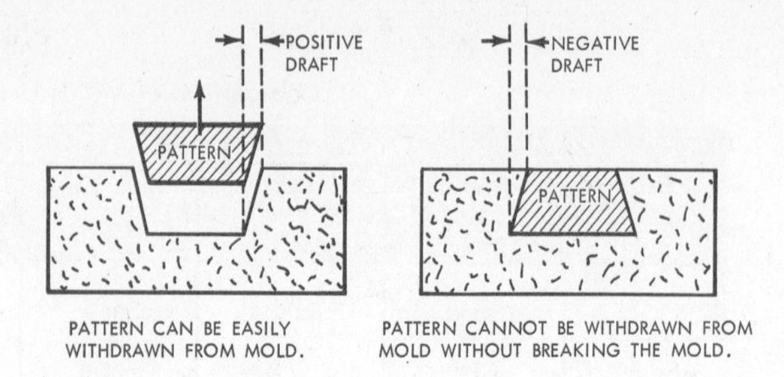

Fig. 39-7. Positive Draft and Negative Draft

A **follow-board pattern** is a one-piece pattern of somewhat irregular shape which is used with a **follow board.** Mold-making with an irregularly shaped pattern is greatly speeded if a cavity is cut into the mold board to hold the pattern in proper position for molding. The cavity is cut so that the parting line of the pattern matches the parting line of the mold halves, Fig. 39-5.

A **sweep** may be used for producing molds having cylindrical cross sections, Fig. 39-6. This method eliminates the need to construct expensive three-dimensional patterns for large castings.

Draft

In order that a pattern may be lifted easily from the mold without breaking the mold, the sides of the pattern must be **tapered.** The taper is also called **draft.** When the pattern is tapered so that it can be removed from the mold, it has **positive draft.** See Fig. 39-7. If, however, the pattern is tapered so that it cannot be withdrawn from the mold without breaking the mold, it has **negative draft.** When the sides of the pattern are straight, the pattern has **zero draft.**

A pattern with negative draft cannot be used to make a mold because it cannot be withdrawn. When a pattern has negative

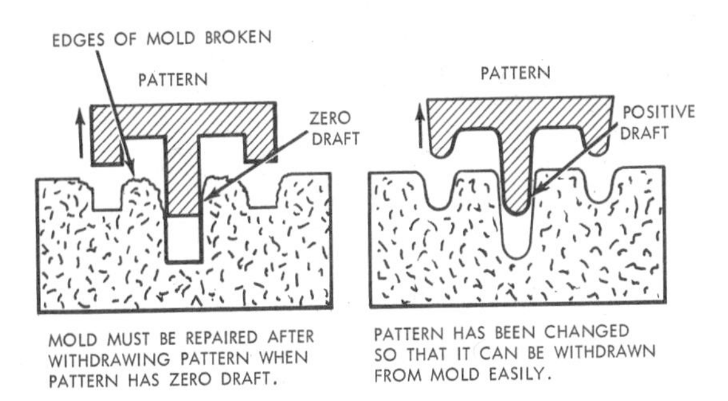

Fig. 39-8. Changing the Shape of a Pattern to Give It Positive Draft

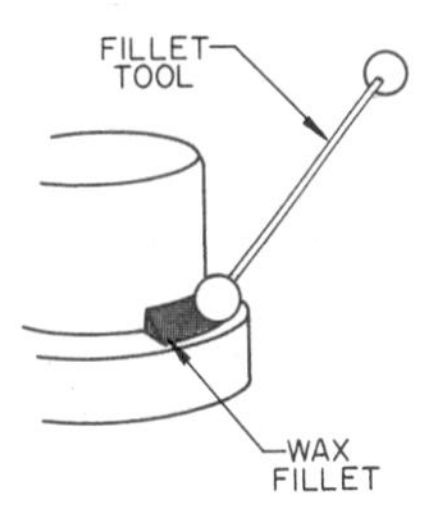

Fig. 39-9. Method of Applying Wax Fillets

draft, it can usually be repositioned, Fig. 39-7. Sometimes, however, the pattern must be reshaped, Fig. 39-8.

Shrinkage

Metal shrinks as it cools. Thus, a casting is larger when it is hot than after it has cooled. When making a pattern, the patternmaker makes the pattern a little larger than the finished casting to allow for shrinkage. He measures with a special ruler that is 1/8″ to 1/4″ longer **per foot** than the standard rule. It is called a **shrink rule.**

The amount of allowance for shrinkage depends on the kind of metal being cast and on whether the mold will cool quickly or slowly. For small castings, the allowances for shrinkage are: iron, 1/8″ per foot; steel, 1/4″ per foot; aluminum, 5/32″ per foot; and brass, 3/16″ per foot.

Fillets

Whenever two surfaces come together on a pattern, they should be connected with a **radius** instead of a sharp line. This radius is called a **fillet.** Fillets help avoid shrinkage cracks at intersections. The size of the fillet should be as large as is compatible with the pattern. Commonly used fillet sizes are 1/8″, 3/16″, and 1/4″. Fillets are made of wax, leather, or wood. Leather and wood fillets are glued to the pattern, while wax patterns are set in place with a warm fillet tool, Fig. 39-9.

Green-Sand Molds

Molds made from a mixture of moist sand and clay are called green-sand molds. A large percentage of the castings made in the United States are made from these molds.

Small sand molds can be made by hand on the bench; this is called **bench molding.** Large sand molds are handmade on the floor; this is called **floor molding.**

Sand for Green-Sand Molds

Sand for green-sand molds is not sand alone, but a mixture of sand, clay, and water. A **green-sand foundry** makes **molding sand** by mixing sand with special clays. Enough water is added to make the mixture moist. The damp, sticky clay makes the grains of sand cling together. The mixture is called "green" because the mold is not dried, baked, or cured before it is used. Other ingredients are also mixed with the clay, sand, and water. Some of these make the mixture black.

In green-sand molding, **tempering** means to mix sand with water to a certain dampness. When the proper amount of moisture has been added, the sand is said to have good temper. It will pack or squeeze in the hand somewhat like snow and break with

Fig. 39-10. Testing Temper of Green Sand

even, square edges, Fig. 39-10. Also, the sand will not cling to the hand when pressed into a lump. It will fall away, leaving the hand almost clean.

If the sand is too dry, some of the sand will be "washed" away from the inside of the mold by the melted metal and the casting will not be the right size. Also, the casting may have some of the washed-away sand imbedded in the metal. If the sand is too damp, the hot metal changes the excess moisture into steam. The steam may cause holes in the casting or may cause the entire mold to explode.

Waterless sand is another type of molding sand. It is used to make molds in the same manner as the water-moistened sand in green-sand molding. It is called waterless sand because a special oil and a formulated bonding material is mixed with pure silica sand to hold it together.

Waterless sand has many advantages and some disadvantages for use in schools. It may be used in casting aluminum, magnesium, bronze, and brass. Its advantages include greater precision, less gases formed, the use of finer sands with lower permeability, finer finishes, molds that may be set aside for several days without evaporation before pouring the casting, molds that may be rammed tighter and with less even pressure, gates that may be made smaller, and fewer vent holes and risers are needed in the mold.

Waterless sand must be conditioned and thoroughly mixed in a **mulling machine** before its initial use. If the sand is riddled carefully and turned over frequently with a shovel it will only need to be reconditioned and remulled about once a year for school use. This service may be performed by a nearby foundry or by the dealer. Casting in waterless sand molds requires good ventilation to carry away fumes and odors.

One should follow the manufacturers' recommendations carefully in mixing the ingredients and in mulling waterless sand. Information concerning these materials and procedures is available from many foundry supply dealers.

Flask

A **flask** is a frame for a mold made from wood or metal, Fig. 39-2. Flasks are made in two halves. The top is called the **cope,** and the bottom is called the **drag.** The halves are held together by **pins** and **sockets.** When they are put together, a **parting plane**[1] separates the cope and drag.

Wood flasks are inexpensive, but they char easily from the heat of hot melted metal. They must be replaced often. Metal flasks last longer.

Molding Board

The board or plate upon which the pattern is laid while pressing the sand around the pattern is called a **molding board.** See Fig. 39-2. It should be slightly larger than the flask and strong enough that it will not bend while the mold is made.

Parting Compound

When packing sand around the pattern to make a mold, the sand sometimes sticks or clings to the pattern. When this happens, it is difficult to remove the pattern from the

[1]In geometry, a plane is an imaginary flat surface that has length and breadth (two dimensions) but zero thickness. In molding and casting, the parting plane is sometimes called a **parting line.**

mold. **Parting compound** is a fine sand or powder used to keep the molding sand from sticking to the pattern. It must be kept dry to work well. **Charcoal dust** or **brick dust** may be used for parting compound. It is usually put into a bag and "dusted on." See Fig. 39-19.

Riddle

A **riddle** (see Fig. 39-12) is like a kitchen flour sieve except that it is larger. It is used to sift sand. It breaks up lumps and leaves the sand "fluffy" so that it will pack properly around the pattern. The **meshes** of some riddles are very fine; others have coarse meshes. Sifting sand with a riddle is called **riddling.**

39-3. Mold Vents

Air, steam, and gas must be allowed to escape from the mold. These cause blowhole defects in the casting. **Blowholes** are holes in castings caused by small explosions of steam and hot gases from the molding sand. **Vent holes** are made over the pattern by pushing a wire or rod into the mold. They allow the steam and gases to escape. See Fig. 39-22.

39-4. Making a Green-Sand Mold, Using a Solid Pattern

Suppose that a mold for a **bookend** with a flat back is to be made. A **solid pattern** is to be used and the flat back of the pattern will be at the **parting line.** The surfaces of the pattern must be clean, smooth, and dry; otherwise, sand will stick to them. The steps for making a sand mold are as follows:

1. Put the flat side of the pattern on the **molding board,** Fig. 39-11. Put the **drag** around the pattern with the **pins** pointing down, and shake some **parting compound** over the pattern.
2. Set the **riddle** on top of the drag, fill it with sand, and sift the sand over the pattern, Fig. 39-12, until it is at least 2″ thick. Press the sand around the pattern with the fingers.

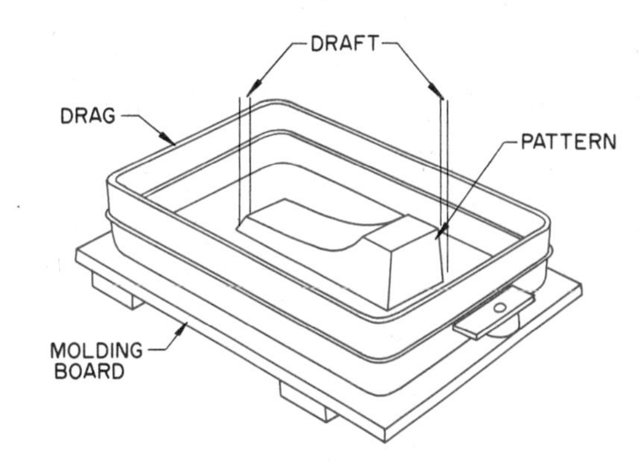

Fig. 39-11. Pattern and Drag Placed on Molding Board

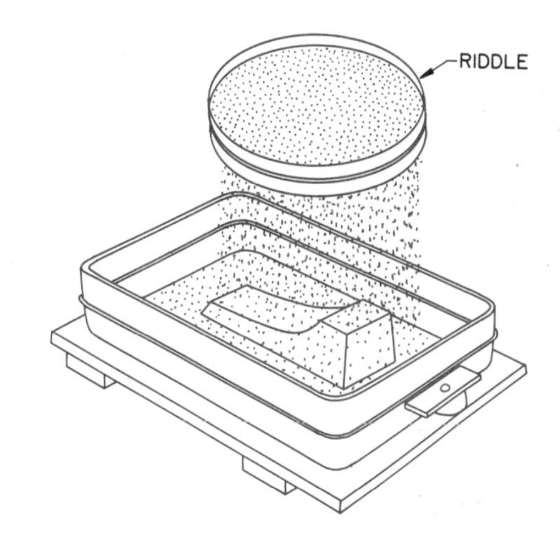

Fig. 39-12. Riddling Sand Over Pattern

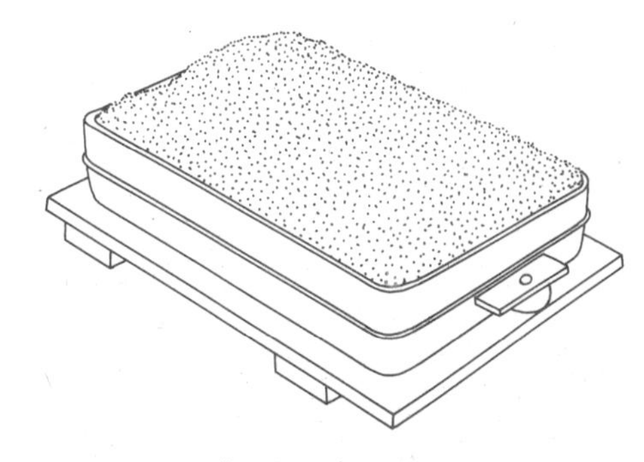

Fig. 39-13. Sand Heaped in Drag

3. Dump the rest of the sand that is in the riddle into the drag. Fill the drag with sand until it is piled up in a heap above the drag, Fig. 39-13. Pack the sand down in the drag and around the pattern with a rammer, Fig. 39-14, alternately using the peen and butt ends. Be careful not to strike the pattern or the sides of the flask. Continue to add sand and ram until the drag is completely full.
4. Smooth the top of the drag with a **strike bar,** which is a board or strip of iron with a **straight edge,** Fig. 39-15.
5. Sprinkle a handful of molding sand over the top, lay the **bottom board** on, and move it back and forth until it rests firmly on the drag.
6. Hold the drag, molding board, and bottom board tightly together at the sides. Slide them to the front of the bench until they begin to drop toward you. At this moment, turn it over to rest on the bottom board, Fig. 39-16. The pins are now pointed up, the molding board is on top, and the bottom board is at the bottom.
7. Remove the molding board; the flat side of the pattern is now up. Make the surface smooth with the **slick and spoon** or **trowel,** Fig. 39-17, and see that the sand is packed around the edges of the pattern.
8. Blow off all loose sand with a **bellows,** Fig. 39-18. Shake some **parting compound** over the pattern and mold to keep the two halves from sticking together, Fig. 39-19.
9. Set the **cope** part of the flask on the drag, Fig. 39-20. Also, set the **sprue pin.** This is a **tapered** wooden or

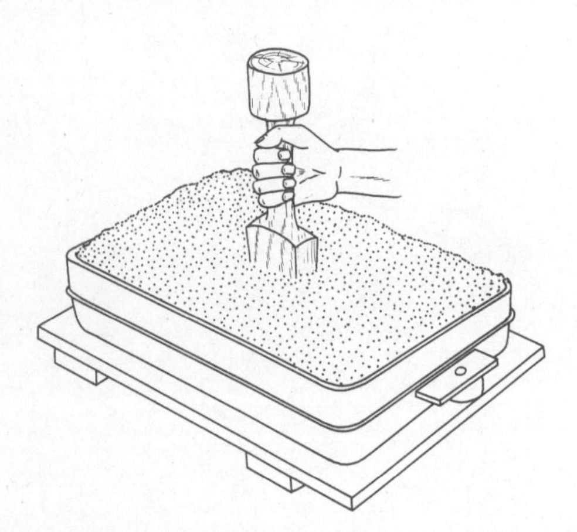

Fig. 39-14. Ramming

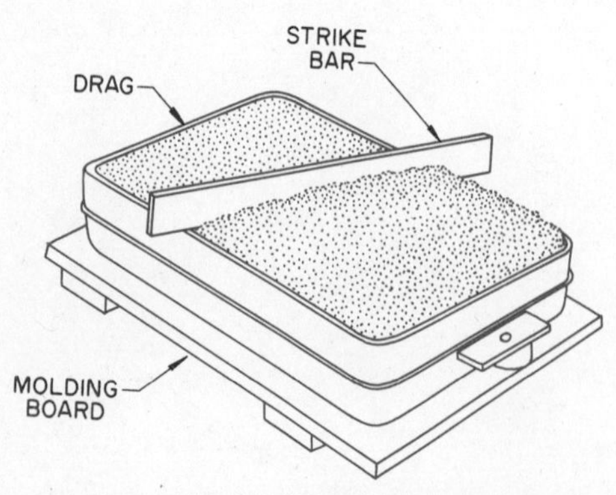

Fig. 39-15. Striking Off Sand with Strike Bar

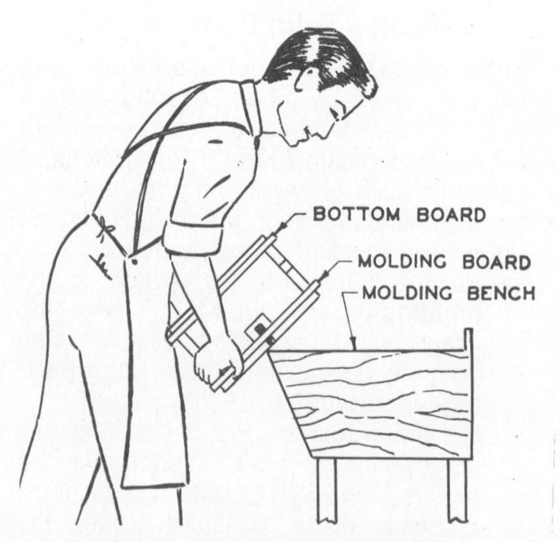

Fig. 39-16. Turning Drag Upside Down

metal pin which is used to make a hole in the cope through which the metal is poured into the mold. Place it on the drag about 1″ from the pattern. The hole is called a **sprue.**

Also set the **riser pin** in place; it is used to make a hole to tell when the mold is full of melted metal. The hole, which is called a **riser,** also allows the air and gases to escape and the dirt from the melted metal to

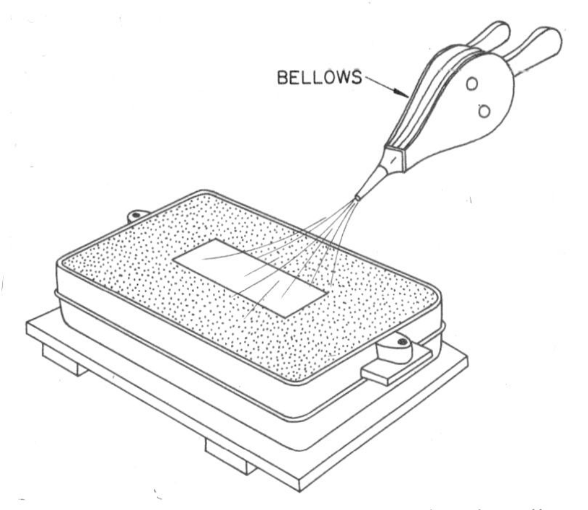

Fig. 39-18. Blowing Off Loose Sand with Bellows

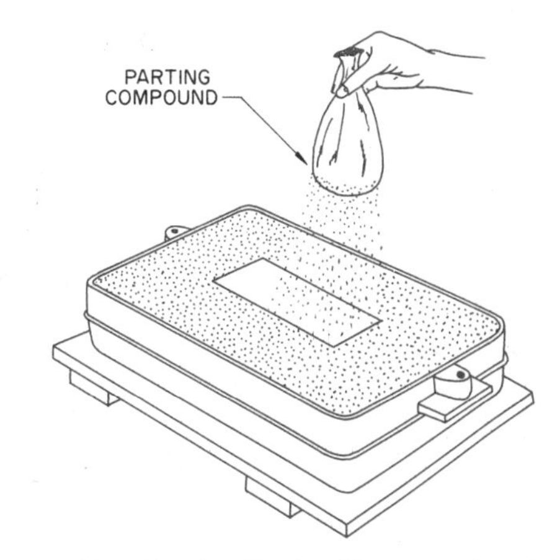

Fig. 39-19. Dusting Parting Compound on Mold

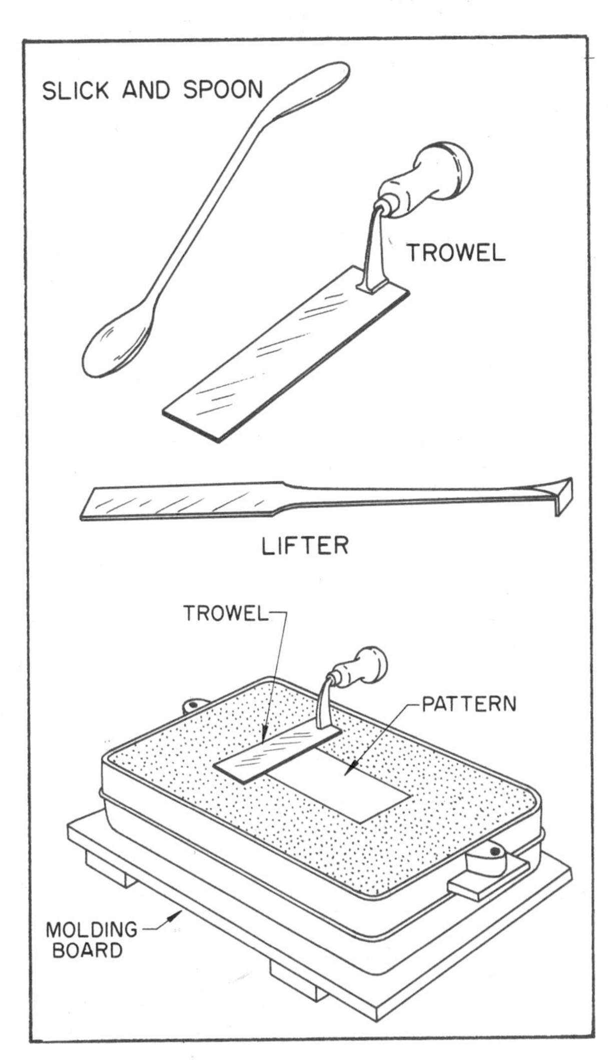

Fig. 39-17. Smoothing Surface

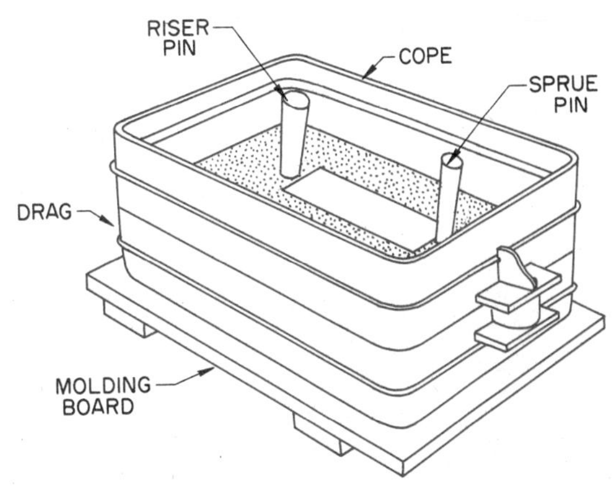

Fig. 39-20. Cope Set on Drag with the Sprue Pin and Riser Pin in Place

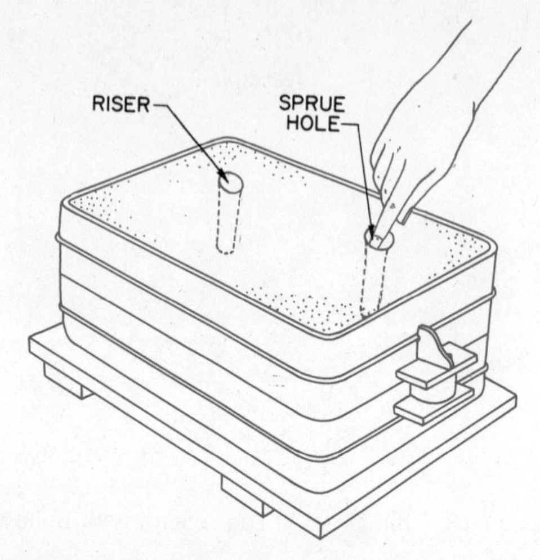

Fig. 39-21. Enlarging Top of Sprue

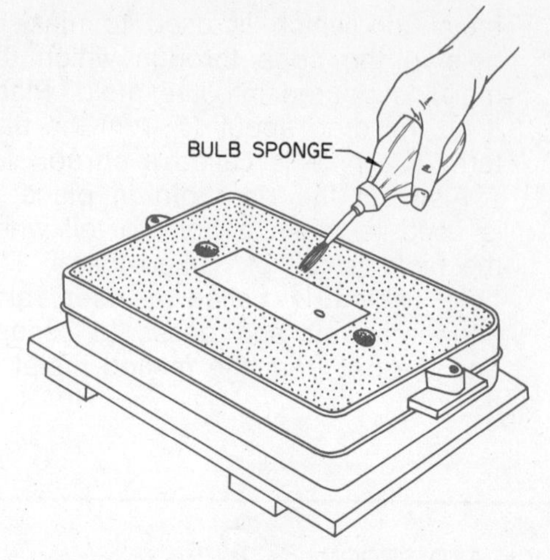

Fig. 39-23. Wetting Edges of Mold with Sponge

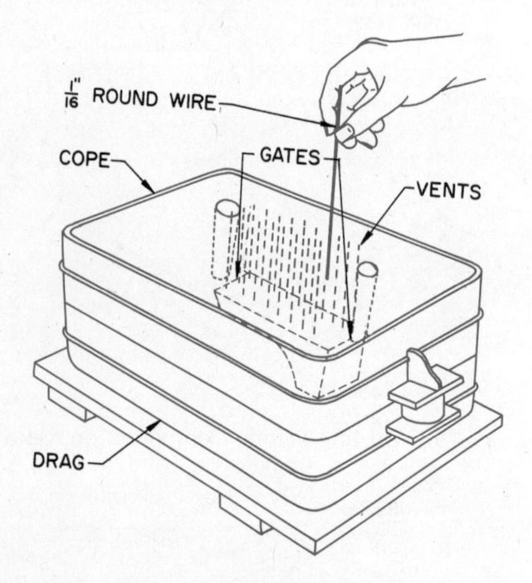

Fig. 39-22. Making Vents in Mold

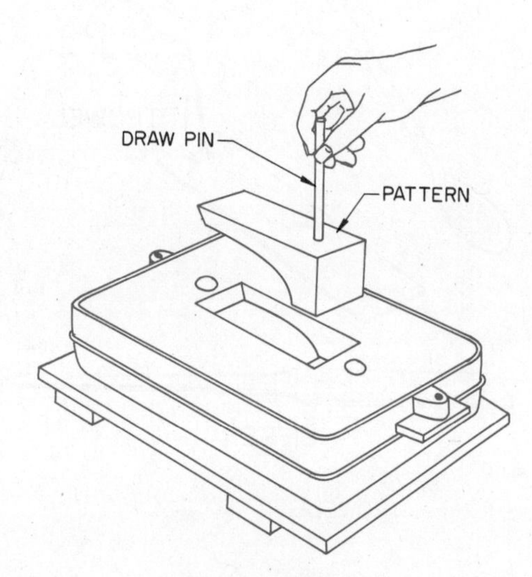

Fig. 39-24. Lifting Pattern Out of Mold

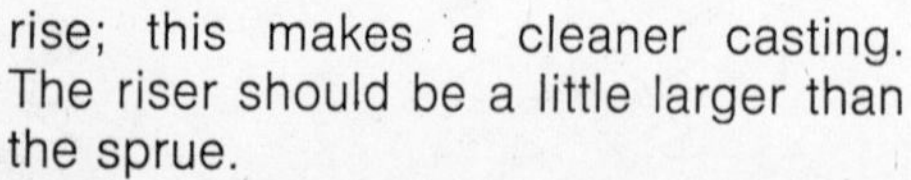

rise; this makes a cleaner casting. The riser should be a little larger than the sprue.

10. Riddle sand into the cope and **ram** as before. The sand in the cope should not be packed as tightly as in the drag in order to allow the gases to escape more easily.
11. Strike off the top. Lift out the sprue pin and riser pin. Round off the top of the sprue hole into the shape of a funnel with the fingers, Fig. 39-21. Make **vents,** which are explained in section 39-3 and Fig. 39-22. These holes are made by pushing a $\frac{1}{16}''$ round wire into the mold to about 1″ from the pattern.
12. Lift the cope off, lay it on its side on

Fig. 39-25. Molds Ready for Melted Metal

the bench, and dampen the edges of the mold next to the pattern with a **bulb sponge.** Dampening makes the sand around the pattern firmer so that the pattern can be lifted out without breaking the mold. See Fig. 39-23.

13. Lift the pattern out with a **draw** pin, which is a piece of steel with threads on one end; the pin is screwed into a hole in the pattern. The draw pin can be **rapped** lightly on all sides to loosen the pattern so it can be lifted easier, Fig. 39-24.
14. Cut a small channel, called a **gate,** in the sand of the drag from the hole made by the pattern to the place where the **sprue** is located. It should be almost ¼″ deep and 1″ wide. This may be done with a **gate cutter,** made of 4″ x 4″ **sheet metal** bent into a U. It should be a little deeper under the sprue than at the other end. Also, cut a gate between the riser and the hole made by the pattern. Patch up small breaks in the mold. Blow off all loose sand with the bellows. Replace the cope on the drag.

Fig. 39-26. Polystyrene Pattern with Sprue, Gate and Runner, and Riser System Attached

15. Place a **flask weight** on top to keep the melted metal from lifting up the cope during pouring. All these are clamped together, if necessary, and set on the floor ready to receive the melted metal, Fig. 39-25.

After the mold is poured and cooled, the casting is removed from the mold. It looks like Fig. 40-10. The sand from a green-sand mold can be reused if it is reconditioned with more clay, water, and other ingredients.

Full Mold Process

The **full-mold casting process** uses patterns of **polystyrene foam,** Fig. 39-26. This material can be shaped very easily and complex patterns can be quickly made by gluing together several simple shapes. The sprue, riser, and gate and runner system are also made of polystrene and are glued to the pattern. Sand is rammed around the polystyrene pattern in the usual manner, except that the pattern is not removed; hence, the name full mold. Because the pattern is not removed, pattern draft can be completely ignored. When the mold is poured, the heat of the metal vaporizes the polystyrene almost instantly and the metal fills the space occupied by polystyrene. Because of the low pattern cost, full-mold casting is economical when only one or a small number of castings are to be made.

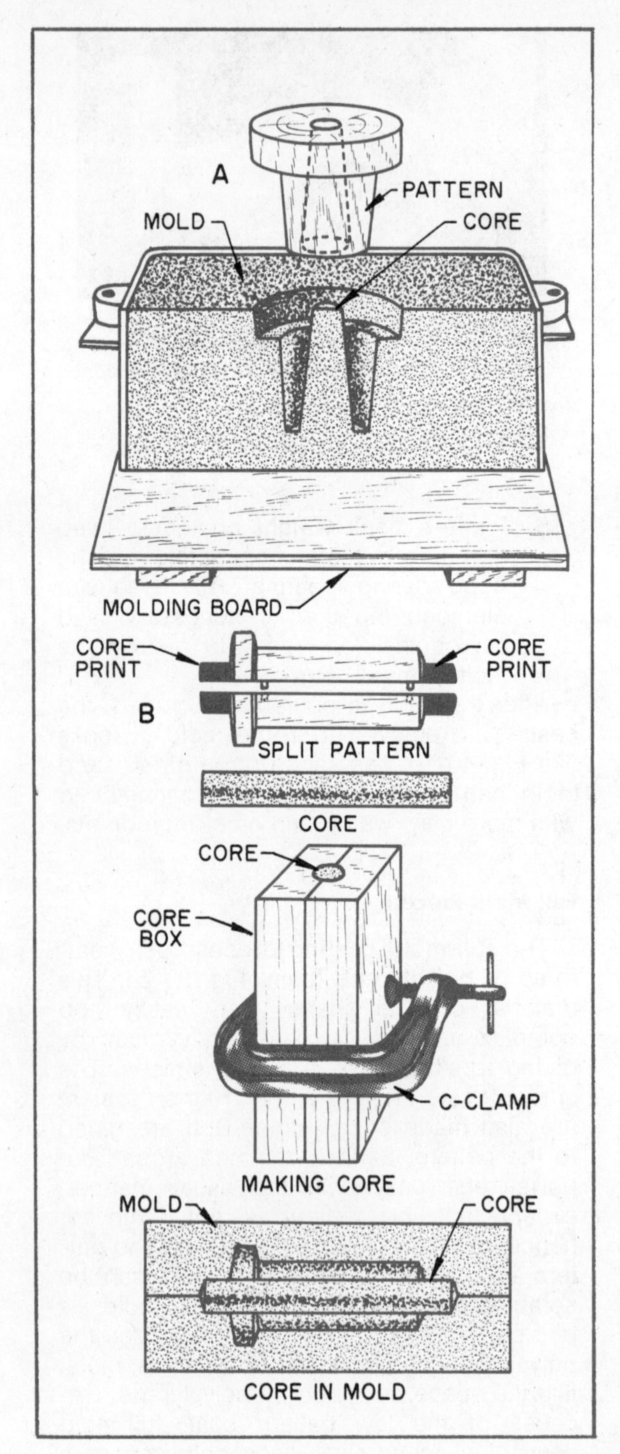

Fig. 39-27. Cores in Molds

39-5. Cores

In molding, a **core** is a separate hardened sand shape put into a mold to form a hole or hollow in a casting, Fig. 39-27. This makes the casting lighter and means that less metal needs to be cut away later. One kind of core may be formed in the mold by simply lifting the hollow pattern out of the mold. See A in Fig. 39-27. The split pattern (B in Fig. 39-27) is made without a hole. Instead, it has **core prints** on the ends of the pattern that form hollows in the mold for the core. After the pattern is lifted out of the mold, the core is set in place. When melted metal is poured into the mold, it flows around the core. After the casting has cooled and been removed from the mold, the core is broken away, leaving a hole in the casting.

A **core box** (Fig. 39-27) is a wooden or metal mold in which a core is shaped from sand. After removing the core from the core box, it is baked in a **core oven** to make it hard. A person who makes cores is a **coremaker.**

CO_2 Process

Another type of sand core (used both in schools and in industry) is made by the **CO_2 process** instead of being baked in an oven. Core sand is mixed with a commercial binder designed for use with the CO_2 (carbon dioxide) process. The core sand is pressed or blown into the core box. It is then hardened by blowing CO_2 gas through the sand for several seconds. The core is then ready for use without baking.

39-6. Dry-Sand Molds

A **dry-sand mold** is stronger than a green-sand mold. It will not break as easily and can be handled more readily. It does not have to be used right away and can be stored. The surface of a dry-sand mold is hard and smooth. The sand will not wash away when the hot melted metal is poured against it as easily as in a green-sand mold.

To make a dry-sand mold, special foundry oils are mixed with sand. These oils coat the grains of sand and make them stick to each other. The sand is **rammed** into the cope and drag to shape the mold, just as with green sand. The flasks are removed and the mold is baked in an oven at 300° to 600° F. When the mold has finished baking, it is removed from the oven and allowed to cool. The parts of the mold are put together and the mold is ready for the melted metal. The casting cannot be spoiled by pockets of steam, which can form in a green-sand mold, since there is no moisture in a dry-sand mold.

39-7. Shell Molding

Another type of sand mold, called a **shell mold,** was developed in Germany in the 1940's. A fine powdered **resin**[2] is mixed with dry molding sand. The sand-resin mixture is poured onto a metal pattern that is already heated to 400° to 600° F. The resin melts and coats the grains of sand. The resin is sticky and causes the grains of sand to stick to each other. At the proper time, the pattern is turned upside down and the excess sand mixture falls off. Only ¼" to ½" of the sand-resin mix sticks to the hot pattern. The pattern is turned upright again and then put into an oven to cure (allow the resin to harden). When the sand mold is removed from the oven, it is hard and thin, thereby the name **shell** mold, Fig. 39-28.

The shell mold is usually made in two halves, Fig. 39-29. When the halves are put together and the complete shell supported in a bed of sand, melted metal can be poured into the mold. Castings made in a shell mold are smooth and require less machining than castings made from a green-sand mold. Shell molds are being used more and more where accuracy and good finish on castings are important. The sand from shell molds cannot be easily reused.

39-8. Permanent Molds

The **permanent mold**[3] is usually made from iron or steel. The cavity of the mold is cast or machined into the metal. Since the permanent mold cannot be destroyed

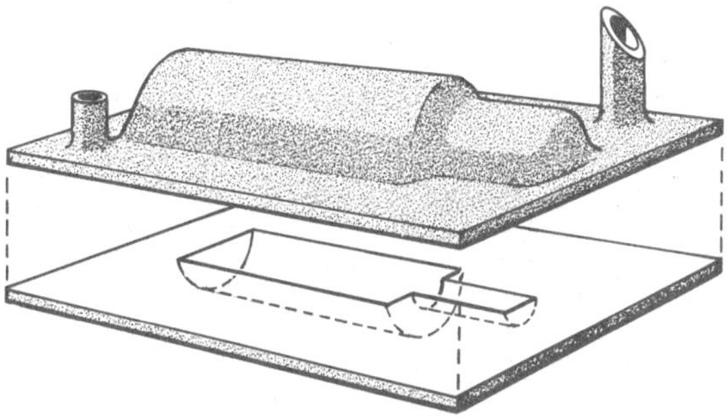

Fig. 39-28. Shell Mold

Fig. 39-29. Shell Molding Machine—Forming Half Shell (Right) and Assembling Pairs (Left)

[2]This **resin** is much the same as one type of those materials generally known as **plastics.** It softens when heated, then hardens with continued heating. Once it has been melted, it will not melt again; therefore, this is called a thermosetting resin.

[3]**Permanent molds** are usually made from metal, but not always. They are called metal molds here to simplify the classification of molds. For special reasons, permanent molds are occasionally cut from nonmetals like block graphite.

Fig. 39-30. Assortment of Small Die Cast Parts

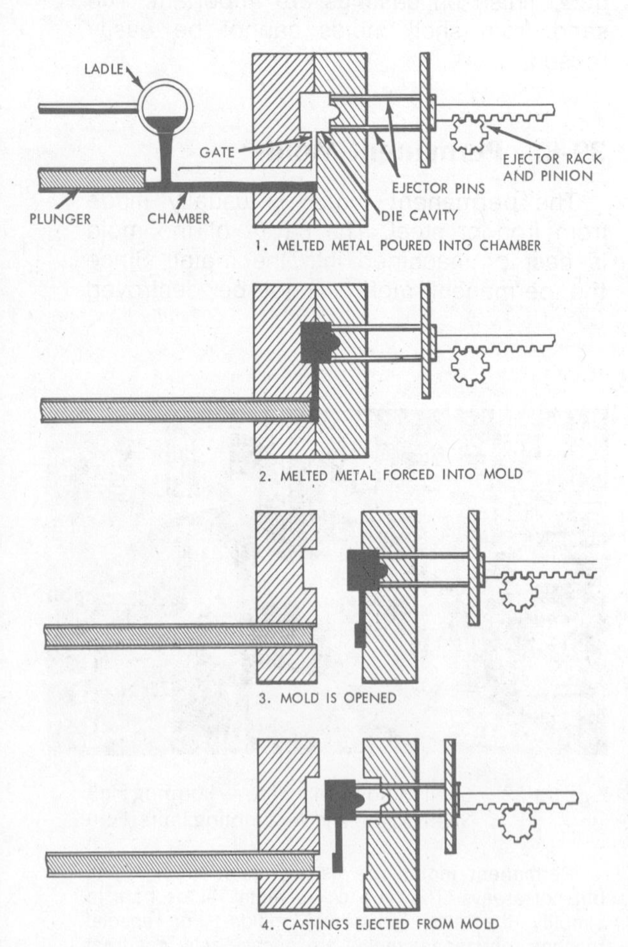

Fig. 39-31. Die-Casting Process

to get the casting out, like the sand mold, only simple shapes can be made from the permanent mold.

Each time melted metal is poured into the permanent mold, it "washes" a little bit of the mold away and the mold becomes slightly larger. When the castings made from this mold become too large, the mold must be thrown away and a new one made to replace it.

The surface of castings made from permanent molds is very smooth. See Fig. 39-30.

39-9. Die Casting

Casting by forcing melted metal into a **die** is called **die casting.** A die is a mold made of metal. It is like a permanent mold except that the metal is forced into the mold under pressure. The metal is said to be **injected** into the mold. Figure 39-31 illustrates the die-casting process. Die-casting machines, Fig. 39-32, are made in many sizes and are capable of high production rates. Both the machine and the dies are very expensive. This expense is justified, however, because many castings can be made from one die.

After being taken from the mold, the die casting needs very little work to make it a

Fig. 39-32. Crane Setting a Die for Aluminum Die Casting

finished casting. Often, only trimming the flash and buffing are required. Many die castings are plated with nickel and chrome.

Much hardware is made by die casting. Automobile door handles and hood ornaments, kitchen cabinet handles, and some lamp bases are examples of die castings.

39-10. Investment Molds

Most molds are made from a pattern that has positive draft so that it can be withdrawn from the mold. An **investment mold** is made from a wax pattern. The wax pattern does not have to be tapered because it is melted out of the mold after the mold is made. Therefore, almost any shape of casting can be made from an investment mold. Often, some parts of the wax pattern have straight sides or sides that have negative draft.

Figure 39-33 shows how to make an investment mold. A wax pattern is molded or cut with a warm knife. The pattern is put into a steel flask that looks like a can with its top and bottom removed. A slurry (watery paste of **silica**[4] and hardener is poured around the wax pattern. The flask is put on a vibrating table which packs the slurry against the wax pattern and removes bubbles of air. Then it is set aside to dry and harden.

Several hours before the mold is to be used, it is turned upside down in a furnace and heated to about 1500° F. The wax melts and runs out, leaving the shape of the wax pattern in the mold. Because the wax is melted out of the mold, the mold is sometimes called a lost-wax mold and the process called the **lost-wax process.**

Investment mold comes from an old English word, "invest," meaning to enclose or surround. Thus, the wax pattern is **invested** in a slurry of silica.

The investment mold makes a casting that is very accurate and has a fine smooth surface. Since the mold is used while still hot from the furnace, it does not chill or cool the melted metal quickly. Therefore, the melted metal can flow into fine cracks and very thin parts of the mold before it becomes solid.

Industry uses investment molds to make castings of many sizes and shapes. The dentist casts gold into investment molds to make fillings and other parts for teeth. The jeweler uses investment molds to cast gold, silver, and platinum into rings, bracelets, pins, trophies, and parts for jewelry.

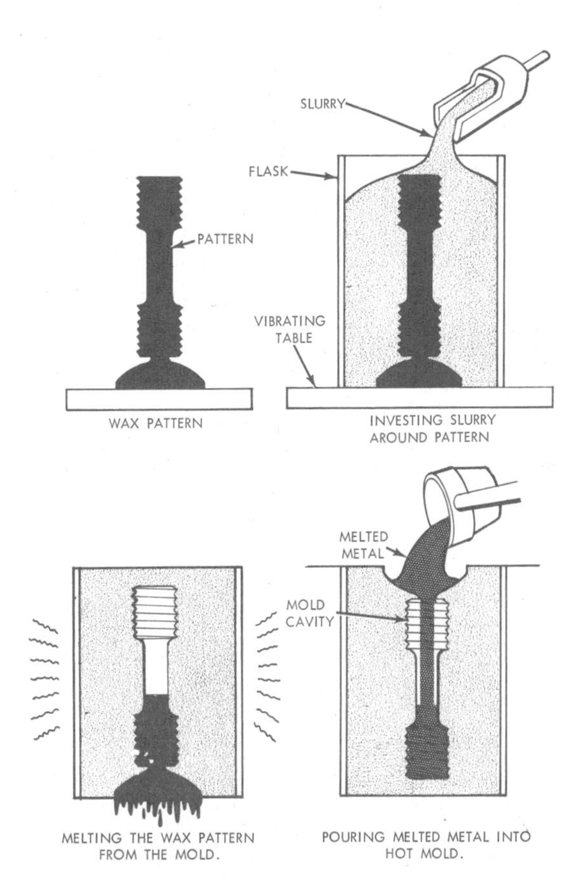

Fig. 39-33. Making and Pouring an Investment Mold

[4]**Silica** is silicon and oxygen combined. It is called silicon dioxide by the chemist. Silicon dioxide is found almost everywhere as sand, quartz, flint, etc. It is used to make molds because it melts at a high temperature and does not change shape or crack when heated. A slurry of silica handles and pours like newly mixed cement.

Fig. 39-34. (Top) Pouring Plaster into Core and (Bottom) Assembling Plaster Cores for Casting an Aluminum Tire Tread Mold

39-11. Plaster-Mold Casting

Plaster molds are sometimes used to make castings from alloys of copper, aluminum, and other metals that melt at low temperatures (400° to 1700° F.). Since they are easy to make, plaster molds are especially useful when only a few castings are to be made, Fig. 39-34.

The plaster is made by mixing water with plaster of paris. The mixture is a thick, creamy paste that stirs and pours like hot, cooked breakfast cereal. The wet plaster is poured over a pattern and set aside to harden. The pattern is then taken out of the mold. The mold must be thoroughly dried before melted metal is poured into it. Otherwise, the moisture will turn into steam that cannot get out of the mold. When this happens, the casting will have holes or a rough surface.

Wood patterns can be used to make plaster molds, but they are not as good as patterns made from metal or plastic. Wood patterns will absorb moisture and swell.

Castings made from plaster molds have a very smooth surface. The sculptor uses plaster molds to cast bronze into statues and fine ornamental work. Special foundries also use plaster molds. The procedure for making a plaster mold is shown in Fig. 39-35.

39-12. Centrifugal Casting

In **centrifugal casting,** the mold is rotated on its longitudinal axis at speeds from 300 to 3000 rpm while the metal is being poured. Centrifugal force causes the molten metal to conform to the mold. The process makes possible the rapid production of cast iron pipe, large-diameter gun barrels, brake drums, and similar hollow objects.

39-13. Extrusion

Extrusion is a process of shaping solid metal by forcing it under high pressure through an opening in a die. Extrusion may be forward or reverse, Fig. 39-36. Extruded shapes may be solid or hollow, and the process may be done hot or cold.

In **hot extrusion,** the metal blank is softened by preheating, effectively lowering the pressure required for forming and making possible extrusions of complex shapes and considerable length. Most hot extrusions are made on hydraulically powered horizontal presses and are long pieces of

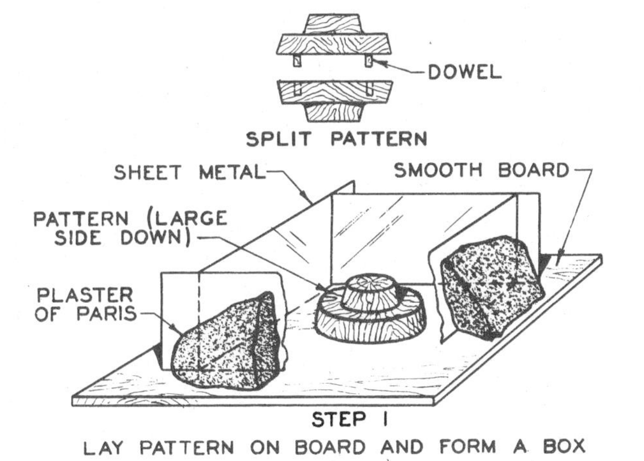

STEP 1
LAY PATTERN ON BOARD AND FORM A BOX

PLASTER OF PARIS
1"
1"

STEP 2
POUR PLASTER OF PARIS INTO MOLD

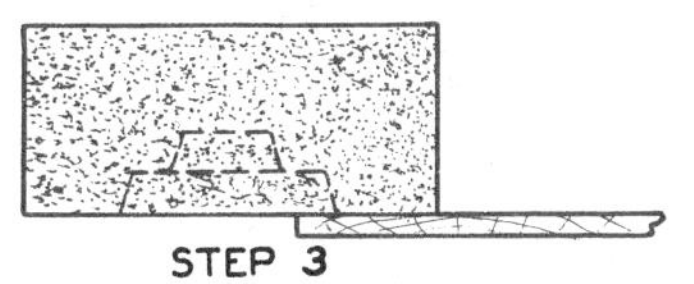

STEP 3
REMOVE SIDES AND SLIDE MOLD OFF BOARD

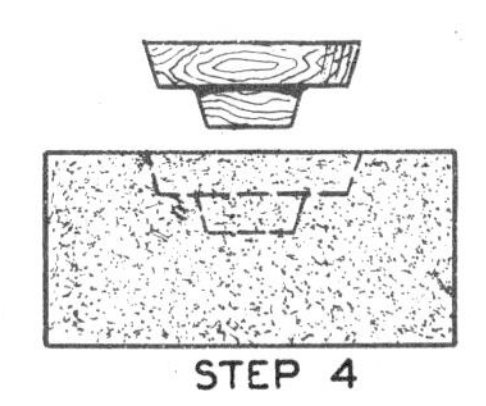

STEP 4
TURN MOLD UPSIDE DOWN, LIFT PATTERN OUT, AND REPAIR MOLD

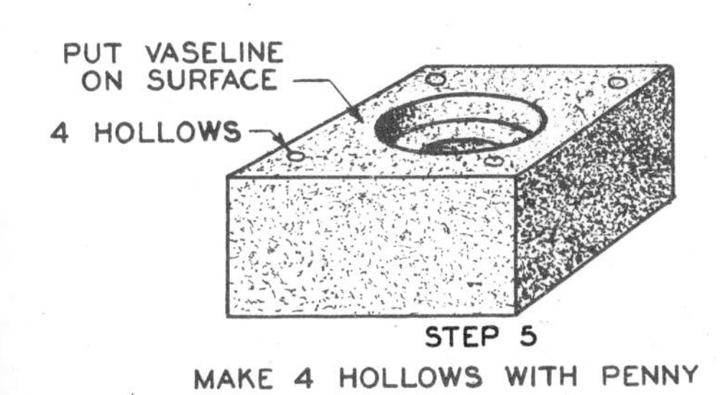

STEP 5
MAKE 4 HOLLOWS WITH PENNY

SHEET METAL

STEP 6
PLACE WHOLE PATTERN IN MOLD, BUILD HIGHER BOX, AND FILL WITH PLASTER OF PARIS

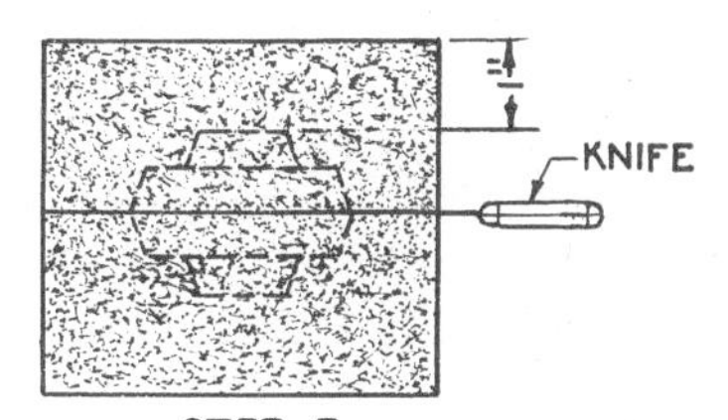

STEP 7
REMOVE BOX AND SEPARATE HALVES

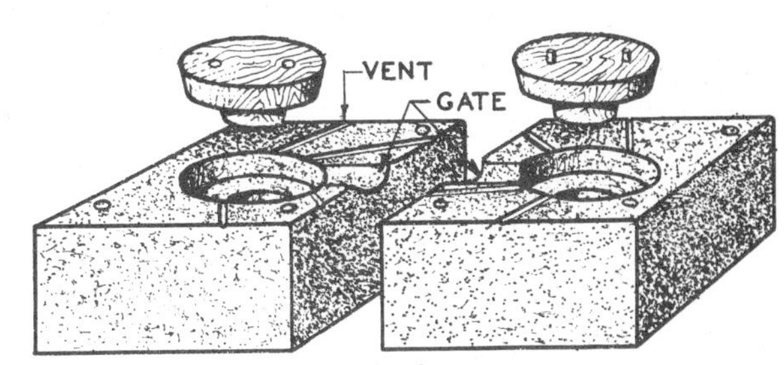

STEP 8
REMOVE PATTERN, REPAIR MOLD, CUT GATE, AND MAKE VENTS

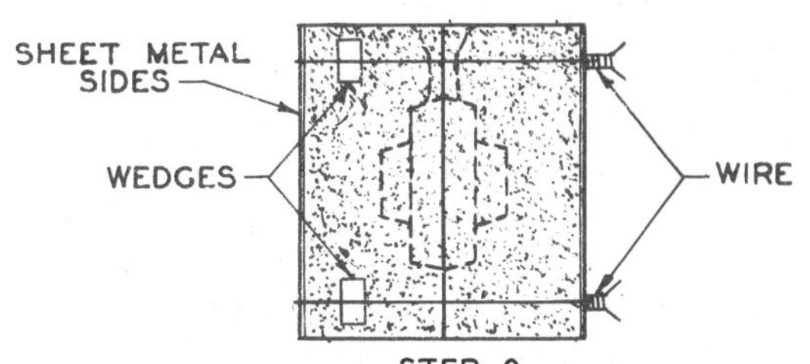

STEP 9
MOLD READY FOR MELTED METAL
(BE SURE THAT THE PLASTER OF PARIS IS DRY BEFORE POURING THE METAL IN)

Fig. 39-35. Making Plaster of Paris Mold

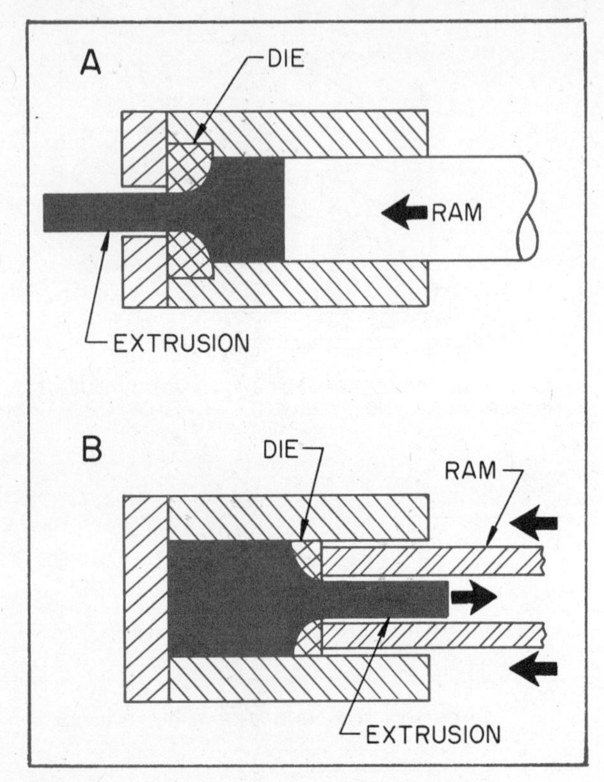

Fig. 39-36. The Two Basic Methods of Extrusion
A. Forward Extrusion
B. Reverse Extrusion

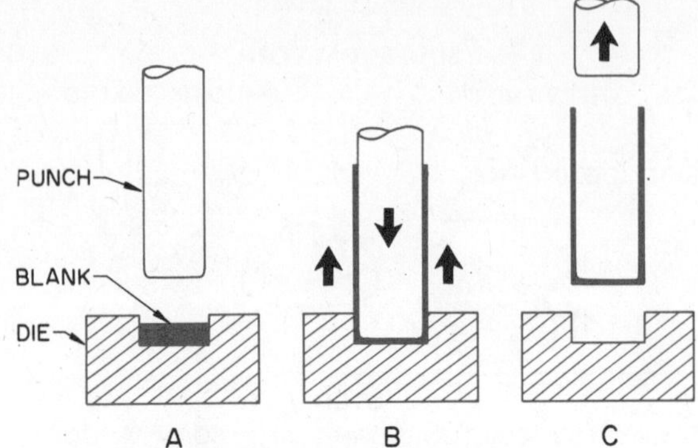

Fig. 39-37. Impact Extrusion Sequence
A. Blank Inserted in Die
B. Extrusion
C. Separation of Tube from Punch

uniform cross section. Extrusion lengths of 20 feet are common, and some machines are capable of making 50-foot lengths. Typical hot extruded products are moldings of brass and aluminum, rods, bars, tubes, and structural shapes of aluminum and steel. Aluminum extrusions may be as large as 2 feet in diameter, but steel extrusions are limited to about 6 inches in diameter.

Cold extrusion is usually done with the metal at room temperature; although some metals, such as magnesium, are heated a few hundred degrees Fahrenheit. Cold extrusion pressures are several times higher than for hot extrusion, thus limiting cold extrusion to small- and medium-size parts. Cold extrusions must also have a uniform wall thickness throughout and be of balanced or symmetrical cross section.

Parts may be cold extruded from aluminum, steel, brass, copper, tin, magnesium, lead, zinc, and titanium. Examples include tubular containers of aluminum for food, beverages, photographic film, electrolytic condensers, ammunition projectile casings, hydraulic cylinders, and wrist pins of steel.

Impact extrusion is a form of cold extrusion used mainly for the manufacture of soft metal tubes for packaging of toothpaste and similar materials, Fig. 39-37. The tube begins as a solid disc of metal at the bottom of a closed die cavity. A punch strikes the disc a sharp blow, causing metal to squirt backwards up the punch, thus forming the tube. The blank thickness determines the length of the tube. After the punch is withdrawn from the die cavity, the tube is blown off the punch by air pressure. The process is very fast, with production rates reaching as high as 80 tubes per minute.

39-14. Powder Metallurgy

Powder metallurgy (P/M) is the technology of producing metal parts by compressing metal powders in precision dies or molds, usually at room temperature. This produces a fragile **green compact** which

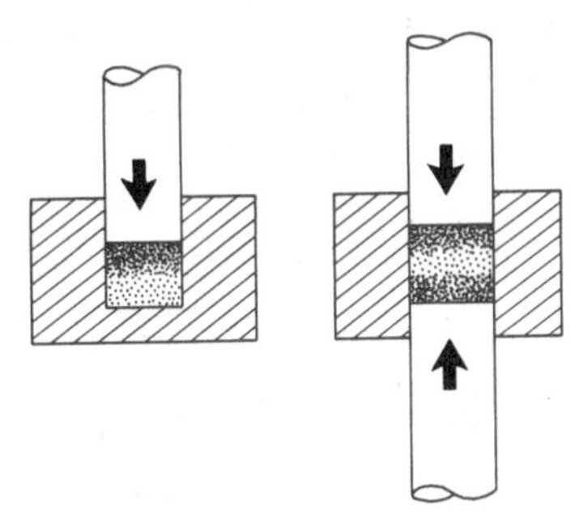

Fig. 39-38. Compaction with Single Punch Produces Less Uniform Density Than When Both Upper and Lower Punches Used

must then be heated in an atmospherically controlled (oxygen-free) **sintering furnace.**

Sintering (heating at elevated temperatures but not to the melting point) causes the green compact to acquire strength through the metallurgical bonding of the metal particles. Parts requiring high strength may be further compressed after sintering and may also be heat treated. When so treated, part strength and density approaches that of solid metal parts.

The quality of P/M parts depends largely on compacting the powders to a uniform density. Compacting pressures from 5 to 50 tons per square inch are used, but most parts are compressed with between 10 and 30 tons per square inch. Since metal powders do not flow readily, even when mixed with lubricants, a die with a single punch is satisfactory only for parts with the simplest of shapes. In order to obtain uniform density, most parts require dies with one or more upper and lower punches, Fig. 39-38. Due to their complexity, P/M dies are expensive, and the process is generally not economical for less than 20,000 parts. P/M is therefore almost exclusively a mass production process. Some presses designed especially for powder metal compacting produce green compacts in excess of 300 parts per minute.

The great bulk of powder-metal parts are made of iron-base or copper-base materials. The four main types of iron-base powders used are straight iron, iron/carbon, iron/carbon/copper, and iron/copper. Copper-base powders most widely used are straight copper, copper/zinc, copper/tin, and copper/nickel/zinc. P/M parts are also made of many other metals such as aluminum, nickel, chromium, stainless steel, beryllium, titanium, and cobalt. Some metals can only be processed by P/M methods; among these are tungsten, tungsten carbide, tantalum, and molybdenum.

Most P/M parts are functional in nature. Typical examples include bearings, gears, cams, and levers; other drive mechanisms and mechanical parts; filters of various types; and magnets and magnetic cores for electrical and electronic applications. P/M parts generally weigh less than a pound, but parts up to four pounds are commonly made, and 100-pound parts are possible.

Two unusual features make the P/M process especially important. It enables materials to be combined which previously could not be combined, and it permits parts of controlled density or porosity to be made. For example, steel parts can now be made with enough copper content so that when assembled and heated to the melting point of the copper the copper flows between the parts and brazes them together. Copper is combined with carbon to make long-wearing commutator brushes for electric motors and generators. Ceramics (nonmetallic materials) are combined with metals to create "cermets," a new cutting tool material.

Control of porosity is of particular advantage in making self-lubricating bearings and in the manufacture of filters. Porosity of P/M bearings is controlled so that they may be **impregnated** with from 10 to 40 percent of oil by volume, an amount usually calculated to lubricate the bearing for life. Filters can be made to trap particles of selected size or to separate liquids of different densities, Fig. 39-39.

Porosity in P/M parts may be removed by filling the pores with a metal of lower

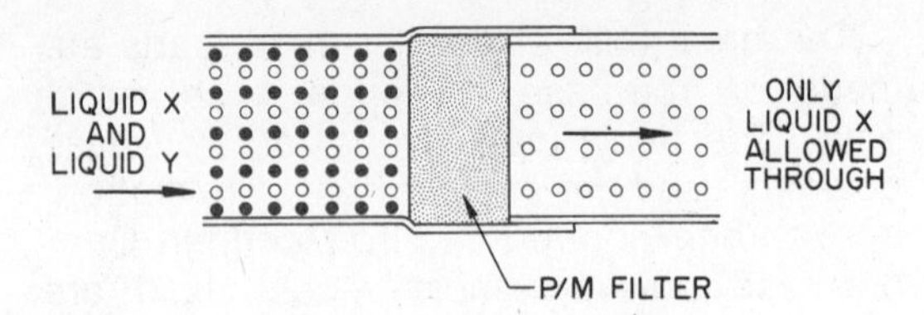

Fig. 39-39. Liquids of Different Densities Can Be Separated by P/M Filters

melting point, a process know as **infiltration.** Infiltration may be needed to effectively seal a part from transmitting fluids or gases or to improve the mechanical properties of the part.

The largest use of P/M parts in the United States is by the automotive industry, which uses over half of all P/M production. An average automobile contains more than 100 P/M parts.

The small appliance and portable power tool industries are also large consumers of P/M parts, accounting for almost 20 percent of P/M production.

Words to Know

bellows
bench molding
blowholes
bottom board
bulb sponge
centrifugal casting
CO_2 process
cope
cope-and-drag pattern
core box
core oven
core prints
die casting
draft
drag
draw pin
dry-sand mold
extrusion
fillet
flask
follow board
follow-board pattern
foundry
full-mold process
gate
gate cutter
green compact
green-sand mold
impact extrusion
impregnation
infiltration
investment mold
loose-piece pattern
lost-wax process
match-plate pattern
mold
molding
molding board
molding sand
mulling machine
negative draft
one-piece pattern
P/M
parting compound
parting line
pattern
permanent mold
plaster mold
polystyrene foam
positive draft
powder metallurgy
resin
riddle
riser
shell mold
shrink rule
sintering
slick and spoon
solid pattern
split pattern
sprue
sprue pin
striker bar
sweep pattern
taper
temper
vent holes
waterless sand
wax pattern
zero draft

Review Questions

1. What is metal casting?
2. What is a foundry?
3. What is a foundry pattern?
4. Name some materials that are used to make patterns.
5. List the different kinds of patterns and give the main advantage of each kind.
6. What is a fillet?
7. Of what materials are fillets made and how are they applied?
8. What is draft?
9. What is shrinkage? How does the patternmaker allow for shrinkage?
10. What is bench molding?
11. What is a flask? Name its two major parts.
12. Why is green sand called green?
13. Name the materials used to make green sand.
14. What does sand temper mean? How is it tested?
15. Why is parting compound used?
16. What is a foundry riddle? For what is it used?
17. Why are molds vented?
18. What hazard is associated with using green sand which is too wet?
19. Describe the full-mold casting process.
20. Why are cores used in molds?

21. What advantages do dry-sand molds have over green-sand molds?
22. Explain shell molding.
23. What is a permanent mold?
24. What is die casting? What are its major advantages?
25. Describe how investment molds are made.
26. Describe how plaster molds are made.
27. What is meant by centrifugal casting?
28. Explain the process of extrusion. How does it differ from impact extrusion?
29. List three advantages of hot extrusion over cold extrusion.
30. Name several products familiar to you which were made by various extrusion processes.
31. Explain how parts are made by powder metallurgy.
32. What are two unique advantages of powder metallurgy? Suggest several products which are made possible by these advantages.
33. Explain the meaning and purpose of impregnation and infiltration in powder metallurgy.
34. What kinds of metals are used in most P/M parts?
35. What industries are the largest consumers of P/M parts?

Mathematics

1. Two hundred and fifty pounds of dry sand are to be mixed to make green-sand molds. If the moisture content of the sand is to be 3%, how many pounds of water must be added?

Career Information

1. Write a report on the composition, mixing, and testing of green sand.
2. Make a list of industries which might mold and cast metal.
3. Compare the work of a patternmaker who makes patterns for sand casting with that of a diemaker who makes dies for die casting.

UNIT 40

Melting and Pouring Metal

40-1. Furnaces for Melting Metal

There are several kinds of furnaces for melting metals. Some foundries make many large castings and so must have large furnaces. One kind, a cupola, is used in an **iron foundry** (a foundry that specializes in making castings from iron). The cupola (see § 17-4) is like a blast furnace (see § 17-3 and Fig. 17-5), but it is smaller. Other foundries specialize in making castings from other metals and use furnaces suitable for those metals.

Fig. 40-1. Soldering Furnace with Provision for Small Melting Pot

Small quantities of low-melting-point metals can be melted in some soldering furnaces, Fig. 40-1. For larger quantities of metal, a gas-fired **crucible furnace,** like the one shown in Fig. 40-2, might be used. With this type of furnace, natural or manufactured gas[1] and air are mixed and burned

Fig. 40-2. Gas-Fired Furnace with Ultraviolet Safety System and Automatic Spark Ignition

[1]Natural gas is found in the earth, usually in areas which also have oil wells. Manufactured gas is made from coal, coke, or petroleum products.

inside the furnace. The hot gases flow around the crucible containing the metal to be melted. A gas furnace of this type is good for melting such metals as lead, brass, aluminum, and zinc, which can be cast at relatively low temperatures.

Where higher temperatures are needed to melt metal, electricity is sometimes used. One type of electric furnace is called an **arc furnace,** Fig. 40-3. It operates, in principle, like a giant electric arc-welding machine. Inside the furnace, electricity is made

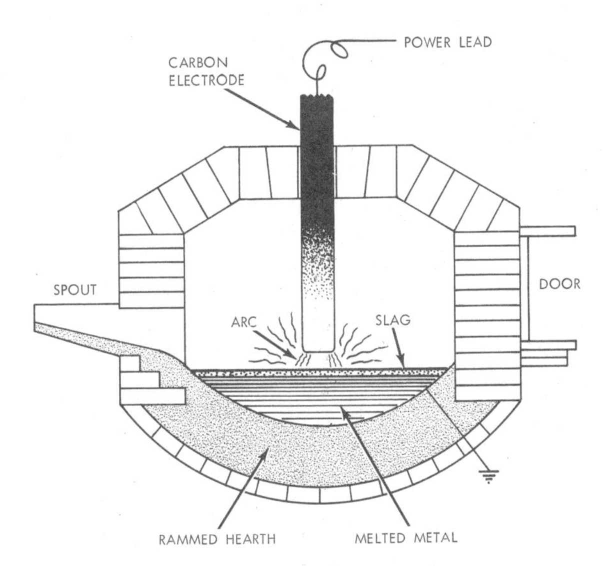

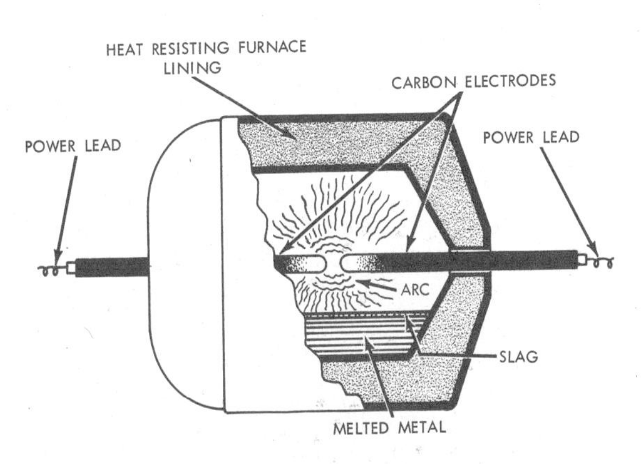

Fig. 40-3. Two Types of Electric Arc Furnace — Direct and Indirect Arc

to jump from a rod made of carbon, called an "electrode," and through an air space, called "the gap." When the electricity jumps this gap, an arc is made which gives off a brilliant white light and large amounts of heat.

40-2. Clothing for Melting and Pouring

Without the proper clothing, melting and pouring can be dangerous. Metal may be spilled from the ladle. Also, gas pockets in the mold and moisture in the ladle or mold can cause explosions. These explosions can spray melted metal and cause body burns and injury to the eyes.

A cap, goggles, heavy leather or asbestos apron, asbestos leggings which cover the shoes, and heavy asbestos gloves should be worn to guard against burns. See Fig. 40-4.

40-3. Melting of Metals

New furnace linings and new **crucibles** should be carefully and slowly **preheated** to minimize the danger of their cracking.

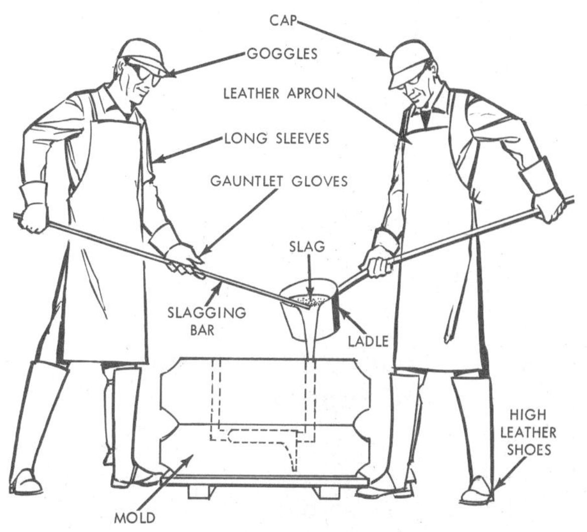

Fig. 40-4. Clothing for Melting and Pouring

Furnaces and crucibles which have already been used can be charged without preheating. Crucibles should not be packed tightly because expansion of the charge may cause the crucible to crack. The crucible should be placed in the center of a pot-type furnace so that it is uniformly heated.

Light the furnace and allow it to preheat with a reduced flame. Then bring the flame up to full heat. After the initial charge of metal has melted, add as much clean, dry metal as is needed for the pour. **Never allow moisture to enter a pot of liquid metal.** The moisture rapidly turns to steam and can cause the molten metal to explode. When the desired amount of metal has melted, check its temperature frequently to guard against overheating.

Just before pouring, **flux** should be added, stirred into the molten metal, and the **slag** should be skimmed from the top with a **slagging bar.** Fluxes are compounds which help purify the molten metal by combining with the **suspended oxides** or **dissolved gases.**

Slag is the waste material that floats on the melted metal. It contains impurities from the pieces of metal that were put into the furnace to be melted. These impurities are sand, scale, and dirt. The slag also includes the metal that has combined with oxygen and nitrogen from the surrounding air during melting. Slag looks like scum and has a dull appearance. It acts like a blanket and keeps more of the melted metal from combining with the oxygen and nitrogen in the air. It must not be poured into the mold. It can be scooped off the top of the melted metal before pouring or held in the ladle with the slagging bar.

40-4. Pouring Temperature and Superheat

If melted metal is poured into the mold before it has reached the proper temperature, it may solidify before it has completely filled the mold cavity. When this happens, the metal is said to have been "poured cold." This is more likely to happen with molds that have thin or small sections through which melted metal must flow to fill the mold. On the other hand, if the metal is heated above the proper temperature, time and heat are wasted, the mold is more likely to be damaged when the metal is poured, the metal becomes more dangerous to handle, and coarse-grained castings often result.

The proper temperature to pour metal is called the **pouring temperature.** The pouring temperature depends on the kind of metal being used, the shape of the mold, and how many molds are to be poured. If several molds are to be poured, the melted metal may be hot enough for the first molds, but too cold for the last mold. If the last mold is poured cold, it will "freeze off," and the mold will be wasted.

The pouring temperature of a metal is often given in terms of its **superheat** or **degrees of superheat.** Superheat is the number of degrees above the melting temperature. For example, aluminum melts at 1218° F. If it is poured into the mold when it reaches 1418° F., it is said to have 200° of superheat. In general, low pouring temperatures produce better castings. Table 16 gives the melting points of various metals.

40-5. The Ladle

A **ladle** is used to transfer melted metal from the furnace to the mold and to pour the melted metal into the mold. Some ladles look like big water dippers, Fig. 40-4. Unlike a water dipper, care must be used to keep the ladle free from moisture. If there is water or moisture in the ladle, it will suddenly turn to steam when melted metal touches it. When moisture **flashes** to steam in the presence of melted metal, an explosion results and the melted metal splatters. The proper clothing should always be worn by the melter and pourer.

Sometimes the ladle is heated separately in an oven or with a gas flame. This is done to dry out the ladle and be sure all moisture is driven away. Heating the ladle also reduces the amount the melted metal will cool as it is poured from the furnace into the ladle.

40-6. Preparation for Pouring

The mold should be placed near the melting furnace so that the melted metal can be poured quickly and will not have to be carried far. Molds should be placed on a bed of dry sand at least an inch thick, the bed extending several inches beyond the molds on all sides. If there are several molds to be poured, place them in a row for convenience and safety of pouring. Place a **pig mold** after the last mold to accept the excess liquid metal.

Be sure that all persons directly involved in the pouring are dressed with the proper safety equipment. Set up a barrier to keep observers a safe distance away from the pouring area. See that there is a clear path from the furnace to the mold area. There should be nothing to stumble over or bump into while carrying the hot metal.

Some means to measure or gage the temperature of the melted metal should be available. An instrument that measures the temperature of melted metal is called a **pyrometer,** Fig. 40-5. **Pyro** means heat; thus a pyrometer is a heat or temperature meter. There are a number of kinds of pyrometers. Your instructor will show you how to determine the temperature of the metal.

Get a metal bar at least 3′ long to use as a **slagging bar.** The **slagging bar** is used to keep slag from running into the mold while the metal is being poured, Fig. 40-4.

Before the melted metal reaches the pouring temperature, it should be decided who will read the pyrometer, who will operate the furnace, who will handle the ladle, and who will direct the pourer. It is desirable to make one or more practice runs to be sure that everyone knows his job and is able to carry it out safely.

40-7. Pouring the Mold

When the melted metal reaches the pouring temperature, the furnace is shut off and the crucible is removed from the furnace with crucible tongs and placed in the crucible pouring tool. Flux is stirred in and the slag is skimmed off. The person directing the pouring should "talk" the pourer over to the mold which is to be poured first. The pourer should pour the melted metal smoothly and steadily, keeping the **pouring cup** of the mold full. The stream of metal should be continuous and uninterrupted until the mold is full. The pourer can tell when the mold is almost full by watching the metal fill up the **riser** of the mold.

During the pouring, someone should keep the slag in the ladle with the slagging

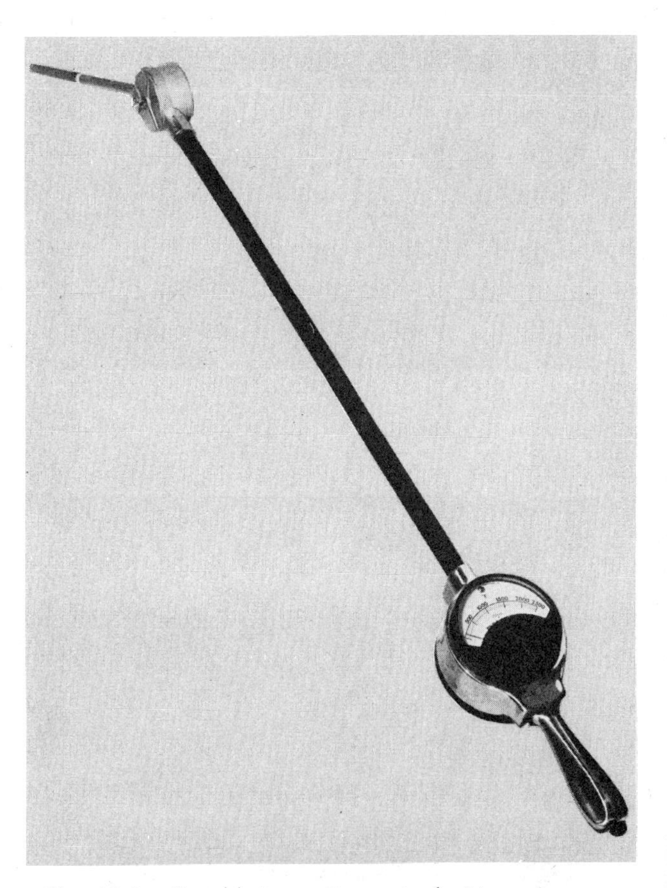

Fig. 40-5. Portable Lance Pyrometer for Measuring Temperature of Molten Nonferrous Metals

bar (see Fig. 40-4) so the slag will not run into the mold and cause a defect in the casting. Several persons should be waiting with shovels of sand to throw onto any melted metal that spills on the floor. Also, there are often special tasks that come up that these persons can do. Should the mold break and the melted metal run out onto the floor (called a **runout**), enough sand should be on hand to cover ail the metal that was in the mold.

Figure 40-6 shows the pouring of aluminum from a large crucible handled by two workers. Figure 40-7 shows molds on a conveyor line being poured by a single worker operating a ladle supported by an overhead crane. The mechanism to which the ladle is attached enables the operator to easily raise or lower the ladle, position the pouring spout over the mold, and tilt it to control the rate of pouring.

Fig. 40-6. Hand Pouring Aluminum into a Sand Mold

40-8. Cleaning and Finishing the Casting

After the mold has cooled so that it can be handled with gloves and **tongs,** it can be broken apart and the casting removed. This is called the **shakeout operation.** In large foundries, the mold is put on a vibrating conveyor. The sand falls through the conveyor, leaving the casting on the conveyor, Fig. 40-8.

When the casting has been shaken out of the mold, it still has much sand and dirt clinging to it. Before the casting can be machined, the sand must be removed or the cutting tools will become dull very quickly. This is called the **cleaning operation.** Several methods are used to clean castings. Among them are **sand blasting, shot blasting,** and **tumbling.**

In sand blasting, sand is blown at high speed against the casting. The sand, dirt,

Fig. 40-7. Sewing Machine Castings Move on Conveyor into Cooling Tunnel

Fig. 40-8. Automatic Casting Shake-Out from Conveyor Trucks

and scale on the casting are rubbed and blown away.

In shot blasting, pieces of metal about the size of an air rifle BB, called **shot,** are hurled against the casting. Again, the sand, dirt, and scale on the casting are rubbed and blown away.

In tumbling, the castings if they are small, are put into large metal barrels and the barrel is turned and tumbled. The castings knock against and rub each other removing the sand, dirt, and scale, Fig. 40-9.

Before the casting is a **finished casting,** the **sprues, gates,** and **risers** (see Fig. 40-10) are sawed off or knocked off with a sledge. Also, the **flash**[2] and sharp edges that occur at the parting line must be removed with a grinder or file. The casting is now a finished casting. It is ready for machining.

Fig. 40-9. Castings Emerging After Being Cleaned by Tumbling

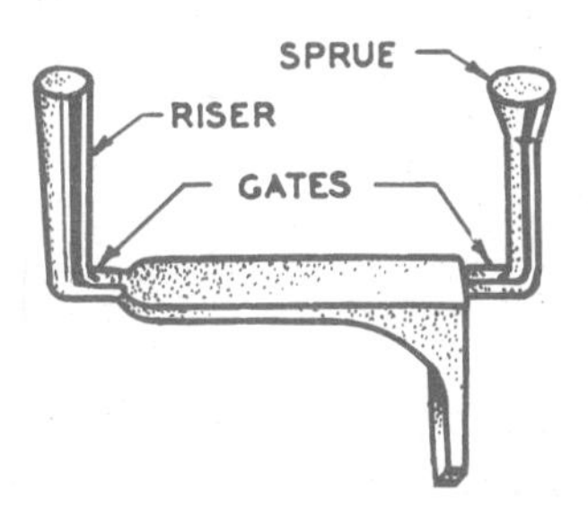

Fig. 40-10. Castings as Taken from Mold

Words to Know

arc furnace
crucible
crucible furnace
cupola
finished casting
flash
flux
frozen metal
iron foundry
ladle
oxides
pig mold
pouring cup
pouring temperature
preheating
pyrometer
raw casting
riser
runout
sand blasting
shakeout
shot blasting
slag
slagging bar
sprue
superheat
tongs
tumbling

Review Questions

1. What is a cupola?
2. What is a crucible furnace?
3. Describe the operation of an electric furnace.
4. List the special clothing which should be worn when melting and pouring metal.
5. What is flux? How is it used?
6. What is slag? How is it useful?
7. Why is it important to pour metal at the proper temperature?
8. What is superheat?
9. What is a ladle? Why should it be dry when used for pouring metal?
10. Describe how a pourer should fill a mold with melted metal.
11. What is a runout?
12. What is a shakeout?

[2]A small open crack occurs at the parting plane of a mold because the cope and drag do not fit together perfectly. When the mold is poured, melted metal fills the open crack. As the metal becomes solid, a thin sharp edge of metal **flash** occurs all around the casting at the parting plane.

13. Name several ways castings are cleaned.
14. How is a raw casting made into a finished casting?

Mathematics

1. Seventeen pounds of sand are used in a mold to make a casting that weighs 5 pounds. How much sand must a foundry handle to make 600 castings? What is the weight ratio of casting to mold?
2. A certain casting is 30″ long at room temperature (70° F.). How long will it be when it is heated to 1270° F. if it enlarges $\frac{1}{16}$″ per foot for each 100° of heating?

Career Information

1. Describe the kind of work a metallurgical engineer or a metallurgical technician would do at a foundry.
2. What sort of education and training would be needed in order to become a salesperson for a foundry?

PART XII

Abrasive Materials and Products, Tool Sharpening

UNIT 41

Abrasives

41-1. What Does Abrasive Mean?

Abrasive comes from the word **abrade,** which means to rub off. An abrasive substance is a very hard, tough material which has many sharp cutting edges and points when crushed and ground into grains like sand. Abrasives must be harder than the materials they cut. Several common forms in which abrasives are used in metalworking include the following:

1. Abrasive cloth (also called coated abrasive)
2. Loose grain and powder abrasive
3. Abrasive compounds (in the form of paste, sticks, or cakes)
4. Grinding wheels
5. Sharpening stones

This unit is concerned principally with the selection and use of abrasives for hand polishing. Most hand polishing is done with abrasive cloth. Occasionally, hand polishing is done with loose-grain abrasive powders or abrasive compounds. The use of abrasive polishing compounds for machine buffing and polishing is included in Unit 35. Grinding wheels and their selection are included in Unit 42. Sharpening stones are included in Unit 44.

In selecting abrasives, one should be familiar with the most common kinds of abrasive material, their properties, grain size, and their uses. These factors apply to abrasive cloth, grinding wheels, and to loose-grain abrasives. They are included in this unit.

41-2. Pride in Work

Things must be beautiful as well as useful. Finely polished plain surfaces have a beauty all their own. One who takes pride in his personal appearance also takes pride in his work. The surface should be filed carefully until satin smooth and free of machine marks. It is then polished smooth and bright with an abrasive. A **craftsman** must be able to produce a piece of finely polished work.

41-3. What Does Polishing Mean?

Polishing means to make smooth, bright, and glossy by rubbing. To polish is to change a rough, uneven, dull surface with irregular scratches to a surface with very fine, uniform, **parallel** cuts or grooves that cannot be seen with the naked eye. **Hand polishing** means to rub the surface of the metal by hand, back and forth, with an **abrasive** or with **steel wool.**

41-4. Properties of Abrasives

The characteristics of a material are called its properties. Abrasives must possess three common properties: (1) hardness, (2) fracture resistance, (3) wear resistance.

Hardness means the ability of the abrasive to scratch the surface of the material to be polished or ground. The abrasive must be harder than the material to be cut.

Fracture resistance of an abrasive means its toughness, that is, the ability to resist breaking or crumbling when pressed hard against the work during polishing or grinding. The fracture resistance should neither be too high nor too low. It should be such that when the abrasive grains become dull they will fracture away, thus exposing new sharp cutting edges. Fracture resistance of grinding wheels is related to the kind of bonding material which binds the grains together.

Wear resistance of an abrasive refers to its ability to resist wear and stay sharp longer. It is related to the hardness of the abrasive material. Thus, harder abrasive materials generally are more wear resistant than softer. However, there are some exceptions to this general rule.

41-5. Kinds of Abrasives

The common abrasive materials are classified as either natural or artificial abrasives. **Natural** abrasives generally are minerals which come from nature. They occur either in the form of grains, like sand, or in the form of large rocklike chunks. The large chunks must be crushed or ground into small abrasive grains. Some common natural abrasives include flint, garnet, emery, corundum, crocus, and diamond. These will be explained in greater detail.

The **artificial** man-made abrasives are also known as synthetic or manufactured abrasives. With the exception of diamond, artificial abrasives are harder than the natural abrasives. Diamond is the hardest abrasive material. The artificial abrasives have largely replaced the natural abrasives in the metalworking industry because of their hardness and wear resistance.

Natural Abrasives

The following are some common natural abrasives.

Flint comes from the mineral quartz — a crystalline rocklike material. It is used in making the familiar yellowish colored abrasive paper called **sandpaper.** It is one of the oldest kinds of abrasive paper and is used in woodworking.

Garnet is a reddish colored glasslike mineral which is crushed into fine abrasive grains. It is harder and sharper than flint and is widely used for woodworking.

Emery is one of the oldest kinds of natural abrasives used for metalworking. It is black in color and is composed of a combination of corundum and iron oxide. **Corundum** is aluminum oxide, Al_2O_3. Prized gems such as emerald and ruby are the purest form of corundum. Emery, used for making an emery cloth for polishing metals, is about 60% corundum. Emery grains are not as sharp as artificial abrasives. The cutting action of emery is slight; therefore, it is used largely as a polishing abrasive.

Crocus is a fine, soft, red abrasive of iron oxide, or iron rust. It may be produced artificially or naturally and is used to clean and polish metal surfaces to a high gloss. It is available in the form of **crocus cloth** or as a polishing compound.

Diamond is the hardest substance known. It is used in the form of abrasive grains which are bonded together to form a thin layer of abrasive. The layer of abrasive is bonded to a wheel, thus forming a grinding wheel. Diamond grinding wheels are used for grinding very hard materials, such as cemented-carbide cutting tools, ceramic cutting tools, glass, and stone. The diamonds used for this purpose are industrial diamonds in the form of chips or grains. They are much less expensive than the diamonds used for jewelry.

A diamond chip may be brazed on the end of a soft steel bar to make a tool for dressing or truing softer grinding wheels. Diamonds used in industry have been produced artificially during recent years. However, they are still almost as expensive as natural diamonds. Fine diamond dust is also used in making lapping compound for lapping hardened steel and other very hard materials.

Artificial Abrasives

The artificial abrasives were developed during the latter part of the 19th century. They are a result of society's effort to produce abrasives which are harder, tougher, and more wear resistant than the natural abrasives.

The artificial abrasives are manufactured at extremely high temperatures in electric furnaces. When first produced, they were very expensive because of the shortage and high cost of electric power at that time. Today, however, the artificial abrasives have largely replaced the natural abrasives in the metalworking industry. Three common artificial abrasives include silicon carbide, aluminum oxide, and boron carbide.

Silicon carbide is made by heating a mixture of powdered sand, coke, sawdust, and common salt in an electric furnace. It comes from the furnace in masses of beautiful, bluish crystals like diamonds. The crystals are crushed into fine abrasive grains which are used in making grinding wheels, abrasive stones, and coated abrasives (abrasive cloth).

Silicon carbide is harder and more brittle than aluminum oxide abrasive. It is hard enough to cut aluminum oxide. It is generally used for polishing or grinding materials of low tensile strength, including the following: cast iron, aluminum, bronze, tungsten carbide, copper, rubber, marble, glass, ceramics pottery, magnesium, plastics, and fiber. Silicon carbide is known by trade names such as Carborundum, Crystolon, Carbolon, and Carbonite.

Aluminum oxide is produced by heating bauxite ore in an electric furnace at extremely high temperatures. With the addition of small amounts of titanium (a lustrous, hard, lightweight, metallic element), greater toughness can be imparted to the aluminum oxide. The center of the solid mass formed in the furnace is aluminum oxide. It is broken up and crushed into fine grains for making grinding wheels, abrasive stones, and coated abrasives.

Aluminum oxide has properties somewhat different than silicon carbide. It is not as hard, but it is tougher and does not fracture as easily. Aluminum oxide abrasives are recommended for grinding and polishing materials of high tensile strength, including the following: carbon steels, alloy steels, hard or soft steels, malleable iron, wrought iron, and tough bronze. Approximately 75% of all grinding wheels in use today are made of aluminum oxide. Aluminum oxide is known by trade names such as Alundum, Aloxite, Borolon, Exolon, and Lionite.

Boron carbide is produced from coke and boric acid in an electric furnace. It is known by the trade name **Norbide,** produced by the Norton Company. It is harder than either aluminum oxide or silicon carbide and can cut either of them. However, it is not as hard as diamond. It is used in stick form to dress or true grinding wheels 10″ or less in diameter. It is also used in powder form, instead of diamond dust, for lapping hardened steel or other very hard materials.

41-6. Grain Size of Abrasives

Grain size refers to the size of the abrasive grains used in the manufacture of abrasive materials. Most abrasive manufacturers produce abrasives for grinding wheels according to the Standard Abrasive Grain Sizes shown in Table 28.

The grain sizes given in Table 28 are the number of meshes or holes per inch in the

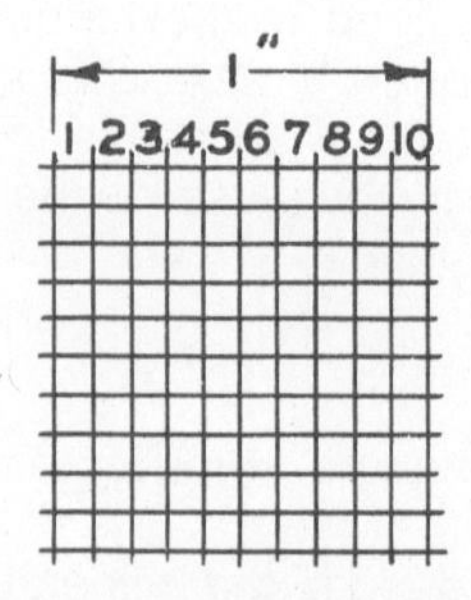

Fig. 41-1. A 10-Grain Screen (Actual Size)

screen, Fig. 41-1. Thus, a 10-grain abrasive is one which will just pass through a 10-mesh screen; that is, a screen which has meshes 1/10 of an inch square (less the width of a screen strand), 10 meshes per inch, or 100 meshes per square inch.

In Fig. 41-2:

(A) is a 60-grain screen;
(B) is the 60-grain screen magnified 16 times;
(C) is the actual size of the grain which passes through the 60-grain screen; and
(D) is the grain magnified 16 times.

41-7. Coated Abrasives

A coated abrasive is composed of a flexible backing material to which abrasive grains are bonded with cement or glue. Hence, **abrasive cloth** and **abrasive paper** are classified as **coated abrasives.** The backing material may be paper, cloth, fiber, or a combination of these materials. Coated abrasives used in the metalworking industry include emery, aluminum oxide, silicon carbide, and crocus. Coated abrasives are available in many forms for both hand and machine polishing or grinding. They are

A. 60-Grain Screen

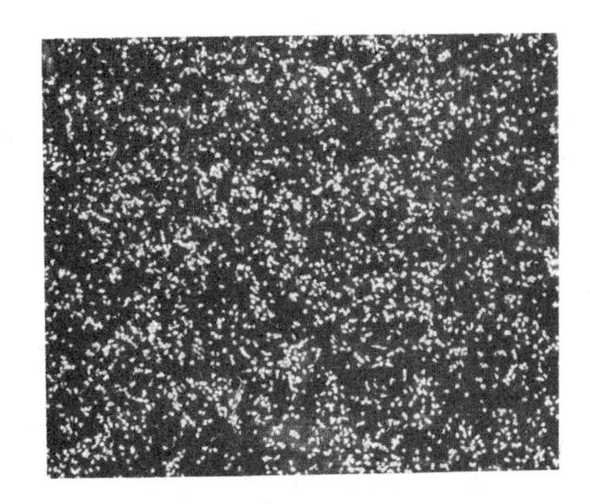

C. 60-Grain Abrasive

B. 60-Grain Screen Enlarged 16 Diameters

D. Abrasive Grains Enlarged 16 Diameters

Fig. 41-2. A 60-Grain Screen and Grain Which Passes through It

Fig. 41-3. Variety of Commonly Used Coated Abrasives

available in the form of belts, rolls, sheets, discs, spiral points, and cones, Fig. 41-3.

Grades of Coated Abrasives

The relative coarseness or fineness rating of coated abrasives is somewhat different than the standard rating given to abrasive grains used for grinding wheels in Table 28. The following are the common abrasive grain sizes used for coated abrasives:

Extra-Coarse — 12, 16, 20, 24, 30, 36
Coarse — 40, 50
Medium — 60, 80, 100
Fine — 120, 150, 180
Extra-Fine — 220, 240, 280, 320, 360, 400, 500, 600

Backing materials on coated abrasives include the three items listed below:

1. **Paper.** Used for hand applications for woodworking and fine finishing abrasives.
2. **Cloth.** Two weights of cloth are used as backing. The lightweight cloth is called **jeans** and is flexible for hand polishing. The heavyweight cloth is called **drills.** It is more stretch resistant and is used for machine buffing.
3. **Fiber.** This type of backing is extra strong, durable, and used for tough machine polishing and grinding applications.

41-8. How Is Abrasive Sold?

Loose-grain abrasive is sold by the pound, usually in cans. Abrasive cloth is sold by the sheet and in packets of 25, 50, and 100 sheets. The sheets are usually 9″ x 11″. It is also sold in rolls ½″ to 6″ wide by 25 or 50 yards long and in endless belts, Fig. 41-3.

41-9. Holding Work for Hand Polishing

To keep from scratching a polished surface with the **hardened steel** jaws of the vise, the work should be placed between **vise jaw caps.** See § 25-3. Jaw caps are usually made from soft brass, copper, or aluminum.

41-10. Choosing Grade of Abrasive

To hand polish a large, roughly filed surface, first rub it with a medium grade of abrasive, such as No. 80 or 100 abrasive cloth; then continue with finer grades until you get the depth of cuts or grooves you want. For polishing a smooth filed surface, use Nos. 120 or 150 abrasive cloth and finish with No. 180 or a finer grade, if desired.

41-11. How to Polish

It has been explained that the reason for polishing is to change a rough, uneven, dull surface with irregular scratches to a surface with very fine, uniform, **parallel** (see § 7-15) cuts or grooves which cannot be seen with the naked eye.

After filing, but before polishing is begun, the surface in Fig. 41-4 contains many tiny, irregular grooves of various depths which

Fig. 41-4. Rough, Uneven Surface (Enlarged)

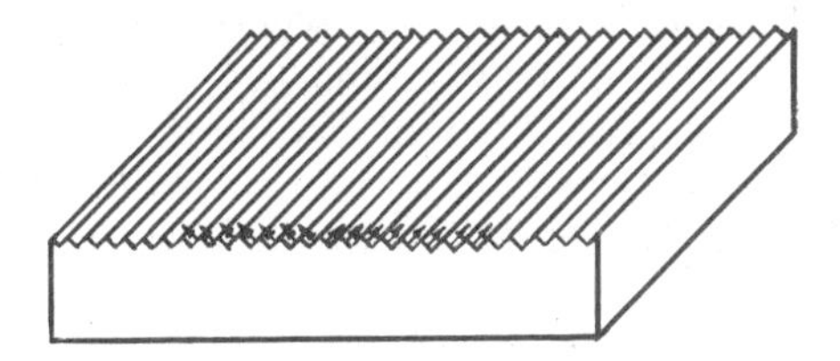

Fig. 41-5. Surface with Uniform and Parallel Abrasive Cuts or Grooves (Enlarged)

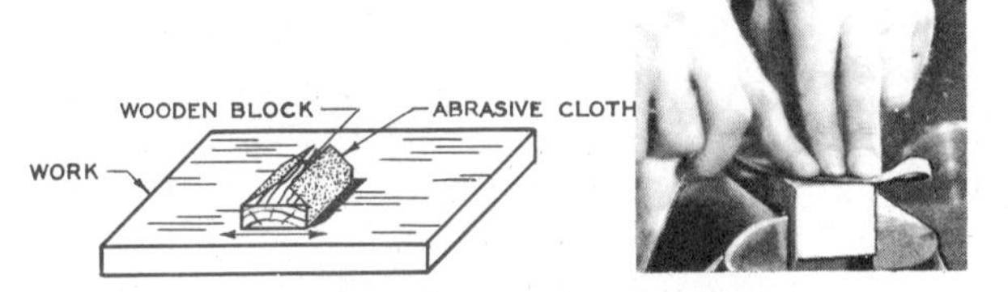

Fig. 41-6. Polishing Flat Work

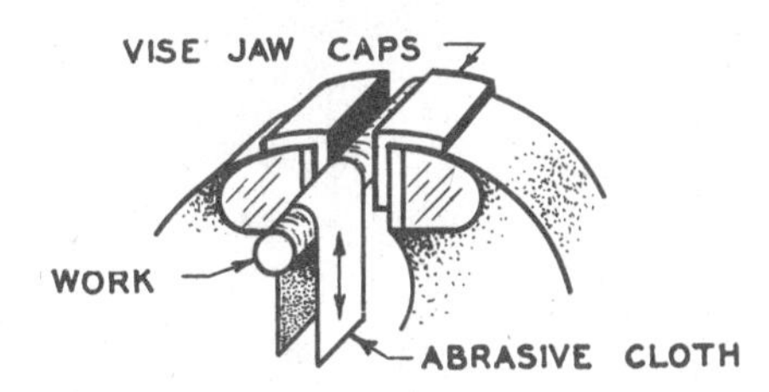

Fig. 41-7. Polishing Round Work

TABLE 28

Abrasive Grain Sizes for Grinding Wheels

Coarse	Medium	Fine	Very Fine
10	30	70	220
12	36	80	240
14	46	90	280
16	54	100	320
20	60	120	400
24		150	500
		180	600

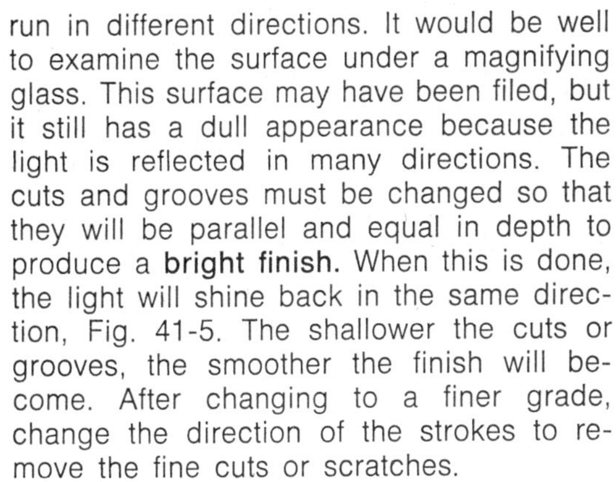

run in different directions. It would be well to examine the surface under a magnifying glass. This surface may have been filed, but it still has a dull appearance because the light is reflected in many directions. The cuts and grooves must be changed so that they will be parallel and equal in depth to produce a **bright finish.** When this is done, the light will shine back in the same direction, Fig. 41-5. The shallower the cuts or grooves, the smoother the finish will become. After changing to a finer grade, change the direction of the strokes to remove the fine cuts or scratches.

A piece of worn-out abrasive cloth, moistened with a drop of oil, makes a smoother finish on steel. This oil-polished surface will not rust as quickly as it will if it is polished dry.

41-12. Polishing Flat and Round Work

When flat work is polished, the abrasive cloth may be rubbed back and forth upon the metal, as shown in Fig. 41-6. When round work is polished on the bench, the work may be held in a vise and the abrasive cloth worked back and forth around the work, as shown in Fig. 41-7. Polishing round work is much more efficient if the workpiece can be rotated in a machine.

41-13. What Does Lapping Mean?

Lapping is the fine removal of small amounts of metal from a **hardened steel** surface that is slightly oversize. It is done where and when ordinary grinding would remove too much metal, especially on measuring tools where smooth surfaces and exact sizes are needed. The tool used for this purpose is called a **lap;** it is made of

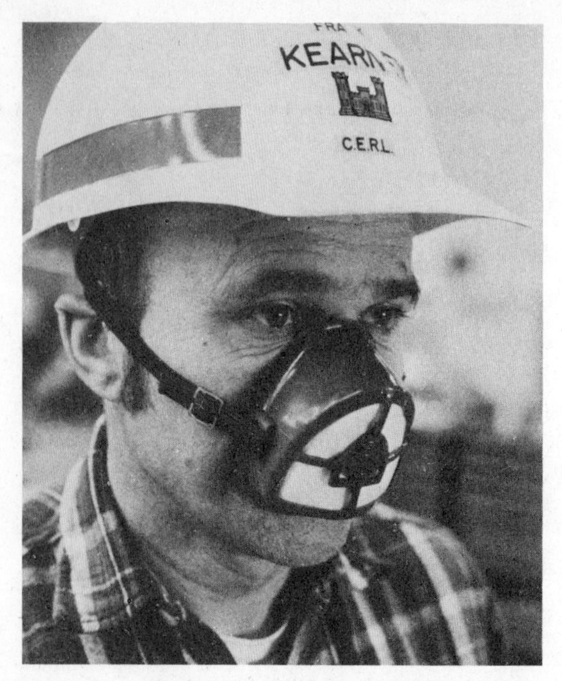

Fig. 41-8. Respirators Filter Out Abrasive Dust Particles

copper, brass, soft cast iron, or lead; it must be of a material softer than the metal to be lapped.

A fine **powder abrasive,** known as **lapping powder,** is used for lapping; the pulverized sizes of the grains of the abrasives are numbered as given in section 41-6. The lapping powder is made into a paste by mixing it with **Vaseline** or **lard oil;** it is then called **lapping compound** or **grinding compound.** The lap is coated with the paste and then rubbed against the surface of the part that is to be slightly ground to size. **Lapping** can be a hand or machine operation.

41-14. Polishing with Steel Wool

Polishing or cleaning of metal by hand may be done with fine steel wool, which is made of **steel shavings.** The fine, sharp edges of the steel shavings scratch the metal and thus polish it. Steel wool is made in seven grades ranging from fine to coarse as follows: 0000, 000, 00, 0, 1, 2, and 3. The fine grades, 0000 and 000, will provide a low-lustre finish on copper, aluminum, and other nonferrous metals.

41-15. Working Safely with Abrasives

Eye protection should always be worn when working with abrasives, especially when using any kind of grinding machine. Whenever possible, grinding should be done wet so as to prevent abrasive dust from contaminating the air breathed. Long exposure to abrasive dust particles can cause a lung disorder known as silicosis. In situations which produce high concentrations of abrasive dust, the lungs can be protected by wearing an effective respirator, Fig. 41-8.

Words to Know

abrasive
abrasive cloth
aluminum oxide
artificial abrasive
boron carbide
coated abrasive
corundum
crocus
diamond
emery
flint
fracture resistance
garnet
grain size
grinding compound
lapping
lapping compound
mesh
natural abrasive
silicon carbide
steel wool
wear resistance

Review Questions

1. What is meant by hand polishing?
2. What is an abrasive?
3. List three properties which abrasives must possess.
4. Explain the principle difference between natural and artificial abrasives.
5. List five natural abrasive materials.
6. List three artificial abrasives.
7. Name five forms in which abrasives are used in metalworking.

8. List several kinds of materials which are ground or polished with aluminum oxide abrasives.
9. List several kinds of materials which are ground or polished with silicon carbide materials.
10. For what purpose is boron carbide used?
11. Explain how abrasive grain size is determined.
12. What is the meaning of the term "coated abrasive"?
13. What kinds of coated abrasive materials are used for metalworking?
14. In what forms are coated abrasives available?
15. What is lapping? For what it is used?
16. What is steel wool? For what is it used?

Mathematics

1. How many openings are there in a piece of 36-mesh screen that is 2″ square?

Social Science

1. What are the dangers of working in a place where there is dust as a result of polishing, especially that done by power?

Career Information

1. What is silicosis?
2. Choose a natural abrasive and write a story telling how it is produced.
3. Choose an artificial abrasive and write a story telling how it is manufactured.

UNIT 42

Grinding Wheels

42-1. What Are Grinding Wheels?

Grinding wheels are abrasive tools used for shaping and finishing metals. They are made by mixing abrasive grains with various bonding materials, pressing this mixture into grinding wheel shapes, and then baking the wheels until they are **cured** (hard). **Mounted wheels** are small grinding wheels which are made with a steel shaft permanently attached. Figure 42-1 gives some indication of the wide range of grinding wheel sizes and shapes which are available.

42-2. Reasons for Grinding

Grinding is used principally for the following:

1. To make cutting edges on chisels, drills, knives, and other cutting tools made from hardened steel.
2. As a machining process to cut metal to its desired shape and size.
3. To make smooth, polished surfaces such as are required on bearing surfaces and rolls for processing various materials.

42-3. How Does a Grinding Wheel Cut?

The **grinding wheel** is made of abrasive grains which have small cutting edges and points. Examine a grinding wheel with a magnifying glass and you will discover many small cutting edges just like the teeth of a file. Thus, the grinding wheel is a **cutting tool**. The **cutting edge** on each grain **cuts** a tiny chip from the metal. These chips are very small indeed, but the wheel turns at a high speed and many small cutting edges cut many small chips in a very short time; thus, much metal is cut away quickly.

42-4. Abrasives

Most grinding wheels are manufactured from two kinds of artificial abrasive materials: aluminum oxide and silicon carbide. The properties and kinds of abrasives used in manufacturing grinding wheels and the materials which may be ground with each kind of abrasive are explained in section 41-5. If you have not read these sections, you should read them before studying this unit further.

42-5. Grain Sizes of Abrasives

All grinding abrasives are first crushed and ground, then passed through sieves of different mesh size, and finally graded accordingly. The grains are numbered according to their sizes, Fig. 42-6. Grinding wheels with coarser grain sizes are used for fast cutting and where quality of finish is not important. Wheels with finer grain sizes are used when only a small amount of metal is to be removed and when a fine finish is required. The grain sizes of abrasives are explained in section 41-6.

Fig. 42-1. Common Types of Tool-Grinding Wheels

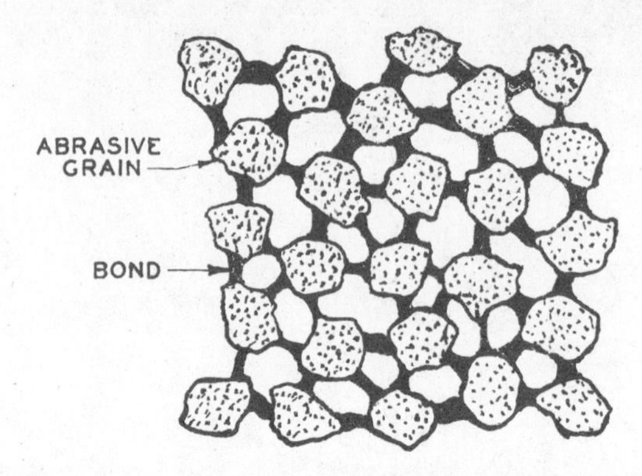

Fig. 42-2. Grains Held Together by Bond (Enlarged)

42-6. Bonds

The bond is the cement which binds or holds the abrasive grains together in the form of a grinding wheel, Fig. 42-2. Several of the different kinds of basic bonding materials that are used in manufacturing grinding wheels are listed below:

earth, or clay (vitrified)	resinoid
	resinoid reinforced
silicate[1]	shellac
rubber	oxychloride
rubber reinforced	

A number of additional bonding materials, which are modifications or combinations of the above bonding materials, are also available. The type of bonding material used in a grinding wheel is identified by a code letter or letters, see Fig. 42-6.

The best bond is one which is not softened by heat when grinding and which holds the cutting points of the abrasive until they are dull. The dull grains are then pulled away from the bond because more pressure is used to grind, and new sharp points are uncovered to begin cutting.

42-7. Vitrified Wheels

To **vitrify** means to make glossy by burning. The **bond** used in vitrified wheels is a kind of earth or clay. The wheels are baked in an electric furnace at about 3000° F. for about 100 hours. Most wheels are made this way. They have large pores, cut easily, and do not **glaze** (dull) easily when grinding is done. Heat, cold, water, oil, or acids do not hurt them.

About 75% of all grinding wheels produced are the vitrified or modified-vitrified types. These wheels, however, are not **elastic**; thin wheels made this way break very easily. Vitrified wheels are made up to 36″ in diameter.

42-8. Silicate Wheels (Semivitrified)

The bond of silicate wheels is **silicate** or water-glass bond. The manufacturing time is less and a thinner-sized wheel can be constructed as compared to the vitrified process. Hard silicate wheels are used for grinding fine edges on tools and knives. They are not recommended for rough grinding. Silicate wheels are made up to 60″ in diameter.

42-9. Elastic Wheels

The bond of elastic wheels is **rubber, shellac,** or **Bakelite.** Very thin, strong wheels are made this way. They can be run faster than vitrified wheels. Elastic wheels can be made as thin as $\frac{1}{64}$″ and are used for sharpening saws, grinding in narrow spaces, making narrow cuts, and cutting off metal, Fig. 42-3.

42-10. Grades of Wheels

Grinding wheels are available in a variety of grades ranging from soft to hard. On a soft wheel, the dull grains are easily released or torn off the wheel while grinding. On a hard wheel, the grains are held tightly and do not tear off easily. Hence, the **grade** of a grinding wheel refers to the looseness or tightness with which the abrasive grains are held together.

[1]**Silicate bond,** also known as **water-glass bond,** is made by melting sand, charcoal, and soda.

Fig. 42-3. Cutting Off the End of a Drill with an Elastic Grinding Wheel

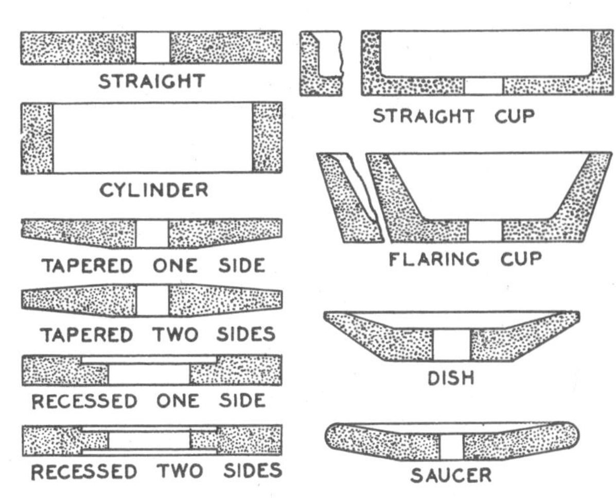

Fig. 42-4. Grinding Wheel Shapes

Softer grades of grinding wheels are generally used for grinding very hard materials, such as hardened tool steel. The abrasive grains dull more easily on hard materials, and they must, therefore, be torn away more easily. When they tear away, new sharp grains are exposed. Harder grades are used for grinding softer materials. Hence, the terms **hard** or **soft,** as related to the **grade** of grinding wheels, have no relationship to the abrasive grain itself. One kind of abrasive grain, such as aluminum oxide, may be used to manufacture either hard-, medium-, or soft-grade grinding wheels. With any given bonding material, the amount of bonding material determines the grade. The grade of a grinding wheel is identified by a letter of the alphabet, as shown in Fig. 42-6.

42-11. Structure of Wheels

The structure of a grinding wheel refers to the spacing between the abrasive grains. Some wheels have abrasive grains which are more **dense** or closely spaced. **Open-grain** wheels have grains spaced farther apart or less dense. Open-grain wheels grind more rapidly than close-grain wheels. Manufacturers identify the structure of grinding wheels by numbers from 1 to 15. The number 1 is the most dense while the number 15 is the least dense, Fig. 42-6.

42-12. Shapes of Grinding Wheels

Grinding wheels are made in a wide variety of shapes chosen by the Grinding Wheel Manufacturers Association of the United States and Canada. Their names describe their shapes. Some common shapes are shown in Fig. 42-4.

42-13. Wheel Faces

The shapes of grinding wheel faces are indicated by letters. See Fig. 42-5.

42-14. Wheel Holes

The hole in a grinding wheel should be about .002″ larger than the diameter of the **shaft** on which it is mounted. This allows the wheel to slide freely, but not loosely, on the shaft. If the hole is so small that the wheel has to be forced on the shaft, there is danger of the wheel cracking.

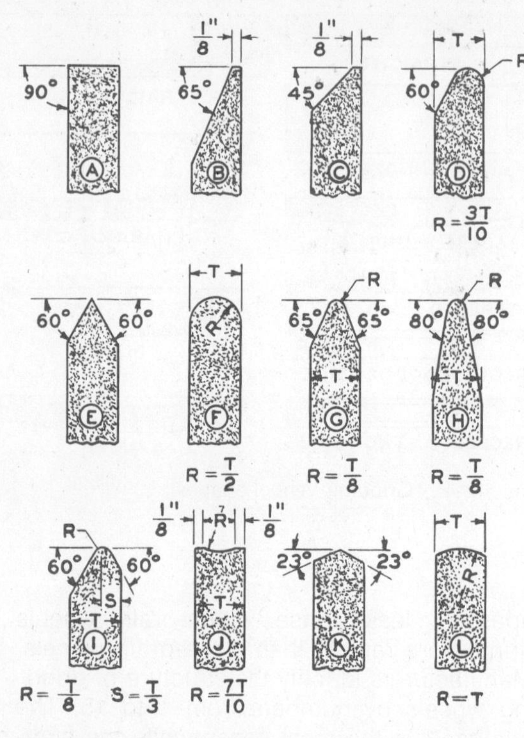

Fig. 42-5. Grinding Wheel Faces

42-15. Wheels for Different Kinds of Work

All materials cannot be ground equally well with the same wheel. The shape of the work and the kind of metal determine the **cutting edge** needed on the grinding wheel. Therefore, different **grain sizes** and **grades** of wheels are recommended for the various kinds of work. It is best to refer to manufacturers' catalogs for the kind of wheel to use. Recommended grinding wheels for various kinds of grinding operations, on various types of materials, are also listed in handbooks for machinists.

42-16. Grinding Wheel Marking System

Most grinding wheel manufacturers use a standard marking system for identifying the following characteristics of grinding wheels:

1. Abrasive type
2. Grain size
3. Grade
4. Structure
5. Bond type

The characteristics of a grinding wheel identified with the code number 51A36-L5V23 are shown in Fig. 42-6. The prefix in the code number is the manufacturer's symbol to indicate the exact kind of abrasive. (This is optional and may not be given.) The suffix is the manufacturer's private code to identify the wheel (optional).

42-17. Ordering Grinding Wheels

Selection

The following factors must be considered in recommending or selecting a grinding wheel for a specific job:

1. Type of grinding operation: hand grinding, surface grinding, tool grinding, cylindrical grinding, etc.
2. Material to be ground: steel, cast iron carbide tools, etc.
3. Amount of stock to be ground: heavy or light rate of stock removal.
4. Quality of finish desired: rough or smooth finish.
5. Area of wheel contact: a wheel with a wide face may require a softer grade.
6. Wheel speed: wheel must be rated at or above the maximum rpm of the grinding machine.
7. Whether grinding is done dry or with a cutting fluid.
8. Abrasive type, grain size, grade, structure, and bond type.

Ordering the Wheel

When ordering a grinding wheel, always provide the following information:

1. Shape of wheel.
2. Type of wheel face.
3. Diameter of wheel.
4. Width (or thickness) of wheel.
5. Diameter of hole.

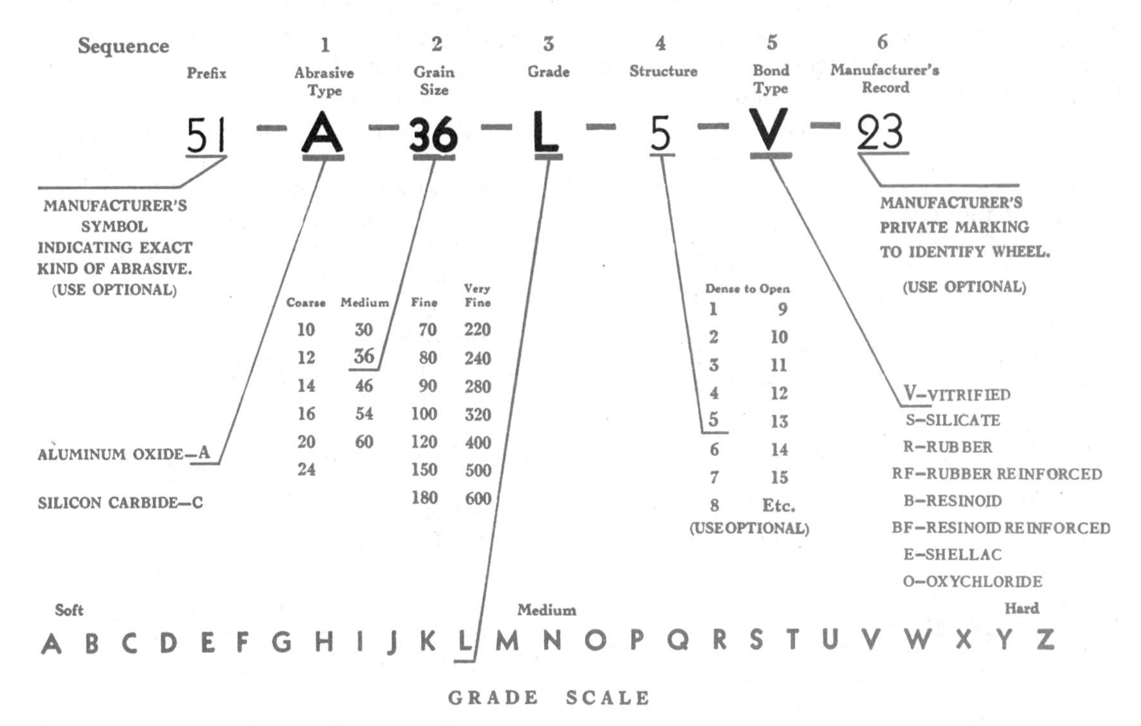

Fig. 42-6. Standard Marking System Used for Identifying Grinding Wheels and Other Bonded Abrasives

6. Speed (rpm) of machine.
7. Identify the following information by using the standard marking system, Fig. 42-6:
   a. Abrasive-type: use a prefix number, if known.
   b. Grain size.
   c. Grade.
   d. Structure.
   e. Bond type.
   f. May specify the manufacturer's record number if desired.

42-18. Diamond Wheels

Diamond wheels are special grinding wheels which have a layer of diamond grains bonded to the operating face of an otherwise nonabrasive **wheel core.** Wheel cores are usually made of metal or Bakelite, a hard plastic.

Diamond wheels are used for cutting and grinding of materials that are very hard or abrasive. They are used primarily for sharpening carbide cutting tools, shaping and sharpening carbide dies, and for cutting and grinding concrete, glass, and other ceramic materials.

For information on diamond grinding wheel shapes, composition, sizes, and wheel markings, consult a manufacturer's catalog or a handbook for machinists.

Words to Know

Bakelite
bond
cured
elastic wheel
glaze
grade
grinding wheel
mounted grinding wheel
semivitrified wheel
oxychloride bond
resinoid bond
rubber bond
shellac bond
silicate bond
structure
vitrified bond
wheel core
wheel face
wheel hole

Review Questions

1. What does grinding mean?
2. How are grinding wheels made?
3. What are the reasons for grinding?
4. Name the two abrasives most used for making grinding wheels.
5. How is the grain size of an abrasive measured?
6. What is meant by the bond of a grinding wheel?
7. Name the kinds of bonding materials used in grinding wheels.
8. What does vitrify mean?
9. What is meant by the grade of a grinding wheel?
10. What should be the relation between the size of the hole in the grinding wheel and the diameter of the shaft on which it is to be mounted?
11. What is meant by the structure of a grinding wheel?
12. Name several common shapes of grinding wheels.
13. Where can you find a list of recommended kinds of grinding wheels for different jobs?
14. Explain the kind of standard marking system which is now used by most grinding wheel manufacturers.
15. Make out an order for a grinding wheel for your shop.
16. For what kinds of work are diamond grinding wheels used?
17. How does the construction of a diamond grinding wheel differ from the construction of a conventional grinding wheel?

Career Information

1. List several occupations which require a knowledge of grinding wheel construction, selection, and use.

UNIT 43

Utility Grinders

43-1. What Are Utility Grinders?

Utility grinders are general-purpose grinding machines. They are used for non-precision grinding operations in which either the grinder or the work being ground is hand held. Examples of work done with utility grinders include removal of burrs and other sharp edges; removal of flash and other casting imperfections (called **snagging**); and preparation of joints for welding, smoothing welds, and off-hand sharpening of cutting tools.

43-2. Types of Utility Grinders

A **bench grinder,** Fig. 43-1, is a small grinder which can be bolted to the top of a bench. It usually has a wheel on each end of a **shaft** which extends through an **electric motor.** This grinder is used to grind tools and for general light grinding. Bench grinders are made for use with grinding wheels up to about 1″ in width and 8″ in diameter.

A **pedestal grinder** is a utility grinder mounted on a free-standing base or pedestal, Fig. 43-2. Various sizes of pedestal grinders are made, the heavy-duty ones using wheels 3″ or more in width and 20″ or more in diameter. The smaller pedestal grinders are used for the same class of work as bench grinders, while the larger machines are generally used for a rougher class of work, such as snagging of castings.

The **wet grinder,** Fig. 43-3, has a pump to supply a flow of water, or **coolant,** to the wheel. The water runs back into the tank and is used over and over again. It carries off the heat caused by grinding and washes away bits of metal and abrasive. If not carried away, these bits of metal would fill up the pores of the wheel and cause what is known as a **loaded wheel.**

Portable grinders are powered by electricity, Fig. 43-4, or compressed air, Fig. 43-5. Some portable grinders are equipped with a **flexible shaft** for versatility and ease of handling. These are often used for delicate work such as the grinding of dies, Fig. 43-6.

Hand-powered portable grinders, Fig. 43-7, are used where electrical and air

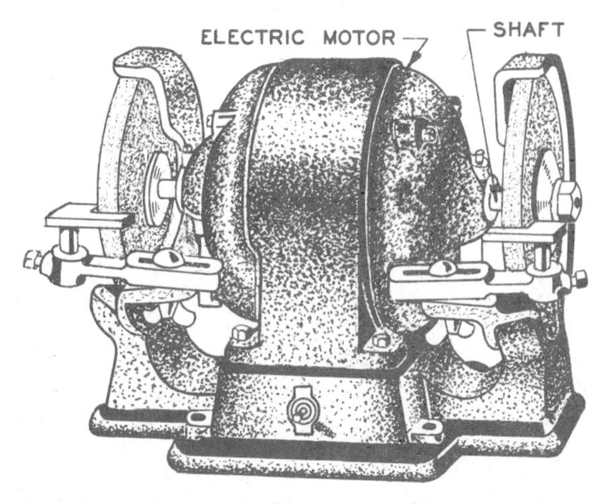

Fig. 43-1. Bench Grinder

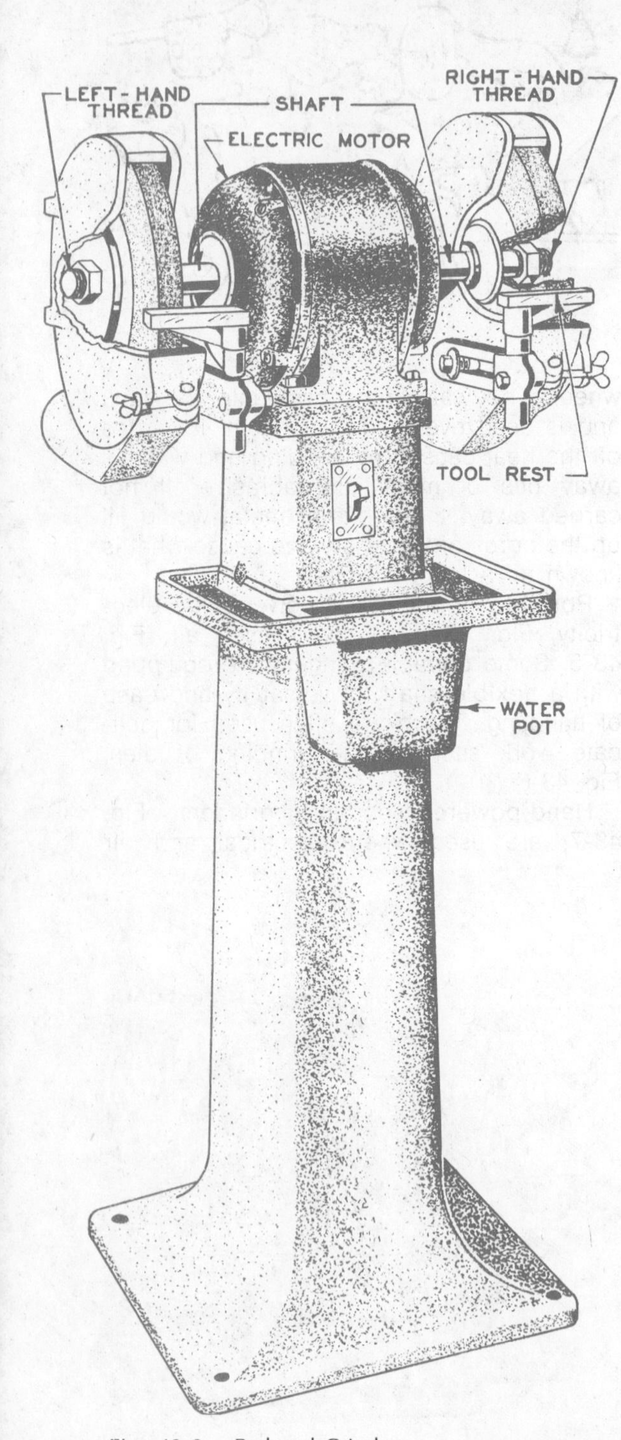

Fig. 43-2. Pedestal Grinder

Fig. 43-3. Wet Grinder

Fig. 43-4. Using Portable Electric Grinder to Smooth a Weld on a Truck Chassis

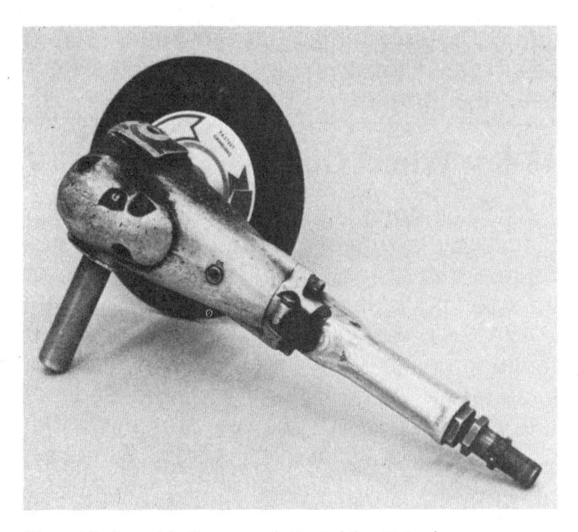

Fig. 43-5. Air Powered Portable Grinder

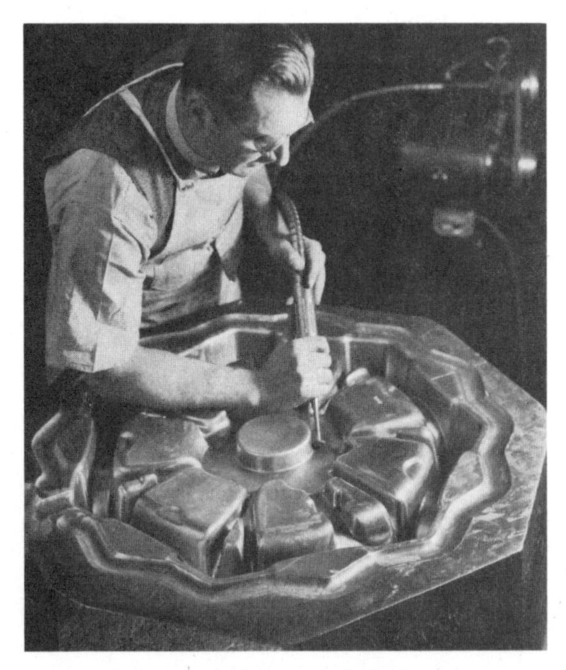

Fig. 43-6. Using a Small Grinding Wheel on a Flexible Shaft

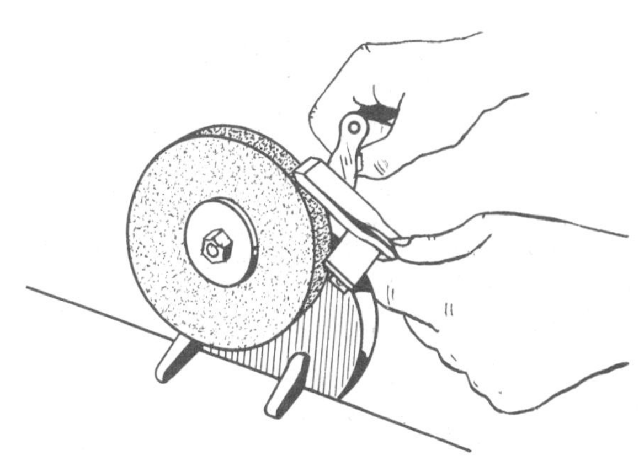

Fig. 43-7. Using a Portable Hand Grinder

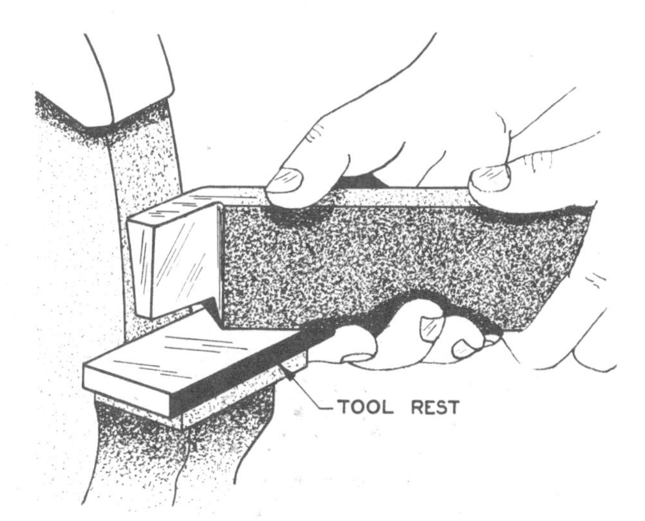

Fig. 43-8. Use of Tool Rest

power are not available. They are designed with a built-in clamp so they can be readily attached to a bench or table top.

43-3. Threads on Grinder Shaft

The ends of the **grinder shaft** on which the wheels are fastened are always **threaded** so that the **nuts** which tighten the wheels to the shaft will tighten as the shaft turns. For this reason, the left side of the grinder has **left-hand threads** on the shaft and nut, while the right side of the grinder has **right-hand threads** on the shaft and nut. See Fig. 43-2.

Be careful to turn the nuts in the right direction when removing or replacing them. To remove the nuts, turn them in the direction that the wheels turn when grinding.

43-4. Tool Rest

Every grinder should have a **tool rest** (see Fig. 43-2) upon which the work is rested while grinding, Fig. 43-8. The tool rest should be set as close to the wheel as possible without touching it. This is done to keep the work from catching between the wheel and the tool rest, Fig. 43-9. Most grinding accidents are the result of the tool catching. The tool rest must be moved closer as the wheel wears smaller to keep the proper space. This must be done while the wheel is at a standstill.

43-5. Water Pot

Some grinders have a **water pot,** Fig. 43-2. It is filled with water and is used to keep the work cool (by dipping it into water often). If the grinder has no water pot, a small pail filled with water should be kept near the grinder.

43-6. Wheel Guards

The operator is protected from flying pieces by a **wheel guard** that nearly surrounds the wheel, Fig. 43-9. Just enough of an opening is left to do the grinding.

Proper wheel guards give complete protection against broken grinding wheels. In a series of tests, not once did a piece of the wheel leave the guard in a manner that could have caused injury to the operator, Fig. 43-9. The wheels in the tests were broken by dropping a steel **wedge** between the **tool rest** and the wheel. This is one of the most common causes of accidents.

43-7. Safety Flanges

Safety flanges are large metal washers, Fig. 43-11, placed on each side of the grinding wheel. They clamp the wheel in place on the shaft and also hold the parts of the wheel together if it breaks, Figs. 43-10 and 43-11. They should be at least one-third of the diameter of the wheel; one-

Fig. 43-9. Wheel Guards Protect against Danger from Broken Grinding Wheels

Fig. 43-10. Flanges Hold Broken Parts of the Grinding Wheel Together

half is better. The flanges should be **recessed;** that is, the side that fits against the wheel should be cut deeper in the middle so that only the outer edge presses against the wheel. The **inside flange** should be **keyed** (see § 26-26) or otherwise fastened to the shaft.

43-8. Safety Washers

Soft washers made of blotting paper, leather, or rubber should be placed between the wheel and the **flanges.** See § 43-7 and Fig. 43-11. These washers should be a little larger than the flanges. The soft material is forced into the pores of the wheel, thus locking the wheel in place.

43-9. Glass Eye Shield

Some grinders have **glass eye shields** through which to look while grinding, Fig. 43-12. The shield, together with safety goggles, provides safe eye protection.

If a grain of **abrasive,** which has rough, sharp edges and points, gets into the eye, it often has to be removed by a doctor. The eye may be swollen and very sore after the grain is removed and the worker may not be able to work for several days.

43-10. Why Do Wheels Break?

Many grinding wheels break due to the following circumstances:

1. Flaws in the wheel.
2. Wrong placing and fastening of the wheel on the shaft.
3. Too much speed.
4. Work getting caught.

43-11. Inspecting Grinding Wheels for Cracks

Before a new wheel is used, it should be carefully inspected for cracks. Strike the

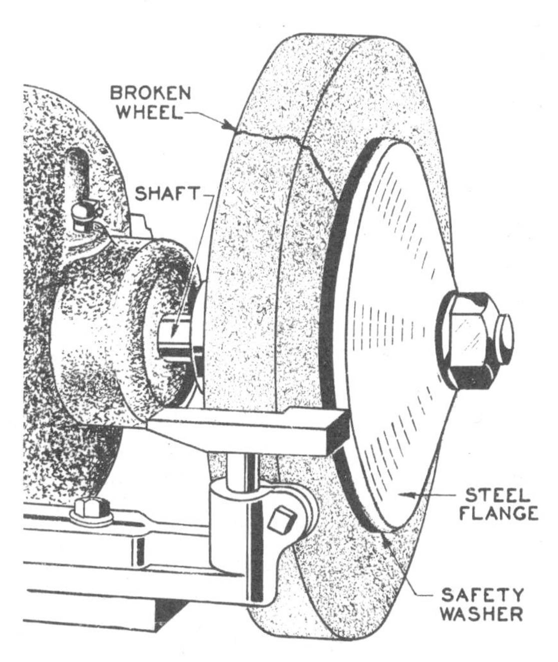

Fig. 43-11. Broken Wheel Held Together by Safety Washers and Flanges

Fig. 43-12. Reinforced Glass Eye Shield for a Grinder

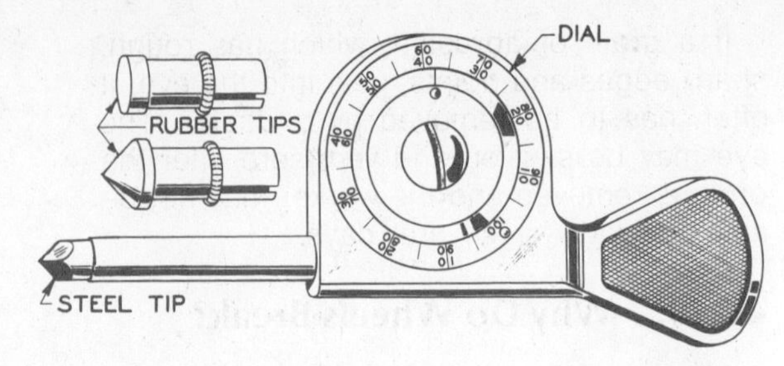

Fig. 43-13. Speed Indicator

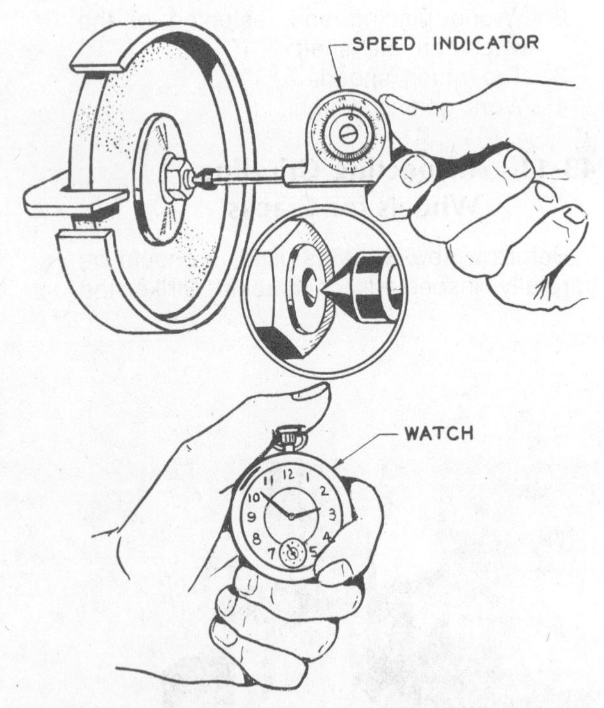

Fig. 43-14. Finding Speed with Speed Indicator

wheel gently with a light object, such as the handle of a screwdriver for light wheels and a **mallet** for heavy ones. Wheels must be dry and free from the sawdust in which they were packed when they are being tested in this way. The sound will indicate if the wheel is cracked. A good wheel will ring clearly when struck.

43-12. Mounting the Grinding Wheel

Grinding wheels sometimes break because of wrong **mounting** (placing and fastening on the shaft). The wheel should not be forced on the shaft as this may cause it to crack. **Safety washers** a little larger than the diameter of the **flanges** should be placed between the sides of the wheel and the flanges. The nut should be just tight enough to hold the wheel firmly.

After **mounting** a wheel, stand to one side and let the wheel run at full speed for at least one minute. It should then be **trued** with a **grinding wheel dresser.** See § 44-4.

43-13. Speeds of Grinding Wheels

Grinding wheels are usually run at a **surface speed** of 4000′ to 6500′ **per minute** (fpm). Surface speed is the speed of the rim of the wheel and the distance it would travel if rolled on the floor for one minute. Note that this is about a mile (5280 feet) a minute or about 33 revolutions per second for a 10″ wheel. At the same rpm, the surface speed of the wheel gets slower as the diameter gets smaller. Make sure that the speed of the wheel is right before grinding. Many electric motors operate at 1725 rpm. This speed gives nearly 6000 fpm for a 13″ wheel and just over 4000 fpm for a 9″ wheel.

To find the surface speed, multiply the **revolutions per minute** (rpm) of the grinder by the circumference of the grinding wheel in feet. The following formula may be used:

$$\text{Surface speed (fpm)} = \frac{\text{rpm} \times 3.14 \times \text{wheel dia. in inches}}{12}$$

43-14. Speed Indicator

The rpm of a grinding wheel may be found with an instrument called a **speed indicator,** Fig. 43-13. The point of the instrument is pressed against the **center-**

drilled holes in the end of the shaft; the number of revolutions made by the shaft and grinding wheel shows on the **dial** of the indicator, Fig. 43-14. One hundred revolutions of the shaft make one revolution on the dial. The revolutions per minute can be found by timing with a watch. Rubber tips, which may be used on different shafts, are furnished with the speed indicator.

Words to Know

bench grinder
coolant
flexible shaft grinder
hand-powered grinder
loaded wheel
pedestal grinder
portable grinder
revolutions per minute
safety flange
safety washer
snagging
speed indicator
surface speed
tool rest
utility grinder
water pot
wet grinder
wheel guard

Review Questions

1. What are utility grinders? For what are they used?
2. Name three ways portable grinding machines are powered.
3. Why are hand-powered portable grinders still being made?
4. How does a wet grinder differ from other pedestal grinders? Name several advantages of wet grinding.
5. Of what advantage is a grinder with a flexible shaft?
6. What is a grinding wheel flange? What is its purpose?
7. What is a safety washer's purpose? Of what is it made?
8. Why does a grinding wheel break?
9. How should a grinding wheel be inspected for cracks?
10. Of what use is a tool rest? How close should it be set to the wheel?
11. What is meant by the surface speed of a grinding wheel? Why is it important?
12. For what is a speed indicator used?

Mathematics

1. What is the surface speed of a 10″ grinding wheel turning at 1800 rpm?
2. How many rpm's is a 6″ grinding wheel turning if its surface speed is a mile a minute?

UNIT 44

Sharpening Tools by Hand Grinding

44-1. Importance of Tool Sharpening

Properly shaped and sharpened tools are essential to the production of high quality work. They require less force to do the work for which they were designed, they produce higher quality finishes, their cutting edges last longer between sharpening, and they are safer to use because they perform in an expected way. Like other tools, the grinding wheels used for tool sharpening must themselves be sharp and run true in order to do their best work.

Fig. 44-1. Loaded Grinding Wheel — Before and After Dressing

44-2. Cause of Loaded and Glazed Wheels

A **loaded** grinding wheel is one which has become clogged with bits of metal, much as a file becomes clogged with filings. See Fig. 44-1. This clogging occurs when soft materials such as lead, copper, brass, aluminum, and rubber are ground. A loaded wheel cuts very poorly, if at all, so the clogged surface of the wheel must be removed with a **grinding wheel dresser.**

A **glazed wheel** is one which has become dull through normal use or because the wheel is not suited to the work. If a wheel loads or glazes quickly, it is probably too hard for that particular task. Use of loaded

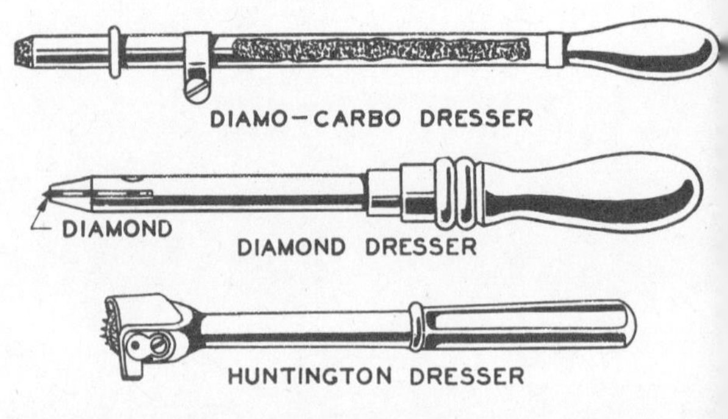

Fig. 44-2. Grinding Wheel Dressers

and glazed wheels causes excess heat which often **burns the temper** out of a tool. See Section 38-7.

44-3. Grinding Wheel Dressers

A **grinding wheel dresser** is a tool used for sharpening and shaping grinding wheels, Fig. 44-2. There are five kinds of grinding wheel dressers commonly used:

1. Diamo-carbo
2. Diamond
3. Huntington
4. Abrasive stick
5. Abrasive wheel

Diamo-Carbo Dresser. The diamo-carbo dresser is a tube filled with a very hard **abrasive,** Fig. 44-2. When held against the grinding wheel, it breaks off the dull abrasive particles and in this way sharpens the wheel, Fig. 44-3.

Diamond Dresser. The diamond dresser has one or more diamonds set in the end of a holder, Fig. 44-2. It is the best, but also the most expensive, of all dressers. It lasts much longer than the others and in the long run is the cheapest if it is properly cared for.

Huntington Dresser. The Huntington dresser uses hardened steel cutters, Fig. 44-2. The cutters consist of alternate star and disc shape wheels which are free to rotate on a shaft. When the dresser is pressed against the face of the grinding wheel, the cutters turn with the grinding wheel and knock off the dull abrasive grains. Replacement cutters are inexpensive and are easy to replace, Fig. 44-4.

Abrasive Stick. The abrasive stick dresser uses a replaceable cutting element of silicon carbide or any other hard abrasive held in a steel tube.

Abrasive Wheel. The abrasive wheel dresser employs a hard abrasive wheel mounted at a slight angle in a hooded holder. It functions in a manner similar to the Huntington dresser.

44-4. Dressing and Truing Grinding Wheels

When the grinding wheel becomes dull, loaded, or out of shape, it must be **dressed** and **trued. Dressing** means sharpening a wheel. **Truing** means cutting the wheel so that there will be no high spots when the wheel is running. Every new wheel should be trued after mounting.

Close-fitting goggles should be worn when the operator is dressing or truing the

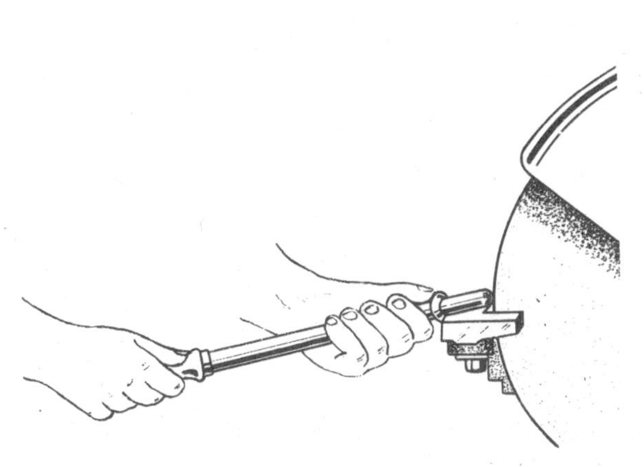

Fig. 44-3. Dressing and Truing a Grinding Wheel and a Diamo-Carbo Dresser

Fig. 44-4. Using a Huntington Dresser

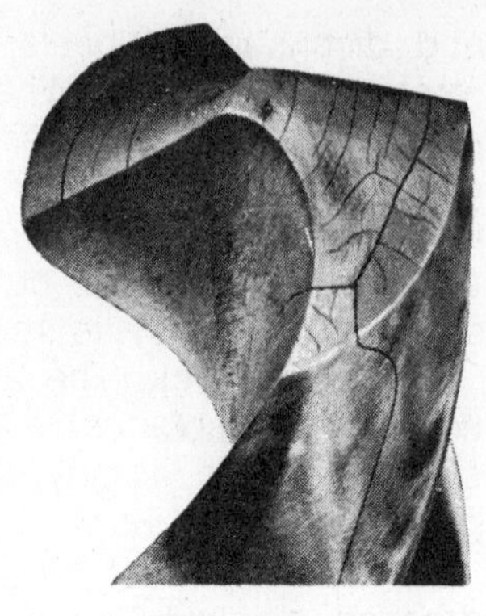

Fig. 44-5. Cracks in a Steel Drill from Improper Cooling

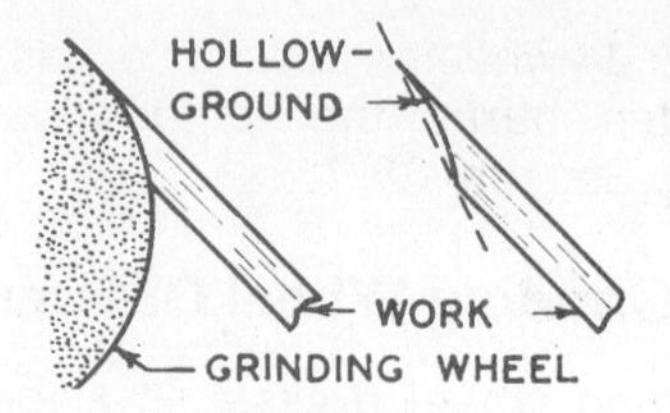

Fig. 44-6. Hollow-Grinding

grinding wheels. This operation produces a cloud of sharp abrasive particles which can be very painful and difficult to remove if they are allowed to enter the eye.

The dresser should be grasped firmly with both hands, brought into light contact with the wheel, and moved slowly and steadily across the wheel face. This process is repeated until the face of the wheel is straight, the wheel runs true, and it is no longer loaded.

44-5. How to Grind Safely

Always wear close-fitting goggles when grinding, and grind only on a grinder which has wheels that are sharp and are properly enclosed with wheel guards. Keep the tool rest as close to the wheel as possible without touching it.

Grind only on the face of the wheel. Grinding on the side of the wheel spoils its shape. Because the side is not properly trued or dressed, it tends to burn the tool quickly. Special wheels are made for side grinding.

Grinding a tool should require only light pressure on the wheel. Avoid applying sudden, heavy pressure which will cause the wheel to wear rapidly and become out-of-round.

44-6. What Does Burning the Temper Mean?

When grinding such tools as chisels, punches, etc., be careful not to **burn** the thin edges or points. The tool is being burnt when it turns to a purple or blue color. Burning a tool causes the steel to lose its **temper**; that is, it loses some of its hardness. Merely grinding off the blue color does not bring back the temper.

Keep the tool cool by dipping it in water often. A **water pot** is handy for this purpose. The work should be moved across the whole width of the wheel to prevent wearing grooves into the wheel or burning the tool.

44-7. Grinding Tools of High-Speed Steel

Tools made of **high-speed steel** need special care in grinding. If such a tool is overheated by grinding and then dipped in cold water, it may crack. Figure 44-5 shows a drill that was overheated during grinding and then cooled in water that was too cold. Most of the cracks could not be seen until the drill was dipped in an acid which ate into the cracks and made them more visible. Cracking can be avoided by grinding only on sharp wheels and by keeping the tool cool by frequently dipping it in water.

44-8. Hollow Grinding

A tool sharpened on the face of a grinding wheel has a curved surface called **hol-**

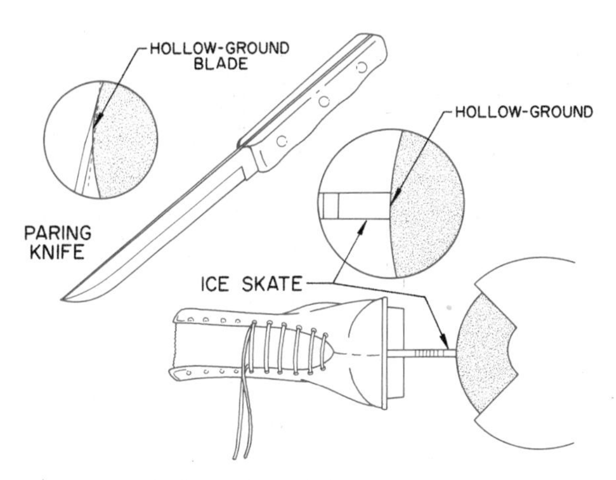

Fig. 44-7. Hollow Ground Carving Knife

low ground which is behind its cutting edge. See Fig. 44-6. The curved sides of a carving knife are an excellent example of hollow grinding, Fig. 44-7. All sharp edge tools such as chisels, plane irons, and knives do their best work when they are hollow ground.

44-9. Cutting Edge Angles

For proper cutting efficiency, the recommended **cutting edge angle** for each tool should be maintained. Gages are available for checking the angles of some tools. A **protractor** can be used in place of a gage.

The harder the material to be cut, the larger should be the angle of the cutting edge; the softer the material to be cut, the smaller should be the cutting edge angle. As an example, compare the 60° cutting angle on a cold chisel used for cutting metal to the angle on a pocket knife or bread knife used to cut softer materials.

44-10. Sharpening a Cold Chisel

A right-handed person should hold the blade end of a chisel in the left hand, which rests on the tool rest, Fig. 44-8. The right hand should hold the head end of the chisel. Press the chisel lightly against the face of the wheel and move it back and forth.

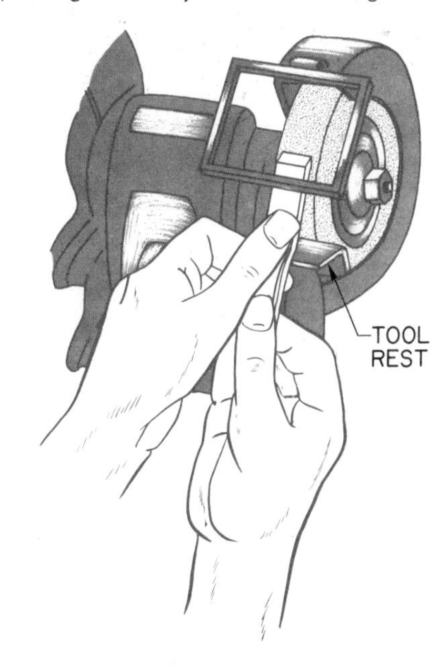

Fig. 44-8. Sharpening a Cold Chisel

Keep the chisel cool by frequently dipping it in water. Examine the chisel often to make sure the same amount of metal is ground off both sides and that the same angle is ground on both sides. As grinding progresses, check the cutting edge angle with a center gage. See Fig. 13-4.

44-11. Drill Sharpening

Drills are sharpened manually on utility grinders or on special purpose drill grinders. Because of the relative complexity of drill grinding, it is treated separately in Unit 45.

44-12. Grinding Screwdrivers

Screwdrivers often have flat tapered blade faces which tend to climb out of the screw slot during use. Hollow grinding of screwdrivers provides parallel instead of tapered faces which make the screwdriver

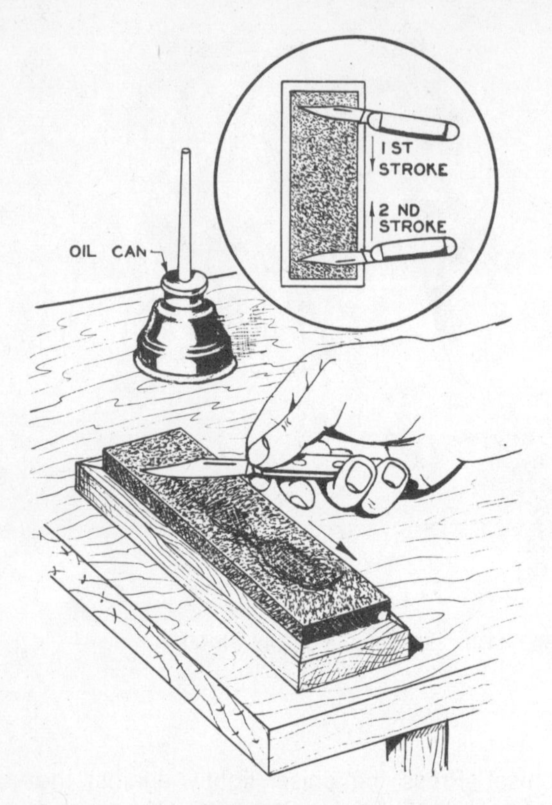

Fig. 44-9. Sharpening a Knife on an Oilstone

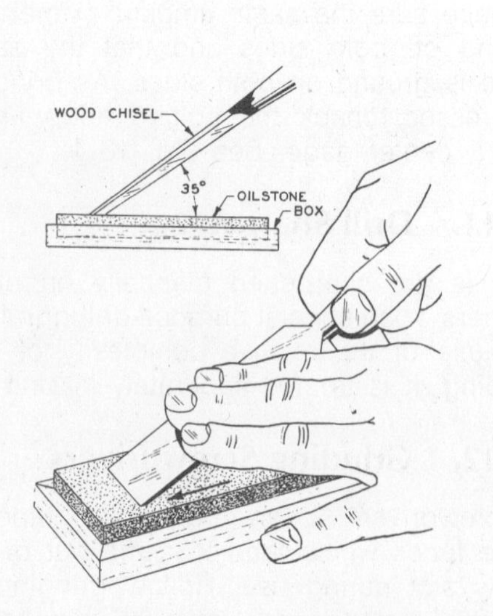

Fig. 44-10. Sharpening a Wood Chisel on an Oilstone

stay in the screw slot much better, thus making the screwdriver a safer tool to use. See section 25-11 and Fig. 25-13.

44-13. Oilstones

Oilstones, also called **hones,** are smooth abrasive stones made in many shapes and sizes. They are used for sharpening, chiefly to put the finishing touches on cutting edges of tools, Figs. 44-9 and 44-10. When using the oilstone, put oil on it to wash away the bits of metal that are ground off the work.

Oilstones should be kept clean and moist. An oilstone which is kept in a dry place should be kept in a box with a cover; a few drops of clean oil should be left on the stone.

A slipstone is a small, wedge-shaped oilstone, Fig. 44-11. It is used to sharpen cutting edges of irregular shapes which cannot be sharpened on the flat oilstone, Fig. 44-12.

Fig. 44-11. Slip Stones

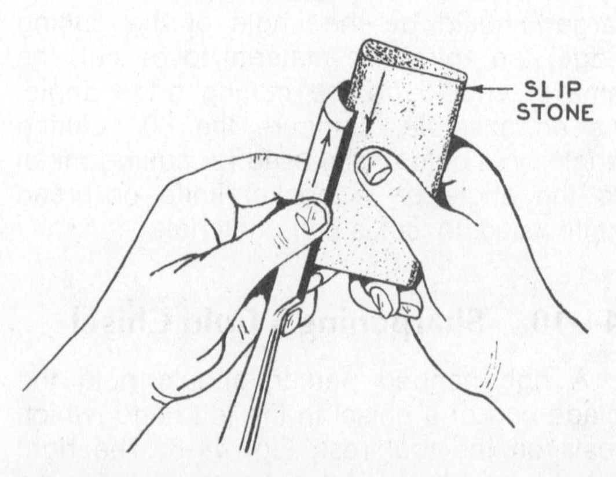

Fig. 44-12. Using a Slip Stone

44-14. Sharpening Cutting Tools on an Oilstone

The sharpening of cutting tools on oilstones is called **honing** or **whetting.** Knives, razors, scrapers, and wood chisels may be sharpened this way.

The oilstone must lie flat on the bench. Put a few drops of oil on the stone. This will keep the stone from **glazing.** All cutting tools should be sharpened with the edge of the tool working against the stone as shown by the arrows in Figs. 44-9 and 44-10. Straight strokes will sharpen a tool quicker than will circular strokes, as shown by the arrows.

Wipe all oil and grit off the oilstone with a rag when you are finished with it. Dirty oil left on the stone dries and carries the steel dust into the pores of the stone.

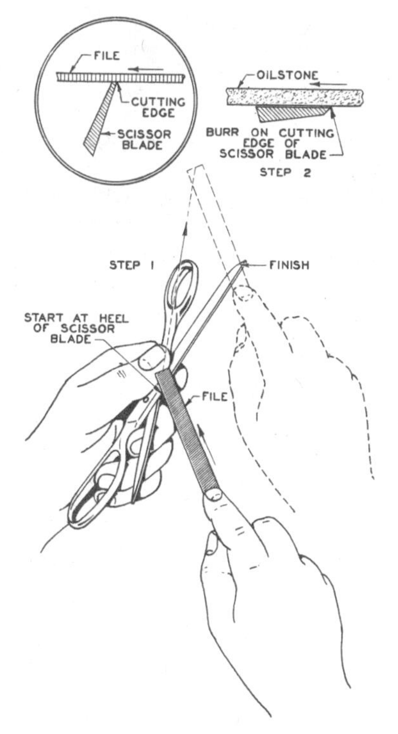

Fig. 44-13. Sharpening Scissors with a File

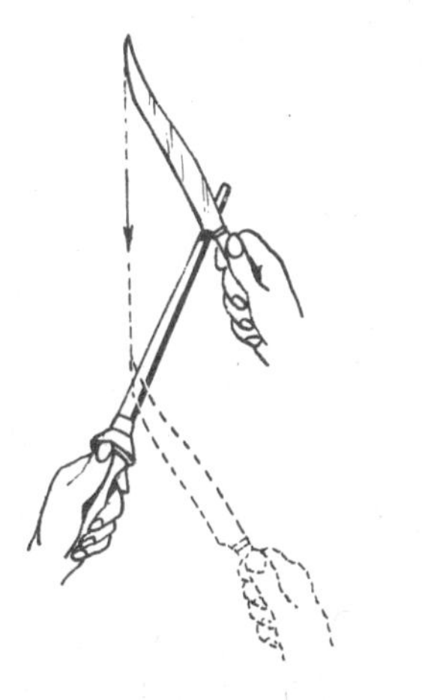

Fig. 44-14. Using a Knife Sharpening Steel

44-15. Sharpening Scissors

Scissors may be sharpened with a small oilstone or with a fine, double-cut file, Fig. 44-13. File or stone lightly against the cutting edge of the scissors. After filing, do not close the scissors until the burr is removed with an oilstone. See Fig. 44-13.

44-16. Steel for Knife Sharpening

Kitchen knives may be sharpened on a **sharpening steel,** a long, tapered rod of hardened steel with a handle on one end, Fig. 44-14. The cutting edge of the knife should be toward the handle of the steel. Stroke from the point of the steel toward the handle and from the handle of the knife toward the point, stroking first on one side of the steel, then on the other side.

Words to Know

abrasive stick dresser
burning the temper
diamo-carbo dresser
diamond dresser
glazed wheel
hollow grinding
hone
Huntington dresser
knife-sharpening steel
loaded wheel
oilstone
slip stone
truing
wheel dressing
whetting

Review Questions

1. Why is tool sharpening important?
2. What is a loaded wheel?
3. What is a glazed wheel?
4. Describe the five kinds of grinding wheel dressers.
5. Why should goggles be worn during the grinding procedure?
6. What is meant by dressing a wheel?
7. What is meant by truing a wheel?
8. What is meant by burning the temper? How can it be avoided?
9. Why should only the face of the grinding wheel be used for grinding?
10. Why should wood and soft metals such as lead, copper, brass and aluminum not be ground on a grinding wheel?
11. What is meant by hollow grinding?
12. For what are oilstones used?
13. Describe the kind of care an oilstone should have.
14. What is a slipstone? How is it used?

Career Information

1. What are some of the dangers involved in using a grinder?
2. In what trades is a knowledge of tool sharpening important? List several tools that are sharpened in each trade.

UNIT **45**

Drill Sharpening

45-1. Importance of Drill Sharpening

Nearly all drilling troubles are caused by wrong sharpening. The results of a badly sharpened drill may be:

1. A broken drill.
2. A hole which is the wrong diameter.
3. A hole which is not perfectly round.
4. A rough finish, Fig. 45-1.

45-2. Drill Sharpening

Four things must be watched when sharpening a drill:

1. Lip clearance
2. Length of lips
3. Angle of lips
4. Location of dead center

If the first three are correct, the fourth will also be correct.

Fig. 45-1. Rough Hole Made by a Dull Drill

45-3. Lip Clearance

Lip clearance is made by grinding away the metal behind the cutting edges, Fig. 45-2, so the cutting edges can cut into the metal.

If there were no lip clearance as on **A** in Fig. 45-2, it would be impossible for the drill to cut into the metal; the bottom of the drill would rub, but not cut. The metal behind the cutting edges must be ground away as on **B** and **C**. Note in **B** and **C** how

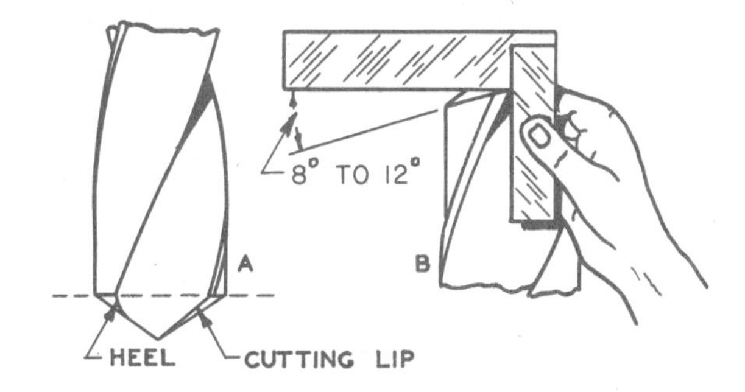

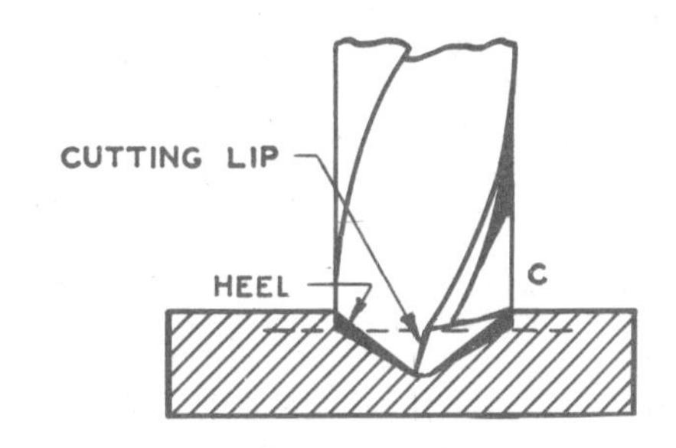

Fig. 45-2. Lip Clearance

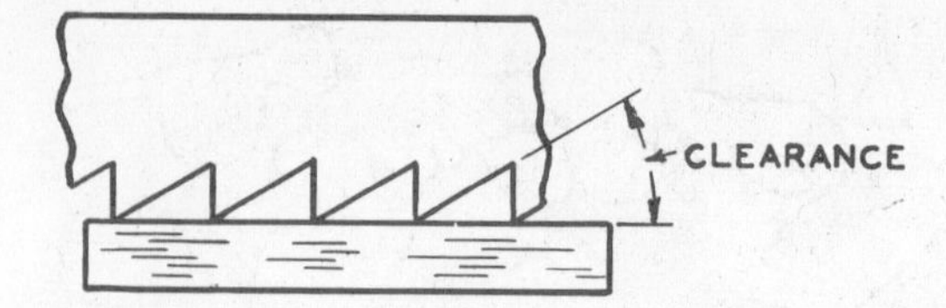

Fig. 45-3. Clearance on Saw Tooth

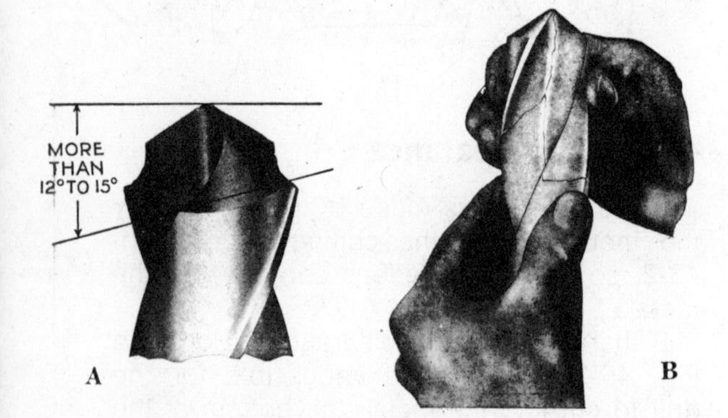

Fig. 45-4. Results of Wrong Lip Clearance

much lower the **cutting lip** is than the **heel;** also note that in **C** the cutting lip has already removed metal ahead of the heel as shown by the dark part of the hole on each side of the drill.

Lip clearance is much like **clearance** of a saw tooth, Fig. 45-3.

45-4. Angle of Lip Clearance

The correct **lip clearance** for a regular point in general-purpose drilling of most steels should be from 8° to 12°. For drilling soft materials under heavy **feeds** the angle of lip clearance may be increased to 12° to 15°. If it is more than 15°, the corners of the cutting edges are too thin and may break off as shown in **A**, Fig. 45-4.

If the angle of lip clearance is much less than 8°, it acts the same as when there is no clearance, and the drill cannot cut into the metal. It may then break in the center along the web as **B** in Fig. 45-4.

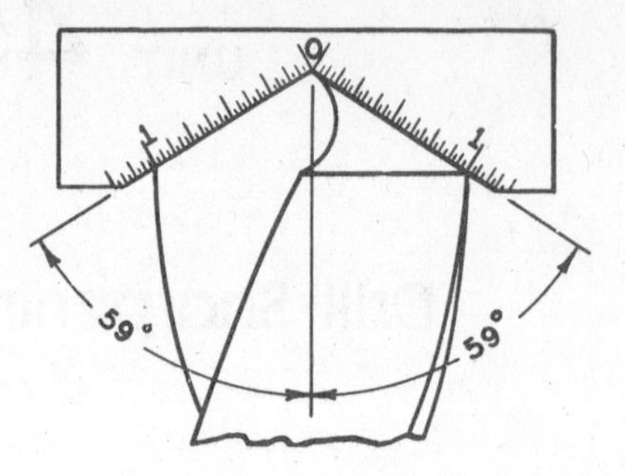

Fig. 45-5. Lips are Same Length with Equal and Correct Angles

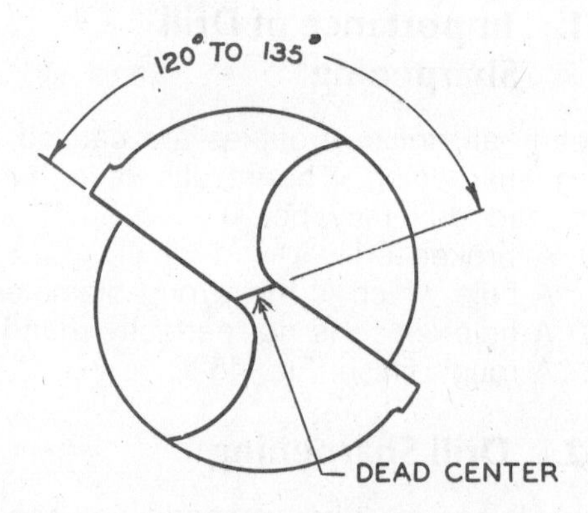

Fig. 45-6. Angle of Dead Center with Cutting Edge

45-5. Length and Angle of Lips

The two lips must be the same length, and their angles must be equal, Fig. 45-5. For ordinary work, 59° is recommended. If the two lips are the same length and at equal angles as shown in Fig. 45-5, the **dead center** will also be centrally located, Fig. 45-6. The line across the dead center should be between 120° and 135° with the cutting edge.

If the angles of the cutting edges are more than 59°, the drill will not cut easily into the metal and, therefore, will not hold its position centrally since it is too flat. If the angles of the cutting edges are less than 59°, more power will be needed to turn the drill, or the drill will cut slower because of the longer cutting edges.

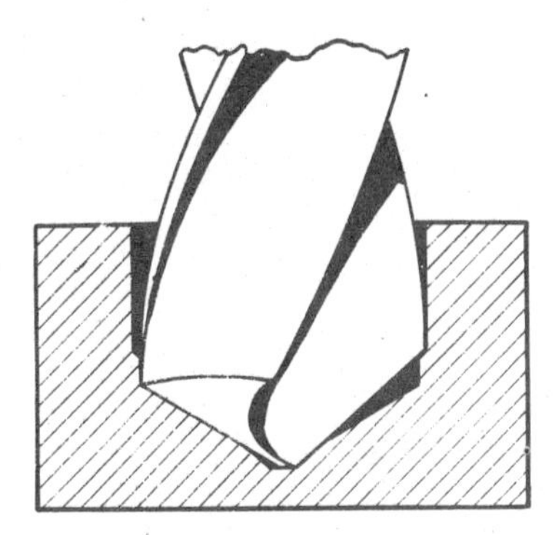

Fig. 45-7. Angles of Cutting Edges Are Unequal

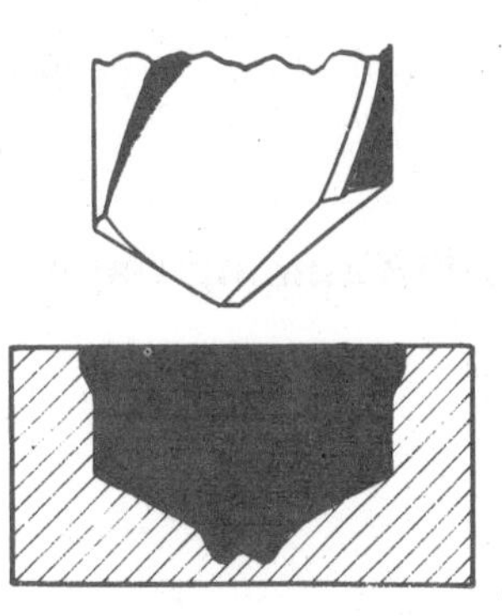

Fig. 45-9. Lips of Unequal Lengths and Unequal Angles on Cutting Edges

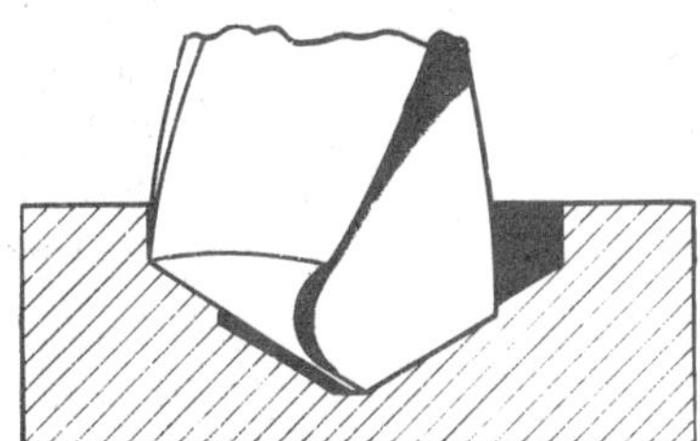

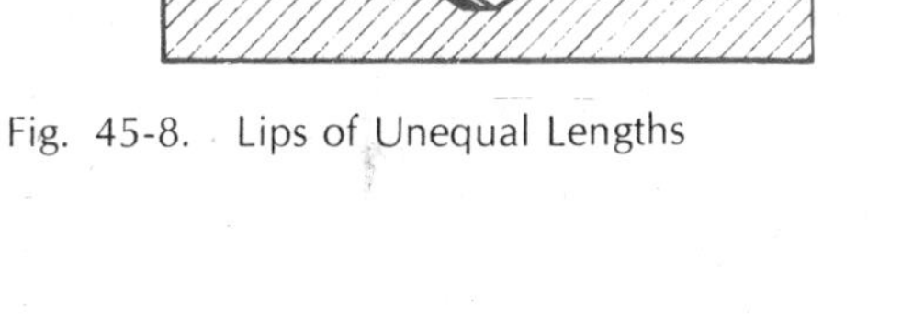

Fig. 45-8. Lips of Unequal Lengths

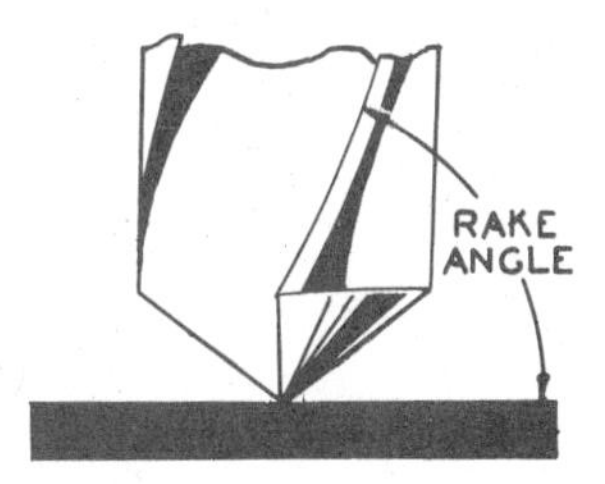

Fig. 45-10. Rake Angle

Only one lip will cut if the dead center is in the center but the angles are different as in Fig. 45-7. One cutting edge will wear quickly and the hole will be oversize.

If the angles on the cutting edges are equal but the lips are of different lengths, as in Fig. 45-8, the dead center will not be in the center. The hole will then be larger than the drill. It may be compared with putting the hub of a wheel anywhere except the exact center. The strain on the drill press is thus very great, wobbling of the spindle results, the drill wears down rapidly, and breakdowns often follow.

The worst conditions occur when the lips are of unequal lengths and the angles on the cutting edges are also unequal, Fig. 45-9. The hole is then larger than the drill, and the drill and drill press are strained. Note that the short lip cuts a smaller hole than the long lip.

45-6. Rake Angle

The rake angle of the drill is the angle between the **flute** and the work, Fig. 45-10. If the rake angle were 90°, the drill would have **neutral** or **zero rake.** If the rake angle is more than 90°, the drill will have **negative rake.**

If the rake angle is too small, the cutting edge is too thin and it breaks under the strain of the work. The rake angle also helps to curl the chips; a large rake angle rolls the chips tightly, while a small rake angle rolls them loosely. The rake angle, as made by manufacturers, should not be changed for ordinary drilling.

However, when drilling **free machining** brass and some types of plastics, drills with standard rakes tend to screw themselves into the material, sometimes breaking the drill, the material, or both. This can be avoided if the drills are reground to provide a slightly negative rake, Fig. 45-11.

45-7. Drill-Grinding Gage

A drill-grinding gage should be used when grinding drills, Fig. 45-12. Such a gage measures the lengths and angles of the **cutting lips.**

45-8. Grinding Drills by Hand

In shops without drill grinding machines, drills must be sharpened by hand, Fig. 45-13. It is best to have the instructor demonstrate how to hold and grind a drill by hand. You will have to do much practicing before you can expect to do a good job.

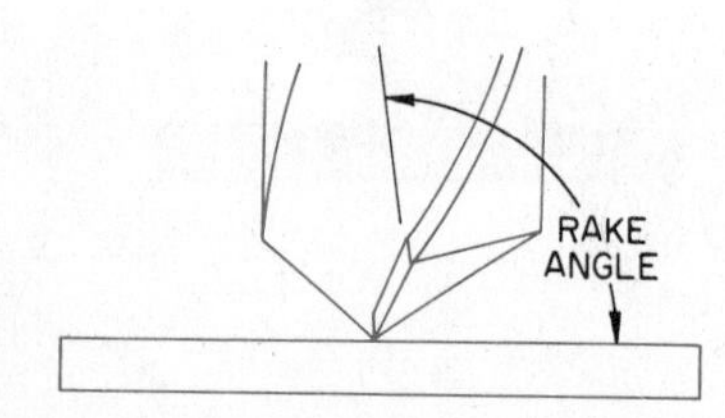

Fig. 45-11. Drill Point Modified to Provide Negative Rake

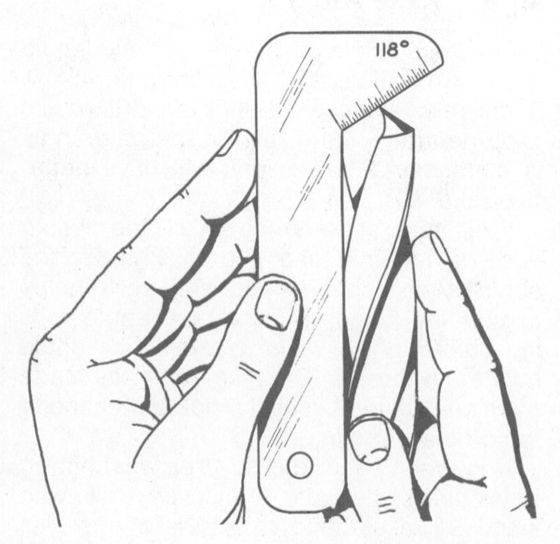

Fig. 45-12. Measuring Lengths and Angles of the Cutting Lips with a Drill-Grinding Gage

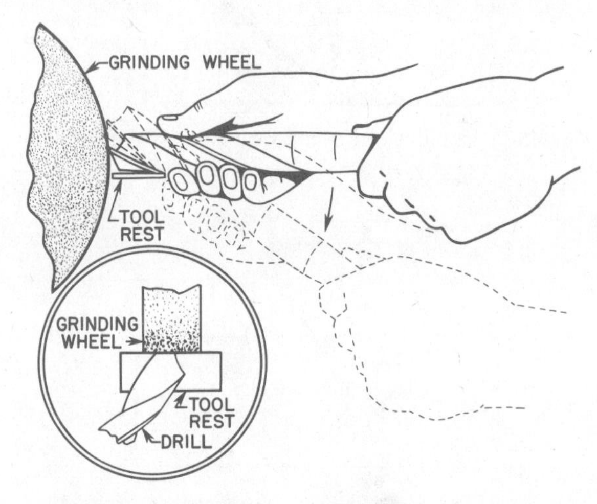

Fig. 45-13. Grinding Drill by Hand

Fig. 45-14. Grinding a Drill with a Drill-Grinding Attachment

Fig. 45-15. Special Purpose Drill-Grinding Machine

45-9. Grinding Drills by Machine

Drills can be ground better and faster on a drill grinding attachment, Fig. 45-14, or on a special purpose drill grinding machine, such as shown in Fig. 45-15.

Drills sharpened by machine are **precision ground** to correct angles and clearances and produce consistently better holes than drills ground by hand.

Words to Know

cutting lip
drill grinding attachment
drill grinding gage
lip clearance
negative rake
positive rake
rake angle
zero rake

Review Questions

1. What three things should be watched when sharpening a twist drill by hand?
2. What is lip clearance?
3. What is the result of too much lip clearance?
4. What is the result of too little lip clearance?
5. How does a drill behave when the angles of the cutting edges are unequal?
6. How does a drill behave when the lips of a drill are of unequal lengths?
7. How can a drill be made to cut oversize?
8. Explain the difference between positive, negative, and zero rake.
9. Drills with negative rake are recommended for drilling what materials?
10. How does a drill grinding gage aid hand sharpening of drills?
11. Is it better to grind a drill by machine than by hand? Why?

Career Education

1. List the tradespeople who must know how to sharpen drills.
2. Make an investigation to find out how many different kinds of drills there are.

PART **XIII**

Controlling Quality

UNIT 46

Quality Control: Measurement, Gaging, and Inspection Tools

46-1. Meaning and Reasons for Inspection

The **quality** of manufactured products is controlled through inspection at various times. **Inspection** is the official checking or examination of materials, parts, or articles at different times while they are being made, or immediately after they are made. The following are thus inspected:

1. Measurements
2. Finish
3. Materials
4. Performance

Parts which do not pass inspection are either reworked or scrapped. Measuring instruments have to be inspected regularly by **inspectors** in **inspection departments,** Fig. 46-1.

46-2. Inspecting Measurements

Measurements must be accurate, which means that they must be exact or correct. Automobile parts, for example, are inspected after every operation during their manufacture.

Figure 46-2 shows an automobile **chassis** being inspected. The **surface plate** is 8′ wide, 18′ long, and is perfectly level and

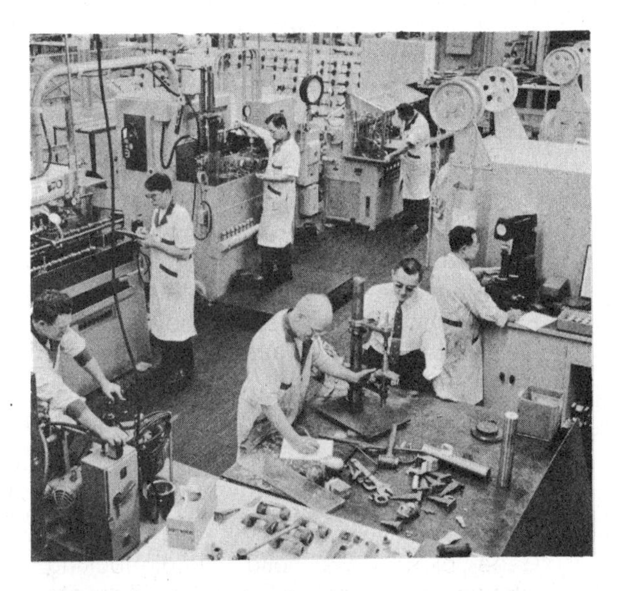

Fig. 46-1. Inspection Room

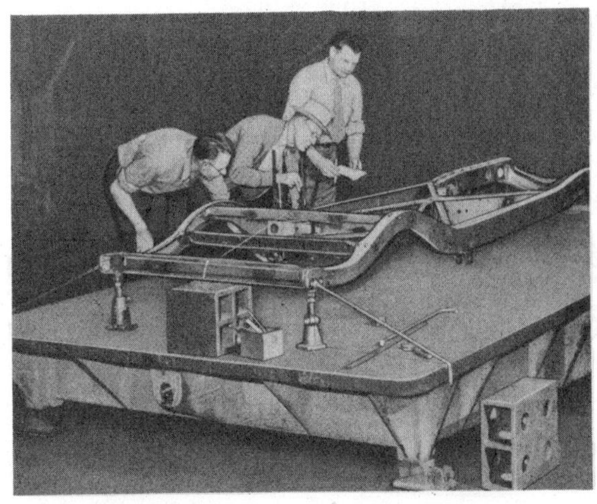

Fig. 46-2. Inspecting an Automobile Chassis

accurate to within 0.005″ from one end to the other.

46-3. Standards for Measurement

Without a standard for linear (straight line) measurement, parts such as automobile parts could not be produced to fit interchangeably on the same make and model automobile. Replacement parts for the same make and model must fit interchangeably anywhere in the world. A **standard** is necessary so that an inch or a metre is the same length in all parts of the world.

The International Bureau of Weights and Measures, established in 1875 near Paris, France, represents most nations of the world, including the United States and Canada. It keeps standards for metric measurement, including a standard for the metre. The international standard for the metre is the distance between two finely scribed lines on a platinum-iridium alloy metal bar at a temperature of 32° F. (0° centigrade). This metal bar was declared the **International Prototype Meter,** and the various member nations received an exact duplicate copy of it. In 1889, the United States received its copy, which is now at the Bureau of Standards at Washington, D. C.

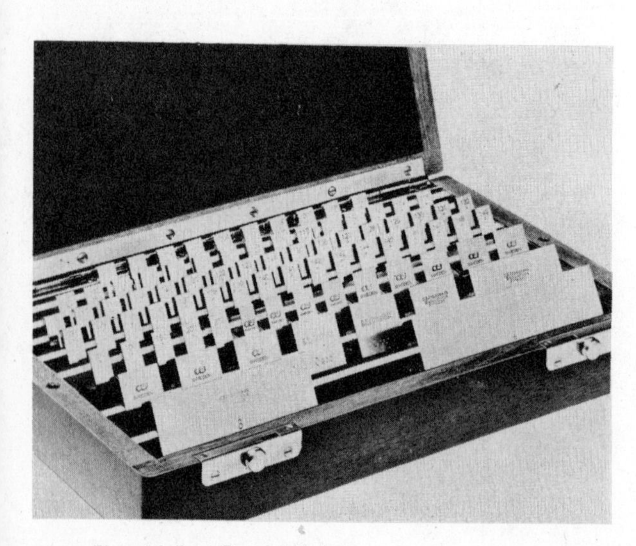

Fig. 46-3. Gage Blocks

In 1893 the U. S. Bureau of Standards adopted the metric system as a **standard** for legally defining the pound and the yard. The length of the U. S. yard was defined as $\frac{3600}{3937}$ metre. One inch was defined as 2.54 centimetres, exactly. Therefore, the units of linear measurement in the English system are now defined in terms of metric units.

In 1959 the English speaking countries, including the United States, Great Britain, and Canada, accepted the **International Inch** by general agreement, without specific legislation. Before that date, the inch in Great Britain was defined in terms of the British Imperial Yard which was several millionths of a millimetre shorter than the U. S. yard.

The International Bureau of Weights and Measures also defined the length of the metre in terms of light wavelengths. One metre equals 1,650, 763.73 wavelengths of orange light emitted by Krypton-86 atoms in an electrical discharge. Since no standard of length is maintained for the English system, the International Inch, as defined in terms of metric units (2.54 centimetres), can be stated in terms of wavelengths of Krypton light as follows:

1″ = 0.0254 metres × 1,650,763.73 wavelengths per metre
1″ = 41,929.3987 wavelengths
1 wavelength = 0.0000238″

Light waves do not vary significantly with temperature changes and changes in atmospheric conditions. Therefore, measurement by this exact method may be duplicated in any part of the world.

Precision **gage blocks,** as shown in Fig. 46-3, are used as a **practical standard** for measurement in machine shops and industrial plants the world over. (Gage is also spelled gauge.)

46-4. Gage Blocks

Gage blocks are solid, simple-looking **hardened steel** blocks, Fig. 46-3. They are ground and finished to different sizes and have extremely flat, precision ground, lapped, and polished surfaces. Gage blocks are exact within a few **millionths of an inch.** Very thin tissue paper is one thousandth of an inch (0.001″) thick; dividing this thickness into a thousand parts gives one millionth of an inch (0.000001″).

The best known gage blocks were invented by Carl E. Johansson in Sweden in 1895. They became known the world over as **Johansson gage blocks** or **Jo-blocks.** Today similar gage blocks are manufactured by several companies according to three grades of accuracy: AA, A, and B. These are U. S. Federal quality classifications based on allowable tolerances for length, flatness, parallelism, and surface finish. The length tolerances for blocks up to one inch, measured at a temperature of 68° F. are as follows:

Class AA:	±0.000 002″
Class A:	+0.000 006″
	−0.000 002″
Class B:	+0.000 010″
	−0.000 006″

Fig. 46-4. Master Planer and Shaper Gage Used with Gage Blocks for Setting Up Work on Surface Plate

Class AA gage blocks are **laboratory** or **master** gages usually used in temperature controlled laboratories as references to check the accuracy of other gages. Class A gage blocks, called **inspection grade,** are used for inspection of finished parts and may also be used for setting working gages. Class B gage blocks are often called **working blocks.** They are used by machinists and tool and die makers in the normal course of their work for such things as accurate layout of lines and hole centers, setting cutting tools accurately, and for checking the accuracy of their other measuring tools, Fig. 46-4.

Gage blocks are assembled by squeezing them together with a twisting motion called **wringing**. This forces all the air from between the blocks. Because the blocks are so flat and smooth, the blocks stick tightly together because of **adhesion.** The adhesion is strong enough to support considerable weight, Fig. 46-5.

46-5. Interchangeability

Interchanageability is the manufacture of parts to such a degree of accuracy that any given part of a product can be exactly replaced with a like part taken from another of the same make and model product. It also means that **replacement parts,** sometimes manufactured years after the product was made, will fit and perform as well as the original parts.

With this degree of accuracy in manufacturing, parts of a product can be made at widely different locations with the assurance that when shipped to an assembly point, all parts will fit together properly. For example, automobile parts are often made hundreds and even thousands of miles away from the assembly plant. Bodies are made in one plant, frames in another, engines in another, wheels in another, and so on. Therefore, it is necessary to make quantities of each part enough alike so that they can be interchanged with each other,

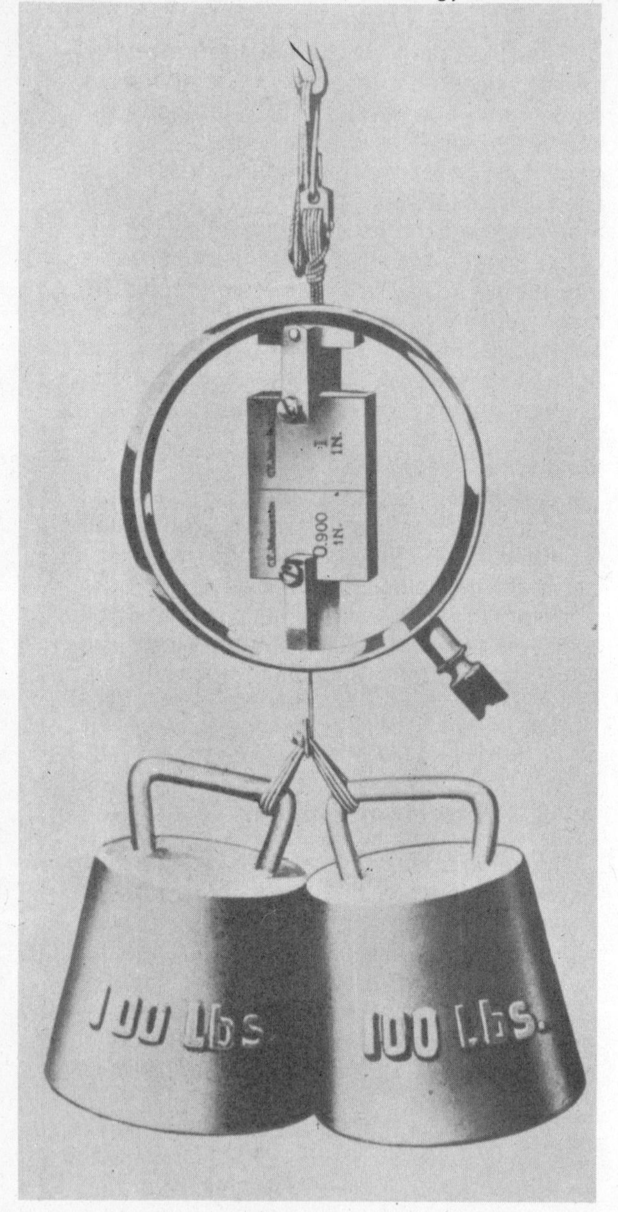

Fig. 46-5. Adhesion of Gage Blocks

and also so they will assemble properly with other parts.

46-6. Dimension Limits

Suppose that a dimension on a drawing is 1″. It is impossible to make the size on the object **exactly** 1″. It may be 1″ when measuring with a **steel rule;** but when measuring with a **micrometer,** it would be found that it is a little over or under 1″. Now suppose it is made so that it will be 1″ when measured with a micrometer, which measures to the nearest 0.001 inch. Yet when it is measured with a **vernier micrometer,** which measures to the nearest 0.0001 inch, it is again a little over or under 1″ because of finer calibrations to 1/10,000ths.

Then suppose that again it is made so that it will be 1″ when measured with a vernier micrometer, but when it is measured with a still finer measuring instrument, it is found to be a little over or under 1″. Each time that a finer measuring instrument is used, it is found that the dimension is either oversize or undersize instead of being **exactly** 1″. Hence, no two parts can be made **exactly** the same size. For this reason, **working drawings** (see § 4-2) often give **double dimensions,** Fig. 46-6. The double dimensions specify the limits of size, including the upper and lower limits.

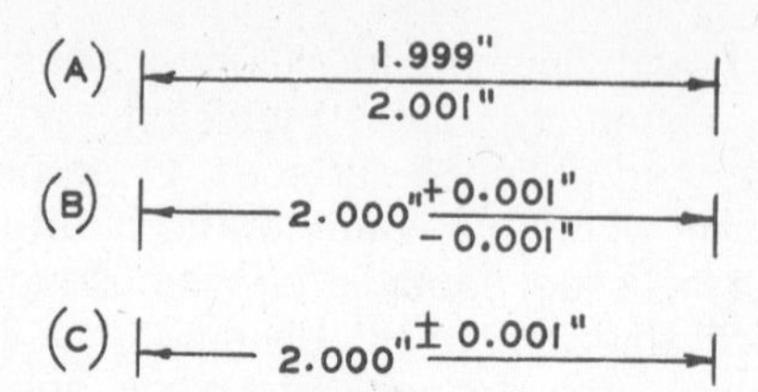

Fig. 46-6. Dimension Limits
In each instance the tolerance is 0.002″

One dimension is sometimes placed above the other as (A) in Fig. 46-6. Thus, the article can be made any size between 1.999″ and 2.001″. If the size is under 1.999″ or over 2.001″, the article cannot be used.

Another way is to use a **plus sign** and a **minus sign** as (B) in Fig. 46-6. Sometimes the plus and minus signs are joined into one sign, as ±. If 0.001″ over or under were allowed, it would appear on the drawing as (C) in Fig. 46-6.

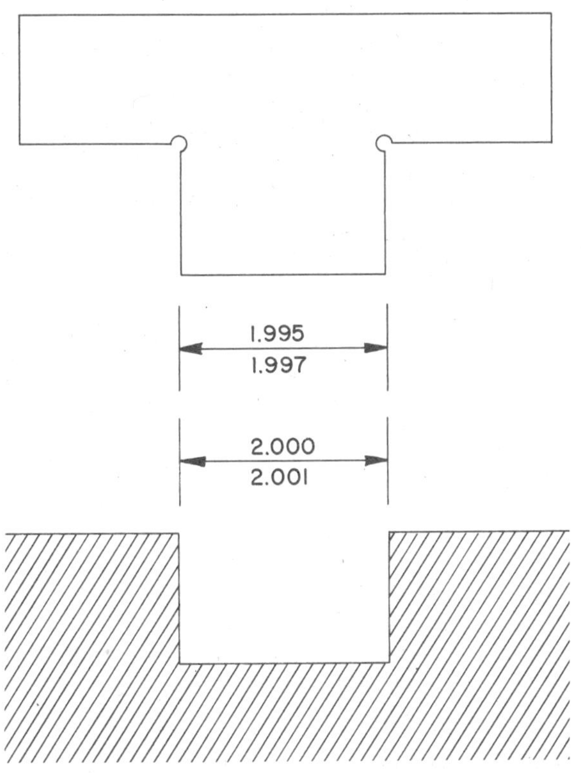

Fig. 46-7. Tongue and Groove Designed with Allowance (Positive) for a Sliding Fit Tolerance on Tongue, 0.002"; Tolerance on Grove 0.001"
These tolerances provide 0.003" positive allowance.

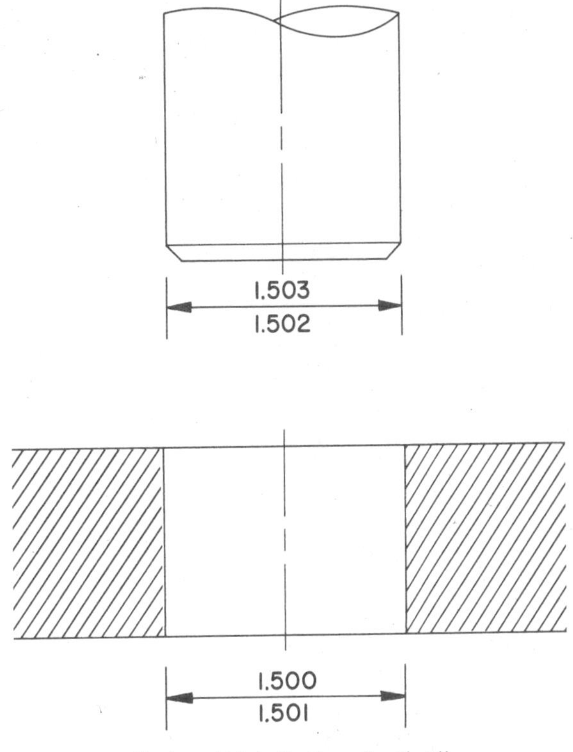

Fig. 46-8. Shaft and Hole Designed with Allowance (Negative) for Force Fit
Loosest fit is 0.001" interference; tightest fit is 0.003" interference.

This small amount oversize and undersize set the **limit.** (See § 24-7.) It tells the **mechanic** at the bench or machine and the **inspector** just how exact the dimension on the work must be. Read also section 46-10.

46-7. Tolerance

Tolerance[1] is the **difference** between the largest and lowest **limits** of a dimension. Thus:

The tolerance of $\frac{1.999}{2.001}$ is .002".

[1]**Tolerance** comes from the word **tolerate,** to endure, put up with, or to allow a certain amount oversize or undersize.

It may be necessary to make the tolerance as small as 0.0001" or as much as ⅛", depending on the use of the part. The smaller the tolerance, the greater must be the care in making the article and the greater the cost of making it. It is a waste of time, money, tools, and energy to make the tolerance any smaller than necessary.

46-8. Allowance

An **allowance** is an intentional difference between the maximum material size limits of mating parts. An allowance may be either positive or it may be negative. A **positive allowance** is the minimum clearance between mating parts, Fig. 46-7. A

negative allowance is the maximum interference between mating parts, Fig. 46-8. A **positive allowance** provides clearance for a running or sliding fit, see Fig. 46-7. In Fig. 46-7, the difference between the maximum material size of the tongue (1.997″) and the maximum material size of the groove (2.000″) provides 0.003″ positive allowance.

A **negative allowance** provides interference between mating parts, thus producing a force fit. Parts which fit together with a force fit must be assembled by driving, pressing, or shrinking them together. The shaft and hole in Fig. 46-8 are designed with a negative allowance. This causes interference between the parts, and they must be forced together. The tightest possible fit (shaft size 1.503″ and hole size 1.500″) provides 0.003″ interference. The loosest possible fit (shaft size 1.502″ and hole size 1.501″) provides 0.001″ interference.

46-9. Master Gages

Master gages are gages that are used to check or inspect other gages. All other gages are compared with them. Thus, **reference gages**, Fig. 46-9, and **gage blocks** are master gages. There are many other kinds.

46-10. Gages for Measuring Different Materials

Gaging is done to determine whether a part or piece of material is produced within specified size limits. Gaging does not determine the actual size of a part but determines whether the part is inside specified dimension limits. The tools, instruments, or devices used to do gaging are called **gages.**

In metalwork, a gage is used to test or check a manufactured part or piece of material. A **double gage** is used to check that the part is inside dimension limits. A gage such as the **limits snap gage** in Fig. 46-11 is a double gage. It is used for inspecting large numbers of parts rapidly. The part must pass through the large opening but not pass through the small opening.

Any part which can be gaged can also be measured with conventional measuring tools such as micrometers. Generally, less

Fig. 46-9. Reference Disks

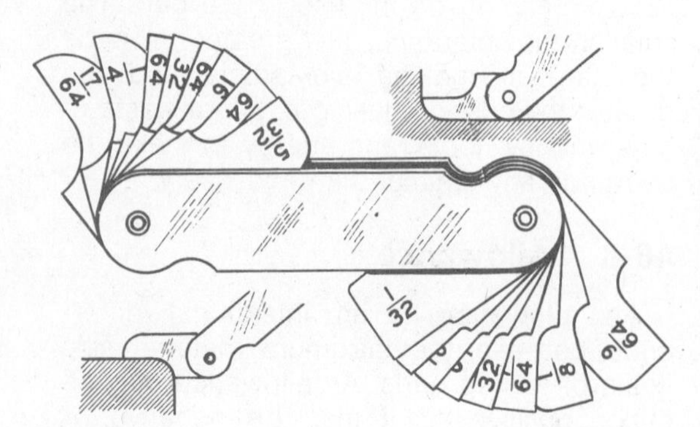

Fig. 46-10. Radius Gage for Measuring Fillets and Rounded Edges

Fig. 46-11. Limit Gages

TABLE 29

Gages, Materials, Gage Numbers, and Size in Decimals

	Manufacturer's Standard Gage[1]	U.S. Standard[2] Gage	Galvanized Sheet Gage	American Standard Wire Gage or Brown & Sharpe Gage	U.S. Steel Wire Gage or American Steel & Wire Co.	Twist Drill and Steel Wire Gage	Am. Steel and Wire Co.	Birmingham or Stubs' Iron Wire Gage	American (National) Standard Screw Gage	
	1	2	3	4	5	6	7	8	9	10
Gage number	For Iron and Steel Sheets	For Iron and Steel Sheets and Plates	For Galvanized Steel Sheets	For Wire & Sheet Metal Except Iron, Steel & Zinc[3]	For Steel Wire (Not Music Wire or Drill Rod)	For Twist Drills and Drill Rod	For Music (or Piano) Wire[4]	For Iron Telephone and Telegraph Wire & Tubing Walls	For Machine Screws	For Wood Screws
0	....	.3125	....	.3249	.3065	....	.009	.340	.060	.060
1	....	.2813	....	.2893	.2830	.2280	.010	.300	.073	.073
2	....	.2656	....	.2576	.2625	.2210	.011	.284	.086	.086
3	0.2391	.2500	....	.2294	.2437	.2130	.012	.259	.099	.099
4	0.2242	.2344	....	.2043	.2253	.2090	.013	.238	.112	.112
5	0.2092	.2188	....	.1819	.2070	.2055	.014	.220	.125	.125
6	0.1943	.2031	....	.1620	.1920	.2040	.016	.203	.138	.138
7	0.1793	.1875	....	.1443	.1770	.2010	.018	.180	...	.151
8	0.1644	.1719	0.1681	.1285	.1620	.1990	.020	.165	.164	.164
9	0.1495	.1563	0.1532	.1144	.1483	.1960	.022	.148	...	.177
10	0.1345	.1406	0.1382	.1019	.1350	.1935	.024	.134	.190	.190
11	0.1196	.1250	0.1233	.0907	.1205	.1910	.026	.120	...	.203
12	0.1046	.1094	0.1084	.0808	.1055	.1890	.029	.109	.216	.216
13	0.0897	.0938	0.0934	.0720	.0915	.1850	.031	.095	...	...
14	0.0747	.0781	0.0785	.0641	.0800	.1820	.033	.083	...	.242
15	0.0673	.0703	0.0710	.0571	.0720	.1800	.035	.072	...	...
16	0.0598	.0625	0.0635	.0508	.0625	.1770	.037	.065	...	.268
17	0.0538	.0563	0.0575	.0453	.0540	.1730	.039	.058	...	...
18	0.0478	.0500	0.0516	.0403	.0475	.1695	.041	.049	...	.294
19	0.0418	.0438	0.0456	.0359	.0410	.1660	.043	.042	...	...
20	0.0359	.0375	0.0396	.0320	.0348	.1610	.045	.035	...	.320
21	0.0329	.0344	0.0366	.0285	.0317	.1590	.047	.032	...	...
22	0.0299	.0313	0.0336	.0253	.0286	.1570	.049	.028	...	...
23	0.0269	.0281	0.0306	.0226	.0258	.1540	.051	.025	...	...
24	0.0239	.0250	0.0276	.0201	.0230	.1520	.055	.022	...	.372
25	0.0209	.0219	0.0247	.0179	.0204	.1495	.059	.020	...	...
26	0.0179	.0188	0.0217	.0159	.0181	.1470	.063	.018	...	...
27	0.0164	.0172	0.0202	.0142	.0173	.1440	.067	.016	...	...
28	0.0149	.0156	0.0187	.0126	.0162	.1405	.071	.014	...	...
29	0.0135	.0141	0.0172	.0113	.0150	.1360	.075	.013	...	...
30	0.0120	.0125	0.0157	.0100	.0140	.1285	.080	.012	...	...

[1]The **Manufacturers' Standard Gage** for steel sheets is now being used for carbon steel and alloy steel sheets.

[2]The **United States Standard Gage** was established by Congress in 1893 for measuring sheet and plate iron and steel. It was the standard gage used for many years.

[3]**Zinc sheets** are measured by a **zinc gage** which is not given in the above table.

[4]The **Music Wire Gage** of the American Steel and Wire Company is used in the United States. It is recommended by the **United States Bureau of Standards.**

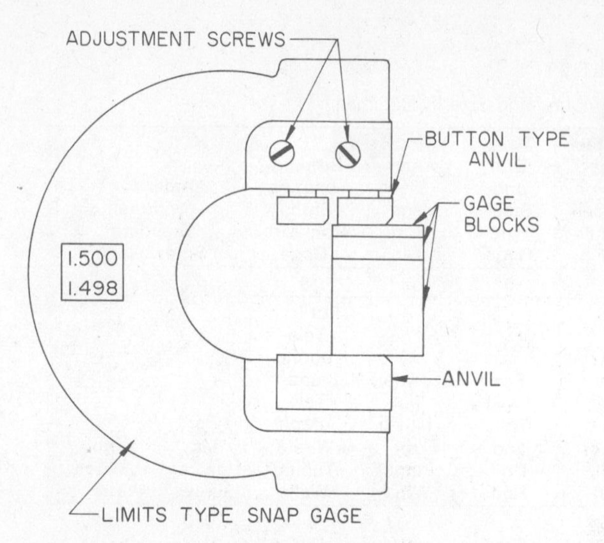

Fig. 46-12. Testing Size of Adjustable Limits Snap Gage with Gage Blocks

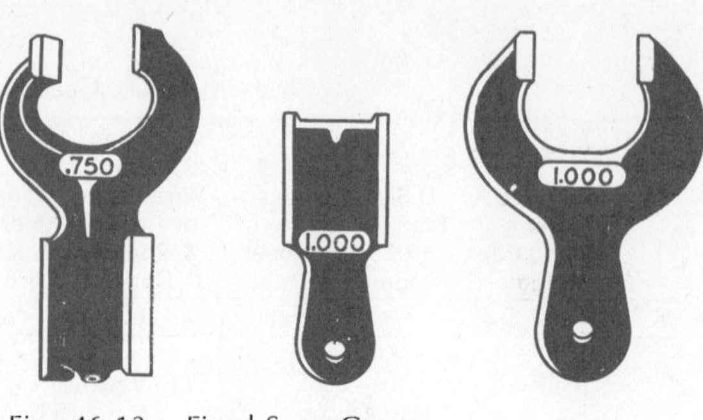

Fig. 46-13. Fixed Snap Gages

training and skill are required for inspecting parts or materials with gages, and there is less possibility for human error.

Gage numbers are used to indicate sizes of wires, drill rod, small drills, **machine screws,** wood screws, the wall thicknesses of seamless tubing, and the thickness of sheet metal and plates. Each size is given a certain number. There are, however, different gages for different materials, and the gage numbers are different for different materials.

Table 29 gives the thicknesses of the gage numbers of different materials. For example, No. 10 sheet steel is 0.1345″ thick, No. 10 sheet copper is .1019″ thick, No. 10 steel wire is .1350″ in diameter, a No. 10 drill is .1935″ in diameter, and a No. 10 machine screw is .1900″ in diameter.

Each **gage system** has a different name. When buying any of these materials, the name or kind of material should be given in addition to the name of the gage, the gage number, and the size in **thousandths of an inch** so that mistakes cannot be made.

46-11. Inspection Tools and Gages

At some time, **inspectors** use most of the **layout tools** described in Unit 6 and the micrometers described in Unit 9.

Many kinds of inspection gages are available for a variety of purposes. Some are designed for general use and can be used for a number of measuring and gaging purposes within a broad range of sizes; some are designed for special gaging purposes within a very narrow size range; while others are designed to gage only one size.

Adjustable gages are designed for general use. They generally can be adjusted for gaging different materials or parts within a certain range of sizes. Several gages of this type include the dial indicating-type gages in Figs. 46-23, 46-30, 46-31, and 46-32. The snap gage in Fig. 46-12 can be adjusted for gaging various dimensions within its size range.

Fixed gages are designed for special measuring applications involving specific size limits. Some are used for gaging only one size. A few of the common fixed gages include caliper-type snap gages, Fig. 46-11; ring gages, Fig. 46-14; plug gages, Fig. 46-15; thread gages, Fig. 46-17; and reference gages, Fig. 46-9. Gages are generally named after their most distinguishing feature, such as their shape, form, or use. In-

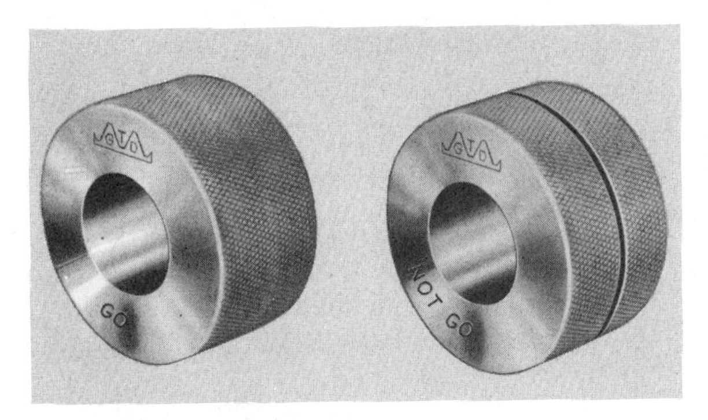

Fig. 46-14. Plain Ring Gage

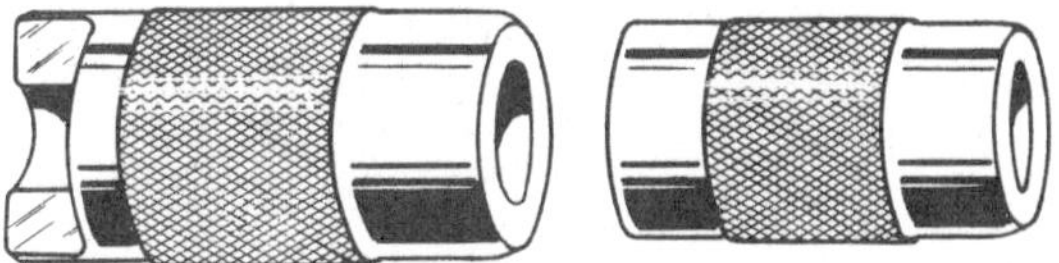

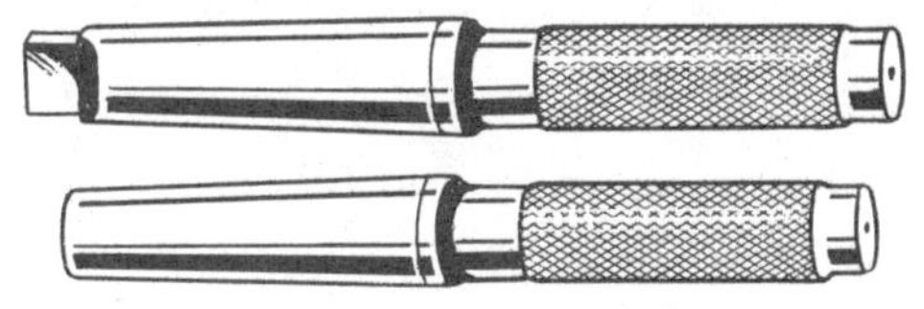

Fig. 46-15. Tapered Plug Gages and Tapered Ring Gages

spection workers and machinists should be familiar with the common kinds of gages and their applications.

Reference gages, also called **reference disks** (see Fig. 46-9) or **master gages,** are made of hardened steel and are used to test or check inspection gages. They are checked to determine whether they are worn, damaged, or out of adjustment. Micrometers can also be checked in this manner.

The size should always be stamped on fixed gages and on reference gages. Gages are available in many sizes. They should be checked from time to time to determine their accuracy.

46-12. Drill Gage, Screw-Pitch Gage, Thickness Gage, Sheet Metal and Wire Gages, Center Gage, and Radius Gage

Drill gages are described in section 52-8.

The **screw-pitch gage** is described in section 21-10.

The **thickness gage** is described in section 24-8.

Sheet metal and wire gages are described in section 31-5.

Two uses of the **center gage** are given in sections 13-11 and 55-9.

The **radius gage,** also known as a **fillet gage,** has a number of blades which fold into a handle like the blades of a pocket knife, Fig. 46-10. It is marked in **fractions** of an inch, as for example, ¼″, and is used to measure rounded corners, or **radii** (plural of **radius,** explained in section 4-5).

46-13. Go and No-Go Gages

Go and no-go gages are often called **limits gages.** They are **double gages** because they have two gaging points or surfaces. One tests the upper size limit and the other tests the lower size limit of a part being gaged. The size limits are those which are indicated on the drawing for the part. The sizes generally are stamped on the gage (see Fig. 46-12). There are several types of go and no-go gages. The common types include snap gages, ring gages, and plug gages.

The principles involved in testing parts with go and no-go gages can be understood by studying Fig. 46-11. The figure shows a cylindrical part being tested with a limits **snap gage.** The upper gaging point is the **go** point, while the lower one is the **no-go** or not-go point. If the part is inside the specified limits on the drawing (the size limits frequently are stamped on the gage also), it will pass through the **go** point or surface, but it will not pass through the **no-go** point or surface.

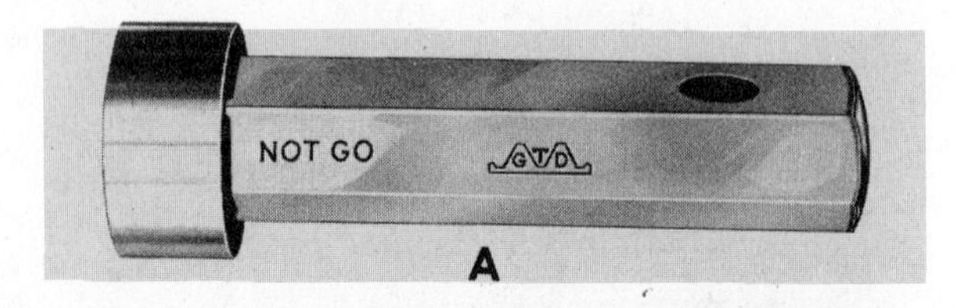

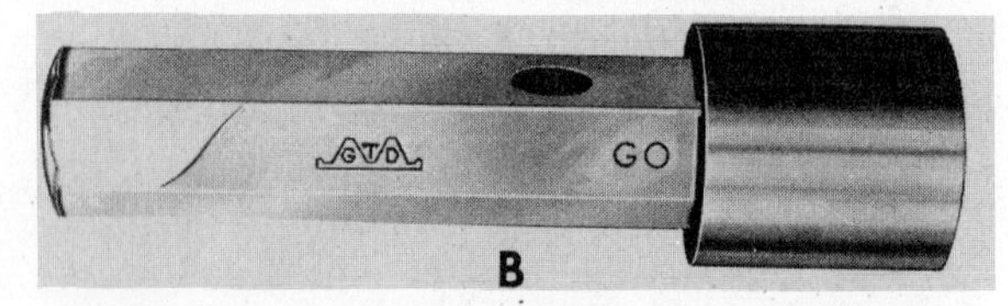

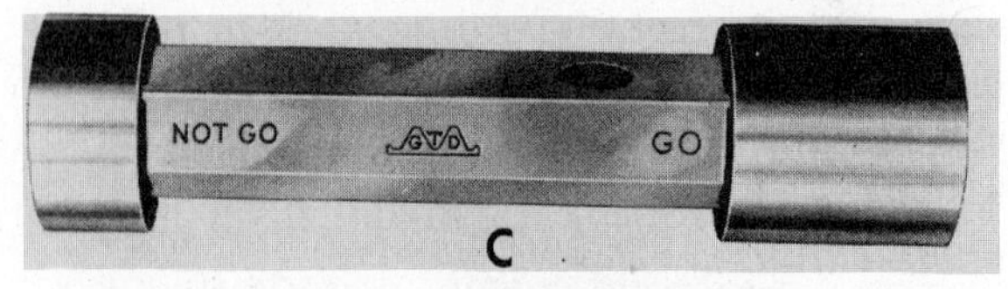

Fig. 46-16. Plain Cylindrical Plug Gages
A. No-Go Gage
B. Go Gages
C. Dougle-End Gage

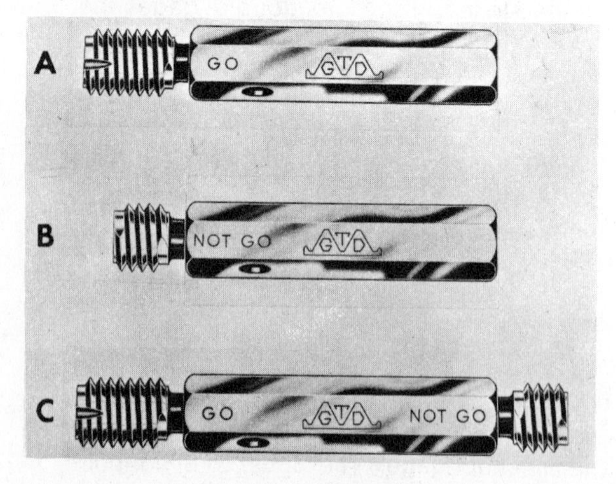

Fig. 46-17. Thread Plug Gages
A. Go Gage
B. No-Go Gage
C. Dougle-End Gage

46-14. Snap Gages

Snap gages are used for checking the outside diameter, length, or thickness of parts. They are used in a manner similar to the way a caliper is used. Hence, snap gages are often called **caliper gages.** Snap gages are available in a wide variety of styles and sizes. They may be the fixed, Fig. 46-13; adjustable, Fig. 46-12; or indicating, Fig. 46-31.

Snap gages of the adjustable type, Figs. 46-11 and 46-12, are widely used limits gages of the go and no-go type. They have one stationary anvil and two adjustable button anvils, Fig. 46-12. The outer button is set at the upper-size limit, and the inner one is set at the lower-size limit. The sizes may be tested for accuracy with **gage** blocks.

Adjustable snap gages may be supplied **set** and **sealed** at specific limits by the manufacturer. The sizes are then stamped on the gage. They are also available **unset** and **unsealed.** They may then be adjusted to the desired size with gage blocks, as in Fig. 46-12.

Limits snap gages are available with gaging rolls, instead of anvils, for testing the pitch diameter of screw threads, Fig. 46-20. Dial indicating snap gages are also available, Fig. 46-31.

46-15. Ring Gages

A ring gage is a hardened steel ring or collar. Three kinds of ring gages are used:

1. Plain ring gages, Fig. 46-14.
2. Tapered ring gages, Fig. 46-15.
3. Thread ring gages, Fig. 46-18.

Plain ring gages (see Fig. 46-14) are used to test the external dimension limits of straight round parts. The **no-go** ring, identified by the groove around the outside diameter, is used to check the minimum size limit. The **go** ring is used to check the maximum size limit. The go ring should pass over a part which is inside specified size limits, with little or no interference. The no-go ring should not pass over the part.

If both rings pass over the part, it is undersize. If neither does, it is oversize.

Tapered ring gages (see Fig. 46-15) have a tapered hole and are used for testing the size and fit of a taper, such as the tapered shank on a drill or reamer.

Thread ring gages, of the go and no-go type in Fig. 46-18, are used for checking the fit and the pitch diameter limits of external screw threads.

46-16. Plug Gages

Three kinds of plug gages are in common use: (1) plain-cylindrical plug gages, Fig. 46-16; (2) cylindrical-taper plug gages, Fig. 46-15; and (3) thread plug gages, Fig. 46-17. Plug gages of special design are also made for checking square holes or holes of special shape.

Plain-cylindrical plug gages, Fig. 46-16, are accurate cylinders which are used to check the size limits of straight cylindrical holes. The go gage should enter the hole with little or no interference. If great pressure is necessary, the hole is too small. The no-go gage should not enter the hole. If it does, the hole is too large.

Cylindrical-taper plug gages, Fig. 46-15, are used for checking the size, amount of taper, and the fit of tapered holes. This type of gage is used to test the tapered hole in drill sleeves, in machine tool spindles, and in various kinds of tool adaptors.

Thread plug gages, Fig. 46-17, are used for checking the size limits and the fit of internal screw threads.

46-17. Thread Measurement and Thread Gages

The emphasis in screw thread measurement is always on the pitch diameter, explained in section 21-8. The tolerances and the limits on the pitch diameter largely determine the class or fit of the thread. The size and accuracy of screw threads usually are measured or tested with the following:

1. Screw-pitch gage (see Fig. 21-8 and § 21-10).
2. Thread micrometer (see Fig. 9-5).
3. Thread plug gage (see Fig. 46-17).
4. Thread ring gage (see Fig. 46-18).
5. Roll-thread snap gage (see Fig. 46-20).
6. Other methods.

Thread Micrometer

The pitch diameter of external 60° V-threads, including Unified and American (National) form threads, can be measured

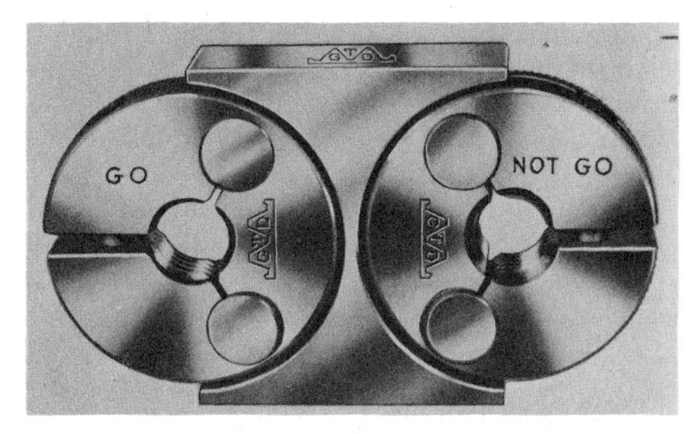

Fig. 46-18. Go and No-Go Thread Ring Gage with Holder

Fig. 46-19. Inspecting Threads with a Ring Gage

directly with a thread micrometer, Fig. 9-5. The micrometer spindle has a 60° conical point. The anvil has a 60° V-groove. The anvil swivels so that it can be held at a convenient angle for reading the micrometer. The micrometer should always be checked for a zero reading in the closed position before using it.

The number of threads per inch which can be measured with a specific thread micrometer is limited. One inch capacity thread micrometers are designed for each of the following ranges of threads per inch: 8 to 13, 14 to 20, 22 to 30, and 32 to 40. Larger capacity thread micrometers are also available. Hence, care should be taken to select the right thread micrometer for the thread to be measured. For example, a 1″ capacity micrometer designed for the thread range from 8 to 13 threads per inch should be selected for measuring the pitch diameter of a ¾-10 UNC thread.

The pitch diameter for each diameter, pitch, and fit or class of thread may vary within certain size limits. The size limits are listed in handbooks for machinists. For example, the pitch diameter for an external ¾-10 UNC class 2A thread may range from 0.6773 to 0.6832″. Hence, if the pitch diameter is found to be inside this size range, the thread will fit properly.

Thread Plug Gage

Limits type thread plug gages, Fig. 46-17, are used for checking internal threads for the proper fit or class of thread. The gages include a **go** gage and a **no-go** gage. The go gage is the longer gage, and it has a chip groove for cleaning the threads. The go gage is the minimum pitch diameter, and the no-go gage is the maximum pitch diameter of the internal thread.

In using thread plug gages, the go gage should enter the tapped hole for the entire length of the gage. The no-go gage may or may not enter. If it enters the hole, it should fit snugly on or before the third thread, thus showing that the thread has the maximum pitch diameter for the specified fit or class of thread. If the gage enters farther, the thread is oversize and will fit too loosely.

Thread plug gages are made for each size, pitch, and fit or class of thread. They are made for all standard Unified and American (National) form threads. They are also made for pipe threads.

Thread Ring Gage

The accuracy and the fit of external threads may be tested with thread ring gages, Figs. 46-18 and 46-19. These are limits-type gages which include both **go** and **no-go** gages. The **go** gage checks the maximum pitch diameter, flank angle, lead, and clearance at the minor diameter. The **no-go** gage checks only the pitch diameter to determine whether it is below minimum size limits.

Both ring gages are used in checking a thread. If the **go** gage does not turn on freely, one of the thread elements is inaccurate, and the thread will not fit the mating internal thread properly. The **no-go** gage should not turn on. If it does, the pitch diameter is under the specified minimum size limits and the thread will not fit properly with the mating thread. Thread ring gages are made for each size, pitch, and fit or class of thread.

Roll-Thread Snap Gage

External Unified and American (National) screw threads can be tested rapidly and accurately with a roll-thread snap gage of the limits type. The gage in Fig. 46-20 is the open-face type which can be used close to shoulders.

The outer or **go rolls** check all thread elements at one time. They are set at the maximum pitch diameter limit. The inner or **no-go rolls** are set at the minimum pitch diameter limit. They check only the pitch diameter to determine whether it is below the minimum pitch diameter specified. Threads which are accurate and within the proper pitch diameter limits will pass through the go rolls and are stopped by the no-go rolls. Roll-thread snap gages are

Fig. 46-20. Roll-Thread Snap Gage

made for each size, pitch, and fit or class of thread.

Other Methods

Screw threads may be inspected or measured by other methods. A method called the **three-wire method** may be used for measuring the pitch diameter of external screw threads. This method is more complex and is explained in handbooks for machinists. Thread measuring and inspection instruments of special design are also available for checking threads.

Fig. 46-21. Electronic Gage Being Used for Inspecting

46-18. Air and Electronic Gages

The mechanical gages previously discussed are of a type known as **attribute** gages. Attribute gages are only capable of determining whether a part size is within preset limits; they do not measure the actual size of the part. When it is desirable to classify parts into groups within size limits, size indicating gages must be used. **Air** and **electronic gages** are essentially **comparators** which are preset, using master gages, to measure a particular size. They are designed, however, to indicate within a limited size range the exact amount the part is over or under size.

Air gages operate by either measuring the volume of air or the drop in pressure resulting from the air escaping between the gaging head and the workpiece. The gage is calibrated to read directly in thousandths, ten thousandths, or even millionths of an inch. Air gages use either a dial gage or a glass tube in which either the height of a cork ball or the height of a column of fluid indicates the size.

Electronic gages use a **stylus** or probe in a gage head to contact the workpiece, Fig. 46-21. Mechanical motion of the stylus is converted to an electric signal which is amplified and used to operate a needle on an electric meter. The electric meter is calibrated to read directly to the desired degree of accuracy. Electronic gages are capable of detecting size differences as small as one millionth of an inch.

Air and electronic gages are also useful for checking flatness, parallelism, straightness, roundness, and concentricity. Mechanical dial-indicating gages are used for some of the same purposes as air and electronic gages.

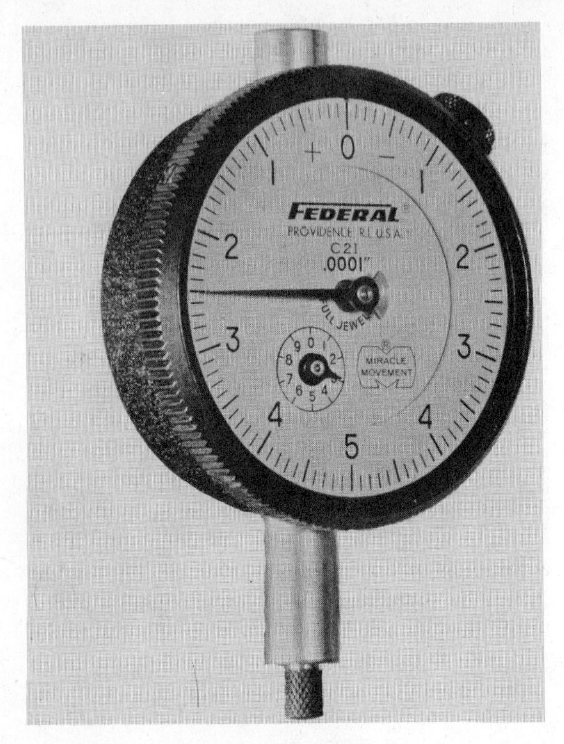

Fig. 46-22. Dial Indicator (Balanced Type)

Fig. 46-23. Checking Height of Machined Part with Dial Indicator

46-19. Dial Indicator Gages

Inspectors, machinists, and toolmakers use a variety of dial-indicating gages and measuring instruments. A **dial indicator,** also called a **dial gage,** looks somewhat like a watch, Fig. 46-22. The dial indicator shows visually the amount of error, in size or alignment, for a part being measured or gaged, see Fig. 46-23.

The graduations on dial gages vary in size. They may be indicated in thousandths (0.001), in ten-thousandths (0.0001), or to the nearest 0.00005″. The dial indicator in Fig. 46-22 has 0.0001″ graduations. The numbered graduations are in thousandths, and the shorter graduations between are ten-thousandths.

Two types of dial gages are in common use. The **balanced type,** as in Fig. 46-22, is numbered in both directions starting with zero. This type is most common on inspection-type gages. The **continuous-reading** type, as in Fig. 46-25, is numbered clockwise, continuously starting at zero. The range of graduations on small gages may vary from 0.010″ to 0.050″.

Dial Indicator Set

The dial indicator set shown in Fig. 46-23 is mounted on a column which is clamped in the T-slot in the steel base. It may be used as a gage for checking the thickness of parts at an inspection bench. The gage also may be swiveled on its column to test the thickness of a workpiece mounted on a machine tool table, see Fig. 46-23.

In using a dial indicator set, the indicator must first be set to **gaging height** or **gaging thickness;** this is the basic thickness of the part to be gaged or tested. The gaging height may be set with a planer gage, with gage blocks, or with other available gaging tools or devices.

The gaging height is established between the dial indicator contact point and the surface on which the part rests. For testing the thickness of parts on a machine tool table, the gaging height is between the dial

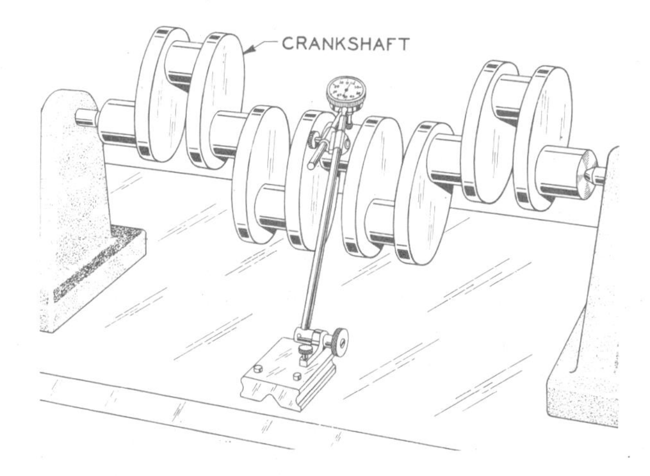

Fig. 46-24. Using a Dial-Indicating Gage for Testing the Straightness of a Part

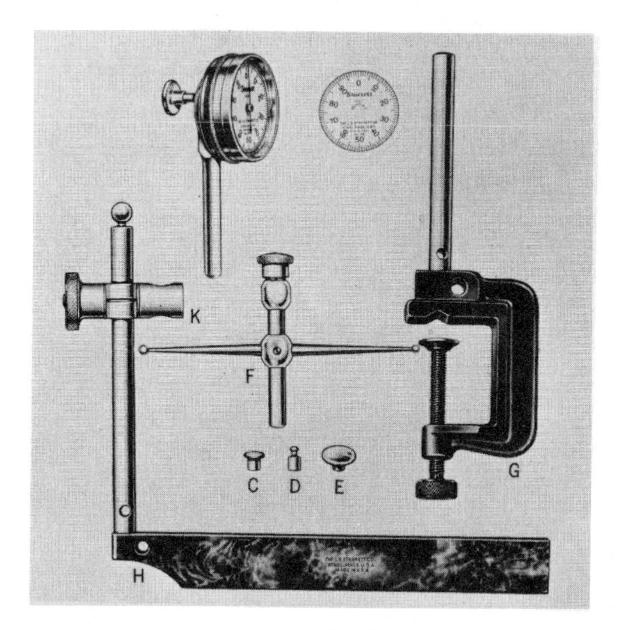

Fig. 46-25. Universal Dial Indicator Set
C. D. & E. Contact Points
F. Hole Attachment
G. Clamp
H. Tool Post
K. Sleeve

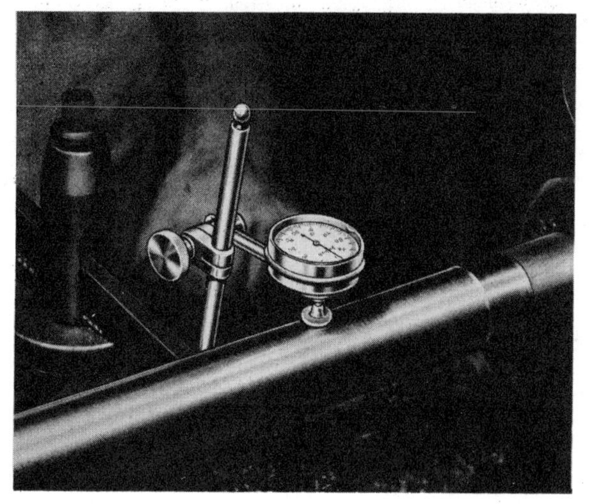

Fig. 46-26. Checking Runout on a Lathe

contact point and the table surface, see Fig. 46-23. In setting the gaging height the dial should be set at **zero,** and it should be under enough spring tension to enable the dial hand to rotate in either direction through the desired measuring or gaging range.

A **universal dial indicator set,** Fig. 46-25, may be used for many kinds of testing and measuring applications. With the variety of accessories provided, it may be mounted on a surface-gage base. It may then be used on a machine table or on a surface plate, see Fig. 46-23. It may be mounted in the tool post on a lathe, as shown in Fig. 46-26. With the hole attachment, holes can be accurately aligned or tested in a lathe chuck or on other machine tools, see Fig. 46-27.

Universal dial indicator sets such as shown in Fig. 46-28 are small, convenient to use, and very versatile. They are often used in conjunction with a vernier height gage for inspection work, Fig. 46-29.

Dial-indicating depth gages, Fig. 46-30, are used for gaging the depth of grooves, shoulders, keyways, holes, and similar recesses. Extension points make it possible to increase the measuring depths.

Dial-indicating snap gages, Fig. 46-31, are used for gaging the diameters of parts. The gage shows whether the parts are with-

Fig. 46-27. Hole Attachment Permits Accurate Internal Tests with Dial Indicator

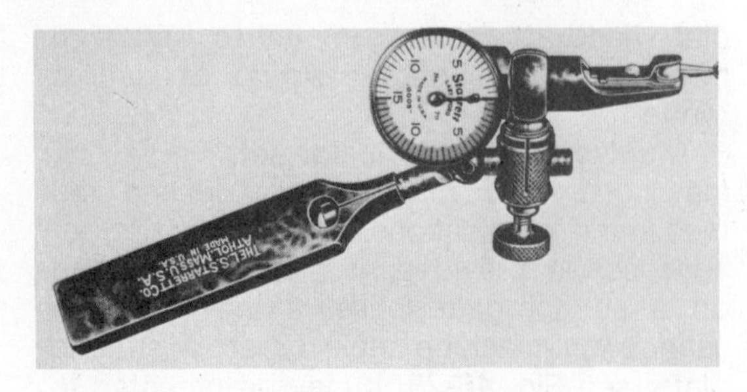

Fig. 46-28. Universal Dial Indicator Set Used for Setup and Inspection

Fig. 46-29. Dial Indicator Used with Vernier Height Gage

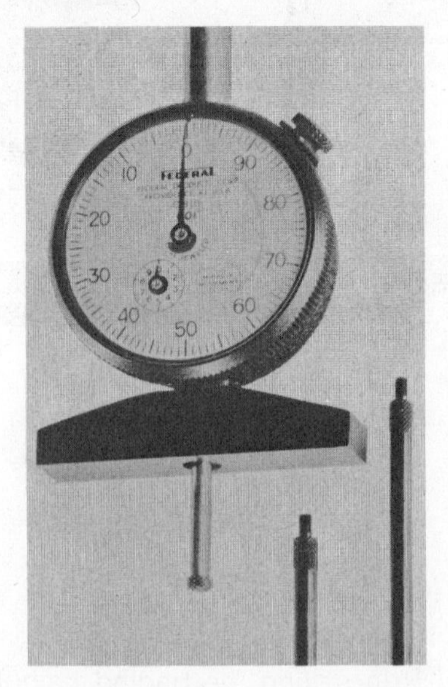

Fig. 46-30. Dial-Indicating Depth Gage with Extension Points

in the size limits specified. The gage may be set for any **basic size** within its capacity, and it shows the amount which the part is over or under the basic size. These gages are available in several sizes. Thus, a gage with a capacity from 0″ to 1″ can be set to gage the thickness of any part within this size range. A dial-indicating snap gage may be used for measuring parts at a bench. It also may be used for measuring parts which are mounted in a machine such as a lathe or cylindrical grinding machine.

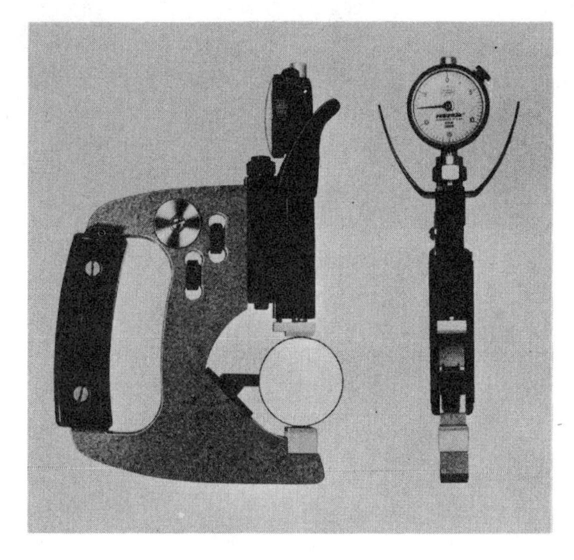

Fig. 46-31. Dial-Indicating Snap Gage

Dial Comparator

A dial comparator, Fig. 46-32, is used for gaging thickness to determine whether parts are within the limits specified. The dial in Fig. 46-22 has 0.001″ graduations and a dial range of 0.100″. It is also equipped with a revolutions counter. The sliding table on the column may be raised or lowered for parts of various thicknesses. The contact point is raised or lowered through its range with the lifting lever at the top. The gaging height is established with either a planer gage or with gage blocks.

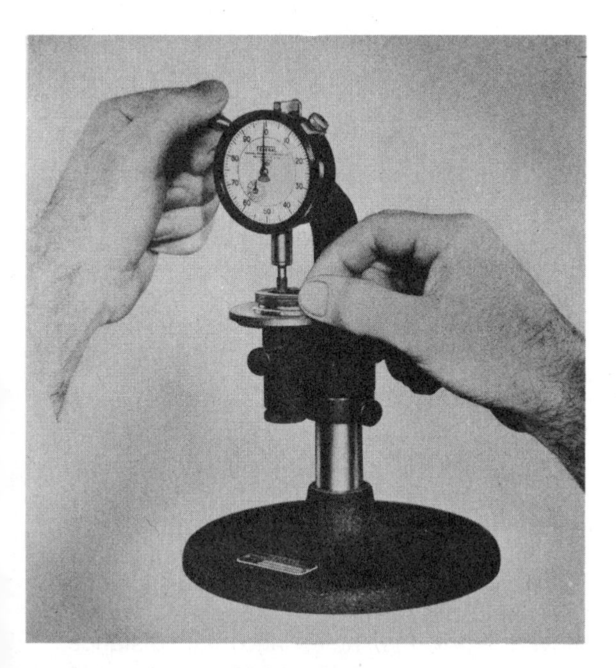

Fig. 46-32. Dial Comparator

Fig. 46-33. Master Planer and Shaper Gage Used to Set Height of Cutting Tools

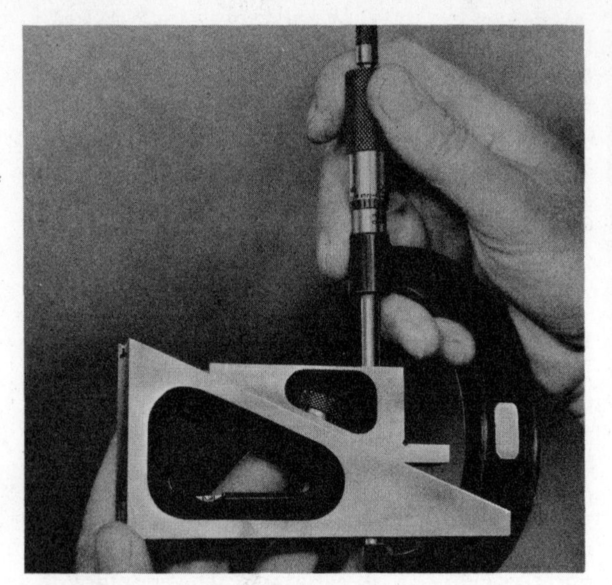

Fig. 46-34. Gage Height or Width Is Measured and Set to Micrometer Accuracy

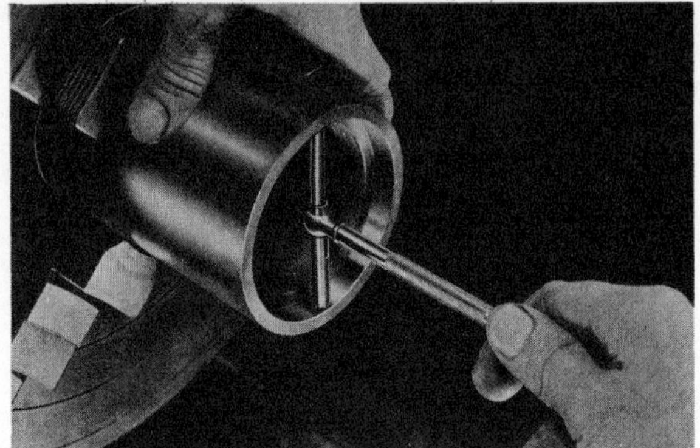

Fig. 46-35. Using a Telescoping Gage

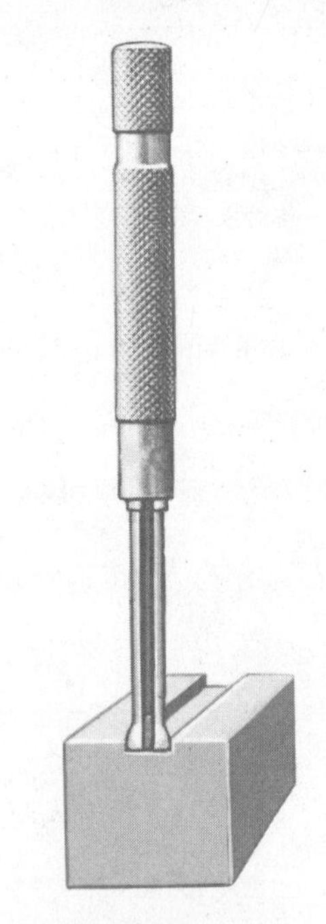

Fig. 46-36. Small Hole Gage

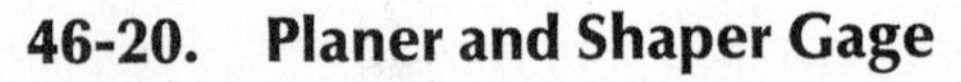

46-20. Planer and Shaper Gage

A **planer and shaper gage,** Fig. 46-33, is often used for setting the height of the cutting tools on machines such as planers and shapers. It also may be used for setting the tool height on milling machines, surface grinding machines, and other machine tools.

The height or width of the gage may be adjusted and set to micrometer accuracy, as shown in Fig. 46-34. Cylindrical extension parts can be screwed into the gage for added length (see Fig. 46-33).

Other uses of a planer and shaper gage include the following: in conjunction with gage block for establishing height on a surface plate (see Fig. 46-4); for establishing gaging heights for gaging tools such as dial-indicating snap gages (Fig. 46-31), dial-indicators (Fig. 46-30), and dial-indicating comparators (Fig. 46-32).

46-21. Telescoping Gage

A **telescoping gage** is used for measuring or gaging the size of holes. The end of the gage has a plunger which is under spring tension when retracted. The gage is used by retracting the plunger and inserting the gage in a hole as in Fig. 46-35; the knurled nut on the handle is then tightened, thus locking the plunger in position. The gage is then removed from the hole, and the distance across the ends of the gage is measured with a micrometer. The telescoping gage may be used for measuring grooves as well as holes. These gages are available in sizes with measuring distances from 5/16″ to 6″.

Small-hole gages, Fig. 46-36, as the name implies, are used for measuring small holes, grooves, and recesses from ⅛″ to ½″ width. When the gage is used, the ball end is inserted in the hole or groove to be measured. The knurled screw on the handle is turned until the ball end expands enough to cause a slight dragging pressure. The gage is then extracted from the hole, and the distance across the ball end is measured with a micrometer.

46-22. Optical Comparators

An **optical comparator** projects an enlarged shadow-like profile of the object being inspected onto a screen. Here both its size and shape are compared to a master drawing. The optical comparator is especially useful for checking small, irregularly shaped objects which cannot easily be measured with conventional tools. Flexible parts — such as springs and soft rubber or plastic objects, which would distort under the pressure of ordinary measuring tools — can easily be inspected with the optical comparator. The accuracy of gear tooth shape and screw thread form and pitch are easily checked in this manner.

46-23. Optical Flats

An **optical flat** is a disc or rectangle of polished quartz with precision flat surfaces. Used in conjunction with a monochromatic (one color) light, usually helium light, the light passes through the optical flat onto the surface of the workpiece being inspected. The light produces dark band patterns on the surface of the workpiece which tell how flat the workpiece is, how parallel it is to another surface, or what its comparative size is. When optical flats are used with helium light, each dark band becomes a measurement of 11.6 millionths of an inch.

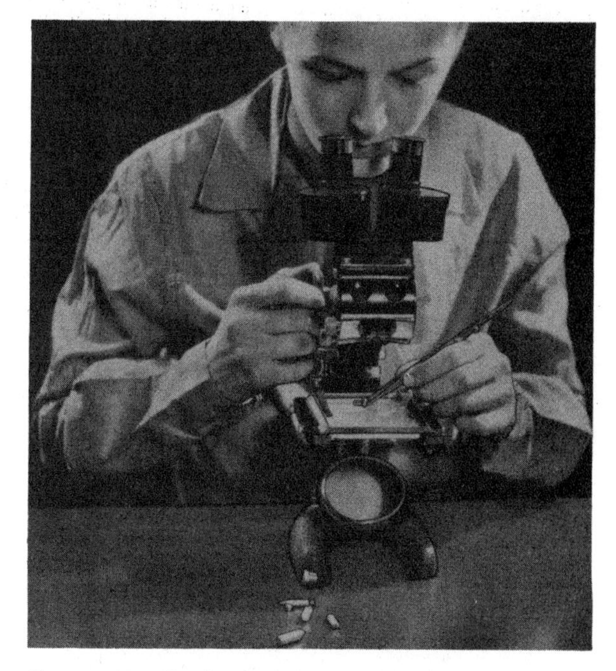

Fig. 46-37. Toolmaker's Microscope

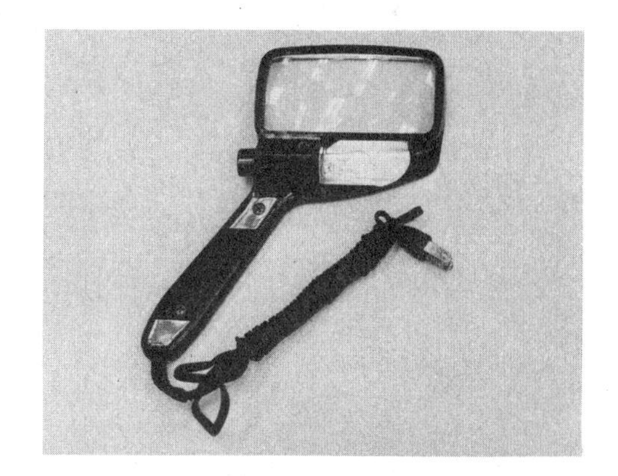

Fig. 46-38. Illuminating Magnifying Glass Used in Assembly and Inspection Work

46-24. Microscopes and Magnifying Glasses

Microscopes are used for a variety of inspection jobs including evaluation of material samples, measurement of surface imperfections, and measurement of miniature parts, Fig. 46-37.

Magnifying glasses aid in the accurate reading of vernier calipers, scales, and other finely graduated instruments. Free standing, illuminating magnifying glasses, Fig. 46-38, are often used to aid assembly and inspection of small components.

46-25. Surface Finish Measurement

The surface finish quality of machined surfaces is extremely important to the service life of parts which are in sliding or rolling contact with each other. If a surface on a bearing is too rough it will rapidly wear out. If its surface is too smooth, it cannot hold sufficient lubricant to keep it from rapidly wearing out. It is the responsibility of design engineers to specify the finish quality necessary for the correct functioning of parts. Machinists and inspectors, then, must have equipment for checking the quality of surface finish.

One method of checking surface finish quality is comparison with **standard specimens** such as shown in Fig. 46-39. Comparison is made by sense of touch. A fingernail is dragged first over the standard specimen and then over the surface being checked. This provides a quick approximation of surface finish quality.

Accurate measurement of surface finish quality is made with an electrical instrument called a **surface finish indicator,** Fig. 46-40. With this instrument, a tracer head that houses a diamond-tipped stylus is drawn over the surface being measured. The mechanical movement of the stylus is converted to an electrical signal which is amplified and used to actuate a needle on an electric meter. The meter scale is marked to read directly in micro-inches (millionths of

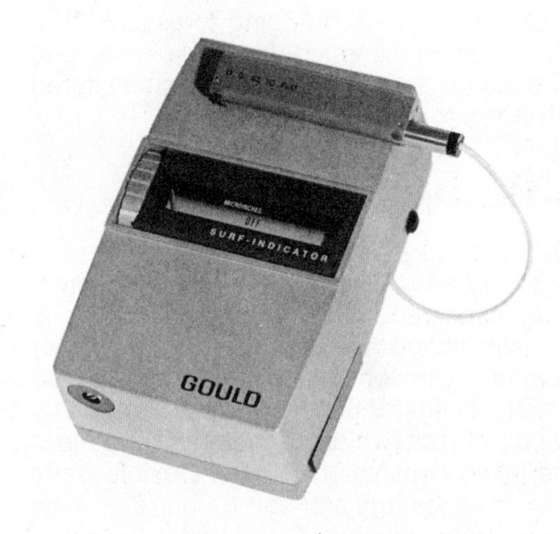

Fig. 46-40. Surface Finish Indicator

Fig. 46-39. Standard Specimen for Surface Finish

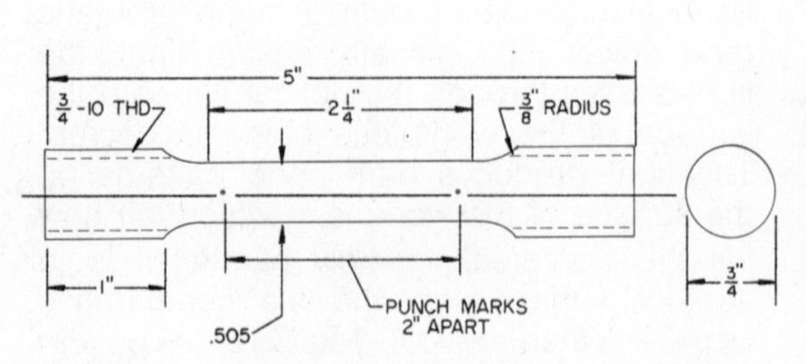

Fig. 46-41. Dimensions of One Size of Standard Tensile Test Specimen

an inch), the unit of measure used for indicating surface finish quality.

46-26. Tensile and Impact Testing

Two tests which are often made on metals to see whether they meet strength specifications are **tensile** and **impact tests.** Tensile strength is the ability of a material to resist forces tending to pull it apart. The tensile strength of materials is expressed as the number of pounds of pulling force required to break a sample having a cross section of one square inch. Tensile tests are usually made with samples made to standard specimen sizes of less than a square inch, Fig. 46-41. A typical tensile testing machine is shown in Fig. 46-42.

Impact tests are made to determine the shock resistance or toughness of a metal. The impact resistance of a metal is expressed in terms of the number of foot-pounds of force required to break a sample of standard size. There are two standard tests, **Charpy** and **Izod,** which use test specimens of the same size, Fig. 46-43. The method of testing differs, however. In the Izod test, the specimen is held vertically at the bottom end, and the direction of the hammer blow is against the notched side of the specimen. For Charpy testing, the specimen is supported at both ends in a horizontal position, and the direction of the hammer blow is from the side opposite the notch.

46-27. Care of Inspection Equipment

Gages are very expensive. They should be kept very clean and handled and stored away with the greatest care and skill.

Many gages have to be protected from heat and cold. For example, the warmth of the hand will **expand** or warp the gage and thus change its size. The change in size may be only .0001″ to .0002″, more or less. The size of a gage lying in the sunlight will change the same way. Rubber or wooden handles are often put on gages to protect them from warmth of the hand while **gaging** a piece of work. Rooms in which gaging is done and in which gages are stored should be kept at 68° F.

A gage which has been dropped on the floor, or otherwise bumped, should be checked or inspected before using it again. Gages become worn after they have been used to inspect many pieces and should, therefore, be inspected from time to time.

Fig. 46-42. Tensile Testing Machine

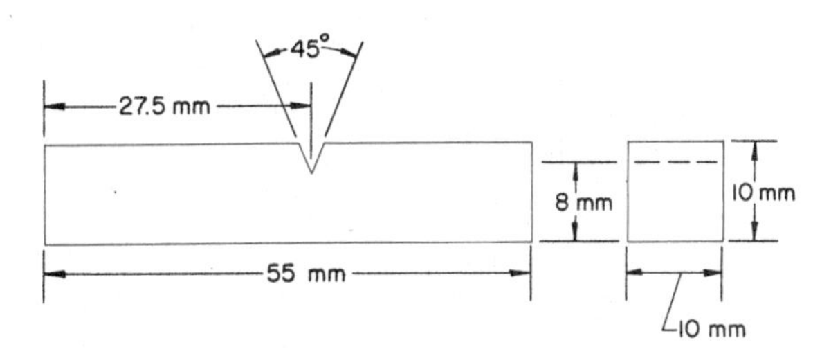

Fig. 46-43. Charpy V-Notch Test Specimen

Words to Know

air gage
adhesion
allowance
attribute gage
Bureau of Standards
caliper gage
Charpy
dial comparator
dial indicator
dimension limit
double dimension
electronic gage
gage or gauge
gage blocks
go or no-go gage
impact testing
inspector
interchangeable part
International Bureau of Weights and Measures
International Inch
International Prototype Meter
Izod
Jo blocks
limit gage
magnifying glass
master gage
master planer and shaper gage
negative allowance
optical comparator
optical flat
positive allowance
radius gage or fillet gage
reference gage
ring gage
roll-thread snap gage
small hole gage
snap gage
standard specimen
surface finish measurement
tapered plug gage
tapered ring gage
telescoping gage
tensile testing
thread gage
thread micrometer
tolerance
wringing

Review Questions

1. What factors are inspected for the purpose of controlling the quality of manufactured metal products?
2. What is the object that is used as the international standard for the meter?
3. What is the length of the International Inch in terms of metric units? Also, in terms of wavelengths of krypton light?
4. What is used as the practical standard of precision measurement in machine shops and industrial plants all over the world?
5. List the three classes of gage blocks, give the accuracy limits for each class, and state briefly the kind of work done by each class.
6. Why should gage blocks be used at 68° F. if possible?
7. Describe how gage blocks are assembled for use.
8. For what are master gages used? What is another name for them?
9. What advantage does the use of gages have over measuring tools?
10. What is meant by interchangeability of parts? Why is it important? How is it made possible?
11. What is meant by dimension limits?
12. What is meant by tolerance on parts?
13. What is meant by allowance on mating parts?
14. Explain the difference between positive and negative allowance.
15. Describe how go and no-go gages are used.
16. What is a snap gage?
17. What is a ring gage? Name three kinds.
18. What is a plug gage? Name three kinds.
19. Explain how a thread micrometer is selected and used to measure a screw thread.
20. Describe how air gages differ from mechanical gages.
21. How does an electronic gage work?
22. List two types of dial gages, and explain how their graduations are numbered.
23. List several uses for a dial indicator set.
24. List two ways in which the gaging height may be established for a dial comparator.
25. Name several kinds of dial-indicating gages.
26. List several uses for a planer and shaper gage.
27. For what purposes is a telescoping gage used?
28. What is a small hole gage?
29. For what kind of inspection are optical comparators well suited?
30. Describe three properties of a surface that can be checked with an optical flat.
31. For what kinds of inspection are microscopes used?

32. How do magnifying glasses aid in inspection work?
33. Why is surface finish quality important?
34. Describe two ways surface finish quality can be inspected.
35. What is tensile testing, and why is it done?
36. What is impact testing, and why is it done?

Career Information

1. Who makes the precision measuring tools and gages used in inspection work?
2. Find out how much education and training is required to become an inspector in an industry of your choice.

UNIT 47

Nondestructive Testing and Inspecting

47-1. Need for Nondestructive Inspection

Nondestructive inspection techniques enable inspectors to check on properties critical to the safe performance of metal parts without causing damage to the parts themselves. These tests are made on parts at certain stages during their manufacture so that defective parts can be rejected as early as possible. Parts are also tested after a period of time in service as a means of discovering and eliminating those with **fatigue cracks.** The aircraft industry does a great deal of nondestructive inspection of engine and airframe parts, both during new construction and during overhaul. Nondestructive testing is concerned with **hardness testing** and **testing for cracks and flaws.**

Fig. 47-1. Rockwell Hardness Tester

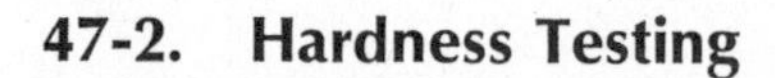

47-2. Hardness Testing

The **hardness** of metals can be determined with several different types of hardness testing instruments. The hardness is designated by a **hardness number,** from a **hardness scale,** which is based on the kind of hardness testing instrument used. The following are three of the most common types of hardness testing instruments:

1. Rockwell Hardness Tester
2. Brinell Hardness Tester
3. Scleroscope Hardness Tester

47-3. Rockwell Hardness Tester

Rockwell hardness tests are based on the depth of penetration made in metal by a specific kind of penetrator point under a specific load. The hardness is indicated directly by a hardness number which is read on a dial, see Fig. 47-1. The hardness number is based on the difference in depth of penetration caused by a **minor** load and a **major** load applied to the penetrator. Deep penetration indicates a softer metal.

Rockwell hardness testers of several types are available. They may be the stationary-type, Fig. 47-1, or the portable-type, Fig. 47-2. The load generally is applied through a system of weights, levers, screws, or a combination of these devices. Testers are available for testing according to the **standard Rockwell Hardness Scales** only or for testing according to the **Rockwell Superficial Hardness Scales** only. Some testers can be used for testing hardness according to either scale.

The **Rockwell-C** (RC) Scale requires use of a diamond-point penetrator called a **brale.** A minor load of 10 kilograms (22 lbs.) and a major load of 150 kilograms (330.8 lbs.) is used. The RC Scale is used for testing the hardness of heat-treated or hardened steels which are harder than Rockwell-B 100.

The **Rockwell-B** (RB) Scale requires use of a 1/16″ diameter ball penetrator made of

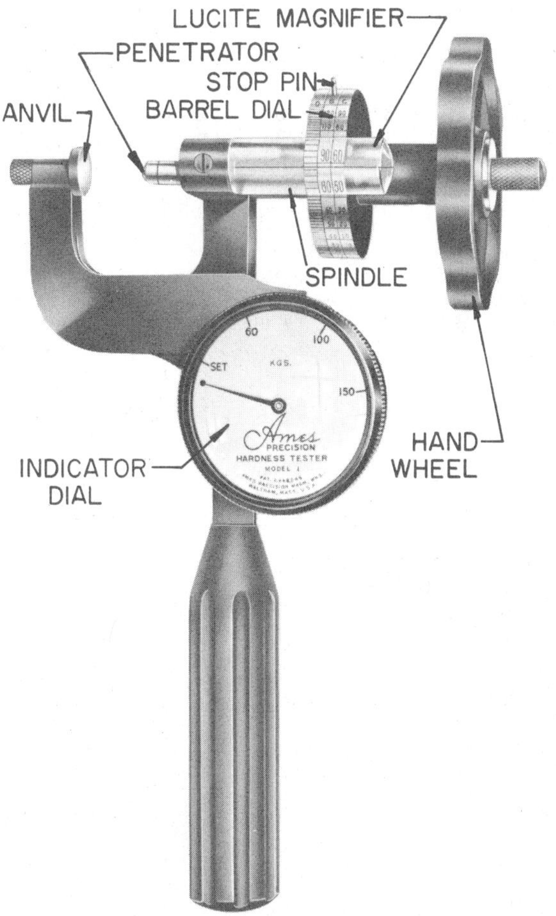

LUCITE MAGNIFIER
PENETRATOR
STOP PIN
BARREL DIAL
ANVIL
SPINDLE
HAND WHEEL
INDICATOR DIAL
Ames PRECISION HARDNESS TESTER

Fig. 47-2. Portable Hardness Tester

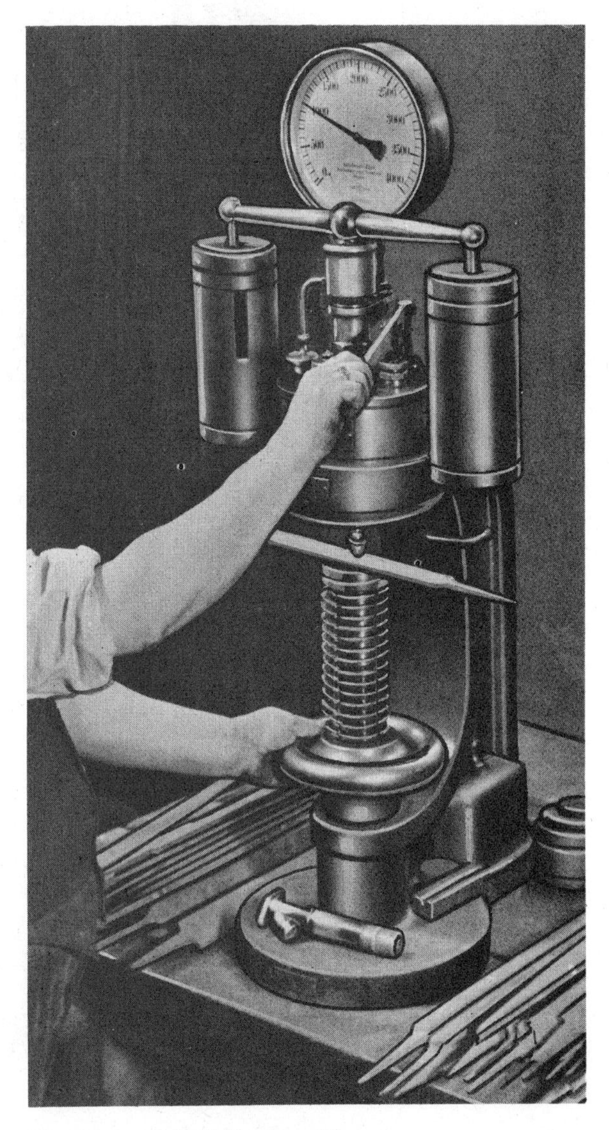

Fig. 47-3. Brinell Hardness Tester

TABLE 30

Hardness Numbers for Steel Approximately Equivalent to Rockwell C Scale

Rockwell C-Scale Hardness No.	Diamond Pyramid Hardness No.	Brinell Hardness No. (10-mm Ball, 3000-kg load)			Rockwell Hardness No.			Rockwell Superficial Hardness No. (Superficial Brale Penetrator)			Shore Scleroscope Hardness No.	Tensile Strength (Approx.) 1000 psi
		Standard Ball	Hultgren Ball	Carbide Ball	A Scale (60-kg Brale)	B Scale (100-kg 1/16" Ball)	D Scale (100-kg Brale)	15-N Scale (15 kg)	30-N Scale (30 kg)	45-N Scale (45 kg)		
68	940	...	...	...	85.6	...	76.9	93.2	84.4	75.4	97	...
67	900	...	...	...	85.0	...	76.1	92.9	83.6	74.2	95	...
66	865	...	...	...	84.5	...	75.4	92.5	82.8	73.3	92	...
65	832	...	...	(739)	83.9	...	74.5	92.2	81.9	72.0	91	...
64	800	...	...	(722)	83.4	...	73.8	91.8	81.1	71.0	88	...
63	772	...	...	(705)	82.8	...	73.0	91.4	80.1	69.9	87	...
62	746	...	...	(688)	82.3	...	72.2	91.1	79.3	68.8	85	...
61	720	...	...	(670)	81.8	...	71.5	90.7	78.4	67.7	83	...
60	697	...	(613)	(654)	81.2	...	70.7	90.2	77.5	66.6	81	...
59	674	...	(599)	(634)	80.7	...	69.9	89.8	76.6	65.5	80	326
58	653	...	(587)	615	80.1	...	69.2	89.3	75.7	64.3	78	315
57	633	...	(575)	595	79.6	...	68.5	88.9	74.8	63.2	76	305
56	613	...	(561)	577	79.0	...	67.7	88.3	73.9	62.0	75	295
55	595	...	(546)	560	78.5	...	66.9	87.9	73.0	60.9	74	287
54	577	...	(534)	543	78.0	...	66.1	87.4	72.0	59.8	72	278
53	560	...	(519)	525	77.4	...	65.4	86.9	71.2	58.6	71	269
52	544	(500)	(508)	512	76.8	...	64.6	86.4	70.2	57.4	69	262
51	528	(487)	494	496	76.3	...	63.8	85.9	69.4	56.1	68	253
50	513	(475)	481	481	75.9	...	63.1	85.5	68.5	55.0	67	245
49	498	(464)	469	469	75.2	...	62.1	85.0	67.6	53.8	66	239
48	484	451	455	455	74.7	...	61.4	84.5	66.7	52.5	64	232
47	471	442	443	443	74.1	...	60.8	83.9	65.8	51.4	63	225
46	458	432	432	432	73.6	...	60.0	83.5	64.8	50.3	62	219
45	446	421	421	421	73.1	...	59.2	83.0	64.0	49.0	60	212
44	434	409	409	409	72.5	...	58.5	82.5	63.1	47.8	58	206
43	423	400	400	400	72.0	...	57.7	82.0	62.2	46.7	57	201
42	412	390	390	390	71.5	...	56.9	81.5	61.3	45.5	56	196
41	402	381	381	381	70.9	...	56.2	80.9	60.4	44.3	55	191

40	392	371	371	371	70.4	...	55.4	80.4	59.5	43.1	54	186
39	382	362	362	362	69.9	...	54.6	79.9	58.6	41.9	52	181
38	372	353	353	353	69.4	...	53.8	79.4	57.7	40.8	51	176
37	363	344	344	344	68.9	...	53.1	78.8	56.8	39.6	50	172
36	354	336	336	336	68.4	(190.0)	52.3	78.3	55.9	38.4	49	168
35	345	327	327	327	67.9	(108.5)	51.5	77.7	55.0	37.2	48	163
34	336	319	319	319	67.4	(108.0)	50.8	77.2	54.2	36.1	47	159
33	327	311	311	311	66.8	(107.5)	50.0	76.6	53.3	34.9	46	154
32	318	301	301	301	66.3	(107.0)	49.2	76.1	52.1	33.7	44	150
31	310	294	294	294	65.8	(106.0)	48.4	75.6	51.3	32.5	43	146
30	302	286	286	286	65.3	(105.5)	47.7	75.0	50.4	31.3	42	142
29	294	279	279	279	64.7	(104.5)	47.0	74.5	49.5	30.1	41	138
28	286	271	271	271	64.3	(104.0)	46.1	73.9	48.6	28.9	41	134
27	279	264	264	264	63.8	(103.0)	45.2	73.3	47.7	27.8	40	131
26	272	258	258	258	63.3	(102.5)	44.6	72.8	46.8	26.7	38	127
25	266	253	253	253	62.8	(101.5)	43.8	72.2	45.9	25.5	38	124
24	260	247	247	247	62.4	(101.0)	43.1	71.6	45.0	24.3	37	121
23	254	243	243	243	62.0	100.0	42.1	71.0	44.0	23.1	36	118
22	248	237	237	237	61.5	99.0	41.6	70.5	43.2	22.0	35	115
21	243	231	231	231	61.0	98.5	40.9	69.9	42.3	20.7	35	113
20	238	226	226	226	60.5	97.8	40.1	69.4	41.5	19.6	34	110
(18)	230	219	219	219	...	96.7	...	...	...	...	33	106
(16)	222	212	212	212	...	95.5	...	...	...	...	32	102
(14)	213	203	203	203	...	93.9	...	...	...	...	31	98
(12)	204	194	194	194	...	92.3	...	...	...	...	29	94
(10)	196	187	187	187	...	90.7	...	...	...	...	28	90
(8)	188	179	179	179	...	89.5	...	...	...	...	27	87
(6)	180	171	171	171	...	87.1	...	...	...	...	26	84
(4)	173	165	165	165	...	85.5	...	...	...	...	25	80
(2)	166	158	158	158	...	83.5	...	...	...	...	24	77
(0)	160	152	152	152	...	81.7	...	...	...	...	24	75

The values in boldface type correspond to the values in the joint SAE-ASM-ASTM hardness conversions as printed in ASTM E140-65, Table 2. Values in parentheses are beyond normal range and are given for information only. Data from Metals Handbook 8th Edition, American Society for Metals. (Reprinted with permission.)

Fig. 47-4. Brinell Microscope for Measuring Diameter of Impression Made by Brinell Hardness Tester

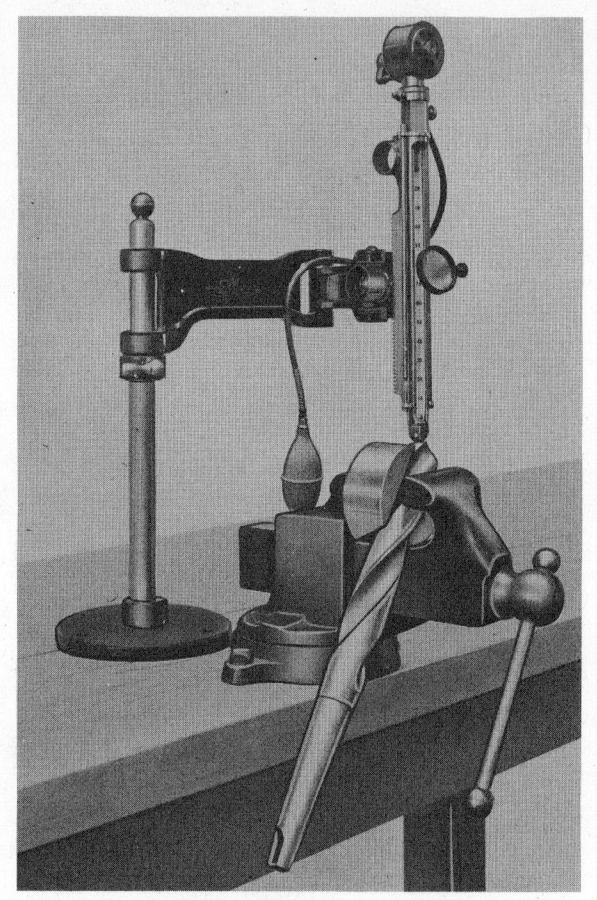

Fig. 47-5. Testing Hardness with Scleroscope

hardened steel. It is used with a minor load of 10-kg and a major load of 100-kg (220.5 lbs.). The RB Scale is used for testing hardness of unhardened steel, cast iron, and nonferrous metals.

Of the various Rockwell Hardness Scales, the RC and RB are standard and are the most widely used. (See Table 30.) With the Rockwell Superficial Hardness Scales, lighter loads are applied and a smaller dent is made in the surface of metal being tested. However, these tests generally are not as accurate as the RC or RB tests.

47-4. Brinell Hardness Tester

Brinell Hardness Tests are made with a testing machine, Fig. 47-3, which forces a hard ball of a specific diameter under a specific load into a smooth metal surface. The ball is 10 millimetres (mm) in diameter. For standard Brinell (BHN) Hardness Tests on steel, a load of 3000-kg (6600 lbs.) is applied. The load is applied steadily and is maintained for a minimum period of 15 seconds for steel and 30 seconds for nonferrous metals.

The diameter of the dent made by the ball determines the Brinell Hardness Number. The diameter of the dent is measured with a microscope, Fig. 47-4, which has a special calibrated measuring lens. The diameter of the dent is converted to a hardness number by using a comparison chart which is supplied with the testing machine.

Brinell testers generally work best on metals which are not extremely hard. These include nonferrous metals, soft steels, and medium-hard steels. The Brinell hardness

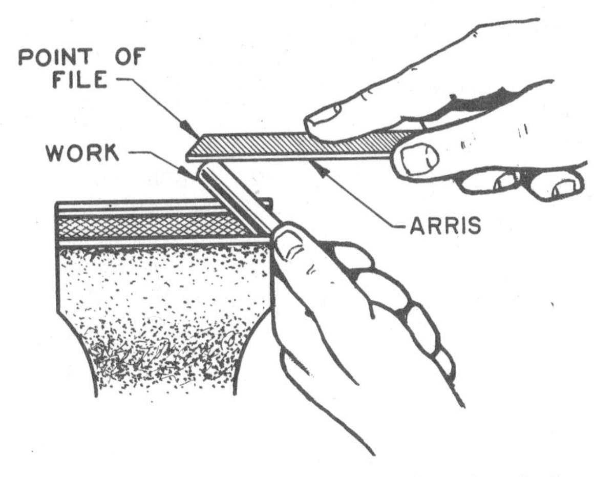

Fig. 47-6. Testing the Hardness of Steel with the Arris of a File

of steel generally ranges from about BHN 150 for soft, low-carbon steel to BHN 739 for hardened, high-carbon steel. On very hard steels, the dent is so small that it is difficult to see or measure.

47-5. Shore Scleroscope Tester

The **Shore Scleroscope Tester,** Fig. 47-5, operates on the rebound principle. It measures the height to which a diamond-tipped hammer rebounds after being dropped on a metal surface to be tested. Harder metals cause the hammer to rebound higher, thus indicating higher hardness values.

Scleroscope Testers are essentially **nonmarring,** particularly on harder metals. A small dent may appear on softer metals. The hardness number is read directly on the vertical column, Fig. 47-5. On some Scleroscope Testers, the hardness numbers are read directly on a dial instead. Small parts may be held in a clamping stand for testing.

47-6. Testing Hardness with a File

The hardness of steel may be tested with the **arris** of a file, see Fig. 47-6. If the file cuts, the steel is soft. If the file does not cut but slips over the steel, it is hard. Do not

TABLE 31

Data for Estimating Hardness of Steel with a File

Rockwell C Hardness Number	Action of File on Steel
20	File removes metal easily with slight pressure
30	File starts to resist cutting metal
40	File cuts metal with difficulty
50	File barely cuts metal with great difficulty
57	File glides over metal without cutting

attempt to test the hardness of hardened steel with the flat surface of the file, or you will dull and ruin the teeth. By studying Table 31 carefully you can make a rough estimate of the hardness of steel by using the arris of a file.

47-7. Tests for Detecting Cracks and Flaws

Several techniques are used to reveal surface cracks and hidden flaws in castings, welds, and even rolled or forged metal. These are listed below:

1. Fluorescent penetrant inspection
2. Magnetic particle inspection
3. Radiographic inspection
4. Ultrasonic inspection

47-8. Fluorescent Penetrant Inspection

This is a technique whereby **fluorescent penetrating oil** is applied to the part by brushing, spraying, or dipping. After being wiped dry, the part is dusted with an **absorbent powder** which draws the fluorescent penetrating oil out of any cracks that are present. When placed under ultraviolet light, the cracks appear in bright fluorescent color.

A similar method, called Spotcheck (a Trademark of Magnaflux Corp.), also uses a dye penetrant but uses a developer which causes the cracks to appear without the need for ultraviolet light. Fluorescent penetrants and dyes cannot detect flaws which do not break through the surface.

47-9. Magnetic Particle Inspection

In **magnetic particle inspection,** the part is first magnetized. It is then either dusted with fine iron powder or coated with a solution in which the magnetic particles are suspended. Flaws in the workpiece cause the lines of magnetic force to become distorted and break through the surface. Here they attract concentrations of the magnetic particles, thereby revealing defects in the metal. This method can only detect flaws which break through or are just below the surface.

47-10. Radiographic Inspection

Radiographic inspection consists of exposing the workpiece to **X-ray** or **gamma ray** radiation and then viewing the resultant image on a fluoroscope or on film. X-rays are very sensitive and are capable of inspecting any thickness of almost any kind of material. Internal flaws of three-dimensional nature such as voids or sand and slag inclusions in castings are readily detected. Cracks, however, are not readily revealed by radiography.

Gamma radiation, because of shorter wavelengths, is more effective in inspecting thick sections. It is also less expensive, and its equipment is more portable than X-ray.

47-11. Ultrasonic Inspection

In **ultrasonic inspection** high frequency sound waves are introduced into the workpiece. The sound waves reflect from the workpiece surfaces and from internal defects as well. This enables both the workpiece thickness and the exact depth of flaws to be measured. For convenience, the sound beam is converted to an electrical signal and made to appear graphically on an oscilloscope screen. Considerable skill is required in the application and interpretation of ultrasonic inspection techniques.

Words to Know

absorbent powder
arris
brale
Brinell Hardness Tester
fatigue cracks
flaw
fluorescent penetrant inspection
gamma ray
hardness number
hardness scale
kilogram
major load
minor load
oscilloscope screen
penetrator
radiographic inspection
Rockwell Hardness Tester
Rockwell Superficial Hardness Tester
Shore Scleroscope Hardness Tester
Spotcheck
ultrasonic inspection
ultraviolet light
void
X-ray

Review Questions

1. Tell what is meant by nondestructive inspection, and give several reasons why it is necessary.
2. How is the hardness of metals specified or designated?
3. List three common types of hardness testing instruments or machines.
4. List two types of Rockwell hardness testers.
5. On what kinds of metal are Rockwell-C tests generally made?
6. What kinds of metal are generally tested according to the Rockwell-B scale?
7. List one advantage and one disadvantage in using the Rockwell Superficial Hardness Scales.
8. Explain how a Brinell hardness test is made.
9. On what kinds of metal do Brinell hardness testers work best?
10. Explain how a Shore Scleroscope hardness tester operates.
11. What is the major advantage in using the Shore Scleroscope tester?
12. What is the Shore Scleroscope equivalent of Rockwell C-60?
13. Describe the fluorescent penetrant in-

spection process. How does it differ from the Spotcheck process?

14. How does the magnetic particle inspection process work? What limits the kind of materials which can be inspected by this process?
15. Explain the radiographic inspection process.
16. Describe the ultrasonic inspection process.
17. What kind of flaws can be detected by dye penetrant and magnetic particle inspection?
18. Is there one best method for detecting imperfections in metal? Explain why or why not.

XIV PART

Introduction to Powered Machining

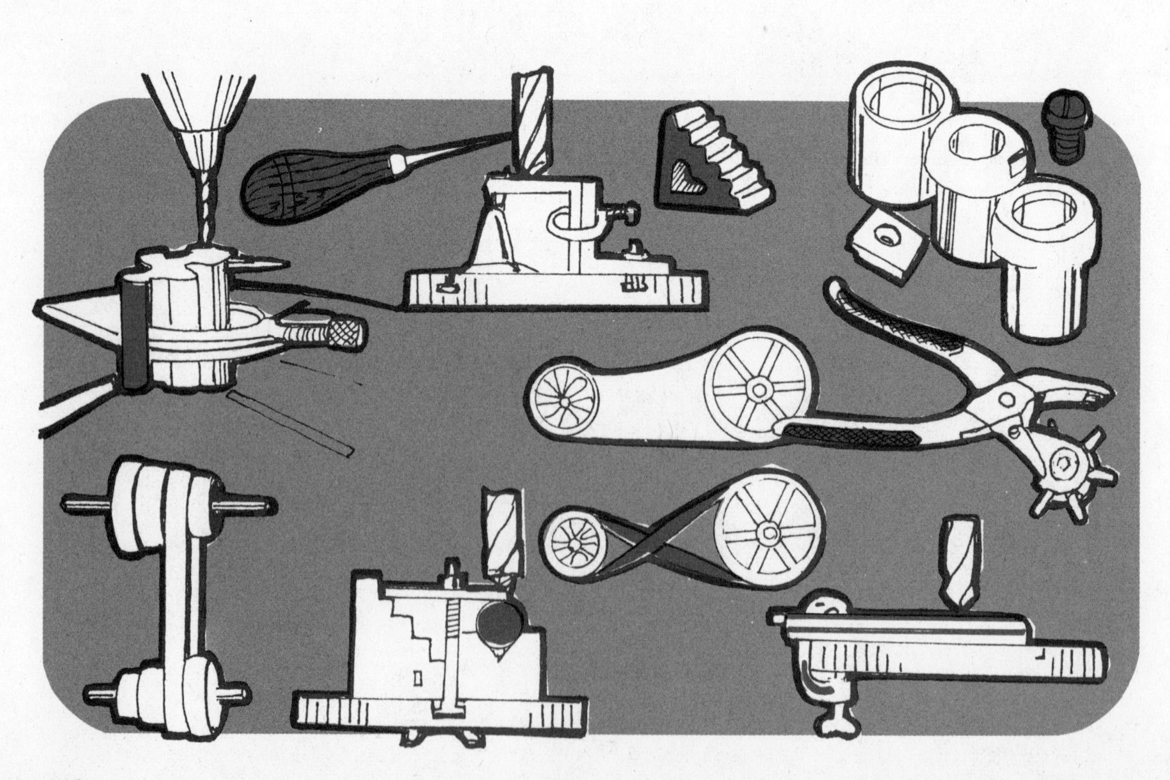

UNIT 48

Lubricants and Cutting Fluids

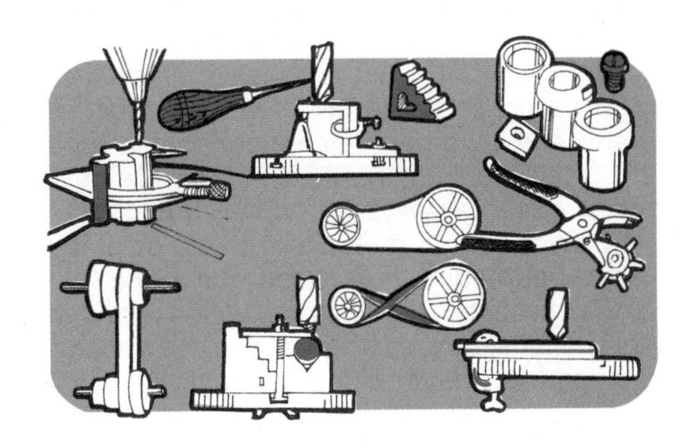

48-1 What are Lubricants?

Lubricants are materials that make surfaces smooth and slippery. They are used to reduce the heat, friction, wear, and vibration caused when surfaces rub together. Over the years, chemical technology has expanded the range of materials available as lubricants to include petroleum oils, animal oils, vegetable oils, greases, metallic films, mineral films, plastics, synthetic fluids, and gases.

48-2. Lubricating Oils

There are three main kinds of lubricating oils:

1. Mineral oils
2. Animal oils
3. Vegetable oils

Mineral Oils

Mineral oils are by far the most versatile and best known lubricants. Petroleum is the leading source of mineral oils.

Lubricating oil, or **machine oil**, is obtained from **petroleum.** It is used to oil the rubbing surfaces, called **bearing surfaces**, of machinery. Its stiffness or thickness, viscosity, can be measured by the length of time in seconds that a standard amount of oil at determined temperatures can flow through a hole. Stiff or thick oil drips slowly.

A medium-heavy oil is used for general machine oiling. It has a Saybolt universal viscosity rating of 250 to 500 seconds at 100° F., and may be called Type C. A thinner oil (70 to 100 second rating) is used in reservoirs such as at lathe spindles and may be called spindle oil or Type A.

TABLE 32

Typical Lubricating Oils for Machines

Company Name	Saybolt Universal Viscosity Rating in Seconds at 100° F.		
	100 Second: Type A LIGHT (SPINDLE OIL)	150-240 Sec.: Type B MEDIUM LIGHT	250-500 Sec.: Type C MEDIUM HEAVY
Mobile	Velocite Oil 10	Gg. Vactra Oil Light	Gg. Vactra Hvy. Med.
Pure	Spindle Oil D	Puropale Medium	Puropale Hvy. Med.
Shell	Vitrea Oil 923	Vitrea Oil 27	Vitrea Oil 33
Sinclair	Cadet Oil A	Warrior Oil	Commander Oil B
South Bend Lathe	CE 2017	CE 2018	CE 2019
Standard (Ind.)	Spindle Oil C	Indoil #15	Indoil #31

Note: This partial listing of typical brands is given as an aid in procuring comparative grades from the company of your preference.

Bed Way Lubricant is a heavy oil (300 to 500 seconds) for sliding surfaces such as cams and lathe beds. Gear Lubricant is used on gears not running in oil to reduce gear noise. (See Table 32 for typical brands.)

Cylinder oil, or **motor oil,** is also obtained from **petroleum.** It is used for oiling hot parts, such as the **pistons** which slide in the **cylinders** of an engine. It is not recommended for machine tool lubrication.

Animal Oils

Animal oils are obtained from the fats of animals. Fish oils are obtained from such fish as herring, salmon, sardine, cod, and the sperm whale. Sperm oil is rare and expensive. It is thinner than the other animal oils and is used to oil fine, delicate instruments and machinery.

Glycerin is a thick, oily, syrupy, colorless, odorless liquid obtained from fats or produced synthetically from alcohol. It has a sweet taste and is used as a medicine. Glycerin freezes at a very low temperature and is thus used as a base for anti-freeze solutions for automobile radiators. It is also sometimes used for oiling machines, especially engines.

Vegetable Oils

Vegetable oils are oils obtained from plants and vegetables, especially from their seeds. Most vegetable oils are used in the manufacture of food products. **Castor oil,** made from the castor bean, is a very good lubricant. It is often used in racing cars, but since it solidifies on cooling, it must be drained immediately after the engine is stopped.

48-3. Greases

Greases are obtained from animal fats and petroleum. Most lubricating greases are petroleum oils which are thickened with one or more of a variety of soaps, clays, carbon, graphite, lead, zinc oxide, and other materials having lubricating properties.

Greases are often used in place of oils because they can be sealed better against loss, they protect against entrance of dirt and moisture, and they tend to cling better to some surfaces.

Greases are classified from 0, softest, to 6, stiffest, by the National Lubricating Grease Institute. Grease selection should follow the machine manufacturer's recommendations.

48-4. Other Lubricants

Graphite, a form of carbon, is black, very soft and slippery, and is used for lubricating some machine parts, especially where there is low speed and moderate heat. It lasts longer than oil, does not get gummy, and does not attract dust.

White lead, a compound of lead, is a white powder which is mixed with linseed oil. It was extensively used as a paint pigment and is a good high-pressure lubricant. When turning work between centers in a metal lathe, the tailstock or dead center is often lubricated with white lead. White lead is poisonous and should be handled with due care.

48-5. Methods of Applying Lubricants

Oil is supplied to lubrication points in a variety of ways.

An **oil hole** is a small hole through which oil flows to a surface which must be oiled. An **oil groove,** which is usually connected to an oil hole, is a small groove by means of which oil is spread evenly over a surface to be oiled.

An **oil cup,** Fig. 48-1, is a small, covered glass or brass cup screwed into a machine near a surface which must be oiled. The oil, which is put into the cup, flows through a hole to the surface to be oiled.

An **oil tube,** or **oil pipe,** is a tube or pipe through which oil flows to rubbing parts of machinery that cannot be reached directly.

Oil tube systems which provide lubrication to an entire machine from one central reservoir are called "one shot" lubrication sys-

Fig. 48-1. All Bearings Should Be Oiled Regularly

tems. These systems are either fully automatic or are manually operated, Fig. 48-2.

Grease is often packed directly into bearings and grease cups by hand. A grease cup is a small, covered cup installed near a surface which must be greased. The grease is forced out of the cup by screwing down its cap. Grease is also forced to lubrication points through **grease fittings** with a grease gun, Fig. 48-3.

Fig. 48-2. One Shot Lubrication System

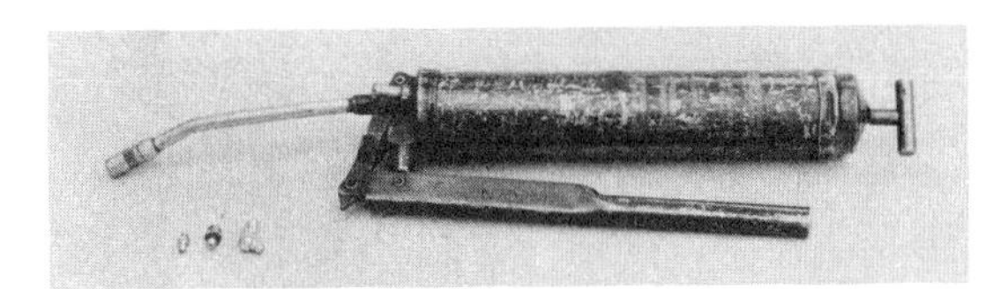

Fig. 48-3. Grease Gun and Grease Fittings Often Used on Machines

48-6. Cutting Fluids

The term **cutting fluid** applies to various types of **cutting oils, cutting coolants, cutting solutions,** and **cutting compounds.** Cutting fluids are applied to metal cutting tools such as drills, taps, dies, reamers, power saws, lathe cutting tools, and milling cutters to make them cut easier.

Cutting fluids improve the cutting or machining of metals in the following ways:

1. Carry away heat.
2. Cool the cutting tool and the work.
3. Lubricate the face of the cutting tool and the chip.
4. Prevent the **adhesion** or pressure welding of a **built-up edge** on the cutting tool. A built-up edge is caused by a small metal chip sticking to the cutting edge of a cutting tool.
5. Aid in flushing away chips.
6. Improve the quality of the machined surface.
7. Increase tool life by reducing tool wear.
8. Permit higher cutting speeds than those used for dry machining.

Most commercially available cutting fluids can be classified under three groups which include **cutting oils, emulsifiable oils,** and **chemical cutting fluids.** However, from the standpoint of composition, there are, in general, six types of cutting fluids:

1. Lard oil.
2. Mineral oil.
3. Mineral and lard oil combinations.
4. Sulfurized and chlorinated mineral oils.
5. Emulsifiable (soluble) oils.
6. Chemical cutting fluids.

Lard oil is an animal oil obtained from lard or hog fat. It is an excellent cutting oil

TABLE 33
Cutting Fluids for Cutting Common Metals

Metals	Power Sawing	Drilling	Reaming	Threading	Turning	Milling	Grinding
Carbon Steels Malleable Iron	EO, MO, ML	EO, Sul, ML	ML, Sul, EO	Sul, ML, EO	EO, Sul, ML	EO, Sul, ML	EO
Wrought Iron Stainless Steels Tool Steels High-Speed Steels	EO, ML, MO	EO, Sul, ML	ML, Sul	Sul, ML, EO	EO, Sul, ML	EO, Sul, ML	EO
Gray Cast Iron	Dry, EO	Dry, EO	Dry, EO	Dry, EO ML	Dry, EO	Dry, EO	EO, MO
Aluminum Alloys	Dry, EO MO	EO, MO, ML	ML, MO, EO	ML, MO, EO, K	EO, MO, ML, K	EO, MO, ML	EO, MO
Copper Base Alloys Brass Bronze	Dry, MO, ML, EO	EO, MO, ML	ML, MO, EO	ML, MO, EO	EO, MO, ML	EO, MO, ML	EO
Magnesium Alloys	Dry, MO	Dry, MO	Dry, MO	Dry, MO	Dry, MO	Dry, MO	Dry, MO

Key:
K—Kerosine
L—Lard Oil
MO—Mineral Oils
ML—Mineral-Lard Oils
Sul—Sulfurized Oils, with or without chlorine
EO—Emulsifiable (soluble) Oils and Compounds
Dry—No cutting fluid

but is so expensive it is usually mixed with mineral oils. Pure lard oil also tends to become rancid and develop a bad odor. Bacteria which breed in lard oil can also cause **dermatitis** and skin irritation among machine operators. Lard oil is useful for machining metals which would otherwise be stained by use of sulfurized mineral oils.

Mineral oils are used in light-duty cutting operations, especially with nonferrous metals and free cutting steel. Mineral-lard oils provide better lubricating properties and are used in medium-duty cutting operations. **Sulfurized and chlorinated** mineral oils are recommended for cutting tough metals and for performing severe machining operations such as broaching and tapping.

Emulsifiable oil is often called **soluble oil** or **water-soluble oil.** It is a special type of mineral oil which mixes or disperses evenly in water. Hence, an emulsifiable oil (soluble oil) cutting fluid is made by **adding the oil to the water** (never water to oil), thus forming a milky-white colored solution. The proportion of water and oil varies according to the severity of the machining operation. For average severity machining operations on ferrous metals, 1 part oil is added to 20 parts water. For average grinding operations 1 part oil is added to 40 parts water. Emulsifiable oil solutions have excellent lubricant and coolant qualities, and they are relatively inexpensive.

Chemical cutting fluids are solutions of **organic** and **inorganic** compounds in water. They generally do not contain petroleum products. Two types are made. Those with lubricants and wetting agents added are used for a wide range of machining operations including severe machining. Plain fluids are used mainly for surface grinding operations.

Kerosene, a thin oil obtained from petroleum, is used mostly as a fuel, but is some-

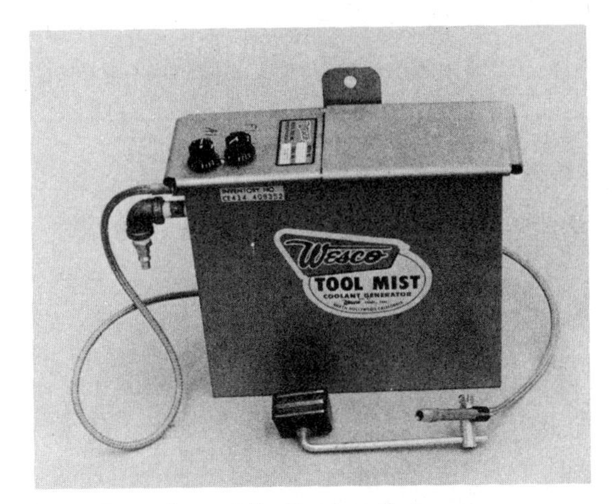

Fig. 48-4. Spray-Mist Coolant System

times used as a lubricant in cutting metals (see Table 33). It is also useful for cleaning oily, greasy, and dirty machine parts. Because it is **flammable,** due care should be taken to use approved storage containers. It should never be used near an open flame or other heat source sufficiently hot to ignite it.

48-7. Cutting Fluid Selection

For efficient cutting or machining of metals, a recommended cutting fluid should be used. Metals such as cast iron and magnesium, however, are often cut dry. Water-based cutting fluids should never be used on magnesium because of the fire hazard. Cast iron contains graphite which lubricates the tool as it cuts. Lard or mineral-lard oil should be used for all hand tapping, threading, and reaming of steel. See Table 33 for more complete information.

48-8. Methods of Applying Cutting Fluids

Cutting fluids may be supplied to taps, dies, and drills with an oil can or a brush. Production machine tools and many general purpose machines have built-in coolant systems which pump the cutting fluid to the point of cutting through a system of pipes. (See Figs. 12-7 and 53-16.)

In factories with many machines using the same cutting fluid, it is often pumped from one large tank to the several machines. It is pumped back to the tank where it is strained and used over and over again.

Spray-mist coolant systems use a water base fluid which is delivered to the cutting tool under pressure of compressed air. The compressed air **atomizes** (breaks into tiny particles) the cutting fluid, providing considerable cooling but little lubrication. The fluid is lost in the process. (See Fig. 48-4.)

Words to Know

animal oil
atomize
castor oil
chemical cutting fluid
chlorine
coolant
cutting oil
cylinder oil
dermatitis
emulsifiable oil
fish oil
glycerin
graphite
grease cup
grease fitting
grease gun
inorganic
kerosene
lard oil
lubricant
machine oil
mineral oil
oil cup
oil groove
organic
petroleum
soluble oil
sperm oil
spray-mist coolant system
sulfur
vegetable oil
viscosity

Review Questions

1. What are lubricants?
2. Name the three main kinds of cutting oils.
3. What type of oils are the most versatile and best known?
4. What kind of oil should be used for lubricating machines?
5. For what is cylinder oil used?
6. What is meant by the viscosity of oil? How is it measured?

7. What is sperm oil? For what is it used?
8. From what materials are lubricating greases made?
9. Why are greases used instead of oils in some cases?
10. How are greases classified according to stiffness?
11. What is graphite? For what kind of lubrication is it used?
12. For what kind of lubrication is white lead used?
13. Why should care be taken in handling white lead?
14. Explain how a "one shot" lubrication system works.
15. List several ways of getting grease to the point of lubrication.
16. List the ways in which cutting fluids improve the machining of metals.
17. What are the principal ingredients used in cutting oils?
18. List the six types of cutting fluids.
19. How is emulsifiable cutting fluid prepared?
20. What kind of cutting fluid is recommended for use in hand reaming, tapping, and threading operations?
21. What are chemical cutting fluids?
22. How are cutting fluids applied on production machines?
23. How does a spray-mist coolant system work?
24. Why are cast iron and magnesium often machined dry?
25. Why should water-based fluids never be used for machining magnesium?

UNIT **49**

Drive Belts and Chains

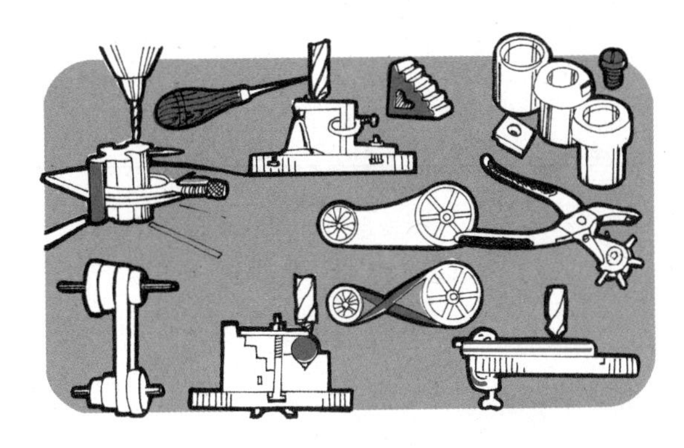

49-1. Why Are Belts Used?

Belts are used to:

1. Carry power from one **pulley** to another (see § 49-2).
2. Change the speed of a pulley.
3. Change the running direction of a pulley (see § 49-7).
4. Carry materials, as a **conveyor.**

49-2. Relation Between Belt and Pulley

A **pulley** is a wheel on which a belt runs. **Cone pulleys,** also called **step pulleys,** have several steps and are used in pairs, Fig. 49-1, to get different speeds. The pulley from which power is taken is called the **driving pulley,** or **driver;** the pulley to which power is carried is called the **driven pulley.** Flat pulleys are larger in diameter at the center of the rim than at the edges. This provides a **crown** in the center of the pulley which helps the flat belt to stay centered on the pulley.

When the driving pulley is enlarged or the driven pulley is decreased in size, the **speed** of the machine is increased. The increase in speed also means less **power** just as an automobile has less power in high gear than in low. A **variable speed pulley,** Fig. 49-2, is one where the effective size of the pulley can be varied throughout a continuous range of sizes without steps.

CROWN OF PULLEY
SHAFT
4-STEP CONE PULLEY
BELT
4-STEP CONE PULLEY

Fig. 49-1. Cone Pulleys

Fig. 49-2. Variable Speed Pulley

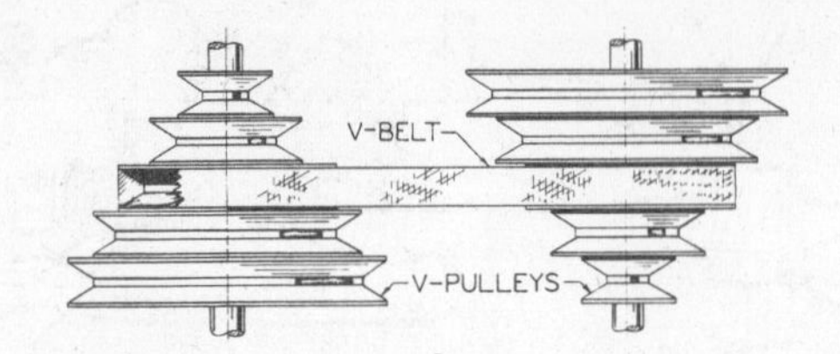

Fig. 49-3. V-Belt

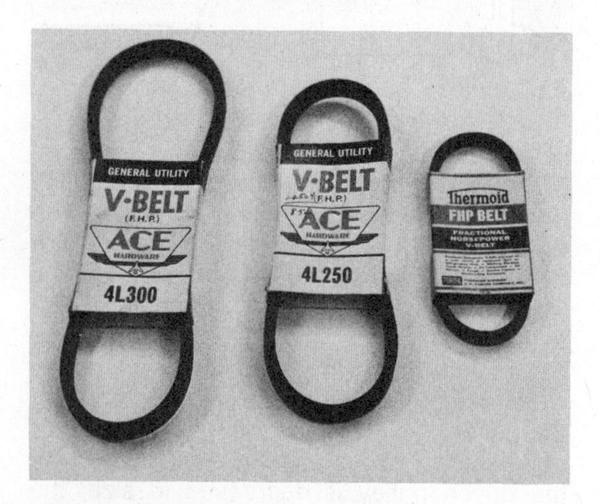

Fig. 49-4. V-Belt Sizes for Fractional Horsepower Motors

As more tension is placed on the V-belt (see § 49-4), it is pulled closer to the center of the variable pulley, making its effective size smaller. **Always adjust a continuously variable speed pulley while it is running or the belt may be damaged.**

49-3. Materials Used For Belts

Belts are made of leather, rubber, canvas, metal, and special composition materials. Leather belts are made from the hides of bulls, cows, and steers. Rubber belts may be solid rubber or be reinforced with fabric, steel, or both. Canvas belts are made from cotton or other fabrics and are generally less expensive than other belts. Metal belts may be a continuous sheet or strip of metal, formed into chains, or made of wire mesh. Wire mesh belts are used to carry materials

Fig. 49-5. Three Popular V-Belt Sizes Often Found on Machine Tools
Each size is identified by the letter designation shown.

through ovens and furnaces where the heat would destroy belts of other materials.

49-4. V-Belts

V-belts run on V-pulleys, Fig. 49-3. V-belts are most often made of a combination of rubber, fabric, and steel wire. Most V-belts are endless. A common use of V-belts and V-pulleys is the fan belt on automobile engines where they drive the fan, alternator, water pump and other accessories. V-belts are used where pulleys are close together. For heavy loads, several V-belts are often used side by side on a multiple pulley called a **sheave.**

V-belts are made in a number of standard sizes. Standard sizes of V-belts for fractional horsepower motors are shown in Fig. 49-4.

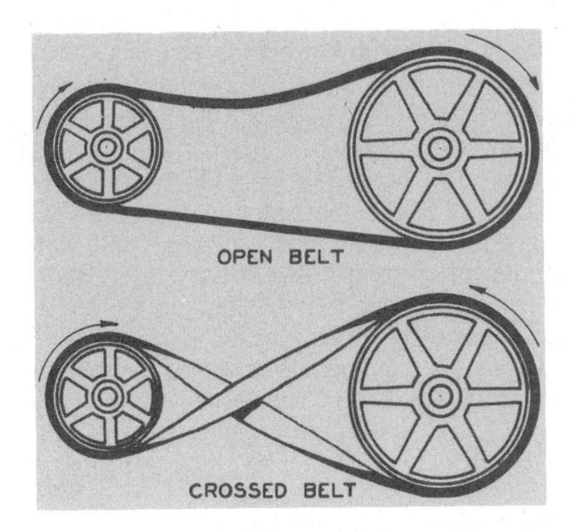

Fig. 49-6. Open and Crossed Belts

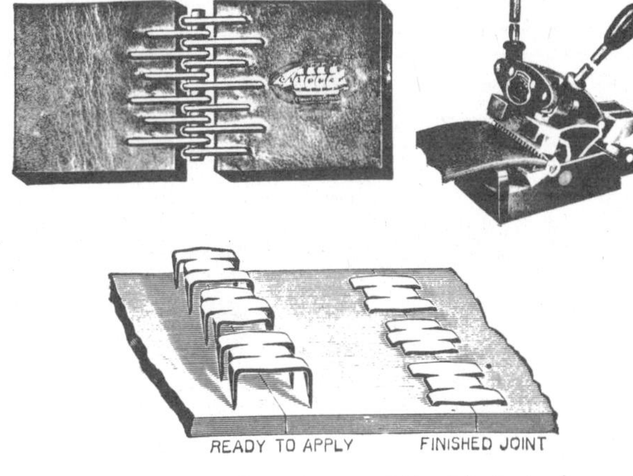

Fig. 49-7. Wire-Belt Lacings and Steel-Belt Hooks

49-5. Cog Belts

Cog belts provide positive, no-slip power transfer such as is required for timing auto engine ignition. They are made of reinforced rubber and are very strong and durable.

49-6. Chain Belts

Chain belts or **drive chains** are used in applications where heavy loads are transmitted, such as **chain hoists,** and where slippage cannot be tolerated as in bicycles and some automatic machine drive systems. They are often used in place of gears.

49-7. Flat Leather Belts

A **single-ply belt** is one thickness of material, usually leather. A **two-ply belt** is two thicknesses cemented together. (See Fig. 49-10.) A **three-ply belt** is made of three thicknesses. Single-ply belts are used on pulleys up to 12″ in diameter, two-ply belts on 12″ to 20″ pulleys, and three-ply belts on 20″ to 30″ pulleys.

Open and Crossed Belts

An **open belt** connecting two pulleys makes them run in the same direction, Fig. 49-6. A **crossed belt** changes the direction.

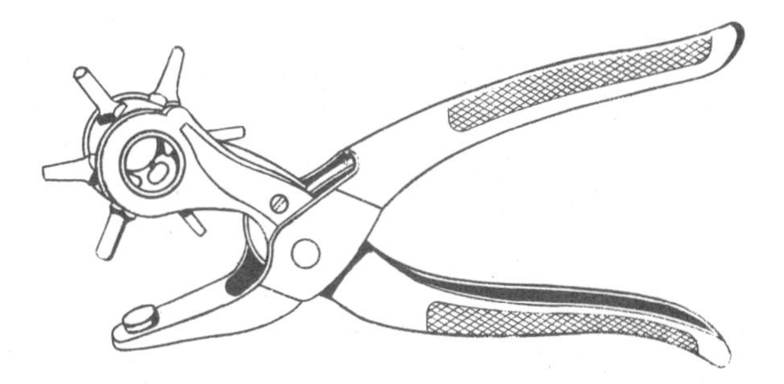

Fig. 49-8. Leather Punch

Methods of Splicing Flat Leather Belts

There are many ways of fastening the two ends of a flat belt together. They may be fastened with **rawhide**[1] **lacing, steel hooks** or **wire hooks, cement,** or **rivets.** Many patented **metal belt fasteners** are sold and the manufacturer's directions should be followed in using them. **Wire-belt lacing** and **steel-belt lacing** are shown in Fig. 49-7.

[1]**Rawhide** is untanned, dressed skin of cattle; it is very hard and tough when twisted into strips and dried.

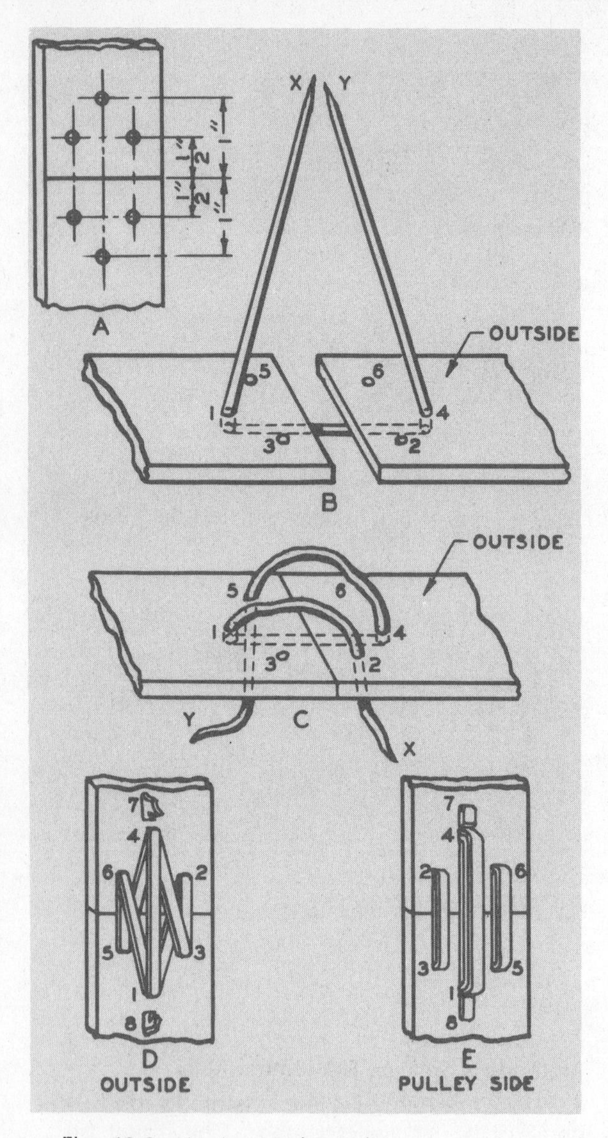

Fig. 49-9. Lacing Leather Belt

Splicing a Belt with Rawhide Lacing

The steps in lacing a leather belt with rawhide lacing are shown in Fig. 49-9. Holes should be at least ¾″ apart from center to center. The holes in one end should be directly opposite the holes in the other end to make the ends match. Belts up to 2″ wide should have three holes in each end. Wider belts should have more holes, but always an odd number of holes, as 3, 5, 7, 9, etc.

Cross the lacing on the outside of the belt, Fig. 49-9 (D). The pulley side of the belt should have no crossed lacing, Fig. 49-9 (E).

Splicing Belts by Cementing

The best way to fasten the ends of a leather belt together is by **cementing**. It is the safest way because nothing can catch or cut the hands. A **cemented belt** is also stronger than a **laced belt** which is weakened by the punched holes. A cemented belt runs smoother than a laced belt on which the lacing forms a hump which thumps against the pulleys. A joint which begins to separate should be recemented at once.

Good **glue** is a fine cement for leather belts, although special **belt cements** are sold. Measure the length of the belt and add enough for the **lapped joint** shown in Fig. 49-11. To cement a belt, first shave down the ends of the belt so that when they are lapped on each other, the joint will be the same thickness as the rest of the belt; then cement the ends together by clamping between two boards until the cement dries. Put

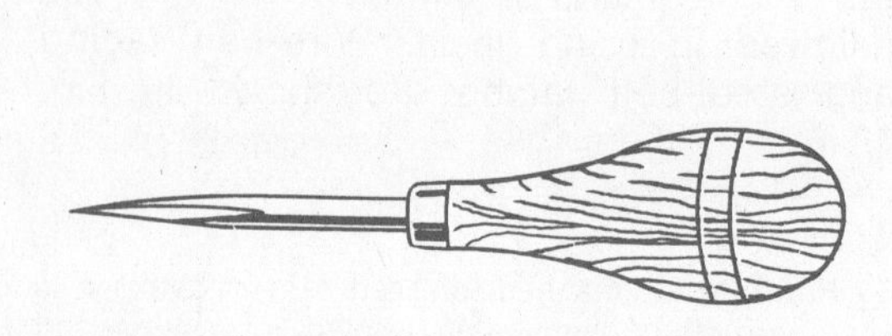

Fig. 49-10. Belt Awl

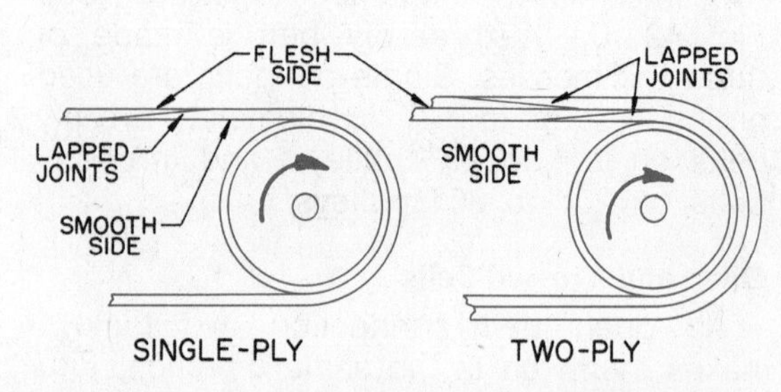

Fig. 49-11. Cemented Belt Joints for Endless Belts

paper between the belt and the boards to keep the boards from sticking to the belt.

The Running Direction of a Cemented Belt

Belts should run with the smooth side, which is called the **grain side** or **hair side,** next to the pulley (see Fig. 49-11). The grain side of the belt will carry more power than the **flesh side** which is the rough side. **Two-ply belts** can run only with the grain side next to the pulley because the flesh sides are cemented together. The cemented **laps** will not be as likely to loosen or curl up when run in the direction shown in Fig. 49-11 as they might if run in the opposite direction.

Care of Leather Belts

Lubricating oil should be cleaned off belts with gasoline. After it is cleaned, a leather belt may be wiped with a cloth moistened with **neat's-foot oil** to keep it from drying and cracking. **Belt dressing** is a sticky mixture which contains asphalt, **pitch,** or **rosin.** It is sometimes put on a belt to make it sticky and to keep it from slipping on the pulleys. Such belt dressing is harmful to the belt and should be used only when it is more costly to shut down the machine than to ruin the belt.

49-8. Shifting Belts

To **shift** a belt means to change it from one pulley to another. Because of the danger of personal injury, never attempt to shift V-belts or flat belts while they are running. Always turn off the machine and wait until the belt comes to a complete stop. Release the belt tension if possible, shift the belt, and reapply the belt tension.

Belt driven machines with variable speed drives **must be running** when speed changes are made. (See Fig. 49-2 and section 49-2.)

Words to Know

belt awl
belt cement
chain belt
chain hoist
cog belt
cone pulley
crossed belt
drive chain
endless belt
flesh side
grain or hair side
lapped joint
leather punch
multiple-ply belt
open belt
pulley crown
rawhide lacing
sheave
single-ply belt
splicing
steel belt lacing
step pulley
variable speed pulley
V-belt
V-pulley
wire mesh belt

Review Questions

1. Give three reasons for using drive belts and chains.
2. Describe a cone pulley. Why are they used?
3. What is meant by the crown of a pulley? What is its purpose?
4. Describe the variable speed pulley. Of what advantage are they?
5. Name five materials used for making drive belts.
6. Describe a V-pulley.
7. When two or more V-belts run together on the same pulley, what is the pulley called?
8. For what are cog belts used?
9. Why are drive chains used in preference to belts sometimes?
10. What is meant by an open belt?
11. Why are drive belts sometimes crossed?
12. Name five ways of splicing flat belts.
13. Which splicing method is best for leather belts? Why?
14. What is rawhide? For what is it used?
15. Which side of a leather belt is placed next to the pulley? Why?
16. What is a single-ply belt? Multiple-ply?
17. What is neat's-foot oil? For what is it used?
18. Describe the safest way to shift belts.

Mathematics

1. How long a belt (open belt) is needed to connect two 14″ diameter pulleys which are 6′ apart from center to center?
2. What is the speed of a 12″ driven pulley if the speed of its 16″ driving pulley is 450 rpm?
3. The speed of a 10″ driving pulley is 650 rpm. What size driven pulley is needed if it is to run at 800 rpm?

Career Information

1. Make a list of the occupations which require a working knowledge of drive belt and chain systems.

UNIT 50

Work Holding Devices and Techniques

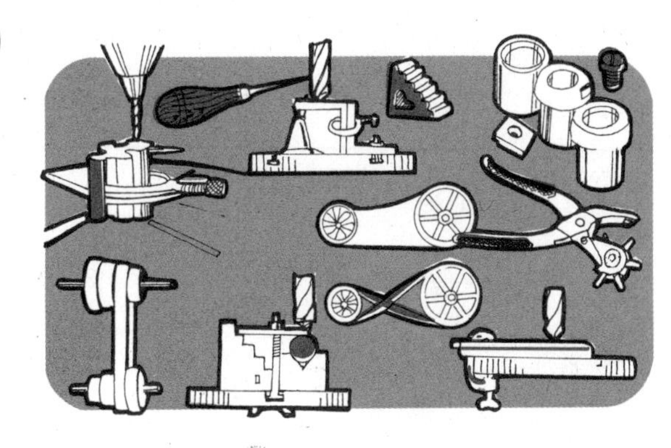

50-1. Setting Up Work

Setting up and clamping a workpiece in a machine vise, holding fixture, or directly to a machine tool table is known as **setting up.** The setup tools and the procedures used are similar for setting up workpieces on such common machine tools as drill presses, milling machines, shapers, and planers. (See Figs. 50-2-50-9.) For safe and accurate machining operations such as drilling, shaping, and milling, the workpiece must be set up accurately and held securely. A variety of machine tool vises and setup tools is used for this purpose.

50-2. Importance of Workpiece Fastening

To keep the workpiece from moving or turning under the pressure of cutting, always clamp it securely in a vise or to the machine table. If the workpiece springs or moves while it is being machined, the cutting tool may break, the workpiece may be thrown from the machine, and injury may result to the operator, workpiece, or machine. Figure 50-1 shows the results of workpiece spring when drilling.

Milling machines, shapers, some drill presses, and other machines have tables with **T-slots** through which bolts may be slipped for clamping vises and workpieces. Some devices and methods of supporting and holding work on machine tables are shown in Figs. 50-2-50-11. Even when work seems well clamped, it often pulls loose. When this happens, shut off the machine and step away from it.

50-3. Strap Clamps

Six kinds of **strap clamps** are shown in Fig. 50-2. They are **plain clamp** or **strap, U-clamp, gooseneck clamp, screw-heel clamp, finger clamp,** and **double-finger clamp.** They are used to hold down the work. (See Figs. 50-5, 50-10, and 50-11.) These clamps are often made in the shop but may be bought just like any other tool. The **U-clamp** can be removed without removing the nut from the bolt. The **goose-**

Fig. 50-1. Drill Broken as a Result of Work Springing

Fig. 50-2. Tools for Holding Work

Fig. 50-3. T-Nut and Stud Set

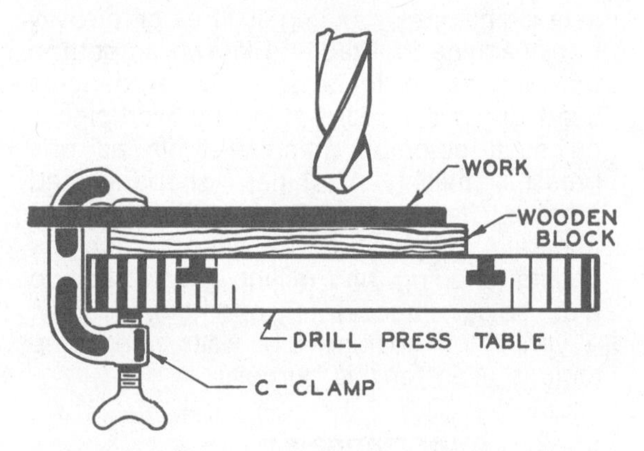

Fig. 50-4. Holding Flat Work with a C-Clamp

neck clamp can be used with a shorter bolt than the other clamps.

Sometimes the work is too high to be clamped in the ordinary way. In such cases, one or more holes are sometimes drilled in the sides of the work, and **finger clamps** are placed in these holes to clamp the work to the table.

50-4. T-Slot Bolts

T-slot bolts are usually used with the strap clamps. They may also be used to **bolt** the **vise** to the machine table. The head of the bolt is made to fit the **T-slot** in the table. (See Figs. 50-2, 50-5 - 50-7, and 50-9).

50-5. T-Nuts and Studs

T-nuts and studs serve the same purpose as T-bolts but are more versatile. If necessary, studs of several lengths can be combined to get the required length. A typical set of T-nuts and studs is shown in Fig. 50-3.

50-6. C-Clamp

A C-clamp, shaped like the letter C, is measured by the greatest distance it can be opened between the jaw and the end of the screw. It is made in many sizes and is useful when clamping work to the table of a machine. (See Figs. 50-2, 50-4, and 50-7.)

50-7. Parallel Clamp

A parallel clamp, also called a **toolmaker's clamp** or **machinists' clamp,** may be used to clamp work in place. (See Figs. 50-2 and 50-8.)

50-8. Jackscrew

The jackscrew, also known as a **planer jack,** ranges in height from two inches upward. It may be used for leveling or supporting odd-shaped work. (See Figs. 50-2, 50-7.)

50-9. Step Block

The step block is used to support and **block up** one end of a **strap clamp.** This levels the clamp so that both ends are the same height from the table. (See Figs. 50-2 and 50-6.)

Step blocks may also be purchased in sets, with or without matching strap clamps.

50-10. Parallels

Parallels are strips of cast iron or hardened steel with opposite sides ground perfectly parallel. They come in pairs exactly alike in size. Parallels are used in leveling and supporting the work. A parallel is placed under each end of the work. They are expensive precision tools and deserve the best of care. (See Figs. 50-2, 50-5, and 50-8-50-10.)

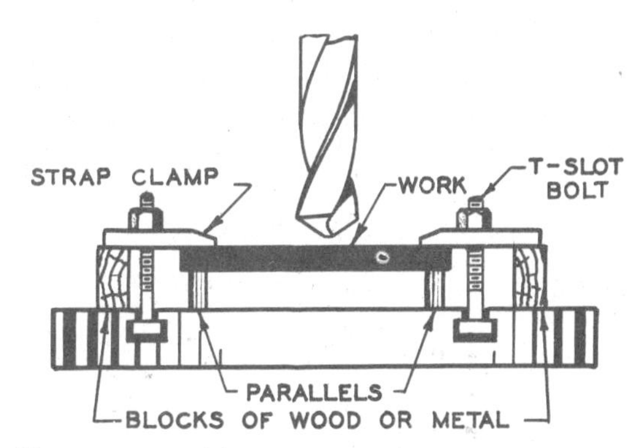

Fig. 50-5. Holding Flat Work on Parallels

50-11. Vises

Vises come in many sizes and types. For light drilling, reaming, tapping, and similar operations, vises are often hand-held. Most vises have provisions for being fastened to the machine table and should be so fastened when accurate or heavy machining is involved. (See Figs. 50-2 and 50-9.)

50-12. V-Blocks

A V-block gets its name from its **V** shape (see Fig. 50-2). The angle of the **V** is usually 90°. V-blocks are used to hold round work, Fig. 50-6. They should be made or purchased in pairs so that one can be placed under each end of long work. (See Fig. 6-27.) They are made with or without clamps.

50-13. Angle Plate

Work is sometimes clamped to an **angle plate,** also called a **toolmaker's knee** which is clamped to the machine table. Such work usually cannot be held otherwise. Figures 50-7 and 50-8 illustrate the use of angle plates in machining setups.

50-14. Holding Flat Work

Flat work, such as a **plate,** may be clamped to the table as shown in Figs. 50-4 and 50-5. The drill can be kept from cutting into the table by placing a piece of wood under the work as in Fig. 50-4, by setting the work so that the drill will pass into the hole in the center of the table, or by using **parallels** as in Fig. 50-5.

50-15. Holding Round Work

Round work such as rods may be held in **V-blocks** as shown in Fig. 50-6.

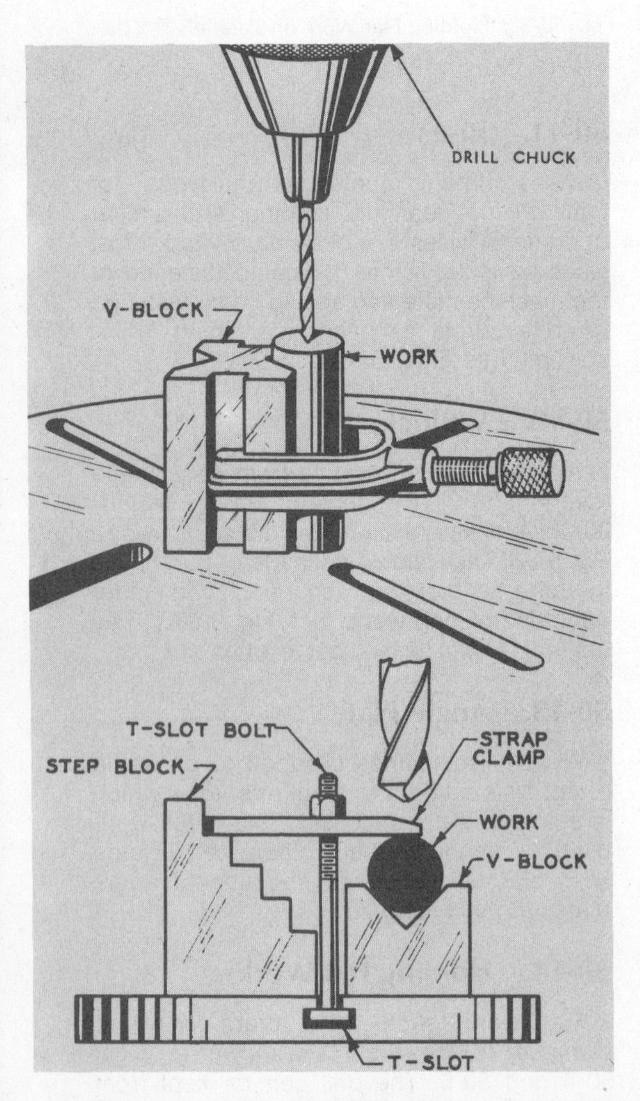

Fig. 50-6. Holding Round Work

50-16. Holding Odd-Shaped Work

Special holding tools may be needed to hold odd-shaped work. Some work may be fastened to an **angle plate** as shown in Figs. 50-7 and 50-8. Odd-shaped work held with **U-clamps** and **adjustable bolts** is shown in Figs. 50-10 and 50-11.

50-17. Fixtures

Fixtures are work-holding devices used in production work which allow a workman to quickly support and clamp a workpiece

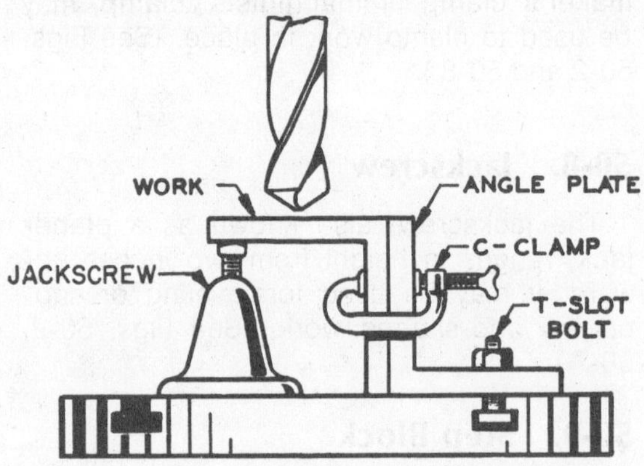

Fig. 50-7. Using Angle Plate and Jackscrew to Hold Work

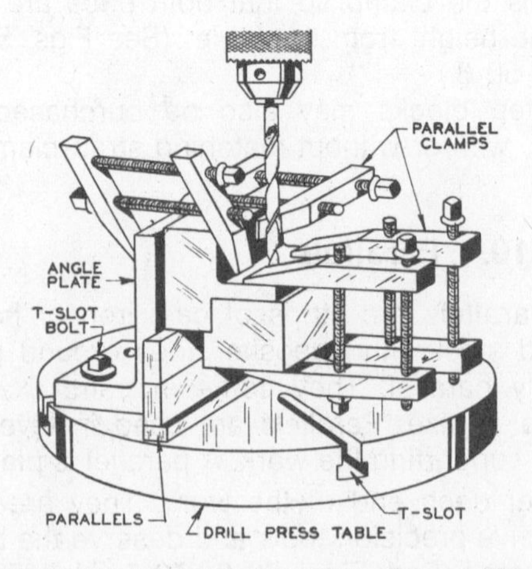

Fig. 50-8. Using Angle Plate, Parallels, and Parallel Clamps to Hold Work

for machining. Fixtures have no provision for guiding the cutting tools. They are usually custom-made to hold a specific part. Fig. 50-12 shows a typical fixture.

50-18. Drill Jigs

If many pieces of the same kind have to be drilled, time and money can be saved by designing and making a **drill jig**, Fig. 50-13, which is a tool for holding the work while it is being drilled. There are many forms of drill jigs. The time which would be necessary to lay out the holes to be drilled in every piece is saved by using a drill jig. Of course, it takes time and costs from a dollar to hundreds and even thousands of dollars to

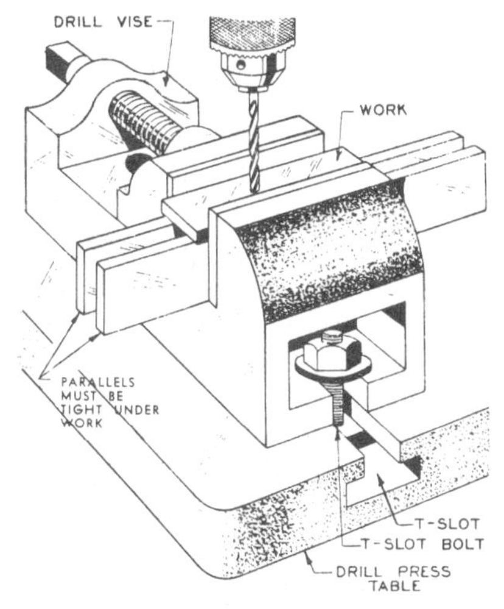

Fig. 50-9. Using Drill Vise and Parallels to Hold Work

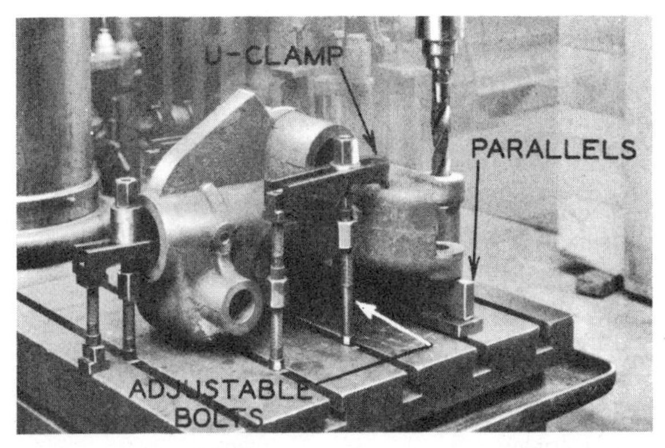

Fig. 50-10. Holding Odd-Shaped Work with U-Clamps and Adjustable Bolts

Fig. 50-11. Holding Spherical Work with U-Clamps and Adjustable Bolts

Fig. 50-12. Angle Plate Setup for Drilling Perpendicular to an Angled Surface

Fig. 50-13. Drill Jigs

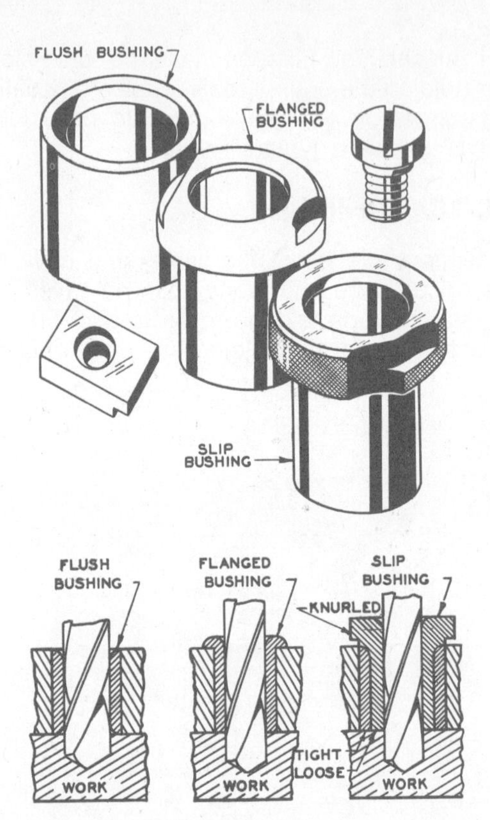

Fig. 50-14. Drill Bushings

make a jig. Costs must, therefore, be studied carefully before starting a job to see whether a jig would save time and money. A jig also helps to drill the holes more exactly than any other way.

50-19. Drill Bushings

Drill jigs are fitted with hardened steel bushings, called **drill bushings.** Figure 50-14 shows three kinds of drill bushings, namely, **flush bushing**, **flanged bushing,** and **slip bushing.** The drill bushing guides the drill.

Slip bushings are used at locations where reaming, tapping, or counterboring must take place. The hole is drilled with the slip bushing in place. Then the slip bushing is removed, providing clearance for tools of larger diameter to be used at the same location.

Words to Know

angle plate
double-finger clamp
drill bushing
drill jig
finger clamp
fixture
flanged bushing
flush bushing
gooseneck clamp
jackscrew
parallel clamp
parallels
screw-heel clamp
setting up
slip bushing
step blocks
strap clamp
stud
T-bolts
T-nuts
T-slot
U-clamp
V-blocks

Review Questions

1. What is meant by setting up a workpiece for machining?
2. Why is it important for workpieces to be accurately positioned, solidly supported, and clamped?
3. What are T-slots? What is their purpose?
4. Describe how strap clamps are used.
5. How do T-bolts differ from T-nuts and studs?
6. Name two types of clamps which can aid workpiece setup.
7. What are step blocks and how are they used?
8. Tell what parallels are and how they are used.
9. Describe how V-blocks are used in setup work.
10. What is an angle plate, and how is it used?
11. What are fixtures? In what way do they differ from jigs?
12. Describe how drill jigs work.
13. Name three types of drill jig bushings.
14. Of what advantage are slip bushings?

PART XV

Machine Tool Processing

UNIT 51

Drilling Machines

51-1. What is a Drilling Machine?

A drilling machine is a machine which holds and turns a drill to cut holes in metal. It is also called a **drill press**, or just **drill**.

51-2. Kinds of Drilling Machines

Among the kinds of drilling machines are:

1. **Hand drill** (see § 51-3).
2. **Portable electric drill** (see § 51-4).
3. **Handfeed drill press** (see § 51-5).
4. **Back-geared upright drill** (see § 51-6).
5. **Gang drill** (see § 51-7).
6. **Multiple-spindle drill** (see § 51-8).
7. **Radial drill press** (see § 51-9).
8. **Turret drill press** (see § 51-10).

51-3. Hand Drill

A common drilling tool which is used for very light work is known as the **hand drill,**

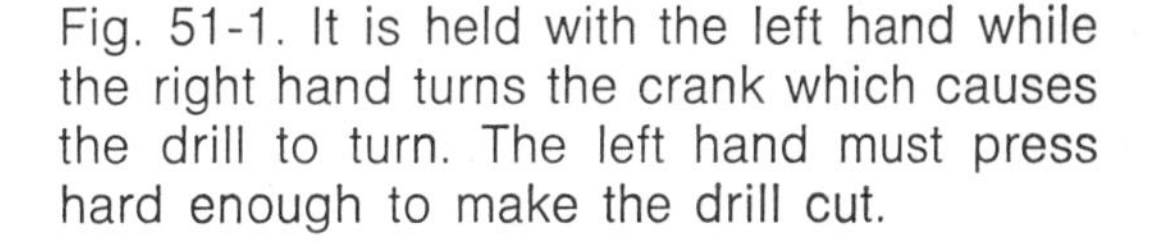

Fig. 51-1. It is held with the left hand while the right hand turns the crank which causes the drill to turn. The left hand must press hard enough to make the drill cut.

51-4. Portable Electric Drill

A small drilling machine which can be carried from job to job is the **portable electric drill,** Fig. 51-2. It can be run from a 115V electric outlet. Figure 51-3 shows how it is used. It is important that portable electric tools be grounded.

51-5. Handfeed Drill Press

Small, light work may be drilled on the **handfeed drill press**. This is a small press in which only the smaller drills are used. It may be one which sets on the bench, called

Fig. 51-1. Hand Drill

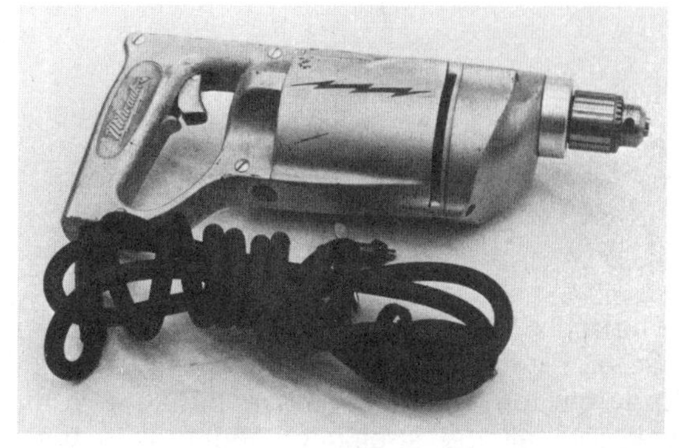

Fig. 51-2. Portable Electric Drill

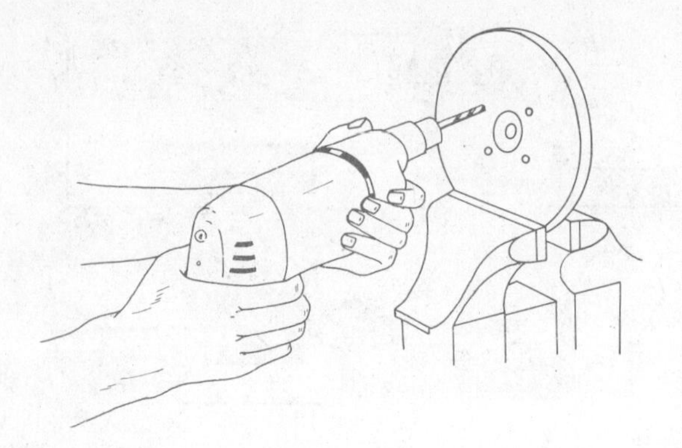

Fig. 51-3. Using a Portable Electric Drill

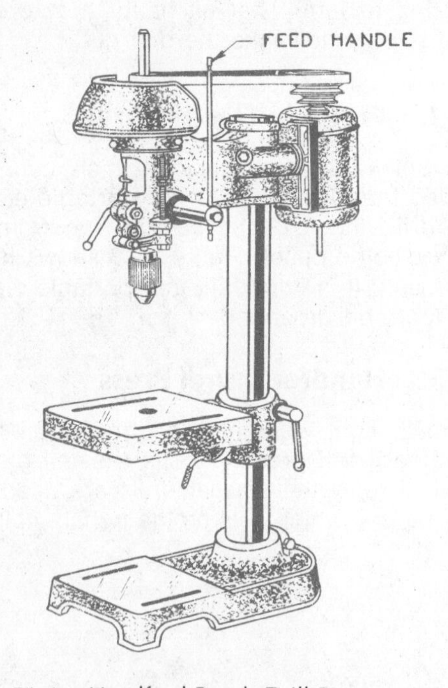

Fig. 51-4. Handfeed Bench Drill Press

a **bench drill**, Fig. 51-4, or it may be a floor model, Fig. 51-5. It is the simplest drill press and may be called **sensitive** because you can **feel** all the strains on the drill in the **feed handle.**

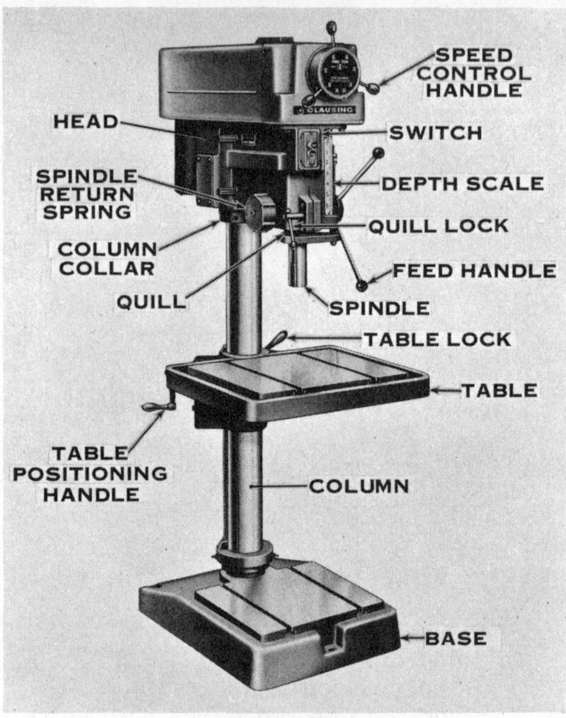

Fig. 51-5. Handfeed Floor Drill Press

51-6. Back-Geared Upright Drill Press

This machine is like the sensitive drill press except that it is larger and more powerful, Fig. 51-6. It has **gears** for changing the speeds; these are called **back gears.** Also, besides **feeding** by hand as on the sensitive drill press, this machine has an **automatic feed,** that is, it uses power to lower the drill. Larger drills can be used than in the sensitive drill.

51-7. Gang Drill

The gang drill, Fig. 51-7, is a drilling machine in which two or more drill presses are **ganged** or made into one machine. Each **spindle** may be run alone. This ma-

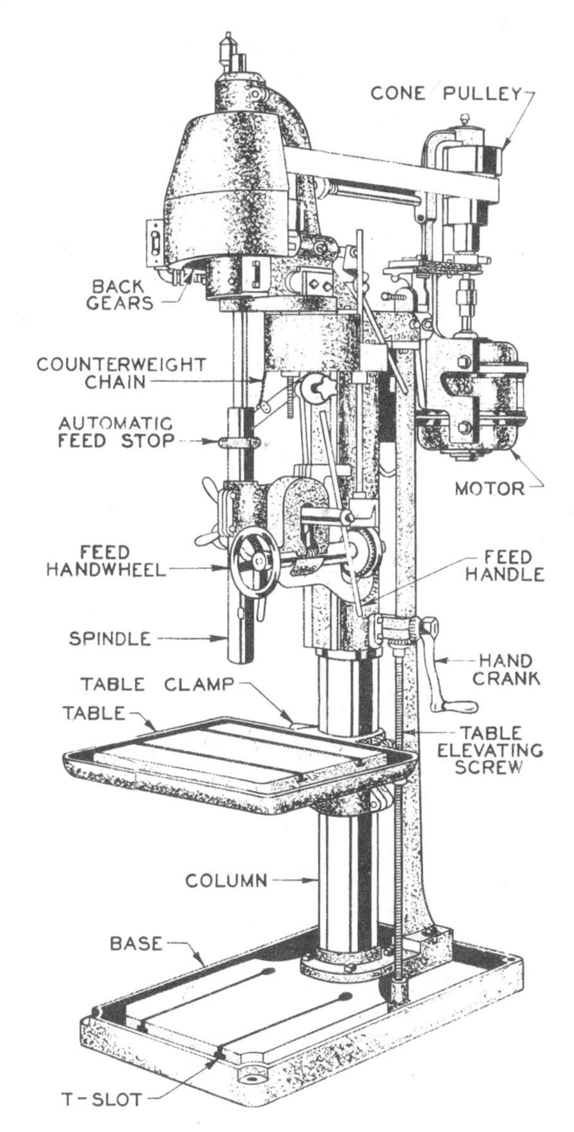

Fig. 51-6. Back-Geared Upright Drill Press and Power Feed

chine is used mainly in **mass production** where different drill press operations are done, one after another. Some work may be done by one spindle and then passed to the next spindle for the next operation, and so on. One spindle may hold a small drill, a second may hold a large drill, a third may

Fig. 51-7. Gang Drilling on Automobile Parts

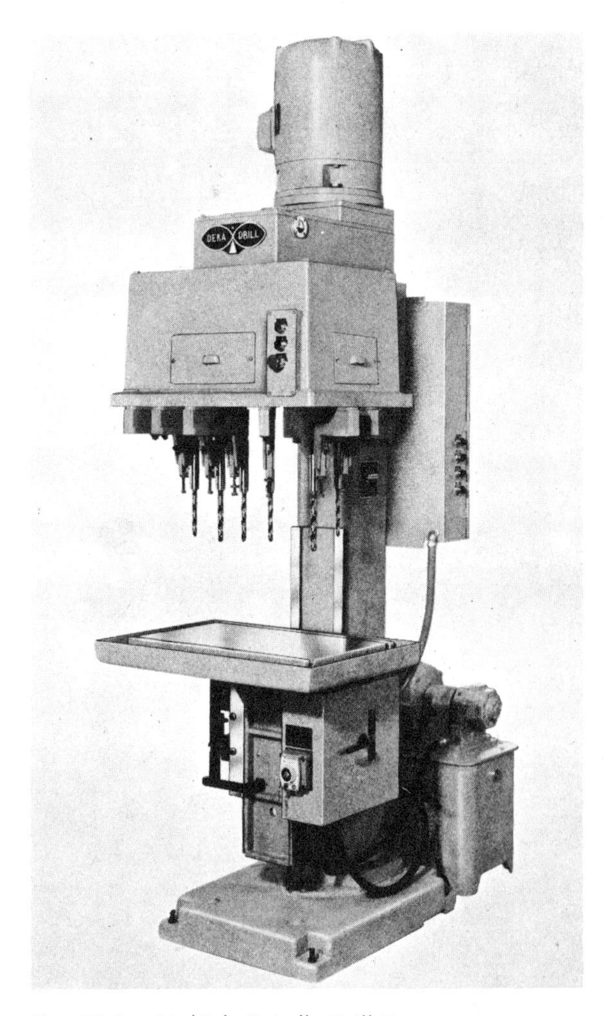

Fig. 51-8. Multiple-Spindle Drill Press

hold a **countersink** (see Fig. 54-6), and so on; the work is thus passed along the table from one spindle to the next.

51-8. Multiple-Spindle Drill Press

This drilling machine has a number of spindles fastened to the main spindle with **universal joints,** Fig. 51-8. Each of these spindles holds a drill, and all the drills run at once, thus drilling at one time as many holes as there are drills, Fig. 51-9. This machine is also used in **mass production.** It is a great timesaver where many pieces, each having a number of holes, have to be drilled. It does away with setting up the job many times. One machine does the work of many machines, and the space that these machines would otherwise occupy is saved.

51-9. Radial Drill Press

This machine has a movable **arm** on which the spindle is mounted. The spindle may be moved and set at different distances from the post or **column.** The arm may also be swung around to the left or right, Fig. 51-10. The arm may be raised or lowered so that work of different heights may be drilled. The radial drill press is used for large, heavy work (such as machine frames) which cannot be moved easily to drill several holes. It is, therefore, necessary to shift the position of the drill. The radial drill makes this possible.

51-10. Turret Drill Press

The **turret drill press,** Fig. 51-11, has a **multiple-spindle turret** with 6, 8, or 10 spindles. A tool is placed in each spindle, and the desired tool is brought into operating position by **indexing** or revolving the turret. Like the gang drill, the turret drill press is a production machine designed for rapidly performing several drill press operations in rapid sequence. The turret drill press shown in Fig. 51-11 can perform up to 6 different drilling operations automatically. Small hand-operated turret drill presses are available also.

51-11. Parts of the Drill Press

The names of drill press parts are the same for all drill presses (Figs. 51-5 and

Fig. 51-9. Multiple-Spindle Drill Presses

Fig. 51-11. Tape-Controlled Drill Press

Fig. 51-10. Drilling Hole in Large Casting with Radial Drill Press

51-6). The **base** of the drill press is the support for the machine; it is bolted to the floor and has **T-slots** so that large work may be bolted to it for drilling. The **column** is the post to which the **table** is fastened; the table holds the work in place while it is being drilled. It also has **T-slots.** By turning the **hand crank** which causes the **table elevating screw** to turn, the table may be raised or lowered and then clamped in place by tightening the **table clamp.** Every drilling machine has a **spindle** which holds and turns the drill. It is upright on most machines and is held by an arm which is fastened to the column. To keep the spindle from dropping, it is balanced by a **spindle return spring** which supports the **quill** — a housing around the rotating spindle moved up and down by the **feed handle.** Some spindles may be counterbalanced by a weight which moves up and down inside the column. This weight is fastened to the spindle with a chain called the **counter-weight chain.**

One **cone pulley** is fastened to the spindle and another is connected to the **motor.** The steps of the cone pulley give as many speeds as there are steps. The **back gears,** located inside the front pulley, give additional speeds. Stop before shifting gears. A variable speed drive gives a continuous range of speeds shown on a dial (Figs. 49-2 and 51-5). It is adjusted only while the drill is running.

The feed stop may be set so the drill will stop cutting when the desired depth is reached. This is useful for repetitive drilling.

51-12. How Are the Sizes of Drilling Machines Measured?

Drilling machines range widely in sizes. The size of an **upright drill press** is measured by the distance from the drill to the column multiplied by two. Thus, a 20-inch drill press can drill the center of a 20-inch circle.

The size of a **radial drill press** is measured from the drill to the column. Thus, a 6-foot radial drill press can drill the center of a 12-foot circle.

Words to Know

automatic feed
back-geared upright drill press
bench drill press
drill press
- automatic feed stop
- back gears
- base
- column
- cone pulley
- counterweight chain
- feed handle
- hand crank
- spindle
- table
- table clamp
- table elevating screw

gang drill press
hand drill
multiple-spindle drill press
portable electric drill
radial drill press
sensitive drill press
turret drill press

Review Questions

1. Name and briefly describe the six kinds of drill presses.
2. Name the main parts of a drill press.
3. What is a sensitive drill press?
4. What is meant by automatic feed?
5. What is the purpose of back gears?
6. How is the size of a drill press measured? A radial drill press?
7. What do gang drill presses and turret drill presses have in common?

UNIT 52

Drills, Sleeves, Sockets, and Chucks

52-1. Kinds of Drills

The common kinds of drills are **twist drill, straight-fluted drill**, and **spade drill**, Fig. 52-1.

52-2. Spade Drill

Spade drills with replaceable blades are used for drilling large holes. They are available in sizes from 1″ to 5″ and are much cheaper than twist drills. (See Fig.52-1.)

52-3. Straight-Fluted Drill

The straight-fluted drill or **farmer drill**, named after its inventor, is used for drilling brass, copper and other soft metals (see Fig. 52-1). It may also be used to drill thin metal.

A variation of the straight-fluted drill, called a **gun drill**, is used for drilling very deep holes.

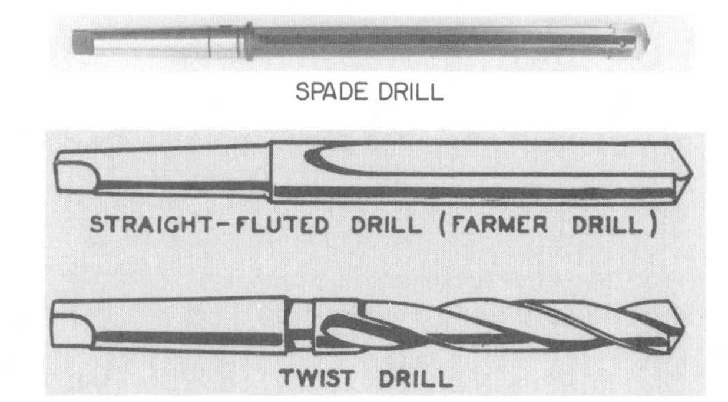

Fig. 52-1. Kinds of Drills

52-4. Twist Drill

The twist drill is the one that is most often used in metalwork. Twist drills are made with two, three, or four **cutting lips**, Fig. 52-2.

The **two-lip drill** is used to drill holes into solid metal. Three- and four-lipped drills, called **core drills**, are used for drilling out **cored holes** in castings. (See Unit 39.)

An **oil hole twist drill** feeds cutting fluid to the cutting edges through holes in the drill, Fig. 52-2.

52-5. Carbon-Steel, High-Speed Steel, and Carbide Drills

Twist drills are made of **carbon steel, high-speed steel**, and **tungsten carbide**. If

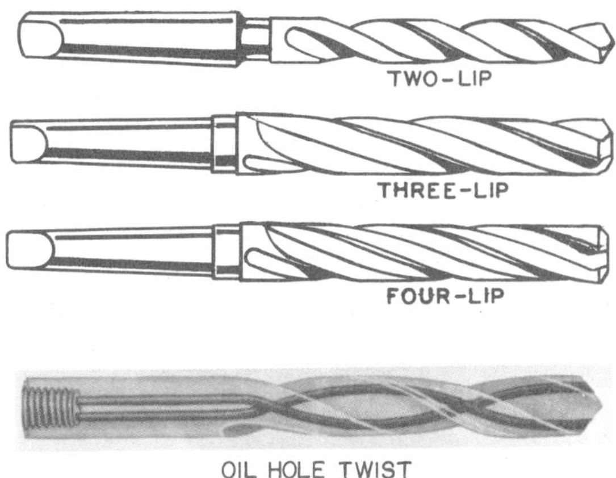

Fig. 52-2. Kinds of Twist Drills

TABLE 34

Drill Sizes

Letter drills begin where **number drills** end.

Number and Letter Drills	Fractional Drills	Decimal Equivalents	Number and Letter Drills	Fractional Drills	Decimal Equivalents	Number and Letter Drills	Fractional Drills	Decimal Equivalents	Number and Letter Drills	Fractional Drills	Decimal Equivalents
80	...	.0135	42	...	.0935		13/64	.2031		13/32	.4062
79	...	.0145		3/32	.0937	6	...	.2040	Z	...	.4130
	1/64	.0156	41	...	.0960	5	...	.2055		27/64	.4219
78	...	.0160	40	...	.0980	4	...	.2090		7/16	.4375
77	...	.0180	39	...	.0995	3	...	.2130		29/64	.4531
76	...	.0200	38	...	.1015		7/32	.2187		15/32	.4687
75	...	.0210	37	...	.1040	2	...	.2210		31/64	.4844
74	...	.0225	36	...	.1065	1	...	.2280		1/2	.5000
73	...	.0240		7/64	.1094	A	...	.2340			
72	...	.0250	35	...	.1100		15/64	.2344		33/64	.5156
71	...	.0260	34	...	.1110	B	...	.2380		17/32	.5312
70	...	.0280	33	...	.1130	C	...	.2420		35/64	.5469
69	...	.0292	32	...	.1160	D	...	.2460		9/16	.5625
68	...	.0310	31	...	.1200	E	1/4	.2500		37/64	.5781
	1/32	.0312		1/8	.1250	F	...	.2570		19/32	.5937
67	...	.0320	30	...	.1285	G	...	.2610		39/64	.6094
66	...	.0330	29	...	.1360		17/64	.2656		5/8	.6250
65	...	.0350	28	...	.1405	H	...	.2660			
64	...	.0360		9/64	.1406	I	...	.2720		41/64	.6406
63	...	.0370	27	...	.1440	J	...	.2770		21/32	.6562
62	...	.0380	26	...	.1470	K	...	.2810		43/64	.6719
61	...	.0390	25	...	.1495		9/32	.2812		11/16	.6875
60	...	.0400	24	...	.1520	L	...	.2900		45/64	.7031
59	...	.0410	23	...	.1540	M	...	.2950		23/32	.7187
58	...	.0420		5/32	.1562		19/64	.2969		47/64	.7344
57	...	.0430	22	...	.1570	N	...	.3020		3/4	.7500
56	...	.0465	21	...	.1590		5/16	.3125			
	3/64	.0469	20	...	.1610	O	...	.3160		49/64	.7656
55	...	.0520	19	...	.1660	P	...	.3230		25/32	.7812
54	...	.0550	18	...	.1695		21/64	.3281		51/64	.7969
53	...	.0595		11/64	.1719	Q	...	.3320		13/16	.8125
	1/16	.0625	17	...	.1720	R	...	.3390		53/64	.8281
52	...	.0635	16	...	.1770		11/32	.3437		27/32	.8437
51	...	.0670	15	...	.1800	S	...	.3480		55/64	.8594
50	...	.0700	14	...	.1820	T	...	.3580		7/8	.8750
49	...	.0730	13	...	.1850		23/64	.3594		57/64	.8906
48	...	.0760		3/16	.1875	U	...	.3680		29/32	.9062
	5/64	.0781	12	...	.1890		3/8	.3750		59/64	.9219
47	...	.0785	11	...	.1910	V	...	.3770		15/16	.9375
46	...	.0810	10	...	.1935	W	...	.3860		61/64	.9531
45	...	.0820	9	...	.1960		25/64	.3906		31/32	.9687
44	...	.0860	8	...	.1990	X	...	.3970		63/64	.9844
43	...	.0890	7	...	.2010	Y	...	.4040		1	1.0000

the drill shank is not stamped **HS**, meaning **high speed,** it is made of carbon steel. Section 19-12 explains how to tell whether a drill is carbon steel or high-speed steel by the **spark test.** High-speed steel drills cost two or three times as much as carbon-steel drills. Note in Table 35 on page 478 that high-speed drills may be run twice as fast as carbon-steel drills.

For drilling very hard or very abrasive materials, either tungsten carbide-tipped or solid tungsten carbide drills are available.

The **web** is the metal in the center running lengthwise between the flutes. It is the backbone of the drill; it gets thicker near the shank and makes the drill stronger. Section **A** in Fig. 52-3 was cut from a drill near the shank while section **B** was cut near the point. Study the difference in the thickness of the web in these two sections.

The **tang** is the flattened end of the shank. It fits into the slot in a **drill sleeve** (see § 52-11) or the **drill press spindle**. It drives the drill and keeps the shank from slipping, especially on large drills. The **lips** are the **cutting edges** of the drill. The **dead center** is the end at the point of the drill; it should always be in the exact center of the point. The **heel** is the part of the point behind the cutting edges.

The flutes are shaped to:

1. Help to form the cutting edges at the point.
2. Curl the chips into small spaces.
3. Form passage for the chips to come out of the hole.
4. Allow the **cutting fluid** to travel to the cutting edges of the drill.

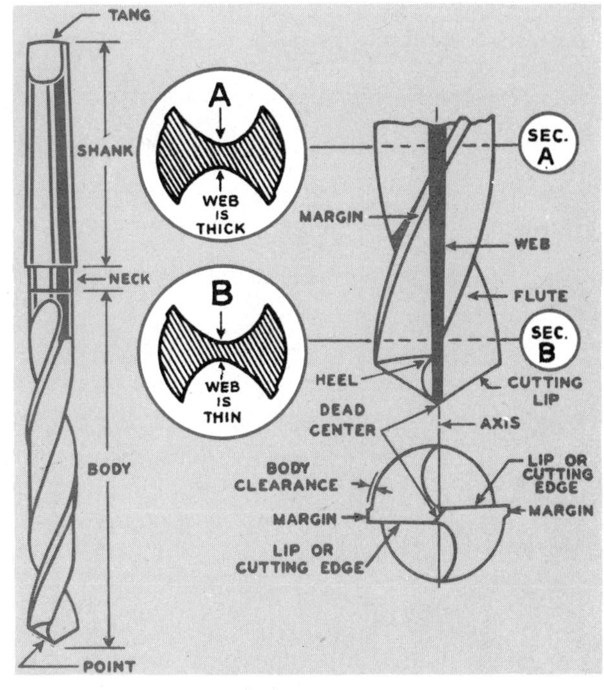

Fig. 52-3. Parts of a Twist Drill

52-6. Parts of Twist Drill

The parts of a twist drill are shown in Fig. 52-3. The **body** is the part in which the grooves are cut. The grooves which run along the sides of the drill are called **flutes.** While drilling, the drill is held by the **shank.** The cone-shaped cutting end is the **point**. The **margin** is the narrow edge alongside the flute. **Body clearance** is the part that has been cut away between the margin and the flute; the rubbing between the drill and the walls of the hole is reduced, and less power is needed to turn the drill. It also keeps from **drawing the temper** out of the drill by overheating.

52-7. Kinds of Drill Shanks

Drill shanks are either **straight** or **tapered,** Fig. 52-4. Taper-shanked drilling tools use the American Standard (Morse) Taper. There are No. 1, No. 2, No. 3, and larger Morse tapers, with No. 1 being the smallest. It is helpful to know the sizes well enough so they can be recognized at a glance.

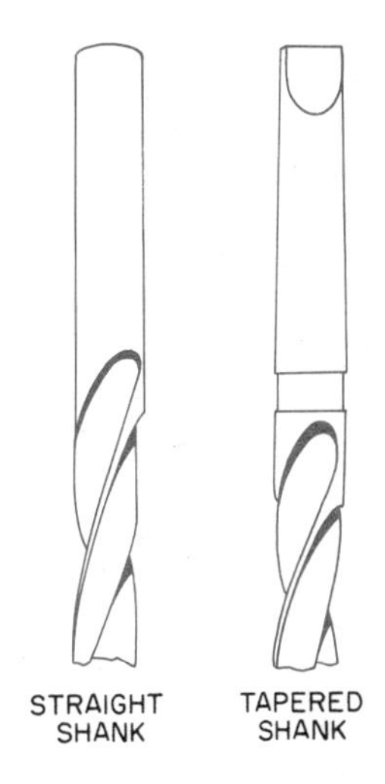

Fig. 52-4. Kinds of Drill Shanks

52-8. Sizes of Drills

Small drills are usually purchased in sets, Fig. 52-5. The size of a drill is known by its

TABLE 35

Drilling Speeds and Feeds
(For Use with High-Speed Steel Drills)

Diameter of Drill	Speed in Revolutions Per Minute (rpm) for High-Speed Steel Drills (Reduce rpm One-half for Carbon-Steel Drills)				Feed Per Revolution Inches
	Low-Carbon Steel Cast Iron (soft) Malleable Iron	Medium-Carbon Steel Cast Iron (hard)	High-Carbon Steel High-Speed Alloy Steel	Aluminum and its Alloys Ordinary Brass Ordinary Bronze	
	80-100 Ft. Per. Min.	70-80 Ft. Per. Min.	50-60 Ft. Per. Min.	200-300 Ft. Per. Min.	
1/8″	2445-3056	2139-2445	1528-1833	6112-9168	0.002
1/4″	1222-1528	1070-1222	764-917	3056-4584	0.004
3/8″	815-1019	713-815	509-611	2038-3057	0.006
1/2″	611-764	534-611	382-458	1528-2292	0.007
3/4″	407-509	357-407	255-306	1018-1527	0.010
1″	306-382	267-306	191-229	764-846	0.015

Fig. 52-5. Set of Number Drills in a Drill Stand

diameter, which may be a number gage, a letter, or a fractional size. Table 34 shows the sizes from the smallest twist drill, which is No. 80, up to 1″ in diameter.

Gage Numbers

Number drills are made in sizes from No. 80 to No. 1 (0.0135″ to 0.228″ diameters). Note that the larger the number, the smaller the drill.

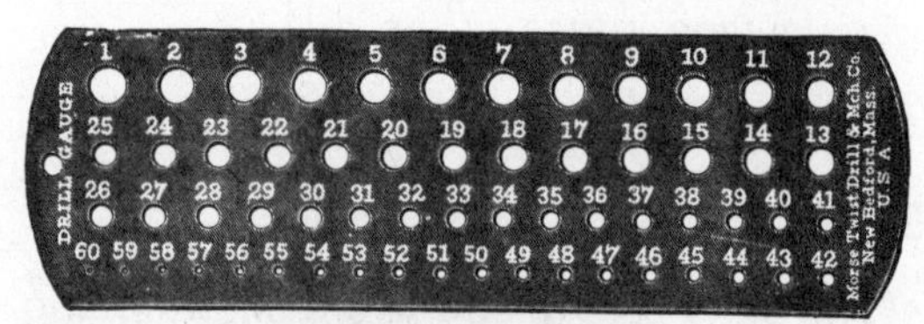

FOR NUMBER DRILLS SIZES NO. 1 TO NO. 60

FOR NUMBER DRILLS SIZES NO. 61 TO NO. 80

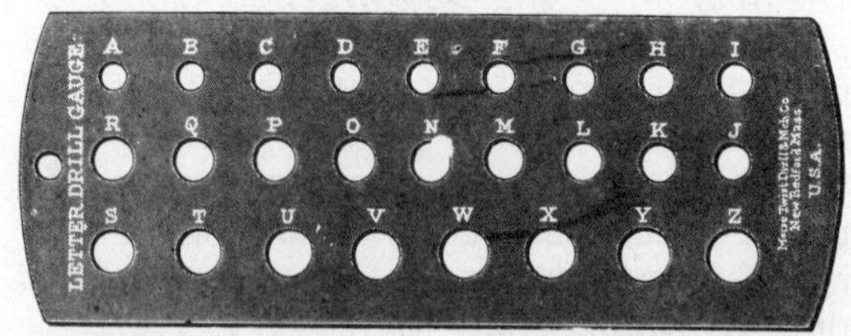

FOR LETTER DRILLS SIZES A TO Z

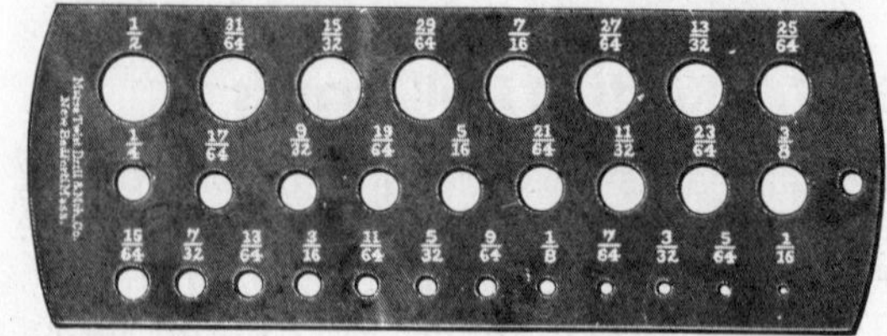

FOR FRACTIONAL DRILLS SIZES 1/16″ TO 1/2″

Fig. 52-6. Drill Gages

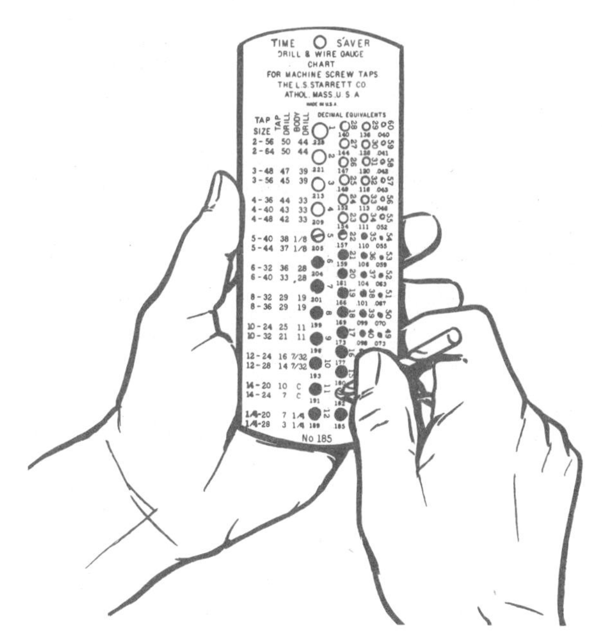
Fig. 52-7. Measuring a Drill with a Drill Gage

Letter Drills

Letter drills are labeled from A to Z (0.234″ to 0.413″ diameters). Note that the letter drills begin where the number drills end.

Fractions

Fraction drills range from 1/64″ to 4″ in diameter or larger. The sizes increase by 64ths of an inch in the smaller sizes and by 32nds and 16ths in the larger sizes. Note that the number and letter drill sizes fall between the fractional drills.

If a drill of a certain size is needed but is not in the shop, sometimes the next size smaller or larger drill can be used instead.

52-9. How to Measure a Drill

The diameter of a drill is stamped on the drill near the shank. Very small drills are not stamped and must be measured with **drill gages,** Figs. 52-6 and 52-7. Occasionally a drill may be incorrectly stamped; therefore, it is always best to measure it with a drill gage or with a **micrometer,** Fig. 52-8. A new drill may be measured across the **margins** at the point; a worn drill must be measured at the ends of the flutes near the **neck.**

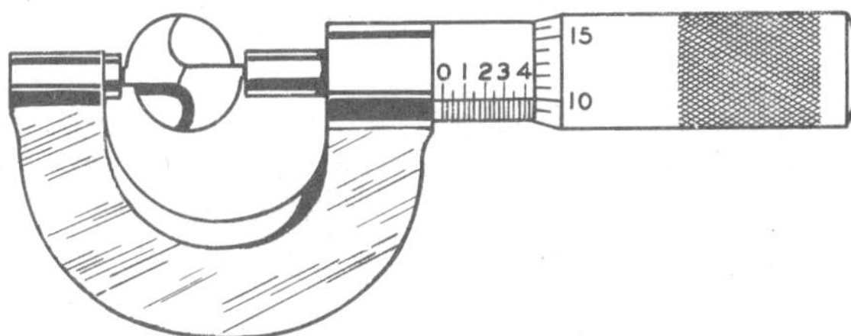
Fig. 52-8. Measuring a Drill with a Micrometer

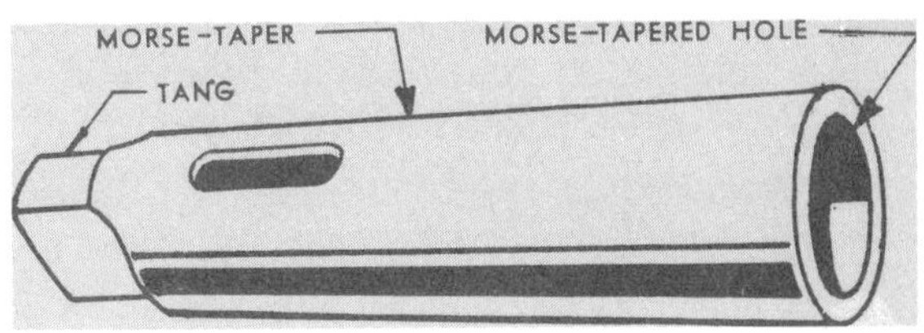

Fig. 52-9. Drill Sleeve

52-10. Ordering Twist Drills

When ordering a twist drill give the:
1. Diameter of drill
2. Shape of shank
3. Kind of material; that is, whether **carbon steel, high-speed steel,** or **carbide.**

For example:
1. ¾″ diameter
2. Taper shank
3. High-speed steel

52-11. Drill Sleeves

Taper shanks on drills come in several sizes, such as No. 1, No. 2, and No. 3 **Morse tapers** (see § 52-7). The **drill press spindle** has a hole with a No. 2 or a No. 3 Morse taper.

A drill with a No. 1 taper will not fit into a spindle with a No. 2 or No. 3 taper. Therefore, to make the drill fit, a **sleeve** like the one shown in Fig. 52-9 with a No. 1 **tapered hole** is placed on the drill shank. The outside of this sleeve has a No. 2 taper. If this taper is still too small to fit into the drill-press spindle, the sleeve with a No. 2 tapered hole and a No. 3 outside taper is also used, thus enlarging the drill shank until it fits snugly into the drill press spindle. The first sleeve is known as a **No. 1 to No. 2 Morse-taper sleeve** while the second is a **No. 2 to No. 3 Morse-taper sleeve.**

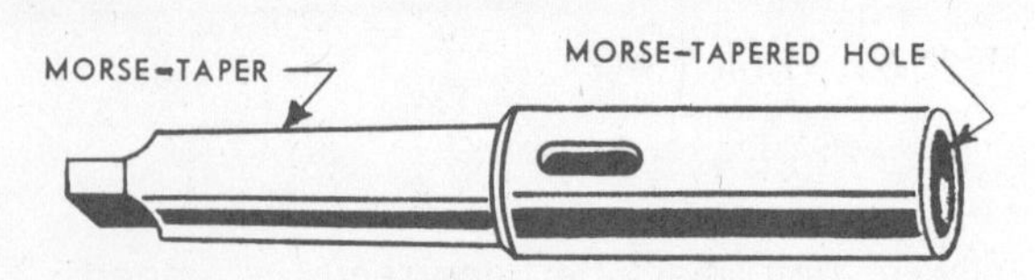

Fig. 52-10. Drill Socket

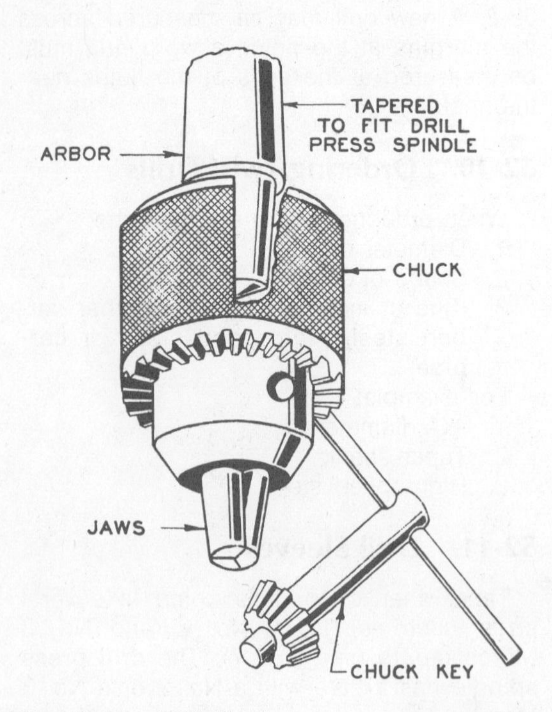

Fig. 52-11. Drill Chuck

The drill sleeve has a flattened end called a **tang** which fits a slot in the drill-press spindle and keeps the drill and drill sleeve from slipping.

52-12. Drill Sockets

Sometimes it is necessary to use a drill with a No. 3 taper shank in a drill press spindle with a No. 2 tapered hole. The drill socket, Fig. 52-10, makes this possible. It has a large hole and a small Morse-taper shank.

52-13. Drill Chuck

Straight shank drills must be held in a drill chuck, Fig. 52-11. This is fitted with a **Morse taper shank,** called an **arbor,** which fits into the **drill press spindle.** The drill chuck shown here has three **jaws.** It grips the drill between the jaws and is adjusted by turning the outside with a **chuck key.**

Words to Know

arbor
carbide drill
carbon-steel drill
chuck key
drill chuck
drill gage
drill press spindle
drill shank
drill sleeve
drill socket
drill stand
farmer drill
four-lip drill
fractional drill
gun drill
letter drill
Morse taper
number drill
oil tube drill
spade drill
straight-fluted drill
straight shank drill
tang
tapered shank
three-lip drill
twist drill
two-lip drill

Review Questions

1. Name three kinds of materials from which drills are made.
2. Name the three kinds of drills. Which is used most for metalwork?
3. How can you tell whether a drill is made of carbon steel or high-speed steel?
4. Describe an oil hole drill.

5. For what is a gun drill used? What kind of a drill is it?
6. Of what purpose is the tang on a drill?
7. What purpose do the flutes of a drill serve?
8. What is the name of the taper used on taper-shanked tools?
9. What part of the drill should be measured with a micrometer to find its size?
10. What is a drill gage? For what is it used?
11. What information must be given to order a twist drill?
12. For what is a drill sleeve used?
13. How does a drill socket differ from a drill sleeve?
14. What kinds of tools are held in a drill press chuck?

Mathematics

1. Is a 17/64″ drill larger than an 11/32″ drill?
2. What is the difference in size between a No. 35 drill and a "D" drill?
3. What is the difference in size between a No. 18 drill and a 3/64″ drill?

UNIT 53

Drilling

53-1. What Is Drilling?

Drilling means to make a hole with a drill. It is one of the basic methods of machining solid materials.

53-2. Drilling Speed

Drilling speed or **cutting speed**, is the distance a drill would travel in 1 minute if it were laid on its side and rolled. Thus a 1″ drill, turning 100 **revolutions**[1] per minute, would roll 1 × 3.1416 × 100 ÷ 12 = 26.18′ per minute. (**Circumference** of a circle = diameter × 3.1416; dividing by 12 changes the inches to feet.)

The speed at which a drill may be turned depends upon:

1. Its diameter.
2. Whether it is made of **carbon steel, high-speed steel,** or **tungsten carbide.**
3. The hardness of the metal that is being drilled.

The smaller the diameter of the drill, the greater should be the speed; the larger the diameter, the slower should be the speed. Too slow a speed in drilling small holes is inefficient and can cause the drill to break. Table 35 shows that high-speed drills can be run twice as fast as carbon-steel drills. It also shows the speeds at which different metals should be drilled. Use a slow speed to drill hard metal and a fast speed to drill soft metal. If the corners of the cutting edges wear away quickly, Fig. 53-1, the speed should be reduced. The speed of a drill may be increased until it begins to show signs of wear; it should then be reduced a little and run regularly at this reduced speed.

Rules concerning speeds cannot be strictly followed because of the many different conditions. The operator's experience and judgment must help him decide at what speed a drill should be run to obtain the best results.

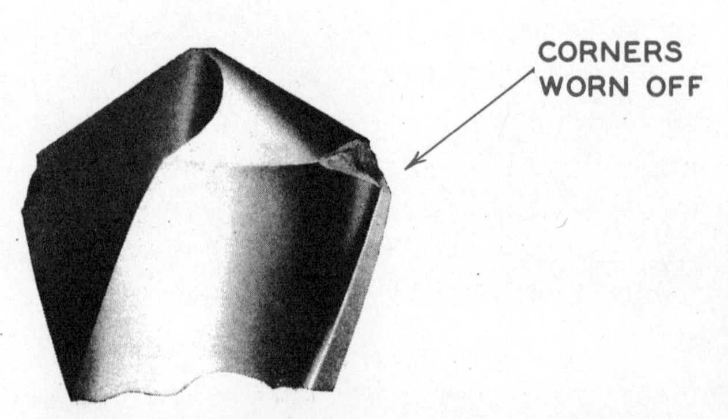

Fig. 53-1. Result of Too Much Speed in Drilling

[1]**Revolution** means one turn completely around.

53-3. Feed

The **feed** of a drill is the distance it cuts into the metal in one turn. It is different for each size of drill and the various materials to be drilled. The drill should be fed into the metal just as fast as it can cut.

As with speed, rules about feed cannot be strictly followed. Experience and judgment must again help to decide how fast the drill should be fed into the metal. The sizes of drills and the general feeds are given in Table 35. The feed should be the same for high-speed steel drills as for carbon-steel drills.

53-4. Putting Drill in Drill Press Spindle

A **tapered-shank drill** must be well cleaned before putting it in the **drill-press spindle**, Fig. 53-2. Examine the **shank** of the drill to make sure that it is not scratched or nicked. A scratched or nicked shank will not fit perfectly in the spindle and will cause the drill to wobble.

The **tang** of the drill should be in the same position as the **slot** in the spindle; a quick upward push fastens the drill in the spindle.

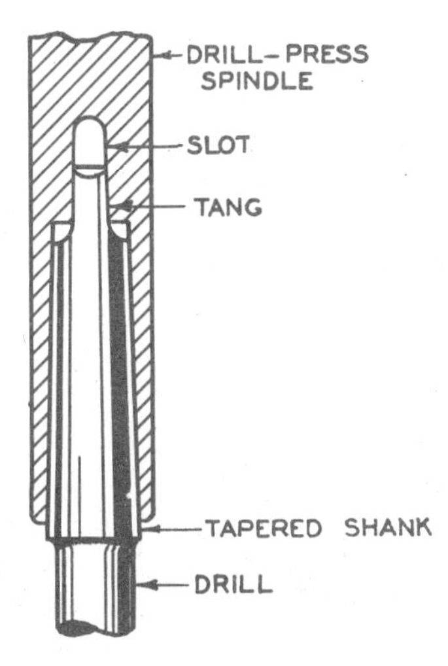

Fig. 53-2. Putting a Drill in a Drill-Press Spindle

The slot holds the tang and turns the drill with the spindle so that the drill cannot slip.

If the taper shank of the drill is too small to fit the spindle, place a **drill sleeve** (see Fig. 52-9) over the shank of the drill. Be sure that the drill sleeve is not scratched or nicked.

53-5. Removing Drill from Drill Press Spindle

The drill or drill chuck should be removed from the drill press spindle with a **drill drift**, Fig. 53-3. Note that one edge of the drill drift is rounded while the other edge is flat.

To remove the drill or drill chuck, put the end of a drill drift into the slot of the spindle with the rounded edge of the drill drift against the upper, rounded part of the slot (see Fig. 53-2). Hold the drill with one hand so that it will not drop and nick the drill press table. Strike the wide end of the drill drift lightly with a **lead hammer**.

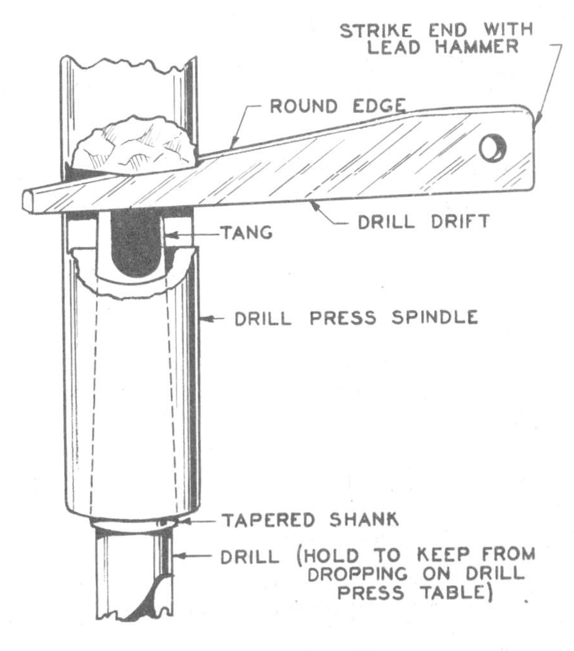

Fig. 53-3. Removing a Drill from the Drill-Press Spindle with a Drill Drift

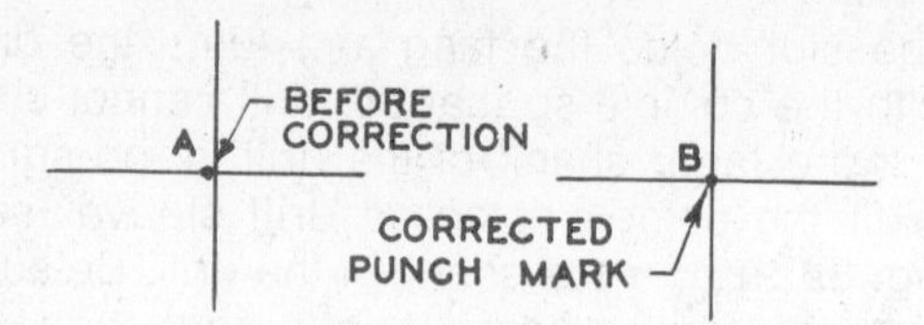

Fig. 53-4. Punch Mark Before and After Correction

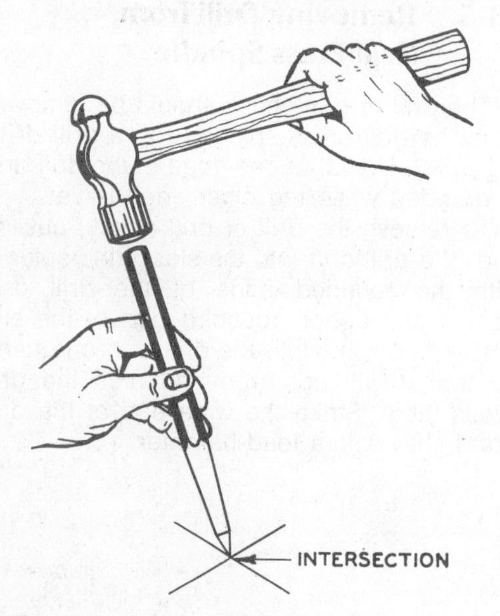

Fig. 53-5. Moving a Punch Mark

A drill drift should also be used to remove a drill from a **drill sleeve** or a **drill socket**.

53-6. Putting Drill in Drill Chuck

A **straight-shank drill** (see Fig. 52-4) may be held in a **drill chuck** (see Fig. 52-11) which is held in the drill press spindle. Tighten the chuck as much as you can with the **chuck key** so that the drill will not slip. Many drill shanks get chewed up and ruined because they slip in the chuck.

Be sure to remove the chuck key from the drill chuck before turning on the drill press.

53-7. Laying Out Hole for Drilling

A layout should be made to show where the hole should be before it is drilled. The first step in laying out a hole to be drilled is to find the **center**. Next, put a small punch mark exactly at this center by tapping the **prick punch** very lightly with the hammer.

If the punch mark is to one side, as **A** in Fig. 53-4, it is necessary to slant the prick punch to move the mark exactly to the center, Fig. 53-5. Finish up with a light blow on the prick punch held in an upright position. When completed, the punch mark should be exactly in the center. (See **B** in Fig. 53-4.)

After enlarging the prick-punch mark with a center punch, the workpiece is ready for drilling.

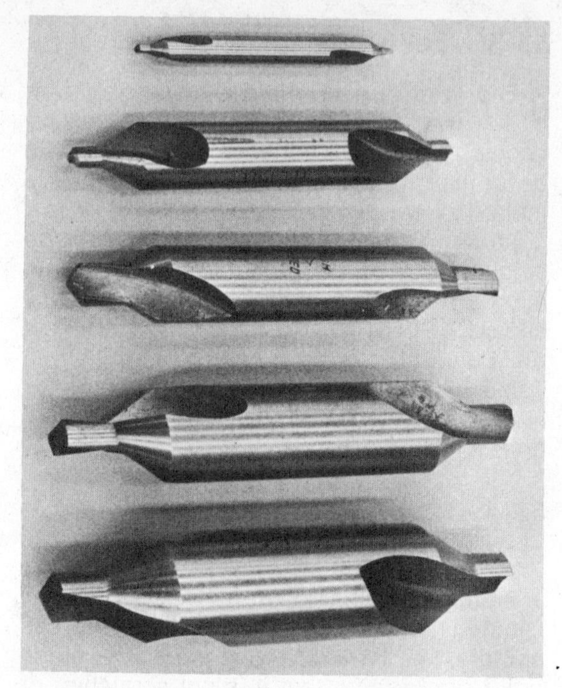

Fig. 53-6. Center Drills

53-8. Center Drilling

Center drilling is done with a tool called a **center drill** or **combination drill and countersink**, Fig. 53-6. They are made in several sizes and are used for the purpose of establishing a hole center for guiding

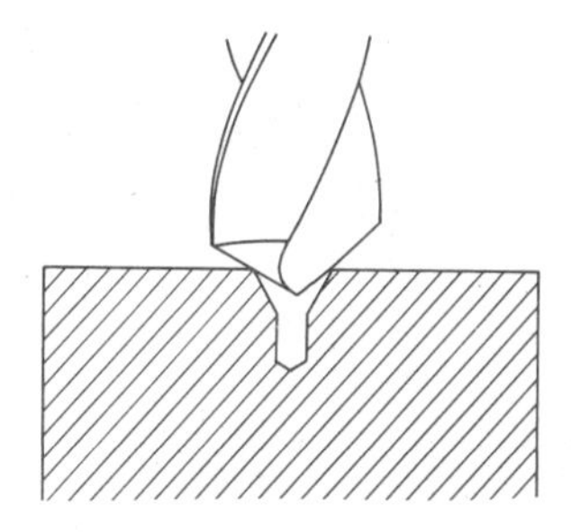

Fig. 53-7. Center Drilled Hole Helps Provide Accurate Location for Drills

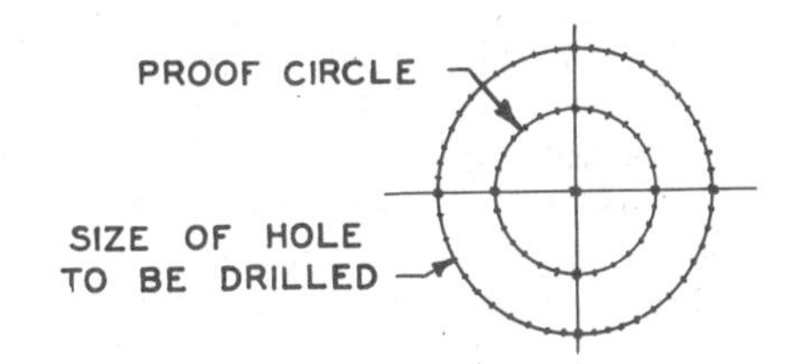

Fig. 53-8. Layout of a Hole for Exact Drilling

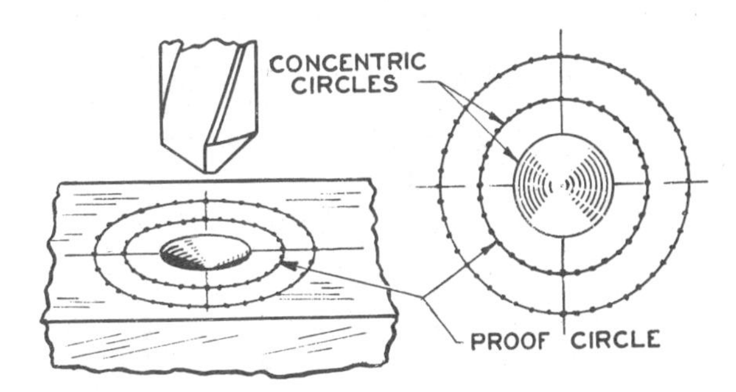

Fig. 53-9. Checking the Accuracy of the Hole

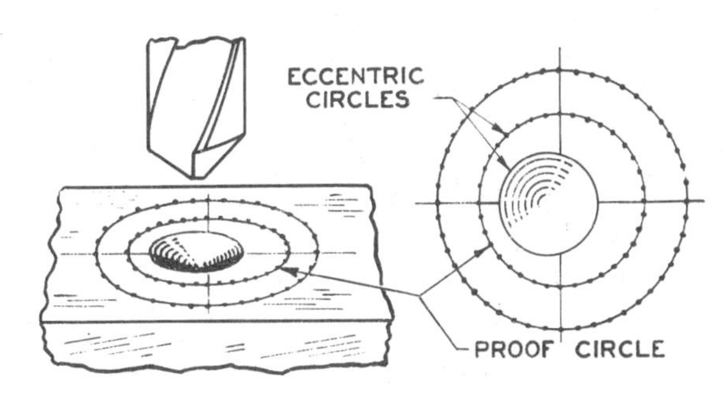

Fig. 53-10. Circle Made by the Drill Is Eccentric to the Layout

drills, Fig. 53-7. They are also used to provide a properly shaped bearing surface for metal lathe centers. (See Fig. 55-41). When the workpiece cannot be rigidly clamped to the machine table, the hole center should be laid out and center punched. The center-punch mark provides a starting place for the center drill. When the workpiece can be rigidly clamped, however, center drilling can be done without the aid of a center-punch mark.

53-9. Laying Out Hole for Exact Drilling

If a hole must be drilled in an exact location on a drill press, the following procedure is recommended.

Find the center of the hole and put a small prick-punch mark at this center. Then, with the divider, scribe a circle the same size as the hole to be drilled. Scribe a smaller circle inside the first one. This is called a **proof circle**, Fig. 53-8; it helps to check that the drill is cutting in the center as will be seen when the hole is drilled. Next, put small prick-punch marks on both circles. Finally, enlarge the prick-punch mark in the center of the circles with a **center punch**. The layout is now ready for drilling.

53-10. Drilling Hole to Layout

After the hole is laid out as shown in Fig. 53-8, it is ready for drilling. To drill the hole, place the point of the drill, while it is turning, exactly over the center-punch mark and enlarge it a little with the drill, Fig. 53-9. Then raise the drill from the work and see if the circle made by the drill is **concentric**[2] with the **proof circle**. (See § 53-9.) If it is concentric, then the drilling may be continued until the hole is finished.

If the circle made by the drill is not concentric, but **eccentric**[2] as in Fig. 53-10, the

[2]**Concentric** means having the same center. **Eccentric** means having different centers.

drilled circle must be drawn back so that it will be concentric to the proof circle. This is known as **drawing the drill** and is explained in section 53-11. **Eccentricity** is caused by:

1. Drill sharpened incorrectly.
2. Punch mark made with center punch not in exact center.
3. Drill wobbling.
4. Hard spots in metal.

53-11. How to Draw the Drill

The way to draw the drill back to the center of the circle is to cut a groove down the side of the hole made by the **point** of the drill with a **round nose chisel**; this must be done on the side farthest from the **proof circle**, Fig. 53-11. This groove is then drilled out; the drill bites into the edge of the groove, makes an egg-shaped hole, and thus shifts the center of the drill to the center of the proof circle.

The drill can only be shifted as long as the two lip corners A and B shown in Fig. 53-12 do not touch the metal. After the groove has been drilled out, see if the new circle made by the drill is **concentric**, Fig. 53-9, with the proof circle. If not, another groove must be cut and the steps repeated until the circle made by the drill is concentric with the proof circle, after which the hole may be drilled until finished.

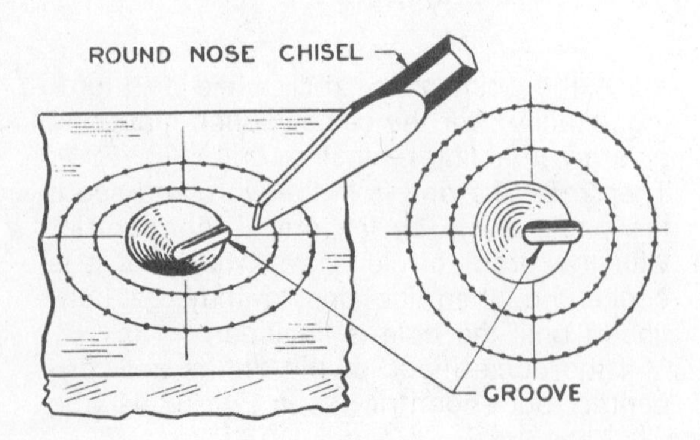

Fig. 53-11. Groove is Cut to Draw the Drill to Proper Position

53-12. Cautions When Drilling a Through Hole

A twist drill that is just breaking through the other side of the hole acts just like a corkscrew. This is when drilling a hole with a twist drill is the most dangerous. The drill may grab and break off or the work may be torn loose from its holdings. To prevent this, hold back on the feed as the drill point begins to break through the underside of the work. **Straight-fluted drills** do not act this way.

53-13. Drilling Large Holes

The larger the drill, the thicker is the **web** between the **flutes** and the wider is the **dead center** (see Fig. 45-6). This dead center does not cut; it interferes with the drill cutting into the metal. To overcome this when drilling a large hole, a small hole (a little larger than the thickness of the web of the large drill) is first drilled through the metal. This small hole

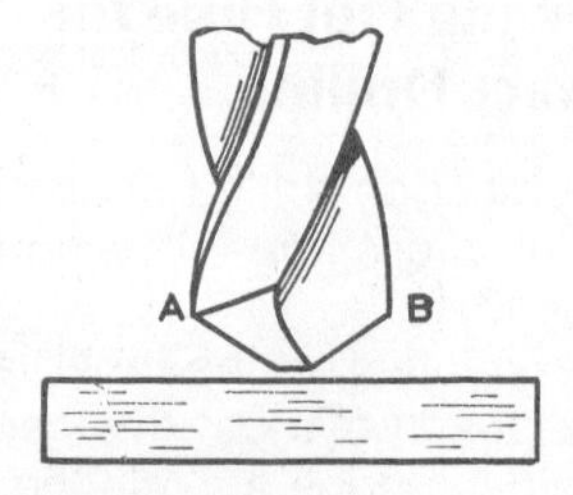

Fig. 53-12. Drill Can Shift Only Until Corners A and B Start Cutting

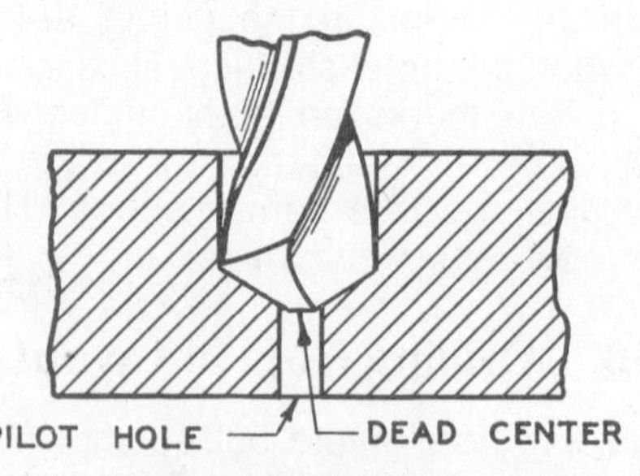

Fig. 53-13. Drilling a Large Hole

is called a **pilot hole**, Fig. 53-13. The large drill may then be used to enlarge the hole. The pilot hole must be drilled exactly to the layout because the large drill will follow the pilot hole exactly. Thus the pilot hole leads or steers the large drill.

53-14. Drilling Accurately Sized Holes

All drills cut **oversize**, that is, the holes they make are slightly larger in diameter than the drills themselves. Larger drills and poorly sharpened drills tend to cut more oversize than small drills or correctly sharpened drills.

The amount a drill will drill oversize can be minimized by a procedure known as **step drilling**. Step drilling simply involves using one or more smaller drills first to remove the bulk of the material; leaving only a small amount of material for the final drill to remove, Fig. 53-14. For drill sizes 3/8″ and smaller, the diameter of the next-to-last drill should be about 1/64″ smaller than the final drill diameter. Above 3/8″, allow 1/32″ size difference.

53-15. Drilling Holes to Depth

Most drilling machine spindles or quills are marked in inches and have a **feed stop** (see Figs. 51-5 and 51-6). To drill a hole to a certain depth, this stop may be set, making use of the depth scale, so that cutting will stop when the desired depth is reached. This is especially useful when many holes of the same depth have to be drilled. If there are no marks on the spindle, a mark may be made with a pencil to show the depth of the hole.

53-16. Chips

Note the difference between cast iron and steel **chips** or **shavings**, Fig. 53-15. When the drill is correctly sharpened, **cast iron chips** are small, broken pieces of metal, while **steel shavings** are long curls.

Chips and shavings have sharp edges and points and may cause bad cuts if picked up with the fingers. They should be removed from the work with a blunt tool or a brush.

53-17. Keep Flutes Clean of Chips

During the drilling, be sure that the flutes of the drill are kept open. Clogging of the flutes may cause the drill to break. To prevent this, the drill may have to be pulled out of the hole a number of times and the chips cleared away.

The flutes are more likely to clog when drilling cast iron than when drilling steel. This is also more likely to happen when drilling deep holes than when drilling shallow ones. When drilling deep holes it may be

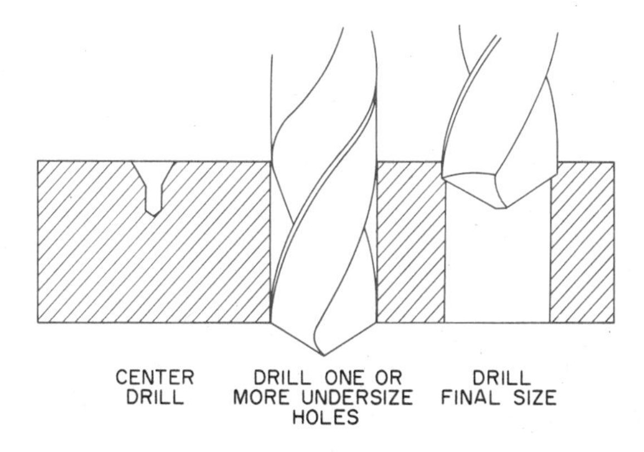

Fig. 53-14. Step Drilling Procedure

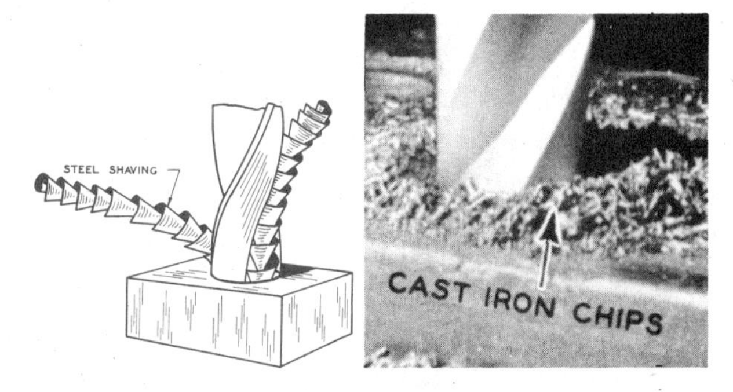

Fig. 53-15. Difference between Steel Chips and Cast Iron Chips

necessary to take the chips out of the hole with a long **magnet** or to blow them out through a long tube. When a tube is used, it should be bent so that the chips will not fly into the operator's face.

53-18. Drilling with Cutting Fluids

During the drilling process, the metal chips rub over the lips of the drill and cause heat. The curling of the chips over the cutting edges of the drill also causes heat, much the same as when a piece of wire is bent back and forth quickly. This heat may **draw the temper** and soften the drill. A **cutting fluid** should, therefore, be used to take the heat away from the lips of the drill and to wash away the chips. The drill will thus last longer, and the hole will be smoother.

Copper is a soft, ductile metal which is also a good heat conductor. For this reason, a drill may be fed into copper with a high rate of feed. However, the lip clearance angle of the drill must be greater than when drilling steel. This is true because the high feed rate will cause the lip clearance angle surface to rub as it advances into the metal. Excessive rubbing will result in a **glazed**, hard, shiny, surface in the hole. The proper cutting action must take place or the drill will be ruined.

Cast iron and copper do not require a cutting fluid. However, an emulsifiable oil solution is sometimes used for cooling purposes when drilling these materials.

Fig. 53-16. Drill Press with Piped Lubricant and Air Feed Increases Production 100% in Appliance Manufacturing Concern

53-19. Cooling Systems

In shops where quantities of lubricant are needed, large systems of tanks and pumps are set up and the **cutting fluid** is forced through pipes to the work on the machines, Fig. 53-16. Machines on which much cutting fluid is used have troughs around the tables to keep the fluid from running to the floor, Fig. 53-16.

Many machines have their own built-in cutting fluid systems.

Words to Know

center drill	drill drift
combination drill and countersink	drilling speed
	eccentric
concentric	pilot hole
cutting speed	proof circle
drawing the drill	step drilling

Review Questions

1. What is meant by drilling speed or cutting speed? Why is it important?
2. What is meant by the feed of a drill? Why is it important?
3. Should a small drill run slower than a large drill if they are drilling the same material?
4. How can you tell when a drill is running too fast?
5. Why must the hole in the drill press spindle be clean before putting a drill, sleeve, or chuck into it?

6. How should a drill chuck or drill sleeve be removed from the drill press spindle?
7. Should a taper-shanked drill be used in a drill chuck? Why?
8. What is involved in making a layout for drilling a hole?
9. What is a center drill. For what is it used?
10. What is a proof circle?
11. What is the meaning of concentric? Eccentric?
12. Why are twist drills somewhat dangerous to use when drilling holes that go completely through a workpiece?
13. Tell what step drilling is and why it is done.
14. How may chips be removed from a deep hole?

Mathematics

1. At what rpm should a 7/8″ diameter carbon-steel drill turn for drilling low-carbon steel?
2. At what rpm should a 1/4″ high-speed steel drill turn to cut medium-carbon steel?

UNIT 54

Other Drill Press Operations

54-1. Other Drill Press Work

Other work than drilling may be done on the drill press. Six basic kinds of **hole machining operations**, Fig. 54-1, are commonly performed on a drill press:

1. **Drilling**
2. **Reaming**
3. **Countersinking**
4. **Counterboring** and **Spot Facing**
5. **Tapping**
6. **Boring**

Another kind of operation, **spot-finishing,** can also be done on a drill press. This is explained in section 35-12.

The basic hole-machining operations listed above can also be performed on many other kinds of machine tools. They can be performed on such basic machine tools as the metal-turning lathe (see Unit 55) and on vertical and horizontal milling machines (see Unit 57). These operations can also be performed on special mass-production machine tools such as turret lathes, screw machines, tapping machines, boring machines, and multipurpose machine tools. The same principles involved in doing the six hole-machining operations on a drill press also apply when performing these operations on other kinds of machine tools.

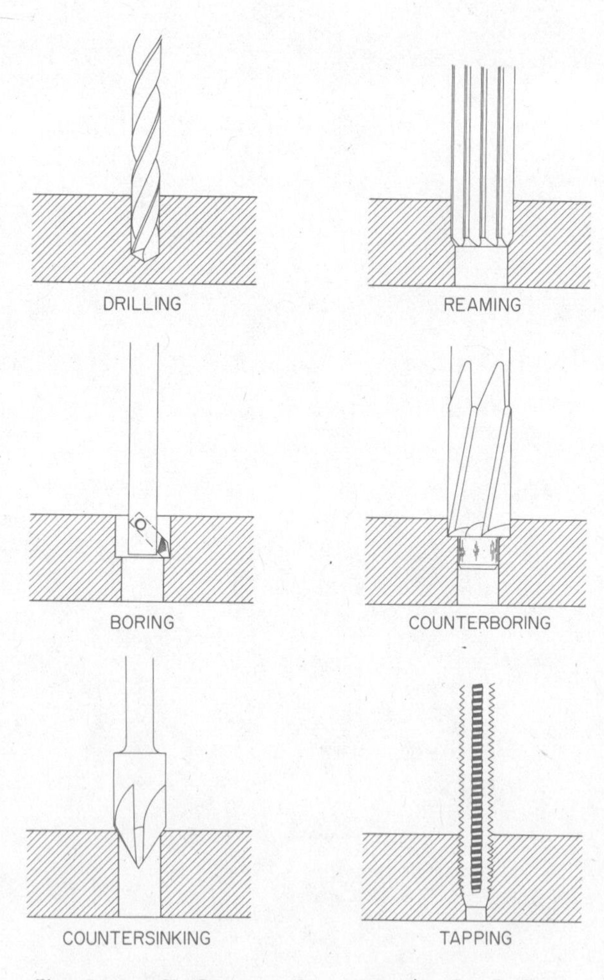

Fig. 54-1. Six Common Operations that Can Be Performed on a Drill Press

54-2. Reasons for Reaming

When a perfectly round hole of a certain diameter with straight and smooth walls is desired, the hole should first be drilled 1/64″

or 1/32″ smaller than the final diameter of the hole and then **reamed**.[1]

54-3. Kinds of Reamers

Reaming is done with a multiple-tooth cutting tool called a **reamer**, Fig. 54-2. The reamer is a **finishing tool** which machines holes straight and true to the exact size desired.

Reamers are available in a wide variety of types and sizes. They generally are made of carbon-tool steel or high-speed steel. They are also available with cemented carbide-tipped cutting edges. Carbon-tool steel reamers generally are satisfactory for hand reaming applications. High-speed steel reamers and reamers with carbide-tipped cutting edges wear longer and stay sharp longer for machine reaming applications.

Principal Kinds

The many types of reamers can be classified in two ways: those which are turned by hand, called **hand reamers**, and those used on machines, called **chucking** or **machine reamers**, Fig. 54-3. **Hand reamers** have a straight shank with a square end. They are turned with a tap wrench which fits over the square end, Fig. 54-4. **Chucking reamers** may have either tapered shanks with a standard Morse taper, or they may have plain, straight shanks which fit into collets or adapters on special production machines. Taper-shank reamers generally are used on drill presses and metalworking lathes. A taper-shank reamer, like a taper-shank drill, can also be installed in the tailstock of a metalworking lathe, Fig. 55-56.

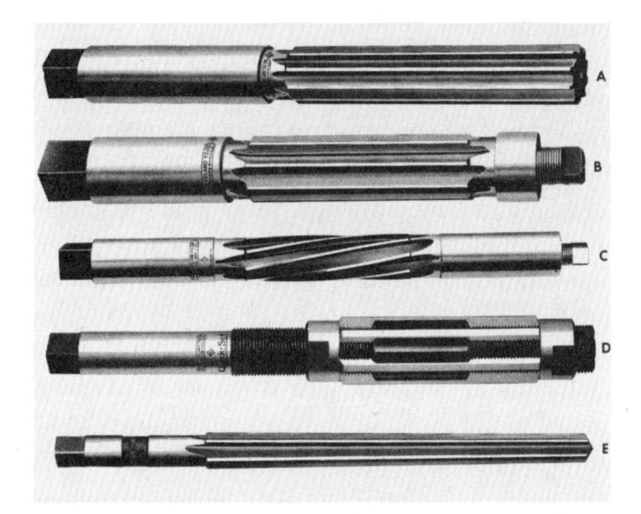

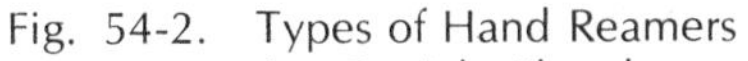

Fig. 54-2. Types of Hand Reamers
A. Straight-Fluted
B. Expansion, Straight Flute
C. Expansion, LH Helical Flute
D. Adjustable
E. Taper Pin

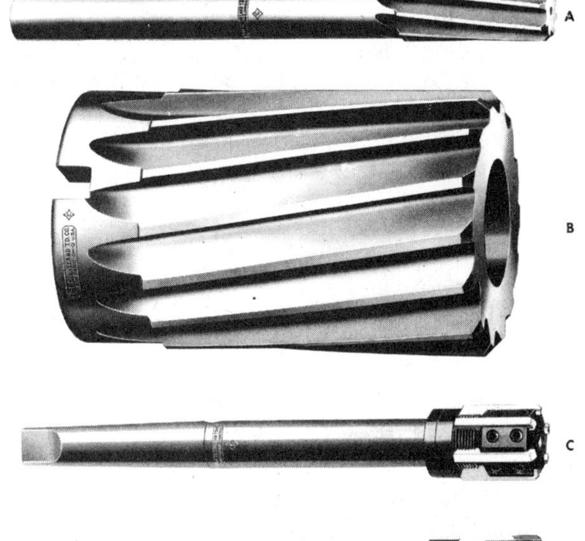

Fig. 54-3. Types of Machine Reamers
A. Straight-Shank
B. Helical-Fluted Shell
C. Adjustable
D. Carbide-Tipped Adjustable

[1]**Ream** comes from the German word **raumer** which means a person, tool, or instrument that cleans and makes room. Hence, the tool for cleaning out and enlarging a drilled hole is called a **reamer**.

Cutting Action

A drill cuts with its cutting edges on the end of the drill. Machine reamers are **end-cutting** reamers which have their cutting edges on the end, and they cut in much the same manner as a drill. The cutting edges are beveled at a 40° to 50° angle, usually 45°, at the end of the reamer.

Hand reamers, however, cut in a manner quite different from machine reamers. Hand reamers cut on the periphery (outside) at the tapered portion of the reamer. The flutes on solid-hand reamers are ground straight for the entire length, except for a portion which is ground with a **starting taper** at the end. The tapered portion does the cutting with a **scraping** action, very much like a hand-scraping tool. The length of the tapered portion of the reamer is generally about equal to the diameter of the reamer.

Flutes

The flutes on reamers may be **straight** or **helical**. Straight-fluted reamers work well for reaming average materials. Helical-fluted reamers are designed for reaming materials which are considered difficult to ream. The helical flutes aid in producing smoother and more accurate holes.

Solid Reamers

The solid reamers of both the hand and chucking types, Figs. 54-2 (**A**) and 54-3 (**A**), are made from one solid piece of steel. They are designed to produce a hole of one specific size. They cannot be adjusted to cut oversize or undersize.

Expansion reamers, Fig. 54-2 (**B**) and 54-2 (**C**), are designed so that the diameters can be expanded slightly to produce a hole slightly oversize for a desirable fit. This kind of operation is often necessary in assembly and maintenance work. The maximum amount of expansion varies according to the diameter of the reamer. It may range from 0.006″ for a ¼″ reamer to 0.012″ for a 1½″ reamer. The reamer is provided with an adjustment screw which is used for expanding the diameter of the reamer. It cannot be adjusted to produce undersize holes. An undersize pilot is provided on the end of some expansion reamers, Figs. 54-2 (**B**) and 54-2 (**C**), to aid in aligning the reamer for straight cutting. Expansion reamers may be either the hand or chucking type.

Adjustable reamers, Figs. 54-2 (**D**), 54-3 (**C**), and 54-3 (**D**), can be adjusted to produce holes of any size within the size adjustment range of the reamer. These reamers are available in either the hand or chucking type, in size ranges from ¼″ to about 3-11/32″ diameter. Hand reamers from ¼″ to 15/16″ diameter are available in size ranges in steps of 1/32″. A ¼″ reamer may be adjusted for holes from ¼″ to 9/32″, a 9/32″ reamer for holes from 9/32″ to 5/16″, and so on. Reamers larger than 15/16″ have adjustment size ranges in steps of 1/16″ or larger. The adjustable hand reamer in Fig. 54-2 (**D**) is provided with cutting blades which slide in precise, tapered slots. The diameter of the reamer is adjusted by loosening the nut on one end of the blades and tightening the nut on the other end, thus sliding the blades in the slots. When the blades become worn out they may be replaced with new blades without grinding.

Shell reamers are primarily machine reamers. They are available with straight flutes or with helical flutes as shown in Fig. 54-3 (**B**). They have a tapered hole which fits tightly on an arbor. Shell reamers of several sizes may fit the same arbor.

Other Reamers

See **taper-pin reamer** in section 26-25, **pipe-burring reamer** in section 27-9, and **pipe reamer** in section 27-12.

54-4. Care of Reamers

The **reamer** is a very fine tool and should be handled with the greatest of care. Particular pains must be taken to protect the **cutting edges**, for a small nick or **burr** will cause the reamer to cut oversize and also to scratch the inside wall instead of making a smooth finish. It is best to store each reamer in a separate wooden box or space in the tool cabinet.

54-5. Hole Size for Machine Reaming

The reamer should be used to make only a light, **finishing cut**. If a drawing calls for a 7/8″ reamed hole, then a drill 1/32″ smaller, or 27/32″ must first be used and the hole then reamed with a 7/8″ reamer. The difference between the sizes of the reamer and the drill, which is 1/32″, is the **allowance** for reaming.

The usual allowance for reaming is 1/64″ for reamed holes less than 1/2″ and 1/32″ for holes 1/2″ and over.

54-6. Hand Reaming

Hand reaming is performed when extreme accuracy is required. It is done with **hand reamers** which are specially designed for hand reaming. A cut of 0.002″ is usually recommended. **Never take cuts greater than 0.005″ with a hand reamer.** Hence, never allow more than 0.005″ material allowance. Holes which are to be hand reamed are generally drilled about 1/64″ to 1/32″ undersize first. They are then bored about 0.002″ to 0.005″ under the desired size of the reamed hole. Another method involves drilling and rough reaming to about 0.002″ undersize, followed by hand reaming. Rough reaming is done with a reamer intentionally ground about 0.002″ undersize, thus leaving an allowance of 0.002″ for hand reaming.

It helps, when hand reaming, to put a lathe center (see § 55-19) into the drill-press spindle and to hold it tightly against the small hole in the end of the reamer while turning the reamer with a tap wrench. It steadies the reamer and keeps it straight with the hole until it is started. A slow, steady, screwlike motion by the reamer with fast **feed** gives the best results. The reamer should never be forced or strained and should always be turned **clockwise**[2] even when removing it from the hole.

[2]**Clockwise** means the direction in which the hands of a clock turn.

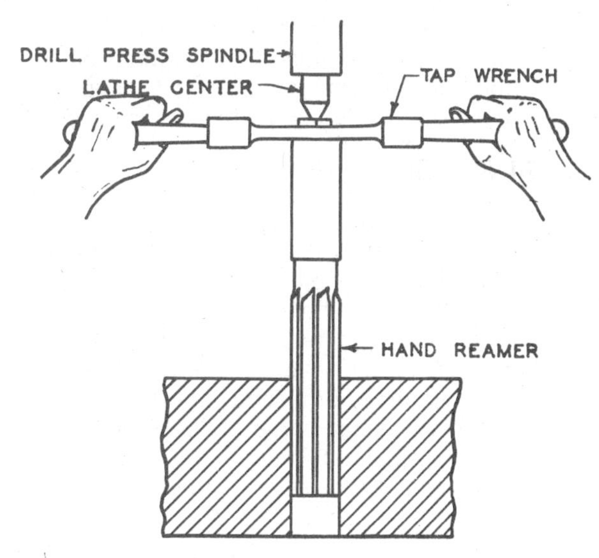

Fig. 54-4. Steadying a Hand Reamer with a Lathe Center in the Drill-Press Spindle (Hand Reaming)

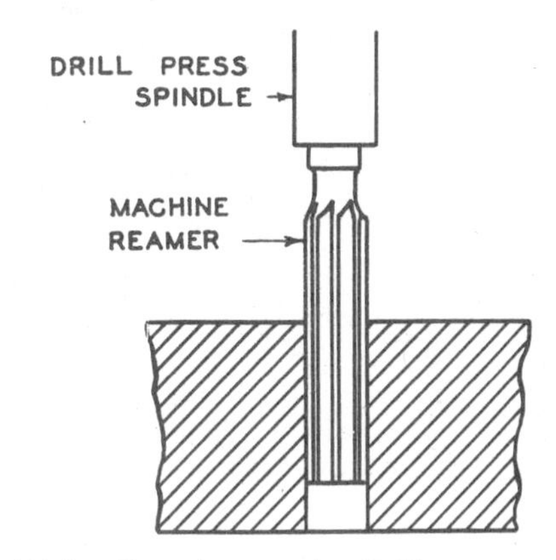

Fig. 54-5. Reaming on the Drill Press (Machine Reaming)

54-7. Machine Reaming

The **machine reamer** is used in the **drill-press spindle** just like a drill, Fig. 54-5. It must always be in a straight line with the hole. For this reason, it is best to remove the drill and insert the reamer without disturbing the setup of the work.

The **speed** for reaming should be much slower than for drilling, while the feed is usually faster than for drilling. Cutting fluid should always be used when reaming steel.

54-8. Countersinks

Countersinking is done with a **countersink.** There are several forms of countersinks. (See Fig. 54-6.) The difference in countersinks is the **angle** of the cutting edges. One kind has an angle of 82° for **flat-head screws**; the other kind, known as a **combination drill and countersink,** or **centerdrill**, has an angle of 60°. The centerdrill has a small drill point and is used mainly to countersink the ends of the work to be held between **centers** on the **lathe** (see § 55-19 and Fig. 55-40).

54-9. Countersinking

Countersinking on the drill press is shown in Fig. 54-6. The 82° **countersink** should run at a slow speed to avoid **chattering**, while the **centerdrill** should run at a high speed. Use cutting oil when countersinking steel.

54-10. Counterboring

Counterboring is done with a cutting tool known as a **counterbore** (see Fig. 54-7). It has a small end which leads or steers the tool into the hole and keeps it central; this small end is called a **pilot**.

Counterboring is done after the hole is drilled. (See Fig. 54-7.) The **speed** for counterboring should be slower than for drilling. The **pilot** should be oiled before entering the hole to keep it from getting rough. Use cutting oil when counterboring steel.

54-11. Reason for Counterboring

The heads of **fillister-head screws** are usually set down into the work. Making the top of a hole larger to receive the head of a fillister-head screw is called **counterboring.**

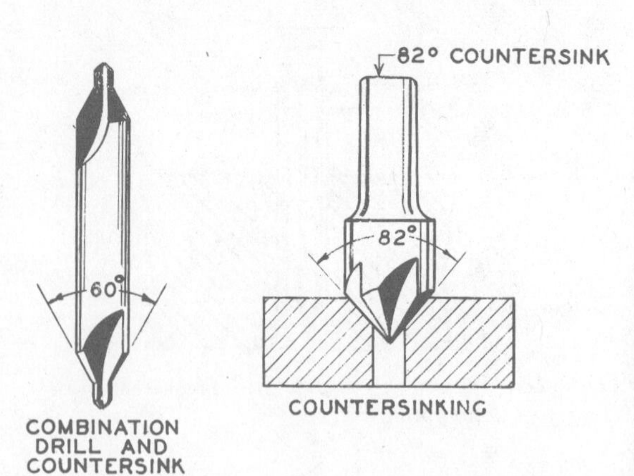

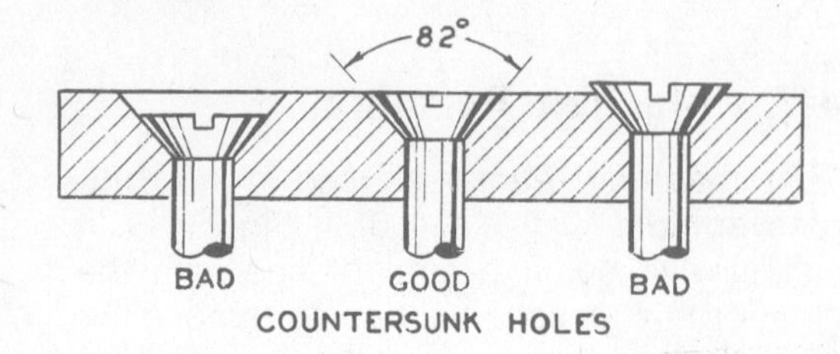

Fig. 54-6. Countersinking

Fig. 54-7. Counterboring for a Fillister-Head Screw

54-12. Spot Facing

Spot facing is somewhat like counterboring, except that in spot facing only a little metal is removed around the top of the hole. Only the surface is made smooth and square to form a flat **bearing surface** for the head of a **cap screw** or for a nut, Fig. 54-8. Spot facing is done with a **spot-facing tool** after the hole is drilled, Fig. 54-9.

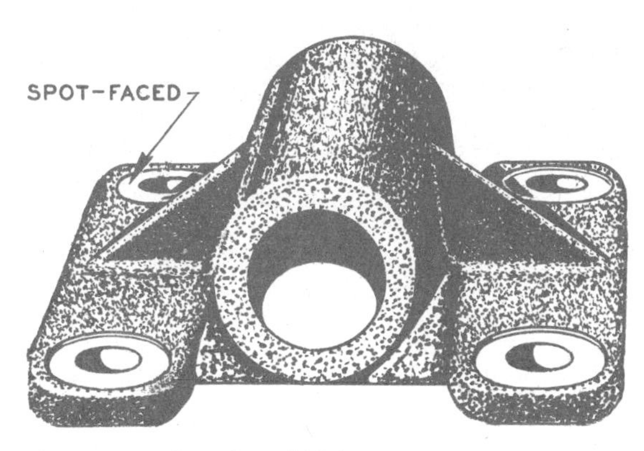

Fig. 54-8. Spot-Faced Holes

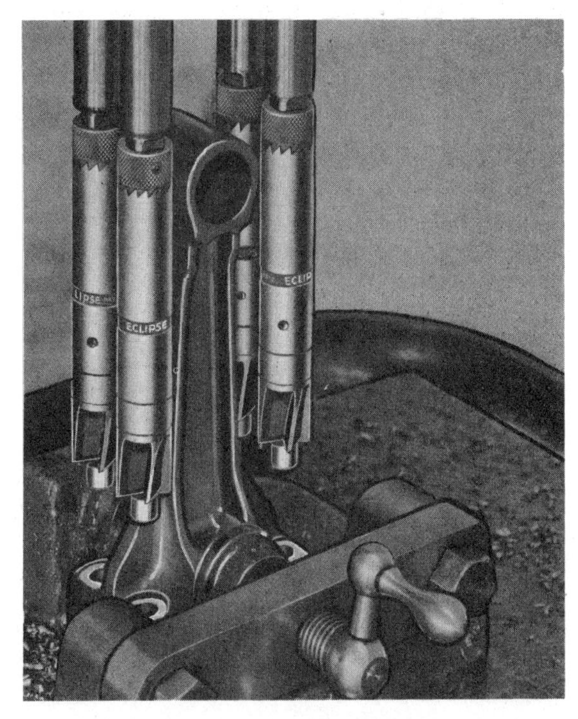

Fig. 54-9. Spot-Facing to Form a Bearing Surface for a Cap Screw

54-13. What Does Tapping Mean?

The forming of **screw threads** on the inside of a hole, as the threads in a **nut**, is called **tapping**. Since tapping is explained in Unit 23, only tapping in the drill press is explained here.

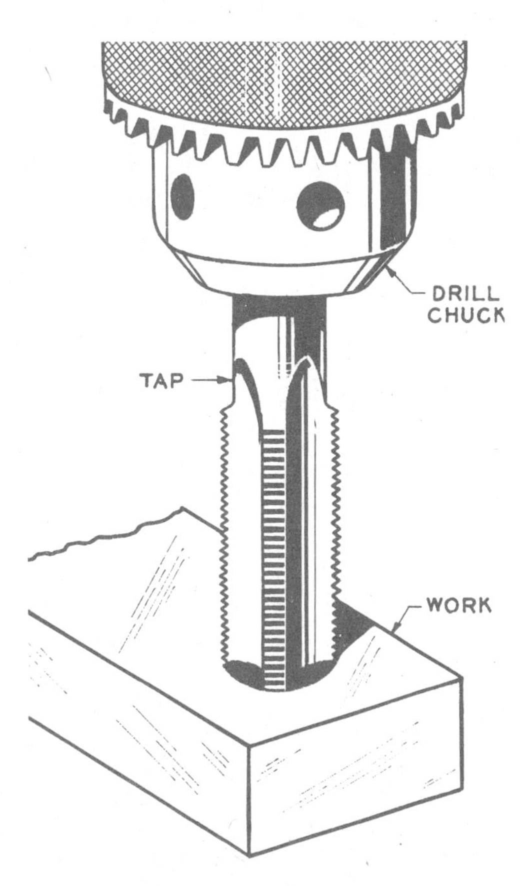

Fig. 54-10. Tapping on the Drill Press
This is an accurate method to use for alignment.

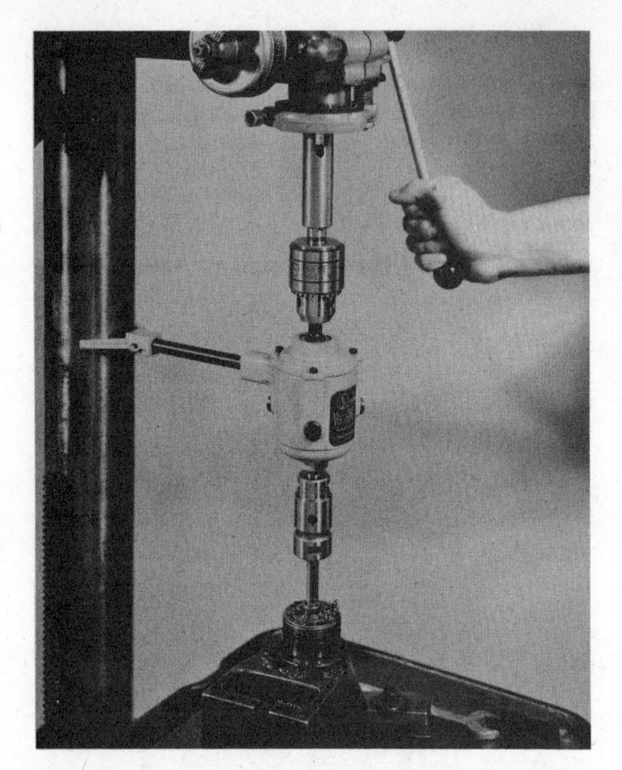

Fig. 54-11. Tapping Attachment Mounted on Drill Press

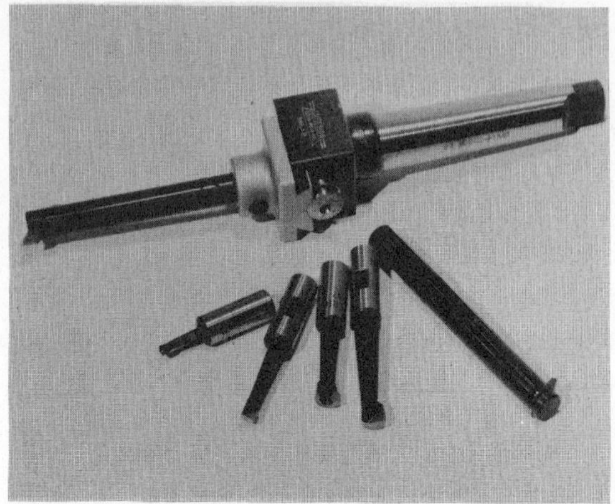

Fig. 54-12. Adjustable Boring Head and Boring Tools

54-14. Tapping in the Drill Press

Tapping in the drill press may be done by hand or with a **tapping attachment**.

For **hand tapping**, drill the required hole for the size tap to be used. Refer to Table 19 on tap drill sizes for this information. Without moving the workpiece, remove the tap drill from the drill chuck and replace it with the tap. Be sure the tap is either a starting tap or a plug tap. Apply a suitable cutting fluid and rotate the drill chuck **by hand** while applying pressure with the drill press feed lever.

After three or four threads have been cut, release the tap from the drill chuck, attach a tap wrench, and finish tapping the hole. This method of hand tapping assures good alignment of the tapped threads with the hole.

Tapping with Power

A **tapping attachment** may be mounted on a drill press as shown in Fig. 54-11. The tap is mounted in a collet chuck which is provided in the tapping attachment. The workpiece to be tapped must be mounted securely in a vise. The vise should also be clamped or bolted down to the drill press table. The tap should be carefully aligned with the hole so that the thread will be tapped straight.

The tapping attachment in Fig. 54-11 is nonreversing. **It does not require that the drill press spindle be reversed to extract the tap** after tapping the hole. With this type of attachment, the tap enters the hole with right-hand rotation as pressure is applied downward on the drill-press feed handle. When the hole is tapped to depth, pressure on the feed handle is released and the tap stops rotating, even though the drill press spindle continues to rotate. The tap is extracted from the hole by applying upward pressure on the spindle with the drill-press feed handle. The upward pressure causes the tap to rotate in a reverse direction, opposite to the spindle rotation.

Some tapping attachments used on drill presses are reversing. With this type, the drill press spindle must be reversed in order to extract the tap after tapping to the desired depth.

Only spiral ground and helical-fluted taps should be used with tapping attachments.

54-15. Boring

Boring, Fig. 54-1, is performed to produce a very straight hole to accurate size. When a drill or reamer of the proper size is not available, a hole may be bored to accurate size instead. For ordinary boring, the hole is drilled from 1/16″ to 1/8″ undersize and is then bored to the desired size.

Boring is performed with a single-point cutting tool as in Figs. 54-1 and 54-12. In Fig. 54-12 a **boring bar** is inserted in a **boring head** which may be inserted in the tapered spindle of a drill press or milling machine. To be done successfully, boring must be done at low rpm with automatic feed, and the workpiece must be clamped solidly to the machine table. The boring head can be adjusted to bore holes accurately to 0.001″ (one thousandth of an inch), or closer. The **cutting-tool bit** mounted in the boring bar in Figs. 54-1 and 54-12 is held in place by a setscrew. The tool bit can be resharpened or replaced when worn out. Holes may also be bored with similar boring tools on a metalworking lathe, as shown in Figs. 55-57 and 55-58.

Words to Know

adjustable reamer
cap screw
chucking reamer
counterbore
counterbore pilot
countersink
expansion reamer
fillister-head screw
finishing cut
hand reamer
machine reamer
reaming
reaming allowance
shell reamer
spot facing
tap drill
taper pin reamer
tapping attachment

Review Questions

1. Name several reasons why reamed holes are sometimes more desirable than drilled holes.
2. Name several kinds of hand reamers.
3. Name several kinds of machine reamers.
4. What is the difference in construction between a machine reamer and a hand reamer?
5. How much metal should be left for the hand reamer to cut away? Machine reamer?
6. Explain how a machine reamer cuts. Does a hand reamer cut the same way?
7. In what way is an expansion reamer different from an adjustable reamer?
8. What is meant by countersinking?
9. What are the included angles of countersinks?
10. What is the difference between countersinking and counterboring?
11. What is the difference between counterboring and spot facing?
12. Explain how hand tapping is done in the drill press.
13. How can you find out what size tap drill to use?
14. Explain what boring is and how it is done on a drill press.

UNIT 55

The Lathe and Lathe Operations

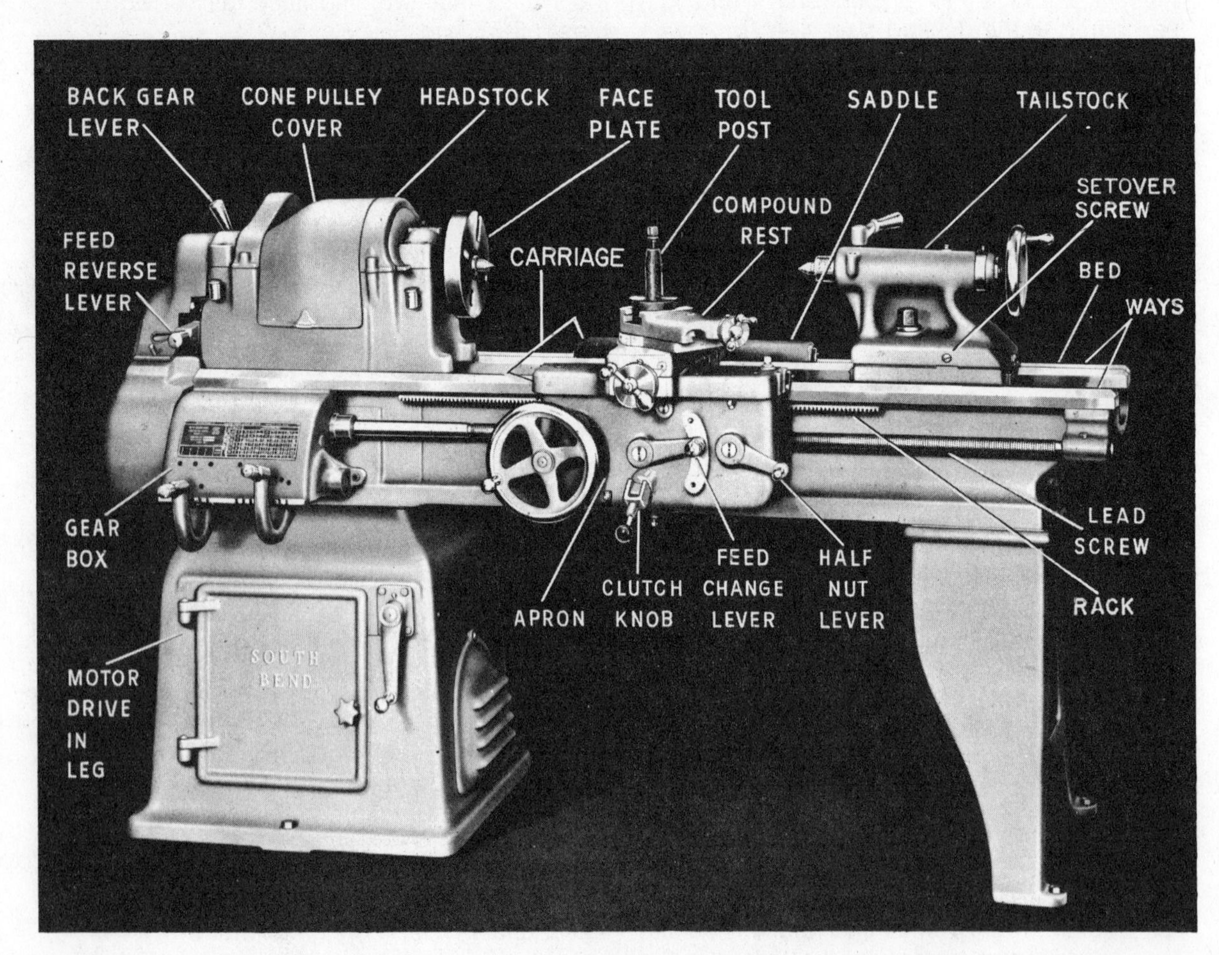

Fig. 55-1. Lathe and Its Parts

55-1. The Lathe as a Machine Tool

The metalworking lathe, Fig. 55-1, is a **machine tool**. A **machine tool** is a machine which is used for cutting metal; it holds both the workpiece and the cutting tool. There are many kinds of machine tools. The most common kinds of machine tools are often called **basic machine tools**. Five of the most basic kinds of machine tools include drill presses, lathes, shapers and planers (Fig. 56-1), milling machines (Fig. 57-1), and grinding machines (Figs. 58-2, 58-14 and 58-18.)[1]

Most of the many other kinds of machine tools are specialized, mass production machine tools. These are modifications or adaptations of one or more of the basic machine tools listed above. For example, specialized production machine tools which are adapted from the lathe include turret lathes, hand screw machines, Fig. 55-2, automatic screw machines, Figs. 55-3 and 55-4, and chucking machines. Hand screw machines are small, hand-operated turret lathes. Chucking machines, some of which are automatic, have no tailstock and are made for turning parts of short length which must be held by a chuck. Each of these machines can perform only some of the operations which can be performed on a lathe. However, the specialized production machines perform these operations more rapidly and efficiently.

It is important that you learn about the basic kinds of machine tools, including the drill presses, lathes, shapers, milling machines, and grinding machines. When you understand the kinds of setups, the kinds of cutting tools, and the kinds of operations which can be performed on these machines, you can apply this knowledge and experience to specialized production machine tools. The principles involved in performing operations on the basic kinds of machine tools also apply to performing these same operations on modern, specialized mass-production machine tools.

55-2. Why the Lathe Is Important

We are living in the age of the machine. The airplane, steamship, locomotive, electric motor, computer, transfer machine, and automobile are among the outstanding inventions. All these wonders were made possible through the development of the lathe. It is the oldest and most important machine tool in industry.

Fig. 55-2. Hand Screw Machine Used for Short Production Runs of Small Parts

Fig. 55-3. Closeup of Machining Area of a Turret-Type Single-Spindle Automatic Screw Machine

[1]Some illustrations in this unit are reprinted by arrangement with South Bend Lathe, Inc. from **How to Run a Lathe.** (copyrighted, all rights reserved)

Fig. 55-4. Eight-Spindle Automatic Screw Machine Used for Very High Production Volume

The modern lathe can perform many different operations. In studying the history of other machines, we find that the idea of their operation comes from the lathe. There are more lathes in this country than any other kind of metalworking machine tool.

The lathe, Fig. 55-1, performs many kinds of external and internal machining operations. **Turning**, Fig. 55-50, is the most common external machining operation performed on a lathe. Turning can produce either straight, curved, or irregular cylindrical shapes. (Also see Fig. 55-29.) Other external machining operations performed on a lathe include knurling, Fig. 55-61, and thread cutting, Fig. 55-67. The lathe can also perform most of the internal hole-machining operations which are normally performed on a drill press. These include drilling, centerdrilling, countersinking, counterboring, boring, and reaming. (See Figs. 54-1, 55-40, 55-56, and 55-58.)

Only a few of the most common turning, boring, hole machining, and thread machining operations are explained in this unit.

55-3. Who Runs the Lathe?

A person who earns his living by doing different operations on a lathe is a **lathe operator**. The **machinist** makes parts for machines on the lathe. The **toolmaker** makes parts of tools on the lathe. The **jeweler** makes small parts of jewelry on a small lathe which is called a **jeweler's lathe**. The **metal patternmaker** makes parts for metal patterns on a lathe.

55-4. Lathe and Its Parts

Before attempting to operate a lathe you should become familiar with the principal component parts, units, and controls on the lathe. These are named in Figs. 55-1 and 55-6. By studying Fig. 55-5 carefully, you can understand the function of each major assembly unit in relation to other major parts on the lathe. The major parts may be designed somewhat differently by individual manufacturers; however, the basic principles which apply to their function are much the same for all lathes.

The controls on different lathes may also be designed somewhat differently. However, all lathes are equipped with similar controls, and these controls perform similar functions. When you have learned the names of the principal parts, parts units, and controls, you can learn to operate the lathe without damaging the machine or injuring yourself.

LINE of POWER

Electrical energy is turned into working power by the motor, then transferred efficiently through the V-belt drive to

(1) Rotate work

(2) Move the cutting tool

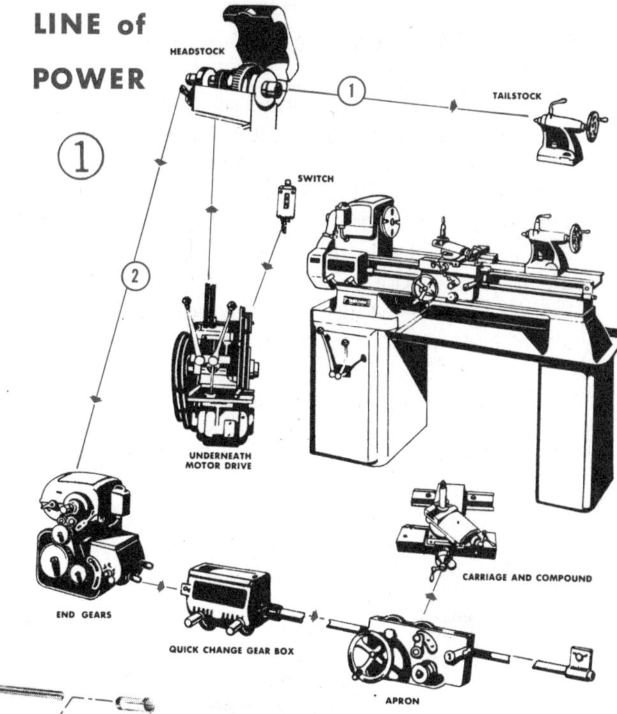

② **HOLDING and ROTATING WORK**

HEADSTOCK

Headstock supports spindle which rotates on "Zero Precision" tapered roller bearings. Work holders are mounted on spindle nose.

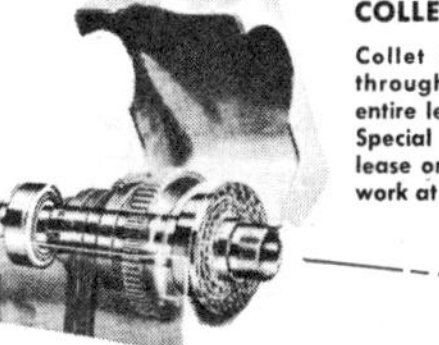

COLLET ATTACHMENT

Collet attachment passes through the hole through entire length of the spindle. Special jaws on collet, release or grip small diameter work at the spindle nose.

LATHE DOG, DOG PLATE AND CENTERS

The lathe dog clamps around the work piece. The dog plate mounts on the spindle. As work piece is placed on the spindle nose center, the tail of the lathe dog is slipped into a slot in the dog plate so that when dog plate revolves it turns the work piece.

TAILSTOCK

Tailstock center supports right end of work held "between centers." It can be offset to cut tapers, locked in any position along lathe bed, and has hand-wheel feed for tailstock tools.

CHUCK

Chuck mounts on spindle nose. Adjustable jaws permit holding of larger diameter, odd-shaped or stub-end work.

③ **HOLDING and MOVING TOOL**

END GEARS

Outboard gear on spindle drives end gear train which operates lead screw through gear box.

QUICK CHANGE GEAR BOX

Double tumbler levers permit rapid selection of desired ratio (Pitch and feed) between spindle r.p.m. and lead screw r.p.m.

CARRIAGE AND COMPOUND

Carriage provides rigid support for cross-slide and travels — either to the right or left along the bed. Cross slide moves compound in or out with power feed or handwheel. Compound swivels to provide angular feeds.

APRON (FRONT VIEW)

Apron controls are centrally grouped with selector lever for power longitudinal and cross feeds, friction clutch for engaging feeds, half-nut lever for thread cutting, and hand wheel for hand traverse of carriage. Built-in safety mechanisms prevent engaging half-nuts and power feeds at the same time.

DETAIL FROM APRON (REAR VIEW)

Lead screw transmits power through apron by (1) spline drive for power feeds and (2) by half nuts for thread cutting. Precision lead screw threads are used only for thread cutting.

Fig. 55-5. How a Modern Lathe Operates

The **lathe bed** is the long part which rests on four legs. The **headstock** is fastened to the left end of the bed, and the **tailstock** can be clamped at any point along the bed. The **carriage** is the part which slides back and forth on the bed between the headstock and the tailstock. The V-shaped tracks of the bed upon which the carriage and tailstock slide are called **ways**.

55-5. Size of Lathe

Lathes range in size from tiny **jeweler's lathes** to large machines. A small lathe that sets on a bench is a **bench lathe**.

Figure 55-7 shows that the size of a lathe is measured by:

1. **Swing**
2. **Length of bed**

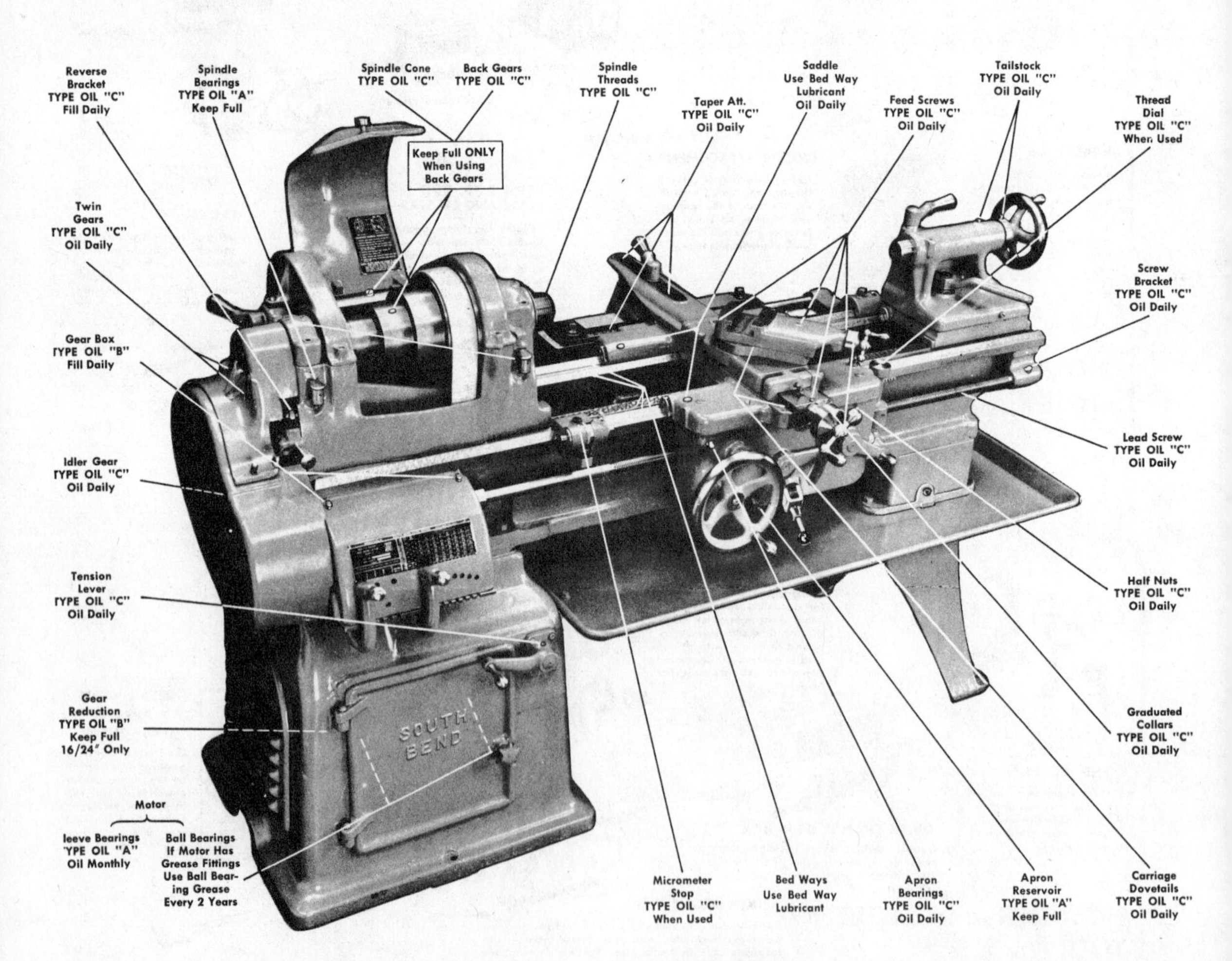

Fig. 55-6. Oiling the Lathe

The **swing** is measured by the largest diameter of work that can be turned in the lathe. For example, a piece 10″ in diameter is the largest work that can be turned in a 10″ lathe. Thus a 10″ × 5′ lathe has a 10″ swing and a bed 5′ long.

55-6. Getting Acquainted with the Lathe

Get acquainted with the lathe by first cleaning it thoroughly and then oiling it. The power should be turned off before this is done so there will be no danger of getting caught in a running lathe.

Move the different handles and levers without power to find out what happens. Parts should move easily, without force. Slide the tailstock to the end of the **lathe bed**. Then move the carriage by hand to about the center of the bed. Next, turn the lathe spindle, Fig. 55-6, by hand to see if any part of the lathe is locked.

Safety Note

If the lathe is turned on while the spindle or any part of the lathe is locked, the lathe may be damaged.

The spindle may have been locked by the previous operator in order to install or remove a face plate or chuck. Ask your instructor to explain how to unlock the spindle, or free any other part of the lathe which is locked, before you start the lathe.

Next, request permission from your instructor to turn on the power. Then, carefully put the lathe through its movements.

Safety Note

Most lathes must be stopped while shifting, adjusting, or changing any of the controls on the headstock end of the machine.

Check with your instructor to see if any of the controls on the headstock end of your lathe may be changed while the lathe is running. The controls on the apron or carriage generally can be moved while the machine is turned on and running. The proper methods for determining the cutting speed, determining the rpm, and setting the rpm of the spindle are explained in section 55-8.

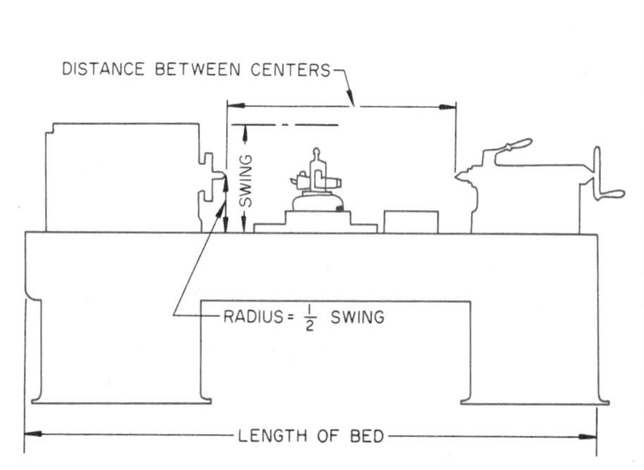

Fig. 55-7. Size of a Lathe

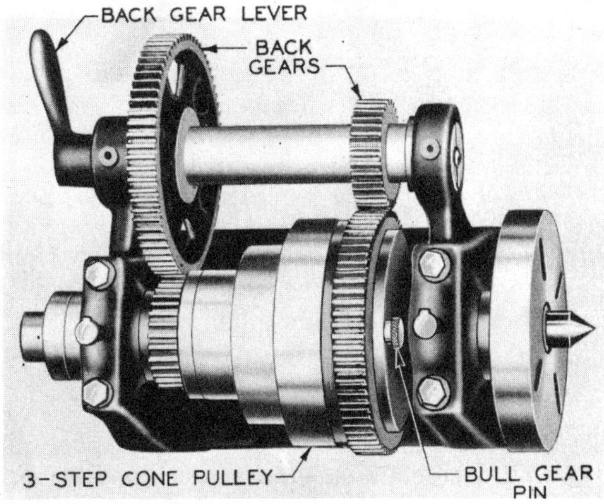

Fig. 55-8. Cone Pulley and Back Gears

Fig. 55-9. Lathe with Flat-Belt Drive

The depth of cut, suggested feeds, and the procedures for using automatic feeds are explained in section 55-9.

55-7. Oiling the Lathe

The lathe should be oiled every day with good **lubricating oil**. (See § 48-2 and Table 32 for grades and comparative brands.) The places where the lathe should be oiled are shown in Fig. 55-6. Consult the chart for your lathe, since requirements may vary. Some lathes have lubrication charts mounted on the machine. Only a drop or two of oil in each **oil hole** is necessary. Keep the **ways** clean and oiled. Wipe off all wasted oil.

55-8. Cutting Speed and rpm

Cutting speed on a lathe is the speed at which the circumference of the work passes the tool bit. It is expressed as surface feet per minute (sfpm). If a continuous chip were cut on a lathe for one minute, the length of the chip would be the cutting speed.

Cutting speed is related to **rpm** (revolutions per minute). On a lathe, rpm means the number of revolutions of the work in one minute. For a stated cutting speed, work of large diameter should run at a lower rpm than work of small diameter.

Soft metal generally should be machined at a higher cutting speed than hard metal. A **rough cut** is made at a lower cutting speed than a **finish cut**. Different metals are cut at different cutting speeds. Suggested cutting speeds for various metals are listed in Table 36.

The cutting speed or rpm for lathe work can be determined with the following formulas:

D = Diameter of work in inches
Pi = Constant 3.1416 (Round off Pi to 3 for quicker calculations)
CS = Cutting speed in surface feet per minute
rpm = Revolutions per minute

$$CS = \frac{D'' \times Pi \times rpm}{12}$$

$$rpm = \frac{CS' \times 12}{D'' \times Pi}$$

Recommended cutting speeds for each kind of steel are included within a range. For example, a recommended cutting speed for turning ordinary low-carbon steel is from about 90′ to 100′ per minute, see Table 36. Hence, it is not necessary to calculate cutting speeds or rpm to exact decimal figures. Therefore, most mechanics substitute the figure 3 instead of 3.1416 for Pi when calculating approximate cutting speeds or rpm for machining metals.

Example: Calculate the approximate cutting speed for stock 1″ diameter in a lathe, revolving at 344 rpm.

$$CS = \frac{1 \times \overset{1}{\cancel{3}} \times 344}{\underset{4}{\cancel{12}}} = \frac{\overset{86}{\cancel{344}}}{\underset{1}{\cancel{4}}}$$

$$CS = 86 \text{ feet per minute}$$

Example: Calculate the approximate rpm for a piece of steel 1¼″ diameter which is to be machined at 90 feet per minute.

$$rpm = \frac{90 \times \overset{4}{\cancel{12}}}{1.25 \times \underset{1}{\cancel{3}}} = \frac{360}{1.25}$$

$$rpm = 288$$

The above formulas also may be used for calculating the cutting speeds for drilling and milling. In these cases, **D** is the diameter of the drill or the milling cutter.

The cutting speed or rpm may also be determined by consulting a **table of cutting speeds**. (See Table 39, p. 585.) These tables are available in technical handbooks and wall charts.

The **cone pulley** and **back gears**, Fig. 55-8, on the headstock make it possible to run the lathe at different speeds. The back gears are fastened to a shaft behind the cone pulley. A lathe with a **step cone pulley** and back gears has eight different speeds. (See Fig. 55-9.) Each of the four steps of the cone pulley gives a different speed, making four speeds; by using the back gears, four slower speeds are obtained, making eight different speeds in all. When the back gears are not used, the **bull-gear pin** (see Fig. 55-8) must be engaged. When the back gears are used, the bull-gear pin must be disengaged. The back gears can be connected with the cone pulley by shifting the **back-gear lever** while the lathe is stopped.

Some lathes have V-type step pulleys which are driven by a V-belt instead of a flat belt, Fig. 55-10. When spindle speeds are changed with this type of pulley, the tension on the belt is first released with a release lever provided for this purpose. The belt then is moved manually to the desired step on the pulley.

Fig. 55-10. Manual-Shift V-Belt Drive

The lathe in Fig. 55-5 has a lever-shift V-belt drive system. Speed changes with this drive are made by shifting either or both of the speed change levers to the desired position. This system provides four speeds in direct drive and four speeds in back-geared drive.

Some lathes have a **variable-speed drive** system. This kind of belt drive system also is used on drill presses and some other kinds of machines. (See Fig. 49-2 and § 49-2). Drive systems of this type are infinitely variable. This means that any desired rpm within the speed range of the lathe may be selected.

Safety Note

With most V-belt variable speed-drive systems changes in speed can be made only while the lathe is running.

With a **geared-head drive** system, as on the lathe in Fig. 55-11, changes in speed are made by shifting the speed-change levers. This involves shifting gears, not belts, see Fig. 55-12. On most head-geared drive systems, however, the lathe must be stopped in order to change speeds.

Safety Note

If the gears are shifted while the lathe is running, serious gear damage may result.

55-9. Carriage Feed, Depth of Cut, and Threading Mechanism

The lathe is designed so that the cutting tool may be fed manually or automatically along the work while machining. The feed is called **longitudinal feed** when the tool is fed along the work, parallel to the lathe bed. The longitudinal feed is used for operations such as **turning**, Fig. 55-50, and boring, Fig. 55-58. The feed is called **cross feed** when the tool is fed across the end of the workpiece, as in facing operations, Fig. 55-38.

Fig. 55-11. Modern-Geared Head Lathe

Fig. 55-13. End-Gear Train

Fig. 55-12. Headstock with All-Geared Drive

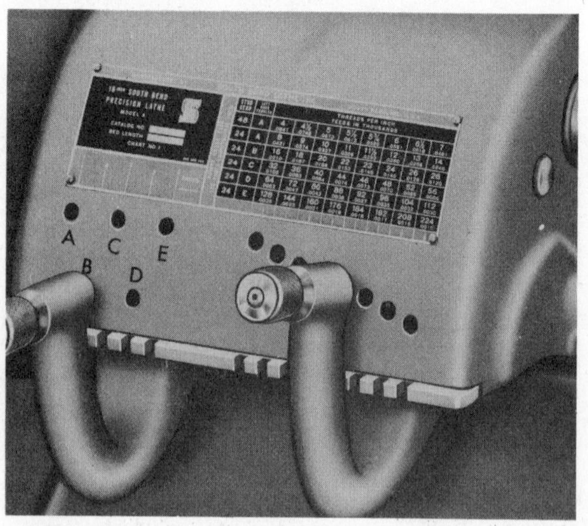

Fig. 55-14. Quick-Change Gear Box with Chart

The amount of feed is the distance that the tool moves along the workpiece during one revolution of the work.

The amount of feed, including both longitudinal and cross feed, is controlled through the use of the feeding and threading mechanisms on the lathe. These mechanisms include the end gears, Fig. 55-13; the quick-change gear box, Fig. 55-14; and the **carriage and apron assembly,** as shown in Fig. 55-15.

The **end gears** transmit power from the lathe spindle to the lead screw through the **gear box**. The lead screw transmits power to the carriage and apron assembly. (See Fig. 55-5.)

The controls on the carriage and apron, Fig. 55-15, affect all movements of the tool and the carriage. The **apron-handwheel** is used for manual, longitudinal movement of the carriage. The **cross-feed knob** is used for manual cross feed of the tool. The **compound-rest knob** is used to feed the tool manually with the compound rest. The **carriage-lock screw** may be tightened to lock the carriage and thus prevent longitudinal movement for operations such as facing.

The compound rest is normally set at an angle of about 29° from the crosswise position, as in Fig. 55-15. This is the angle at which it is set for cutting Unified and American (National) Screw threads (Fig. 55-72). It also is a convenient angle for performing a majority of other lathe operations. The compound rest may be swiveled and set at any desired angle for turning or boring tapers. (See Fig. 55-52.)

The **feed-change lever** is used to select any of the following feeds: longitudinal feed, cross feed, and feeds for threading. For longitudinal or cross feeding, the lever is moved to the upper or lower position as desired. For thread-cutting feeds, the feed change lever is located at the center position as shown in Fig. 55-15.

The **clutch knob**, Fig. 55-1, (also called the **automatic-feed knob**, Fig. 55-15) is used to engage or disengage automatic longitudinal or cross feed. The **half-nut lever,** Fig. 55-15, is used to engage the feed for thread-cutting operations only. The **feed-reverse lever**, Fig. 55-1, is used to reverse the direction of the lead screw, and thus reverse the direction of either longitudinal or cross feeds.

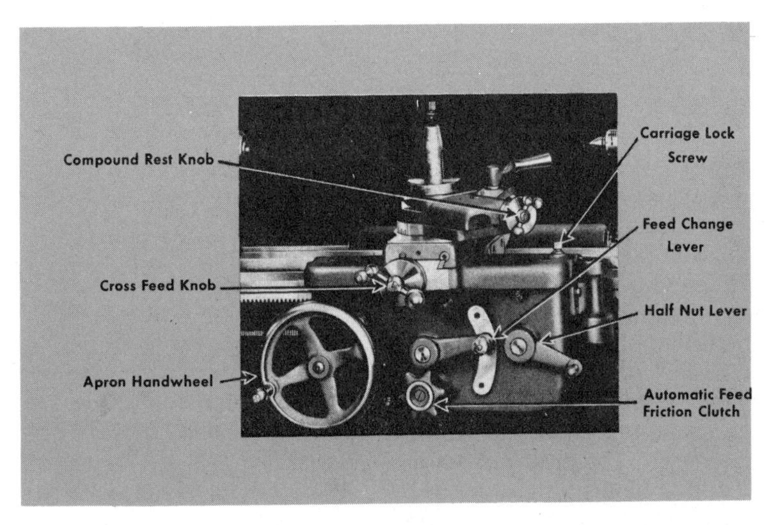

Fig. 55-15. Parts of Lathe Carriage and Apron Assembly

SOUTH BEND LATHE WORKS — SOUTH BEND, IND. U. S. A.

POWER CROSS FEED .375 TIMES LONGITUDINAL FEED

STUD GEAR	LEFT HAND TUMBLER	THREADS PER INCH FEEDS IN THOUSANDTHS							
48	A	4 .0841	4½ .0748	5 .0673	5½ .0612	5¾ .0585	6 .0561	6½ .0518	7 .0481
24	A	8 .0421	9 .0374	10 .0337	11 .0306	11½ .0293	12 .0280	13 .0259	14 .0240
24	B	16 .0210	18 .0187	20 .0168	22 .0153	23 .0146	24 .0140	26 .0129	28 .0120
24	C	32 .0105	36 .0093	40 .0084	44 .0076	46 .0073	48 .0070	52 .0065	56 .0060
24	D	64 .0053	72 .0047	80 .0042	88 .0038	92 .0037	96 .0035	104 .0032	112 .0030
24	E	128 .0026	144 .0023	160 .0021	176 .0019	184 .0018	192 .0017	208 .0016	224 .0015

Fig. 55-16. Index Chart for Quick-Change Gear Lathe

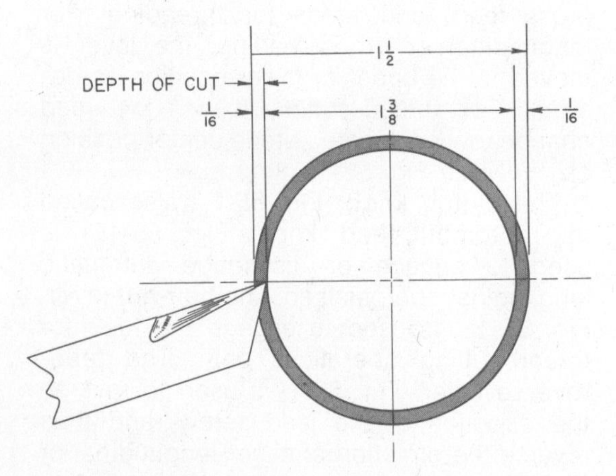

Fig. 55-17. The Diameter Is Reduced Twice the Depth of the Cut When Turning

The levers on the quick-change gear box, Fig. 55-14, are used for the selection of feeds. An **index plate**, Fig. 55-16, is located on the quick-change gear box. Note that there are large numbers and small decimal numbers on the index plate. The large numbers indicate the pitch or number of threads per inch for threading operations only. Hence, if the large number 16 were selected, the lathe would feed longitudinally 1/16″ for each revolution of the workpiece. The carriage would travel 1″ while the workpiece revolved 16 times. Thus, 16 threads would be cut for 1″ of length. The **half-nut lever** must be used for threading operations, and it will function only when the **feed-change lever** is in the center position.

The small decimal numbers on the index plate indicate the amount of feed per revolution for longitudinal and cross feeds. It should be noted that the amount of cross feed is not always the same as the amount of longitudinal feed. This varies on different lathes. For the lathe represented in Fig. 55-14, the cross feed is equal to 0.375 times the longitudinal feed. Thus, if a coarse feed of 0.0105″ were selected, the cross feed would equal 0.375 × 0.0105″, or 0.004″ per revolution of the workpiece. The ratio of cross feed to longitudinal feed generally is indicated on the index plate located on the quick-change gear box, Fig. 55-16.

The wheels, levers, and knobs used for the selection and control of feeds and speeds are designed somewhat differently for different lathes. Therefore, it is always best to have your instructor show you how these controls work before operating the lathe. This information is often available in the **maintenance manual** which generally is delivered by the manufacturer with each new lathe.

Feed Selection

The amount of feed varies with the type of metal and the **depth of cut**. Soft metals are generally cut with coarser feeds than hard metals. Finishing cuts are made with finer feeds than roughing cuts. Suggested feeds for machining steel would be .005″ for a rough cut and .002″ for a finish cut.

Depth of cut means the distance the tool is advanced into the work at a right angle to the work. It is the depth measured between the machined surface and the work surface, Fig. 55-17. This definition is true for all machine tools. A deeper cut may be taken on soft metals than on hard metals. The depth of a finishing cut is less than the depth of a roughing cut. On small lathes a common rough cut depth in steel would be .060″ to .125″. Finishing cuts are usually .010″ to .020″ on any lathe. Cuts less than .010″ deep generally are too small to be accurate. A .010″ cut removes .020″ from the diameter. A 1/16″ cut removes ⅛″ from the diameter, see Fig. 55-17.

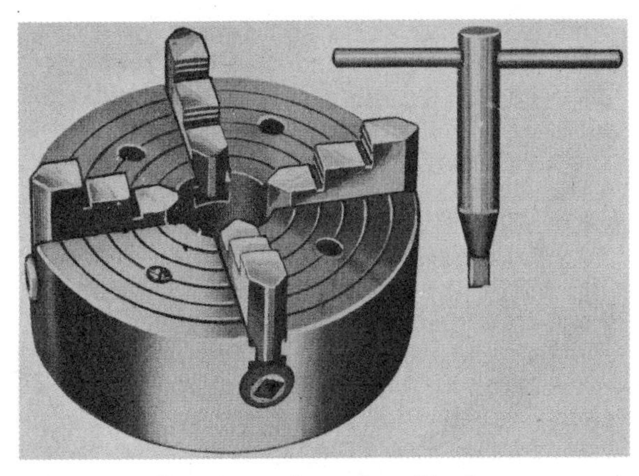

Fig. 55-18. Four-Jaw Independent Chuck

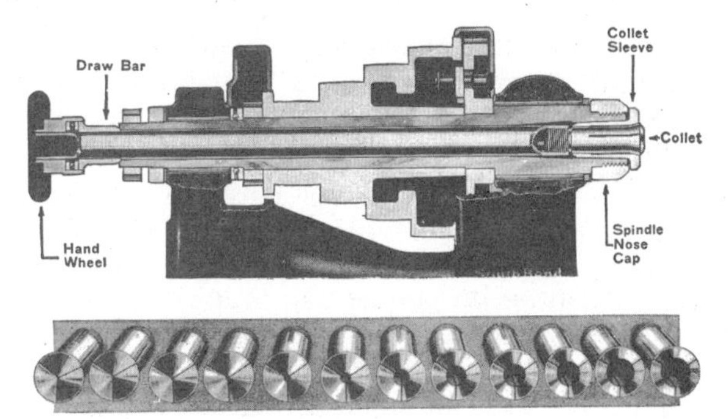

Fig. 55-20. Cross Section of Collet Chuck and Set of Collets

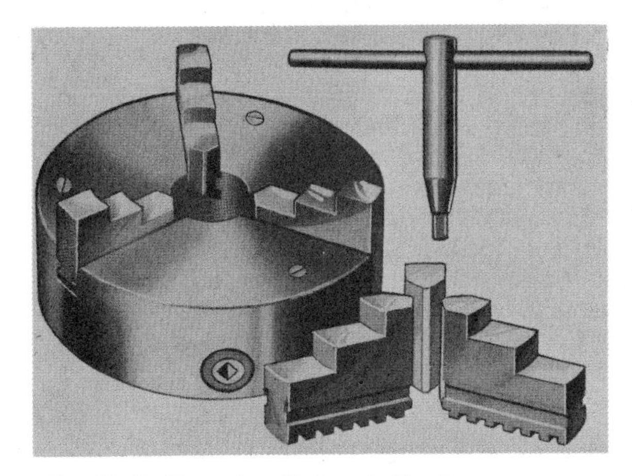

Fig. 55-19. Three-Jaw Universal Chuck

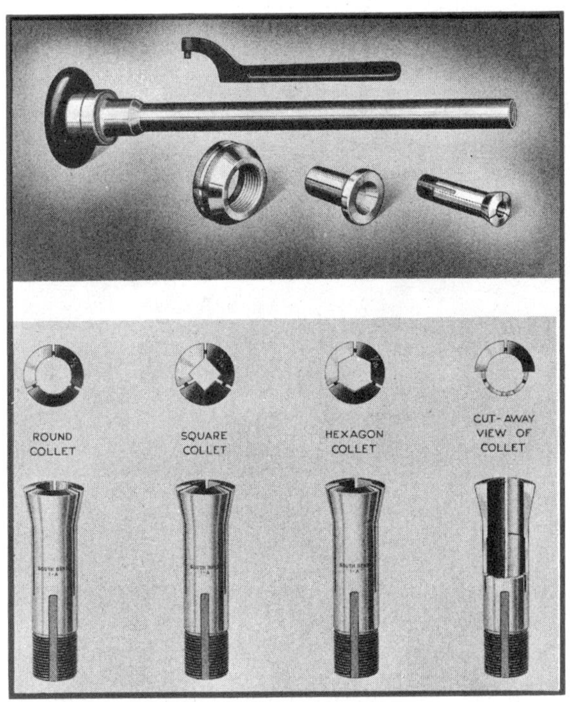

Fig. 55-21. Collet Chuck and Types of Collets

55-10. Lathe Chucks

A lathe chuck is used to hold work. It should be screwed on or off the **headstock spindle** (see Fig. 55-44) while the lathe is stopped. There are three principal kinds of lathe chucks: the **four-jaw independent chuck,** Fig. 55-18; the **three-jaw universal chuck**, Fig. 55-19; and **collet chucks**, Figs. 55-20 and 55-21.

The **four-jaw independent chuck** has four **jaws** and each must be moved separately with a **chuck key**, also called a **chuck wrench**; it is used chiefly to hold work that is not perfectly round. It may be used to hold work which is round, square, rectangular, or irregular in shape. The jaws on this kind of chuck are **reversible**; that is, they can be taken off and put on again in the opposite direction. They have steps so that different sizes of work can be held as shown in Figs. 55-52 and 55-58.

The four-jaw chuck is more accurate than a three-jaw chuck because the work can be centered in the chuck exactly with a dial indicator, Fig. 55-26. However, even with considerable experience, it takes much more time to center work accurately in a four-jaw chuck than in a three-jaw universal chuck.

The **three-jaw universal chuck** has three jaws that work at the same time. Thus when one jaw is screwed down upon the work with the chuck key, the other two jaws move also. This kind of chuck usually has two sets of jaws; one set holds the larger work while the other set holds rings and small work (see Figs. 55-38, 55-40, and 55-61). Three-jaw universal chucks center work accurately to within 0.002″ to 0.003″. Generally, they retain this accuracy until either the jaws become worn or the threads on the spindle nose become nicked or damaged.

Collet chucks of the spring collet-type, Fig. 55-21, are made to hold work which is close to a specific diameter. A spring collet should be used only for holding work which is within about 0.005″ of the designated size of the collet. They are commonly used for work smaller than 1″ in diameter.

There are several types available. A common one is the **draw-in type** which fits into the nose of the lathe spindle with a handwheel draw bar. Collets are available in sizes by 1/64ths, up to 1-1/16″ diameter. Collet chucks are very accurate, save time in mounting work, and speed up production.

There are two other kinds of spindles: the long, taper-key drive spindle shown in Fig. 55-22 and the cam-lock type spindle shown in Fig. 55-23.

Chucks should be handled carefully. They should be oiled regularly, kept clean, and not

Fig. 55-22. Spindle Nose with Long-Taper Key Drive

Fig. 55-23. Cam-Lock Spindle

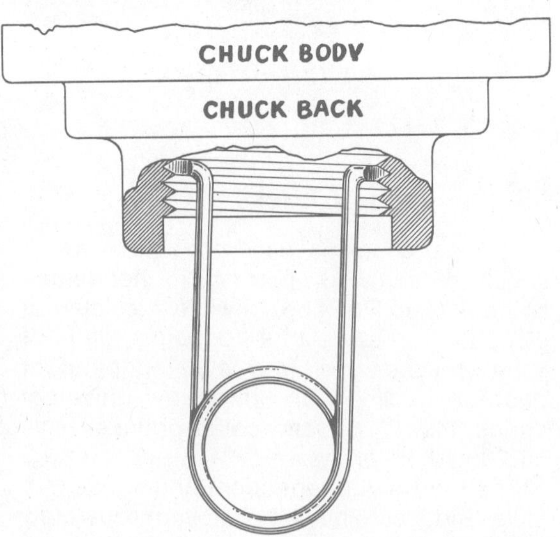

Fig. 55-24. Cleaning Threads in Chuck

TABLE 36

Lathe Tool Cutting Angles[2] and Cutting Speeds

Material	Side Relief	End Relief	True Back Rake	Side Rake	Suggested Cutting Speeds in Feet Per Minute[3]
Aluminum	10°	10°	35°	15°	200-1500
Brass	10°	8°	0°	0°	150- 300
Bronze	10°	8°	0°	0°	90- 100
Cast Iron, Hard	8°	8°	5°	8°	30- 50
Cast Iron, Malleable	8°	8°	8°	10°	80- 100
Cast Iron, Soft	8°	8°	8°	10°	50- 80
Fiber	15°	15°	0°	0°	80- 100
Free Machining Steel	10°	10°	16°	10°	150- 350
High-Carbon Steel	8°	8°	8°	8°	50- 70
Low-Carbon Steel	10°	10°	16°	10°	90- 100
Medium-Carbon Steel	10°	10°	12°	10°	70- 90
Plastics, Acrylics	15°	15°	0°	0°	60- 70
Plastics, Molded	10°	12°	0°	0°	150- 300

[2]All angles are true working angles measured from horizontal and vertical planes.

[3]Use the lower speeds on roughing cuts and when machining dry. Use higher speeds when using cutting fluids and for finishing cuts. See Table 33 for selection of cutting fluids.

overloaded or abused. The threads in threaded nose chucks should be cleaned with a thread cleaning tool as in Fig. 55-24.

55-11. Chucking

Chucking means to fasten work in a **lathe chuck** so that it can be machined.

A **four-jaw independent chuck** will hold work that is not perfectly round. Each jaw works separately. Place the work in the center of the chuck between the four jaws and tighten them. The circles on the chuck (see Fig. 55-25) help to locate the work in the center. Then run the lathe at a medium speed, rest the hand on the **carriage** or on the **compound rest**, and hold a piece of chalk so that it touches only the high spots of the work as it turns, Fig. 55-25. The **chalk mark** tells which jaws have to be reset. It usually takes a few trials of marking with chalk to get the work to turn without wobbling.

The **chalk method** of centering work in a four-jaw chuck is accurate enough for many lathe operations. It is satisfactory for jobs which can be completed in one setup. For operations requiring more accurate centering, a dial indicator is used. However, the workpiece should always be centered as

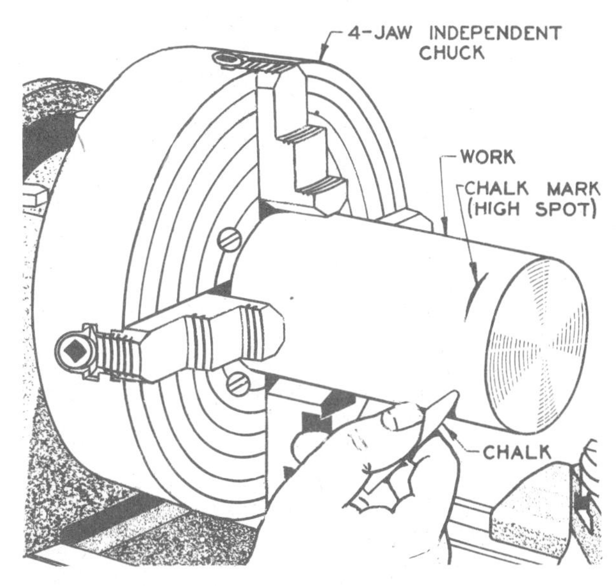

Fig. 55-25. Centering the Work in a Four-Jaw Chuck

Fig. 55-26. Centering Work with a Dial-Test Indicator

Fig. 55-27. Testing Face of Work with a Dial Indicator

close as possible by the chalking method described above, before using the dial indicator.

When centering work in a four-jaw chuck with the use of the dial indicator, it is best to center the work between one pair of jaws at a time. For example, center the work between the jaws numbered 1 and 3, and test with the dial indicator until the same reading

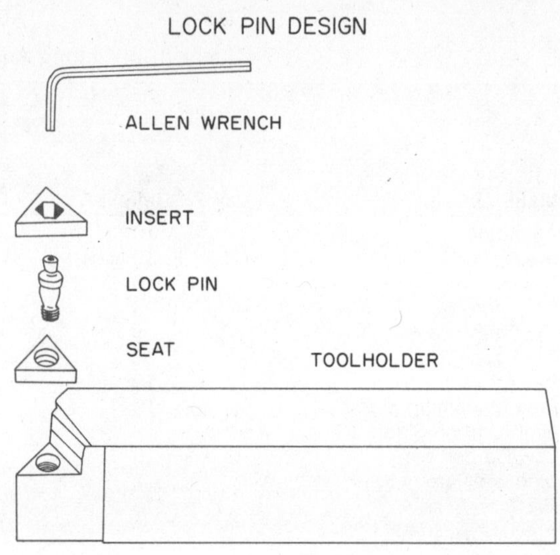

Fig. 55-28. Special Toolholder System for Holding Throw-Away Insert Cutting Tools

is obtained at these locations on the workpiece. When the workpiece has been centered accurately between the first pair of jaws, it can be centered between the second pair easily and rapidly.

To chuck in a **three-jaw universal chuck** or a **collet chuck**, it is only necessary to insert the workpiece and tighten the chuck; the workpiece is centered automatically.

55-12. Lathe Cutting Tools

Cutting in the lathe is done with **cutting tools**, also called lathe tools. The **cutting edges** are shaped differently depending on the type of toolholder in which they will be mounted, the type of metal to be machined, and the type of cut to be made, Fig. 55-29.

Lathe cutting tools are commonly made of the following materials:

1. High speed steel
2. Cast alloys
3. Cemented tungsten carbide
4. Ceramic

Lathe cutting tools are made into small pieces called **tool bits** which are held in

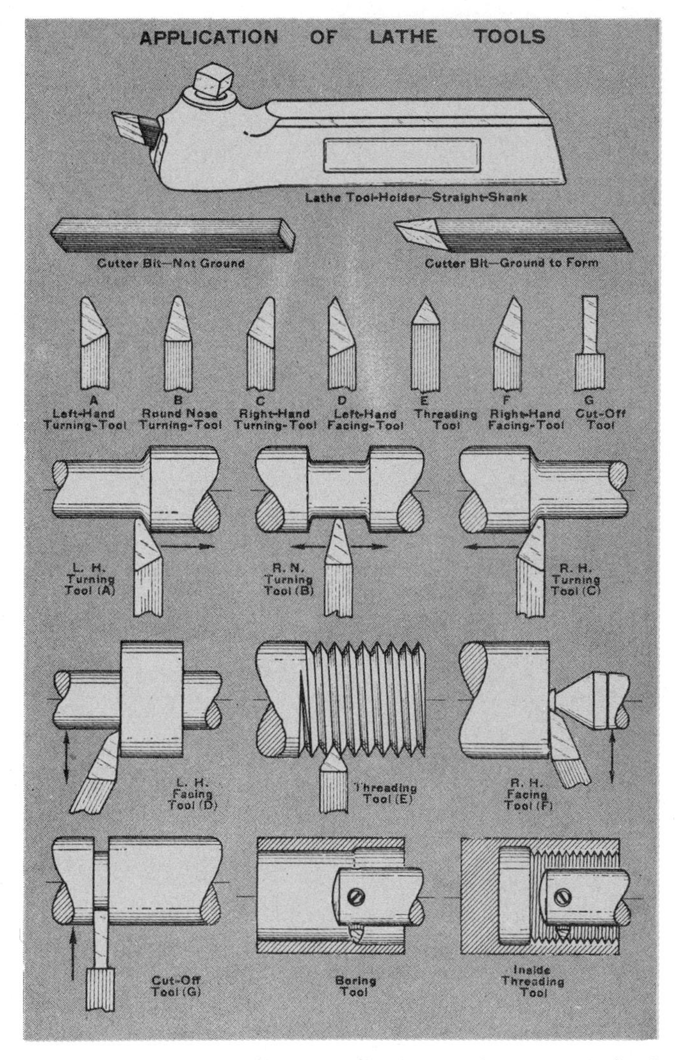

Fig. 55-29. Lathe Tool Bits and Their Applications with Various Lathe Operations

toolholders. High-speed steel, cast alloy, and carbide-tipped tool bits are small square bars which are held in conventional toolholders, Figs. 55-29 to 55-32. Tungsten carbide and ceramic tool bits are made as **throw-away inserts** and require special tool holders, one type of which is shown in Fig. 55-28. Inserts are made in several sizes and shapes.

A different toolholder is required for holding each different insert size and shape. Throw-away inserts are so called because they are thrown away, instead of being resharpened, after all their cutting edges become dull. This is not as wasteful a practice as it might at first seem. Note that the inserts, except the diamond shape, have cutting edges completely around the top and bottom surfaces. When one cutting edge becomes dull, the tool is unclamped and **indexed** or rotated to the next cutting edge and reclamped. When all the cutting edges

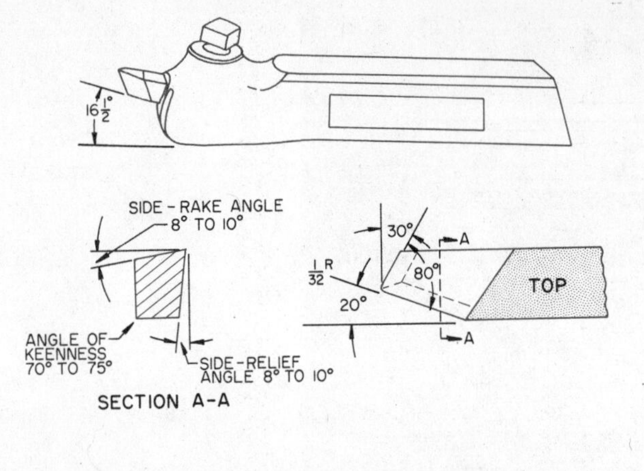

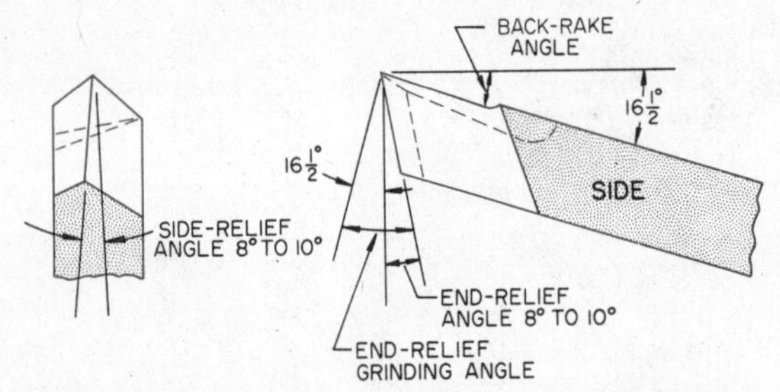

Fig. 55-30. (Top) Standard 16½° Toolholder for High-Speed Tool Bits; (Below) R. H. General-Purpose Turning Tool with Angles Given

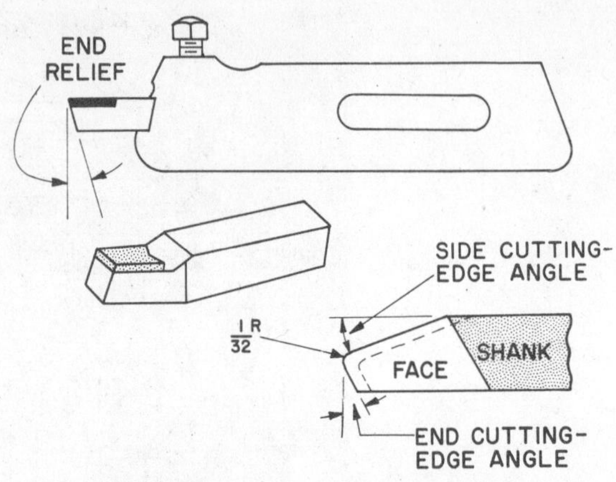

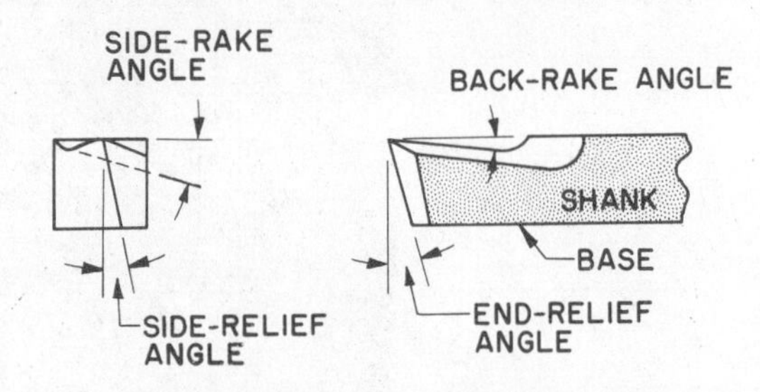

Fig. 55-31. (Top) Zero-Degree Toolholder; (Below) L. H. General-Purpose Turning Tool
Note grinding angles are the same as relief angles.

on the top surface are dull, the insert is turned over, exposing a fresh set of cutting edges. Some of the used carbide material is now being recycled; it is mixed with new material to make new inserts.

55-13. Conventional Toolholders

The toolholder supports the **tool bits.** There are two common classes of turning toolholders: **16½° toolholders and zero-degree toolholders.** When the tool bit is mounted in a 16½° toolholder, it is tilted upward 16½°, increasing the back rake angle, Fig. 55-30. When a tool bit is mounted in the zero-degree toolholder, no back rake is provided by the holder, and any necessary back rake must be ground on the tool bit, as in Fig. 55-31.

The zero-degree toolholder is designed for mounting tungsten carbide-tipped tools. See Fig. 55-31. As the carbide tip is brittle, it must have a minimum **grinding angle**. This type of toolholder also may be used for tool bits made of high-speed steel or cast alloys, as in Fig. 55-31. The grinding angle for tools in the 16½° toolholder is greater than for zero-degree toolholders. Compare Figs. 55-30 and 55-31.

The toolholder is mounted in the tool post, see Fig. 55-35. There are three types of turning tool bit holders: the **left-hand**, the **straight**, and the **right-hand** toolholders, see Fig. 55-32.

The **left-hand toolholder** is bent so that cutting can be done close to the **chuck**, Fig. 55-49, without the chuck striking the **carriage** or **compound rest.**

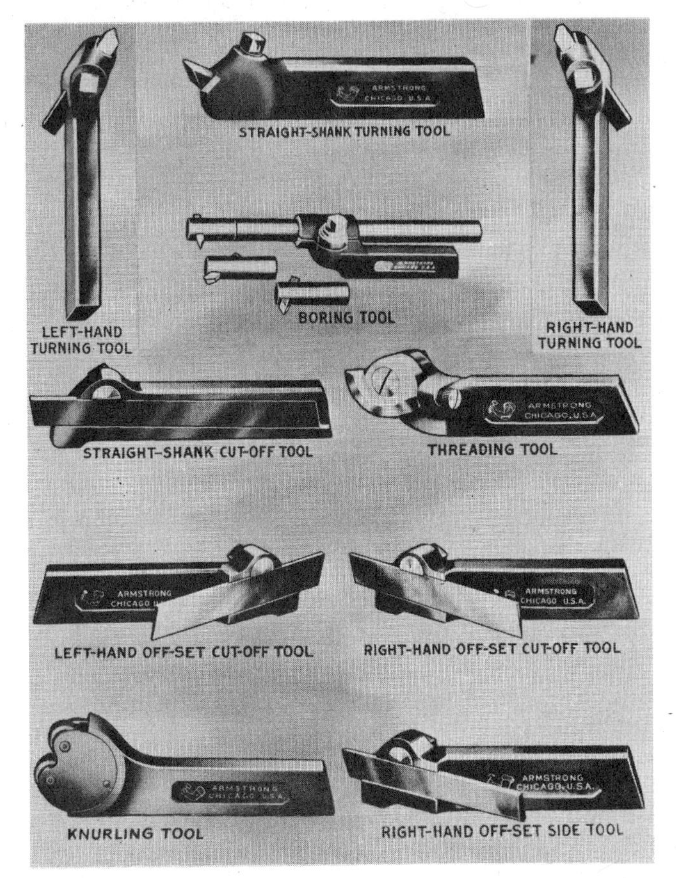

Fig. 55-32. Commonly Used Lathe Toolholders and Cutting Tools

The **right-hand toolholder** is bent so that cutting can be done close to the **tailstock**.

The **straight toolholder** is straight and is used for most **straight turning** on long workpieces. (See § 55-22 and Fig. 55-50.)

In addition to turning toolholders, the following kinds of toolholders with cutting tools also are used: cut-off tools, boring tools, knurling tools, and threading tools (see Fig. 55-32.)

55-14. Grinding Lathe Turning Tools

Tool bits may be ground to many different shapes (see Fig. 55-29) with good machining results. However, the basic relief angles and rake angles must be provided.

Locate each of the terms below in Fig. 55-33. Also, locate some of these terms in Figs. 55-30 and 55-31.

Face: The face is the top ground portion of the tool from which the chips slide off.

Flank: The flank is the ground surface below the cutting edge of the tool.

Nose angle: The angle formed by the side-cutting edge and the end-cutting edge.

Nose radius: The rounded portion of the nose of the tool. The nose radius for a finishing tool is greater than for a roughing tool (see Fig. 55-34).

Side-relief (formerly side clearance) is the angle formed between the flank and a vertical plane or the work being cut.

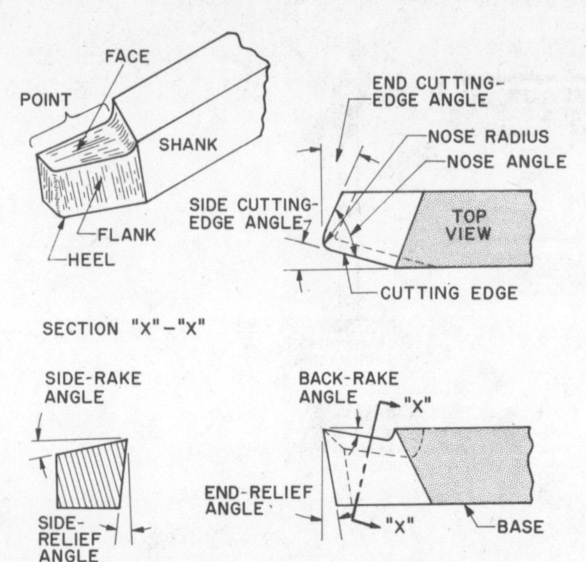

Fig. 55-33. Cutting Tool Terms Applied to Single Point Tools Used on Lathes, Shapers, and Planers

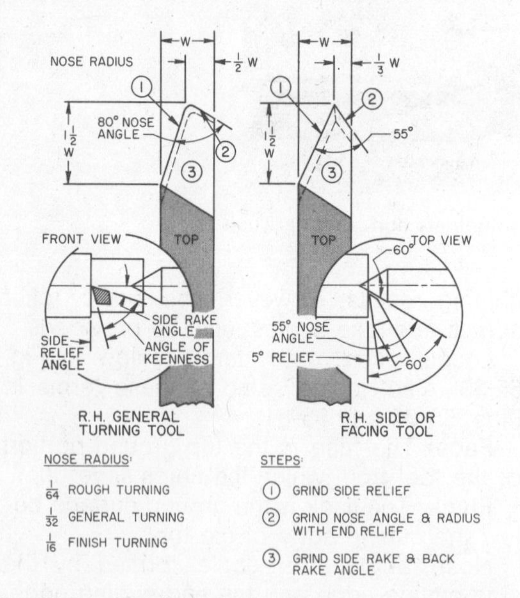

Fig. 55-34. Grinding the Two Most-Used Tool Bits

End-relief (formerly front clearance) **angle** is the angle formed at the nose of the tool bit between the ground end of the tool and the work.

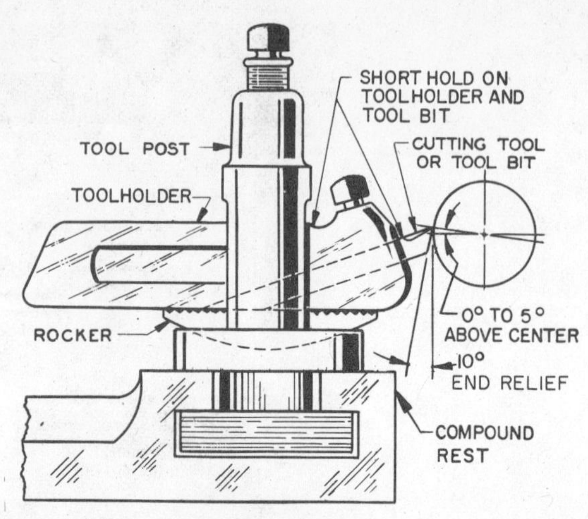

Fig. 55-35. Setting the Cutting Tool

End-relief grinding angle is the angle which must be ground on the end of the tool bit in order to produce the actual end relief when the tool is mounted in the lathe. For tools mounted at the level of the lathe center in a zero-degree toolholder, as in Fig. 55-36, the end grinding angle is the same as the end relief. Carbide-tipped tools, Fig. 55-31, generally are set at the level of the center line of the work, as in Fig. 55-36. For tools mounted in a 16½° toolholder, the end-grinding angle is greater than the actual end-relief angle (see Fig. 55-30). The end-relief angle is checked with the tool mounted in the toolholder, as shown in Fig. 55-37.

Side-rake angle is the angle formed between the face of the tool and a horizontal plane. When a continuous, long wirelike chip is formed, a decreased side-rake angle will frequently cause the chip to coil up and break off.

Back-rake angle is the angle formed between a horizontal plane and a line sloping back from the cutting edge at the end of the tool. A back-rake angle of 16½° is automatically provided for tools mounted in a 16½° toolholder. Tools mounted in a zero-degree toolholder have no back rake provided and any required back rake must be ground on

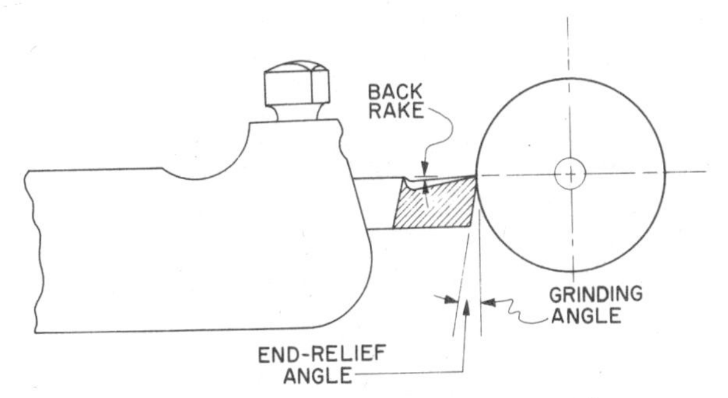

Fig. 55-36. Cutting Tool Set on Center with Zero-Degree Toolholder

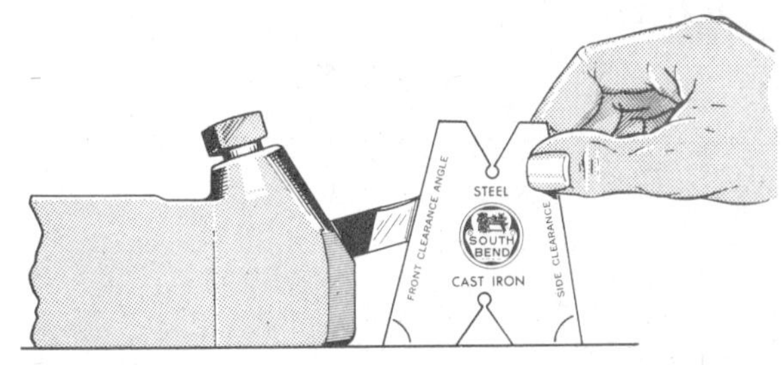

Fig. 55-37. Checking Relief (Clearance) Angles on Tool Bit

the tool. The back-rake angle influences the direction in which the chip leaves the nose of the tool. A decreased back rake frequently causes the chip to break off more readily. The **side-cutting edge angle** and the **end-cutting edge angle** are shown in Fig. 55-33.

A **right-hand** (**RH**) **lathe tool bit** has the cutting edge on the left side. The tool cuts from right to left, or from the tailstock toward the headstock of the lathe (see Fig. 55-30).

A **left-hand** (**LH**) **lathe tool bit** has the cutting edge on the right side. The tool cuts from left to right, or from the headstock toward the tailstock (see Fig. 55-31).

In sharpening a lathe tool bit, first determine the grinding angles (see Table 36) for the type of tool bit desired. The steps for grinding a RH general turning tool and a RH facing tool are indicated in Fig. 55-34. The angles should be checked carefully with a tool-angle gage at the angle the tool will be held in a holder, Fig. 55-37. A tool-angle gage can be made very simply from a strip of sheet metal 2″ × 3″. The required side-relief angle and end-relief angle may be laid out with a bevel protractor (see Fig. 7-4) and cut off. Finally, the ground surfaces should be whetted on an oilstone to remove any roughness on the cutting edges or nose radius. Keep the tool cool while grinding by dipping it frequently in water.

55-15. Setting the Cutting Tool

The **toolholder** should have a short hold on the **tool bit**, and the **tool post** should

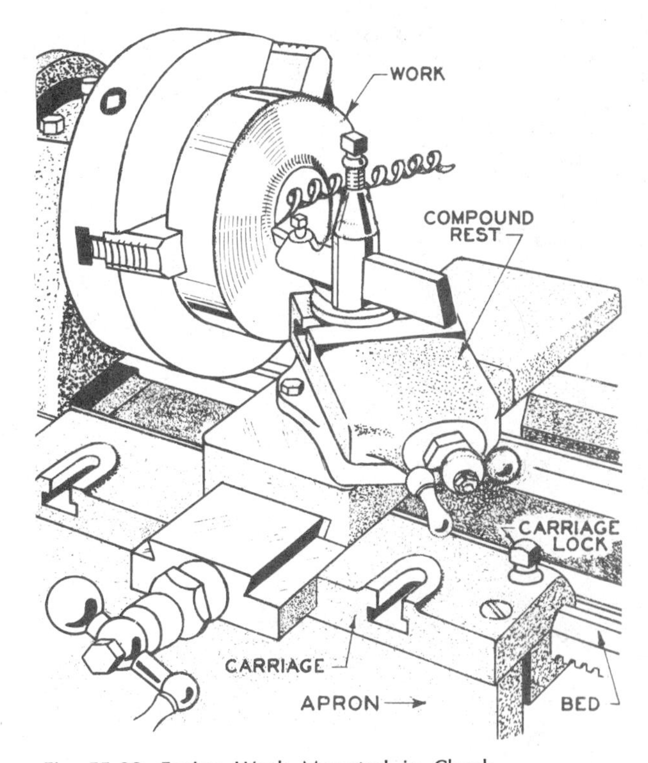

Fig. 55-38. Facing Work Mounted in Chuck

have a short hold on the toolholder, Fig. 55-35. If the toolholder extends too far out of the tool post or if the tool bit extends too far out of the toolholder, the springing or vibrating of the cutting edge will cause **chatter marks**.

The tool post should be set to the left on the **compound rest** (see Fig. 55-49).

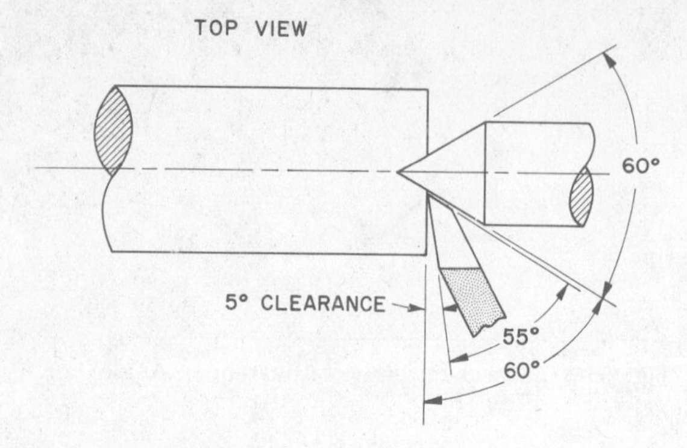

Fig. 55-39. Right-Hand Side Facing Tool, Facing End of Work between Centers of a Lathe

The **cutting edge** of the tool should pass through the center of the work. A good way is to set the cutting edge of the tool at the height of the **lathe center.** High-speed steel tools may be set a little above as in Fig. 55-35, if desired. Modern carbide-tipped tools (or those ground similarly) are set at the height of the lathe center, as in Fig. 55-36.

55-16. Facing

Facing is the cutting or **squaring** of the end of a piece of work, Fig. 55-38. Set the

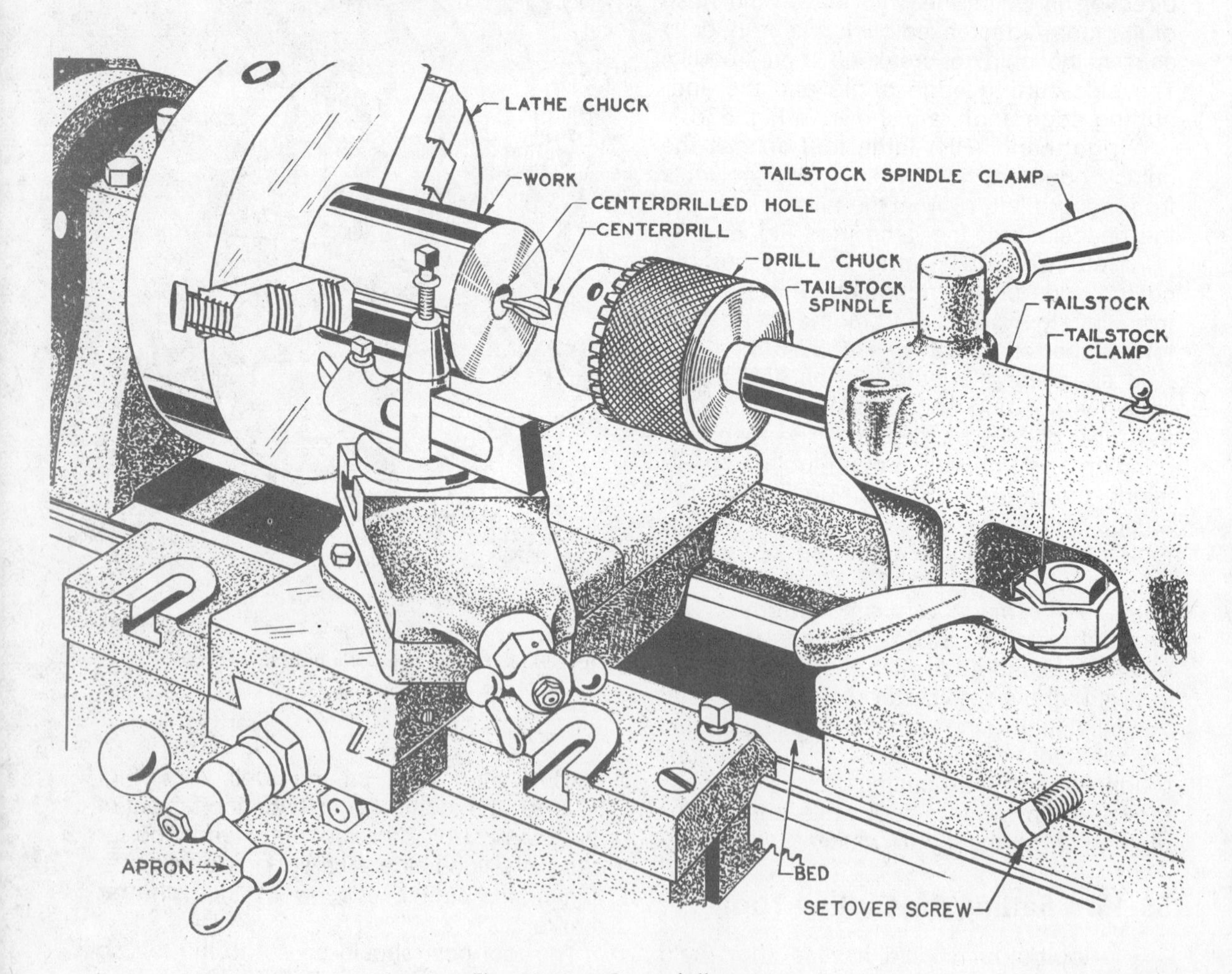

Fig. 55-40. Centerdrilling

cutting tool so that the cutting edge passes through the center and is the only part that touches the work. Then lock the **carriage** to the **bed** by tightening the **carriage lock**, Fig. 55-38. This keeps the tool against the work.

The type of tool and its setting for facing long work between centers of the lathe are shown in Fig. 55-39 and at the right in Fig. 55-34.

55-17. Centerdrilling

A piece of metal that is to be **turned** in the lathe must have small, **centerdrilled** holes in both ends. The small holes form **bearing** surfaces so that the work can be held between **lathe centers**, called **centers** for short. (See Fig. 55-48.)

The drilled part of the hole must be deeper than the **countersunk** part to make room for the sharp point of the lathe center. It also holds a small amount of center lubricant to prevent the dead center from overheating. Centerdrilling is done with a combination drill and countersink. The centerdrilled hole must be drilled to the proper depth, as shown at the top on Fig. 55-41. If the hole is either too shallow or too deep, the lathe center will be damaged. Centerdrilling may be done on the drill press or on the lathe as in Fig. 55-40.

Safety Note

Before centerdrilling on a lathe, the headstock and tailstock must be in accurate alignment.

The alignment may be checked by aligning the centers as shown in Fig. 55-44. If the centers are not accurately aligned, the point of the centerdrill will break off.

To centerdrill on the lathe, fasten the **centerdrill** in a **drill chuck** which is held in the **tailstock**. Fasten the work in the **lathe chuck**. Then, slide the tailstock so that the centerdrill is near the work, and clamp the tailstock to the **bed**. The centerdrilling is then done by turning the **handwheel** of the tailstock while the work is turning.

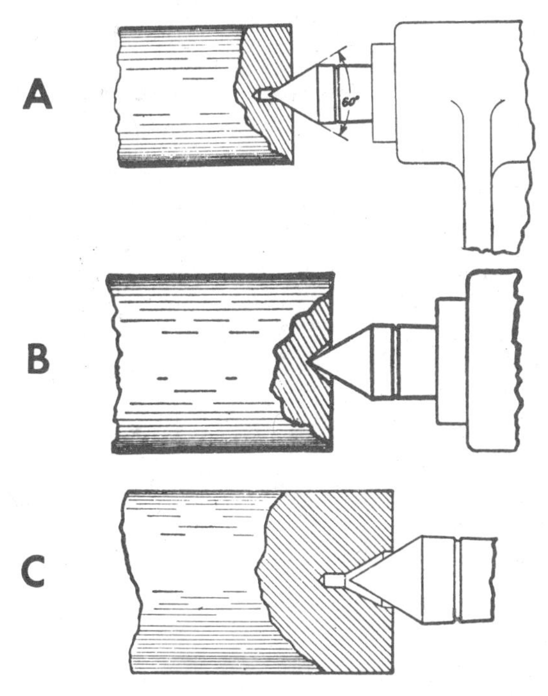

Fig. 55-41. Good and Poor Center Drilled Holes
A. Good
B. Hole Too Shallow and Wrong Angle
C. Hole Too Deep

55-18. Faceplate

The faceplate, Figs. 55-48, 55-50, is used to drive the lathe dog (see section 55-20) when turning between centers. (See Fig. 55-46). Workpieces, or workpiece holding fixtures, are also sometimes bolted directly to the faceplate for machining operations.

55-19. Lathe Centers and Setover Screws

The two lathe centers, called **centers** for short, are used to support the work for turning between centers. (See Fig. 55-48). The one in the headstock is called the **headstock center** or **live center** because it turns with the **headstock spindle**. The center in the **tailstock spindle** is called the **tailstock center** or **dead center** because it does not turn. Both centers have a **Morse taper shank** (see Fig. 55-42).

The point of the dead center is **hardened** to make it tough so that when the work rubs on it and causes heat, the point will not wear off. Dead centers are also available with carbide tips for greater wear resistance. **Live tailstock centers**, Fig. 55-43, revolve with the workpiece and thus eliminate the problem of dead center wear and lubrication.

The tailstock center may be removed from the tailstock spindle by turning the **handwheel**. This screws the spindle into the tailstock and knocks out the center. To loosen the center in the headstock spindle, put a rod called a **knockout rod** in the hole at the other end of the spindle. Hold the center with the right hand, and with the left hand tap lightly with the bar.

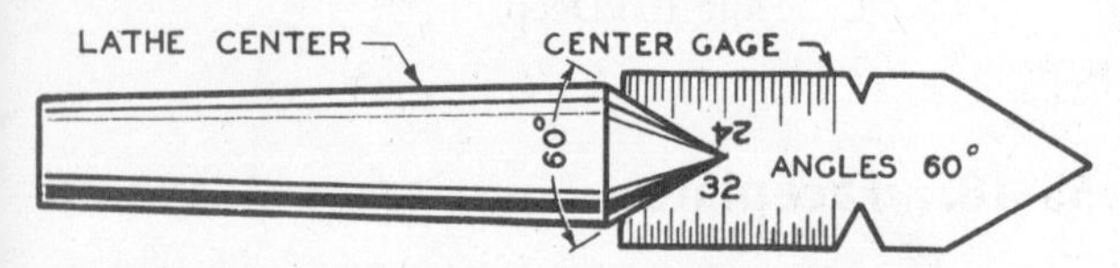

Fig. 55-42. Testing the Angle of a Lathe Center

Fig. 55-43. Live Tailstock Center

Good work depends to a great extent upon the condition of the lathe centers. Each point must have an angle of 60° which may be tested with a **center gage**, Fig. 55-42. The lathe centers and the **spindle holes** must be very clean, because the smallest nick or speck of dirt will cause trouble with the work at a later time. They must fit the spindle holes exactly so that there will be no looseness.

The centers must also be tested for **alignment,** which means that the two points must meet when the tailstock with its center is slid up to the headstock center, Fig. 55-44. Run the lathe at a medium speed to see if the live center wobbles. If it wobbles, a new point must be cut or ground on it.

For straight turning, the centers must be perfectly aligned. For taper turning, the tailstock may be intentionally moved off center a calculated amount. If the dead center is to one side of the live center, it is aligned by turning the **setover screws** on the tailstock, Fig. 55-45. The alignment of the centers can best be tested by taking a straight cut (see § 55-22) and then measuring the diameter at both ends of the work with a micrometer. If both measurements are the same, the centers are **aligned** and the turning may go on. If the measurements are not the same, the setover screws have to be set again and another cut must be made for another test. This must be repeated until both measurements are the same. Small movements can be determined with a dial indicator.

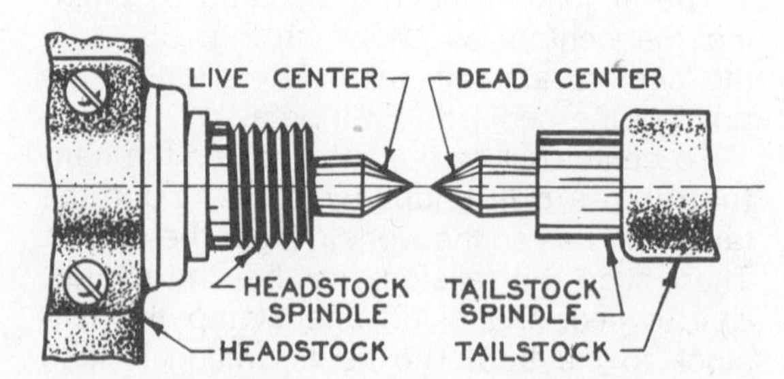

Fig. 55-44. Proper Alignment of Lathe Centers, Viewed from Above

55-20. Lathe Dogs

A lathe dog, Fig. 55-46, is used to keep work which is to be turned between centers from slipping (see Fig. 55-48). It is clamped on the end of the work, Fig. 55-47, which is then placed between centers. The **tail** of the dog slips into the slot in the faceplate.

There are many sizes of dogs. Use the smallest one that will slip over the work. The **clamp dog** is used for large work.

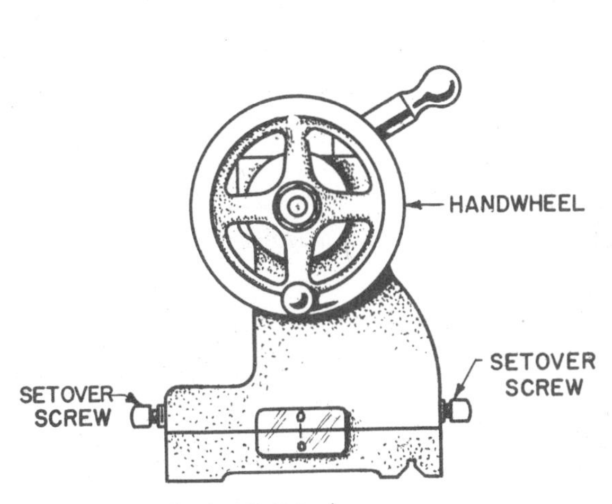

Fig. 55-45. Setting Tailstock

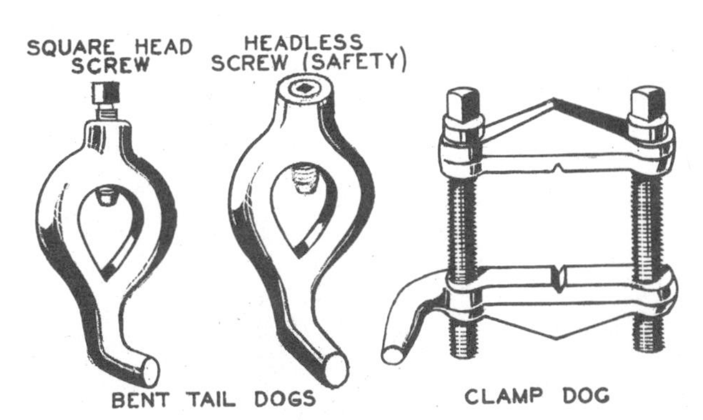

Fig. 55-46. Lathe Dogs

55-21. Mounting Work between Lathe Centers

To place work between centers for **straight turning** as in Fig. 55-50:

1. **Face and centerdrill** the work.
2. Screw the **faceplate** on the headstock spindle.

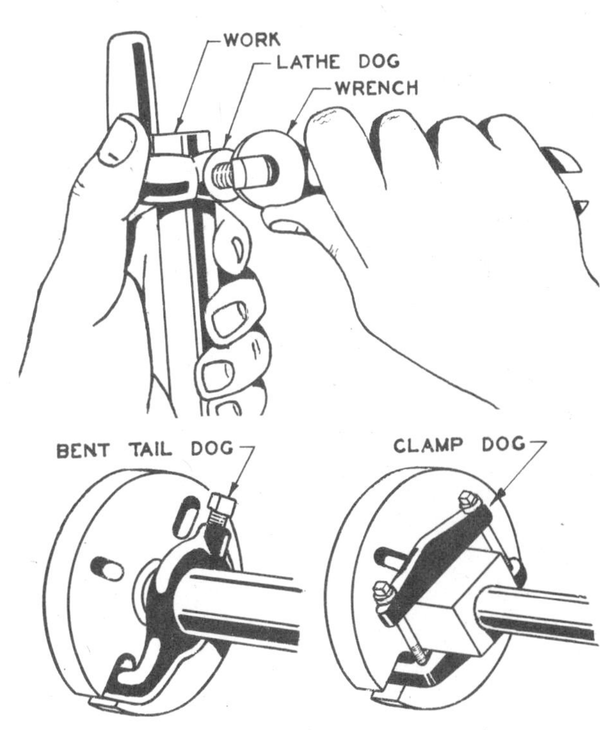

Fig. 55-47. Fastening Lathe Dogs to the Work

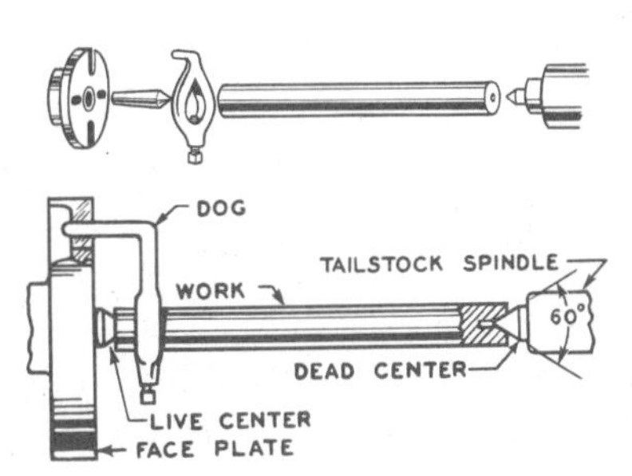

Fig. 55-48. Work Held between Lathe Centers

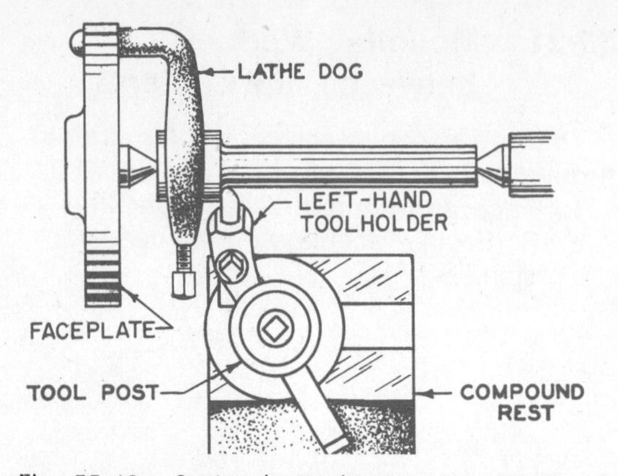

Fig. 55-49. Setting the Tool Post and Toolholder to Cut Close to Faceplate

3. Clamp a **lathe dog** on the work.
4. Be sure the lathe centers are in good condition and the centerdrilled holes in the ends of the work are clean.
5. Put a drop of center lubricant or a mixture of oil and **white lead** in the centerdrilled hole which goes on the tailstock center.
6. Place the work between the lathe centers by sliding the **tailstock** up to the work.
7. Clamp the tailstock to the **bed**.
8. Turning the **handwheel** on the tailstock will set the **dead center** so that the work will move freely on the centers,

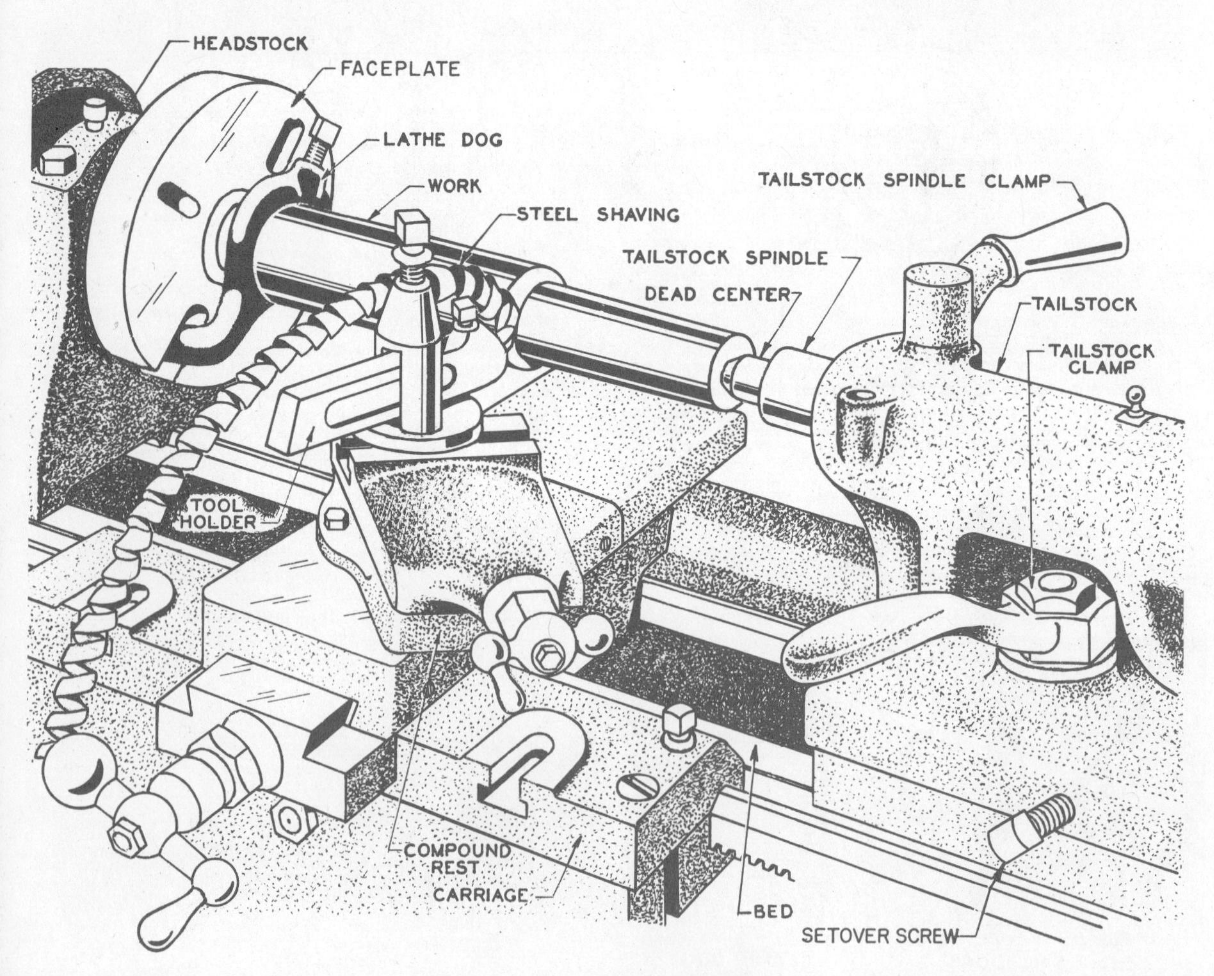

Fig. 55-50. Straight Turning

yet be tight enough to allow no looseness endwise. The **tail** of the dog must fit loosely in the slot of the faceplate.
9. Clamp the **tailstock spindle clamp**.

55-22. Straight Turning

For straight turning, the work is first placed on **centers**, and the cutting tool is set for cutting. The tool should be set on the left side on the **compound rest** so that the **lathe dog** will not strike it, Fig. 55-49. For most turning, the compound rest is set at an angle of about 29° or 30° from the crosswise position, as shown in Fig. 55-50. However, when it is necessary to turn very close to the lathe dog or chuck, the compound rest may be set at the crosswise position, as shown in Fig. 55-49.

The cutting should be from the tailstock toward the headstock. The reason for this is that when the cutting is toward the headstock, the pressure is on the headstock center which turns with the work. Should the tool feed toward the tailstock, then the pressure is on the dead center. This increases the rubbing, and the point of the dead center may be damaged.

A squeak is a sign that something is wrong. Either the centers are too tight against the work or the dead center needs lubrication.

55-23. Taper Turning

The machining of tapers is an important operation on the lathe. There are three methods of turning tapers. The one chosen depends on the angle of the taper, the length of the taper, and the number of pieces to be turned. The **offset tailstock method** is most commonly used. In this method, the right end of the turning will be smaller when the tailstock is moved toward the operator and larger when moved away from the operator. Tapers can also be turned by using a **taper attachment** as in Fig. 55-51.

Compound Rest Method

If the taper is short and the angle of taper is known, the taper can be turned by setting the compound rest for this angle and turning the compound rest feed screw by hand. Both internal and external tapers can be turned by the **compound rest method**. An internal taper can be bored as shown in Fig. 55-52.

When the angle of taper is not given, it can be found by using the mathematics of **trigonometry**. For example, to find the angle of taper for the part shown in Fig. 55-53:

1. Construct the line A-B parallel to the center line so that it intersects the corner of the workpiece and forms the right triangle A-B-C.

Fig. 55-51. Turning a Taper with a Taper Attachment

Fig. 55-52. Boring an Internal Taper with Compound Rest-Set at Desired Angle

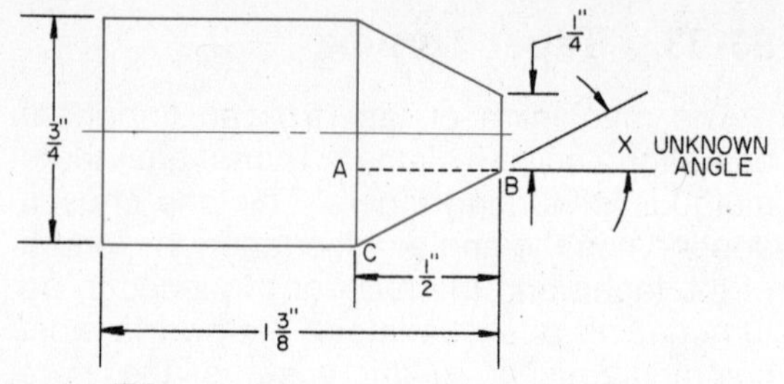

Fig. 55-53. Problem in Calculating Taper Angle

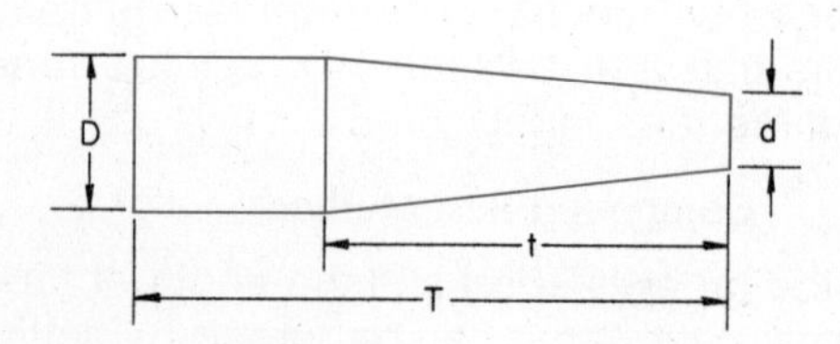

Fig. 55-54. Symbols Used in Formulas for Calculating Tapers

2. The length or base of the right triangle A-B-C is ½″ as read from the drawing.
3. By subtraction we find that the height of the triangle A-B-C is ¼″.
4. Using the tangent function of trigonometry;

$$\text{Tangent } \angle x = \frac{\text{side opposite}}{\text{side adjacent}} = \frac{1/4''}{1/2''} = \frac{.250}{.500} = .500$$

Consulting a table of **trigonometric functions**, we find that .500 in the tangent column corresponds to an angle of approximately 26° 35′. Since the compound rest is marked in whole degrees only, the fractional part of degrees can only be estimated.

Tailstock Offset Method

The amount of taper may be specified either by the taper in inches per foot or by the length of the taper and the diameters at the ends of the taper. Two formulas will enable you to calculate the tailstock offset for

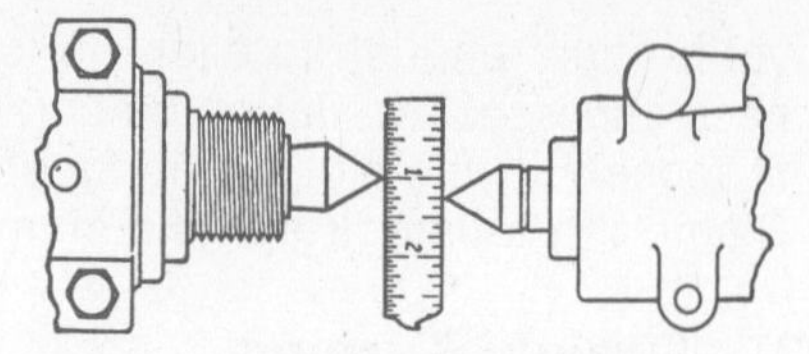

Fig. 55-55. Checking Amount of Tailstock Setover

either specification. These formulas use the following symbols, as illustrated in Fig. 55-54.

T = Total length of stock in inches
t = Length of portion to be tapered in inches
D = Large diameter
d = Small diameter
Offset = Setover of tailstock in inches
tpf = Taper per foot

Taper given in inches per foot: Divide the total length of the stock in inches by 12 and multiply this quotient by one-half the taper per foot specified. The result is the offset in inches:

$$\text{Offset} = \frac{T}{12} \times \tfrac{1}{2}\ \text{tpf}$$

For example, to calculate the amount of tailstock offset for turning work 4″ long with a taper of .600″ per foot:

$$\text{Offset} = \frac{4}{12} \times \tfrac{1}{2} \times .600'' = \frac{1}{3} \times .300'' = .100''$$

Diameters given at ends of taper: Divide the total length of the stock by the length of the portion to be tapered and multiply this quotient by one-half the difference in diameters. The result is the offset in inches:

$$\text{Offset} = \frac{T}{t} \times \tfrac{1}{2}\ (D\text{-}d)$$

For example, to calculate the amount of tailstock offset for turning work with a length of 6″, a tapered length of 3″, a large diameter of ½″ and a small diameter of ¼″:

$$\text{Offset} = \frac{6}{3} \times \tfrac{1}{2}\ (.500\text{-}.250)$$
$$= .250''$$

To move the tailstock, use the setover screws shown in Fig. 55-45. The most accurate way to determine the amount of the setover is to measure the offset between the live and dead centers, Fig. 55-55. A dial indicator (see Fig. 46-25) may be used to observe accurately the amount of setover. The cutting edge of the tool must be exactly at the level of the center of the work when cutting tapers.

55-24. Drilling, Reaming, and Counterboring

Drilling, reaming, countersinking, and counterboring — commonly performed with a drill press — can also be performed readily in a lathe. When they are performed in a lathe, the workpiece is held in an appropriate lathe chuck. The drill, reamer, countersink, or counterboring tool is held in the tailstock. The tailstock should always be accurately aligned with the headstock for all hole-machining operations.

For drilling and reaming with small diameter, straight-shank drills, countersinks, and reamers, the tool is held in a drill chuck. The drill chuck is mounted in the tailstock, as in Fig. 55-40. The tailstock is clamped to the lathe bed so that it does not move. The tool is then fed to the desired depth with the tailstock handwheel.

Larger drills with taper shanks are mounted directly into the tailstock, as shown in Fig. 55-56. Taper-shank reamers, counterbores, and other special taper-shank cutting tools also may be mounted directly in the tailstock for hole-machining operations.

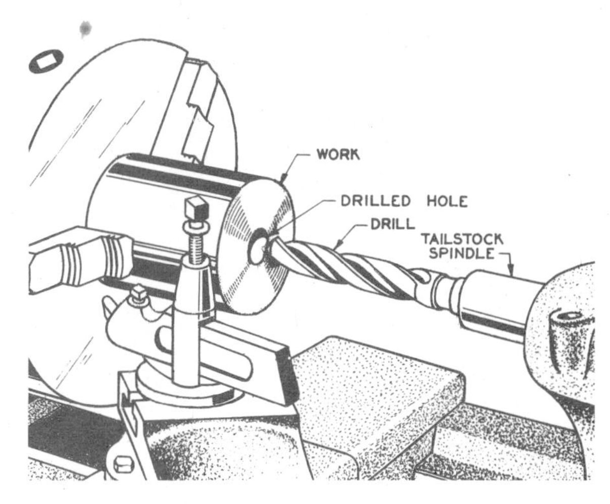

Fig. 55-56. Drilling

55-25. Boring in a Lathe

It is impossible to make an exact or perfect hole with a drill. A hole may be needed that is larger than an available drill. A hole may be made exactly round (**true**) by **boring**. There is no limit to the size of a bored hole.

Boring is the cutting and enlarging of a round hole to make:

1. A more exact size.
2. A hole that will not wobble.
3. The hole accurate with its **axis**.

Boring is done with a **boring tool**. There are two kinds of boring tools: the kind that is shown in Figs. 55-57 (A) and 55-58 and the smaller, **forged** boring tools shown in Fig. 55-57 (B). The **tool bit** is held in a **boring bar** which is held in the **tool post** of a lathe, and the cutting is done as shown in Fig. 55-58. The work turns, and the tool is held in a fixed position by the tool post and the carriage moves parallel to the axis of the hole.

Boring can also be done on a **drill press**, **milling machine** (see Fig. 57-51), or on a specially built **boring machine**; in these machines the work is held still and the boring tool turns. (See § 54-15 and Figs. 54-1 and 54-12.)

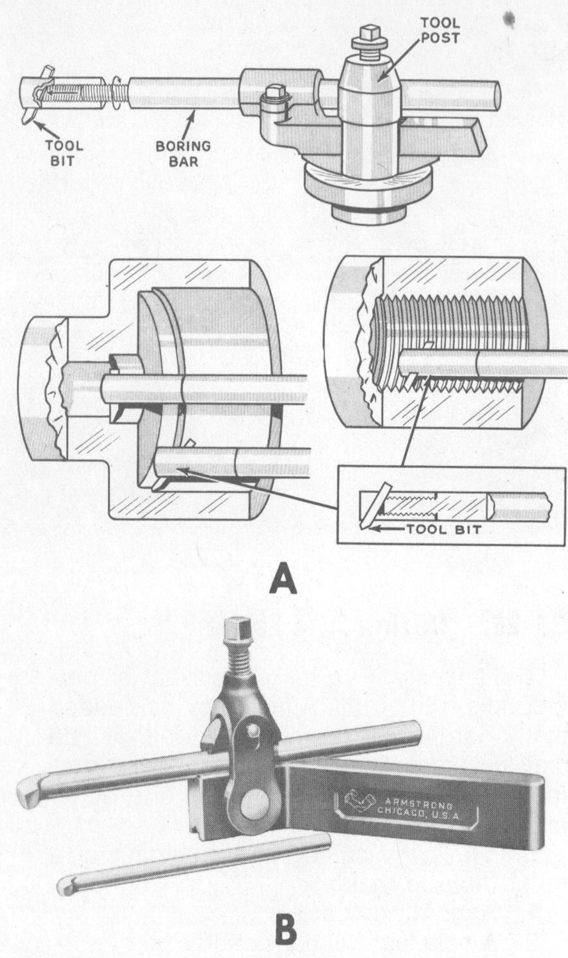

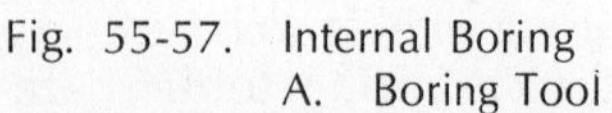
Fig. 55-57. Internal Boring
A. Boring Tool
B. Forged Boring Tool in Boring Toolholder

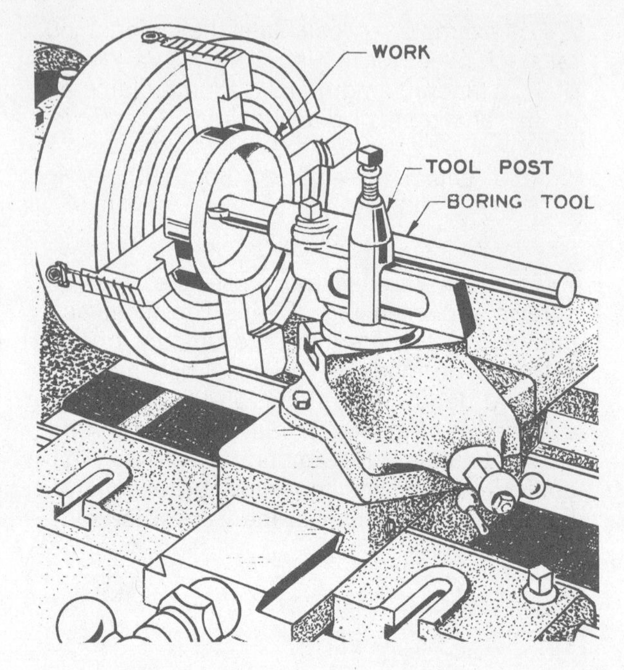

Fig. 55-58. Boring

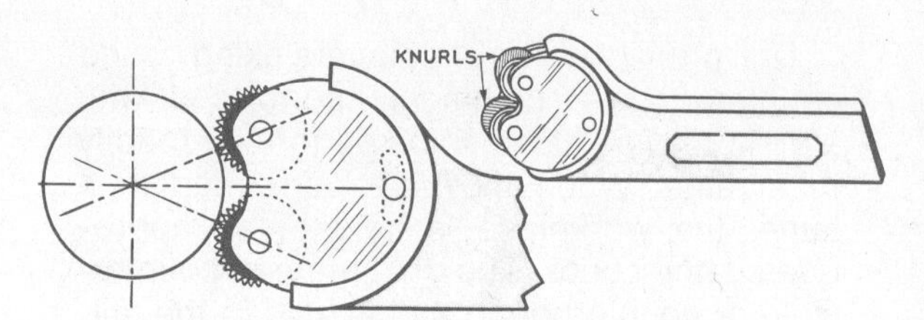

Fig. 55-59. Knurling Tool

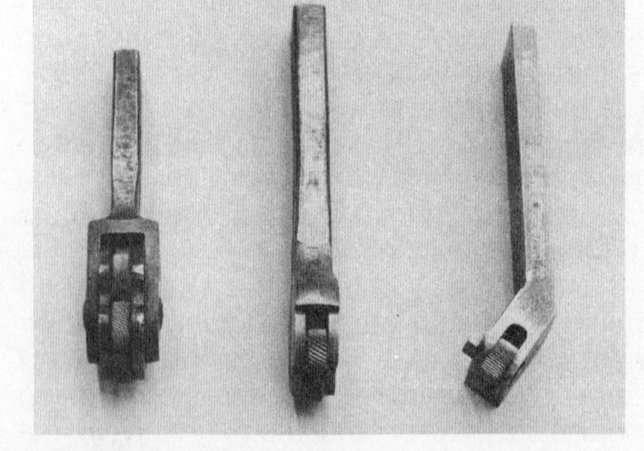

Fig. 55-60. Knurling Wheels

55-26. Knurling

Handles on some tools and screws are made rough in order to give a better grip as the handles on **scribers** in Fig. 6-3. This is called **knurling**[4] and is done with a **knurling tool** in the lathe, Figs. 55-59 and 55-61. Two small hardened steel wheels or rolls called **knurls**, turn in the knurling tool when pressed

[4]Pronounced **nurling**; also spelled nurling.

Fig. 55-61. Knurling

into the rotating work. There are coarse, medium, and fine knurls in diamond and straight line patterns, Fig. 55-60.

Procedure

1. Set the knurling tool at a right angle to the work, Fig. 55-61.
2. Set the lathe for a slow back geared speed see section 55-8.
3. Set the lathe for a feed of about 0.020″ to 0.035″, see section 55-9.
4. Force the tool slowly and firmly into the workpiece, starting at the right end.
5. Start lathe, and apply cutting fluid liberally.
6. Engage the longitudinal feed, and let the tool feed across the workpiece to the desired length.
7. When the tool reaches the left end, disengage the longitudinal feed and stop the lathe. Without withdrawing the tool, reverse the direction of longitudinal feed. Cross feed the tool into the work about 0.010″ to 0.015″ more. Start the lathe, engage the longitudinal feed, and allow the tool to travel back to the right end. Repeat this procedure until the knurl is cut to the desired depth.

Fig. 55-62. Mandrel

Fig. 55-63. Work Mounted on a Mandrel

55-27. Mounting Work on Mandrel

A **mandrel**, Fig. 55-62, is a solid steel bar with a slight taper. Sometimes it is necessary to machine cylindrical parts accurately in relation to a drilled or bored hole which is in the part. Gear blanks, V-pulleys, and collars are parts of this type. When these kinds of parts are machined, a tapered mandrel is pressed into the hole in the part. A lathe dog is clamped on the mandrel and it is mounted between centers in the lathe, as shown in Fig. 55-63. The outside diameter and the faces of the part are then machined accurately in relation to the hole.

55-28. Cutoff Tool

Cutoff tools Fig. 55-64, are held in a cutoff-toolholder. Three kinds of cutoff-toolholders are shown in Fig. 55-32; these include straight, right-hand and left-hand toolholders. The cutoff-tool blade should be ground as shown in Fig. 55-64. An end relief of about 5° is adequate. Since the tool is provided with tapered sides which provide side relief, the sides should not be ground. A back rake of 0° to 5° is adequate.

Cutoff operations are performed on workpieces which are mounted in a chuck.

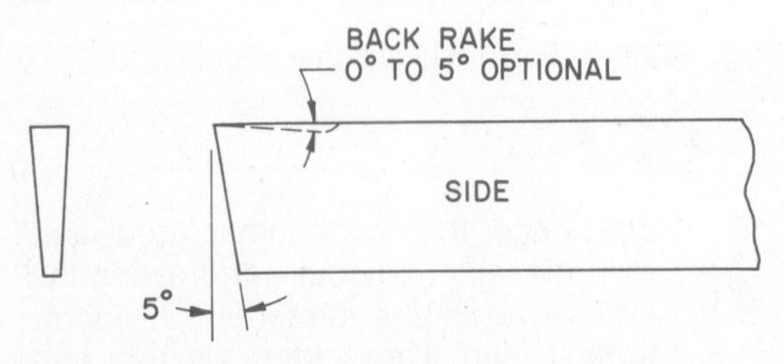

Fig. 55-64. Cutoff Tool Blade

Safety Note

Never attempt to cut off stock mounted between centers.

To do so will cause the workpiece to bend or break and then fly out of the lathe. The cutoff tool may be used for cutting grooves, cutting to a shoulder, or for cutting off stock. The cutoff tool should be mounted at the height of the centerline of the work, see Fig. 55-36. If chatter or vibration develops, it generally can be reduced by reducing the cutting speed. Cutoff tools work best when liberally supplied with a cutting fluid.

55-29. Steady Rest and Follower Rest

A **steady rest** is used to hold long bars rigid or steady while turning, threading, or boring in a lathe. The steady rest can be used to support work mounted in the chuck, as in Fig. 55-65. It also can be used to support workpieces mounted between centers, as shown in Fig. 55-66. The steady rest is clamped to the lathe bed. The jaws in the steady rest are adjusted so that they rub the

Fig. 55-65. Using Steady Rest for Boring or Internal Threading

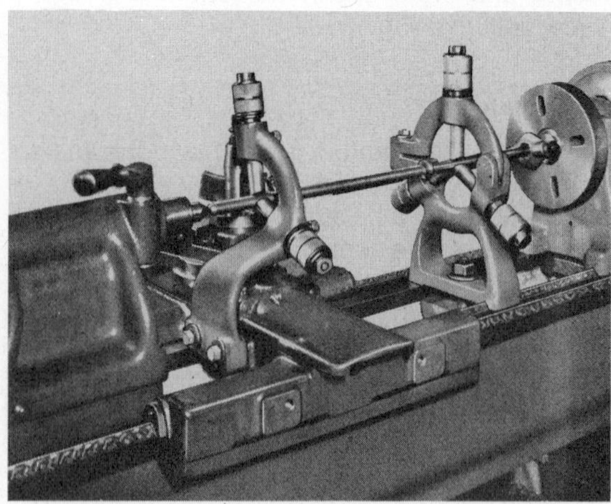

Fig. 55-66. Using Both Steady Rest and Follower Rest for Cutting Threads in a Long Bar

work lightly. Machining operations can be performed with the carriage being used on either side of the steady rest.

A **follower rest** is a supporting device which is attached to the saddle of the lathe, as shown in Fig. 55-66. It is used for either turning or cutting threads on long workpieces. The jaws on the follower rest are adjusted so they touch the work lightly. The follower rest follows along the work, thus holding it steady while the piece is being machined. The follower rest may be used alone, or it may be used in conjunction with the steady rest, as shown in Fig. 55-66.

55-30. Cutting Threads on a Lathe

Screw threads can be cut on work mounted in a lathe, as shown in Fig. 55-67. The workpiece may be mounted in a chuck, or it may be mounted between centers. The threads may be right-hand, left-hand, internal or external threads.

Before cutting screw threads on a lathe, you should be familiar with the kinds of threads, thread fits, classes of threads, and the calculations necessary for cutting threads. This information is included in Unit 21.

Fig. 55-67. Cutting Thread on a Lathe

Thread-Cutting Tools

Two kinds of single-point, thread-cutting tools commonly are used for cutting threads on a lathe. An ordinary lathe tool bit may be ground as shown in Fig. 55-68 for cutting either National Form Threads or Unified (National) Form Threads. If the tool is ground with a relatively sharp point, it may be used for cutting threads with a wide variety of pitches. The tool is set on the centerline of the work for cutting threads, Fig. 55-70. The tool also is set square with the work, Fig. 55-71.

Another kind of thread-cutting tool is shown in Fig. 55-69. The top face of this tool should always be in line with the shank of the toolholder as shown. The tool is then set on the centerline, Fig. 55-70, and is set square with the work, Fig. 55-71. This kind of threading tool also may be used for cutting either National Form Threads or Unified (National) Form Threads. The sides of the

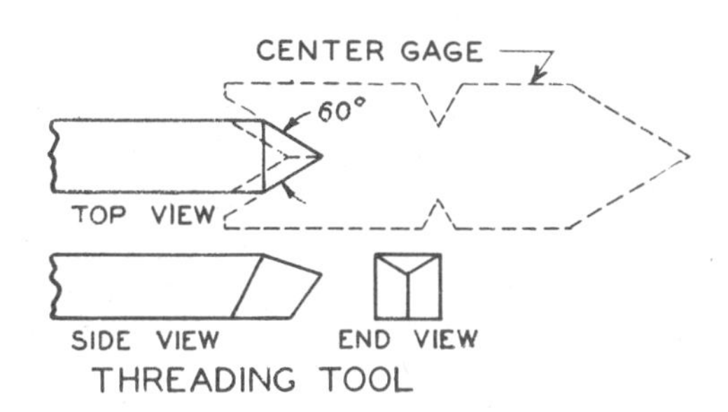

Fig. 55-68. Threading-Tool Bit

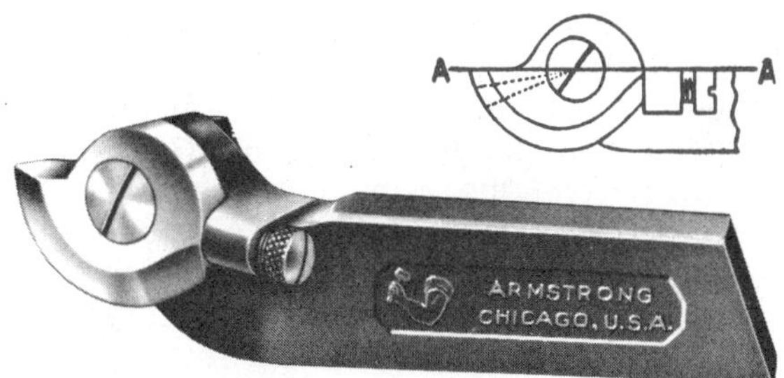

Fig. 55-69. Threading Tool

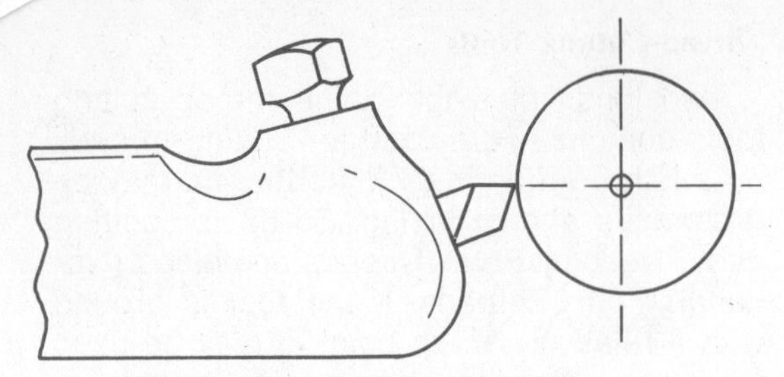

Fig. 55-70. Top of Tool Bit Set on Center for Cutting Screw Threads

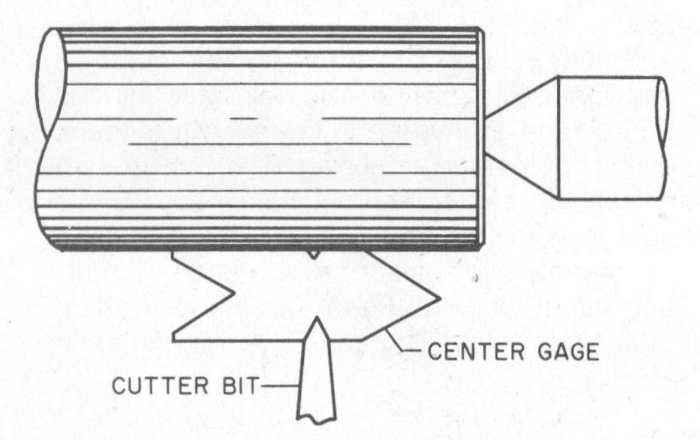

Fig. 55-71. Threading-Tool Set Square with Work

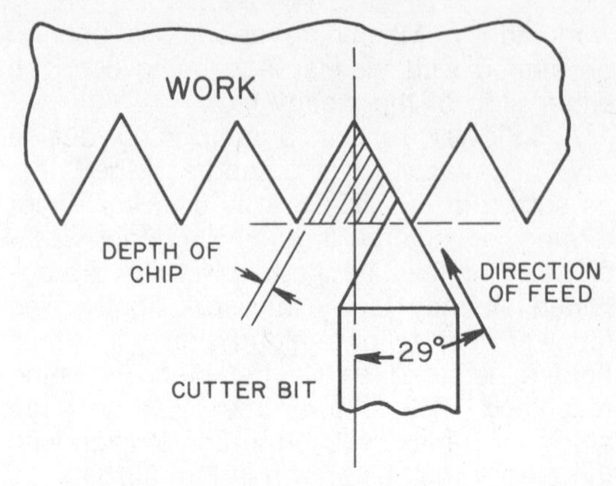

Fig. 55-72. Action of Threading Tool with Compound-Rest Set at 29° Angle

threading tool are designed so that there is always adequate side relief. Hence, the tool may be resharpened by grinding the top face only. The face is ground flat across the top so the tool may be used for cutting either right-hand or left-hand threads. When the cutting tool is worn out, a new one can be installed on the toolholder.

When right-hand external Unified and American (National) Form Threads are cut, the compound rest should be set at a 29° angle to the right, as in Fig. 55-67. The tool travels from the right toward the left for right-hand external threads. A thread with the desired 60° V-form is thus produced. Several cuts are required. The tool is fed into the work with the compound rest for each additional cut until the thread is cut to the desired pitch diameter.

When left-hand external threads are cut, the compound rest is swung 29° to the left. The feed is then reversed, and the carriage and the tool travel from the left toward the right.

The procedure for cutting external Unified and American (National) Form Threads is included in this section. An internal thread is cut with a threading tool bit inserted in a boring toolholder as shown in Fig. 55-57. If you wish to cut internal threads, acme threads, or any other kind of special thread on a lathe, ask your instructor to show you the correct procedure. Procedures for cutting these kinds of threads are included in more advanced machine tool technology books.

Procedure

The following is the procedure for cutting right-hand external Unified and American (National) screw threads on a lathe:

1. Determine the number of threads per inch to be cut. Set the quick-change gear levers to the desired pitch. (See Figs. 55-14 and 55-16.)
2. Determine the depth of the thread. (See § 21-6.)
3. Mount the stock in the lathe. It may be mounted in the chuck or between centers, depending on the length and design of the part to be threaded.

Fig. 55-73. Thread Dial Indicator

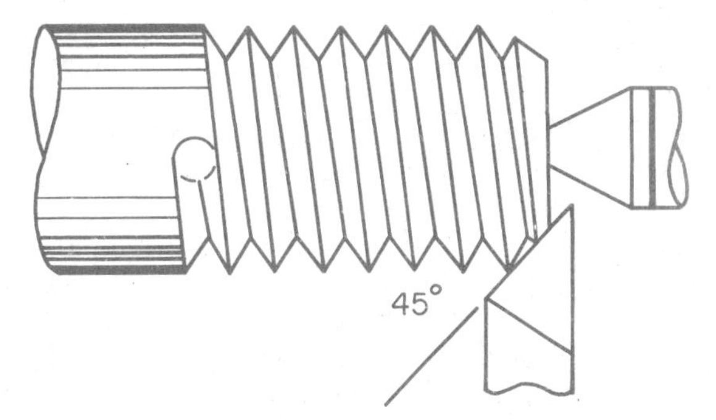

Fig. 55-74. Chamfer End of Thread with 45° Chamfer

4. Lay out the length to be threaded, and mark it with a lead pencil or with the point of a tool bit. If desired, a groove may be cut at the end of the thread, using a cutoff tool. The depth of the groove should be equal to the depth of the thread. The groove makes it easier to avoid breaking off the point of the cutting tool when the tool is withdrawn after each cut.
5. Set the compound rest at a 29° angle to the right. (See Fig. 55-72.)
6. Set the point of the tool bit on the centerline of the work, Fig. 55-70, square with the work, as in Fig. 55-71.
7. Set the lathe at a low, back-geared rpm when cutting your first threads. With experience, you can cut threads at higher speeds.
8. Set the feed-change lever, Fig. 55-15, to the **threading position**. On most lathes this is located in the center position.
9. Advance the tool until it just touches the work. Then set the micrometer collar on the **cross-feed knob** to **zero**, Fig. 55-15. Also set the micrometer collar on the **compound-rest feed knob** to zero, Fig. 55-15.
10. Bring the carriage to the right end of the work, clear of the work.
11. Advance the tool 0.005″ with the **compound-rest feed knob.** NOTE: The depth for each additional cut is set with the **compound-rest feed knob.** As the depth of thread increases with succeeding cuts, reduce the depth of cut to 0.003″ then to 0.002″. The last several cuts should not be deeper than 0.001″.
12. If the **thread-dial indicator** is not engaged, Fig. 55-73, engage it and tighten it in position. The dial should revolve when the carriage is moved.
13. Start the lathe. When a **numbered line on the dial** is in line with the **index line** on the outer ring of the thread-dial indicator, engage the **half-nut lever,** Fig. 55-15.
14. Apply cutting oil to the threads.
15. At the end of the thread, disengage the **half-nut lever**, and at the same time, turn the **cross-feed knob** one complete turn, counterclockwise, to withdraw the tool.
16. Bring the carriage back to the right end of the thread in position to start the next cut.

17. Turn the **cross-feed knob** one revolution to the right to the original **zero** starting position.
18. Repeat steps 12 through 17 until the thread is cut to the desired depth. As the depth of thread is increased, the pitch diameter should be measured with a **thread micrometer** to avoid cutting the thread too deeply. Another way to determine when the thread is cut to approximately the right depth is to test it with a standard-threaded nut with the same kind of thread which is being cut. The nut should turn on the external thread without excess looseness.
19. When the thread has been cut to the desired depth, chamfer the end of the thread at a 45° angle to a depth equal to the thread depth. (See Fig. 55-74.)

Words to Know

apron
automatic screw machine
back-gear lever
boring bar
bull-gear pin
carriage
carriage lock
ceramic
chatter mark
chucking
clamp dog
collet
compound rest
countershaft
cross slide
cutting speed
dead center
depth of cut
leadscrew
left-hand toolholder
faceplate
facing
feed
four-jaw independent chuck
gib
handwheel
headstock
headstock center or live center
headstock spindle
knockout rod
knurling
lathe chuck
lathe dog
lathe tool
power feed
rake
right-hand toolholder
setover screw
sfpm
spindle hole
straight toolholder
straight turning
swing
tailstock
tailstock center or dead center
tailstock spindle clamp
taper turning
threading tool
three-jaw universal chuck
throw-away insert
tool bit
- back-rake angle
- end-relief
- end relief grinding angle
- face
- flank
- left-hand tool
- nose angle
- right-hand tool
- side-rake angle

tool post
turret lathe
ways

Review Questions

1. What is a machine tool?
2. List five of the most basic kinds of machine tools.
3. List several specialized mass-production machine tools which are adapted from the lathe.
4. What kinds of operations can be performed on a lathe?
5. What kinds of workers run lathes?
6. Name the main parts of the lathe.
7. Name the main parts of the headstock.
8. Name the main parts of the tailstock.
9. Where is the leadscrew located?
10. Where is the half nut located?
11. Where is the rack located?
12. Where are the back gears located?
13. How is the size of a lathe measured?
14. Why should one avoid turning the lathe on when the spindle is locked?
15. Why should most lathes be stopped before shifting or changing the controls on the headstock end of the lathe?
16. Can the controls on the carriage and apron generally be adjusted while the lathe is running?
17. How often should a lathe be oiled?
18. Define the meaning of **cutting speed** as it applies to lathes.
19. List the recommended cutting speed range for turning low-carbon steel on a lathe.
20. For what purposes are back gears used on a lathe?
21. How is the speed changed on lathes equipped with a V-belt type variable-speed drive?

22. Can speed changes be made on geared-head lathes while the lathe is running?
23. What is meant by **longitudinal feed** on a lathe?
24. What is meant by **cross feed**?
25. For what purpose is the carriage lock screw used?
26. At what angle is the compound rest set for normal turning operations?
27. What are automatic feeds?
28. What control is used to engage the automatic feed?
29. What control is used to engage the feed for thread cutting?
30. What purpose do the levers on the quick-change gear box serve?
31. Explain the purpose of the large numbers and the small numbers on the index plate on the quick-change gear box on the lathe.
32. Is the amount of cross feed and longitudinal feed per revolution always the same? Explain.
33. How much is the diameter reduced when turning with a 1/16″ deep cut?
34. List three types of lathe chucks and explain their uses.
35. Explain how a workpiece is centered in a four-jaw chuck, using chalk as an aid.
36. Of what materials are lathe cutting tools commonly made?
37. List several different kinds of lathe toolholders.
38. What type of toolholder is used for holding carbide tipped cutting tools?
39. What is meant by **end relief** and **side relief** on a tool bit?
40. What is meant by side rake on a tool bit? Back rake?
41. At what height should the cutting edge of the tool be set?
42. What is meant by **facing**?
43. Why must the tailstock center be in accurate alignment with the headstock center before centerdrilling work mounted in the chuck on the lathe?
44. For what purposes is the faceplate used on a lathe?
45. What is the angle of the point on a lathe center?
46. Why are dead centers hardened?
47. What is meant by a "live" tailstock center?
48. For what purpose are the **setover screws** on the tailstock used?
49. For what purpose are lathe dogs used?
50. Why should turning generally be done from the tailstock toward the headstock?
51. When turning between centers, what does a "squeak" indicate?
52. List three methods which can be used for turning tapers on a lathe.
53. How should the center be removed from the headstock?
54. How should the center be removed from the tailstock?
55. Why should the lathe centers be free from dirt before installing them in the lathe?
56. List several hole-machining operations which can be performed on a lathe.
57. Why are holes bored rather than just drilling them to the desired size?
58. What purpose does knurling serve?
59. What is a lathe mandrel and for what purpose is it used?
60. How can you generally avoid chatter on cutoff operations?
61. For what purpose is a steady rest used on a lathe?
62. For what purpose is a follower rest used on a lathe?
63. Explain, briefly, how threads can be cut on a lathe.
64. Explain two kinds of thread-cutting tools which can be used to cut external threads on a lathe.
65. What kind of cutting tool is used for cutting internal threads on a lathe?

Mathematics

1. Calculate the tailstock offset for turning a cylinder 10″ in length, 6″ to be tapered with a taper of ½″ per foot.
2. Calculate the tailstock offset of a drift punch having a total length of 6″, 3″ to

be tapered, large diameter of 5/8", and small diameter of 1/8".

3. Calculate the angle of taper needed to machine the punch in Problem 2 by the compound rest method of taper turning.

4. At what rpm should a 2½" diameter aluminum workpiece turn to obtain a cutting speed of 300 sfpm?

5. Is 340 rpm a satisfactory speed for machining a 1½" diameter bar of low-carbon steel?

UNIT 56

Shapers, Planers, and Their Operations

56-1. The Shaper

The **shaper**, Fig. 56-1, is one of the common basic machine tools. It uses a single-point cutting tool which is very similar to a lathe tool bit. Shapers are used primarily for machining flat surfaces. The flat surfaces may be horizontal or vertical. Grooves, slots, or keyways can be machined with a shaper. Curved surfaces also can be laid out and

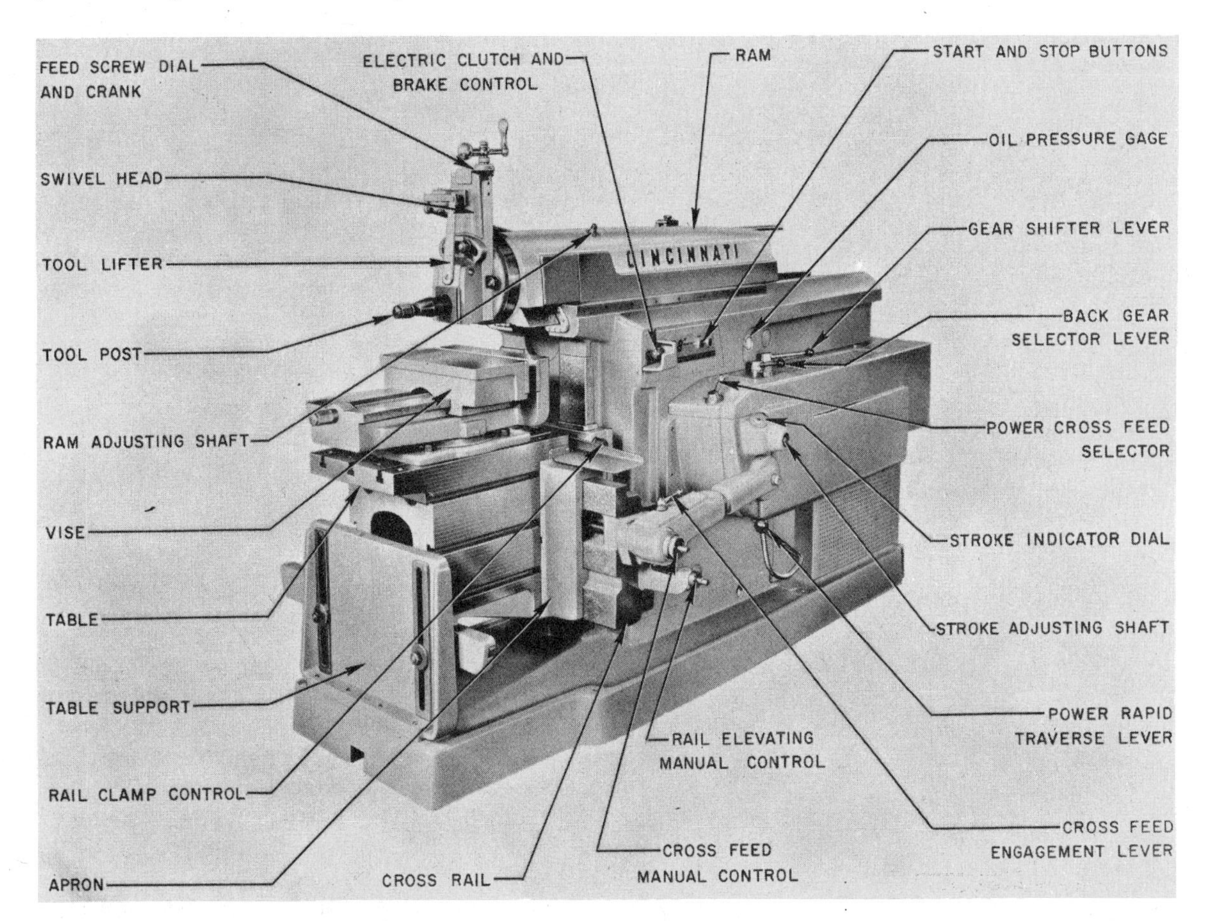

Fig. 56-1. Plain Heavy-Duty Shaper, Showing Principal Parts

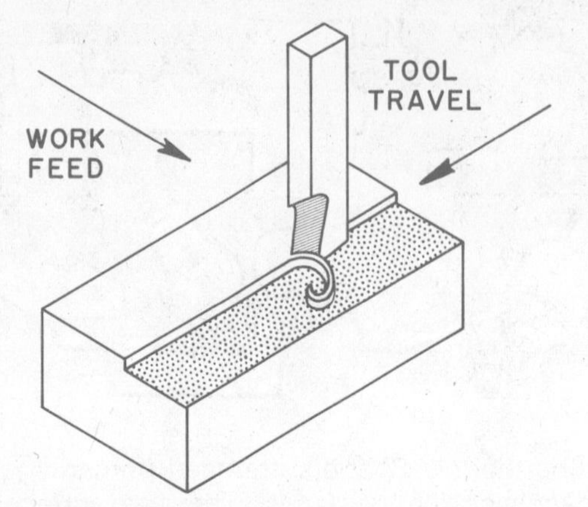

Fig. 56-2. Tool in Relation to Workpiece When Shaping

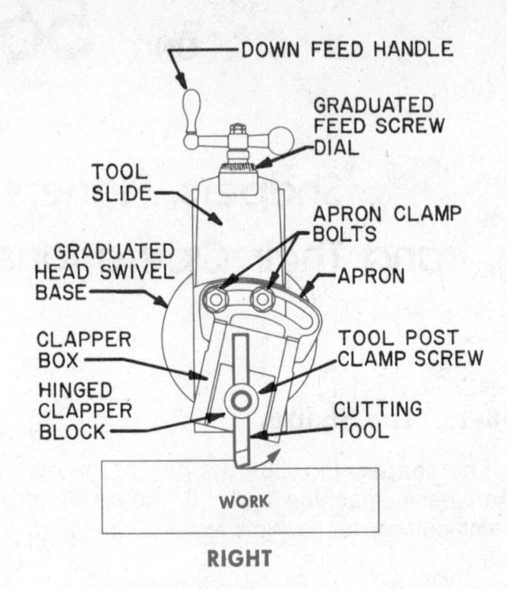

Fig. 56-3. Tool Head Assembly

machined by handfeeding the tool along the layout line.

The workpiece is held tightly in a machine vise or bolted directly to the table. The cutting tool is held in a toolholder which is moved back and forth in a straight line by a **ram**. The cutting tool **peels** off a chip each time the ram moves forward on a **cutting stroke.** The table feed is selected before operating the shaper. The table feeds crosswise during the backstroke of the ram. The machine then makes the next cutting stroke, see Fig. 56-2.

Either a shaper or a milling machine may be used for machining flat, vertical, or angular surfaces. Either machine also may be used for machining grooves. Because of a more rapid rate of metal removal, milling machines are rapidly replacing shapers and planers (see Fig. 56-34) in production machine shops. However, because of the ease of grinding and maintaining shaper tools, the shaper is frequently used in maintenance machine shops, tool and die shops, and school shops.

Two basic types of shapers are available: the **horizontal-type**, as illustrated in this unit, and the **vertical-type**. The horizontal-type is most common and is most widely used. On the vertical-type, the ram moves up and down in a vertical position, instead of the horizontal position.

The basic procedures used and the principles involved in machining with a shaper also apply to operations performed on a planer, Figs. 56-34 and 56-35. Single-point cutting tools, similar to shaper tools, are used on planers. Planers and their use are explained in section 56-23.

56-2. Shaper Size

The size of a shaper generally is designated by the maximum length of **stroke**. The common maximum stroke lengths for shapers of various sizes are 7″, 12″, 14″, 16″, 18″, 24″, and 36″. In each case, the length of stroke may be adjusted for any length from 0″ to the maximum length for the machine.

The maximum stroke length, however, is not the only factor used in designating the physical size and work capacity of shapers. They are sometimes designated for light-duty, medium- or standard-duty, or heavy-duty work. Heavy-duty shapers are heavier and more rigid than the light- or medium-duty types. For example, shapers with a 16″ stroke may range from about 2000 to 6000 pounds in weight.

56-3. Parts of Shaper

The principal parts of the shaper are identified in Figs. 56-1 and 56-3. The names of the principal parts should be learned before attempting to operate the machine. The following are some of the most important parts on a shaper:

Base

The **base** is the large casting which supports the machine. The base of large shapers rests on the floor. The base of smaller shapers often rests on a pedestal or on a bench.

Column

The **column** is an accurately machined vertical surface on the front of the large **base casting** which houses the internal parts of the machine. The complete, large base casting also is sometimes called the column. The accurately machined vertical column provides a bearing surface for the **cross rail**.

Cross Rail

The cross rail supports the **table** at one end. A **table support leg** supports the table at the opposite end. A **swivel vise** is bolted to the table for mounting workpieces to be machined.

Ram

The ram is a heavy casting which slides back and forth in **ways** which are accurately machined in the top of the large base casting.

Tool Head Assembly

The tool head assembly, Fig. 56-3, is mounted on the end of the ram. The **cutting tool** may be mounted in a **toolholder**, or it may be clamped directly in the **tool post** as shown in Fig. 56-3. The tool post is mounted on a hinged **clapper block**, which is mounted in the **clapper box**, which in turn forms part of the **apron**. During the cut, the clapper block is forced back solidly into the base of the clapper box. On the return stroke, the clapper box swings freely, thus preventing damage to the point of the cutting tool.

The apron is swiveled to the desired angle and is clamped to the **tool slide** with **apron-clamp bolts**. The depth of the cut is established with the **down-feed handle**, and it may be read directly to 0.001″ on the graduated **feed-screw dial**. The **tool head** may be swiveled at any desired angle for making angular cuts, as shown in Fig. 56-30.

Drive System

Most shapers are driven by an electric motor. The drive system uses a system of belts, gears, and levers to change speeds and the length of the cutting stroke. A **clutch** is used to engage or disengage the power to the machine. Some large shapers are hydraulically operated.

56-4. Adjustments with Power Off

Become acquainted with the manual adjustments before operating the shaper with power. The operating controls and levers are somewhat different on different makes of shapers. However, the general procedures used in making various adjustments are similar on most shapers. On most kinds of shapers the following adjustments can be made manually, without the machine being turned on:

1. Tool-slide adjustment (see § 56-5).
2. Horizontal table movement (see § 56-6).
3. Table elevation (see § 56-7).
4. Length of stroke (see § 56-8).
5. Position of stroke (see § 56-9).
6. Cutting speed (see § 56-10).
7. Feed adjustment (see § 56-11).

56-5. Tool-Slide Adjustment

Move the tool slide, Fig. 56-3, up or down with the **down-feed handle.** This handle turns the tool-slide feed screw, thus establishing the depth of cut desired. The down-feed handle is used for feeding the tool manually when making vertical cuts, Fig. 56-29, and for making angular cuts, Fig. 56-30. Some shapers are equipped with a **tool-slide lock screw** or **lever**. The locking device prevents the tool from **digging in**, due to backlash in the down-feed screw, while making heavy horizontal cuts.

When horizontal cuts are made, both the tool and the tool slide should have as little overhang as possible, see Fig. 56-4. Check to see that the tool and the toolholder are clamped tightly. Also check to see that the apron is swiveled in a direction opposite to the direction in which the tool is feeding, see Fig. 56-3. This will enable the tool and clapper block to swing up and away from the work on the **return** or backward stroke. When possible, the tool should be in a vertical position, as shown in Fig. 56-4, so that it cannot dig further into the work if it becomes loosened. When the depth of cut has been established and set, the tool-slide locking lever or device should be tightened just enough to hold snugly.

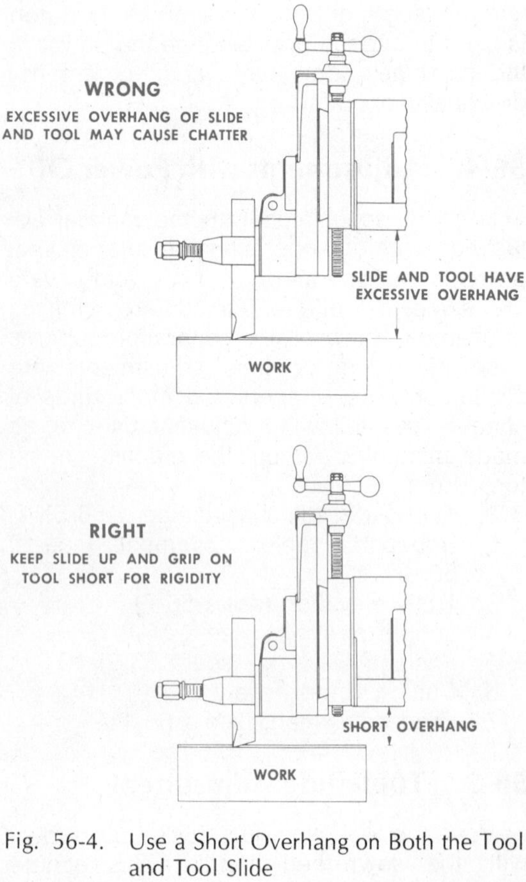

WRONG
EXCESSIVE OVERHANG OF SLIDE AND TOOL MAY CAUSE CHATTER
SLIDE AND TOOL HAVE EXCESSIVE OVERHANG
WORK

RIGHT
KEEP SLIDE UP AND GRIP ON TOOL SHORT FOR RIGIDITY
SHORT OVERHANG
WORK

Fig. 56-4. Use a Short Overhang on Both the Tool and Tool Slide

56-6. Horizontal Table Movement

Place a hand crank on the **longitudinal- or cross-feed screw**, Fig. 56-1, and move the table crosswise in either direction. In order to do this, the **cross-feed engagement lever** or knob must be disengaged. The table then slides freely in either direction.

The **table assembly** on a shaper consists of several major parts, see Fig. 56-1. The parts include the **table**, the **apron** to which

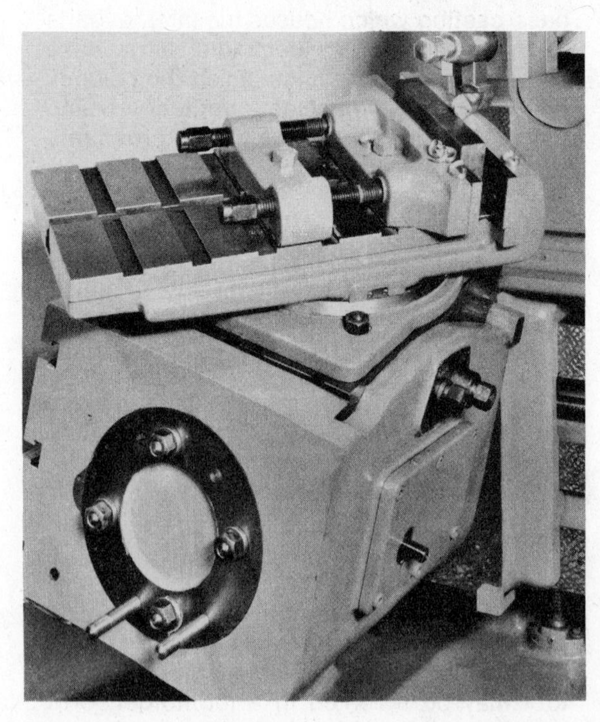

Fig. 56-5. Set Up with Both Universal Table and Vise Swiveled

the table is attached, and the **table support**. A heavy-duty swivel vise is generally bolted to the table.

The table on a shaper may be either the **plain**-type or the **universal**-type. A universal table, as shown in Fig. 56-5, may be swiveled about its axis for making angular cuts.

56-7. Table Elevation

The table can be raised or lowered as required for large or small workpieces. The table should be set at an elevation which permits minimum overhang of the tool and tool slide, as in Fig. 56-4. On most shapers, the cross rail is equipped with several column **clamping nuts** which are carefully adjusted for proper clearance along the machined vertical column bearing surfaces. These clamping nuts may be located on one or both columns. Only **one** of the clamping nuts is designed for clamping the cross rail and table assembly securely to the column while cuts are made. This **clamping nut** must be loosened **slightly** in order to raise or lower the table. Ask your instructor to show you which **column clamping nut** must be loosened in order to raise or lower the table. The clamping nuts on the table support must also be loosened before raising or lowering the table.

The cross rail and table assembly are raised or lowered with a hand crank placed on the **rail elevating manual control**, Fig. 56-1. When the table is at the proper elevation, lower it and again raise it slightly to take out the backlash in the adjustment screw. Then tighten the **column clamping nut** and the clamping nuts on the table support legs. For many jobs with workpieces of similar size, the table may be left at the same elevation. However, it must be raised or lowered for smaller or larger workpieces.

56-8. Length of Stroke

The stroke should be at least 3/4″ longer than the cut. It should start about 1/2″ before the cut and should continue at least 1/4″ beyond the cut. The length of stroke is determined by the position of the crank pin and sliding block. These are located on the **vibrating arm** inside the machine, See Fig. 56-6. The vibrating arm is connected to the main-drive gear. The farther the crank pin is located from the axis of the drive gear, the longer the stroke. On the small shaper in Fig. 56-6, the graduation marks on the vibrating arm indicate the length of stroke.

The length of stroke is adjusted somewhat differently on different shapers. On the large shaper in Fig. 56-1, the stroke-adjusting shaft is turned to the desired stroke length. The length is indicated on the stroke-indicator dial.

On most smaller shapers, first move the ram to the extreme rear position if the handwheel is provided; if no handwheel is provided, the power must be used. On a small shaper, such as the machine in Fig. 56-6, loosen the nut on the crank pin and move the sliding block to the graduation mark which indicates the desired length of stroke. Then tighten the nut on the crank pin.

Some shapers have a **lock nut** located on the stroke selector shaft. On machines so equipped, the lock nut must be loosened before attempting to adjust the length of

Fig. 56-6. Rocker Arm Is Graduated for Stroke Adjustment

stroke. After the length has been adjusted, the lock nut again must be tightened before the machine can be operated.

56-9. Position of Stroke

The position of a given stroke, such as a stroke 6″ long, may be moved forward or backward, depending on the location of the workpiece mounted in the vise or bolted to the table. Thus, the position of the stroke is always adjusted after the workpiece is mounted. First the length of the stroke must be adjusted before attempting to adjust the position of the stroke.

To adjust the position of the stroke, first move the ram to the extreme rear position and adjust the length of stroke desired. Loosen the ram **hand clamp** device on machines so equipped; the ram hand clamp is located on top of the ram or above the stroke indicator. The position of the ram is then adjusted by turning the **ram adjusting shaft**, Fig. 56-1, with a hand crank. On small machines, the ram is pushed manually to the desired position. When the position of the stroke has been adjusted, the ram **hand clamp** again must be tightened before starting the machine.

56-10. Cutting Speed

The cutting speed is designated in surface feet per minute (sfpm). The cutting speed is increased by increasing the number of cutting strokes per minute. The cutting speed for shaper work, like the cutting speed for turning on a lathe, depends upon the following:

1. Kind of material in the cutting tool.
2. Kind of material being machined.
3. Machinability of the material being machined.
4. Depth of cut.
5. Rigidness of the machine and the workpiece setup.

The cutting speeds for shaping with high-speed steel cutting tools are similar to the cutting speeds used for turning on a lathe. The same recommended cutting speeds may be used, (see Table 36). The following formula may be used for calculating the approximate number of strokes per minute required for a given cutting speed:

$$N = \frac{CS \times 7}{L}$$

N = Number of strokes per minute
CS = Cutting speed for metal being cut (in feet per minute)
7 = Multiplier to convert feet to inches (a shaper cuts about two-thirds of the time)
L = Length of cutting stroke in inches

Example: Calculate the approximate number of strokes per minute required for shaping a piece of low-carbon steel with a 6″ stroke at 90 surface feet per minute.

$$N = \frac{90 \times 7}{6}$$

$$N = \frac{630}{6}$$

$$N = 105 \text{ strokes per minute}$$

When the number of strokes per minute is given, the cutting speed can be calculated with the following formula:

$$CS = \frac{N \times L}{7}$$

The method for changing the number of strokes per minute is somewhat different for different makes of machines. On larger machines of the type in Fig. 56-1, the number of strokes desired is selected by shifting the two gear-shaft levers provided. The number of strokes per minute on machines with variable-speed drive is changed by turning the handwheel **while the machine is running.** On some shapers, the number of strokes per minute is changed by changing a V-belt to different steps on the drive pulley.

56-11. Feed Adjustment

The feed can be adjusted for heavy or light feeds. For horizontal cuts, **feed** is the distance the table moves horizontally along the cross rail for each new cut, see Fig. 56-2. The amount of feed depends on the same factors as cutting speed. Lighter feeds generally are used for (1) deep cuts, (2) cuts with light-duty shapers, and (3) producing a smoother surface finish.

For finishing cuts, a feed of 0.010″ to 0.020″ and a depth of 0.005″ to 0.015″ is recommended. A finishing tool with a nose radius of ⅛″ or larger is generally used. Sometimes a finishing tool with a flat nose or a nose with a slightly elliptical shape is used, see Fig. 56-17. With finishing tools of this kind, much coarser feeds may be used.

For deep roughing cuts, a fine feed of 0.010″ to 0.020″ is often used. For shallower roughing cuts, a coarser feed may be used. Larger shapers can take much heavier cuts than small shapers, see Fig. 56-7. As a general rule, the feed and the depth of the cut should not be greater than the machine, the cutting tool, and the setup can stand. If the workpiece is not held tightly in position, it may be pushed off the machine.

On most smaller shapers the feed is changed by sliding the feed connecting rod in the **slotted crank disk**. The feed connecting rod operates the longitudinal-feed screw which feeds the table crosswise. To increase the feed rate, move the feed-adjustment selector toward the outside of the slotted crank.

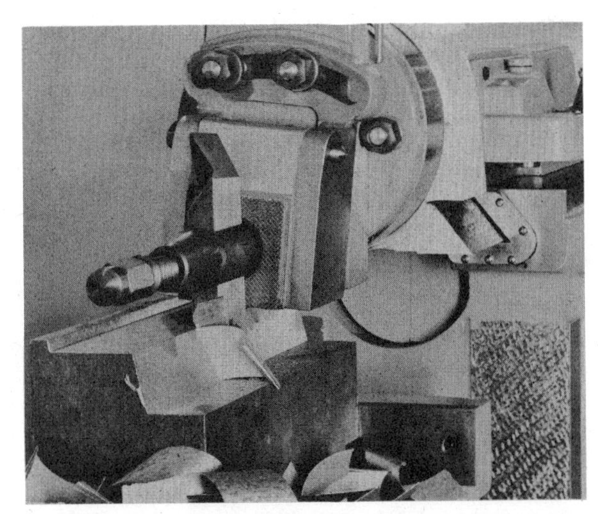

Fig. 56-7. Making a Heavy Cut with a Large Solid Cutting Tool on a Heavy-Duty Shaper

Safety Note

The feed must be set so that the table feeds crosswise on the return or backward stroke.

If the table feeds on the forward or cutting stroke, the feed-adjustment selector should be moved to the opposite side of the slotted crank disk. If the table feeds crosswise during the cutting stroke, the tool will be damaged.

On large shapers, as in Fig. 56-1, the feed is changed by manually turning a dial to the desired feed rate. The feed is engaged on most shapers by moving a **feed-engagement lever,** knob, or ratchet in the desired direction of feed.

56-12. Adjustments with Power On

Many smaller shapers are equipped with a **handwheel** which enables the operator to run the **ram** through the complete stroke without turning on the power; on machines so equipped, this procedure should always be followed before turning on the power. Many newer shapers and larger shapers are not equipped with a handwheel for manual movement of the ram. On these machines power to the ram is controlled with the **clutch lever** or with a **clutch and brake control,** see Fig. 56-1.

Safety Note

Never run the ram back into the column with the tool slide set at an angle. The slide will strike the column and cause serious damage to the tool head assembly. (See Fig. 56-8.)

1. Before starting the machine, observe that the tool slide is in a vertical position (for either horizontal or vertical cuts), and see that the tool will clear the vise, the workpiece, and the setup.

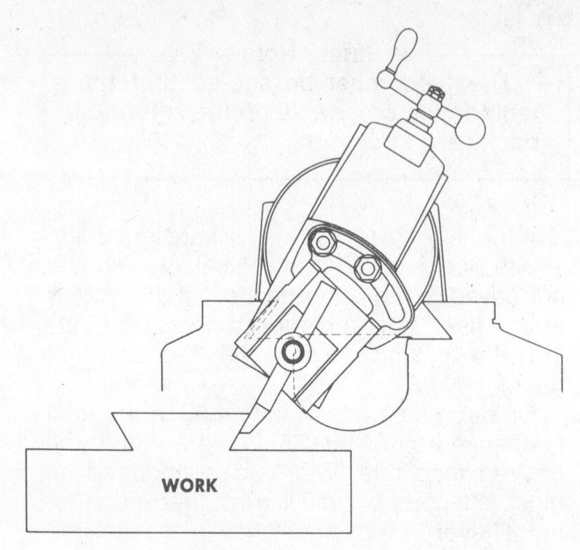

Fig. 56-8. CAUTION: Do Not Run the Ram Back into the Column with the Slide Set at an Angle; Slide Will Strike Column

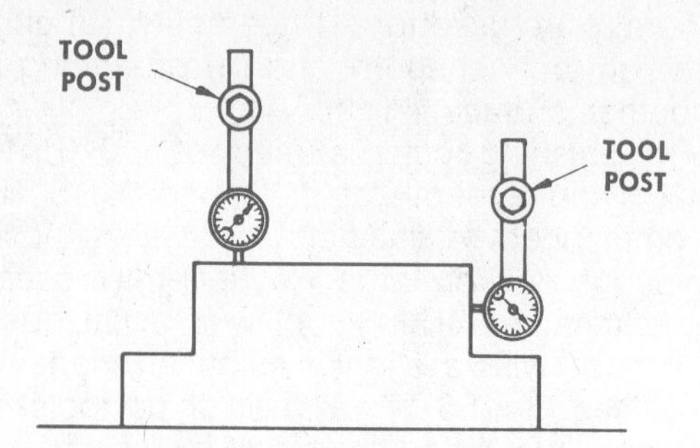

Fig. 56-9. Using Dial Indicator for Setting Work Level and Parallel

2. Turn on the motor, engage the clutch, and observe the stroke.
3. Engage the power automatic cross feed with the **cross-feed engagement lever**, Fig. 56-1. When the lever is swiveled to the right, the table feeds to the right along the table cross rail. When the lever is swiveled to the left, the table feeds to the left. The left side of the machine is the side on which most of the controls are located. The operator operates the machine from the left side. Some shapers have a **ratchet knob**, instead of a **cross-feed engagement lever**, for engaging automatic cross feed.
4. After observing the operation of the stroke and the automatic cross feed, disengage the cross feed, disengage the clutch, and stop the machine.

56-13. Vise Alignment

A swivel vise mounted on the shaper table can be used for holding many kinds of workpieces. The vise generally is bolted on the top surface of the table. However, it may also be bolted on the side of the table for special job setups, see Fig. 56-26. The base of the vise is provided with a keyway and a key which keep the swivel base aligned with the table. The vise jaws can usually be set accurately enough for most jobs by careful alignment of the degree graduation mark and the index mark on the swivel base.

For very accurate work, either the workpiece or the vise may be aligned more accurately with the use of a dial indicator, see Fig. 56-9.

56-14. Safety with the Shaper

1. Always wear approved safety goggles or a safety face shield.
2. Be certain that the vise and the workpiece are fastened securely.
3. Use the right cutting tool for the job.
4. Remove all wrenches or other material from the machine table before starting the machine.
5. Select the proper speed, feed, and depth of cut for the kind of material being machined.
6. Do not run the ram back into the column when the tool slide is set at an angle. To do so will cause serious damage to the tool head assembly. (See Fig. 56-8.) On machines equipped with a **handwheel**, run the ram

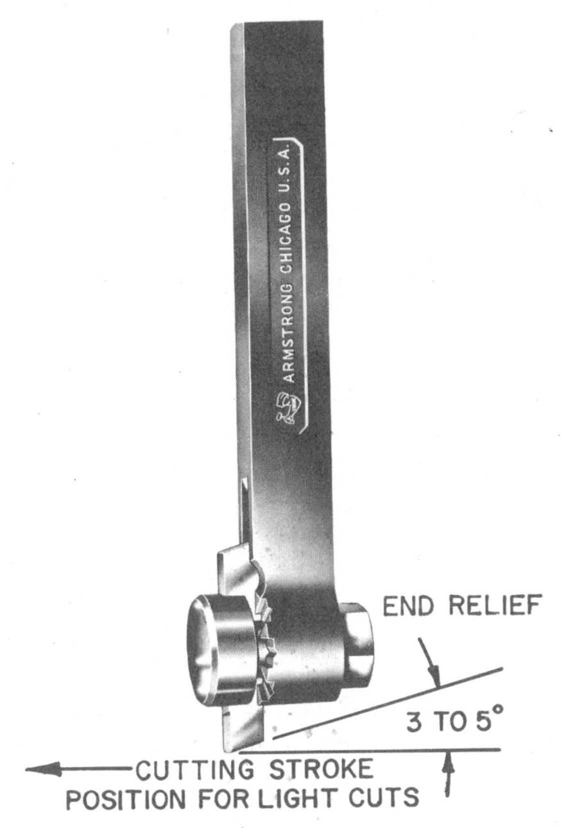

Fig. 56-10. Planer and Shaper Toolholder

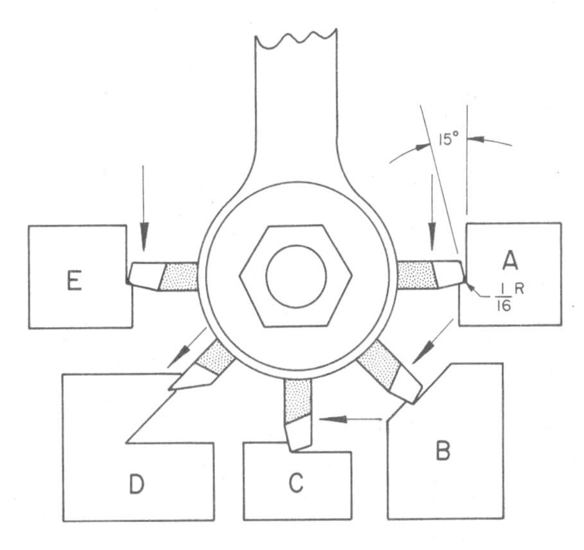

Fig. 56-11. Various Tool Positions in Swivel Head of Planer and Shaper Toolholder
A. Vertical Cut
B. Angular Cut
C. Horizontal Cut
D. Angular Dovetail Cut
E. Vertical Cut

through its full stroke manually before turning on the power to start the machine. This will help determine that the tool clears the work, and the tool slide does not strike the column.

7. Be sure that no one is inside the safety zone for the machine.
8. Check to see that the back of the ram will not strike some object or some person.
9. Stand to the left of the machine, parallel to the stroke of the machine, while operating it. Do not reach across the work or the machine table while the machine is running.
10. Do not touch the tool or the work while the machine is running.
11. Remove chips with a brush only when the machine is stopped.
12. Make adjustments on the machine only while it is stopped.
13. Stop the machine before leaving it.
14. Clean the machine and the area when you complete the job.
15. Remove sharp burrs or edges from machined parts to avoid being cut.

56-15. Tools and Toolholders

Shapers use **single-point** cutting tools which are very similar to single-point lathe tools. The shaper tool generally is held in a standard **planer and shaper toolholder**, as shown in Figs. 56-10 and 56-11. The terms which apply to single-point cutting tools on a shaper or planer are essentially the same as the terms which apply to lathe tools. Thus, cutting-tool terminology for many kinds of single-point cutting tools has been standardized. Cutting-tool terms are shown in Figs. 56-31 and 56-33 and are explained in section 55-14. These terms should be understood before attempting to grind a shaper tool bit.

Fig. 56-12. Using Planer and Shaper Toolholder in Close Corners

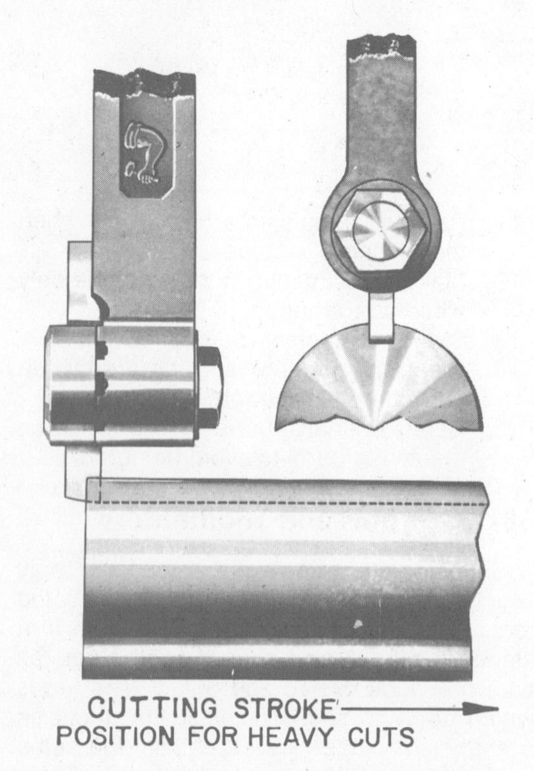

Fig. 56-13. Tool and Toolholder Reversed for Cutting Keyways and for Making Heavy Cuts

Shaper Toolholder

A standard planer and shaper toolholder, Fig. 56-10, holds the cutting tool parallel to the shank of the toolholder and perpendicular to the surface of the work. Note that no rocker block is provided under the tool; therefore, no back-rake angle is provided by the toolholder. The desired back-rake angle must be ground on the tool bit. With the planer and shaper toolholder, the tool may be rotated to five different positions for various kinds of cuts, as shown in Fig. 56-11. The toolholder and tool may be swiveled to various other positions for machining other kinds of surfaces, as in Fig. 56-12.

The position of the tool in relation to tool travel is important when using the planer and shaper toolholder. For **light cuts**, the tool is clamped in the front of the toolholder as shown in Fig. 56-10. For **moderate or heavy cuts,** the toolholder and the tool generally are reversed as shown in Fig. 56-13. This procedure brings the point of the tool back to the shank of the toolholder, causing

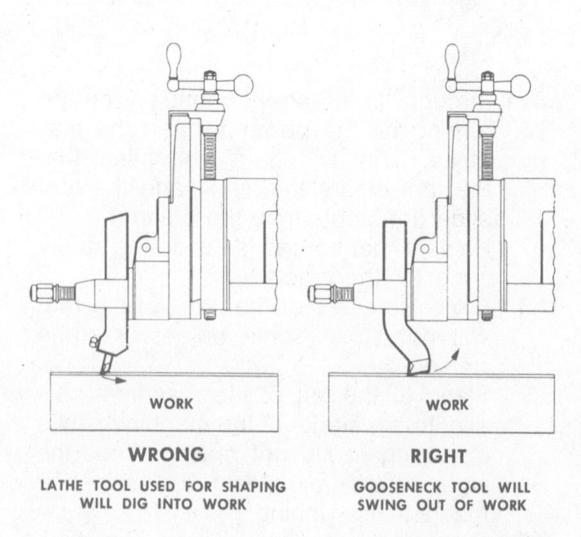

Fig. 56-14. Position of Tool and Angle at Which Tool Is Held Affects Cutting Angle

an effect similar to using a **gooseneck tool.** This reduces the possibility of **tool chatter** or **digging in**. (See Fig. 56-14.)

Lathe Toolholder

A **zero-degree** (straight) lathe toolholder, Fig. 55-31, may also be used in a shaper. It holds the tool bit parallel to the shank of the toolholder and perpendicular to the work in the same manner as the shaper and planer toolholder. However, a 16½° toolholder, Fig. 55-30, should not be used in a shaper since the tool will have more tendency to **dig in**, as in Fig. 56-14.

Extension Toolholder

Extension toolholders, Fig. 56-15, are used for internal machining operations such as cutting keyways, Fig. 56-16, or for machining holes of various shapes. The toolholder holds a bar which is similar to a boring bar used on a lathe.

Shaper Tools

Shaper tool bits are ground in a manner similar to lathe tool bits. On shaper tools, however, smaller relief (formerly called clearance) angles are used. An end-relief angle of 3° to 5° and a side-relief angle of 3° to 5° are appropriate for most shaping operations, see Fig. 56-10. The small relief angles provide greater strength at the cutting edge where high-impact forces are applied during the cutting stroke.

Side-rake and back-rake angles vary with the kind of material being machined. **Side-rake angles** similar to those for lathe tools, Table 36 may be used. An average side-rake angle of 10° is satisfactory for many general-purpose shaping operations on steel or cast iron. Little or no **back-rake** angle is necessary for most shaping operations. However, a back-rake angle of 2° to 5° is often used for shaping soft steel or soft cast iron. Certain special finishing tools are sometimes given a back-rake angle to increase the keenness of the cutting edge.

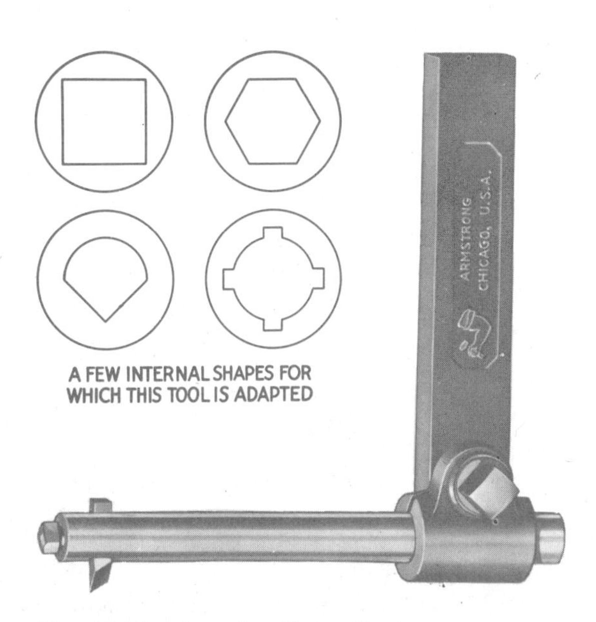

Fig. 56-15. Extension Shaper Tool

Fig. 56-16. Shaping an Internal Keyway

The following ranges of angles are recommended for a general-purpose shaper tool:

End relief, 3° to 5°
Side relief, 3° to 5°
Side rake, 10°
Back rake, 0° to 5°

Large solid-shank cutting tools are often used for heavy-duty shaping operations, as in Fig. 56-7. For the majority of shaping operations, however, tool bits made of high-speed steel are used.

On shapers, a **left-hand** or **left-cut** tool is most often used. The definition of a **right-hand** or **left-hand** single-point tool is the same for all single-point tools, whether they are used on a lathe, shaper, or planer. When the tool is viewed from the point end face up, a left-cut tool has the cutting edge at the left side. Hence, when looking at the cutting tool from the front of the shaper, the cutting edge of the tool is at the left side, and the table and the work feed from the left toward the right. On a lathe, however, a right-cut tool is most often used.

Some of the common shapes of cutting tools used on the shaper include the following:

1. **Roughing Tools**: A left-cut roughing tool is shown in Fig. 56-11 at **A, B,** and **C**. Note the 1/16″ nose radius on the tool. A right-cut roughing tool is shown at **E** in Fig. 56-11. A round-nose tool with a flat face (no side-rake angle) may be fed in either direction for roughing cuts, Fig. 56-19.
2. **Finishing Tools**: Several kinds of finishing tools may be used for finishing cuts. The tool shown in Fig. 56-11 at **A**, **B**, and **C**, but with a nose radius of ⅛″ or larger, will produce good results with feeds of 0.010″ to 0.020″. For finishing cuts with coarser feeds, a square-nose tool or a tool with a slight elliptical - shaped nose is recommended, (see Fig. 56-17). A side-rake angle of about 10° to 15° generally is

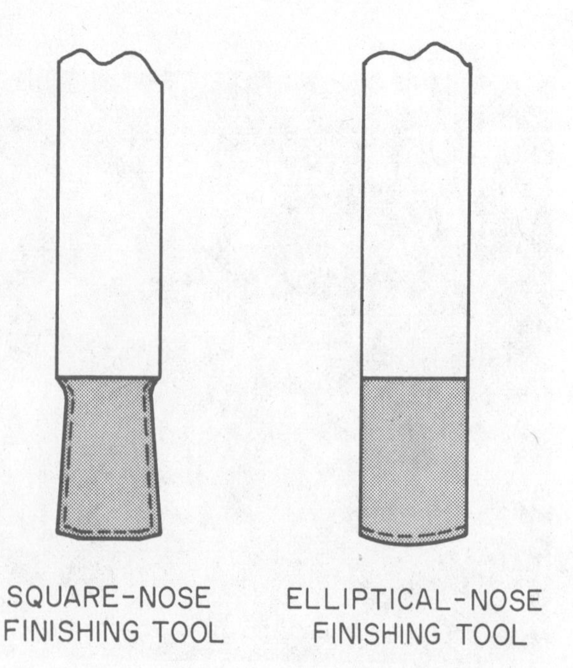

Fig. 56-17. Tools for Finishing Cuts on the Shaper

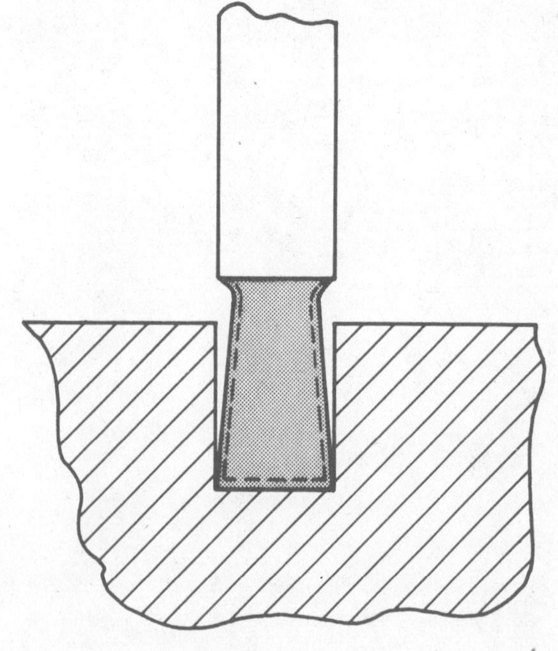

Fig. 56-18. Square-Nose Tool for Cutting Grooves or Keyways

used on the elliptical-nose tool for finishing cuts on steel. The side-rake angle causes a shearing effect which produces an improved surface finish on many kinds of steel.

3. **Square-nose Tool**: A square-nose tool Fig. 56-18, is used for operations such as cutting grooves, shoulders, or keyways.
4. **Dovetail Tool**: A dovetail tool is shown at D in Fig. 56-11. This kind of tool is used for shaping dovetail angles, as shown in Fig. 56-31.

56-16. Holding the Work

The workpiece must be held securely in position for machining.

Machine Vise

Many kinds of workpieces can be held in the vise provided with the machine. The vise may be bolted to the top of the machine table or to the side of the table as in Fig. 56-26. Other kinds of workpieces can be bolted directly to the machine table as in Fig. 56-25. A wide variety of work-holding tools may be used for bolting workpieces directly to the machine table. (Examples are shown and explained in Unit 50.)

The machine table and the vise should be free from nicks or dirt before the vise is bolted to the table. The vise jaws should also be free of dirt or nicks before the work is clamped in the vise. The workpiece should

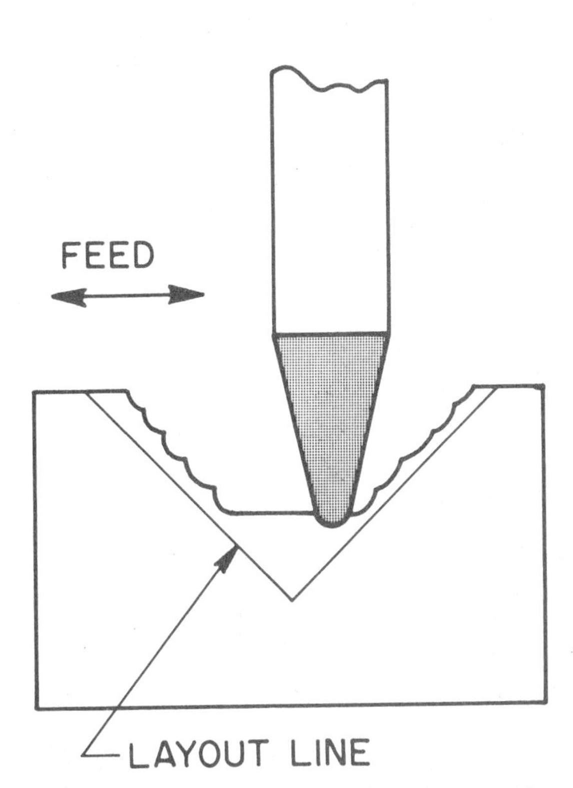

Fig. 56-19. Roughing Out a V-Shaped Area with a Round-Nose Tool

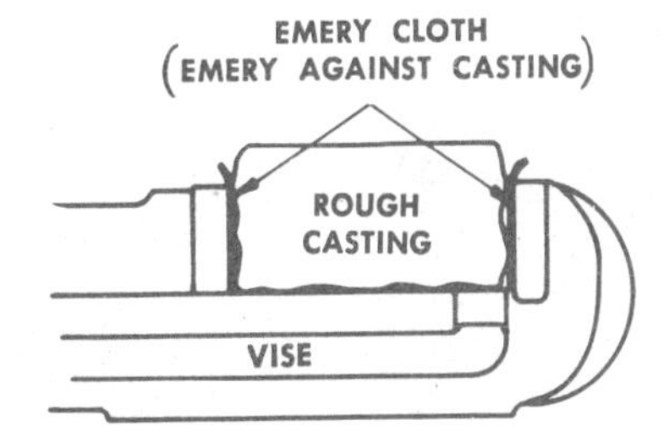

Fig. 56-20. Mounting a Rough Casting in a Vise

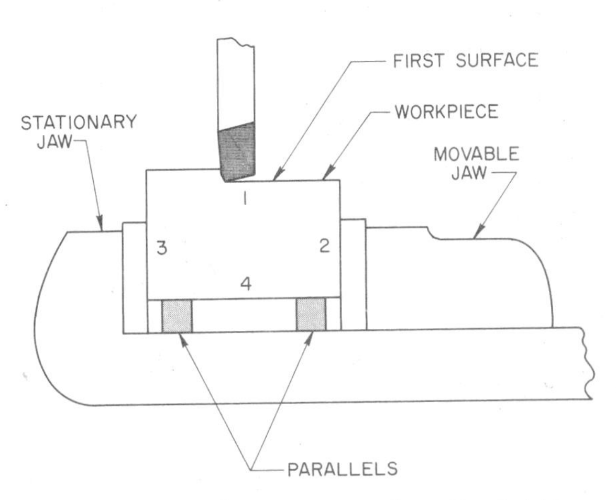

Fig. 56-21. Workpiece Mounted on Parallels in Vise; Machining the First Surface of a Block to Be Shaped Square and Parallel

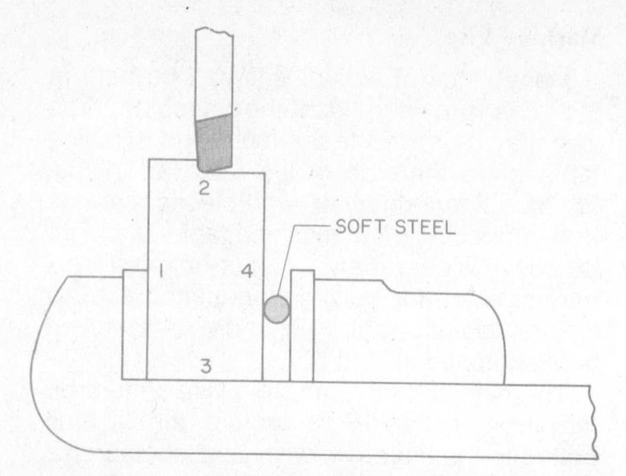

Fig. 56-22. Machining the Second Surface of a Block to Be Shaped Square and Parallel; the Finished Surface Number 1 Is Placed Against the Stationary Jaw

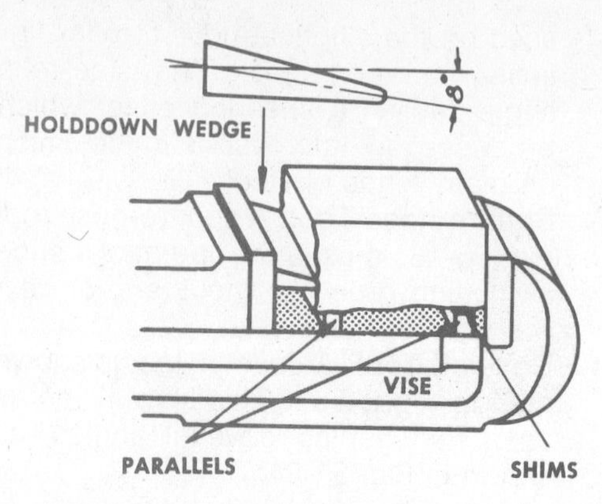

Fig. 56-24. Using Hold-Downs

Fig. 56-23. Hold-Downs

extend above the hardened vise jaws far enough for the cutting tool to avoid striking the jaws. (See Fig. 56-21.)

Emery cloth or soft aluminum sheet may be placed against the rough surface of rough castings to aid in holding them securely in the hardened vise jaws, Fig. 56-20. Sometimes it is necessary to place sheet metal or paper shims underneath a workpiece to raise it to the proper level. (See Fig. 56-24.)

Parallels

Precision parallels are used to raise and seat the workpiece at the proper elevation in the vise, Fig. 56-21. They also hold the workpiece parallel to the bottom of the vise. A round aluminum, brass, or soft steel bar sometimes is used between the movable jaw and the workpiece, as in Fig. 56-22. The round bar holds the finished surface of the workpiece firmly against the stationary vise jaw. This procedure causes the surface being machined to be square with the finished surface which is against the stationary jaw.

Hold-Downs

Hold-downs are wedge-shaped, hardened steel bars with the thick edge beveled at an angle of 2° or 3°, Fig. 56-23. This causes the thin edge of the hold-downs to press the workpiece downward, thus seating it parallel to the bottom of the vise, see Fig. 56-24. Hold-downs will hold the work securely for light and moderate cuts.

Indexing Attachment

Flat surfaces can be machined on round shafts by mounting the shaft between centers on an **indexing attachment**, see Fig. 56-27. Any number of surfaces may be equally spaced or spaced at any desired interval by using this attachment.

56-17. Horizontal Cuts

Shapers are most frequently used for machining plane, flat, true surfaces. One or more roughing cuts, as necessary, are made first. The final cut generally is a light finishing cut which produces a smooth finish. Be certain that you understand the safety practices recommended in section 56-14 before starting the machine. Wear approved safety goggles or an eye shield.

Procedure for Roughing Cut

1. Select the workpiece and remove burrs or bumps. Burrs may be filed off. Larger bumps on castings may be ground off. Make layout lines which indicate the amount of material to be removed.
2. Remove dirt or chips from the vise. See that the vise is bolted securely to its swivel base and to the table.
3. Mount the workpiece in the vise as explained in section 56-16. A large workpiece may be bolted to the table, as shown in Fig. 56-25.
4. Select a left-cut roughing tool and clamp it in the toolholder as shown at C in Fig. 56-11. Clamp the toolholder in the tool post with short tool overhang and with short tool-slide overhang, shown in Fig. 56-4.
5. The clapper box may be clamped in the vertical position for horizontal cuts. However, some operators prefer to swivel the clapper box to the right, shown in Fig. 56-3, for horizontal cuts. With the clapper box swiveled, the tool swings **up and away** from the work on the return stroke. Tighten the apron clamp bolts. Adjust the cutting tool and toolholder so that the tool is in a vertical position or slightly to the right, shown at the right in Fig. 56-3.
6. Adjust the table elevation so that the top of the workpiece is just below and clear of the cutting tool.

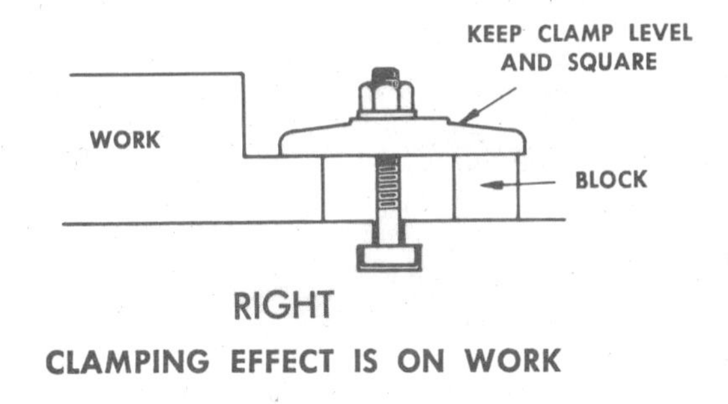

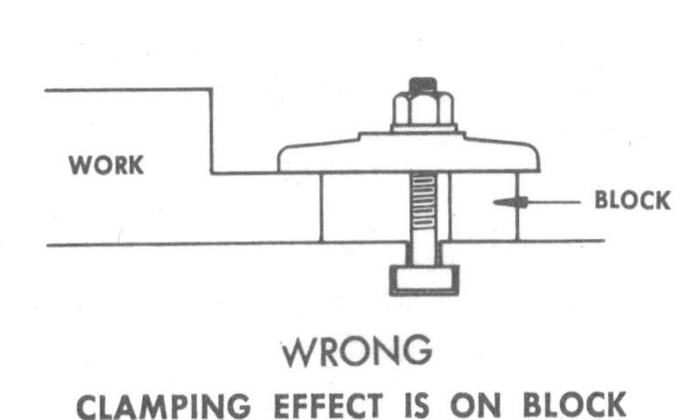

Fig. 56-25. Clamping Work to Table

7. Adjust the length of the stroke. The stroke should be at least ¾″ longer than the length of the workpiece.
8. Adjust the position of the stroke. The stroke should start about ½″ before the cut, and it should end with the cutting tool at least ¼″ beyond the workpiece at the end of the cutting stroke.
9. Set the machine for the number of strokes necessary for the desired cutting speed.
10. Adjust the tool for the desired depth of cut. To do this, feed the work crosswise under the tool with the left hand. With the right hand on the down-feed screw, feed the tool down until it nearly touches the work. With the left hand, feed the work crosswise away from the tool, and with the right hand, feed the tool down to the desired depth of cut.

    On cast iron the cut should be deep enough to cut well under the hard surface scale, usually a minimum of

1/16″ deep. In general, a roughing cut may be as deep as the sturdiness of the machine and setup will permit. On most medium-duty shapers, a cut from 0.060″ to 0.125″ depth is recommended. Always allow enough material for a finishing cut of 0.005″ to 0.015″ depth.

11. Adjust the machine to the desired feed rate. A feed of 0.020″ to 0.030″ is suggested for average roughing cuts.
12. With the handwheel, on shapers so equipped, run the ram through the complete stroke without turning on the power. Check to see that the stroke length and position are correct. Be sure that the tool slide does not strike the vertical column.
13. Turn on the machine, engage the clutch, and observe the stroke for proper clearance of the work.
14. Engage the automatic cross feed and make a horizontal roughing cut. Be sure that the table feeds during the return stroke.
15. When the roughing cut has been completed, disengage the clutch and the automatic cross feed. With a hand crank on the cross-feed screw, feed the table back to the original position, ready for the next cut.
16. If more roughing cuts are required, make them by repeating steps 13, 14, and 15.

Procedure for Finishing Cut

1. Select a finishing tool and mount it in the toolholder, as in step 4 above.
2. Set the finishing tool for the desired depth of cut. With a square-nose tool or an elliptical-nose tool, a cut from 0.005″ to 0.015″ depth is recommended.
3. Set the machine for the desired feed rate. Feed rates of 0.020″ to 0.060″ may be used with square-nose or elliptical-nose finishing tools. With a round-nose tool which has a nose radius of ⅛″ or larger, a feed of 0.010″ to 0.020″ should be used for best results.
4. Start the machine and make the cut as in steps 12 through 15 above.

Fig. 56-26. Vise Mounted on Side of Table

Fig. 56-27. Flat Surfaces Being Machined on Shaft with Indexing Head

56-18. Vertical Cuts

Vertical cuts are made when machining the ends of stock square. They are also used for machining shoulders square, as in Fig. 56-29. Vertical cuts are made by feeding the tool downward with the down-feed handle.

Procedure for Roughing Cut

1. Follow steps 1, 2, and 3 in making horizontal cuts (§ 56-17).
2. Swing the clapper box to the right and clamp tightly, as in Fig. 56-3. The tool will then swing **up and away** from the work on the return stroke.
3. Check to see that the tool slide is in the vertical position, as indicated by the index mark on the graduated swivel base.
4. Select a tool and toolholder and mount in the tool post.
   a. For machining across the end of a workpiece which extends beyond the end of the vise, select a right-cut tool. Position the tool in a shaper toolholder as at **E** in Fig. 56-11. The toolholder should have enough overhang that the tool can cut along the vertical surface without striking the work with the clapper box. Check for clearance of the clapper box by feeding the tool down and up past the surface to be machined.
   b. For machining a vertical surface to a shoulder, as in Fig. 56-29, select a left-cut tool. The tool with the round nose, Fig. 56-28, works best

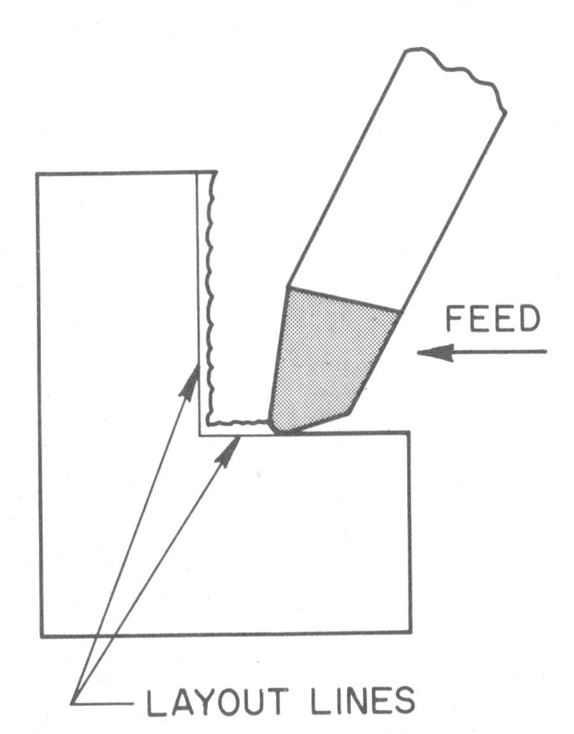

Fig. 56-28. Machining the Horizontal Surface on a Shoulder

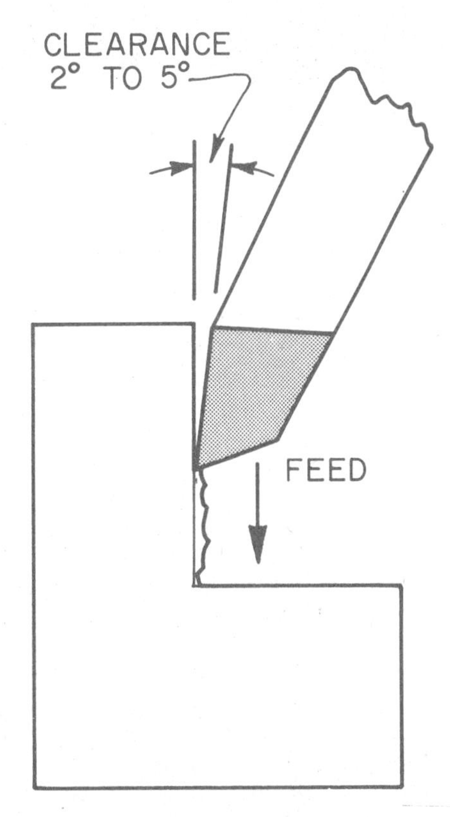

Fig. 56-29. Machining the Vertical Surface on a Shoulder

Fig. 56-30. Making an Angular Cut with a Shaper

for roughing cuts. The tool with the pointed nose, Fig. 56-29, works best for finishing cuts.
Mount the tool in a zero-degree lathe toolholder, the type shown in Fig. 55-31. Mount the toolholder in the tool post at an angle which provides 2° to 5° clearance, as in Fig. 56-29. Allow the toolholder to extend far enough below the clapper box to complete the cut without having the clapper box strike the work. This may be checked by feeding the tool down and up, past the surface to be machined.

5. Adjust the table elevation if necessary. See that the tool is clear of the work and located just above the work.
6. Adjust the length of stroke.
7. Adjust the position of the stroke.
8. Set the machine for the number of strokes required for the desired cutting speed. (See § 56-10.)
9. With the tool above the workpiece, move the table horizontally so that it is in position for the proper **width of cut.**
10. Turn on the power, engage the clutch, and observe that the stroke clears the work properly.
11. Make the vertical cut: With the right hand on the down-feed handle, feed the tool down about 0.005″ to 0.010″ on each return or backward stroke. Continue cutting to the desired depth.
12. Stop the machine, and feed the tool up so it clears the top of the workpiece. Check the machined surface for squareness with a square.
13. If more than one roughing out is necessary, repeat steps 9 through 12.

Procedure for Finishing Cut

1. Select and install a finishing tool of the type shown in Fig. 56-29.
2. Make a fine finishing cut, 0.005″ to 0.015″ depth. Feed the tool down about 0.005″ on each backward or return stroke.

56-19. Machining a Block Square

In many machining applications, it is necessary to machine a block or bar of metal square and parallel. The surfaces must be parallel, and the ends must be parallel. When a block is squared with a shaper, the block is machined by using a series of horizontal and vertical cuts. The top, bottom, and both sides of a block or rectangular bar, as in Fig. 56-21, are machined with horizontal cuts. The work is mounted in the vise with the jaws parallel to the ram. The ends of long blocks are machined with vertical cuts.

A short block may be positioned vertically in the shaper vise for machining the ends square. However, when this procedure is used, the vise should be turned at right angles with the direction of the stroke.

Procedure for Shaping a Block Square and Parallel

1. Machine surface No. 1 with horizontal cuts, as in Figs. 56-21 and 56-20.
2. Machine surface No. 2 with horizontal cuts, as in Fig. 56-22. Place the machined surface No. 1 against the stationary jaw. Insert a round, soft metal rod between the work and the movable jaw, as in Fig. 56-22.
3. Machine surface No. 3 with horizontal cuts, as in Fig. 56-22. Place surface No. 1 against the stationary jaw, as in step 2 above.
4. Machine surface No. 4 with horizontal cuts, as in Fig. 56-21. If the work does not seat solidly against both parallels, use a hold-down wedge, as in Fig. 56-24.
5. To machine the ends, swing the vise so the jaws are at right angles to the ram. Check for right angle accuracy with an indicator, as explained in section 56-13. Mount the workpiece, on parallels if necessary, with one end extending slightly beyond the jaws. Machine the first end with a vertical cut, as explained in section 56-18.
6. Reverse the ends of the workpiece in the vise, and machine the second end to length with vertical cuts.

56-20. Machining an Angle

Angular or bevel cuts may be made on a shaper as shown in Fig. 56-30. When the cut is made, the tool head must be swiveled to the desired angle. The angle is indicated by the degree graduation marks on the swivel base. The cut is made by hand-feeding the tool with the down-feed handle during each return or backward stroke. Note in Fig. 56-30 that the clapper box must be tilted in a direction opposite from the angular surface being machined. Thus the tool swings **up and away** from the cut on the return or backward stroke.

An angular finishing cut is being made in machining a dovetail in Fig. 56-31. Note

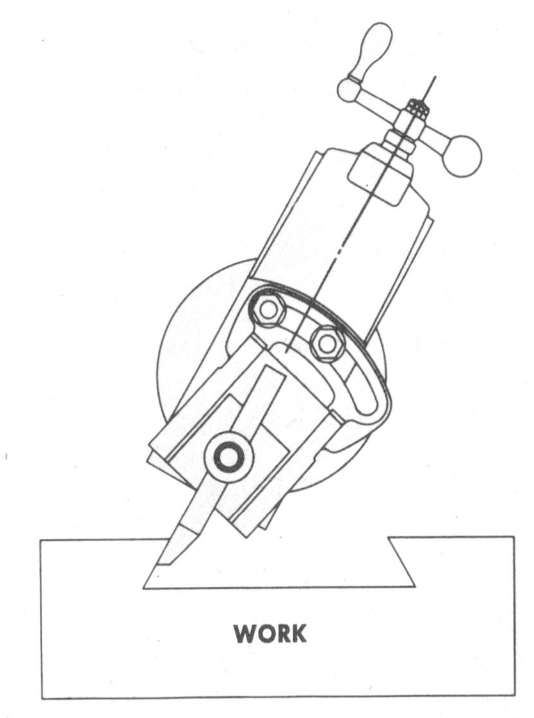

Fig. 56-31. Cutting an Angle with Shaper Head Set at an Angle

again that the clapper box is tilted in a direction opposite from the angular surface being machined.

A necessary precaution always must be taken when shaping an angular surface with the tool head swiveled. **Do not run the ram back into the column with the tool slide set at an angle, or the slide will strike the column.** (See Fig. 56-8.) Thus the length and position of the stroke must be such that the tool head does not go back into the column. Hence, on machines so equipped, the ram should always be run through its complete stroke with the handwheel before turning on the power.

Another way to make angular or bevel cuts is to mount the workpiece in the shaper vise at an angle and make horizontal cuts as in Fig. 56-26. The advantage of this method is that automatic cross feed may be used. When making an angular cut by this method, mount the workpiece in the vise so that the layout line for the finished surface is horizontal, parallel with the table top or vise jaws.

56-21. Making Irregular Cuts

An irregular or curved surface can be machined with a shaper, as shown in Fig. 56-32. The surface to be machined must be laid out first, as shown in Fig. 56-33. If large areas of material are to be removed, they are first removed with horizontal roughing cuts, as in Fig. 56-19. The final finishing cut is made by hand-feeding the table horizontally with the left hand, while hand-feeding the tool to the desired depth with the right hand. Skilled operators sometimes use a combination of automatic cross feed, while hand-feeding the tool to the desired depth.

Fig. 56-32. Shaping an Irregular Surface

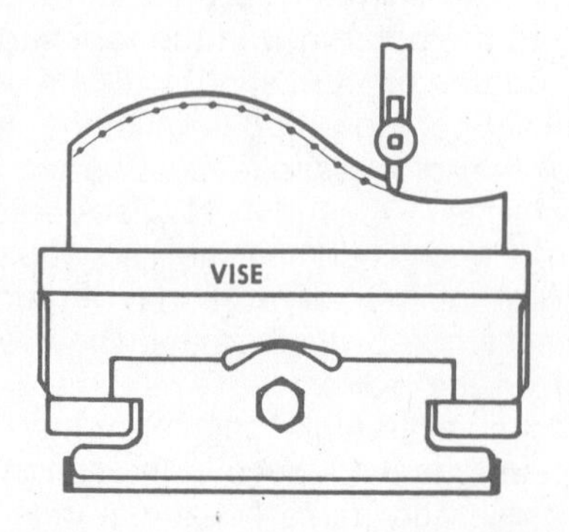

Fig. 56-33. Shaping Irregular Surface to Layout Line

56-22. Cutting Slots or Keyways

Internal slots or keyways may be machined on a shaper by mounting the cutting tool in an extension toolholder as shown in Figs. 56-16 and 56-15. Square or rectangular keyways, as in Fig. 56-16, are cut with a square-nose cutting tool of the type shown in Fig. 56-18. When internal keyways are cut, the length and position of the stroke must be adjusted carefully so that the clapper head or tool post does not strike the workpiece.

External slots or keyways are cut with a square-nose tool of the type shown in Fig. 56-18. The tool is mounted in a shaper toolholder as shown in Fig. 56-13. When a short keyway or slot is machined in a long workpiece, a hole is first drilled at the end of the slot to the depth of the slot. The length and position of the stroke must be carefully adjusted so that the end of the forward stroke stops at the center of the drilled hole.

56-23. Planer

The **planer**, Figs. 56-34 and 56-35, is very much like the shaper, but it is larger and will cut flat surfaces on work that is too large to be handled on the shaper. The work is clamped onto the table, which moves back and forth under the cutting tool. The cutting tools for the planer are single-point tools which have the same general shapes as those for the lathe and shaper, but are usually larger. The cutting tool is held in the **tool head** that is in turn held by the horizontal **cross rail**. The cross rail is attached to

Fig. 56-34. Large Industrial Planer with Two Tool Heads

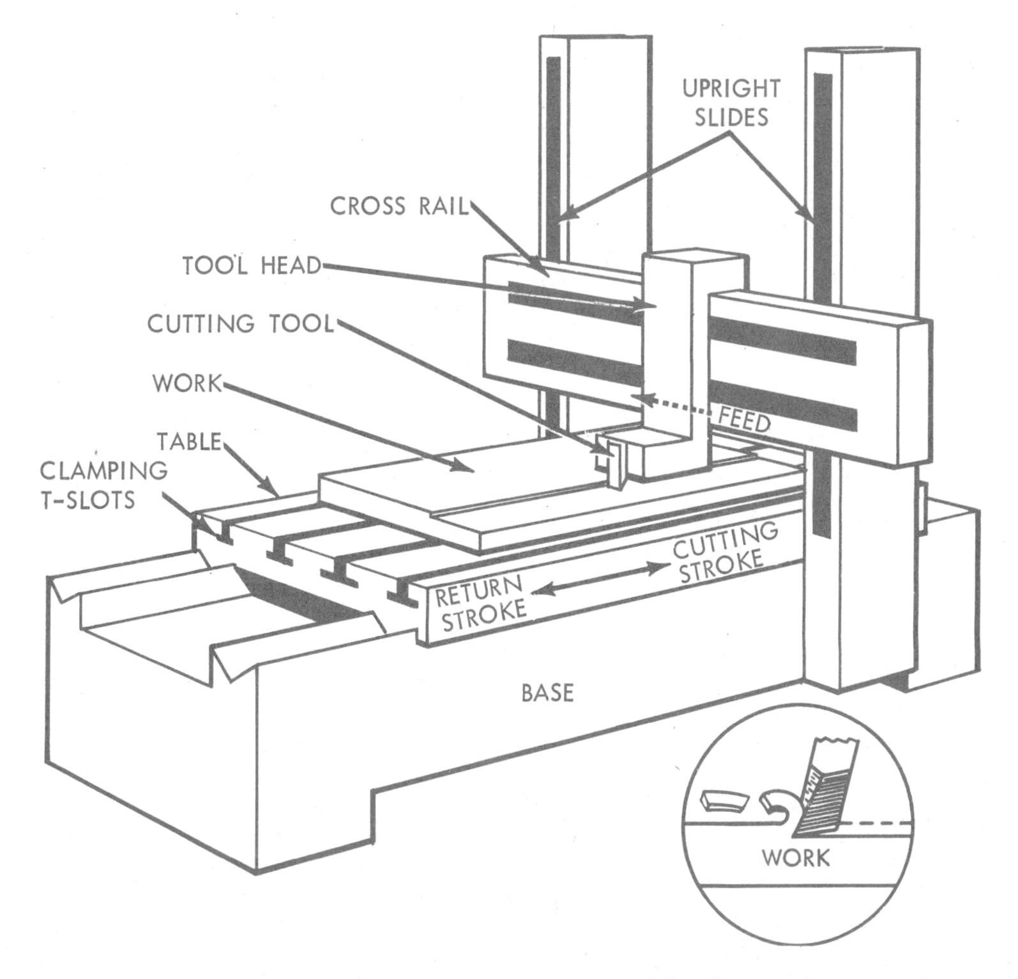

Fig. 56-35. Principles of Operation of the Planer

vertical **uprights** and can be adjusted up or down.

The cutting tool peels off a new chip on each cutting stroke. At the end of the cutting stroke, the table reverses direction and moves back for another cutting stroke. After the table has returned, the tool head moves the cutting tool over for a new cut.

Since the planer is a large and costly machine tool, it is only found in shops and tool rooms that do a great deal of large work. However, the principles involved in setting up and operating a planer are very similar to those which apply to the shaper.

The size of a planer is determined by the largest size of work that it will machine. For example, a planer 26″ (dimension between uprights) × 26″ (maximum between the table and the cross rail) × 7′ (length of the table) will machine a piece of work up to 26″ wide, 26″ high, and 7′ long. Large planers can machine surfaces more than 20′ in length.

Conventional planers, those designed for use with single point cutting tools, have now been largely replaced with planer-type milling machines.

Words to Know

apron
clapper block
clapper box
clutch
column
column clamping nut
cross feed
cross rail
cutting speed
cutting stroke
digging in
dovetail
down-feed handle
end relief
extension toolholder
gooseneck
graduated feed screw dial
hold-downs
indexing attachment
milling machine
parallels
plain table
planer
planer and shaper toolholder
ram
ram positioner
return stroke
shaper
side rake
side relief
single-point tool
slotted crank disk
square nose tool
strokes per minute
table support
tool chatter
tool head assembly
tool post
universal table
vibrating arm
vise alignment

Review Questions

1. What kinds of surfaces can be machined with a shaper?
2. What is the principal **advantage** in machining flat surfaces with a shaper, rather than a milling machine?
3. What is the principal **disadvantage** in machining flat surfaces with a shaper, rather than a milling machine?
4. In what kinds of machine shops are shapers most frequently used?
5. What is the principal difference between horizontal and vertical shapers?
6. How is the size of a shaper generally designated?
7. What is the advantage in having a shaper equipped with a universal table?
8. List the principal parts of a shaper.
9. Name the principal parts of the tool head assembly.
10. What purpose does the tool slide-lock screw serve?
11. Why should the toolholder and cutting tool generally be in a vertical position for making horizontal cuts?
12. Explain the general procedure used for raising or lowering the shaper table.
13. How much longer should the length of stroke be than the length of the surface to be machined?
14. Describe how the length of the stroke can be changed on one kind of shaper.
15. Describe briefly how the position of the stroke is adjusted on one kind of shaper.
16. Describe how the cutting speed for shaping is designated.
17. Explain how the rate of table feed can be adjusted for one kind of shaper.
18. What depth of cut is generally recommended for making finishing cuts on a shaper with a finishing tool?
19. What is the minimum nose radius generally recommended for a finishing tool on a shaper?

20. What rate of feed is often used for very deep, roughing cuts on a shaper?
21. List a general rule which can be used in establishing the maximum feed and depth of cut for a shaper.
22. Does the table feed crosswise during the cutting stroke or during the return stroke?
23. Why should one always run the ram through its complete stroke with the handwheel, if the machine has one, before operating the machine with power?
24. Why should one avoid running the tool head back into the column when the tool slide is swiveled at an angle?
25. Explain how a vise can be accurately aligned with the use of a dial indicator.
26. List several safety practices which should be followed when using a shaper.
27. List three kinds of toolholders which can be used on a shaper.
28. List the recommended angles for a general-purpose shaper tool: (1) end-relief angle; (2) side-relief angle; (3) side-rake angle; (4) back-rake angle.
29. Which is used more frequently on a shaper, right- or left-hand cutting tools?
30. List two common types of finishing tools used on a shaper.
31. What material can be used as an aid in holding rough castings in a shaper vise?
32. Why are parallels often used under workpieces mounted in the shaper vise?
33. Describe how hold-downs are used to mount workpieces in a machine vise.
34. Explain how vertical cuts are made with a shaper.
35. Briefly explain the steps to use in machining a block square and parallel.
36. Explain several ways in which an angular cut can be made with a shaper.
37. Explain how an irregular or curved surface can be machined on a shaper.
38. How can an internal keyway be machined with a shaper?
39. In what ways is a planer similar or different from a shaper?

Mathematics

1. Calculate the number of strokes per minute required for shaping an 8″ long aluminum workpiece at 200 sfpm.
2. How many strokes per minute are required for shaping a high-carbon steel workpiece 2″ long at 60 sfpm?

Career Information

1. What kind of employment do you suppose workers now have who were once shaper and planer machine operators? Explain your answer.
2. Find out what you can about employee retraining programs.

UNIT 57

The Milling Machine and Milling Operations

57-1. What Is a Milling Machine?

A **milling machine**, Fig. 57-1, is a machine tool which cuts metal with a multiple-tooth cutting tool called a **milling cutter**, Fig. 57-6. The workpiece is mounted on the milling machine table and is fed against the revolving milling cutter. The speed of the cutting tool and the rate at which the workpiece is fed may be adjusted for the piece being machined.

With heavy-duty milling machines, several kinds of milling cutters may be mounted on the machine arbor for milling several surfaces at the same time, Fig. 57-6. A wide variety of milling cutters is available for use in machining many kinds of surfaces and for performing many kinds of milling operations. (See Figs. 57-31 to 57-46.)

Fig. 57-1. Universal Milling Machine, Showing Principal Parts

57-2. Why Milling Machines Are Important

The milling machine is important because it is one of the five basic machine tools, and it is capable of performing a wide variety of machining operations. Since they use multiple-toothed cutting tools, milling machines have a higher rate of metal removal and are therefore more efficient than machines such as the shaper and planer which use single-point cutting tools.

The variety of milling operations which can be performed on a milling machine depends on the following:

1. The type of machine.
2. The kind of milling cutter used.
3. The kinds of accessories or attachments available for use with the machine.

Milling machines can machine flat surfaces, including horizontal, vertical, and angular surfaces. (See Figs. 57-11, 57-12 and 57-13.) They machine many kinds of shoulders, grooves, T-slots, dovetails, and keyways. (See Figs. 57-34, 57-35, and 57-36.) Milling machines also machine irregular or curved surfaces with formed-tooth milling cutters. (See Figs. 57-39, 57-40, 57-41, and 57-42.)

Milling machines, particularly vertical milling machines, Figs. 57-9 and 57-10, can perform the basic hole-machining operations commonly performed on a drill press.

A milling machine equipped with a dividing head, Fig. 57-1, can machine equally spaced surfaces, grooves, or gear teeth on cylindrical shaped parts, Fig. 57-19. By swiveling the table about its axis, universal milling machines can machine spirals, Fig. 57-22. Plain milling machines equipped with a universal spiral attachment can also machine spirals, see Fig. 57-21.

Accessories are available that expand the variety of operations which can be performed on milling machines. This unit, however, is concerned primarily with the basic milling operations commonly performed in beginning machine shop classes. Operations of an advanced nature are described very briefly. These descriptions will provide some understanding of the kinds of work which can be done on milling machines.

57-3. Types of Milling Machines

Milling machines can be classified under two basic types. These include **bed-type** and **knee-and-column** milling machines. Bed-type milling machines are manufacturing milling machines which are used for mass-production purposes. The table does not move transversely (crosswise) toward the column. However, the spindle, on which the cutter is mounted, may be adjusted vertically as well as horizontally.

Bed-type machines may be equipped with one, two, three, or more spindles and cutters. The machine in Fig. 57-2 has three spindles and three heavy-duty **face-milling cutters**. Parts are mounted on both sides of the table, and three surfaces are **face milled** at the same time.

Planer milling machines are so called because they are constructed and operated in much the same manner as the older conventional single-point cutting tool planers. (See Figs. 56-34 and 56-35.) They are used for milling operations on very large workpieces.

Fig. 57-2. Bed-Type Special Manufacturing Milling Machine

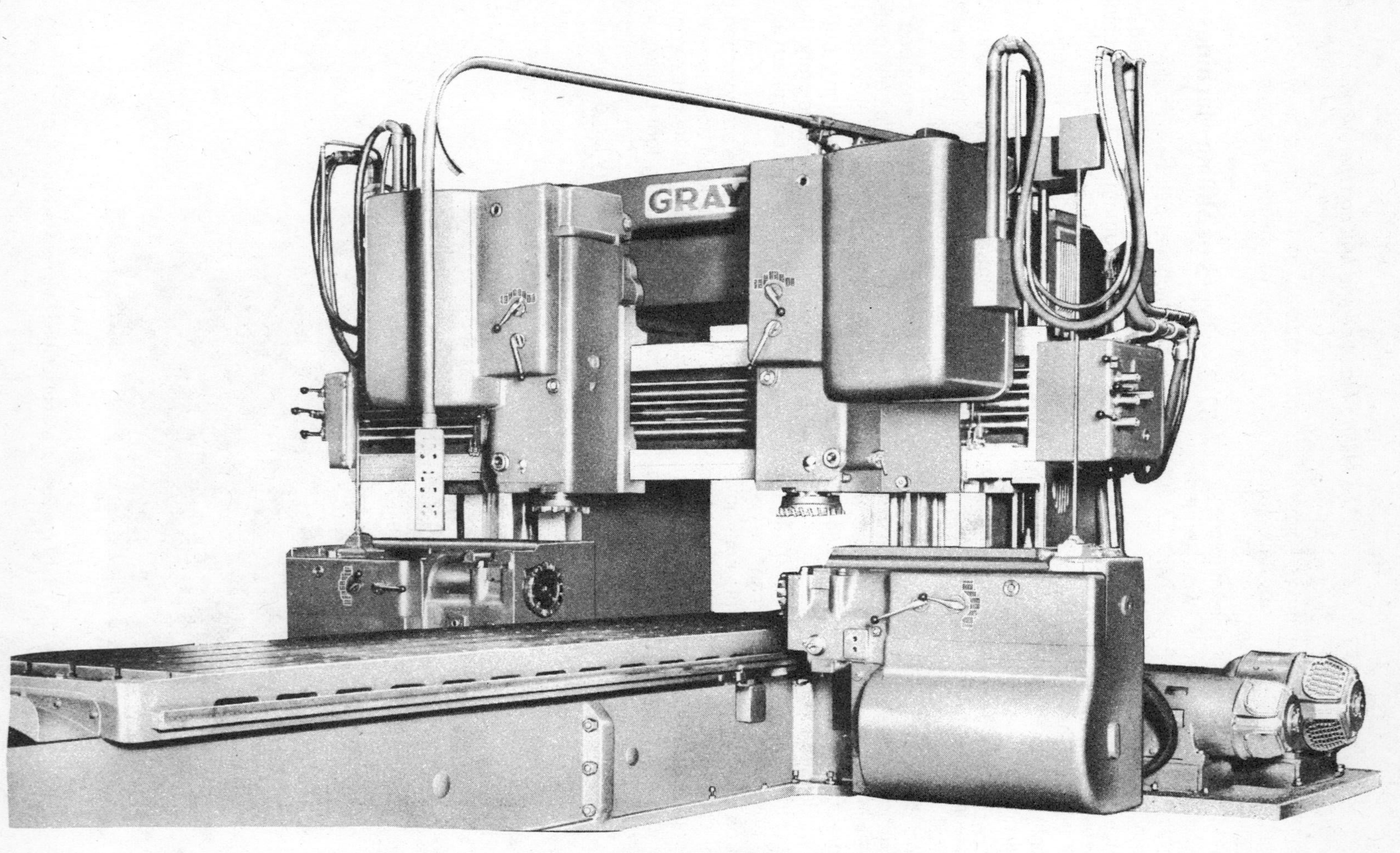

Fig. 57-3. Planer Type Milling Machine

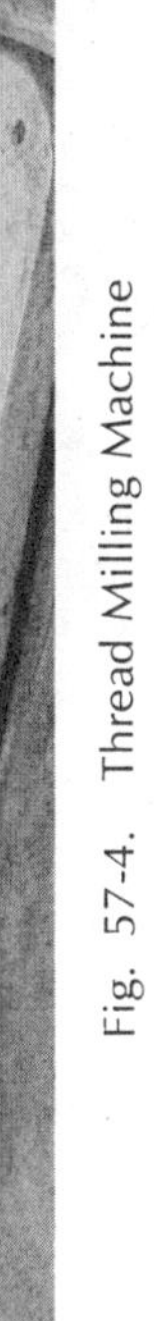
Fig. 57-4. Thread Milling Machine

Special purpose milling machines include **thread milling machines,** Fig. 57-4, and **gear hobbing machines**. Thread milling machines are designed for rapidly machining precision threads on a production basis, since they are capable of cutting several inches of threads simultaneously with only slightly more than one revolution of the workpiece, Fig. 57-5.

Fig. 57-5. Closeup of Thread and Milling Operation

Gear hobbing machines cut gear teeth on gear blanks with a milling cutter called a **gear hob**. Horizontal gear hobbers are used for cutting gears of small diameter. Vertical gear hobbers are also made for cutting large diameter gears.

Milling operations are also performed on horizontal boring mills. These versatile machines are also built for machining large workpieces such as machine bases and are capable of performing all of the precision hole-machining operations as well as milling.

57-4. Knee-and-Column Milling Machines

Knee-and-column milling machines, as the name indicates, are equipped with a **knee** and **column**, see Fig. 57-1. The body or

Fig. 57-6. Milling a Casting with Several Cutters Mounted on the Arbor of a Horizontal Milling Machine

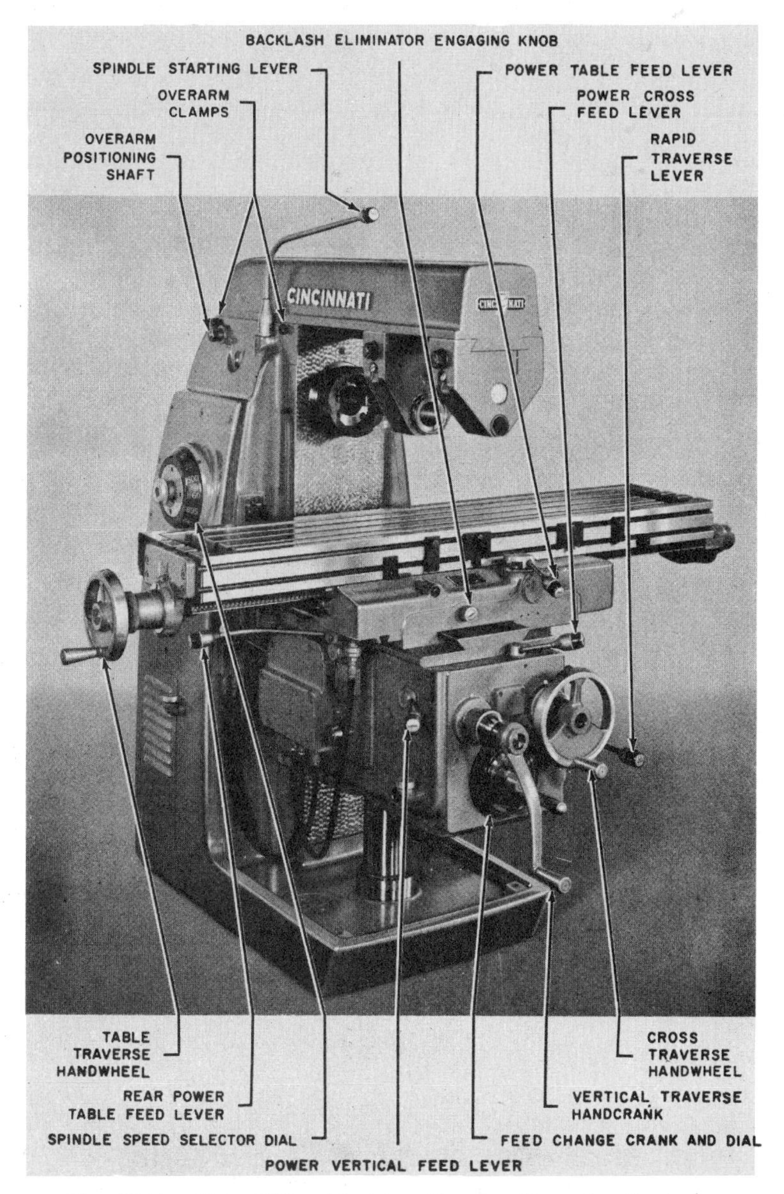

Fig. 57-7. Plain Milling Machine, Showing Operating Controls

frame of the machine is a large casting which includes the **base** and the upright portion called the **column**. The front of the column has accurately machined dovetail ways which are often called the column. (See Fig. 57-1.) The **knee** is mounted on the dovetail ways of the machined column.

The knee can be raised vertically to any desired elevation. The **saddle** and **table** are mounted on top of the knee. The table can feed **longitudinally** (horizontally to the right or to the left), and it can feed **transversely** (crosswise or cross feed in or out from the column). Thus the table on knee-and-column

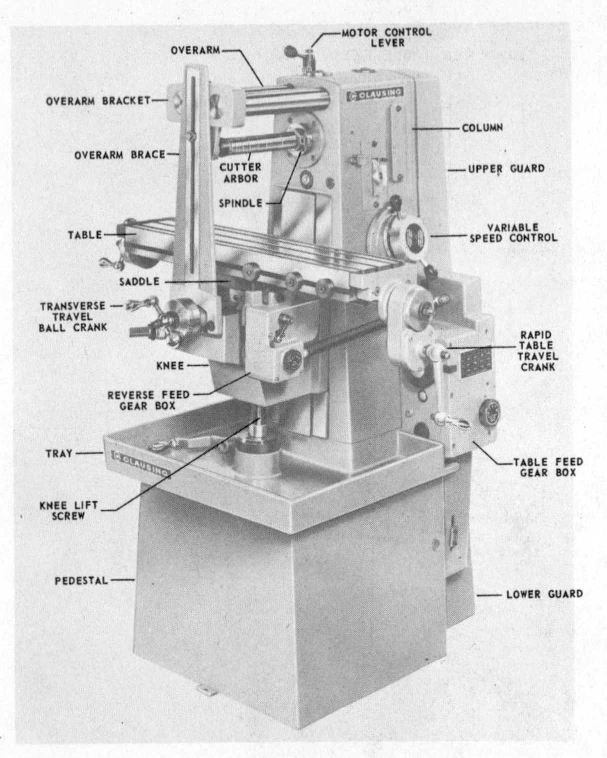

Fig. 57-8. Small Plain Milling Machine-

milling machines can be fed or adjusted in three directions.

The directions are:

1. Vertically
2. Longitudinally
3. Cross feed or transversely

The table can be fed manually or with automatic power feed.

Knee-and-column milling machines can be further classified into the following types:

1. Horizontal milling machines (see § 57-5).
   a. Plain
   b. Universal
2. Vertical milling machines (see § 57-6).
3. Combination horizontal and vertical milling machines (see § 57-7).

Knee-and-column milling machines are manufactured in a wide range of sizes and weights. Some school shops have standard. heavy-duty milling machines as shown in Figs. 57-1, 57-7, and 57-9. Many schools have smaller, lighter weight machines such as those shown in Figs. 57-8 and 57-10. Regardless of the size of the machine, the principles involved in various milling operations are very similar. Heavy cuts on a milling machine require a heavy-duty, sturdy machine. Hence, lighter cuts and lighter feeds must be taken when using smaller, lightweight machines.

57.5 Horizontal Milling Machines

Horizontal milling machines usually have the milling cutter mounted on a horizontal arbor as shown in Fig. 57-6. The arbor fits into the spindle nose which is located on the machined face of the vertical column. The arbor is supported rigidly with an **arbor support** or **overarm support,** Fig. 57-6. Milling cutters used on horizontal milling machine arbors, therefore, have an arbor hole.

End-milling cutters, called **end mills** (Fig. 57-43), can be mounted horizontally in the spindle nose of horizontal milling machines as in Figs. 57-28, 57-44, and 57-18. Thus end-milling operations also can be performed with the cutter operating in a horizontal position on horizontal milling machines. End-milling operations, however, are more commonly performed on vertical milling machines. (See Figs. 57-12 and 57-20).

Horizontal milling machines are of two basic types: **plain** and **universal.**

Plain Horizontal Milling Machine

Plain horizontal milling machines are shown in Figs. 57-7, and 57-8. Machines of this type are in wide use in school shops, tool-and-die shops, maintenance machine shops, and in many industrial machine shops. They are used for all of the common horizontal milling operations. These include milling flat horizontal surfaces, vertical surfaces, angular surfaces, curved surfaces, irregular surfaces, grooves, and keyways.

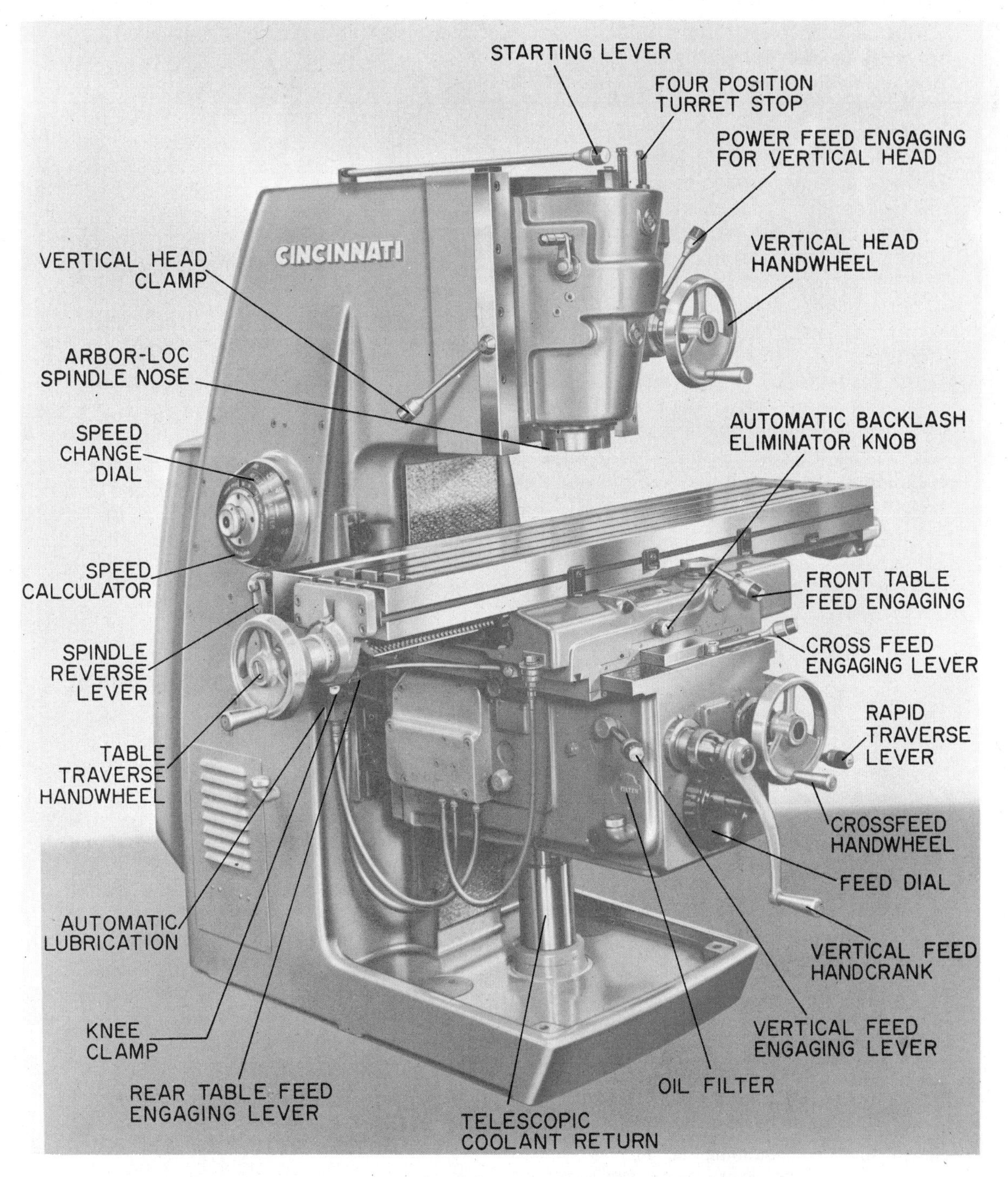

Fig. 57-9. Vertical Milling Machine, Showing Operating Controls

The table on plain milling machines cannot be swiveled for helical milling operations. However, with the use of a dividing head and a universal spiral milling attachment, such operations can be performed. (See Fig. 57-21.) A vertical milling attachment, as in Fig. 57-20, may be used to perform vertical milling operations on a plain milling machine.

Many makes of small plain milling machines are available such as in Fig. 57-8. They are much less expensive than such larger machines as shown in Fig. 57-7. However, they also have less power, weight, sturdiness, and cutting capacity than the larger machines. The smaller machines are in wide use in many school shops.

The plain milling machine in Fig. 57-8 is equipped with power longitudinal table feed. Handfeed is used for transverse (cross) table feed or for changes in table elevation.

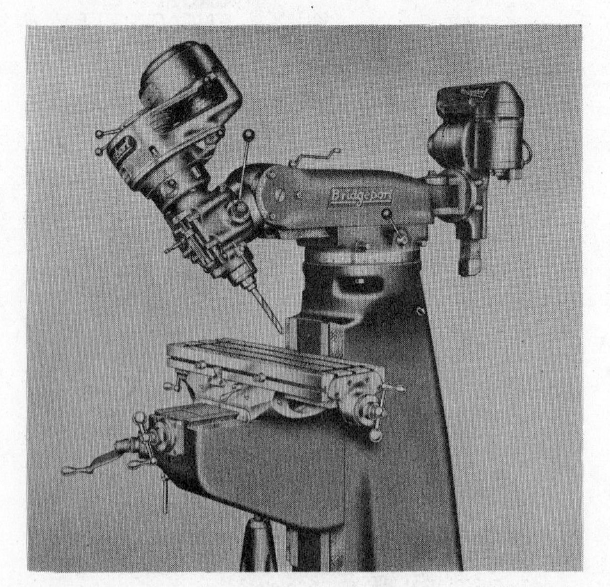

Fig. 57-10. Vertical Milling Machine
The head may be swiveled for drilling or milling at an angle. Vertical shaping attachment is mounted at opposite end of ram.

Universal Horizontal Milling Machine

A universal horizontal milling machine is shown with its principal parts labeled in Fig. 57-1. The distinguishing difference between plain and universal machines is that the table can be swiveled about its vertical axis on the universal machine. Universal machines usually are equipped with a dividing head, tailstock, and a dividing-head lead driving mechanism. These accessories make it possible to perform helical milling operations, as in Fig. 57-22. Helical milling is required for cutting helical gears or for making helical cutting tools such as drills, reamers, or milling cutters.

A vertical milling attachment, Fig. 57-20, can be used on the universal milling machine for performing vertical milling operations. In addition to helical milling operations, the universal milling machine can do all of the operations performed on the plain milling machine.

57-6. Vertical Milling Machines

Vertical milling machines are shown in Figs. 57-9 and 57-10. The spindle on vertical milling machines normally is in a vertical position, similar to the spindle on a drill press. However, the head may be swiveled on some machines for angular milling or hole-machining operations, Figs. 57-10 and 57-13.

Vertical milling machines use cutters called **end mills**, Figs. 57-43 and 57-44. The end mills are mounted in the nose of the vertical spindle, Figs. 57-10 and 57-12. Various kinds of spindle adapters are used for mounting the end mills, Fig. 57-27.

Except for the position of the spindle, vertical milling machines are very similar to plain horizontal milling machines. Compare the two machines in Figs. 57-7 and 57-9; notice the similarity of the principal parts and controls. Many of the basic attachments and accessories can be used interchangeably on horizontal and vertical machines.

Operations

Vertical milling operations can be performed on vertical milling machines or with a vertical milling attachment on a horizontal milling machine. (See Fig. 57-20.) Vertical milling machines can machine horizontal surfaces, angular surfaces, shoulders, grooves, keyways, dovetails, and T-slots. (See Figs. 57-48 to 57-56.) In addition, vertical milling machines can perform hole-machining operations such as drilling, countersinking, boring, counterboring, and reaming. (See Fig. 54-1.)

Heavy-Duty Machines

Heavy-duty vertical milling machines of the type shown in Fig. 57-9 are capable of taking very heavy cuts. They are equipped with power longitudinal feed, power cross feed, and power vertical feed for table elevation. The spindle in the head is also provided with power vertical feed.

Light-Duty Machines

Light - duty and medium - duty vertical milling machines are widely used in tool rooms, small industrial machine shops, and in school shops. Several types are shown in Figs. 57-8 and 57-10. They may be purchased with manual feed only, or with one or more of the following kinds of power feed:

1. Power longitudinal table feed.
2. Power transverse feed (cross feed).
3. Power vertical spindle feed.
4. Power table elevation.

Light-duty machines produce good results when used for cuts which are within their capacity.

Hole Machining

Vertical milling machines are often used for drilling and other hole-machining operations which require very accurate location of the holes. The 0.001″ graduation marks on the table transverse handwheel and the cross-feed handwheel, Fig. 57-9, make it possible to locate the centers of holes accurately. However, always remember to turn the feed wheel or crank in a direction which corrects for wear or backlash in the feed screw before setting the graduated collar at the **zero** index mark. The 0.001″ graduation marks on the vertical head handwheel also make it possible to machine holes accurately to depth.

57-7. Combination Horizontal and Vertical Milling Machines

Some milling machines may be classified as **combination horizontal and vertical milling machines.** This type of machine normally is available with a standard overarm for use as a horizontal milling machine. However, it is also available with the special overarm which has an independent overhead spindle for vertical milling operations. Machines of this type can perform either horizontal milling operations, Figs. 57-11 and 57-18, or vertical milling operations, Figs. 57-12 and 57-13.

Fig. 57-11. Horizontal Milling with a Combination Horizontal and Vertical Machine

Fig. 57-12. Vertical End Milling with a Combination Horizontal and Vertical Machine

Fig. 57-13. Angular-End Milling with a Combination Horizontal and Vertical Milling Machine

57-8. Principal Parts

Before attempting to operate a milling machine, you should know the names of the principal parts and the controls on the machine. The parts and controls on most knee-and-column milling machines, although not exactly alike, are very similar. Many of the basic parts were described in section 57-4. The similarities and differences between plain and universal horizontal machines are explained in section 57-5. The similarities and differences between vertical and horizontal machines are included in section 57-6.

The names of the principal parts and controls of horizontal milling machines can be found in Figs. 51-1, 57-7, and 57-8. The names of the principal parts and controls on vertical machines can be found in Fig. 57-9.

57-9. Milling Machine Controls and Adjustments

Before operating a milling machine, one must know how to make several kinds of adjustments on the machine. These include knee elevation, table adjustments, speed, and feed. It is always best to have your instructor explain the principal parts of the milling machine in your shop. It is also good practice to review the instruction manual provided by the manufacturer of the machine before operating the machine.

Knee Elevation

The knee must be raised or lowered in order to establish the proper elevation of the workpiece under the cutting tool. For **peripheral** (horizontal) milling operations the knee must be raised to establish the **depth of cut.** For **vertical** milling operations, the depth of cut may be established either by raising the knee or by feeding the tool to depth with the **vertical-head handwheel,** Fig. 57-9. Generally, it is best to set the depth of cut by raising the knee.

A **knee clamp** (a lever or a locking nut) locks the knee securely to the column during milling operations. The knee clamp must be loosened before raising or lowering the knee. The knee is then raised or lowered with the **vertical-feed hand crank**. Notice that there is a **micrometer collar** with 0.001″ graduations on this control, and also on the longitudinal-feed and cross-feed controls. Some machines are equipped with **power vertical feed** which is engaged with the **vertical-feed engaging lever.** When the knee is located at the proper elevation, the knee clamp must be tightened. If it is not tightened, vibration, chatter, a rough machined surface, and possible cutter damage may result.

A good way to establish the **depth of cut** is to loosen the knee clamp and start the machine. Raise the knee until the cutter just touches the workpiece. Bring the workpiece out from under the cutter. It should be brought to the correct side of the cutter so that the feed is adjusted for **up milling**, Fig. 57-14 Set the micrometer collar on the vertical-feed crank at zero, and raise the knee for the desired depth of cut. If you raise the table too high, lower it again by turning the handwheel at least ½ turn lower than necessary. Then, raise the table to the desired elevation. This procedure will correct for backlash or wear in the table elevation screw. Be sure to lock the knee clamp before starting the cut.

Transverse Table Movement

Transverse table movement is cross movement of the **saddle** and **table** toward or away from the column. During most milling operations, the saddle is clamped securely to the top of the knee with one or more **saddle-clamp levers.** This reduces table vibration. Therefore, before making transverse table adjustments, loosen the saddle clamp. Then the table can be moved toward or away from the column with the **cross-feed handwheel**. After table adjustment, the saddle clamp again must be tightened.

The saddle clamps must be loosened when performing operations which involve transverse (cross) table feeding. Some milling machines are equipped with power transverse table feed. On machines so equipped, the feed is engaged with the **cross-feed engaging lever,** Fig. 57-9.

Longitudinal Table Movement

Longitudinal table movement is table travel from side to side, either toward the right or the left. The table may be fed manually with the **table traverse handwheel**, also called the **longitudinal feed**. (See Fig. 57-7.) Most machines are equipped with power longitudinal table feed. The table power feed is engaged with the **power table-feed lever**. (See Figs. 57-7 and 57-9.)

Most milling machine tables are provided with a **table-clamp lever** which may be tightened to prevent longitudinal table movement during certain operations. The table-clamp lever should be tightened during hole-machining operations on either horizontal or vertical milling machines.

Rapid Traverse

Larger milling machines are equipped with a **rapid-traverse control**. (See Figs. 57-7 and 57-9.) This control enables the operator to move the knee or table rapidly in either direction with power. Beginners should be very careful in using this control. Serious damage can result if the workpiece should strike the cutter or arbor while the knee is raised rapidly or while the table is traversed rapidly. It is best to ask your instructor to show you how to use the rapid-traverse control before attempting to use it yourself. Skilled operators are able to speed up production through the use of this control.

Spindle Speed Adjustment

The spindle speed is designated in **rpm** (revolutions per minute). On some machines the rpm is selected and set by turning a **speed-change dial** to the desired rpm. (See Figs. 57-7 and 57-9.) Other machines may be equipped with levers which shift gears for the desired rpm. Machines of the type shown in Fig. 57-8 are equipped with a variable-speed drive which must be adjusted while the machine is running. On machines with a step pulley and V-belt drive, the rpm is changed by shifting the belt to a different step on the pulley.

The direction of spindle revolution may be changed on most milling machines. This may be done with a spindle-reversing switch, button, or lever. The machine in Fig. 57-9 has a spindle-reverse lever on the left side of the machine.

Feed Adjustment

The rate of feed and the method for determining the rate of feed are explained in section 57-19. On most machines, the feed rate is changed through a series of change gears in a feed-change gear box. On some machines, the desired feed rate is changed by turning a **feed dial** directly to the desired feed rate indicated on the dial. The machines in Figs. 57-7 and 57-9 are equipped with this type of feed dial located on the front of the knee. The table is fed manually on small machines not equipped with power feeds. The operator then must use his best judgment in feeding the table at the proper rate of feed. Feeding too rapidly can cause cutter breakage.

DIRECTION OF CUTTER ROTATION

WORK FEED

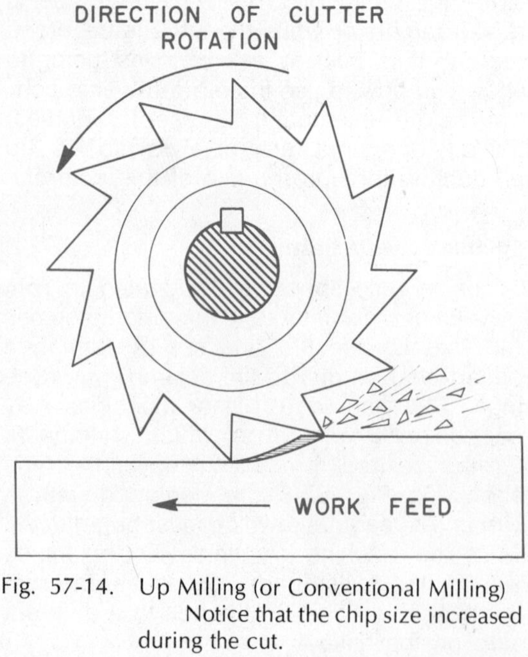

Fig. 57-14. Up Milling (or Conventional Milling) Notice that the chip size increased during the cut.

57-10. Direction of Feed

The direction of feed in relation to the direction of cutter rotation is an important factor in all milling operations. Two methods of feed are possible. When the work is fed against the direction of milling cutter rotation, Fig. 57-14, the method is called **up milling** (formerly called conventional milling). When the workpiece is fed with the direction of the milling cutter rotation, Fig. 57-15, the method is called **down milling** (formerly called climb milling).

Safety Note

Down milling is done only on machines equipped with anti-backlash devices; check your operator's manual.

Smaller milling machines, and many older machines, are not equipped with anti-back-

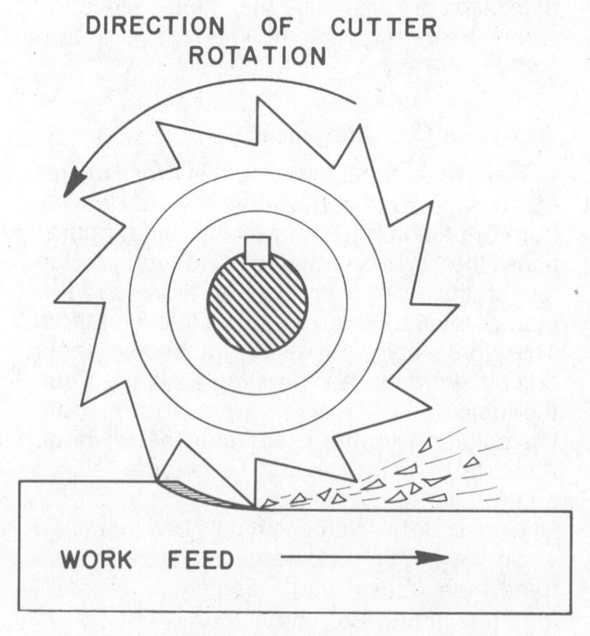

Fig. 57-15. Down Milling (or Climb Milling) Notice that the chip size decreases during the cut.

lash devices. Therefore, only the up-milling method can be used with them. Modern heavy-duty machines of the types shown in Figs. 57-1, 57-7, and 57-9 are equipped with backlash eliminators.

Up (Conventional) Milling

With up milling, the cutter tooth starts into the work with a chip of zero thickness and ends with a thick chip. The cutter starts into clean metal and ends by lifting off the rough surface scale. Thus the cutter stays sharp longer. However, the workpiece must be clamped very tightly in the vise or to the table. The cutting forces tend to lift or pull the work out of the vise. The direction of feed forces the work against the cutter, thus compensating for wear or backlash in the table leadscrew and feeding mechanism.

Down (Climb) Milling

With down milling, the cutter tooth starts into the work with a thick chip and ends with a thin chip, Fig. 57-15. The scraping action of the cutter tooth at the end of the chip tends to produce a smoother surface. The cutting forces tend to pull the workpiece under the cutter. Thus, any backlash in the leadscrew or feeding mechanism can cause vibration, chatter, and possible cutter breakage. A hard surface scale on the workpiece will dull the cutter more rapidly.

Down milling is gaining wider use today on certain production milling operations. It generally produces a better surface finish on harder steels. Small, thin parts and parts which are otherwise difficult to hold can be machined more easily by this method. **But remember that to use this method, the machine must be equipped with an anti-backlash device.**

The principles involved in these methods of milling also apply to vertical end-milling and face-milling operations.

57-11. Accessories and Workholding Devices

A variety of attachments or accessories is available for use on plain, universal, and vertical milling machines. Some of the accessories make it possible to hold the work more effectively. Other accessories make it possible to expand the range of operations which can be done with the machine.

Swivel Vise

The swivel vise, Fig. 57-16, is bolted to the table with T-slot bolts, Fig. 50-2. The vise can be swiveled at any desired angle. The angle is indicated by the degree graduations on the swivel base. A majority of the workpieces or projects machined in beginning machine shop classes can be mounted in a swivel vise for milling. Alignment of the vise is explained in section 57-21.

Universal Vise

The universal vise, Fig. 57-17, can be used for machining angles on workpieces. The

Fig. 57-16. Swivel Vise

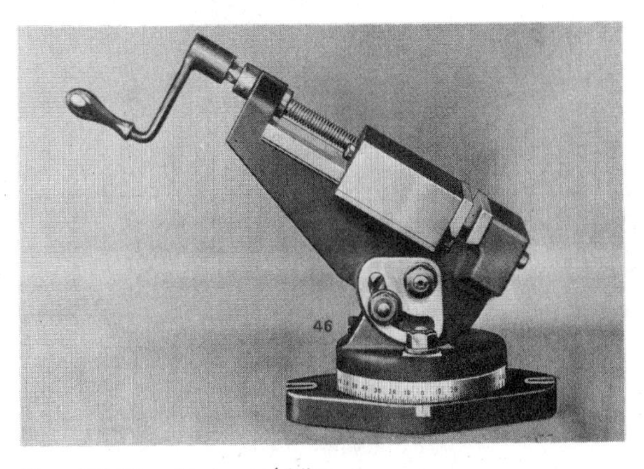

Fig. 57-17. Universal Vise

vise can be swiveled at any desired angle on its swivel base. The vise jaws can be tilted at any desired angle, from 0° to the 90° vertical position.

Dividing Head

A dividing head, Figs. 57-21 and 57-57, is used for holding a workpiece and for dividing it into a number of equally spaced angular divisions. An **index crank** and an **index plate,** which has many circular rows of holes, are located on the front of the dividing head. By **indexing**, explained in § 57-24, a number of equally spaced angular divisions may be machined on a workpiece. (See Fig. 57-19.)

A **tailstock** is bolted to the table and is used with the dividing head for holding workpieces between centers, Fig. 57-19. Both the tailstock and the dividing head are equipped with 60° centers, the same as lathe centers. The tailstock center can be raised above the center point of the dividing head. This is done when tapered grooves or tapered surfaces are machined on a workpiece.

Fig. 57-18. Milling a Shoulder on Workpiece Mounted on a Swivel Table
This shows how horizontal-end milling operations can be performed with a horizontal machine.

A workpiece, such as the gear blank in Fig. 57-19, is pressed on a **lathe mandrel**. The mandrel has a driving dog clamped on the driving end and is mounted between the centers of the dividing head and the tailstock.

A universal 3-jaw chuck may be mounted on the spindle nose of the dividing head for holding short workpieces for many kinds of horizontal or vertical milling operations. One example of its use is shown in Fig. 57-20. The dividing head may be swiveled to any desired angular position, from 0° to the 90° vertical position. Thus, with a round workpiece mounted in the chuck in a vertical position, it is possible to mill any number of equally spaced flat surfaces on its circumference. This procedure can be used for machining a square or hexagonal head on a bolt or on the end of a shaft. This kind of setup also can be used for milling screw slots in the head of a screw.

Lead-Drive Mechanism

The lead-drive mechanism is used with a dividing head for milling helical or spiral

Fig. 57-19. Using the Dividing Head to Cut a Spur Gear

surfaces or grooves, Fig. 57-21. Helical milling is used for milling helical gears, helical milling cutters, helical reamers, drills, and similar items.

Vertical Milling Attachment

This attachment can be mounted to the column and the spindle of horizontal milling machines, Fig. 57-20. The vertical milling attachment makes it possible to perform a wide variety of vertical milling operations on horizontal machines. The attachment may be used in a vertical position or at an angle for milling angular surfaces, Fig. 57-20.

Universal Spiral Attachment

This attachment makes it possible to machine helical or spiral surfaces or grooves with a plain horizontal milling machine. (See Fig. 57-21.) The attachment is mounted on the column and is driven by the machine spindle.

On universal milling machines helical milling is done by swinging the milling machine table on its swivel base, Fig. 57-22. The dividing-head lead driving mechanism causes the work to revolve as the table travels longitudinally, thus causing the helical groove to be machined.

Circular Milling Attachment

Several kinds of circular milling attachments are available for use on both horizontal and vertical milling machines. Hand feeds and power feeds are available. The hand feed attachment, shown in Fig. 57-23, is widely used in many shops. A workpiece can be bolted directly to the attachment table, or it may be mounted in a vise which is bolted to the attachment table.

The circular milling attachment makes it possible to machine circular edges, shoulders, or grooves on vertical-type machines.

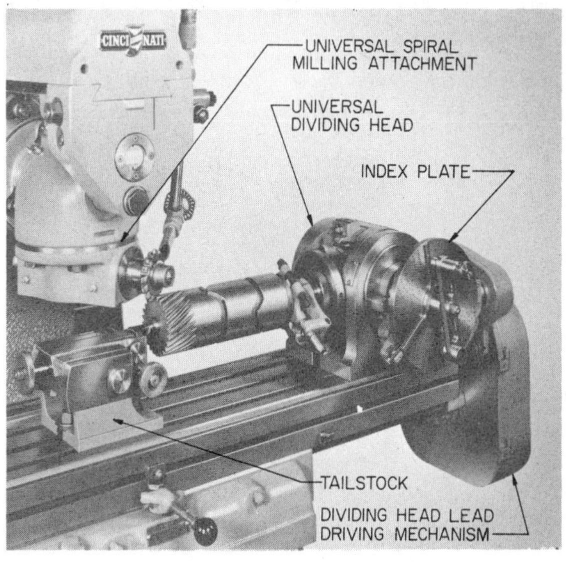

Fig. 57-21. Milling Helical Gear Teeth

Fig. 57-20. Making an Angular Cut with a Vertical Milling Attachment on a Horizontal Milling Machine

Fig. 57-22. Cutting a Left-Hand Helix with the Table Swiveled on a Universal Milling Machine

Fig. 57-23. Circular Milling Attachment, Hand-feed-Type

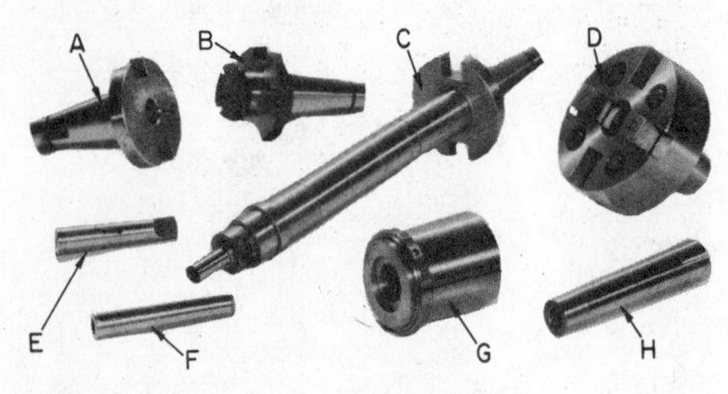

Fig. 57-24. Arbors, Collets, and Adapters
A. Collet Adapter
B. Shell-End Mill Arbor
C. Arbor, Style A.
D. Arbor Adapter
E. Reducing Collet
F. Solid Collet
G. Bushing
H. Split Collet

The swivel base of the circular table is provided with degree-graduation marks. Hence this attachment can be used for milling grooves at any desired angle across the top surface of a workpiece. It also permits machining the vertical edges of a workpiece at an angle. (See Fig. 57-18.)

57-12. Arbors, Collets, Adapters, and Holders

A wide variety of milling machine arbors, collets, adapters, and holders is available for holding milling cutters.

Arbor Shanks

Most manufacturers of standard and heavy-duty milling machines have adapted the **national milling machine taper** for the tapered hole in the machine spindle. Arbors for these machines have the same kind of tapered shank. Standard milling machine tapers are steep tapers with 3½″ taper per foot. Because of the steep taper, they are the **self-releasing** type. Hence they must be held in place with a **draw-in bolt** or a locking device or collar, Fig. 57-28.

Standard milling machine tapers are made in several sizes, designated by the numbers 30, 40, 50, and 60. The No. 50 is the most common and is used on machines of the type shown in Figs. 57-1, 57-7, and 57-9. The No. 40 taper is used on smaller machines.

Some milling machine spindles have a standard shallow taper, such as the **Brown & Sharpe taper** or the **Morse taper**. These are **self-holding tapers**. A few manufacturers use a special taper for the spindle hole on their milling machines.

Style-A arbors, as shown at the top in Fig. 57-25, have a small pilot at the outer end. The pilot fits in a small bearing in a style-A arbor support, Fig. 57-30. The **style-A arbor support** permits the use of small diameter cutters. This arbor support easily passes over the vise when using small cutters close to the vise jaws. Style-A arbors generally are used for light-duty milling operations. An additional inner arbor support can also be used with the style-A arbor for more rigid support. (See Fig. 57-1.)

Style-B arbors, as shown at the bottom in Fig. 57-25, do not have a pilot at the outer end. Instead, they have one or more **bearing sleeves** which are larger in diameter than the arbor **spacing collars**. The bearing

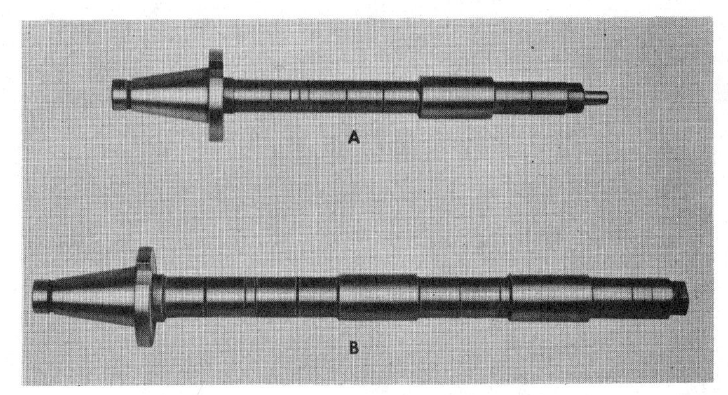

Fig. 57-25. Milling Arbors
A. Style A
B. Style B

sleeves run in the large bearings in the style-B arbor supports, Fig. 57-29. Style-B arbors are used for heavy-duty milling operations. They are desirable for any milling operation so long as the arbor supports clear the vise and the workpiece.

Spacing collars are provided on both style-A and style-B arbors. They hold the arbor rigid and permit spacing the milling cutter at any location along the arbor. The ends of the collars are precision-ground to extreme accuracy. This causes the collars to hold the arbor straight when the cutter is installed and the arbor nut is tightened. A tiny nick or chip between the collars can cause the arbor to bend and the cutter to run untrue. The arbor, cutter, and collars must be wiped clean before the cutter is installed on the arbor. The cutter may be keyed to the arbor. The key should be long enough so that it extends into one collar on each side of the cutter.

Style-C arbors, as shown at B in Fig. 57-24, are used for holding shell-end mills and face mills. (See Fig. 57-44.) Thus style-C arbors are often called **shell-end mill arbors**.

Adapters

Adapters are devices which are used to mount cutters of various kinds on the milling machine spindle. An **arbor adapter** (Fig. 57-24, D) is used for mounting large face mills directly to the machine spindle. The **collet adapter** (Fig. 57-24, A) is used for mounting end mills (Fig. 57-43) on the spindle. The tapered hole in the collet adapter is the **self-holding** type. Usually it has either a Morse taper or a Brown & Sharpe taper. Some machines have collets with a special kind of taper. End mills (Fig. 57-43) with tapered shanks fit into the tapered hole in the collet adapter. A reducing sleeve (Fig. 57-24, E) is inserted in the collet adapter if the tapered hole is larger than the shank on the end mill. Then the end mill is inserted in the reducing sleeve.

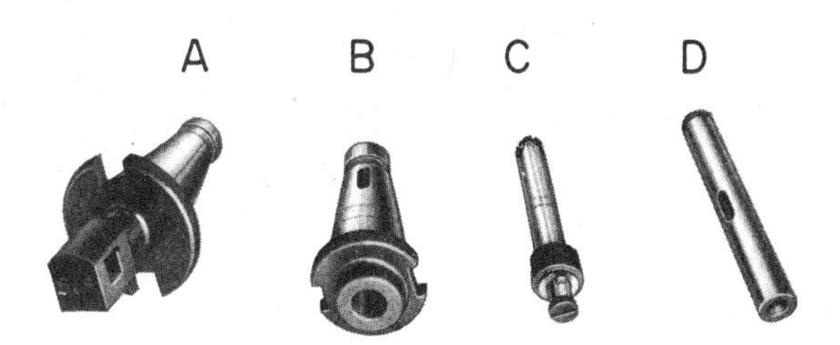

Fig. 57-26. Milling Arbors, Collets, and Adapters
A. Fly Cutter Arbor
B. Adapter for Taper Shank-End Mill
C. Adapter for Small Shell-End Mill
D. Collet for Taper Shank-End Mill

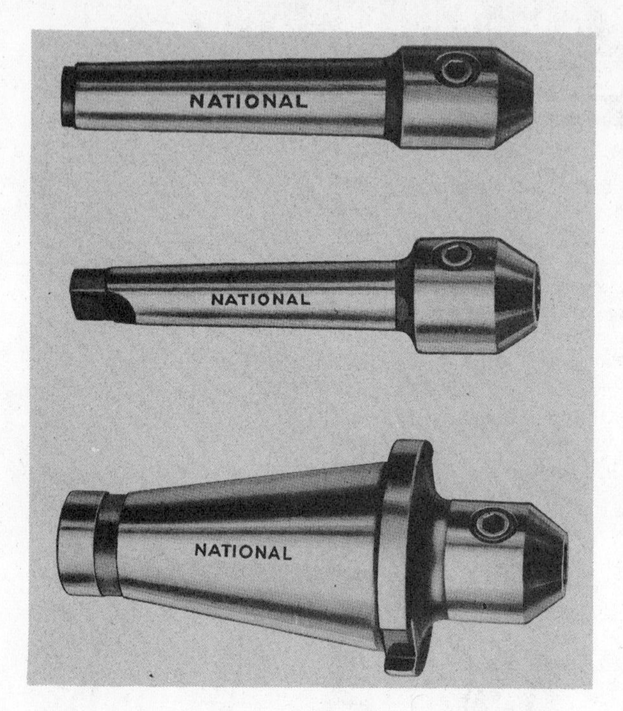

Fig. 57-27. Holders for Straight Shank-End Mill

Fig. 57-28. Mounting Drills for End Mills in the Spindle Nose of Horizontal Milling Machine

Holders

Holders of the types shown in Fig. 57-27 are used for holding straight shank-end mills (Fig. 57-43). The holders are available with holes of various sizes for end mills of different diameters. The setscrew holds the end mill securely in place. A variety of end-mill holders is available. Some are designed to fit directly into the machine spindle; others are inserted in a collett adapter which is installed in the machine spindle.

The holder at the bottom in Fig. 57-27 has a standard milling machine taper and is held in the spindle with a draw bar. The other two holders have self-holding tapers. The one at the top is threaded and is held in the spindle with a draw bar. The holder at the center has a driving tang which prevents it from slipping.

57-13. Removing Standard Arbors

Most milling machines have arbors and collet adapters which have standard, national milling-machine taper shanks. These are held in the machine spindle with either a draw bar or with a locking collar. (See Fig. 57-28.) Arbors or adapters which are held with a draw bar are removed in the following manner:

1. Loosen the draw bar two or three turns. **Do not unscrew the drawbar all the way.**
2. Strike the end of the draw bar with a lead hammer.
3. While holding the arbor with the left hand, unscrew the draw bar with the right hand.
4. Remove the arbor from the machine spindle.

57-14. Removing Self-Holding Arbors

Some older milling machines and many smaller vertical milling machines have arbors or adapters with a Brown & Sharpe taper, Morse taper, or with a special taper. These usually are the self-holding type tapers which also are held in the machine spindle with a draw bar. The draw bar should not be turned up too tightly, or it will be very difficult to remove the arbor or adapter.

To remove the arbor or adapter, loosen the draw bar by turning it in the proper direction. Some draw bars have left-hand threads while others have right-hand threads. First check to see whether the threads are right- or left-hand. As the draw bar is loosened, it presses against a **retaining collar**; this forces the arbor or adapter out of the spindle at the opposite end. If the arbor does not release when reasonable force is applied to the draw bar with a wrench, request assistance from your instructor.

57-15. Milling Cutters

Milling cutters are made in a wide variety of standard shapes and sizes. They can be used for machining flat surfaces, grooves, angular surfaces, and irregular surfaces. Milling cutters of **special** design, for special kinds of operations or for machining surfaces of special shapes, are also available.

Cutter Materials

Milling cutters may be of solid material such as carbon-tool steel, high-speed steel, or tungsten carbide. The majority of the milling cutters used in school shops are solid and are made of high-speed steel.

Milling cutters are also available with **tungsten-carbide teeth**. The carbide teeth are brazed on the tips of the cutter, as shown at the right in Fig. 57-33, and **G, H,** and **J** in Fig. 57-43. Very large cutters often have **inserted teeth**, Fig. 57-2. The inserted teeth may be made of high-speed steel, tungsten carbide, or **cast alloy**. The body of the cutter generally is made of a tough grade of alloy steel, thus reducing the cost of the cutter.

High-speed steel cutting tools rank high in impact resistance, wear resistance, and in general toughness. Hence, they are able to withstand the abuse and vibration which often occur on lightweight milling machines frequently used in school shops.

Carbide-tipped cutters should be used on rigid setups and on heavy machines. Vibration or chatter causes them to fracture quite easily. However, they can be used with cutting speeds two to four times greater than high-speed steel cutters.

Fig. 57-29. Milling a Crankshaft in a Special Fixture
The Style-B Arbor is used in this heavy-duty milling operation.

Fig. 57-30. Straddle Milling

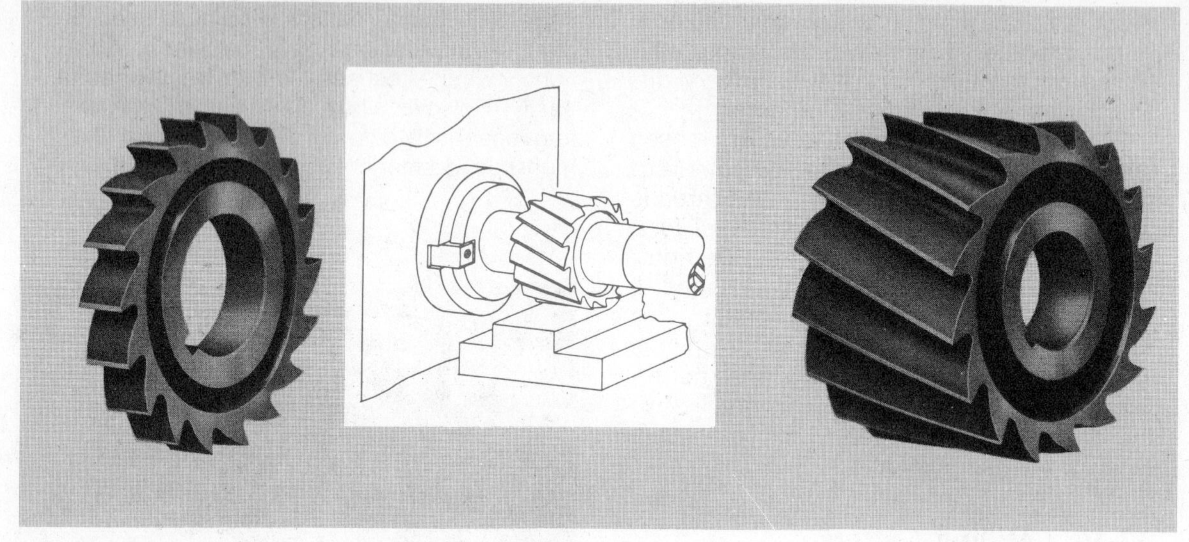

Fig. 57-31. Light-Duty Plain Milling Cutters

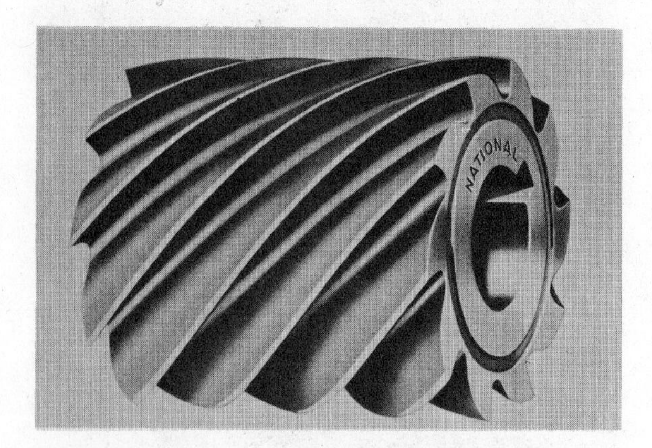

Fig. 57-32. Heavy-Duty Plain Milling Cutter

Plain-Milling Cutters

Plain-milling cutters are cylindrical in shape and have cutting teeth on the periphery (circumference) only, Fig. 57-31. They have an accurately ground hole and are mounted on an arbor for use on horizontal-type milling machines. They are used for machining plain, flat surfaces. Sometimes they are used in combination with other kinds of cutters for milling special kinds of surfaces, Fig. 57-6.

Light-duty plain-milling cutters, Fig. 57-31, have relatively fine teeth. For example, a 2½″ diameter cutter generally has 14 to 18 teeth. Cutters of this type which are less than ¾″ width have straight teeth parallel to the axis of the cutter. Cutters ¾″ and wider have helical teeth with an 18° helix angle. They are available in a variety of widths and diameters.

Heavy-duty plain-milling cutters, Fig. 57-32, are also called **coarse-tooth milling cutters**. They are similar to the light-duty plain cutters except that they have fewer teeth and a steeper helix angle. For instance. a 2½″ diameter cutter generally has 8 teeth, and the helix angle is 45°. Heavy-duty cutters are recommended for heavy cuts wherever large amounts of metal must be removed. They also work well for light and moderate cuts. In fact, lightweight machines generally can take heavier cuts and produce a better surface finish with coarse-tooth plain cutters than with fine-tooth plain cutters.

Side-Milling Cutters

Side-milling cutters, Fig. 57-33, 57-34, and 57-35, are similar to plain-milling cutters in

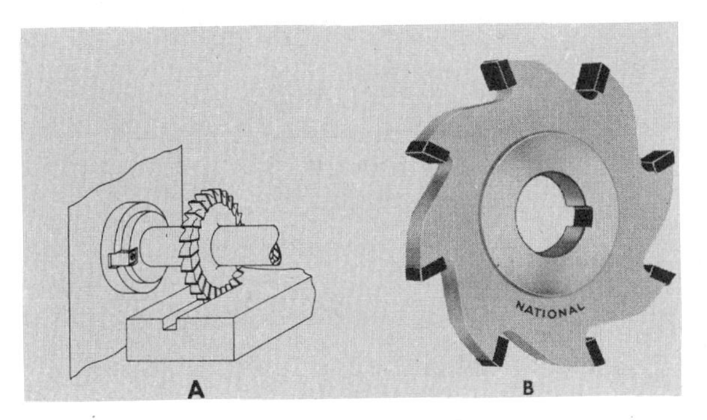

Fig. 57-33. Plain-Side Milling Cutters
A. Solid High Speed Cutter
B. Carbide Tipped Cutter

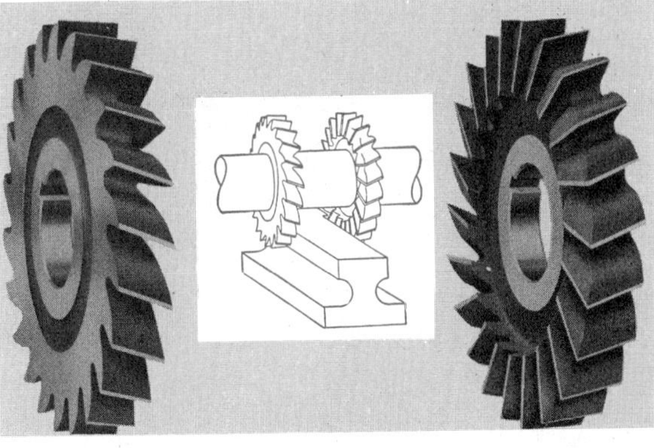

Fig. 57-34. Half-Side Milling Cutters
(Right) Single-Sided Plain Mill Cutter
(Center) Straddle Milling Setup
(Left) Single-Sided Plain Mill Cutter

that they have cutting edges on the periphery. However, they also have cutting edges on the sides. The cutting edges on the periphery do most of the cutting. The teeth may be either **straight** (Fig. 57-33), **helical** (Fig. 57-34), or **staggered** (Fig. 57-35).

Side-milling cutters are used for milling the sides of a workpiece or for cutting slots or grooves. They also can be used for **straddle milling**. This involves cutting two sides at the same time, Figs. 57-30 and 57-34. The width between the two cutters is established with spacing collars and shims.

Plain side-milling cutters, Fig. 57-33, have straight teeth on the periphery and both sides. They are used for moderate-duty side-milling, slotting, and straddle-milling operations.

Half-side milling cutters, Fig. 57-34, have teeth on the periphery and only one side. These cutters are recommended for heavy-duty side-milling and straddle-milling operations.

Staggered-tooth side-milling cutters, Fig. 57-35, are narrow cutters with teeth which alternate to either side. This tooth arrangement provides more chip clearance and reduces scoring on the side surfaces being machined. Cutters of this type are recommended for heavy-duty machining of grooves or keyways.

Fig. 57-35. Staggered-Tooth Side-Milling Cutter

Metal-Slitting Saws

Metal-slitting saws, Fig 57-36, are used for ordinary cutoff operations and for cutting narrow slots. They are available with several kinds of teeth.

Plain metal-slitting saws, Fig. 57-36, have teeth on the periphery only. They are available in widths from 1/32″ to 3/16″ and in diameters from 2½″ to 8″. They have fine teeth, and the sides of the teeth taper toward the hole. The taper prevents the blade from binding in the slots or saw kerf as it rotates. The feed rate should be small (usually about ¼ to ⅛ that used for plain milling cutters). (See Table 38.)

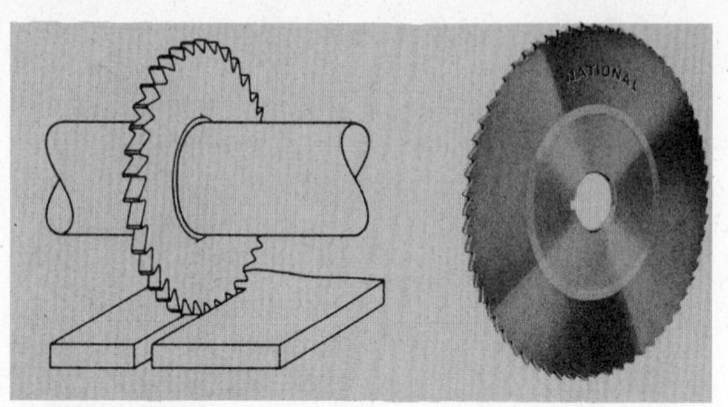

Fig. 57-36. Plain Metal Slitting Saw

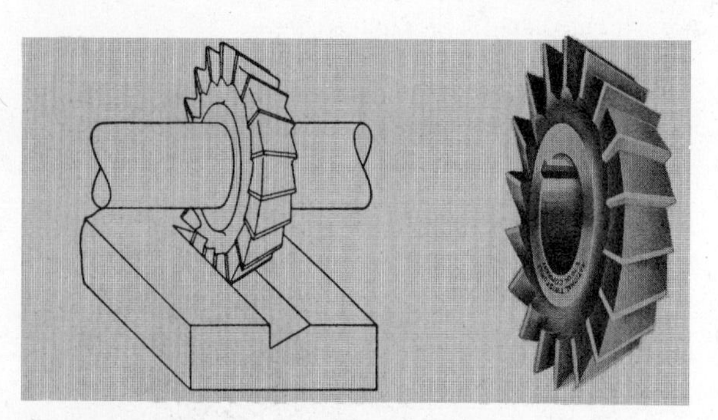

Fig. 57-37. Single-Angle Milling Cutter

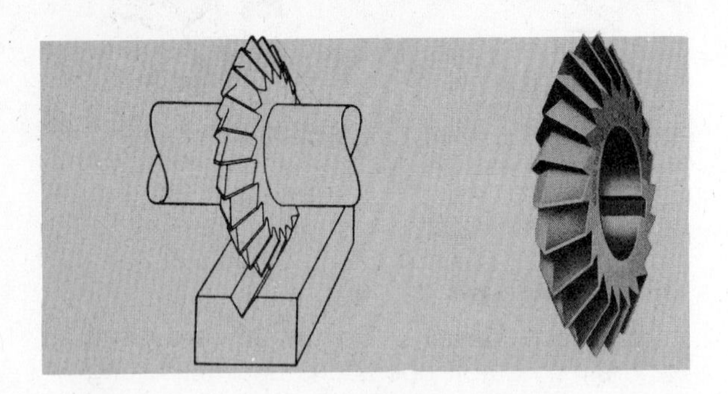

Fig. 57-38. Double-Angle Milling Cutter

Staggered-tooth metal-slitting saws, similar to staggered-tooth side-milling cutters, may be used for wider and deeper cuts. They are available in widths from 3/16″ to ¼″.

Screw-slotting cutters are special, fine-tooth plain-slitting saws. They are used for cutting screw slots and are available in widths from 0.020″ to 0.182″.

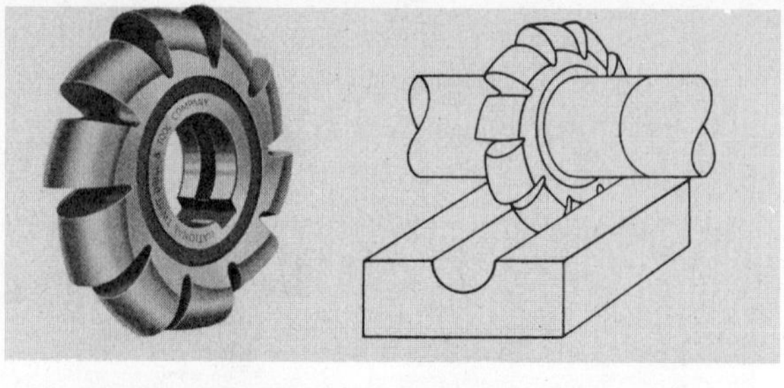

Fig. 57-39. Convex Milling Cutter

Fig. 57-40. Concave Milling Cutter

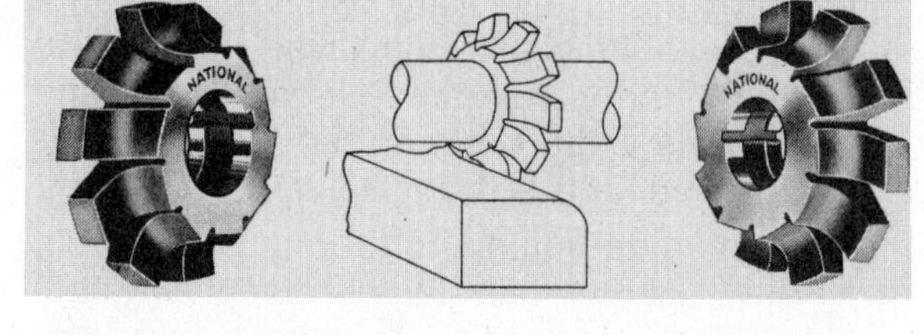

Fig. 57-41. Corner-Rounding Milling Cutters

Angular Milling Cutters

Cutters within this classification, Figs. 57-37 and 57-38, are used for machining V-notches, grooves, serrations, dovetails, and reamer teeth.

Single-angle cutters, Fig. 57-37, have a single angle with cutting edges on both sides of the angle. Generally they are available with either 45°, or 60° angles.

Double-angle cutters, Fig. 57-38, have V-shaped teeth. They usually are available with 45°, 60°, or 90° angles.

Fig. 57-42. (Left) Spur Gear Milling Cutter (Right) Fluting Cutter

Form-Relieved Cutters

Cutters within this classification are used for cutting curved surfaces or surfaces of irregular shape. Several kinds of form-relieved cutters are shown in Figs. 57-39 to 57-42. They are used for cutting curved grooves, rounded corners, or flutes in reamers, milling cutters, or gear teeth.

The "Hand" of Milling Cutters

The term **hand** is used to describe the following factors involved in milling:

1. Hand of the cutter.
2. Hand of the helix.
3. Hand of the cut.

Hand of the cutter refers to the direction in which the cutter must rotate to cut. A cutter may be a right-hand cutter or a left-hand cutter. The hand is determined by looking at the front end of the cutter (toward the spindle nose or column) while the cutter is mounted in the machine spindle. A **right-hand** cutter must rotate **counterclockwise** to cut. A **left-hand** cutter must rotate **clockwise** to cut. Thus, end mills, reamers, drills, and similar cutting tools are designated right hand (RH) or left hand (LH). All of the end mills in Fig. 57-43 are right hand.

Hand of the helix describes the direction of the helical flutes on the milling cutter or on other cutting tools such as drills and reamers. The hand of the helix is determined by looking at either end of the cutting tool and noting the direction in which the helical flutes twist. If they twist away and toward the right, they have right-hand helical flutes, Fig. 57-31. If they twist away and toward the left, they have left-hand helical flutes, Fig. 57-32.

Hand of the cut refers to the direction of cut. A cut may be a **right-hand cut** or a **left-hand cut**. The hand of the cut is also determined by looking at the front end of the cutter (toward the spindle nose or column) while the cutter is mounted in the machine spindle. A **right-hand cut** requires **counterclockwise** rotation of the cutter, Fig. 57-47. A **left-hand cut** requires **clockwise** rotation of the spindle. Arbor-type milling cutters with straight teeth can be mounted for cutting with either a right-hand cut or left-hand cut.

It is also possible to make either right-hand or left-hand cuts with arbor-type helical cutters. However, when possible, helical cutters should be installed so that side thrust on the cutter tends to force the cutter toward the column, Fig. 57-47.

End-Mill Cutters

End-milling cutters commonly are called **end mills**. They are designed for milling slots, shoulders, curved edges, keyways, and pockets where ordinary arbor-type cutters cannot be used. However, end mills can also be used for performing many of the same operations performed by arbor-type cutters.

End mills include two basic types, **solid** and **shell**. With the solid mills, the teeth and the shank are integral parts. A variety of solid end mills is shown in Fig. 57-43.

With shell end mills, the body and the shank are separate parts, Fig. 57-44. Shell end mills are mounted on style-C arbors, as shown at **B** in Fig. 57-24. Small shell-end mills are mounted on an adapter as shown at **C** in Fig. 57-26.

End-milling cutters generally have teeth on the circumference and on the end. On **square-nose** cutters, most of the cutting is done by the teeth on the circumference. On **round-nose** cutters (see **F** in Fig. 57-43), a large portion of the cutting is done at the end. The teeth on the circumference may be straight or helical.

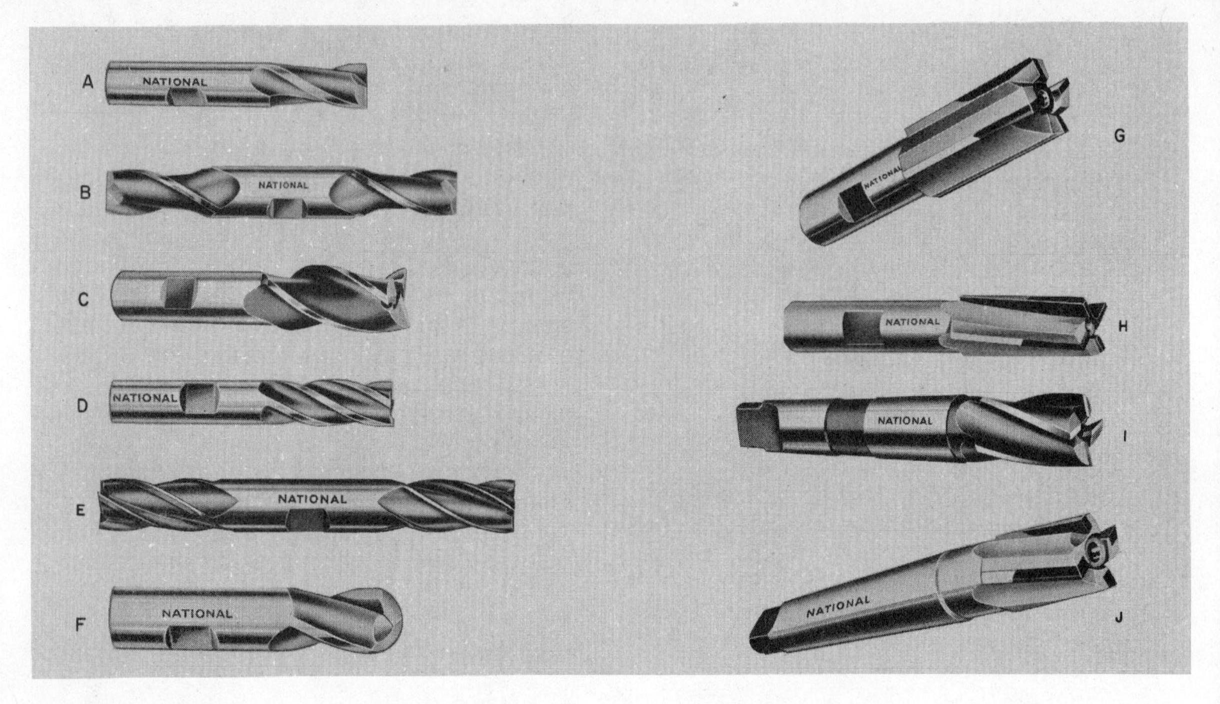

Fig. 57-43. End Mills
A. Two-Flute Single-End
B. Two-Flute Double-End
C. Three-Flute Single-End
D. Multiple-Flute Single-End
E. Four-Flute Double-End
F. Two-Flute Ball-End
G. Carbide-Tipped Straight Flutes
H. Carbide-Tipped RH Helical Flutes
I. Multiple-Flute with Taper Shanks
J. Carbide-Tipped with Taper Shank and Helical Flutes

With the exception of shell-type cutters, the shanks are either **straight** or **tapered**. The flat surface on the shank provides a means for holding the end mill securely in the end-mill adapter with a setscrew, Fig. 57-27. The taper-shank end mills have a flat-drive tang which prevents them from turning. They are mounted in the machine spindle with adapters, as shown at **B** and **D** in Fig. 57-26.

Two-flute end mills are designed with **end-cutting teeth** for **plunge** and **traverse** milling. Hence, this kind of end mill can be fed to depth like a drill. It then can be fed longitudinally.

Multiple-flute end mills have three, four, six, or eight flutes. Some types have end-cutting teeth, as at **C** in Fig. 57-43. These may be used for plunge milling to depth as well as for longitudinal milling.

Ball-end mills, as at **F** in Fig. 57-43, are used for milling pockets in dies. They are also used for milling fillets or slots. They have end-cutting teeth which may be used for drilling to depth (plunge milling) as well as longitudinal milling. **Four-fluted ball-end**

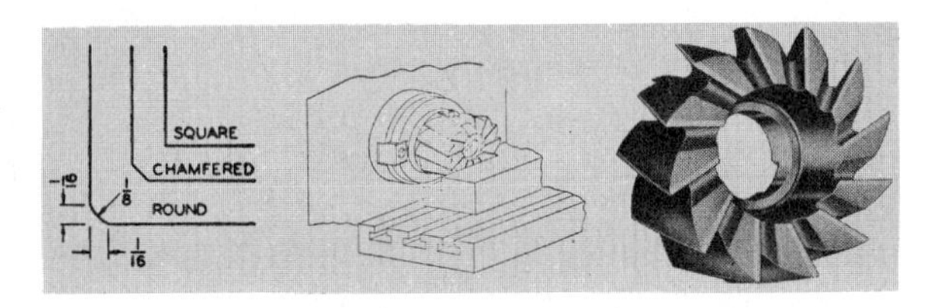

Fig. 57-44. Shell-End Mill

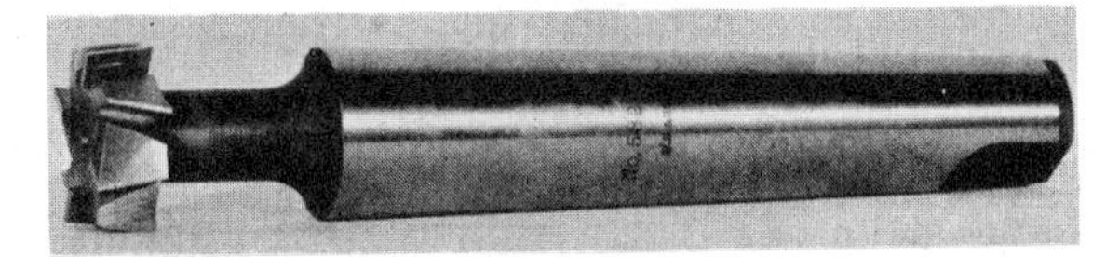

Fig. 57-45. T-Slot Cutter

Fig. 57-46. Woodruff Key Seat Cutter

mills are also available and are used for similar operations.

Shell-end mills, Fig. 57-44, are made in larger sizes than most shank-type end mills. Generally, they are made in diameters from 1¼″ to 6″. They may have either helical or straight teeth. Cutters of this type are used for machining larger shoulders or surfaces. The teeth on all types of milling cutters stay sharp longer if they have a chamfer or a radius ground on the corner of the teeth, Fig. 57-44.

T-slot milling cutters, Fig. 57-45, are used for milling T-slots such as those on milling machine tables. The narrow portion of the T-slot is machined first with a side-milling cutter or an end mill. The wide portion then is cut with the T-slot cutter.

Key-seat cutters, Fig. 57-46, are used for cutting seats for woodruff keys. They are available in sizes for all standard woodruff keys. The end mill is made in sizes from ¼″ to 1½″ diameter. The arbor mill is made in diameters from 2⅛″ to 3½″.

Depth of cut is an important factor in preventing end-mill breakage. As a general rule, **the maximum depth of cut should not be greater than one-half the diameter of an end mill**. On hard, tough steel, the maximum depth should not generally exceed ¼ the diameter of the end mill. Suggested feeds for end mills are indicated in Table 38.

57-16. Importance of Using Sharp Milling Cutters

Milling cutters must be sharp to produce a good surface finish and to cut efficiently. When the cutter becomes dull, extreme forces are exerted on the cutter, the arbor, and the machine spindle, resulting in poor finishes and possible cutter breakage. On horizontal machines, the extreme forces can bend the arbor, and it will no longer run true.

Only experienced persons should grind milling cutters. Cutters are expensive and can be damaged if sharpened improperly. This kind of instruction normally is included in advanced machine shop classes. It is not included within the scope of this book. Information and procedures for sharpening milling cutters generally are included in a manual or handbook provided by the manufacturer of cutter- and tool-grinding equipment.

57-17. Using Cutting Fluids

Cutting fluids should be used for all milling operations on steel, aluminum, and copper alloys. Gray cast iron may be machined dry, or an emulsifiable (soluble) oil solution may be used on it. Cutting fluids and their selection are explained in sections 48-6 and 48-7. (See Table 33, page 452 for selection of recommended cutting fluids for common metals.)

57-18. Cutting Speeds and rpm

For milling, **cutting speed** refers to the circumferential speed of the milling cutter. It is expressed in **surface feet per minute** (sfpm). You can visualize the cutting speed of a milling cutter by imagining it as the distance the cutter could roll across the floor during one minute.

Different cutting speeds should be used when machining different metals. If the cutting speed is too fast, the cutter will become overheated and dull rapidly. If the speed is too slow, time will be wasted and production costs will increase.

TABLE 37

Cutting Speeds for Milling Roughing Cuts with High-Speed Steel Cutters

Material	Cutting Speed Range in sfpm
Low-carbon steel	60-80
Medium-carbon steel, annealed	60-80
High-carbon steel, annealed	50-70
Tool steel, annealed	50-70
Stainless steel	50-80
Gray cast iron, soft	50-80
Malleable iron	80-100
Aluminum and its alloys	400-1000
Brass	200-300
Bronze	100-200

These suggested speeds may be varied as follows:
For finishing cuts ....Increase 25-50%
For carbon-steel cutters ....Decrease about 50%
For cutters with cast-alloy tips ......Increase 50-75%
For cutters with cemented-carbide tips ....Increase 200-400%
Feeds should be as much as the cutter, the setup, and the equipment will safely stand. Recommended cutting fluids should be used, see Table 33, page xxx.

Factors Affecting Cutting Speeds

One of the most important factors affecting cutting speed is the machinability rating of the metal. (See § 19-15.) Metals with high machinability ratings can be machined at higher cutting speeds than those with lower ratings. (See Table 37.)

When the machinability rating is doubled. the cutting speed may be doubled, provided that proper cutting fluids are used. For example, steel with a rating of 100 may be machined at twice the cutting speed of steel with a rating of 50. The machinability ratings for various metals are included in handbooks for machinists.

The following factors affect cutting speeds for milling operations:

TABLE 38

Feeds (Inches per Tooth) for Milling Roughing Cuts with High-Speed Steel Cutters

Material	Plain Mills (Heavy-Duty)	Plain Mills (Light-Duty)	Face Mills	Side Mills	End Mills	Form-Relieved Mills	Slitting Saws
Low-carbon steel, free machining	.010	.006	.012	.006	.006	.004	.003
Low-carbon steel	.008	.005	.010	.005	.005	.003	.003
Medium-carbon steel	.008	.005	.009	.005	.004	.003	.002
High-carbon steel, annealed	.004	.003	.006	.003	.002	.002	.002
Stainless steel, free machining	.008	.005	.010	.005	.004	.003	.002
Stainless steel	.004	.003	.006	.004	.002	.002	.002
Cast iron, soft	.012	.008	.014	.008	.008	.004	.004
Cast iron, medium	.010	.006	.012	.006	.006	.004	.003
Malleable iron	.010	.006	.012	.006	.006	.004	.003
Brass and bronze, medium hardness	.010	.008	.013	.008	.006	.004	.003
Aluminum and its alloys	.016	.010	.020	.012	.010	.007	.004

These feeds are suggested for roughing cuts on heavy-duty machines, and they may be increased or decreased depending on machining conditions. For average conditions, it may be necessary to reduce these rates by 50%. For finishing cuts the rates generally should be reduced 50%.

TABLE 39

Cutting Speeds (sfpm) for Various Diameters

Feet per Min. / Diameter Inches	30′	40′	50′	60′	70′	80′	90′	100′	110′	120′	130′	140′	150′
	Revolutions per Minute												
1/16	1833	2445	3056	3667	4278	4889	5500	6111	6722	7334	7945	8556	9167
1/8	917	1222	1528	1833	2139	2445	2750	3056	3361	3667	3973	4278	4584
3/16	611	815	1019	1222	1426	1630	1833	2037	2241	2445	2648	2852	3056
1/4	458	611	764	917	1070	1222	1375	1528	1681	1833	1986	2139	2292
5/16	367	489	611	733	856	978	1100	1222	1345	1467	1589	1711	1833
3/8	306	407	509	611	713	815	917	1019	1120	1222	1324	1426	1528
7/16	262	349	437	524	611	698	786	873	960	1048	1135	1222	1310
1/2	229	306	382	458	535	611	688	764	840	917	993	1070	1146
5/8	183	244	306	367	428	489	550	611	672	733	794	856	917
3/4	153	203	255	306	357	407	458	509	560	611	662	713	764
7/8	131	175	218	262	306	349	393	436	480	524	568	611	655
1	115	153	191	229	267	306	344	382	420	458	497	535	573
1 1/8	102	136	170	204	238	272	306	340	373	407	441	475	509
1 1/4	92	122	153	183	214	244	275	306	336	367	397	428	458
1 3/8	83	111	139	167	194	222	250	278	306	333	361	389	417
1 1/2	76	102	127	153	178	204	229	255	280	306	331	357	382
1 5/8	70	94	117	141	165	188	212	235	259	282	306	329	353
1 3/4	65	87	109	131	153	175	196	218	240	262	284	306	327
1 7/8	61	81	102	122	143	163	183	204	224	244	265	285	306
2	57	76	95	115	134	153	172	191	210	229	248	267	287
2 1/4	51	68	85	102	119	136	153	170	187	204	221	238	255
2 1/2	46	61	76	92	107	122	137	153	168	183	199	214	229
2 3/4	42	56	69	83	97	111	125	139	153	167	181	194	208
3	38	51	64	76	89	102	115	127	140	153	166	178	191

This table can be used to determine the approximate rpm for drilling, milling, turning, and boring operations for diameters up to 3 inches. It also gives the cutting speeds produced by various rpm with the diameters given.

1. Kind of metal being machined.
2. Machinability rating of the metal.
3. Hardness of the metal (if heat treated).
4. Kind of cutting tool material (high-speed steel, cast alloy, or cemented carbide).
5. Whether proper cutting fluids are used.
6. Depth of cut and rate of feed (roughing or finishing cut).

Suggested cutting speeds for roughing cuts with high-speed steel milling cutters are given in Table 37. There is no one correct cutting speed for milling one kind of metal. A range of speeds generally will produce good results with a specific kind of metal. It is common practice to select an average cutting speed or a speed which is somewhat less than average. With satisfactory results, the cutting speed may be increased.

Revolutions Per Minute (rpm)

After the cutting speed to be used has been determined, the machine spindle must be set at the proper rpm. Cutting speed and rpm have different meanings, and they should not be confused. **A small diameter milling cutter must turn at a higher rpm than a larger diameter cutter in order for both to cut at the same cutting speed.** An example shows this more clearly: A 1″ diameter end mill and a 2½″ diameter arbor cutter are to cut at the same cutting speed. For both to cut at 70 sfpm, the 1″ diameter cutter must turn at 267 rpm, while the 2½″ diameter cutter must turn at 107 rpm. (See Table 39.)

Note that Table 39 can be used for determining the approximate rpm for milling, drilling, turning, and boring operations up to diameters of 3″. It also gives the cutting speeds produced by various rpm when the diameters are given. The machine spindle may be adjusted, as explained in section 57-9, for the rpm which is nearest the desired rpm.

Calculating rpm

The rpm of a given cutting speed for milling and hole-machining operations can be calculated with the following formula:

$$\text{rpm} = \frac{\text{CS} \times 12}{\text{D} \times \pi}$$

Where:
CS = Cutting speed in sfpm
D = Diameter of cutter in inches
π = Pi or 3.1416 (a constant)
rpm = Revolutions per minute

Example: Calculate the rpm for a 3″ diameter cutter which is to mill steel at 90 sfpm.

$$\text{rpm} = \frac{90 \times 12}{3 \times 3.1416}$$

$$\text{rpm} = \frac{1080}{9.4248}$$

$$\text{rpm} = 114.6$$

(Compare this to Table 39.)

Or this approach:

$$\text{rpm} = \frac{90 \times \overset{4}{\cancel{12}}}{\underset{1}{\cancel{3}} \times 3.1416}$$

$$\text{rpm} = \frac{360}{3.1416}$$

$$\text{rpm} = 114.6 \text{ or } 115$$

The rpm selection on the machine may not be accurate enough to set the speed at exactly 115 or 120, so select the closest arbor speed. The number 3 can be substituted for 3.1416 when calculating the **approximate** cutting speed. This procedure is satisfactory for most applications, thus giving 120 rpm.

Calculating Cutting Speed

The cutting speed for milling and hole-machining operations can be calculated when the diameter of the cutting tool and the rpm are known:

$$\text{CS} = \frac{\text{D}'' \times \pi \times \text{rpm}}{12}$$

Example: Calculate the cutting speed of a ½″ diameter end mill which is milling at a speed of 550 rpm.

$$\text{CS} = \frac{0.500 \times 3.1416 \times 550}{12}$$

$$\text{CS} = \frac{863.94}{12}$$

$$\text{CS} = 72 \text{ sfpm}$$

The number 3 may be substituted for 3.1416 for calculating the approximate cutting speed. This procedure is satisfactory for most jobs, in this case, approximately 69 sfpm.

57-19. Rate of Feed

The **rate of feed** for milling is the rate at which the workpiece advances into the milling cutter. It is an important factor in determining the rate of metal removal. The feed rate, together with the width and depth of cut, determines the rate of metal removal.

The tendency for beginners is to use too light a feed and a cutting speed which is too high. This dulls the cutter rapidly and shortens tool life. In general, the feed rate should be as great as the cutting tool, the machine, and the work setup can stand without excessive vibration.

Each tooth on the milling cutter should cut a chip of proper size. On cutters with small, fine teeth, the chip should be small. The feed for each tooth on a milling cutter is indicated in **inches per tooth** for each revolution of the cutter. Suggested average feeds, in inches per tooth, for milling roughing cuts with high-speed steel cutters are listed in Table 38. On light-duty milling machines, the indicated feeds should be reduced. For finishing cuts, the feeds should be reduced by 50%.

Inches Per Minute

The feed rate on most milling machines is set in terms of **inches per minute**. The feed rates may be adjusted for various settings ranging from about ¼″ to 30″ per minute. The feed rates are adjusted with feed-selector dials or levers on the machine.

Calculating Rate of Feed

The following procedure generally is used in calculating the rate of feed:

1. Determine the desired cutting speed. For example, 70 sfpm for low-carbon steel.
2. Determine the rpm of the cutter.
3. Determine the number of teeth on the cutter.
4. Determine the feed in inches per tooth. (See Table 38.)
5. Calculate the feed rate with the following formula:
   $F = R \times T \times rpm$

Where:
F = Feed rate in inches per minute
R = Feed per tooth per revolution
T = Number of teeth on cutter
rpm = Revolution per minute of cutter

Example: Determine the feed rate for milling low-carbon steel at 70 sfpm, 89 rpm, using a heavy-duty plain-milling cutter 3″ in diameter with 10 teeth and a feed of 0.006″ per tooth.

$F = 0.006 \times 10 \times 89$

$F = 5.34$ inches per minute

With the feed-selector dial or levers, adjust the feed rate to the feed closest to 5.34 inches per minute.

57-20. Safety for Milling

1. Wear approved safety goggles.
2. Wipe up any oil on the floor around the machine.
3. Be certain that the table is clean and dry before making a setup.
4. Always be certain that holding devices such as a vise, angle plate, dividing head, or tailstock are fastened tightly to the table.
5. Select the right kind of cutter for the job.
6. Always be sure that the arbor, cutter, and collars are clean before mounting them in the spindle. Use a rag for handling sharp cutters.
7. When seating workpieces in a vise use a lead hammer.
8. Be certain that the vise or other holding devices clear the arbor and overarm supports.
9. Select the proper cutting speed, rpm, and rate of feed for the job.
10. Disengage the control handles when using automatic feeds.
11. Be certain that the column clamps, saddle clamps, and table clamps are loosened when making setup adjustments. Be certain to tighten them after the setup.
12. Keep your hands away from the revolving cutter at all times.
13. Clear chips away from the cutter with a brush, such as a paint brush.
14. Release any automatic feeds after completing the job.
15. Do not allow unauthorized persons within the safety zone of the machine.

16. Clean and wipe the machine when you are finished. Wipe up any oil from the floor. Remove chips with a small shovel or scoop. Never touch them with the fingers.

57-21. Vise Alignment

When a workpiece is mounted in a milling machine vise, the vise must be properly aligned. Generally, the stationary jaw of the vise must be either at **right angles** with the face of the machine column, or it must be **parallel** with the face of the column. For most operations, the stationary vise jaw must be parallel with the machined face of the column.

Before the alignment of the vise on a **universal milling machine** is checked, the table must be swiveled parallel with the face of the column. This can be checked by placing a wide parallel bar between the column and the table. Feed the table transversely (in toward the column), by hand, until it squeezes the parallel lightly against the column. If the table is parallel, no light should show between the parallel and the column. For greater accuracy, a dial indicator can be used to determine if the table is parallel with the face of the column. Correct the alignment of the swivel table if necessary.

Right-angle squareness of the vise with the column can be checked by placing the **blade** of a steel square, Fig. 6-18, against the stationary jaw of the vise. At the same time, place the **beam** of the square against the machined surface of the column. For greater accuracy, a dial indicator can be clamped on the machine arbor, with the indicator plunger against the stationary vise jaw. The table is then fed transversely (crosswise, in and out) by hand-feeding.

Parallelism of the stationary jaw of the vise, with the surface of the machine column, can be checked with a dial indicator. Clamp the indicator to the arbor with the indicator plunger touching the stationary vise jaw. Handfeed the table longitudinally and note the indicator dial. The indicator reading will not vary if the vise jaw is parallel with the face of the column.

57-22. Milling with a Horizontal Milling Machine

The following basic steps of procedure should be used for milling with a horizontal milling machine:

1. Check alignment of the table.

   This step is necessary with universal milling machines only. The side of the table must be parallel with the machined face of the column.
2. Check the alignment of the vise.

   The stationary jaw usually is aligned parallel with the machined face of the column. However, for some operations it should be at right angles with the face of the column. For special angular milling operations, the vise may be swiveled at any desired angle with the face of the column.
3. Mount the workpiece securely in the vise.

   The surface to be machined must extend high enough above the hardened vise jaws that the milling cutter will not cut into the jaws. If the cutter strikes the hardened jaws, it will fracture the cutter teeth and damage the jaws. Seat the workpiece firmly on parallels if it is necessary to elevate the workpiece above the vise jaws.
4. Position the workpiece under the arbor beneath the intended location of the cutter.

   This is done by moving the saddle toward the column with the cross-feed handwheel. The workpiece should be as close as possible to the column.
5. Select the right kind and size of cutter and arbor.

   Always handle sharp cutters with a cloth. Also place a clean cloth on the machine table where the cutter, arbor, and collars will be placed before installing.

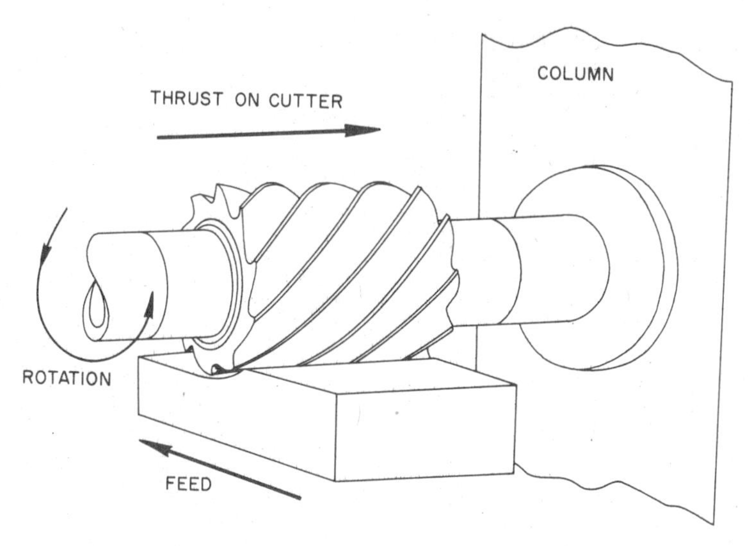

Fig. 57-47. Install Helical Cutter so Thrust Forces the Cutter toward the Column
This is a plain milling cutter which has a left-hand helix. It is mounted on the arbor for right-hand cutting.

For plain milling of flat surfaces, the cutter should be a little wider than the surface to be machined. The cutter should be large enough so that the arbor support bearing will clear the vise.

6. Install the arbor.

   Clean the shank and the spindle hole. Use a draw bar to hold the arbor securely in the machine spindle.

7. Mount the cutter on the arbor, and position the arbor support.

   The arbor, collars, and cutter must be clean and free from nicks before mounting the cutter. When plain helical cutters are used, the cutter should be installed so that **thrust** tends to push the cutter toward the column, as in Fig. 57-47.

   Space the cutter at the desired position with spacing collars. The cutter should be as close to the column as the position of the workpiece will permit. Key the cutter to the arbor. The key should extend into the collars on both sides of the cutter. Place the nut on the arbor loosely.

   Install the arbor support; allow at least ⅛″ space between the end of the arbor threads and the bearing in the arbor support. Finally, tighten the arbor nut securely against the arbor collars and cutter.

8. Adjust the cross-feed location of the workpiece.

   It should be centered, or located in the desired position, under the cutter. Tighten the saddle clamp or clamps.

9. Determine the desired cutting speed. (See Table 37.)
10. Determine the spindle rpm required for the desired cutting speed. (See Table 39.)
11. Adjust the spindle rpm.
12. Calculate the rate of feed required.

    This is explained in section 57-19.

13. Adjust the machine for the proper rate of feed.

14. With the workpiece clear of the cutter, start the machine and observe whether the direction of spindle rotation is correct. (See Fig. 57-47.)

    Reverse the direction of spindle if the cutter is rotating in the wrong direction.

15. Engage the automatic feed and observe whether the direction of feed is correct.

    The workpiece should travel against the direction of cutter rotation for **up cutting** or **conventional milling**, as in Figs. 57-14 and 57-47.

16. Adjust the table elevation for the proper depth of cut.

    This is explained under the heading **knee elevation** in section 57-9. If a finishing cut is to be made after a roughing cut, allow about 0.010″ to 0.020″ material thickness for the finishing cut. If the surface is to be surface ground after milling, allow at least 0.005″ material thickness, per surface, for grinding.

17. Apply cutting fluid, engage the table feed, and make the roughing cut.

    On machines so equipped, cutting fluid may be applied with a pump. On others it may be applied to the cutter with a paintbrush.

    If excessive vibration or chatter takes place, stop the machine. It may be caused by one of the following:

    (a) Depth of cut too deep,
    (b) Rate of feed too fast,
    (c) Cutting speed too high,
    (d) Knee clamp loose,
    (e) Saddle clamp loose, or
    (f) Bolts holding the vise may be loose.

    Make the necessary adjustment.

    Machines not equipped with automatic longitudinal table feed must be fed by hand. Feed the table at a steady rate with the table feed handwheel.

    Try to avoid stopping the table feed before finishing a cut. If the table feed is stopped and started again, a ridge or irregular surface results at the location where the feed was stopped.

18. At the end of the roughing cut, disengage the table feed, stop the machine, and brush the chips away from the machined surface. **Never move the workpiece back under the revolving cutter.**

    It is always good practice to lower the table an amount equal to one turn of the vertical-feed hand crank before bringing the table back to the starting position.

19. If an additional cut must be made, bring the table back to the starting position in preparation for the next cut.

    Measure the workpiece. If another roughing cut is needed, raise the table one turn of the vertical-feed hand crank plus the depth of cut desired.

20. Make a finishing cut if required.

    A cut from 0.010″ to 0.020″ deep generally produces a good surface finish. Make the finishing cut as in steps 15 through 18. Measure the workpiece and take an additional cut if necessary.

21. Remove the workpiece from the machine.

22. Clean the machine.

    Store tools and other machine accessories in the proper place.

23. Remove the sharp arrises or burrs from the workpiece with a file.

57-23. Milling with a Vertical Milling Machine

The principles and procedures involved in most vertical milling operations are similar, even though the particular machine or the controls on the machine are somewhat different in design.

Procedure

The following are basic steps of procedure for milling a flat surface with a vertical milling machine.

1. Select an end mill for the job.

   For machining a flat surface, the cutter should be slightly wider than the surface to be machined. (See Fig.

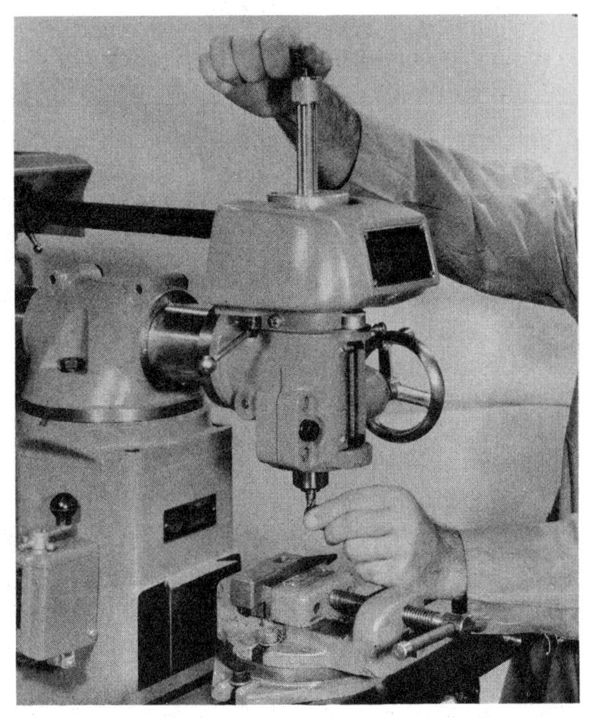

Fig. 57-48. Using a Draw Bar to Fasten End Mill Cutter in the Collet of a Vertical Mill

Fig. 57-49. Machining a Flat Angular Surface with a Shell-End Mill

57-49.) If the surface is wider, several cuts will be required.

2. Mount the milling cutter in the machine spindle. (See Fig. 57-48.)

   Some machines have collets which hold straight-shank end mills. The cutter should be inserted in the collet, and the collet should be drawn up tightly with the draw bar provided.

3. Check to see that the milling head is perpendicular to the table.

   On some machines, the head may be swiveled for milling angular or beveled surfaces. For machining flat horizontal surfaces, the head must be perpendicular to the table.

4. Mount the vise on the table with T-slot bolts, and align the vise.

5. Mount the workpiece in the vise.

   The surface to be machined must extend high enough above the vise jaws so that the cutter will not cut into the hardened jaws. (See Fig. 57-49.) Seat the workpiece firmly on parallels if necessary.

6. Determine the desired cutting speed (See Table 37.)

7. Determine the spindle rpm required for the desired cutting speed. (See Table 39.)

8. Adjust the machine for the desired rpm.

9. Calculate the rate of feed required.

   This is explained in section 57-19.

10. Adjust the machine for the correct rate of feed.

11. With the workpiece clear of the cutter, start the machine and observe whether the direction of spindle rotation is correct.

    Reverse the direction of the spindle if the cutter is rotating in the wrong direction.

When milling shoulders or vertical surfaces with an end mill, use the **up-milling** method. (See Fig. 57-14.) By looking at Fig. 57-14 and imagining it as the top view of an end mill which is cutting along a vertical surface, you can visualize the up-milling method as it applies to end-milling operations.

12. Adjust the table elevation for the proper depth of cut.

    Use the micrometer collar on the knee crank to set the depth of cut accurately. This procedure is explained under the heading, **knee elevation**, in section 57-9.

    The maximum depth of cut for end mills generally should not exceed ½ the diameter of the cutter. On hard, tough steel the maximum depth of cut should not exceed ¼ the diameter of the cutter. On light-duty machines, the maximum depth of cut must be reduced further.

    If a finishing cut is to be made after the roughing cut, allow about 0.010″ to 0.020″ material thickness for the finishing cut. If the surface is to be surface ground after milling, allow at least 0.005″ material thickness, per surface, for grinding.

13. Apply cutting fluid, engage the automatic longitudinal table feed, and make the roughing cut.

    On machines so equipped, cutting fluid may be applied with a pump. On others, it may be applied with a paintbrush.

    If excessive vibration or chatter takes place, stop the machine. It may be caused by one of the following:

    (a) Depth of cut too deep
    (b) Rate of feed too fast
    (c) Cutting speed too high
    (d) Knee clamp loose
    (e) Saddle clamp loose
    (f) Bolts holding the vise may be loose. Make the necessary adjustment.

    Machines not equipped with automatic longitudinal table feed must be fed by hand. Feed the table at a steady rate with the table feed handwheel.

14. At the end of the roughing cut, disengage the table feed and brush the chips away from the machined surfaces. Measure the workpiece.

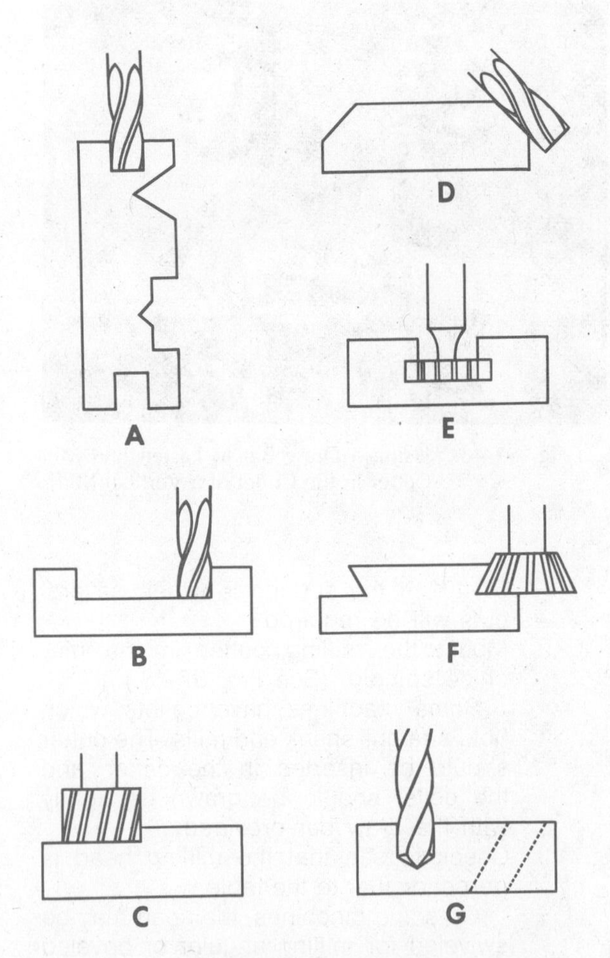

Fig. 57-50. Operations Performed with a Vertical Milling Machine
A. Milling a Groove
B. Milling a Recessed Area
C. Milling a Shoulder
D. Milling a Chamfer
E. Milling a T-Slot
F. Milling a Dovetail
G. Drilling a Hole

15. If additional roughing cuts must be made, bring the table back to the starting position in preparation for the next cut.
16. Use the micrometer collar on the vertical knee crank for setting the depth of cut for additional cuts.
17. Make additional roughing cuts by repeating steps 11 through 16.
18. Make a finishing cut if necessary.

    A finishing cut 0.010″ to 0.020″ generally produces good results on horizontal, flat surfaces.

    A finishing cut 0.002″ to 0.005″ along vertical surfaces, shoulders, and grooves generally produces good results.

Fig. 57-51. Milling a Groove with an End Mill

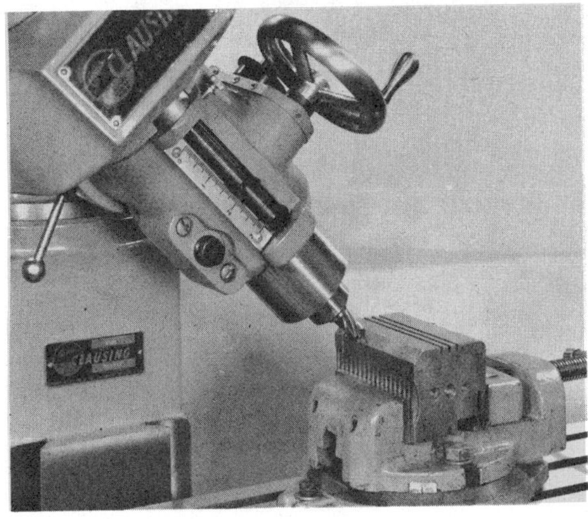

Fig. 57-53. Milling a Bevel with the Head Swiveled

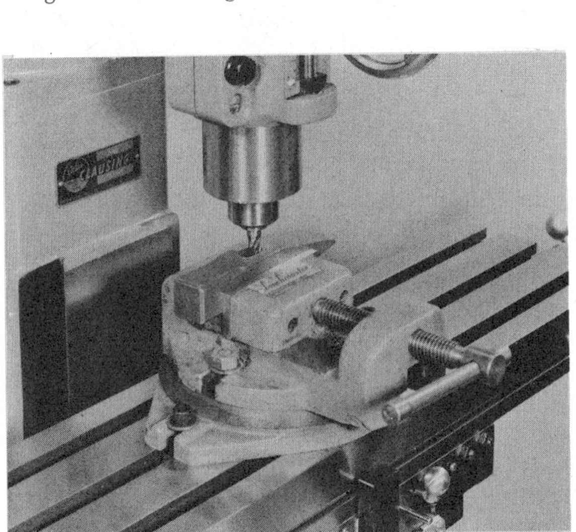

Fig. 57-52. Milling the Eye for a Hammer with an End Mill

Fig. 57-54. T-Slot Machined with a T-Slot End Mill

Fig. 57-55. Boring a Perpendicular Hole at an Angle with Head Swiveled
The boring tool is mounted in a boring head.

Fig. 57-56. Drilling at an Angle
The holes are equally spaced by indexing with a dividing head.

19. Remove the workpiece from the vise, clean the machine, and put all tools and accessories in the proper place.
20. Remove sharp arrises and burrs from the workpiece with a file.

Other Vertical Milling Setups

A variety of vertical milling operations and setups is illustrated in Figs. 57-50 to 57-56.

57-24. Indexing

Indexing is done to move a workpiece so that a series of equally spaced divisions can be machined. Indexing is done with a dividing head. One of the most common types is the **universal dividing head**, Fig. 57-57. Indexing is necessary to machine hexagonal or square bolt heads, spur gears (Fig. 57-19), helical gears (Fig. 57-21), flutes in reamers, keyways, and similar items. Other uses of the dividing head are explained in section 57-11.

The principal parts of a universal dividing head are shown in Fig. 57-57. Indexing is done in a similar manner on most kinds of dividing heads. The **index crank** is attached to the **worm shaft** which has a **worm gear** on the end, inside the dividing head. The worm gear drives the **work spindle**. The work spindle has a **40-tooth worm wheel** (sometimes called a worm gear) on the end, inside the dividing head, which meshes with the worm gear on the worm shaft. Hence, when the index crank is turned 40 complete turns, the work spindle turns one complete turn or 360°. Or, one turn of the index crank causes the work spindle to revolve through 9°. ($40 \times 9° = 360°$.)

The **index-plunger pin** holds the crank in position so that the spindle cannot turn.

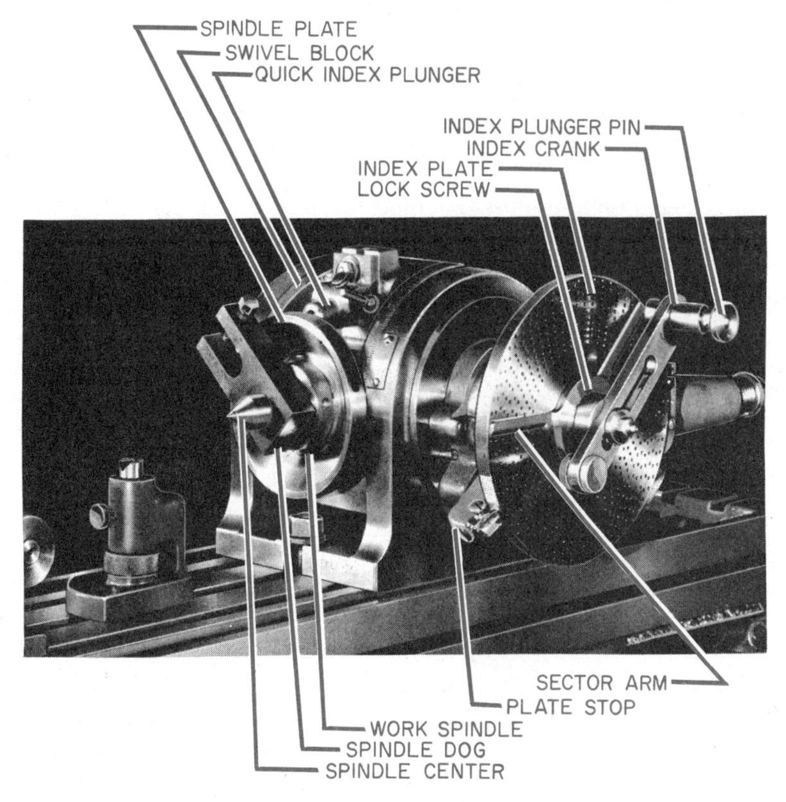

Fig. 57-57. Universal Dividing Head, Showing Principal Parts

Thus, the index-plunger pin must be withdrawn from the hole in the index plate before the crank can be turned.

The **plate stop** holds the index plate in position so that it cannot turn. It does not turn when the index crank is turned. The index plate does not turn, except for operations involving the machining of a helix or spiral. Thus the index plate must be held in a stationary position while bolt heads, keyways, spur gears, and similar jobs are machined.

Rapid Indexing

The rapid method of indexing is often called **direct indexing** and can be done easily and rapidly. The worm shaft (with index crank on the end) must be disengaged from the work spindle. This is done with a thumbscrew or lever provided for this purpose. Then, by withdrawing the **quick index plunger pin** from the hole in the back of the **spindle plate,** one can turn the spindle and plate freely by hand.

There are 24 equally spaced holes and spaces in the back of the **spindle plate**. The holes are spaced 1/24th of a revolution apart. It is possible to index 24, 12, 8, 6, 4, and 2 equally spaced divisions by this method. If 12 divisions are desired, every second space or hole is used. For 8 divisions, index 3 holes or spaces. For 6 divisions, index 4 holes or spaces. For 4 divisions, index 6 holes or spaces. For 2 divisions, index 12 holes or spaces.

Plain Indexing

The plain method of indexing, sometimes called **simple indexing**, makes it possible to index many numbers of divisions not possible by the rapid indexing method. Thus 5, 7, 9, 11, 13, and many other numbers of equally spaced divisions can be indexed by the plain indexing method. This method of indexing generally is required for machining gears.

The **worm shaft** must be engaged with the **work spindle** for plain indexing. Thus the work spindle is turned by withdrawing the **index plunger pin** and turning the index crank.

The **index plate** has several circles of holes. A different number of holes is equally spaced on each **hole circle**. Some dividing heads have holes on each side of the index plate. The plate may be reversed for the desired hole circle. An index plate of the type in Fig. 57-57 has the following numbers of holes in circles and can be considered to be a standard index plate:

One side 24-25-28-30-34-37-38-39-41-42-43
Other side
46-47-49-51-53-54-57-58-59-62-66

Two **sector arms** are located on the front of the index plate. The arms can be swiveled freely around the plate. A **lock screw** is used to lock the two arms securely for a space representing a fraction of a hole circle. Thus it is possible to index a fractional part of a revolution with the index crank.

The following formula can be used to determine the number of revolutions of the index crank for a desired number of equally spaced divisions:

$$T = \frac{N}{D}$$

Where:

T = Number of turns, or fractional part of turns, of index crank

N = Number of turns of index crank for one revolution of work spindle (usually 40 turns)

D = Number of divisions required for workpiece

Example:

Determine the number of revolutions of the index crank for indexing each tooth space for an 18-tooth spur gear.

$$T = \frac{40}{18}$$

$$T = 2\frac{4}{18} = 2\ \frac{2}{9}\ \text{turns}$$

When a fractional part of a revolution of the index crank must be made, the fractional part is established between the two sector arms. To determine a fractional part of a revolution, select a circle with a number of holes which is divisible by the denominator of the fraction for the number of turns. In the above example, the number 54 is divisible by 9. (54 ÷ 9 = 6) or (1/9 = 6/54). Thus 2/9 = 12/54. The sector arms are then spaced with 12 spaces between holes on the 54-hole circle. (**Note**: This is the space between 13 holes. It is the number of spaces between holes which is important. It is like counting the spaces between the fingers of your hand. You have four spaces between five fingers.)

When indexing for the 18-tooth gear above, each gear tooth is indexed 2 12/54 revolutions. That is 2 complete turns of the index crank, plus the 12/54 revolution between the sector arms.

It is always a good idea to practice indexing first without actually machining the part. Start indexing by using the **number one** hole in the hole circle. Mark the hole with a pencil; then index for each space. In the case of the 18-tooth gear above, index 18 spaces. After indexing 18 spaces, the index plunger pin should be in the number **one** starting hole.

Other more complex methods of indexing are sometimes necessary. These are described in handbooks for machinists.

Words to Know

arbor
arbor adapter
bearing sleeves
collets
column
cross feed
cutting speed
dividing head
down milling
end mill
end milling
feed
feed rate
gear hobber

hand of cut
hand of cutter
hand of helix
hole machining
horizontal boring mill
knee clamp
knee elevation
longitudinal feed
micrometer collar
milling cutter
milling machine
  bed-type
  horizontal
  knee-and-column
  plain
  universal
  vertical
planer milling machine
power feed
rapid traverse
saddle
saddle clamp
spacing collar
spiral milling
straddle milling
swivel vise
taper shanks
  Brown & Sharpe taper
  Morse taper
  National milling machine taper
  self-holding taper
  self-releasing taper
thread milling
transverse table feed
T-slot
universal vise
up milling
vertical feed

Review Questions

1. What is a milling machine?
2. Why are milling machines important?
3. What three factors determine the variety of machining operations which can be performed on a milling machine?
4. List several kinds of flat surfaces which can be machined with a milling machine.
5. List several kinds of curved or irregular surfaces which can be machined with a milling machine.
6. List several kinds of hole-machining operations which can be performed with a milling machine.
7. Describe a knee-and-column milling machine.
8. List the three directions in which the table can be adjusted on knee-and-column milling machines.
9. List three different classifications or types of knee-and-column milling machines.
10. List two types of horizontal milling machines.
11. What is the distinguishing difference between plain and universal horizontal milling machines?
12. List the kinds of operations which can be performed on a universal milling machine, but which cannot be performed without special accessories on a plain milling machine.
13. How can vertical milling operations be performed on either plain or universal milling machines?
14. List the kinds of operations which can be performed with vertical milling machines or with vertical milling attachments on horizontal machines.
15. Describe how holes can be accurately spaced and drilled with a vertical milling machine.
16. Why should the knee clamps and saddle clamps be tightened when milling?
17. What is the purpose of a rapid-traverse control on a milling machine?
18. Explain how the spindle rpm can be adjusted or changed on one kind of milling machine.
19. Why would it be necessary to change the direction of spindle rotation on a milling machine?
20. Explain how the power feed rate can be changed on one kind of milling machine.
21. Explain the difference between up-milling and down-milling. Why is up-milling preferred on conventional machines?
22. For what purpose is a vertical milling attachment used?
23. For what purpose is a universal spiral milling attachment used?
24. For what purpose is a circular milling attachment used?
25. Explain the characteristics of the **national milling machine taper.**
26. List two kinds of self-holding tapers used on milling machine spindles.
27. How will nicks on the face of spacing collars affect the straightness of the arbor?

28. Explain the steps of procedure used to remove arbors which have a standard milling machine taper and which are held in the machine spindle with a draw bar.
29. List several kinds of cutting tool materials from which milling cutters may be made.
30. For what kinds of operations are plain milling cutters used?
31. For what kinds of operations are side milling cutters used?
32. List several kinds of side milling cutters and explain their uses.
33. For what operations are metal slitting saws used?
34. List several types of angular milling cutters.
35. For what purposes are form relieved cutters used?
36. How is the **hand** of a milling cutter determined?
37. How is the **hand** of the helix on a cutter determined?
38. For what kinds of operations are end mills used?
39. List two types of shanks on end mills.
40. What type of end mills can be used for plunge milling?
41. List several kinds of end milling cutters and explain the kinds of operations for which they may be used.
42. Explain how a T-slot can be machined.
43. List a general rule concerning the **maximum** depth of cut for an end mill.
44. On what kinds of milling operations should cutting fluids be used?
45. Explain the meaning of **cutting speed** for milling operations.
46. How does the machineability of metal affect the cutting speed which may be used?
47. List several factors which affect the cutting speed used for milling operations.
48. List the formula for calculating the **approximate** cutting speed for milling operations when the rpm is known.
49. List the formula for calculating the approximate rpm for milling when the cutting speed is known.
50. What is meant by the rate of feed for a milling machine?
51. Explain how the **rate of feed per tooth** is selected for a given kind of milling cutter.
52. List several safety precautions to be observed while setting up and operating a milling machine.
53. List six ways in which a workpiece may be mounted on a milling machine.
54. How can the vise be checked for right-angle squareness with the column?
55. How can the vise be checked for parallelism with the column or the T-slots on the table?
56. List several factors which may cause vibration or chatter while milling.
57. What is indexing, and for what kinds of milling operations is it used?
58. Describe a bed-type milling machine.

UNIT 58

Precision Grinding Machines and Grinding Operations

58-1. What Is Grinding?

Grinding is a machining process which removes metal from the workpiece either with a revolving abrasive (grinding) wheel, a moving abrasive belt, a disc, or some other form. When the workpiece is brought into contact with the abrasive tool, tiny chips of metal are removed, Fig. 58-1. Because of frictional heat, the chips may appear as red-hot sparks immediately after leaving the workpiece, but they are rapidly cooled by air. Thus, abrasive tools remove metal in chip form in much the same way as do lathe tools, saw blades, and milling cutters.

The term **grinding** is used when a relatively small amount of metal is removed, as in tool sharpening and in finishing hardened steel workpieces to size. The use of abrasive tools primarily for rapid removal of large amounts of metal in order to produce a workpiece of desired shape and approximate size is referred to as **abrasive machining**.

Grinding operations can be classified under two headings — precision and nonprecision. **Nonprecision** grinding involves the removal of metal which usually cannot be removed efficiently in any other way and which does not require accuracy or close tolerances. Examples include such jobs as reshaping cold chisels and center punches, snagging the rough spots from castings, and reshaping screwdriver blades. Non-precision grinding generally is done on a pedestal grinder or with a portable grinder.

Precision grinding includes many kinds of grinding operations which require grinding accurately to specified size limits. Precision grinding machines are available for many kinds of precision grinding operations. Flat surfaces are ground with surface grinding machines, Fig. 58-14. Round surfaces are ground with cylindrical grinding machines, Fig. 58-2. Milling cutters are ground with tool and cutter grinding machines. (See Figs. 58-9, 58-10, and 58-11.)

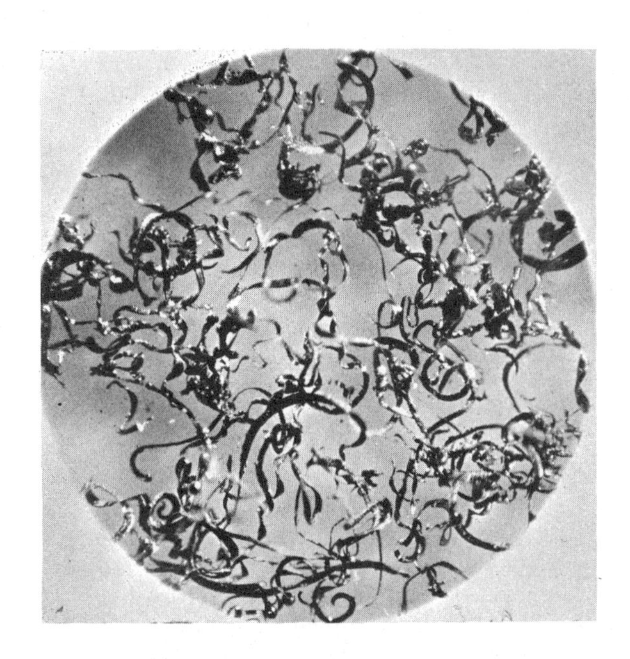

Fig. 58-1. Magnified View of Metal Chips Produced by Grinding

Fig. 58-2. Grinding a Part in a Plain Cylindrical Grinding Machine
Safety goggles are recommended.

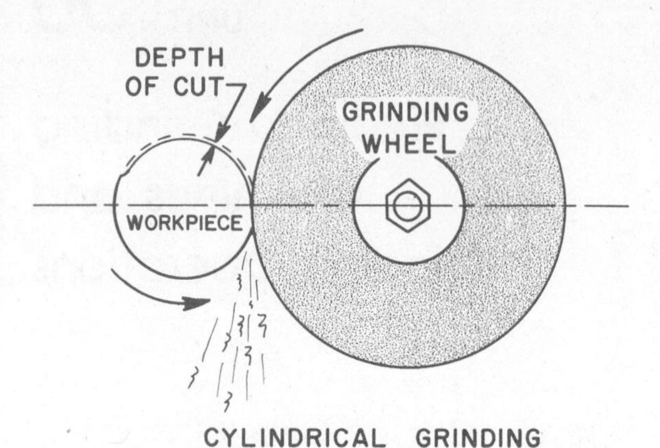

Fig. 58-3. Relationships of Grinding Wheel and Workpiece in Cylindrical Grinding

Grinding Tolerances

Modern industrial practice requires grinding many kinds of machine parts to tolerances of plus or minus 0.0001″. Special parts for precision instruments and gages are sometimes ground to tolerances of plus or minus 0.000020″ (20 microinches, or 20 millionths of one inch). Grinding operations, therefore, make it possible to machine to closer tolerances than with other common chip-machining operations. A second advantage of grinding is that it can be used to machine materials which are too hard to be machined by other common chip-machining operations. For example, metal-cutting tools and heat-treated parts may be machined by grinding.

58-2. Kinds of Grinding Operations

Many parts are rough machined first on a lathe, shaper, milling machine, or special-production machine tool. Then they are machined to finish size by grinding. Parts which are heat treated generally become warped or distorted during heat-treatment processes. Hence, they must be rough machined to an oversize dimension before heat treatment. After heat treatment, they generally are too hard to be machined by other methods. They must, therefore, be machined to final specified size limits by grinding.

The following are several common basic classifications of precision grinding operations:

1. Cylindrical grinding
2. Internal grinding
3. Centerless grinding
4. Tool and cutter grinding
5. Form grinding
6. Surface grinding

Cylindrical Grinding

This classification includes various grinding operations involved in external grinding of parts with a cylindrical or conical shape.

The relationship between the grinding wheel and the workpiece for cylindrical grinding operations is shown in Fig. 58-3. Roughing cuts vary from 0.001″ to 0.004″ deep, depending upon the shape and size of the workpiece and upon the size and rigidness of the machine. Finishing cuts generally vary from 0.0002″ to 0.001″ deep.

Several types of cylindrical grinding machines may be used. **Plain cylindrical grinding** machines can only grind cylindrical surfaces, Fig. 58-2. Universal cylindrical grinding machines allow grinding of parts with

Fig. 58-4. Grinding a Taper with the Table Swiveled on a Universal Cylindrical Grinding Machine

Fig. 58-5. Grinding a Steep Taper with the Head Swiveled on a Universal Cylindrical Grinding Machine

Fig. 58-6. Headstock Swiveled 90° for Face Grinding with a Universal Cylindrical Grinding Machine

either cylindrical or conical (tapered) shapes. The machine table can be swiveled for grinding long tapers, as in Fig. 58-4. On universal cylindrical grinding machines, the headstock may be swiveled for grinding steep tapers, as in Fig. 58-5. On universal cylindrical grinding machines, the headstock also can be swiveled 90° for **face grinding** flat surfaces, as in Fig. 58-6.

For cylindrical grinding of long parts, the workpiece is mounted between the headstock and footstock centers, similar to mounting work between centers on a lathe. Short workpieces can be held in either a universal 3-jaw chuck or in a 4-jaw chuck mounted on the headstock as shown in Figs. 58-6 and 58-7.

Internal Grinding

This kind of grinding operation produces a smooth and accurate surface in a cylindrical hole, Fig. 58-7. The internal surface may be ground straight or tapered. On special jobs the grinding wheel can be cut to a special form for grinding an internal surface of irregular form.

For industrial production purposes, special internal grinding machines are used. However, in most tool-and-die shops, small machine shops, and school shops, internal grinding is performed with an **internal grinding attachment** on a universal cylindrical grinding machine, as in Fig. 58-7.

Internal grinding also can be performed with a tool post grinder mounted on a lathe. With this setup, the workpiece is mounted in the lathe chuck. The tool post grinder is mounted in the tool post T-slot on the compound rest.

Centerless Grinding

This is a form of cylindrical grinding. It is done without using center holes or a chuck for holding the workpiece while grinding. Centerless grinding is done with a **centerless grinding machine**. The relationship of the workpiece, the regulating wheel, and the grinding wheel on centerless grinding machines is shown in Fig. 58-8. Straight or tapered objects such as spindles, piston pins, roller bearings, and lathe centers are ground by this method. Form grinding may also be done with centerless grinders.

Fig. 58-7. Internal Grinding of a Part Mounted in a Chuck on a Universal Grinding Machine

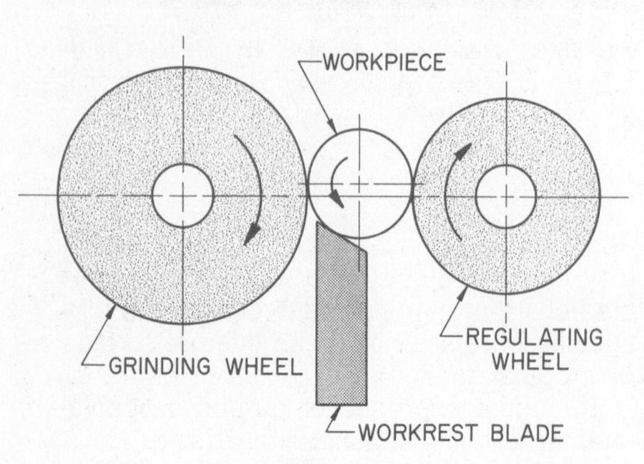

Fig. 58-8. Relationship of Grinding Wheel and Workpiece in Centerless Grinding

Tool and Cutter Grinding

Tool and cutter grinding is the grinding of milling cutters, counterbores, reamers, taps, and similar metal cutting tools on tool and cutter grinding machines specially designed for this purpose. Cutters must be sharp to produce good surface finish and to cut efficiently. When cutters become dull, higher pressures are required to force them to cut. This causes unnecessary strain on the cutter, its holder, and the machine spindle. These forces can break the cutter and damage the cutter holder.

The tool and cutter grinding machine setup for sharpening a plain-tooth helical milling cutter is shown in Fig. 58-9. Figure 58-10 shows the setup for sharpening a form relieved milling cutter, and Fig. 58-11 shows the arrangement for sharpening an end milling cutter.

Fig. 58-9. Grinding a Plain-Tooth Helical Milling Cutter on a Cutter and Tool Grinder

Tool and cutter grinding attachments for sharpening cutting tools are also available for use on some lathes and on certain kinds of surface grinders.

Form Grinding

This refers to grinding surfaces of special form or shape. The face of the grinding wheel must be cut to conform with the shape of the surface to be ground. For example, the grinding wheel in Fig. 58-12 is cut to conform with the shape of the screw thread which it is cutting. The grinding of fillets and rounds is another example of form grinding.

Grinding wheels are available with faces having standard shapes for grinding contours which are often used, see Fig. 42-5. For other nonstandard form grinding operations, the face of a grinding wheel can be cut with a diamond tool to any desired shape. (See Fig. 58-25.) Form grinding can be done on surface grinding machines, cylindrical grinding machines, and on special grinding machines. (See Fig. 58-23.)

Surface Grinding

This kind of grinding produces a smooth, true, flat surface on parts. Surface grinding machines can grind flat surfaces (Fig. 58-13), angular surfaces (Fig. 58-29), and grooves (Fig. 58-27). By shaping the face of the grinding wheel round, curved, or to some special contour, surfaces with a special shape can be ground with a surface grinding machine.

58-3. Surface Grinding Machines

Surface grinding machines can be classified as **horizontal spindle** (Figs. 58-14 and 58-26) and **vertical spindle** (Figs. 58-15 and 58-16).

Fig. 58-11. Setup for Grinding an End Mill with Helical Teeth

Fig. 58-10. Setup for Grinding Face of Teeth on Form-Relieved Cutters

Fig. 58-12. Abrasive Machining a Worm Screw from a Solid Piece

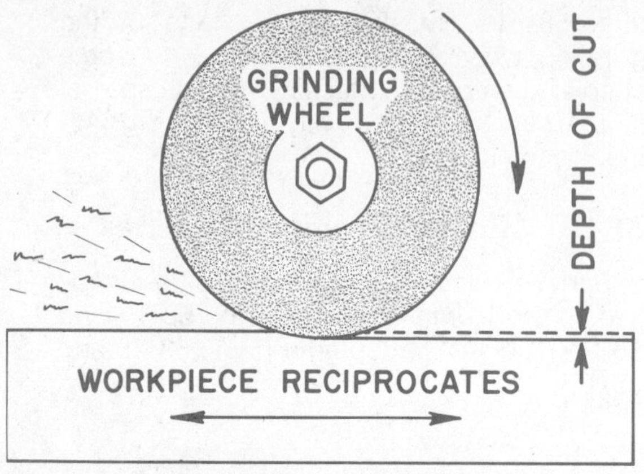

Fig. 58-13. Relationship of Grinding Wheel and Workpiece with Horizontal-Spindle Surface Grinder

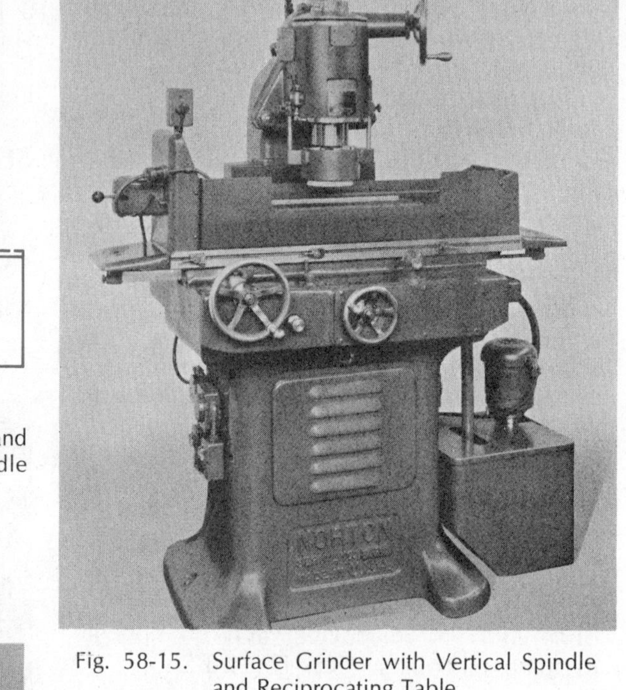

Fig. 58-15. Surface Grinder with Vertical Spindle and Reciprocating Table

Fig. 58-14. Surface Grinder with Horizontal Spindle and Rotary Table
Safety Goggles Are Recommended

Fig. 58-16. Surface Grinding on a Large Vertical Spindle-Type Surface-Grinding Machine with a Rotary Table

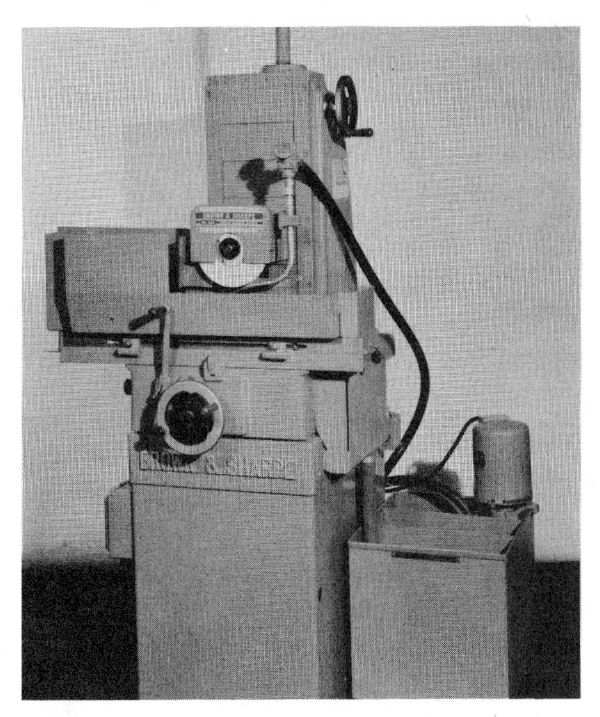

Fig. 58-17. Handfeed Surface Grinder with Wet-Grinding Attachment

Horizontal spindle surface grinders, Fig. 58-26, are most common in small machine shops, tool-and-die shops, maintenance shops, and school shops. With this machine, the workpiece is mounted on the table, usually on a magnetic chuck (Fig. 58-26), in a vise (Fig. 58-28), or bolted to the table (Fig. 58-30). The table reciprocates back and forth under the grinding wheel which remains in the same position. As the table reciprocates longitudinally (back and forth from right to left), it is fed crosswise under the grinding wheel, Fig. 58-13. Thus, the wheel takes a new cut each time the table reciprocates.

Some horizontal spindle surface grinding machines are equipped for handfeeding only. Machines of this type, Fig. 58-17, must be fed by hand both longitudinally and crosswise; a **table hand wheel** and a **cross-feed handwheel** are provided. Many horizontal

Fig. 58-18. 6″ x 18″ Surface Grinding Machine with Hydraulic and Hand Feeds

(1) Cross-feed handwheel; (2) Dial locknut, Cross-feed handwheel; (3) Table handwheel; (4) Set screw; (5) Throttle adjustment bushing; (6) Table throttle lever; (7) Table reversing lever; (8) Dust deflector; (9) Reversing lever contact roller; (10) Tables; (11) Wheel guard; (12) Upright; (13) Table dog; (14) Carrier locknut; (15) Fine-feed adjustment knob; (16) Dial locknut, elevating handwheel; (17) Elevating handwheel; (18) Fine-feed locknut; (19) Cross-feed directional lever; (20) Cross-feed regulating screw; (21) Wheel truing and rapid positioning lever; (22) Oil level sight glass; (23) Base; (27) Bed; and (40) Elevating screw guard, upper.

spindle machines are equipped with automatic power feeds for horizontal and cross feeding, as in Fig. 58-18. Machines of this type also can be fed manually with the **table handwheels.**

A second kind of horizontal-spindle surface grinder is equipped with a **rotary table**, Fig. 58-14. The rotary table is mounted on top of a longitudinal table. The workpiece is mounted on a magnetic chuck which revolves on the rotary table underneath the grinding wheel. As grinding takes place, the longitudinal table is fed in either direction under the grinding wheel.

Vertical spindle surface grinding machines, Figs. 58-15 and 58-16, have the grinding wheel mounted on a vertical spindle. Workpieces to be ground are mounted on the table which may be reciprocating, Fig. 58-15, or revolving, Fig. 58-16. Vertical spindle machines provide a larger area of contact between the grinding wheel and the workpiece and, therefore, grind more rapidly than horizontal spindle machines. For this reason, they are widely used for production grinding where many parts must be ground rapidly. Since more heat is also created, a cutting fluid generally must be used while grinding with machines of this type.

58-4. Features of Horizontal Spindle Surface Grinding Machines

Horizontal spindle surface grinding machines are most commonly used in small commercial shops, tool-and-die shops, and in school shops. Hence this unit is concerned principally with machines of this type. The parts and controls on horizontal spindle grinding machines may be designed somewhat differently by various machine manufacturers. However, the principles involved in the operation of many machines of this basic type are similar.

Principal Parts

The principal parts and controls of a horizontal spindle surface grinding machine, equipped for either handfeeding or power feeding, are shown in Fig. 58-18.

Size

The size and capacity of surface grinding machines generally is designated by the size of the working area of the table. A **magnetic chuck** with approximately the same working area is often mounted on the table for convenience in holding workpieces, as in Fig. 58-27. A size 6″ × 18″ machine, as shown in Fig. 58-18, has a table with working area of 6″ cross travel and 18″ longitudinal table travel. The smaller machine in Fig. 58-17 has a designated size of 5″ x 10″, yet the table work area is 5″ × 11″. The working area of the table is slightly larger than the designated size for some machines. The distance between the center of the grinding wheel and the working surface of the table also is a factor in determining the maximum height of workpieces which may be ground.

Wheel Elevation

The wheel is mounted on a horizontal spindle which may be raised or lowered with the **elevating handwheel**, Fig. 58-18. The handwheel generally has graduations which make it possible to adjust the wheel elevation in increments of 0.0002″. Machines of the type shown in Fig. 58-19 also have an auxiliary **fine-feed adjusting knob** which makes it possible to adjust the wheel elevation in increments of 0.0001″.

Horizontal spindle machines may have **direct drive** or **V-belt drive** from an electric motor. With direct drive, the grinding wheel is mounted on the end of the motor spindle.

Wheel RPM

The cutting speed and the rpm of grinding wheels are explained in Unit 43. The maximum rpm indicated on the grinding wheel always should be equal to or higher than the rpm of the wheel spindle. If the spindle rpm exceeds the maximum indicated rpm for the wheel, the wheel may fly apart and injure someone. On machines with V-belt drive and a step pulley, the rpm can be changed as necessary for smaller or larger grinding wheels. Most surface grinding machines, however, are designed to operate at one standard speed.

For example, a spindle speed of approximately 3450 rpm is often used on machines which use a 7″ diameter grinding wheel. This gives a cutting speed of 6319 surface feet per minute. The grinding wheel should be mounted with cardboard disks on each side of the wheel, as shown in Fig. 58-21.

Table Feed

The table travels longitudinally (to the right or left), and crosswise (towards or away from the column).

Table travel may be handfed with the **table hand crank** or **table handwheel**. The table is fed crosswise by hand with the **cross-feed handwheel.**

On machines equipped with power feed, the rate of longitudinal table travel can be adjusted. Table speed should be slower for finishing cuts than for roughing cuts. The amount of cross feed also can be adjusted. Lighter cross feeds are used for finishing cuts than for roughing cuts.

The controls used for changing the rate of longitudinal table feed and the rate of cross feed are somewhat different on each kind of machine. (See Fig. 58-19.) Therefore, it is always best to have your instructor explain and show you how to use the various controls before using the machine for the first time. The procedures for using the feed controls and other controls are included in the handbook which generally is provided by the manufacturer of the machine.

Table dogs, Fig. 58-19, are provided for setting the length of table travel. As the table travels, the dogs contact the table **reversing lever**, thus causing the table to travel in the opposite direction.

Cross feeds may be set from 0.010″ to 0.250″ on machines equipped with automa-

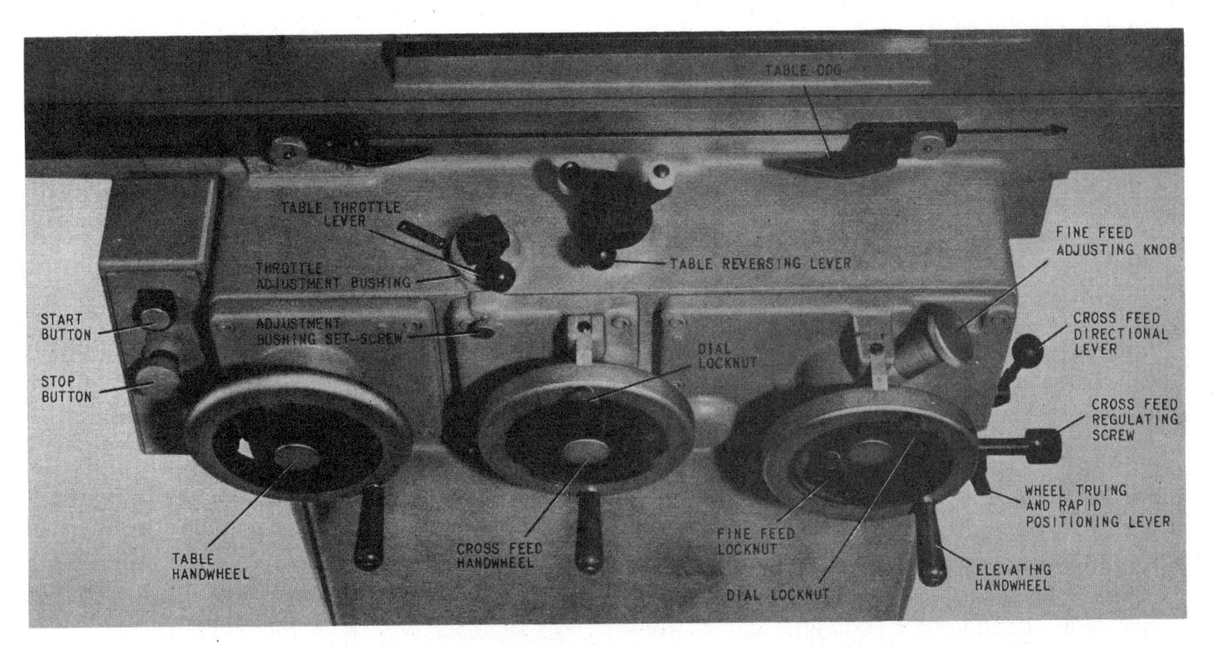

Fig. 58-19. Front Operating Controls on a 618-Surface Grinding Machine Equipped with Power and Handfeeds

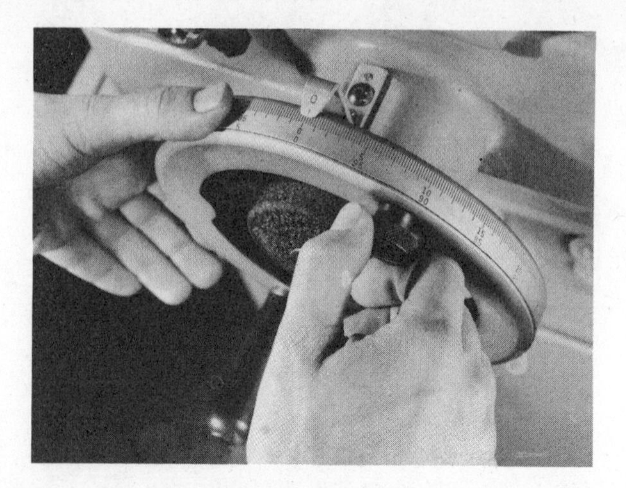

Fig. 58-20. Cross-Feed Handwheel Has 0.0002″ Graduations

tic cross-feed mechanisms. However, for average roughing cuts on smaller machines, a cross feed of 0.050″ to 0.100″ is satisfactory. For finishing cuts, the cross feed should be less, usually from 0.030″ to 0.050″. **Never use cross feeds in excess of one-half the width of the wheel**. The cross-feed handwheel has 0.0002″ graduations for accurately setting the amount of cross feed, see Fig. 58-20.

Depth of Cut

The depth of cut may vary according to the following:

1. Whether the cut is a roughing or a finishing cut.
2. Whether grinding is done wet or dry.
3. The rigidness of the machine and the rigidness of the setup.

Deep roughing cuts generally produce a rougher surface than shallow finishing cuts. Surface finish generally is better with wet grinding than with dry grinding. And heavier cuts generally can be made on heavier machines than on lightweight machines. For average conditions, roughing cuts from 0.002″ to 0.003″ should be used. Finishing cuts of 0.001″ or less produce good results.

Wet-or-Dry Grinding

Abrasive grains fracture from the wheel, and fine metal chips are produced while grinding. For dry grinding a **vacuum exhaust attachment** is recommended. This attachment collects the abrasive grit and dust and it keeps the machine and the work area clean.

A **wet grinding attachment**, Fig. 58-17, is used for grinding with a cutting fluid. The wet attachment consists of a pump, liquid container, and splash guards. The pump provides cutting fluid at the surface area being ground, thus carrying away heat. Hence, heavier cuts generally can be taken. Cutting fluid also improves surface finish, increases grinding wheel life, and carries grit and grinding dust away.

Safety Note

If the cutting fluid is turned on and comes into contact with the porous grinding wheel while it is stopped, the cutting fluid will be absorbed and will collect at the bottom of the wheel, making the wheel out of balance. Should the wheel be turned on in this out-of-balance condition, centrifugal force will cause the wheel to be thrown toward the heavy side. Depending on wheel size, condition, and tightness on the spindle, the wheel could be thrown out of true, or it might crack or even break. The rule, therefore, should always be this: turn the wheel on before turning the coolant on; turn the coolant off before turning the wheel off.

An emulsifiable (soluble) oil solution is recommended for grinding ferrous metals. A solution composed of 40 parts water and 1 part emulsifiable oil is recommended. Cutting fluids for use with other metals are listed in Table 33, page 452.

58-5. Grinding Wheel Selection

Grinding wheels are made of abrasive grains which are held together with a bonding material. Several kinds of abrasive may be used, and the grain size may range from very fine to very coarse. The kinds of abrasive, their properties, and grain sizes are discussed in Unit 41.

Grinding wheels can be made very hard or very soft depending on the kind and amount of bonding material used in their manufacture. Complete information concerning grinding wheels is included in Unit 42. Included in Unit 42 are such factors as bond type, grade or hardness of wheels, grain structure of wheels, the grinding wheel marking system, and the factors to be considered when ordering a grinding wheel.

A standard grinding wheel marking system is now used by most manufacturers for identifying the characteristics of each grinding wheel. The code number generally is located on the cardboard disk on the side of the wheel.

The following kinds of straight grinding wheels are recommended for surface grinding with horizontal spindle, reciprocating surface grinders:

Material	Kind of Grinding Wheel
Soft Steel	23A46 - J8VBE
Cast Iron	32A46 - I8VBE (or)
	37C36 - KVK
Hardened Steel	32A46 - H8VBE
General-Purpose Wheel	23A46 - H8VBE
Nonferrous Metals	37C36 - KVK

For more specific grinding wheel recommendations, consult a standard handbook for machinists, a manufacturer's catalog, or a manufacturer's sales representative.

58-6. Truing and Dressing the Wheel

When a grinding wheel becomes dull, loaded, or out of shape, it must be dressed and trued. **Dressing** means to sharpen a wheel. **Truing** means to cut the wheel so that there will be no high spots when the wheel is running. Truing also refers to forming a wheel to a particular shape, such as a convex or concave shape. Wheels are trued to special shapes for form grinding. A diamond tool, Figs. 58-22 and 58-25, is used for dressing or truing grinding wheels.

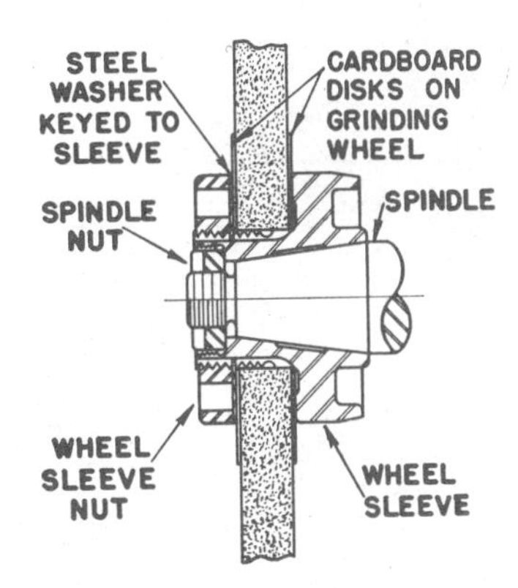

Fig. 58-21. Proper Mounting of Grinding Wheel

Procedure for Truing or Dressing

1. Wear safety goggles.
2. Mount the grinding wheel on the spindle as shown in Fig. 58-21.
3. Select a wheel truing fixture, and place it on the table or on a magnetic chuck. (See Fig. 58-22.) The diamond point tool should be placed ahead of the vertical center line of the wheel, and it should be inclined slightly in the direction of wheel travel. This procedure will prevent the diamond dressing tool from **gouging** or **digging into** the grinding wheel.
4. Clamp the diamond dressing tool in position.
5. Start the machine. With the **elevating handwheel**, lower the wheel until it just touches the diamond point tool.
6. With the **cross-feed handwheel**, move the table so that the diamond cuts across the wheel.

Fig. 58-22. Truing a Wheel on a Surface Grinder

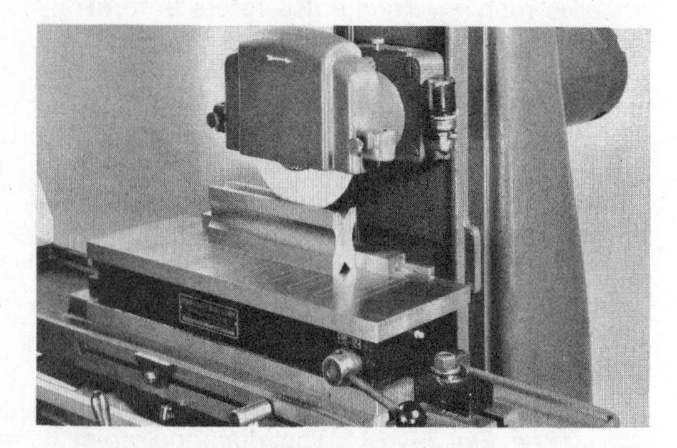

Fig. 58-23. Grinding a V-Shaped Recess

7. Lower the wheel 0.0005″ for each additional cut, as in step 6, until the wheel is true. If the machine is equipped with a wet attachment, a cutting fluid should be used while truing or dressing the wheel.

Shaping the Wheel

It is sometimes necessary to grind a special shape or form such as a rounded fillet, a V-groove, or some irregular surface. In

Fig. 58-24. Radius and Wheel Truing
Note: Tool is mounted at right angles to slide for cutting angular surface on wheel.

such cases, it is necessary to cut or true the grinding wheel to conform to the shape which is to be ground. For example, a V-shaped groove is being ground in Fig. 58-23. For this purpose, the grinding wheel must be ground to the exact shape of the V-groove. A **radius and wheel truing attachment,** Fig. 58-24 and 58-25, can be used for shaping wheels. In Fig. 58-24, the attachment is used for cutting an angular face on the wheel. In Fig. 58-25, the swivel base of the attachment is used for cutting a convex radius on the wheel.

58-7. Holding Work for Surface Grinding

Workpieces of various shapes and sizes can be mounted on the surface grinding machine in a variety of ways, depending on the kinds of accessories available. The following

Fig. 58-25. Shaping Convex Surface on a Grinding Wheel

Fig. 58-26. Workpiece Held on a Permanent Magnetic Chuck

Fig. 58-27. Grinding a Slot with Workpiece Mounted on Magnetic Chuck

Fig. 58-28. Workpiece Held in a Vise

methods are suggested for holding work for surface grinding.

1. **By using a magnetic chuck:**
   Mount the chuck on the table, and bolt it in position. (See Fig. 58-26.) Place the workpiece on the surface of the chuck in the desired position. Turn the magnetic control lever 180° to the

on position. Workpieces made of magnetic materials can be held securely by this method. If the workpiece has only small areas in contact with the chuck, flat pieces of steel should be placed on the chuck against both ends of the workpiece. The extra-flat pieces help hold the work securely in position.

Fig. 58-29. Workpiece in an Adjustable Swivel Vise

2. **By using a machine vise**:
   Bolt a machine vise on the machine table, as in Fig. 58-28. Clamp the work-as in Fig. 58-29. Angular or beveled surfaces can be ground by using this method. A dial indicator or a surface gage can be used for determining whether the surface to be ground is in a horizontal plane.
   piece in the vise. An indicator or surface gage may be used to determine whether the surface to be ground is in a horizontal plane.
3. **By using an adjustable swivel vise**:
   Bolt the vise, Fig. 58-29, to the table and clamp the work in the vise. Then position the vise at the desired angle,
4. **By using an adjustable vise**:
   An adjustable vise, as shown in Fig. 58-32, can be placed on a magnetic chuck. The workpiece is then clamped in the vise. The vise can be swiveled at any desired angle, ranging from the horizontal to the 90° vertical position. This method works well for grinding angular or beveled surfaces. It also works well for grinding vertical surfaces on parts of convenient shape, as in Fig. 58-32.
5. **By using a precision vise**:
   A precision vise, Fig. 58-31, can be placed on a magnetic chuck for hold-

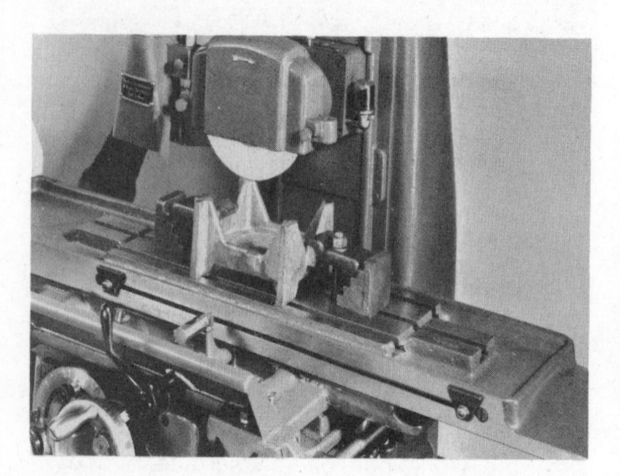

Fig. 58-30. Workpiece Held in Position with Clamps

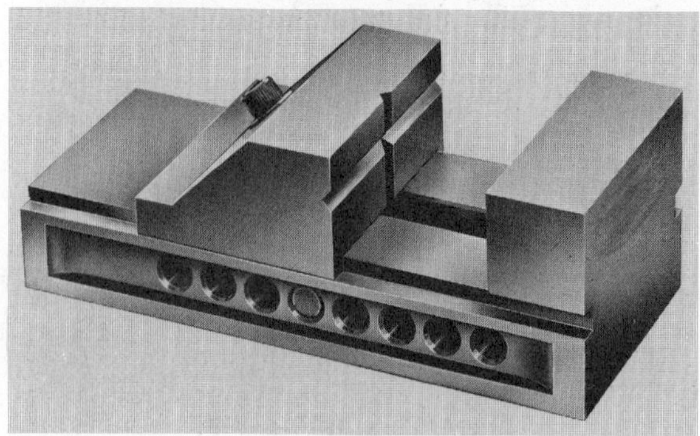

Fig. 58-31. Precision Vise for Use on a Magnetic Chuck

Fig. 58-32. Workpiece Mounted in an Adjustable Vise

ing small workpieces securely and accurately. This kind of vise works well for holding parts with a small cross-sectional area. Round parts can be held in the V-groove in the vise jaw.

6. **By clamping**:
   A workpiece can be positioned, aligned, and clamped to the machine table. (See Fig. 58-30.) Various work-holding tools may be used for setting up the workpiece on the machine table. (See Unit 50 and Fig. 50-2.)
7. **By using V-blocks**:
   Round workpieces can be clamped on V-blocks for grinding. The V-blocks can be clamped to the table, as in Fig. 58-33, or they can be held in position with a magnetic chuck.

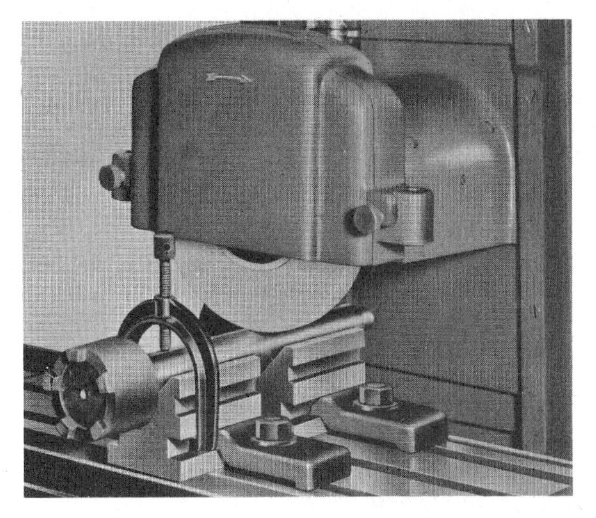

Fig. 58-33. Workpiece Clamped to Machine Table

58-8. Surface Grinding Procedure

The following are the **preliminary steps of procedure** to be followed in setting up for surface grinding. When the preliminary steps have been completed, the specific procedures for **manual-feed machines**, or for **power-feed machines** should be followed.

Preliminary Steps

1. Select a grinding wheel which is recommended for the kind of material to be ground. (See § 58-5.) Test the wheel to see that it is sound. This is done by striking it lightly with a light hammer. A

clear ring indicates that the wheel is sound and has no cracks.

2. Mount the grinding wheel on the **wheel sleeve**, as shown in Fig. 58-21. Cardboard disks should be placed on each side of the wheel, and the **sleeve nut** should be tightened snugly. If the sleeve nut is fastened too tightly, the wheel may crack.
3. Mount the wheel and sleeve on the machine spindle. Wipe any dust or dirt from the spindle and from the hole in the sleeve before mounting the unit on the spindle. Tighten the spindle nut snugly. (The spindle nut threads will most likely be LH.) True the wheel as explained in § 58-6.
4. Remove all burrs or nicks from the workpiece with a file.
5. Wipe dust or dirt from the machine table, magnetic chuck, or other work-holding tools or accessories. Then mount the workpiece on the machine.
6. Lubricate the machine with the proper machine oil before using it.
7. **Protect your eyes** by wearing approved safety goggles or face shield.

Procedure for Manual Feeding

1. Complete the **preliminary steps of procedure**, steps 1 through 7, listed above.
2. With the **table hand crank** and the **cross-feed handwheel**, move the workpiece under the grinding wheel. Turn the machine on. With the **elevating handwheel**, lower the grinding wheel until it just clears the workpiece.

   On machines equipped with power feed, the power-feed mechanism should be disengaged for manual operation. The table handwheel must be engaged for manual operation. It automatically becomes disengaged when the power-feed mechanism is used.
3. With the cross-feed handwheel, feed the workpiece outward until the far side of the wheel extends beyond the work a distance equal to ¾ the width of the wheel face. Thus ¼ the width of the wheel is above the work. Move the table to one side until the wheel clears the end of the workpiece.
4. With the elevating handwheel, lower the grinding wheel 0.002″ to 0.003″ for a roughing cut. Allow sufficient material for a final finishing cut of 0.0005″ to 0.001″ depth. Lock the grinding head in position with the **locking screw**, if the machine is so equipped. Turn on the cutting fluid, or the exhaust, as desired.
5. With the table hand crank or handwheel, feed the table longitudinally under the grinding wheel at a steady rate until the workpiece has passed beyond the grinding wheel at least 1″.
6. With the cross-feed handwheel, feed the work crosswise one complete turn, about 0.100″, for the next cut. On finishing cuts, a cross feed of 0.020″ to 0.050″ is more desirable.
7. Continue grinding as in steps 5 and 6 until the entire surface has been ground.
8. Measure the thickness of the workpiece. If more roughing cuts are necessary, continue grinding as in steps 4 through 7. However, cross feed each additional cut in the direction opposite to the previous cut. This causes the wheel to wear off more evenly.
9. Measure the thickness of the workpiece. in depth, as desired.
10. When the work has been ground to the specified dimension, stop the machine, remove the workpiece, and clean the machine and accessories. Return all tools and accessories to their proper place.
11. Remove sharp burrs and sharp corners with a file so that you do not get cut by them.

Procedure for Power Feeding

1. Complete the **preliminary steps of procedure**, 1 through 7, listed above.

2. Be sure that the power-feed mechanism is disengaged while setting up work, in preparation for automatic feed. On the machine in Fig. 58-18 and 58-19, this is done by turning the table throttle lever to the **off** position. The power cross feed also should be disengaged. The table handwheel and the cross-feed handwheel then can be engaged for manual operation while making the setup.
3. With the **table hand crank** and the **cross-feed handwheel**, move the workpiece under the grinding wheel. Turn the machine on. With the **elevating handwheel,** lower the grinding wheel until it just clears the workpiece.
4. With the cross-feed handwheel, feed the workpiece outward, away from the column, until the far side of the wheel extends beyond the work a distance equal to ¾ the width of the wheel face. Thus ¼ the width of the wheel is above the work.
5. Move the table to the left until the workpiece has passed the grinding wheel a distance of at least 1″.
6. Bring the right-hand **reversing dog** against the right side of the **table-reversing lever**. Fasten the dog in position by tightening the clamping bolt.
7. Move the table to the right until the end of the workpiece has passed the grinding wheel a distance of at least 1″.
8. Bring the left-hand **reversing dog** against the left side of the **table-reversing lever**. Fasten the dog in position by tightening the clamping bolt.

Safety Note

When setting the table dogs for automatic longitudinal table feeding, be sure to allow sufficient over-travel of the work in both directions so that cross-feed action can take place before the work travels back under the wheel.

9. With the elevating handwheel, lower the grinding wheel 0.002″ to 0.003″ for a roughing cut. Allow sufficient material for a final finishing cut of 0.0005″ to 0.001″ depth. Lock the grinding head with the locking screw, if the machine is so equipped. Turn on the cutting fluid, or the exhaust system, as desired.
10. Set the power cross-feed mechanism for a feed of about 0.100″ for roughing cuts. On finishing cuts, a feed of 0.020″ to 0.050″ is more desirable.
11. Engage the automatic longitudinal table-feed mechanism. On the machine in Figs. 58-18 and 58-19 this is done with the **table throttle lever.**
12. Engage the automatic cross-feed mechanism for automatic feed in the proper direction. On the machine in Figs. 58-18 and 58-19 this is done with the **cross-feed directional lever.**
13. When the cut has been completed, disengage the automatic longitudinal table feed and the automatic cross feed.
14. Bring the table to one side, against the table dog, so that the end of the workpiece is at least 1″ beyond the grinding wheel.
15. Measure the workpiece. If additional roughing cuts are necessary, proceed as in steps 9 through 14 above. However, for each additional cut, feed the table crosswise in the direction opposite to the direction of feed for the previous cut. This procedure causes the wheel to wear more evenly.

    If additional roughing cuts are not necessary, make a finishing cut. Finishing cuts generally should be 0.001″ or less in depth.
16. When the work has been ground to the specified dimension, stop the machine, remove the workpiece, and clean the machine and accessories. Place all tools and accessories in the proper place.
17. Remove sharp burrs or sharp arrises with a file so that you do not get cut by them.

Words to Know

abrasive machining
centerless grinding
conical surface
cutting fluid
cylindrical grinding
dressed wheel
dry grinding
face grinding
fillets
form grinding
horizontal spindle
internal grinding
magnetic chuck
nonprecision grinding
precision grinding
reciprocating table
rotary table
surface grinding
table dog
tool and cutter grinding
trued wheel
vertical spindle
wet grinding
wheel elevation

Review Questions

1. Explain the principal difference between grinding and abrasive machining.
2. Explain the difference between non-precision grinding and precision grinding operations.
3. What are the main advantages of precision grinding?
4. To what tolerances or size limits are precision grinding operations often performed?
5. What is meant by surface grinding?
6. What is meant by cylindrical grinding?
7. Describe several ways workpieces are held in a cylindrical grinding machine.
8. List several kinds of cylindrical grinding operations.
9. How do plain and universal cylindrical grinders differ?
10. Describe internal grinding and explain how it is done.
11. Explain how form grinding is done.
12. What is meant by centerless grinding?
13. For what are tool and cutter grinding machines used?
14. List several kinds of surfaces which can be ground on a surface grinding machine.
15. Describe how a horizontal spindle surface grinder operates.
16. Describe the difference between a reciprocating and rotary surface grinding machine.
17. Describe a vertical spindle surface grinding machine and its use.
18. How is the size of a surface grinding machine usually designated?
19. Why is it important to check the operating rpm of the grinding wheel before mounting it on the surface grinder?
20. For what purpose are table dogs used on surface grinding machines?
21. How much cross feed should be used for roughing cuts on horizontal spindle surface grinding machines?
22. How much cross feed should be used for finishing cuts on a horizontal spindle surface grinding machine?
23. List several factors which must be considered in determining the depth of cut for surface grinding.
24. For average conditions, what should be the depth of a rough cut when surface grinding? Finishing cut?
25. What are the advantages of wet grinding?
26. On machines equipped with coolant systems, explain why the coolant should always be turned on after turning on the grinding wheel, and turned off before stopping the grinding wheel.
27. What kind of cutting fluid is recommended for wet grinding of ferrous metals?
28. With the use of a code number, according to the standard grinding wheel marking system, list a recommended grinding wheel for surface grinding hardened steel.
29. Explain the difference in meaning between **truing** and **dressing** a grinding wheel.
30. List four kinds of vises which can be used for holding a workpiece on a surface grinder.
31. List several methods, other than using vises, which may be used for holding a workpiece for surface grinding.

UNIT 59

Automation and Numerical Control

59-1. What is Automation?

The term automation comes from the word **automatic** which means **self-acting** or **self-adjusting**. Automation now generally applies to machines, equipment, or processes which operate without direct control and adjustment by an operator. Automated machines, processes, or systems of manufacturing generally possess the following characteristics:

1. They operate with little or no human help.
2. They detect or sense the need for a corrective adjustment in the process or operation.
3. They make the required corrective adjustments with little or no human assistance.

The true concept of automation is **continuous automatic production**. This method of production, in some applications, can automatically produce the part(s), inspect, assemble, test, and package the product in one continuous flow.

Feedback

Automated machines, processes, or systems utilize a principle called **feedback** as a basis for self-adjustment. Information about the process or operation is detected and reported (fed back) to an adjusting control. An understanding of the following terms or devices used in automated systems will help in understanding the meaning of feedback.

Output. The work produced by a machine operation or process is output. The output may be either a product or a service, such as a machined part or the opening of a door.

Input. The commands, data, or standards specifying the output is called the input. The commands or data may concern such factors as size dimensions, position location, machine speed, weight, temperature, color, pressure, chemical composition, vibration, resistance, or sound.

Sensors. Electronic, mechanical, or other kinds of instruments or devices which detect (sense) and report conditions of the output products or services are sensors. Sensors may measure size, weight, temperature, color, pressure, or chemical composition. Sensors feed information back to the **control center.**

Control center. The control center compares the output with the input commands or data. With complex automated machines or systems, this device is most often an electronic computer which processes information or mathematical data. On simple automatic machines or processes, the control center may be a simple electrical regulator or a mechanical regulator. The information received from the sensors is compared with the input specifications. The control center gives the necessary instructions to the machine — to stop, continue, or change speed or direction. Thus, the control center actually regulates or controls the automated machine or process.

Feedback Control Loop

Engineers use the term "feedback control loop" to describe how an automated machine or process functions. The term **loop** refers to the flow of information through the automated system. A simple automated process utilizing a **closed-loop** system is the heating system in a home. The system includes a furnace, a thermostat, a thermometer and an electrical circuit with a contact switch. See Fig. 59-1. The desired temperature (input), for example 72° F., is set on the thermostat. When the temperature falls below 72° F., the thermostat sensor closes a switch in the electrical circuit to the furnace, thus starting the furnace which produces heat (output). When the temperature in the room rises to the desired input temperature setting (feedback), the thermostat sensor opens the switch in the electrical circuit, thus shutting off the furnace. In this simple system, the thermostat is both sensor and control. The thermometer is only a "readout" device to show the performance of the heating system to the homeowner. Similar **closed-loop systems** are used for the automatic operation of air-conditioning systems, hot-water heaters, and refrigerators.

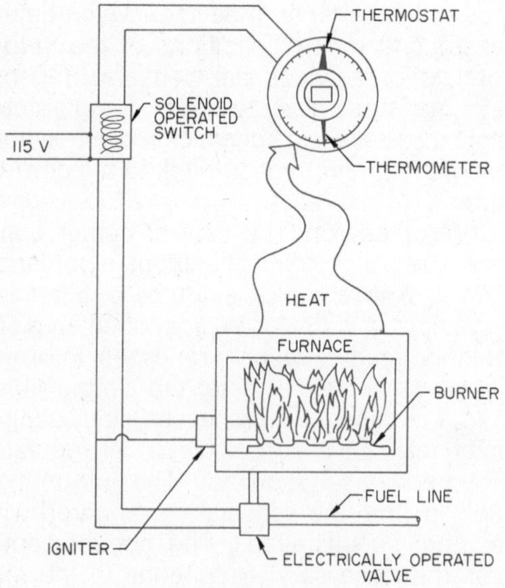

A feedback control loop may be either the closed-loop or open-loop system. The closed-loop system is one in which a measurement of the output is fed back and compared with the input for the purpose of reducing their differences. The open-loop system is one in which no corrective adjustment takes place as a result of the feedback. Instead, a signal such as a light or horn is actuated to warn an operator to make the necessary corrective adjustments. On some open-loop systems, the machine or process is stopped until corrective adjustments are made by the operator. Because closed-loop feedback systems are capable of making required corrective adjustments, they are used on more automated machines or processes.

THERMOSTAT
SOLENOID OPERATED SWITCH
115 V
THERMOMETER
HEAT
FURNACE
BURNER
FUEL LINE
IGNITER
ELECTRICALLY OPERATED VALVE

Fig. 59-1. A Simple Closed-Loop Feedback System
The machine (furnace) produces a product (heat) which is measured by the thermostat. The thermostat controls the furnace by activating the switch, as required for starting and stopping the furnace, thereby maintaining a constant temperature.

Automatic Material Handling

Advanced automatic systems often involve continuous automatic handling of materials. Such systems have been developed in industries which handle large volumes of materials. These systems are used in such **continuous process** industries as steelmaking, papermaking, chemical processing, and petroleum refining. **Unit-processing** also utilizes automatic material handling as in stamping automobile fenders, machining engine blocks, printing sheets of paper, and filling milk bottles.

Data Processing

Most automated machines or systems must process information in one form or another. Even the simple automatic heating system requires the comparison of input information (the desired temperature setting) with output information (the existing temperature).

Offices, schools, banks, and industrial plants use **electronic computers**, Fig. 59-10, for processing mathematical information and many other kinds of data. Computers have complex systems of storing data, or "memory." They store vast amounts of facts, figures, and other information — all in numerical, alphabetical, or symbolic form. When instructed or "programmed" properly, they are capable of solving complex problems.

Businesses use electronic computers for such accounting and bookkeeping tasks as keeping records of customer payments, calculating employee wages, printing their checks, and recording bank deposits and withdrawals. Engineers use computers to make the mathematical calculations required in designing roads, bridges, buildings, and dams. Scientists use computers for making necessary calculations in launching and controlling space vehicles. Industrial plants use computers to control materials, machines, and entire manufacturing systems.

Numerical Control

The electronic computer stores numbers, letters, and other symbolic data. It receives numbers as instructions, manipulates numbers to perform calculations, and transmits numbers to indicate results. Number codes are often used for letters and words. With the wide use of the computer for processing **numerical** information, the application of the computer to control the operation of a machine was a very logical development in industry. Since dimensions are given in numbers, other instructions can be coded into numbers. The use of numbers or numerical data to control the operation of production machines and processes came to be called **numerical control**.

Automated machines or processes, then, involve the application of at least one of the following principles in working with materials and information:

1. Feedback control
2. Automatic material handling
3. Data processing
4. Numerical control

59-2. Importance of Numerical Control

Automatic operation by numerical control is readily adaptable to the operation of many metalworking machines or processes. The use of NC is rapidly increasing for automatic operation of machine tools such as drill presses, milling machines, lathes, boring machines, grinding machines, and punch presses. NC also is coming into use for flame-cutting and welding applications. **Point-to-point** NC is used for spot welding while continuous-path NC is used for continuous-path welding (See § 59-4.)

NC is used with inspection machines which take measurements on parts through the use of sensing probes. The machines record the difference between the actual size and the specified size.

NC is being used for assembly operations such as those involved in wiring complex electronic systems. It is also used for tube-forming machines, wire-wrapping machines, and steel-rolling machines.

A more complex system of NC, using various symbols, is also used for making drawings automatically with special drafting machines. This kind of NC system generally requires the use of an electronic computer. Abbreviated information from a sketch is fed into a computer. The computer makes many calculations which locate the coordinate points and provide other required information for drawing lines and curves. When equipped with the proper accessories, a computer can prepare punched tape which is inserted into the NC system of the automated drafting machine. The NC system reads the tape commands and controls the operation of the drafting machine, thus producing the drawing automatically.

59-3. Numerical Control Machining Systems

Numerical control, abbreviated NC, is a system of controlling a machine or process through the use of numerically coded information which is the **input** for the machine's control system. The instructions to the machine are in the form of coded numbers punched into a ribbon of paper tape. The function of the tape may be compared to functions performed on jigs or fixtures of conventional machine tools. The tape can be stored for future use in repeating the same job or process. With some numerical control systems, the coded numerical information is inserted into the control system on punched cards or on magnetic tape instead of punched tape.

Direct numerical control (DNC) systems now exist which connect the machine tool directly to the computer through a typewriter **computer terminal**. Instructions (programs) for operating the machine(s) are stored in the computer's memory and can be called into use through the computer terminal. This system eliminates the need for tape or card input.

Feedback System

On NC machine tools, the positioning of the tool must take place automatically in accordance with commands on the coded tape or other input medium. The positioning accuracy of the tool generally must be to tolerances of plus or minus 0.001″. Some NC systems can maintain positioning accuracy to tolerances of plus or minus 0.0002″. This generally requires a positioning system which has a **feedback control loop** designed for automatic corrective adjustments.

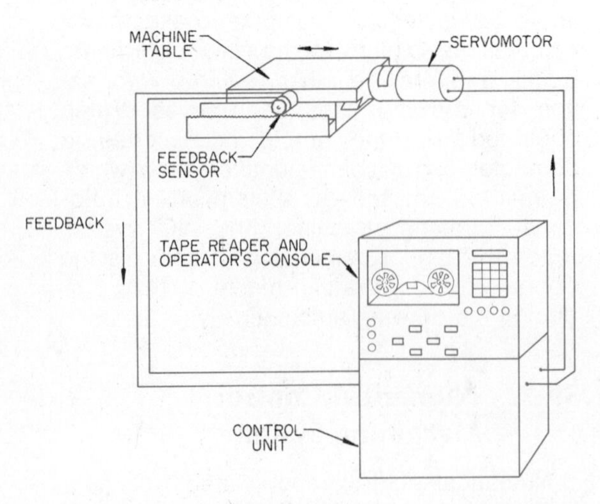

Fig. 59-2. A One-Axis Closed-Loop Numerical Control System

Most NC systems use a **close-loop feedback** system called a **servomechanism**. A servomechanism is an automatic feedback control system which controls the mechanical position of the tool in relation to the work.

A servomechanism may use electric motors or hydraulic cylinders for mechanical positioning of the tool in relation to the workpiece.

The principle involved in the operation of the servomechanism on a machine tool is similar to the principle involved in the operation of the closed-loop feedback system on the furnace in Fig. 59-1. The control system processes the numerical information on the punched tape (input) to create a signal to the servomechanism. A sensing device constantly senses the position of the table or the tool. When the desired position is signaled by the sensor, the control sends a "stop" signal. This prevents movement from the desired position or tool path. The sensing device in a servomechanism generally is an electronic, mechanical, or optical device. (See Fig. 59-2.)

59-4. Basis for NC Measurement

A system of **rectangular coordinates**, called the **cartesian coordinate** system, Fig. 59-5, is the basis for NC measurements. Three-dimensional objects (those with length, width, and thickness) can be described by this rectangular coordinate system. All points of an object are described by imaginary lines perpendicular to the axes. Generally, the horizontal plane includes the X- and Y-axes. In this plane, along the X-axis, all measurements to the right of the origin are in the + X direction and those to the left are in the − X direction. At exactly 90° to the X-axis, and in the same horizontal plane, is the Y-axis with its plus and minus direc-

Fig. 59-3. Numerically Controlled Machining Center
This 2-axis point-to-point NC machine automatically drills, mills, taps, and bores.

Fig. 59-4. Numerically Controlled Turret Drilling Machine
This machine has a 3-axis point-to-point positioning system with all axes tape controlled.

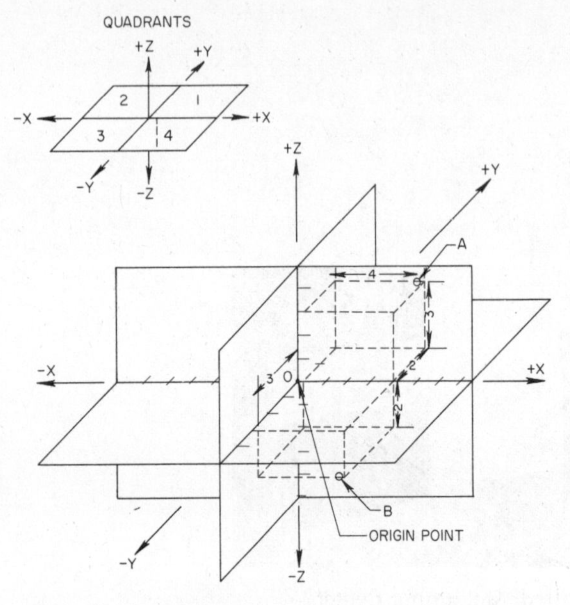

Fig. 59-5. Rectangular Coordinate System (The Cartesian-Coordinate System)
All dimensions are given from the point of origin. Point A has coordinates X = +4, Y = +2, Z = +3. Point B has coordinates X = +4, Y = −3, Z = −2.

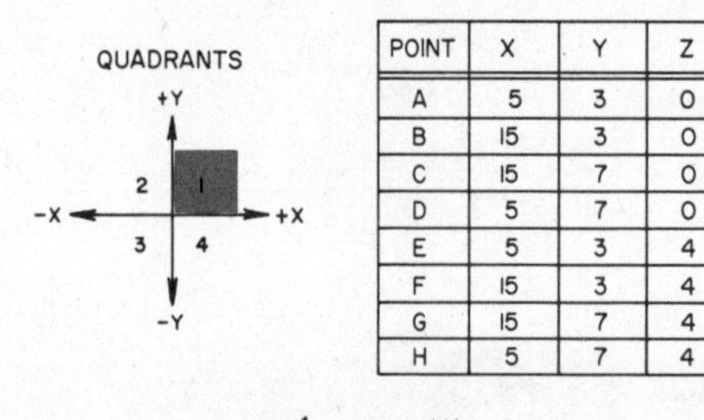

POINT	X	Y	Z
A	5	3	0
B	15	3	0
C	15	7	0
D	5	7	0
E	5	3	4
F	15	3	4
G	15	7	4
H	5	7	4

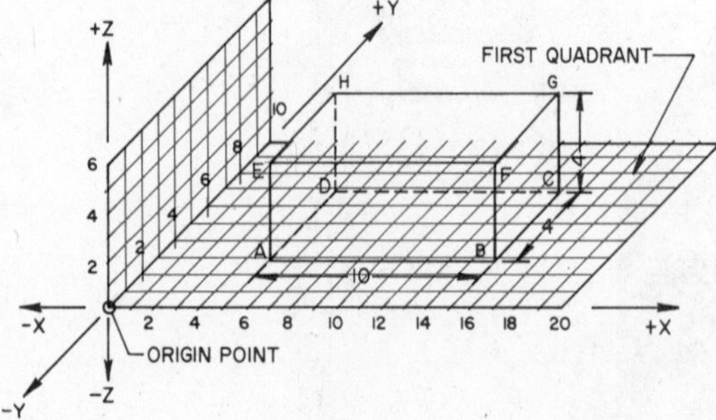

Fig. 59-6. All Coordinate Points are Plus When Located in the First Quadrant with Absolute Dimensioning
The + Sign May Be Omitted.

tions. The Z-axis, with its plus and minus directions, is perpendicular to both the X- and Y-axes.

The rectangular system of coordinates is used to describe the dimensions of all parts for numerical-control programming. The axes in this coordinate system also are used for machine axis designation. (See Figs. 59-7 and 59-8.)

In Fig. 59-5 notice that the X- and Y-plane is divided into quadrants. Many NC systems are designed so that all points on an object are located in the **first** quadrant, as shown in Fig. 59-6. With systems of this type, all positions are designated positive or plus (+). When the object is so located, the positive sign (+) may be omitted when preparing a program manuscript. The machines in Figs. 59-3, 59-4, and 59-21 use the first quadrant system. Study the positions of points **A** through **H** until you understand the method for describing points in the first quadrant of the coordinate system.

59-5. Machine Axis Designations and Movements

The axes of a machine tool correspond to the principal machine movements. The axes

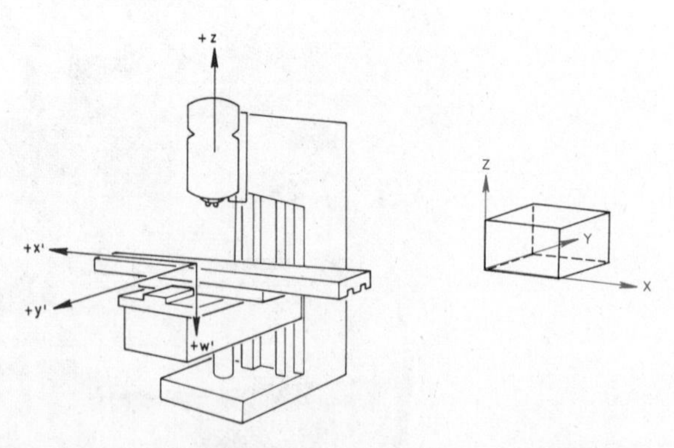

Fig. 59-7. Coordinate Axes for Vertical Knee Mill, Drilling Machine, and Jig Boring Machine
Programmers, setup men, and operators should think only in terms of unprimed numbers.

are designated in accordance with the axes of the rectangular coordinate system. (See Figs. 59-5 and 59-6.) Generally, the longest axis of machine travel is designated the X-axis. Observe that the X- and Y-axes are in the horizontal plane in Fig. 59-5 and the Z-axis in the vertical plane.

Several examples of basic machine tools and their axes designations, according to NAS 983[1], are shown in Figs. 59-7 and 59-8. Programmers, setup men, and operators need to use only the **unprimed** numbers or figures such as X, Y, and Z. The primed numbers, such as X′, Y′, and Z′, are for the machine manufacturer's use for design standardization purposes.

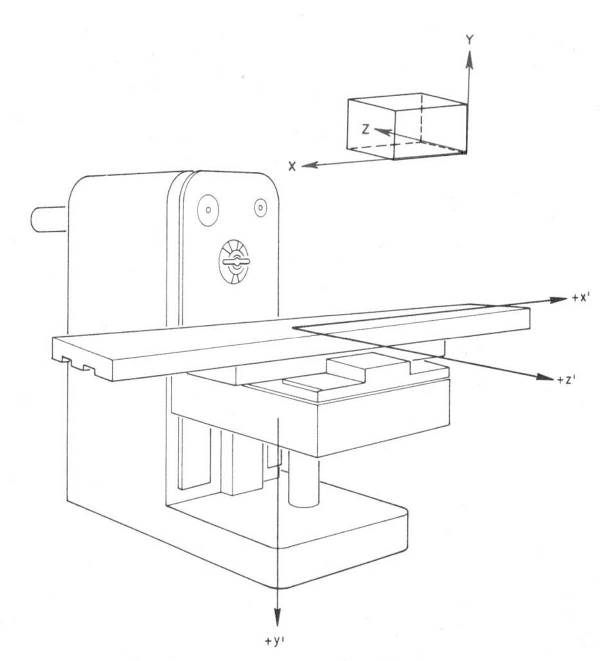

Fig. 59-8. Coordinates Axes for Horizontal-Spindle Lathes and Related Machines
Unprimed numbers apply for programmers and operators.

[1]National Aerospace Standard 983, copyright 1966, Aerospace Industries Association of America, Inc. Reprinted by permission.

Some NC machine tools utilize an **incremental** system for dimensioning parts (§ 59-10) and for indicating the direction of tool travel in relation to the workpiece. Machines with this type of programming system generally use plus (+) and minus (−) signs to indicate direction of tool travel. The + indicates travel in one direction while the − indicates travel in the opposite direction.

On 2-axis machines, the X- and Y-coordinate positions are programmed and included on the control tape. There is no Z-axis program to control tool depth, such as the depth of a drilled hole or the depth of a milled slot. Instead, multiple-tool depth stops are set manually by the operator before machining the part. Each depth stop on the multiple-type depth stop device is numbered. On some systems, the operator changes to the appropriate numbered stop during each manual tool change. On other systems, the tool stop number can be programmed on the control tape, and the machine selects the proper depth stop.

A tool depth stop actually has two stop positions. The first position stops **rapid** Z-axis (tool) travel at **gage height** which is a short distance above the workpiece. A gage height of 0.100″ is satisfactory. The second position stops further Z-axis travel when the tool has fed to depth.

On 3-axis machines, the tool-depth coordinate positions are programmed and included with other information on the control tape. Therefore, the depth of tool is controlled automatically by the tape. The machines in Figs. 59-4 and 59-22 are 3-axis machines with X-, Y-, and Z-axes travel automatically controlled.

59-6. Kinds of NC

There are two principal kinds of NC equipment: (1) point-to-point positioning, and (2) continuous-path contouring equipment. The point-to-point equipment is simpler than the continuous-path type, both in design and in programming of the input information.

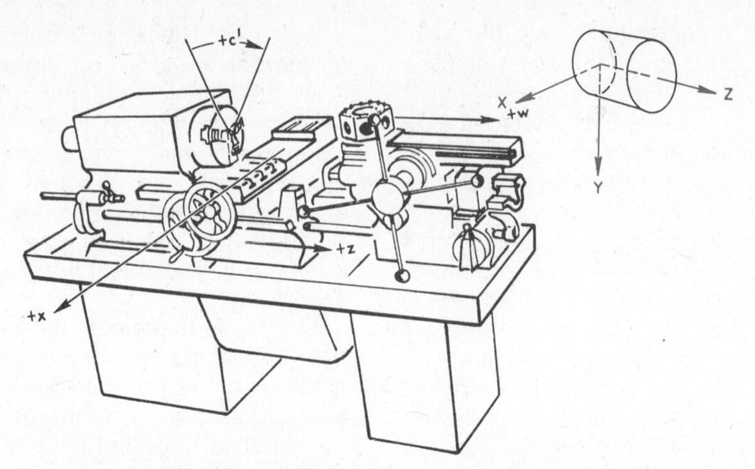

Fig. 59-9. Coordinate Axes for a Horizontal Spindle Lathe
Unprimed numbers apply for programmers and operators.

Fig. 59-10. A General-Purpose Electronic Computer System Aids in Preparation of Control Tapes for Continuous-Path Numerical Control Machine Tools
A. Information is punched at console at right.
B. Details are located in tape storage at rear.
C. Completed information is printed on printer at front.

Point-to-Point NC

A **point-to-point** numerical-control system is basically a positioning system. Its principal function is to move a tool or workpiece from one specific position point to another point. Generally, the actual machine function, such as a drilling operation on a machine tool, also is activated at each position by commands on the control tape.

Point-to-point NC systems also are often called **discrete positioning** systems. They are used for such hole-machining operations as drilling, countersinking, counterboring, reaming, boring, and tapping. Point-to-point NC systems also are used on hole-punching machines, spot-welding machines, and for wiring complex electronic circuits.

Most point-to-point NC systems also can perform **straight-cut** functions such as straight-line milling. The machine tools in Figs. 59-3, 59-4, 59-21, and 59-22 have point-to-point NC systems which also include straight-cut milling capabilities. On 2-axis machines, as in Figs. 59-3 and 59-21, the rate of feed along the axes generally is controlled manually by the machine operator by setting a dial on the control console.

Some straight-cut systems can make cuts at 45° to either the X- or Y-axis. On 3-axis machines, as in Figs. 59-4 and 59-22, the rate of feed for straight milling cuts is automatically controlled by the tape.

Continuous-Path NC

Machining operations which require the cutting tool to move in a curved path require continuous-path NC.

An example of continuous-path contour milling is shown in Fig. 59-11. The cutting tool must cut, within prescribed tolerances, to the curved line specified on the blueprint for the part being machined. Therefore, the center line path of the cutting tool must be described geometrically in terms of its X- and Y-coordinate dimensions.

Most NC continuous-path systems are programmed in terms of straight-line paths. Therefore, the contour generally is broken down into short straight-line **chords**, or segments. The chords must be short enough that the cutting tool path will produce a reasonably smooth contour within the specified tolerance. In Fig. 59-11 the chord lengths are greatly exaggerated so that they can be visualized. Note that curves with a small radius have shorter chord segments than curves with a larger radius. Programming requires that each end of each chord and the center of the arc it represents be specified in terms of X and Y coordinates.

Close-tolerance contour machining requires a greater number of straight-line chord segments. Therefore, a greater number of coordinate positions are also required. The example in Fig. 59-11 prescribes a tolerance of −0.001″.

Since the tolerance is designated minus, the contour cannot be oversize, but may be as much as 0.001″ undersize.

Some continuous-path NC machine tools, such as vertical milling machines, are of the 2-axis type. This type generally cuts continuous-path contours in the X- and Y-plane only. Three-axis machines can cut 3-dimensional contours, such as die cavities, which require cutting movements along the X-, Y-, and Z-axes simultaneously. Continuous-path NC systems are more expensive and more complex to program.

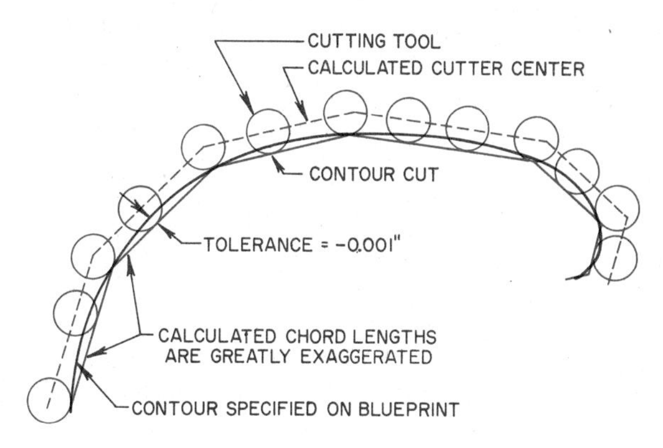

Fig. 59-11. Contour Milling with Continuous-Path Numerical Control
The arc is divided into chord segments (exaggerated example) which are within specified tolerances. The cutting tool travels along the straight line chords, thus generating a curve made of these tiny straight line segments.

59-7. Advantages of Numerical Controlled Production

NC machines and processes are used for many mass-production purposes. However, NC machine tools make their greatest contributions as mass-production tools when they are used for the production of parts which must be made in small quantity. It is estimated that 3/4 of all metal parts are manufactured in lots of fewer than 50 pieces.[2] Because of their close tolerance capability, NC machine tools often can be used at considerable savings in the production of even one part of one kind.

For the production of machined parts in large quantity, mass production with conventional machine tools, dies, and fixtures is frequently more economical and efficient.

[2]U.S. Department of Labor, **Outlook for Numerical Control of Machine Tools,** Bulletin 1437, March, 1965, p. 5. (Bulletin prepared by the U.S. Department of Labor, W. Willard Wirtz, secretary; Bureau of Statistics, Ewan Clague, commissioner; U.S. Government Printing Office.)

An example will illustrate the advantage of conventional mass-production machining methods for parts produced in large lot numbers. Suppose that 10,000 identical parts must be produced, each having 32 drilled holes. These parts can be produced most rapidly with a multiple-spindle drill press which can drill all holes in each part simultaneously. (Drill presses of the types shown in Figs. 51-8 and 51-9 can be used for jobs of this kind.) Of course, expensive jigs and fixtures usually would be required before the parts could be set up and machined. If the cost of tooling were $5000, the tool cost per piece would be $0.50. The machining cost per piece might be as low as $0.33 per piece (60 per hour at $20 machine and labor cost). The total cost per piece would be $0.83.

On the other hand, if an NC machine tool with a single-tool spindle were used to drill the required holes, the job would take much longer. It would require many more man-hours of labor costs. The additional labor costs, together with the relatively high cost of NC machines, would probably result in higher cost per part machined. For example, an NC machine and operator may easily cost $30 per hour to employ. Drilling the same 32 holes would typically require 5 minutes of machine time — 12 parts could be produced per hour at a per-piece cost of $30 ÷ 12 or $2.50. Parts produced in small quantity generally do not justify the high cost of fixtures for each part. Therefore, small quantities of parts generally can be produced at a lower cost per part with NC machines.

There are several advantages in using NC machine tools. The following is a brief summary:

1. NC machines generally obtain closer tolerances without requiring costly jigs for guiding the tool.
2. NC machines do not require storage space for large quantities of jigs and fixtures.
3. Inserting a tape can be done more quickly than positioning jigs and fixtures — a savings in setup time.
4. One NC machine, as in Figs. 59-3 and 59-4, can take the place of several conventional machines such as a drill press, tapping machine, and vertical milling machine. This saves required floor space by machines which would perform the same operations.
5. Material handling is reduced with an NC machine, as material would be moved several times when using conventional machines.
6. Tapes can be prepared more rapidly than fixtures can be made, thus saving **lead** time. Lead time is the time required for planning and tooling up for mass production of a product or part.
7. The tape can be easily stored for use at a later time for another small quantity of parts. This reduces storage space of parts and the cost of keeping large inventories.
8. Design changes can be made easily as a new tape can be produced quickly and at low cost.
9. Duplicate tapes can be made for producing parts on similar machines in the same plant or in plants at other locations.
10. NC is more accurate than a machinist working without jigs and fixtures. Since parts are not spoiled because of human error, waste also is reduced.
11. Tape preparation takes the place of template preparation for continuous-path NC machines.
12. Less inspection cost is required with NC since the machine is more consistently accurate than workers.
13. Management can better control the rate of production to meet the required quantity, quality, and accuracy of customer demands.
14. Small quantities are produced at lower cost, thus reducing the cost of the finished product. Hence, more people can buy the products and the standard of living can be improved.

59-8. Production Steps in NC Machining

The steps involved in manufacturing machined parts with point-to-point and straight-cut NC machine tools are summarized in Fig. 59-12. The **designer**, who frequently is an **engineer**, designs and makes a sketch of the part.

The **parts programmer** performs tasks which were formerly done by the machine operator. He writes instruction in code form on a **manuscript** instead of machining the part. He programs the sequence of operations, selects tools, determines cutting speeds and feeds, and indicates when to apply coolant. He indicates the location of each operation with coordinate dimensions. An example of a program manuscript for a simple part to be machined on a 2-axis machine is shown in Fig. 59-18.

The information on the program manuscript is then punched into a ribbon of 1″ wide paper tape by a typist using a tape-punching machine. (See Fig. 59-14.) The typist types the manuscript on a **manuscript form** sheet, thus checking to see that the coded information is the same as that on the hand written **manuscript** form sheet. At the same time, the machine punches coded holes in the tape which is in the box at the left end of the machine.

The machine control tape is given to the machine operator who inserts it into the **tape reader** in the NC unit of the machine tool. The machine operator mounts the workpiece on the machine table. On the 2-axis point-to-point machine, he changes tools, adjusts feeds, speeds, tool depth stops, and turns on the coolant. On the more complex 3-axis machines, most of these duties are done automatically according to commands on the control tape. On machines of this type, the operator merely loads the part, starts the machine, and unloads the machined part.

On single-spindle machines, the operator generally must make tool changes manually. On machines with turret tool heads, as

DESIGN ENGINEER

1. Designs part.
2. Makes sketch of part.

DRAFTSMAN

1. Makes engineering drawing of part.

PARTS PROGRAMMER

1. Studies part drawing.
2. Describes location of part in relation to zero reference point on machine table. (In some plants this is done by the draftsman.)
3. Determines sequence of operations.
4. Determines tool selection.
5. Identifies position points on parts.
6. Determines coordinate dimensions for position points in relation to zero reference point on machine table.
7. Prepares the program manuscript.

TYPIST

1. Punches tape with tape-punching typewriter.
2. Verifies tape with tape verifier.
3. Produces extra tapes as needed.

NC MACHINE OPERATOR

1. Inserts tape in NC unit.
2. Mounts workpiece in machine.
3. Installs tools in machine spindle.
4. Changes tools as necessary on machines without tool turrets or automatic tool changers.
5. Makes machine adjustments such as feeds, speeds, tool depth stops, and coolant on machines where these are not tape controlled.
6. Removes workpiece from machine.

Fig. 59-12. Steps in NC Production with Point-to-Point Machine Tools

in Fig. 59-4, tool changes take place automatically by rotation of the turret, as commanded by the tape. Large 3-axis NC machines of the type in Fig. 59-22 are equipped with automatic tool changers. In small job shops and in school shops, one person must often perform all of the above operations for NC machining.

59-9. Dimensioning Methods for NC

Drawings for parts which are to be machined on NC machines can be dimensioned by the **incremental** (conventional) method or by the **absolute** method.

Incremental Dimensioning

The part illustrated in Fig. 59-20 is dimensioned by the **incremental** method. This is the usual or conventional method used for dimensioning. With this method, the distance from one point to a second point on a part is given without reference to a common reference point, such as a **zero point**. (See Problem 2 in § 59-11.) Parts which are to be machined on point-to-point NC machines with **incremental control systems** generally are dimensioned by the incremental method.

Absolute Dimensioning

The drawing shown in Fig. 59-19 is dimensioned by the **absolute** method. All points are dimensioned from a common reference point. The **part reference point** should not be confused with the **machine table reference point.** Parts which are to be machined on point-to-point NC machines with **absolute control systems** generally must be dimensioned by the absolute method.

Most absolute NC control systems are designed so that the axis movements of the machine, such as the X- and Y-axes in Fig. 59-19, are located in the first quadrant of the coordinate system. On equipment so designed, all coordinate position points are indicated with positive numbers. Thus the position of point **A** is + X = 04.000 and + Y = 03.000. When coordinate position points are programmed for most NC systems, the plus (+) sign and the decimal points are

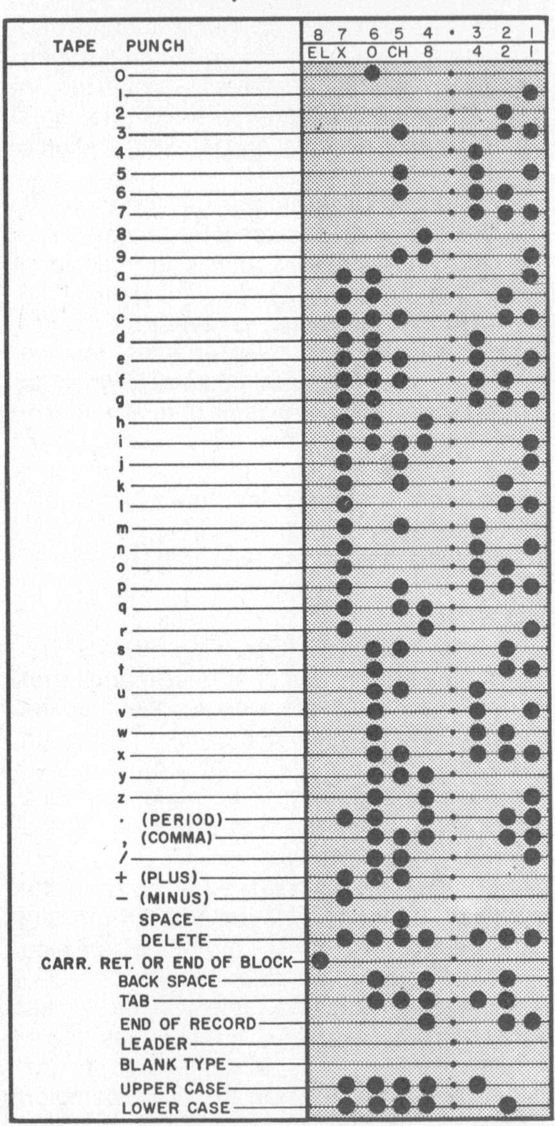

Fig. 59-13. EIA (Electronic Industries Association) Standard Code for 1″ Wide, 8-Track Tape

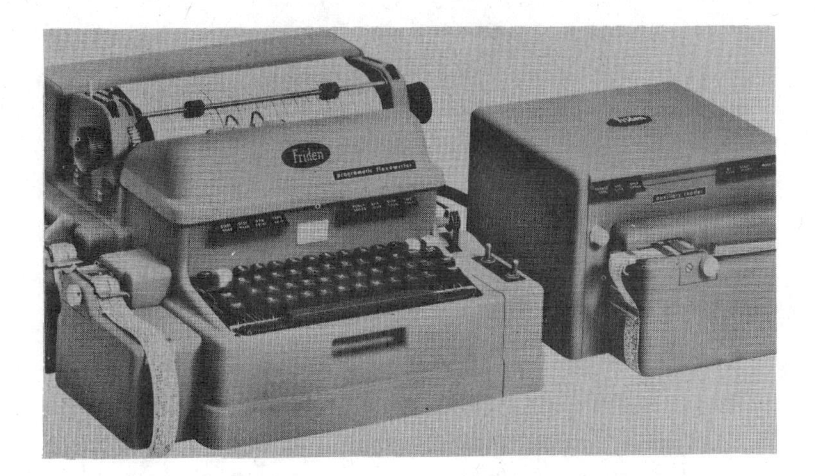

Fig. 59-14. Tape-Punching Machine and Verifier
The typist inserts the original tape in the verifier and proceeds to retype the manuscript copy. If the second tape is not identical to the first tape in the verifier, the machine carriage locks to reveal the location of the error. The typist unlocks the keyboard and corrects the error.

omitted. The coordinate points generally are indicated with five digits. (See Fig. 59-18 and Problems 1 and 2 in § 59-11.)

59-10. NC Tape

The tape used to control an NC machine is 1″ wide, 8-track tape with Electronic Industries Associaiton (EIA) coding in the form of punched holes, as in Fig. 59-13. Each number, letter, and symbol is called a **character**. The characters include the decimal digits from 0 through 9, the letters of the alphabet, and special characters such as **tab** or **end-of-block**. The characters are punched into the tape with a tape-punching machine, Fig. 59-14. The tape generally is made of paper or plastic-laminated paper. The plastic-laminated type is more durable and therefore is used if many parts are needed.

Binary Numbers

The code system of punched holes in the tape is based on the **binary system of numbers.** The binary number system has only

TABLE 40

Example Codes for NC Preparatory Functions

Cycle Code	Cycle Name
G78	Mill stop
G79	Mill
G80	Cancel
G81	Drill
G82	Dwell
G84	Tap
G85	Bore

TABLE 41

Example Codes for Miscellaneous Functions

Function Code	Name of Function
M00	Program stop
M02	End of program
M06	Tool change
M26	Full-spindle retract
M50	No cam
M51-59	Select depth cam (nine depth cams available)

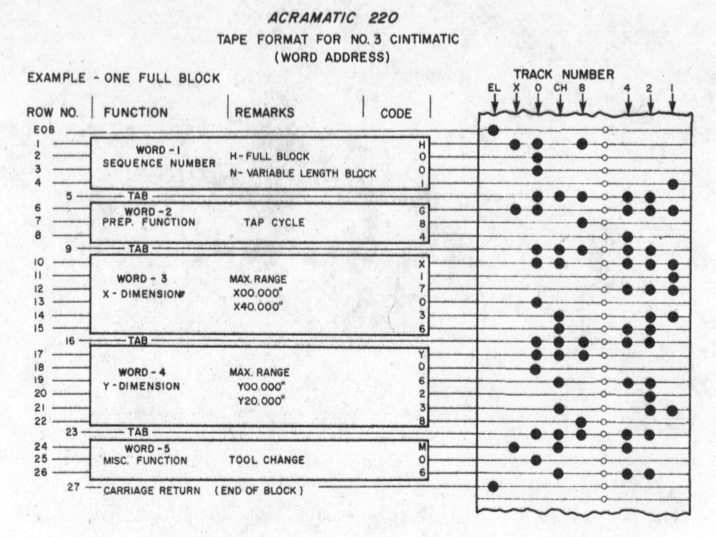

Fig. 59-15. Word-Address Tape Format

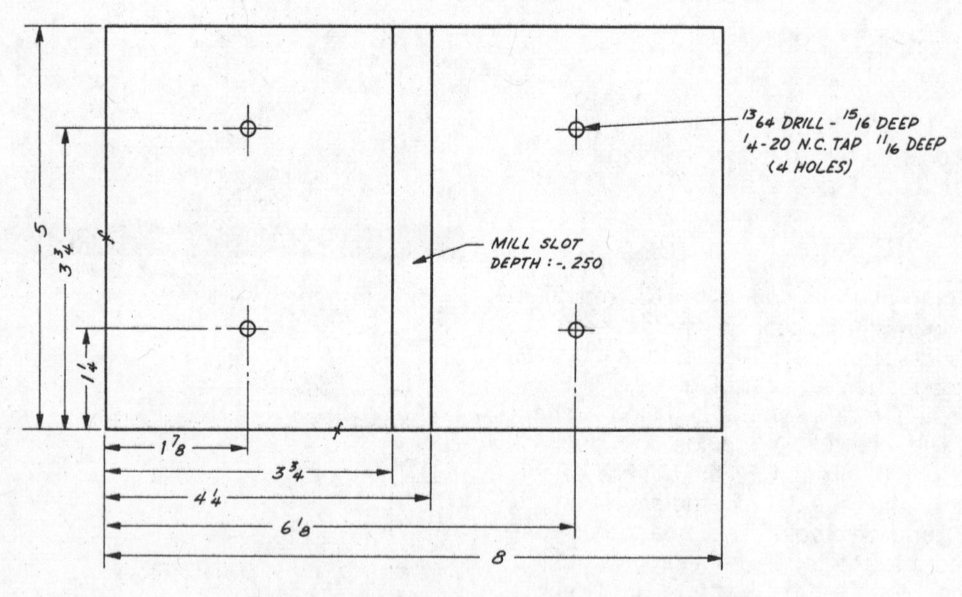

Fig. 59-16. Sample Part, Drawing Number 101

two digits, 0 to 1. (The decimal numbering system has ten digits, 0 to 9.) The binary number system is used in electronic computers and other electronic devices. Electrical circuits can be established on a basis of the two binary numbers, 0 to 1, because these numbers represent the two possible electrical conditions: 1. **on** or **positive** or 2. **off** or **negative**. Thus, when a character is punched in a tape, a hole either **is** or **is not** punched in each track in the tape. The punched holes, therefore, actuate electrical circuits in the NC system. You can learn more about the binary number system in

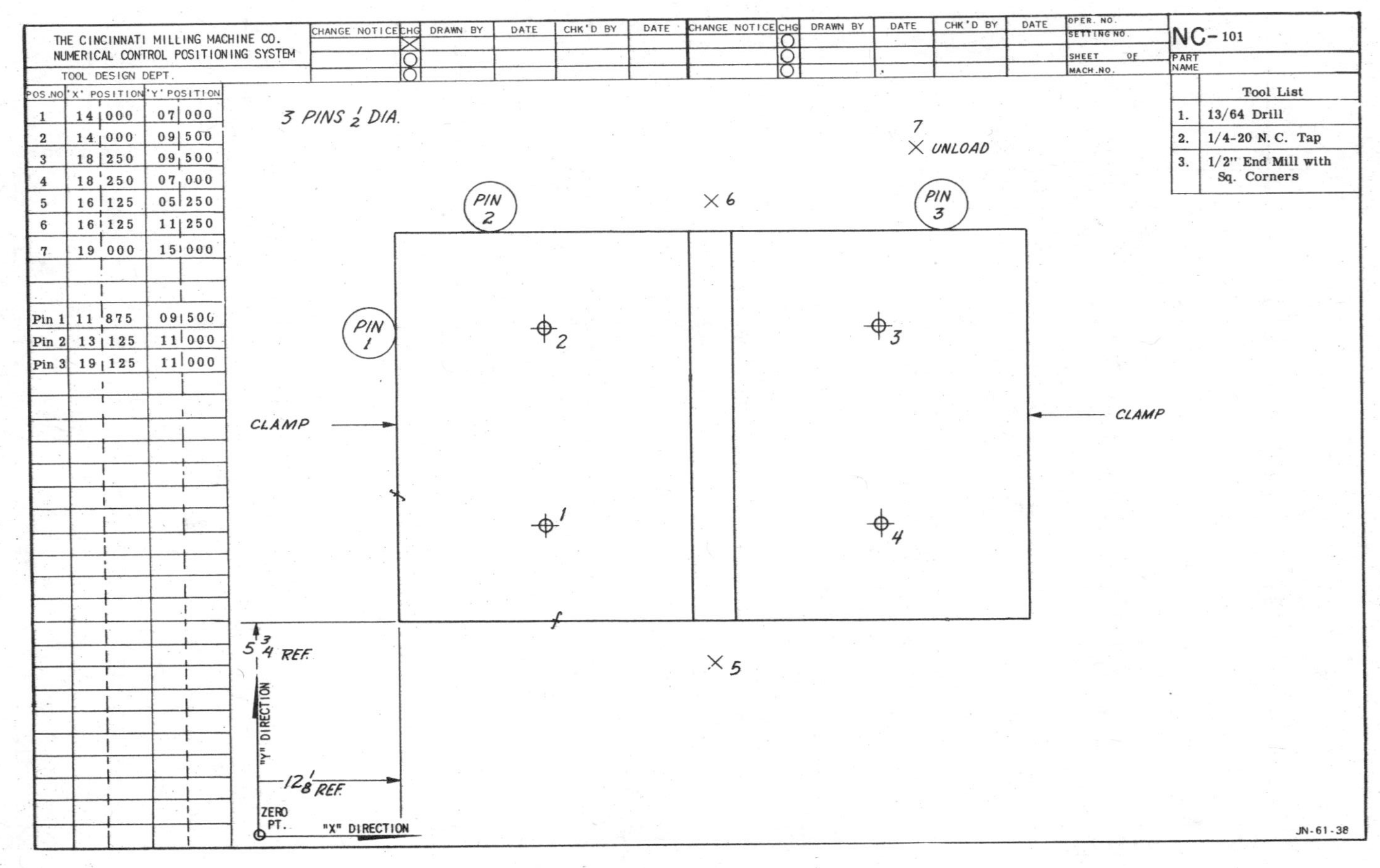

Fig. 59-17. Sample Part Number 101, Showing Location on Machine Table
Note stop pins and clamp location. The X and Y coordinate position is included.

PROGRAM FOR # 3 CINTIMATIC

CINCINNATI
ACRAMATIC POSITIONING
NUMERICAL CONTROL SYSTEM

PART NAME: SAMPLE PART | PART NO.:
DRAWING NO.: 101 | REVISION: | FIXTURE NO.: | PROGRAMMED BY: | PAGE OF PAGES:
SET UP AND TOOL INFORMATION: | DATE:

H or N SEQ. NO.	G PREP. FUNCT.	X POSITION	Y POSITION	M MISC. FUNCT.	POS. NO.	TOOL NO.	HEAD POS.	SPINDLE FEED IN./MIN.	SPEED RPM	DEPTH OF CUT	TABLE FEED IN./MIN.	TOOL REMARKS
H001	G81	X14 000	Y07 000	M51	1							13/64 Dia. Drill x 15/16 Deep
N002			Y09 500		2							
N003		X18 250			3							
N004			Y07 000	M06	4							
H005	G84	X18 250	Y07 000	M52	4							1/4-20 NC Tap x 11/16 Deep
N006			Y09 500		3							
N007		X14 000			2							
N008			Y07 000	M06	1							
H009	G78	X16 125	Y05 250	M53	5							1/2 Dia. End Mill x 0.250 Deep
N010	G79		Y11 250		6							
N011	G78				6							
N012	G81				6							
H013	G80	X19 000	Y15 000	M02	7							
												Unload

Fig. 59-18. Completed Program for Sample Part Number 101 Shown in Fig. 59-17.

books with more complete coverage of NC. Several are listed in the bibliography.

Tape Format

The format is the general arrangement of the information on the control tape. (See Fig. 59-15). It is the arrangement of the punched holes on a paper tape, or the arrangement of magnetized areas on a magnetic tape. Several different formats are used by the different manufacturers of NC machines. The format and the method for programming a part to be machined by NC are explained in the **Programmer's Manual** supplied by the manufacturer of the NC system. Attempts are being made by industry to agree on a standard tape format.

The following formats are used on various point-to-point and straight-cut NC systems:

1. Fixed sequential format
2. Tab sequential format
3. Word address format
4. Word address, variable block format (combines features of the sequential and word address formats)

An example of the **word address, variable block format** is shown in Fig. 59-15. A program manuscript is shown in Fig. 59-18 for the part shown in Figs. 59-16 and 59-17. This tape format is used on the machines in Figs. 59-3 and 59-21. This basic format, with additional information, is used for 3-axis machines of the type shown in Fig. 59-4.

Block

An NC manuscript is made of blocks of coded information. A **block** is a word or group of words that forms a unit. On perforated tape, the block must be separated from other similar units by an **end-of-block** character. (See Fig. 59-13.) An example of one full block of information is shown in Fig. 59-15.

The term **word address** means that each word included in the block is addressed with a letter character identifying the meaning of the word. Thus, the letter character **H** or **N** identifies the sequence number word; the letter X identifies the X-dimension word, etc.

Examples of **full block** and **variable length blocks** are shown in the program manuscript in Fig. 59-18. Sequence numbers 1, 5, 9, and 13 are full blocks; the others are variable length blocks. NC systems designed for variable length formats can store information in the **memory storage** unit when the information is the same as that in previous blocks.

Information in Blocks

A block of information for a 2-axis point-to-point NC program (Fig. 59-19) includes the following information:

1. Sequence number
2. Preparatory function
3. X-dimension
4. Y-dimension
5. Miscellaneous function

Sequence Number

The sequence number indicates the relative location of each block of information on the program manuscript and also on the perforated tape. For some NC programming manuscripts, the sequence number is designated with three digits, as in Fig. 59-18. Sequence 1 is designated as 001, sequence 12 as 012, and sequence 999 as 999. Many NC systems are equipped with a **sequence number readout** which indicates the sequence number of the operation which is being performed. A letter character such as **H** or **N** is used as an address for the sequence number, Fig. 59-18.

X- and Y-Coordinate Dimensions

The coordinate dimensions are indicated with six characters: the letter address and five digits. The letters X, Y, and Z identify the coordinate-axis dimension. Coordinate dimensions are indicated to the nearest one-thousandths of an inch. Thus, an X-coordinate dimension of 4½″ is converted to 04.500. The decimal point is omitted on the program on the tape.

Preparatory Functions

Code numbers are used to prepare an NC system for a particular mode of operation. For point-to-point systems, a preparatory function code commands the machine to prepare for a specific machining cycle. Preparatory functions have a cycle code number which identifies the name of the function. Table 40 gives examples of the code used for the program shown in Fig. 59-18.

Preparatory function code numbers may differ for different NC machines.

POSITION COORDINATES		
POS. NO	X POSITION	Y POSITION
A	04 ¦ 000	03 ¦ 000
B	08 ¦ 000	03 ¦ 000
C	08 ¦ 000	06 ¦ 000
D	04 ¦ 000	06 ¦ 000
LOAD	10 ¦ 000	09 ¦ 000

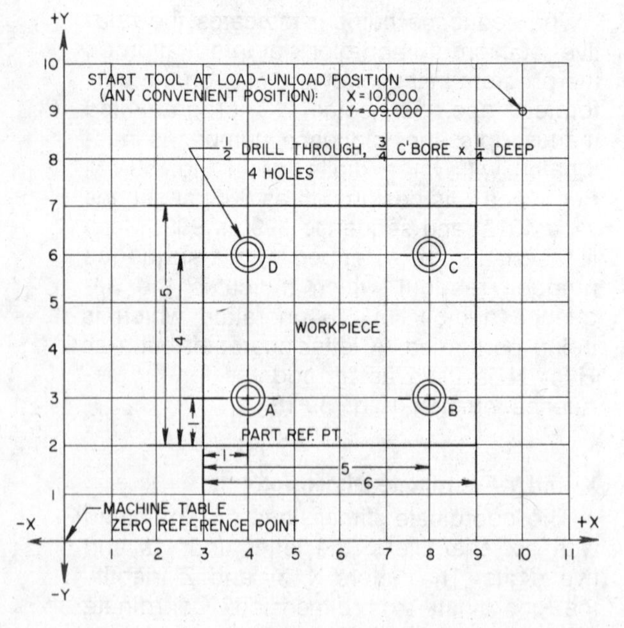

Fig. 59-19. Workpiece with Absolute Dimensioning Located in First Quadrant on Table of Machine with 2-Axis Absolute NC Positioning System (Fixed-Zero Reference)

Miscellaneous Functions

Miscellaneous functions are **on-off** functions of an NC machine. Examples of miscellaneous function codes used for the sample program in Fig. 59-18 are given in Table 41.

Miscellaneous function codes also vary from manufacturer to manufacturer.

Tool Number

The tool number is designated by code number, such as T1, T2, etc. A tool number may or may not be included in the block of information on the program for a 2-axis machine. The tool number is included on programs for 3-axis turret-type machines, and for machines which are capable of automatic selection and installation of different tools.

The kind of tool for each tool change generally is listed on the program manuscript. This procedure is recommended whether tool changes are made manually or automatically. Since the machine operator is given a copy of the program manuscript, he is able to determine, according to the sequence number, which tool should be in use.

59-11. Determining Sequence and Coordinate Dimensions for a Parts Program

The first step in the preparation of the parts program is to study the drawing carefully. There is less chance of error in determining coordinate dimensions if the drawing shows the location of the **part reference point** in relation to the **machine table zero-reference point**, Figs. 59-17 and 59-19. Parts generally are not located so that the part reference point coincides with the machine table reference point. However, the X- and Y-coordinate dimensions in the program must be for the actual coordinates of the machine table.

An example of X- and Y-coordinate dimensions for points A through D is shown in Fig. 59-19. The coordinate dimensions show the positions of points on a part which is located on the table of a machine equipped with an absolute NC positioning system. The machine has a **fixed-zero table reference point.** The table reference point is the reference point from which all X- and Y-coordinate dimensions must be programmed for the machine.

The principles involved in parts programming can be understood more readily by studying the sample programs which follow. The sample programs illustrate the principles involved in determining the **sequence** of operations and the **coordinate dimensions** for the positions at which operations are to be performed. The codes for the **preparatory functions** and for the **miscellaneous functions** are intentionally omitted since these codes vary for different machines. Once again, it must be understood that the **block format** also varies for different machines. Therefore, you should always consult the **Parts Programmer's Manual** for a specific NC machine to determine the block format and the code numbers for preparatory functions, miscellaneous functions, and tool numbers.

Programming Example No. 1

Point-to-point positioning to be performed with an **absolute** NC system.

Table 42 shows the program of coordinate positions and tool changes for machining the holes in the part shown in Fig. 59-19. The part is to be machined on a machine with an absolute NC control system with the following characteristics:

Position system —
Point-to-point (2 axes)
Type of format —
Word address, variable block, (with five digits for X- and Y-dimensions)
Type of dimensioning —
Absolute, located in the first quadrant
Tools —
T_1 = ½″ drill; T_2 = ¾″ counterbore

Programming Example No. 2

Point-to-point positioning with an **incremental** NC system.

With incremental NC systems, the **dimension** and the **direction** of tool travel, **in relation to the workpiece**, must be shown for each movement to a new coordinate position. The direction of tool travel is indicated with positive (+) and negative (−) signs, according to the coordinate system. (See Fig. 59-5.) On some incremental NC systems, the positive sign (+) may be omitted from the program and from the tape. When there is no sign, the system assumes that the tool travel is in the positive (+) direction.

Table 43 gives the program for the sequence, coordinate positions, and tool changes for machining the holes in the part shown in Fig. 59-20. The part is to be machined on a machine with an incremental NC system having the following characteristics:

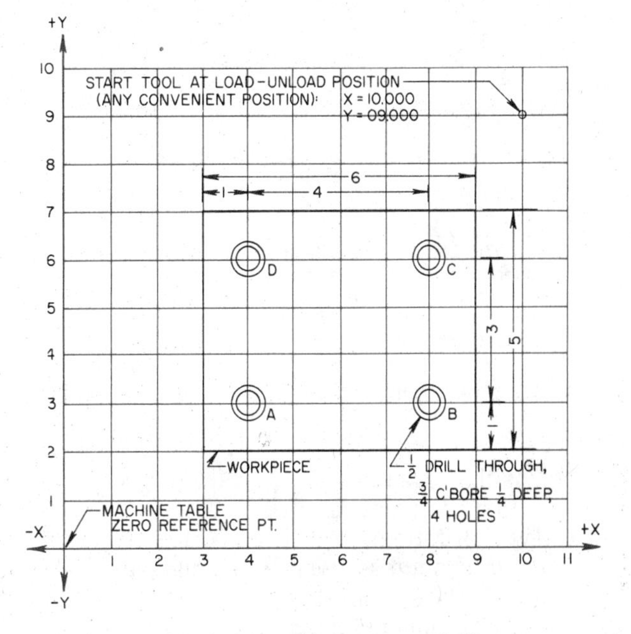

Fig. 59-20. Workpiece with Incremental (Conventional) Dimensioning in First Quadrant on Table of Machine Equipped with 2-Axis Incremental NC Positioning System

TABLE 42

Point-To-Point Positioning with an Absolute NC System

Sequence	X Info.	Y Info.	Tool No.	Position	Notes
001	X10.000	Y09.000	T_1	Load	½" dia. drill, four holes
002	X04.000	Y03.000		A	Same tool
003	X08.000			B	No change in Y-axis, same tool
004		Y06.000		C	No change in X-axis, same tool
005	X04.000			D	No change in Y-axis, same tool
006			T_2	D	Change tool: ¾" c'bore, 4 holes, ¼" deep
007		Y03.000		A	No change in X-axis, same tool
008	X08.000			B	No change in Y-axis, same tool
009		Y06.000		C	No change in X-axis, same tool
010	X04.000			D	No change in Y-axis, same tool
011	X10.000	Y09.000		Unload	Return to load-unload position

TABLE 43

Point-To-Point Positioning with an Incremental NC System

Sequence	X Info.	Y Info.	Tool No.	Position	Notes
001	X10.000	Y09.000	T_1	Load	½" dia. drill, four holes
002	−X06.000	−Y06.000		A	Location of pt. A with reference to loading position
003	X04.000			B	Location of pt. B with ref. to pt. A
004		Y03.000		C	Location of pt. C with ref. to pt. B
005	−X04.000			D	Location of pt. D with ref. to pt. C
006			T_2	D	Tool change: ¾" dia. C' bore, 4 holes, ¼" deep
007		−Y03.000		A	Location of pt. A with ref. to pt. D
008	X04.000			B	Location of pt. B with ref. to pt. A
009		Y03.000		C	Location of pt. C with ref. to pt. B
010	−X04.000			D	Location of pt. D with ref. to pt. C
011	X06.000	Y03.000		Unload	Location of load-unload position with ref. to pt. D

NOTE: The X and Y positions may be checked for errors by algebraically adding the total of each column. The sum of each column should be equal to the coordinates of the starting position. (In the above problem, the starting position is X10.000 and Y09.000.)

Position system —
Point-to-point (2 axes)
Type of format —
Word address (with five digits for X- and Y-dimensions)
Type of dimensioning —
Incremental (conventional) in the first quadrant.
Tools —
$T_1 = ½''$ drill; $T_2 = ¾''$ counterbore

59-12. NC Machine Tools in Industry

Several point-to-point NC machine tools and their applications were mentioned earlier — two-axis machining centers (Fig. 59-3) and three-axis turret-type drilling and milling machines (Fig. 59-4). In addition to these basic kinds of NC machines, several other types are in wide use. These include **hori-**

Fig. 59-21. Horizontal-Spindle NC Machining Center
This point-to-point NC machine automatically drills, mills, taps, and bores.

Fig. 59-22. 3-Axis Numerical Control Machining Center
This machine automatically performs drilling, tapping, boring, and milling operations. It automatically selects and changes tools by tape command.

zontal spindle two-axis machining centers, Fig. 59-21, three-axis machining centers, Fig. 59-22, and vertical and horizontal lathes of two or more axes, Fig. 59-23.

Horizontal Spindle NC Machining Center

The 2-axis point-to-point NC machine shown in Fig. 59-21 automatically drills, mills, taps, and bores. Its operation is very similar to the operation of the vertical drilling and milling machine shown in Fig. 59-3. However, the horizontal spindle machine has several distinct advantages. With the horizontal spindle, the chips fall away from the tool due to gravity. The workpiece may be mounted on an optional revolving index table, so that one setup makes all sides accessible for various machining operations.

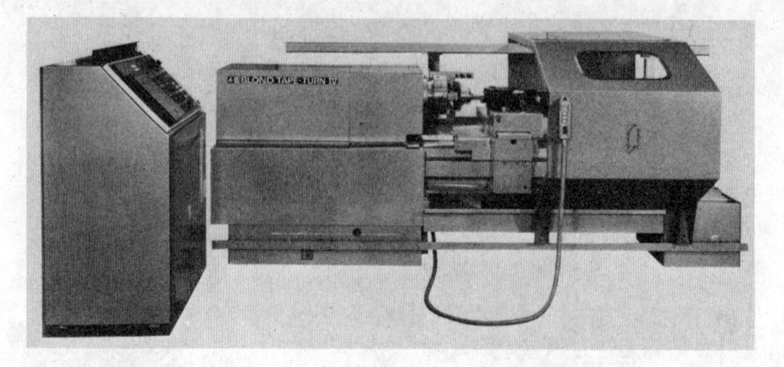

Fig. 59-23. NC Turning Center

The workpiece also can be indexed for machining sides with various angles.

Three-Axis NC Machining Center

The 3-axis point-to-point NC machining center shown in Fig. 59-22 is automatically operated by tape control for all three axes. The input information on the tape includes the following: (1) the sequence of operations, (2) machining functions, (3) coordinate positions for all three axes, (4) tool-selection number, (5) feed rate, (6) spindle speed, (7) coolant flow, and other miscellaneous functions.

Since the Z-axis tape control includes control of tool depths, it is not necessary for the operator to set the depth stops for each tool. This machine selects the proper tools, as commanded by the tape, and controls the depth. It has a heavy-duty boring head and a tool changer with a capacity of 14 tools. The universal head has a tool changer with a capacity for 30 tools, such as drills, taps, reamers, countersinks, and light milling cutters. Thus, the machine can select and use as many as 44 tools automatically for one job. The various operations that are required are completed automatically once the operator mounts the workpiece and starts the machine.

Lathes

Point-to-point and continuous path NC lathes of several types are also in wide use. The conventional single horizontal spindle NC lathe machines parts automatically with the exception of tool changes and workpiece loading and loading which are done manually. (See Fig. 59-23.) The versatility and efficiency of these machines is enhanced either with the use of a turret tool post carrying up to four cutting tools, or with preset quick change tooling.

The conventional lathe operations of straight turning, facing, boring, threading, grooving and cutoff, and profiling with form tools can be done on point-to-point NC lathes. Continuous path NC lathes are also capable of turning any angle of taper and almost any curved profile without the use of form tools.

NC lathes of the type shown in Fig. 59-23 are often called **turning centers**. They are equipped with either a four- or six-station turret tool post, and an optional rear turret of six or eight stations. Tool changes, which are programmed and occur automatically by tape command, are made by rotating the turrets to bring the desired tool into operating position.

Other types of NC lathes include horizontal turret lathes, Fig. 59-9, and vertical turret lathes or boring mills.

NC lathes can be equipped with automatic bar feed or devices for automatic loading and unloading of workpieces, thus making them fully automatic.

Increasing Use of NC Machines

The use of NC machine tools is expected to continue to increase. In recent years, more than 60% of new machine tools sold were NC machines. In addition, many conventional machine tools are being **retrofitted** (converted) to NC.

Persons who are interested in becoming parts programmers or NC machine tool operators should have a good understanding of basic machine tool operations and the speeds and feeds of various metals and tools. This kind of experience can be acquired in metalworking and machine shop classes in high schools, vocational schools, technical schools, and some colleges.

Many employers have programs for training NC machine tool operators and parts programmers. NC machine tools are very expensive, and many schools are unable to acquire them. Persons who have acquired a good understanding of basic machining operations in metalworking classes are frequently employed by manufacturers who are willing to provide further training.

Words to Know

absolute dimensioning
automation
2-axis machine
3-axis machine
binary number
block of information
cartesian coordinate system
closed loop
continuous-path NC
coordinate dimensions
data processing
discrete positioning system
electronic computer
feedback control loop
fixed-zero reference point
incremental positioning
machining center
manuscript form sheet
miscellaneous function
numerical control NC
open loop
part reference point
parts program
parts programmer
point-to-point NC
preparatory function
retrofit
sensor
sequence number
servomechanism
straight-cut function
tape character
tape format
turning center
verifier
word address

Review Questions

1. What is meant by the term automation?
2. List three characteristics which automated machines or systems possess.
3. What is meant by the term **output** as it applies to automated systems?
4. What is meant by the term **input** as it applies to automated systems?
5. What is meant by the term **sensor** as it applies to automated systems?
6. What is the function of a computer in an automated system?
7. What is the meaning of the term **feedback control loop** in an automated system?
8. Explain how the feedback control loop for an automatic heating system functions.
9. Explain the difference between an **open-loop** and a **closed-loop** feedback system.
10. What is meant by the term numerical control?
11. List four ways the coded information can be provided to the NC control system for a machine tool.
12. Of what advantage is direct numerical control?
13. Explain how a servomechanism functions on an NC machine.
14. Name the two main kinds of NC control systems.
15. For what kinds of machining operations are continuous-path NC machines used?
16. What kind of work can be done on a 3-axis continuous-path milling machine which cannot be done on a 2-axis machine?
17. In what industries are 3-axis continuous-path NC machine tools most widely used?
18. What is the principal function of a point-to-point NC system?
19. List several applications, other than machining, for which NC point-to-point systems are used.

20. Most point-to-point NC machine tools also are equipped to perform straight-cut functions. Give an example of a straight-cut operation.
21. List several different types of point-to-point NC machine tools.
22. NC machine tools make their greatest contribution at what production volume?
23. At what production volume are conventional machine tools more efficient than NC machine tools?
24. List several advantages of NC machine tools for production of machined parts.
25. Outline the basic steps involved in NC production of machined parts.
26. Explain the kind of work which is done by a parts programmer.
27. Describe the cartesian coordinate system and explain its three axes.
28. Which quadrant of the coordinate system is the first quadrant?
29. For absolute dimensioning, how are programmed coordinate dimensions designated when all movements of the machine tool NC system are located in the first quadrant?
30. Which axis of a machine tool generally is designated as the X-axis?
31. What type of NC system uses plus (+) and minus (−) signs on programs to indicate the direction of tool travel in relation to the workpiece?
32. Explain the difference between the **part reference point** and the **machine table reference point.**
33. Explain the features of the tape which is used to control NC machine tools.
34. What is a **character** on a tape?
35. What is meant by the binary number system?
36. What is meant by the term **tape format?**
37. List four kinds of tape formats used on various point-to-point and straight-cut NC systems.
38. What is meant by **block** on a program manuscript or on a tape?
39. What is meant by the **word address** as it applies to a tape format?
40. List the kinds of information which generally are included in a block of information for a 2-axis point-to-point NC program.
41. What is a **preparatory function** and how may it be designated on a program?
42. What is a **miscellaneous function** and how can it be designated on a program?
43. How are different tools designated on a program?
44. List as many kinds of metal cutting machine tools as you can which are now available with NC control systems.
45. What is the difference between a machining center and a turning center?
46. Describe how NC machine tools can be made fully automatic.

Career Information

1. What kinds of experience should one have in order to become a parts programmer?
2. Where can one learn to become an NC parts programmer?
3. What is the outlook for employment opportunities in the field of NC programming?

UNIT 60

Nontraditional Machining Processes

60-1. Needs for New Machining Processes

Much research effort has been directed towards finding economical ways of machining extremely hard materials such as carbides, ceramic oxides, and cermets, and such difficult to machine metals as beryllium and titanium. As a result, a number of important new machining processes have been developed, each with its own unique advantages.

60-2. Chemical and Laser Beam Machining

Chemical milling and **chemical blanking**, processes which use strong caustic or acid solutions to dissolve away unwanted metal, are discussed in Unit 33. **Laser beam machining** (LBM) utilizes the same power source as **laser welding**, discussed in Unit 29. For machining instead of welding, different power settings are used. Lasers can machine any known material. The intense heat of the laser beam causes the material to be removed chiefly by vaporization. However, because the rate of metal removal is very small, lasers are only used for jobs requiring a small amount of metal to be removed. Typical jobs include drilling holes, sometimes as small as .001″, in extremely hard materials such as diamond or carbide for wire drawing dies, and removing minute amounts of metal in balancing parts which spin at high speed.

60-3. Electron Beam Machining (EBM)

Electron beam machining (EBM) also uses the same equipment as **electron beam welding**, which was discussed earlier in Unit 29. As with lasers, for machining, the equipment is operated at different power settings than for welding. Machining is accomplished by focusing a high-speed beam of electrons on the workpiece. Sufficient heat is generated to vaporize any known material, but metal removal is very small. The process is used for much the same kind of work as lasers, but it is capable of much greater accuracy.

Disadvantages of the process include the need for skilled operators, high equipment cost, and limits on the workpiece size due to size of the vacuum chamber available. The process also generates X-rays, thus requiring radiation shielding of the work area.

60-4. Electrical Discharge Machining (EDM)

Electrical discharge machining (EDM) is a process which removes metal by controlled electrical arcing or sparking between tool and workpiece while submerged in a **dielectric** (insulating) fluid, Fig. 60-1. A typical EDM machine is shown in Fig. 60-2. EDM is valued for its ability to machine complex shapes in metals of any hardness. It is widely used in making injection and compression molds for rubber and plastic molding, molds for die casting metals, and dies for forging and metal stamping. It is also valuable for its ability to remove broken taps and studs. Because no tool pressures are involved, it is useful for machining delicate workpieces such as metal honeycomb structures. A disadvantage is that only materials which can conduct electricity can be machined.

Electrode (tool) materials, which must also conduct electricity, include high purity graphite, brass, copper, copper-graphite, tungsten and several tungsten alloys, and zinc alloys. The EDM process causes much greater tool wear than conventional machining processes. Often, several identical tools must be made, one or more for roughing cuts, and one or more for making finishing cuts. Sometimes stepped tools can accomplish the same result as several tools, Fig. 60-3. Metal removal rate is directly related to amperage settings. Low amperage settings produce slow rates of metal removal and good finishes, while high amperage settings produce higher rates of metal removal and poorer finishes, Fig. 60-4. Finishes of 30 to 150 microinches are normal, but with care, a 10 microinch finish is possible.

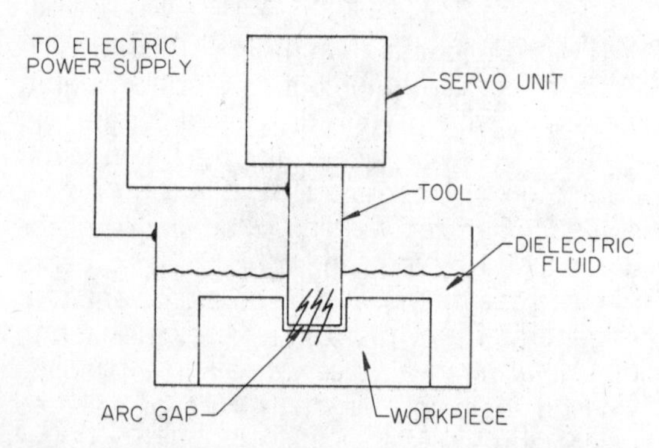

Fig. 60-1. Basic Components of an EDM System

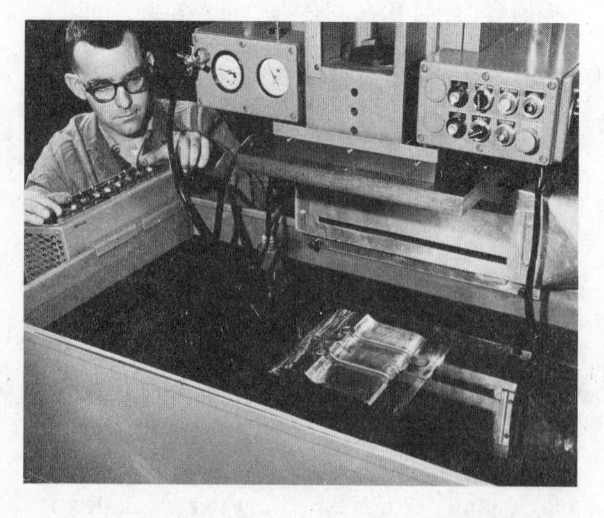

Fig. 60-2. EDM Machine with Electrolyte Surrounding Die Block

Since a gap must exist between the tool and the workpiece, the tool produces a cavity slightly larger than itself. This size difference is referred to as tool **overcut** and generally amounts to only a few thousandths of an inch. Under ideal conditions, the overcut can be as small as .0002″.

In order to obtain uniform cutting action and good finishes, a flow of dielectric fluid must be directed through the arc gap so as to sweep away the chips. The fluid stream may be directed along the tool, through a hollow tool, or through a hole in the workpiece opposite from the point of tool entry.

EDM is a firmly established metalworking process.

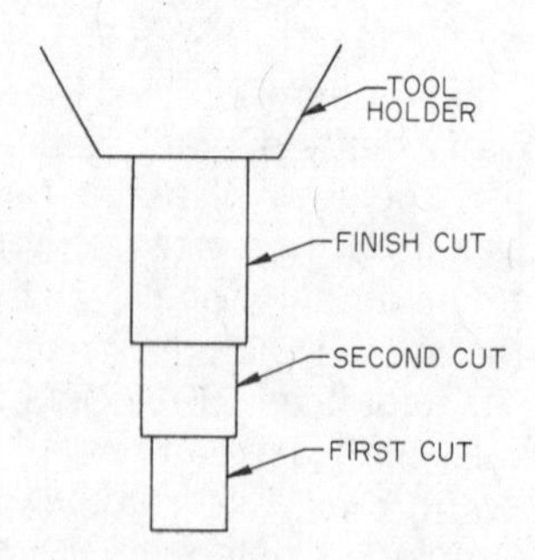

Fig. 60-3. Stepped Cutting Tool

60-5. Electrochemical Machining (ECM)

Electrochemical machining (ECM) is based on the same principles as electroplating. (See Unit 35.) However, instead of depositing metal on the workpiece, ECM reverses the process so that metal is **deplated** or removed from the workpiece. The basic components of an ECM system are shown in Fig. 60-5.

In ECM, the tool becomes the cathode and the workpiece the anode. A gap between the tool and workpiece of from .001″ to .010″ is maintained to provide space for the flow of electrolyte, and to keep the electrical circuit from becoming shorted. A low-voltage, high-amperage direct current passes from the workpiece to the tool through the electrolyte, causing metal particles to be dissolved from the workpiece into the electrolyte by electrochemical reaction.

Since the electrolyte is pumped through the gap between the tool and workpiece at pressures from 200 to 300 psi, the dissolved metal particles are swept away and filtered out, preventing them from being deposited on the tool.

Accuracy of the process and tool life are very good for these reasons: (1) the tool never touches the workpiece, (2) it receives no buildup of metal from the workpiece, and (3) it has negligible wear from the flow of electrolyte.

ECM can machine any metal which conducts electricity, regardless of hardness. Absence of any tool pressure on the workpiece makes the process ideal for machining thin materials and fragile workpieces.

Comparatively low metal removal rates prevents ECM from competing with conventional machining methods when metals with good machineability are involved. It excels in machining difficult to machine metals, especially when holes or cavities of complex shape must be made. No burrs are produced by ECM; the finishes are bright and smooth and ordinarily do not require polishing.

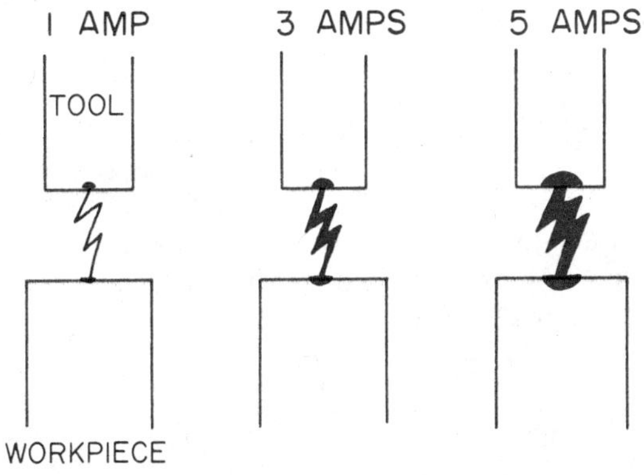

1 AMP 3 AMPS 5 AMPS
TOOL
WORKPIECE

Fig. 60-4. Effect of Amperage on Metal Removal Rate and Finish Quality

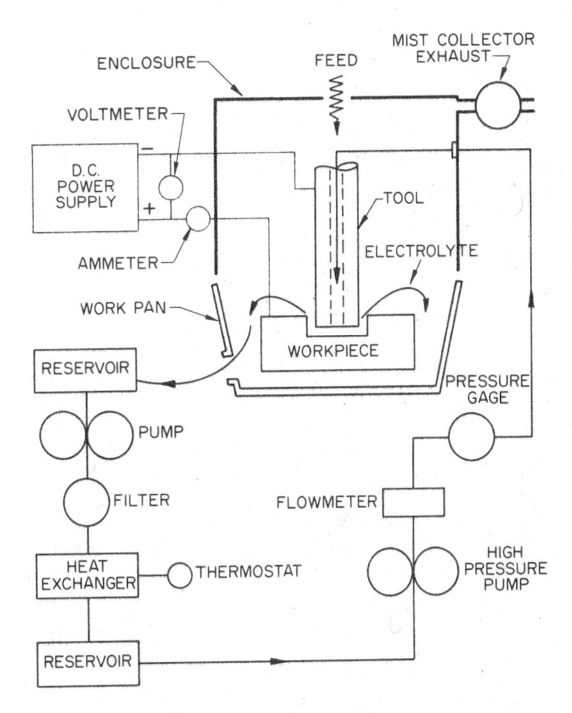

Fig. 60-5. Basic Components of an ECM System

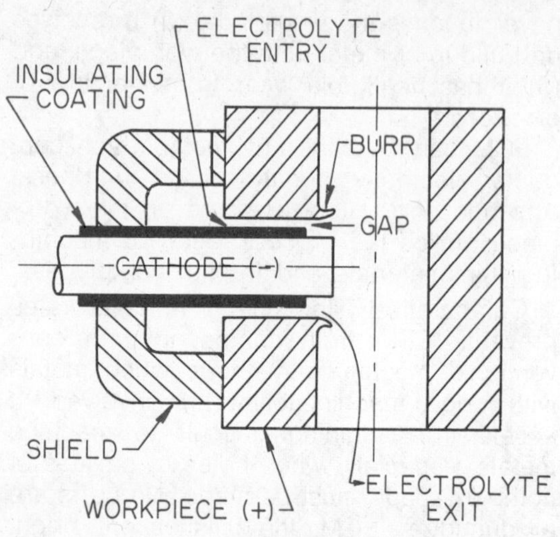

Fig. 60-6. Tooling Arrangement for Electrochemical Deburring

60-6. Electrochemical Deburring (ECD)

Electrochemical deburring (ECD) is an adaptation of ECM, using tooling designed for the removal of burrs and sharp edges from parts machined by conventional methods. (See Fig. 60-6.)

60-7. Electrochemical Grinding (ECG)

Electrochemical grinding (ECG) applies the principles of ECM to the conventional grinding process, Fig. 60-7. Grinding wheels which are used are of conventional abrasive materials but must be made with an electrically conductive bond. ECG removes metal as much as 80% faster than conventional grinding without an appreciable loss in accuracy or surface finish. Grinding operations can often be completed in one pass of the grinding wheel. Only about 10% of conventional grinding wheel pressure is necessary, and, since 90% of the metal is removed by electrochemical reaction (deplating), wheel dressing and wheel wear are sharply reduced.

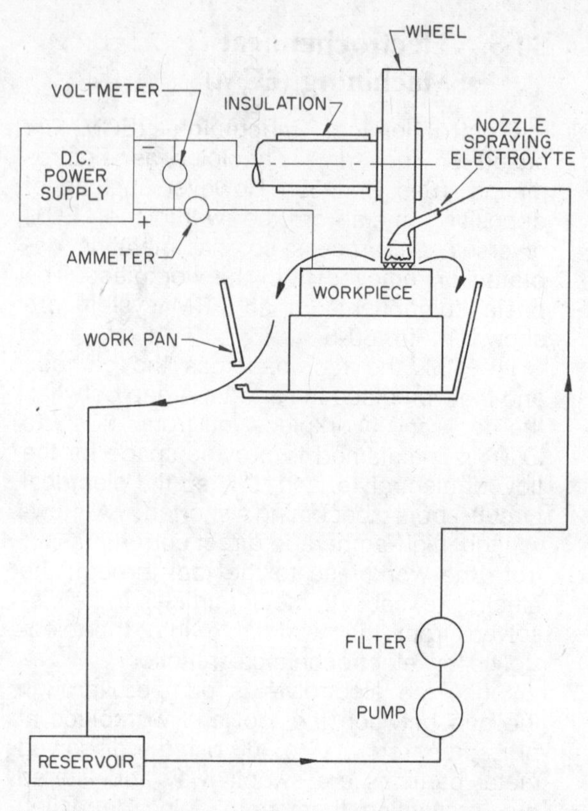

Fig. 60-7. Basic Components of an Electrochemical Grinding System

60-8. Ultrasonic Machining (USM)

In ultrasonic machining (USM) — also known as impact grinding, fine abrasive particles suspended in a fluid (usually water) are directed into the gap between the tool and the workpiece, Fig. 60-8. The tool is vibrated a few thousandths of an inch at an ultrasonic frequency of 20,000 or more cycles per second. This movement causes the abrasive particles to bombard the workpiece with high velocity, thus grinding the workpiece to the shape of the tool.

The abrasive particles may be boron carbide, silicon carbide, or aluminum oxide. Grit sizes used range from 280 to 800 mesh, depending on the desired degree of accuracy and the quality of finish.

Fig. 60-8. Basic Elements of an Ultrasonic Machining System

Fig. 60-9. Ultrasonic Machining of Tungsten Substrate Core
The tools are at upper right.

Tools are usually made of soft steel or brass and must be a mirror image of the shape to be machined. Tool wear is high due to the abrasive cutting action, which usually involves one or more roughing tools and a finishing tool for each job. The tools are usually fastened to their toolholders by brazing. A tolerance of .002″ is easily held and .0005″ is possible by using finer abrasive grits. Finishes are good, being in the vicinity of 20 microinches for ordinary work.

A major advantage of the process is that it can machine materials which cannot conduct electricity and which, therefore, cannot be machined by EDM, ECM, or ECG. Glass, ceramic materials, metal oxides, and precious and semiprecious gems are nonconducting materials which can be readily machined by USM. This process is also valued for its ability to machine carbides, tough alloys, and hardened steels. Soft materials such as carbon, graphite, and plastics are cut as readily as hard materials.

The process is very versatile, enabling any shape to be produced in a workpiece for which a tool can be made. (See Fig. 60-9.)

Words to Know

chemical blanking
chemical milling
deplating
dielectric
EBM
ECD
ECG
ECM
EDM
electrical discharge machining
electrochemical deburring
electrochemical grinding
electrochemical machining
electrochemical reaction
electrode
electrolyte
electron beam machine
impact grinding
laser beam machining
LBM
metal removal rate
overcut
stepped cutting tool
ultrasonic machining
USM

Review Questions

1. Describe the process of laser machining and identify its main advantage.
2. How is laser machining similar to laser welding?
3. Describe the electron beam machining process. How is it similar to electron beam welding?
4. List the advantages and disadvantages of electron beam machining.
5. Explain the process of electrical discharge machining.
6. What is the unique property of a dielectric fluid?
7. For what kinds of work is EDM especially well suited?
8. Why are stepped cutting tools sometimes used in EDM?
9. Name several methods of providing dielectric fluid streams between the tool and workpiece in EDM.
10. How is electrochemical machining similar to electroplating?
11. What prevents the buildup of metal on the tool in ECM?
12. List the main advantages of the ECM process.
13. How is the process of electrochemical deburring related to ECM?
14. How does the electrochemical grinding process differ from conventional grinding?
15. In what way are grinding wheels for ECG different from grinding wheels used for conventional grinding?
16. What are the main advantages of electrochemical grinding?
17. Explain the ultrasonic machining process.
18. What are the main advantages of ultrasonic machining?
19. What is meant by tool overcut?
20. Why does overcut take place in EDM? ECM? USM?

Decimal Equivalents of Common Fractions

Fraction	Decimal
1/64	.01563
1/32	.03125
3/64	.04688
1/16	.0625
5/64	.07813
3/32	.09375
7/64	.10938
1/8	.125
9/64	.14063
5/32	.15625
11/64	.17188
3/16	.1875
13/64	.20313
7/32	.21875
15/64	.23438
1/4	.250
17/64	.26563
9/32	.28125
19/64	.29688
5/16	.3125
21/64	.32813
11/32	.34375
23/64	.35938
3/8	.375
25/64	.39063
13/32	.40625
27/64	.42188
7/16	.4375
29/64	.45313
15/32	.46875
31/64	.48438
1/2	.500
33/64	.51563
17/32	.53125
35/64	.54688
9/16	.5625
37/64	.57813
19/32	.59375
39/64	.60938
5/8	.625
41/64	.64063
21/32	.65625
43/64	.67188
11/16	.6875
45/64	.70313
23/32	.71875
47/64	.73438
3/4	.750
49/64	.76563
25/32	.78125
51/64	.79688
13/16	.8125
53/64	.82813
27/32	.84375
55/64	.85938
7/8	.875
57/64	.89063
29/32	.90625
59/64	.92188
15/16	.9375
61/64	.95313
31/32	.96875
63/64	.98438
1	1.0000

Index

Decimal Equivalents of Common Fractions

64ths	32nds	16ths	8ths / 4ths / 2nds	Decimal
1/64				.01563
	1/32			.03125
3/64				.04688
		1/16		.0625
5/64				.07813
	3/32			.09375
7/64				.10938
			1/8	.125
9/64				.14063
	5/32			.15625
11/64				.17188
		3/16		.1875
13/64				.20313
	7/32			.21875
15/64				.23438
			1/4	.250
17/64				.26563
	9/32			.28125
19/64				.29688
		5/16		.3125
21/64				.32813
	11/32			.34375
23/64				.35938
			3/8	.375
25/64				.39063
	13/32			.40625
27/64				.42188
		7/16		.4375
29/64				.45313
	15/32			.46875
31/64				.48438
			1/2	.500

64ths	32nds	16ths	8ths / 4ths / 1	Decimal
33/64				.51563
	17/32			.53125
35/64				.54688
		9/16		.5625
37/64				.57813
	19/32			.59375
39/64				.60938
			5/8	.625
41/64				.64063
	21/32			.65625
43/64				.67188
		11/16		.6875
45/64				.70313
	23/32			.71875
47/64				.73438
			3/4	.750
49/64				.76563
	25/32			.78125
51/64				.79688
		13/16		.8125
53/64				.82813
	27/32			.84375
55/64				.85938
			7/8	.875
57/64				.89063
	29/32			.90625
59/64				.92188
		15/16		.9375
61/64				.95313
	31/32			.96875
63/64				.98438
			1	1.0000